The Global Nonlinear Stability of the Minkowski Space

Princeton Mathematical Series

1. The Classical Groups *by Hermann Weyl*
3. An Introduction to Differential Geometry *by Luther Pfahler Eisenhart*
4. Dimension Theory *by W. Hurewicz and H. Wallman*
8. Theory of Lie Groups: I *by C. Chevalley*
9. Mathematical Methods of Statistics *by Harald Cramér*
10. Several Complex Variables *by S. Bochner and W. T. Martin*
11. Introduction to Topology *by S. Lefschetz*
12. Algebraic Geometry and Topology *edited by R. H. Fox, D. C. Spencer, and A. W. Tucker*
14. The Topology of Fibre Bundles *by Norman Steenrod*
15. Foundations of Algebraic Topology *by Samuel Eilenberg and Norman Steenrod*
16. Functionals of Finite Riemann Surfaces *by Menahem Schiffer and Donald C. Spencer*
17. Introduction to Mathematical Logic, Vol. I *by Alonzo Church*
19. Homological Algebra *by H. Cartan and S. Eilenberg*
20. The Convolution Transform *by I. I. Hirschman and D. V. Widder*
21. Geometric Integration Theory *by H. Whitney*
22. Qualitative Theory of Differential Equations *by V. V. Nemytskii and V. V. Stepanov*
23. Topological Analysis *by Gordon T. Whyburn* (revised 1964)
24. Analytic Functions *by Ahlfors, Behnke, Bers, Grauert et al.*
25. Continuous Geometry *by John von Neumann*
26. Riemann Surfaces *by L. Ahlfors and L. Sario*
27. Differential and Combinatorial Topology *edited by S. S. Cairns*
28. Convex Analysis *by R. T. Rockafellar*
29. Global Analysis *edited by D. C. Spencer and S. Iyanaga*
30. Singular Integrals and Differentiability Properties of Functions *by E. M. Stein*
31. Problems in Analysis *edited by R. C. Gunning*
32. Introduction to Fourier Analysis on Euclidean Spaces *by E. M. Stein and G. Weiss*
33. Étale Cohomology *by J. S. Milne*
34. Pseudodifferential Operators *by Michael E. Taylor*
36. Representation Theory of Semisimple Groups: An Overview Based on Examples *by Anthony W. Knapp*
37. Foundations of Algebraic Analysis *by Masaki Kashiwara, Takahiro Kawai, and Tatsuo Kimura. Translated by Goro Kato*
38. Spin Geometry *by H. Blaine Lawson, Jr., and Marie-Louise Michelsohn*
39. Topology of 4-Manifolds *by Michael H. Freedman and Frank Quinn*
40. Hypo-Analytic Structures: Local Theory *by François Treves*
41. The Global Nonlinear Stability of the Minkowski Space *by Demetrios Christodoulou and Sergiu Klainerman*
42. Essays on Fourier Analysis in Honor of Elias M. Stein *edited by C. Fefferman, R. Fefferman, and S. Wainger*
43. Harmonic Analysis: Real-Variable Methods, Orthogonality, and Oscillatory Integrals *by Elias M. Stein*
44. Topics in Ergodic Theory *by Ya. G. Sinai*

The Global Nonlinear Stability of the Minkowski Space

Demetrios Christodoulou
and Sergiu Klainerman

PRINCETON UNIVERSITY PRESS
PRINCETON, NEW JERSEY
1993

Published by Princeton University Press, 41 William Street,
Princeton, New Jersey 08540
In the United Kingdom: Princeton University Press, Chichester,
West Sussex

Library of Congress Cataloging-in-Publication Data

Christodoulou, Demetrios, 1951–
The global nonlinear stability of the Minkowski space / Demetrios Christodoulou and Sergiu Klainerman.
p. cm. — (Princeton mathematical series : 41)
Includes bibliographical references.
ISBN 0-691-08777-6 (CL)
1. Space and time—Mathematics. 2. Generalized spaces.
3. Nonlinear theories. I. Klainerman, Sergiu, 1950–.
II. Title. III. Series.
QC173.59.S65C57 1993
530.1'1—dc20 92-15664
CIP

This book has been composed in Times, Perpetua, and Computer Modern using TEX

Princeton University Press books are printed on acid-free paper and meet the guidelines for permanence and durability of the Committee on Production Guidelines for Book Longevity of the Council on Library Resources

Printed in the United States of America

10 9 8 7 6 5 4 3 2 1

Table of Contents

II. Bianchi Equations in Space-Time

III. Construction of Global Space-Times. Proof of the Main Theorem

Acknowledgments

We would like to express our gratitude to C. Fefferman for his interest in our work and the precious time he spent in hearing and reading details of our proof. His comments helped us clarify obscure points and make the book easier to read. We would also like to thank S. T. Yau for his early interest and encouragement. Many thanks to W. T. Shu for reading and correcting parts of our manuscript. Finally, we want to express our gratitude to our families for their loving support given to us during the long years in which our present work has matured.

CHAPTER 1

Introduction

The aim of this book is to provide a proof of the nonlinear gravitational stability of the Minkowski space-time. More precisely, our work accomplishes the following goals:

1. It provides a constructive proof of global, smooth, nontrivial solutions to the Einstein-Vacuum equations, which look, in the large, like the Minkowski space-time. In particular, these solutions are free of black holes and singularities.
2. It provides a detailed description of the sense in which these solutions are close to the Minkowski space-time in all directions and gives a rigorous derivation of the laws of gravitational radiation proposed by Bondi. It also describes our new results concerning the behavior of the gravitational field at null infinity.
3. It obtains these solutions as dynamic developments of all initial data sets, which are close, in a precise manner, to the initial data set of the Minkowski space-time, and thus it establishes the global dynamic stability of the latter.
4. Though our results are established only for developments of initial data sets which are uniformly close to the trivial one, they are in fact valid in the complement of the domain of influence of a sufficiently large compact subset of the initial manifold of any "strongly asymptotically flat" initial data set.

According to Einstein, the underlying geometry of space-time is that given by a pair $(\mathbf{M}, \mathbf{g})$ where $\mathbf{M}$ is a 3+1-dimensional manifold and $\mathbf{g}$ is an Einstein metric on $\mathbf{M}$, that is, a nondegenerate, 2-covariant tensorfield with the property that at each point one can choose 3+1 vectors e_0, e_1, e_2, e_3 such that $\mathbf{g}(e_\alpha, e_\beta) = \eta_{\alpha\beta}$; $\alpha, \beta = 0, 1, 2, 3$ where η is the diagonal matrix with entries $-1, 1, 1, 1$. The Einstein metric divides the nonzero vectors X in the tangent space at each point into timelike, null, or spacelike vectors according to whether the quadratic form $\langle X, X \rangle = \mathbf{g}_{\alpha\beta} X^\alpha X^\beta$ is, respectively, negative, zero, or positive.

The set of null vectors forms a double cone, called the null cone of the corresponding point. The set of timelike vectors forms the interior of this cone. It has two connected components whose boundaries are the corresponding components of the null cone. The set of spacelike vectors is the exterior of the null cone, a connected open set. Any physically meaningful space-time should be time orientable, that is, one can choose in a continuous fashion a future-directed component of the set of timelike vectors. This allows us to specify the causal future and past of any point in space-time. More generally, the causal future of a set $S \subset \mathbf{M}$, denoted by $J^+(S)$, is defined as the set of points q that can be reached by a future-directed causal curve that initiates at S.[1] Similarly, $J^-(S)$ consists of the set of all points q that can be reached from S by a past-directed causal curve.

The boundaries of past and future sets of points in **M** are null geodesic cones, often called light cones. Their specification defines the *causal structure* of the space-time, which, up to a conformal factor, uniquely determines the metric.

A hypersurface M in $\mathbf{M}$ is said to be spacelike if its normal direction is timelike at every point on M. We denote by g the Riemannian metric induced by **g** on M. The covariant differentiation on the space-time **M** will be denoted by **D**, while that on M will be written with the symbols D or ∇. Similarly, we denote by **R**, respectively R, the Riemann curvature tensors of **M**, respectively M. Recall that for any given vector fields X, Y, Z on $(\mathbf{M}, \mathbf{g})$,

$$\mathbf{D}_X\mathbf{D}_Y - \mathbf{D}_Y\mathbf{D}_X Z = \mathbf{R}(X,Y)Z + \mathbf{D}_{[X,Y]}Z,$$

or, in components, relative to an arbitrary frame e_α, $\alpha = 0, 1, 2, 3$,

$$\mathbf{D}_\beta\mathbf{D}_\alpha Z^\gamma = \mathbf{D}_\alpha\mathbf{D}_\beta Z^\gamma + \mathbf{R}^\gamma{}_{\sigma\beta\alpha} Z^\sigma.$$

The extrinsic curvature, or second fundamental form, of M in $\mathbf{M}$ will be denoted by k. Recall that if T denotes the future-directed unit normal to M, we have

$$k_{ij} = -\langle \mathbf{D}_{e_i} T, e_j \rangle = \langle T, \mathbf{D}_{e_i} e_j \rangle$$

with e_i, $i = 1, 2, 3$, an arbitrary frame on **M**.

We will use the notation $\in_{\alpha\beta\gamma\delta}$ to express the components of the volume element $d\mu_{\mathbf{M}}$ relative to an arbitrary frame. Similarly, if e_i, $i = 1, 2, 3$ is an arbitrary frame on M, then $\in_{ijk} = \in_{0ijk}$ are the components of $d\mu_M$, the volume element of M, with respect to the frame $e_0 = T, e_1, e_2, e_3$.

The Riemann curvature tensor **R** of the space-time satisfies the following:

[1] A differentiable curve $\lambda(t)$ whose tangent at every point is a future-directed timelike or null vector.

Bianchi Identities

$$\mathbf{D}_{[\epsilon}\mathbf{R}_{\alpha\beta]\gamma\delta} = \frac{1}{3}(\mathbf{D}_\epsilon\mathbf{R}_{\alpha\beta\gamma\delta} + \mathbf{D}_\alpha\mathbf{R}_{\beta\epsilon\gamma\delta} + \mathbf{D}_\beta\mathbf{R}_{\epsilon\alpha\gamma\delta}) = 0.$$

The traceless part of the curvature tensor is

$$\mathbf{C}_{\alpha\beta\gamma\delta} = \mathbf{R}_{\alpha\beta\gamma\delta} - \frac{1}{2}(\mathbf{g}_{\alpha\gamma}\mathbf{R}_{\beta\delta} + \mathbf{g}_{\beta\delta}\mathbf{R}_{\alpha\gamma} - \mathbf{g}_{\beta\gamma}\mathbf{R}_{\alpha\delta} - \mathbf{g}_{\alpha\delta}\mathbf{R}_{\beta\gamma})$$
$$+ \frac{1}{6}(\mathbf{g}_{\alpha\gamma}\mathbf{g}_{\beta\delta} - \mathbf{g}_{\alpha\delta}\mathbf{g}_{\beta\gamma})\mathbf{R}$$

where the 2-tensor $\mathbf{R}_{\alpha\beta}$ and scalar $\mathbf{R}$ are respectively the Ricci tensor and the scalar curvature of the space-time. We call this the conformal curvature tensor of the space-time. We notice that the Riemann curvature tensor has 20 independent components while the conformal curvature and Ricci tensors have 10 components each.

The conformal curvature tensor is a particular example of a Weyl tensor. These refer to an arbitrary 4-tensors W that satisfy all the symmetry properties of the curvature tensor and in addition are traceless. We say that such W's satisfy the Bianchi equation if, with respect to the covariant differentiation on $\mathbf{M}$,

Bianchi Equation

$$\mathbf{D}_{[\epsilon}W_{\alpha\beta]\gamma\delta} = 0.$$

For a Weyl tensorfield W the following definitions of left and right Hodge duals are equivalent:

$${}^\star W_{\alpha\beta\gamma\delta} = \frac{1}{2} \in_{\alpha\beta\mu\nu} W^{\mu\nu}{}_{\gamma\delta}$$

$$W^\star_{\alpha\beta\gamma\delta} = W_{\alpha\beta}{}^{\mu\nu}\frac{1}{2} \in_{\mu\nu\gamma\delta}$$

where $\in^{\alpha\beta\gamma\delta}$ are the components of the volume element in $\mathbf{M}$. One can easily check that ${}^\star W = W^\star$ is also a Weyl tensorfield and ${}^\star({}^\star W) = -W$. Given an arbitrary vectorfield X, we can define the *electric-magnetic decomposition* of W to be the pair of 2-tensors formed by contracting W with X according to the formulas

$$ii_X(W)_{\alpha\beta} = W_{\mu\alpha\nu\beta}X^\mu X^\nu$$

and

$$ii_X({}^\star W)_{\alpha\beta} = {}^\star W_{\mu\alpha\nu\beta}X^\mu X^\nu.$$

These new tensors are symmetric, traceless, and orthogonal to X. Moreover, they completely determine W, provided that X is not null (see [CH-K]).

Given a vectorfield X and a Weyl field W, $\mathcal{L}_X W$ is not, in general, a Weyl field, since it fails to be traceless. To compensate for this, we define its modified Lie derivative

$$\hat{\mathcal{L}}_X W_{\alpha\beta\gamma\delta} = \mathcal{L}_X W_{\alpha\beta\gamma\delta} - \frac{1}{2}\left(\pi^{\mu}{}_{\alpha} W_{\mu\beta\gamma\delta} + \pi^{\mu}{}_{\beta} W_{\alpha\mu\gamma\delta}\right) + \left(\pi^{\mu}{}_{\gamma} W_{\alpha\beta\mu\delta} + \pi^{\mu}{}_{\delta} W_{\alpha\beta\gamma\mu}\right) + \frac{3}{8}\mathrm{tr}\pi W_{\alpha\beta\gamma\delta}$$

where π is the deformation tensor of X, that is

$$\pi = \mathcal{L}_X \mathbf{g}.$$

One can associate [Be-Ro] to the conformal curvature tensor or, more generally, to any Weyl tensorfield W, a 4-tensor that is quadratic in W and plays precisely the same role for solutions of the Bianchi equations as the energy-momentum tensor of an electromagnetic field plays for solutions of the Maxwell equations.

Bel-Robinson Tensor

$$Q_{\alpha\beta\gamma\delta} = (W_{\alpha\mu\beta\nu} W_{\gamma}{}^{\mu}{}_{\delta}{}^{\nu} + {}^{\star}W_{\alpha\mu\beta\nu}\, {}^{\star}W_{\gamma}{}^{\mu}{}_{\delta}{}^{\nu})1/2.$$

Q is fully symmetric and traceless; moreover, it satisfies the positive energy condition, namely, $Q(X,Y,Z,I)$ is positive whenever X,Y,Z,I are future-directed timelike vectors (see [Ch-K]) for a proof of these properties of Q). Moreover,

$$\mathbf{D}^{\delta} Q_{\alpha\beta\gamma\delta} = 0$$

whenever W satisfies the Bianchi equations. This remarkable property of the Bianchi equations is intimately connected with their conformal properties. Indeed, they are covariant under conformal isometries. That is, if $\phi : \mathbf{M} \longrightarrow \mathbf{M}$ is a conformal isometry of the space-time, that is, $\phi_* \mathbf{g} = \Omega^2 \mathbf{g}$ for some scalar Ω and W is a solution, then so is $\Omega^{-1}\phi_* W$.

It is well known that the causal structure of an arbitrary Einstein space-time can have undesirable pathologies. All these can be avoided by postulating the existence of a Cauchy hypersurface in M—a hypersurface Σ with the property that any causal curve intersects it at precisely one point.[2] Einstein space-times with this property are called *globally hyperbolic*. Such space-times are, in particular, stable causal, that is, they allow the existence of a globally defined differentiable function t whose gradient $\mathbf{D}t$ is timelike everywhere. We call t a *time function*, and the foliation given by its level surfaces a *t-foliation*. We denote by T the future-directed unit normal to the foliation.

[2] In particular, Σ is a spacelike hypersurface

Topologically, a space-time foliated by the level surfaces of a time function is diffeomorphic to a product manifold $\Re \times \Sigma$ where Σ is a 3-dimensional manifold. Indeed, the space-time can be parametrized by points on the slice $t = 0$ by following the integral curves of $\mathbf{D}t$. Moreover, relative to this parametrization, the space-time metric takes the form

$$ds^2 = -\phi^2(t,x)dt^2 + \sum_{i,j=1}^{3} g_{ij}(t,x)dx^i dx^j \tag{1.0.1}$$

where $x = (x^1, x^2, x^3)$ are arbitrary coordinates on the slice $t = 0$. The function $\phi(t,x) = \frac{1}{(-\langle \mathbf{D}t, \mathbf{D}t \rangle)^{1/2}}$ is called the *lapse function* of the foliation; g_{ij} is its first fundamental form. We refer to 1.0.1 as the canonical form of the space-time metric with respect to the foliation.

The foliation is said to be normalized at infinity if

Normal Foliation Condition

$$\phi \longrightarrow 1 \quad \text{as} \quad x \longrightarrow \infty \quad \text{on each leaf } \Sigma_t.$$

The second fundamental form of the foliation, the extrinsic curvature of the leaves Σ_t, is given by

$$k_{ij} = -(2\phi)^{-1}\partial_t g_{ij}. \tag{1.0.2}$$

We denote by ∇ the induced covariant derivative on the leaves Σ_t, and by R_{ij} the corresponding Ricci curvature tensor. Relative to an orthonormal frame e_1, e_2, e_3 tangent to the leaves of the foliation, we have the formulas

$$\begin{aligned}
\mathbf{D}_i e_j &= \nabla_i e_j - k_{ij}T \\
\mathbf{D}_i T &= -k_{ij}e_j \\
\mathbf{D}_T e_i &= \bar{\mathbf{D}}_T e_i + (\phi^{-1}\nabla_i\phi)T \\
\mathbf{D}_T T &= (\phi^{-1}\nabla_i\phi)e_i
\end{aligned}$$

where $\bar{\mathbf{D}}_T e_i$ denotes the projection of $\mathbf{D}_T e_i$ to the tangent space of the foliation. It is convenient to calculate relative to a frame for which $\bar{\mathbf{D}}_T e_i = 0$.[3]

Since Σ_t is three dimensional, we recall that the Ricci curvature R_{ij} completely determines the induced Riemann curvature tensor R_{ijkl} according to the formula

$$R_{ijkl} = g_{ik}R_{jl} + g_{jl}R_{ik} - g_{jk}R_{il} - g_{il}R_{jk} - \frac{1}{2}(g_{ik}g_{jl} - g_{jk}g_{il})R$$

where R is the scalar curvature $g^{ij}R_{ij}$. The second fundamental form k, the lapse function ϕ, and the Ricci curvature tensor R_{ij} of the foliation are connected to the space-time curvature tensor $\mathbf{R}_{\alpha\beta\gamma\delta}$ according to the following equations:

[3] This frame is called Fermi propagated.

The Structure Equations of the Foliation

$$\partial_t k_{ij} = -\nabla_i \nabla_j \phi + \phi(\mathbf{R}_{iTjT} - k_{ia}k^a{}_j) \quad (1.0.3a)$$

$$\nabla_i k_{jm} - \nabla_j k_{im} = \mathbf{R}_{mTij} \quad (1.0.3b)$$

$$R_{ij} - k_{ia}k^a{}_j + k_{ij} trk = \mathbf{R}_{iTjT} + \mathbf{R}_{ij} \quad (1.0.3c)$$

where ∂_t denotes the partial derivative with respect to t, and $\mathbf{R}_{iTjT}$, and $\mathbf{R}_{mTij}$ are the components $\mathbf{R}(\partial_i, \mathbf{T}, \partial_j, \mathbf{T})$ and $\mathbf{R}(\partial_m, \mathbf{T}, \partial_i, \partial_j)$, respectively, of the space-time curvature relative to arbitrary coordinates on Σ. Equation 1.0.3a is the second-variation formula, while 1.0.3b and 1.0.3c are, respectively, the classical Gauss-Codazzi and Gauss equations of the foliation.

In view of 1.0.3c, the equation 1.0.3a becomes

$$\partial_t k_{ij} = -\nabla_i \nabla_j \phi + \phi(-\mathbf{R}_{ij} + R_{ij} + trk k_{ij} - 2k_{ia}k^a{}_j). \quad (1.0.3d)$$

Taking the trace of the equations 1.0.3c, 1.0.3b, and 1.0.3a, respectively, we derive

$$R - |k|^2 + (trk)^2 = 2\mathbf{R}_{TT} + \mathbf{R} \quad (1.0.4a)$$

$$\nabla^j k_{ji} - \nabla_i trk = \mathbf{R}_{Ti} \quad (1.0.4b)$$

$$\partial_t trk = - \triangle \phi + \phi(\mathbf{R}_{TT} + |k|^2) \quad (1.0.4c)$$

where $|k|^2 = k_{ij}k^{ij}$.

In contrast to Riemannian geometry, where the basic covariant equations one encounters are of elliptic type, in Einstein geometry the basic equations are hyperbolic. The causal structure of the space-time is tied to the evolutions feature of the corresponding equations. This is particularly true for the Einstein field equations, where the space-time itself is the dynamic variable.

The Einstein field equations were proposed by Einstein as a unified theory of space-time and gravitation. The space-time $(\mathbf{M}, \mathbf{g})$ is the unknown; one has to find an Einstein metric $\mathbf{g}$ such that

Einstein Field Equations

$$\mathbf{G}_{\mu\nu} = 8\pi \mathbf{T}_{\mu\nu}$$

where $\mathbf{G}_{\mu\nu}$ is the tensor $\mathbf{R}_{\mu\nu} - 1/2\mathbf{g}_{\mu\nu}\mathbf{R}$, with $\mathbf{R}_{\mu\nu}$ the Ricci curvature of the metric, $\mathbf{R}$ its scalar curvature, and $\mathbf{T}_{\mu\nu}$ the energy momentum tensor of a matter field (e.g., the Maxwell equations). Contracting twice the Bianchi identities $\mathbf{D}_{[\epsilon}\mathbf{R}_{\alpha\beta]\gamma\beta} = 0$, we derive

Contracted Bianchi Identities

$$\mathbf{D}^\nu \mathbf{G}_{\mu\nu} = 0,$$

which are equivalent to the divergence equations of the matter field

$$\mathbf{D}^\nu \mathbf{T}_{\mu\nu} = 0.$$

In the simplest situation of the physical vacuum, $\mathbf{T} = 0$, the Einstein equations take the form

Einstein-Vacuum Equations

$$\mathbf{R}_{\mu\nu} = 0.$$

In view of the four contracted Bianchi identities mentioned previously, the Einstein-Vacuum equations, or E-V for short, can be viewed as a system of $10-4=6$ equations for the 10 components of the metric tensor $\mathbf{g}$. The remaining 4 degrees of freedom correspond to the general covariance of the equations. Indeed, if $\Phi : M \longrightarrow M$ is a diffeomorphism, then the pairs $(M, \mathbf{g})$ and $(M, \Phi^* \mathbf{g})$ represent the same solution of the field equations.

Written explicitly in an arbitrary system of coordinates, the E-V equations lead to a degenerate system of equations. Indeed, the principal part of the Ricci curvature $\mathbf{R}_{\mu\nu}$ is

$$\frac{1}{2}\mathbf{g}^{\alpha\beta}(\partial_\mu \partial_\alpha \mathbf{g}_{\beta\nu} + \partial_\nu \partial_\alpha \mathbf{g}_{\beta\mu} - \partial_\mu \partial_\nu \mathbf{g}_{\alpha\beta} - \partial_\alpha \partial_\beta \mathbf{g}_{\mu\nu}).$$

Thus, for a given metric $\mathbf{g}$, the symbol σ_ξ at a point p of the space-time manifold $(M, \mathbf{g})$ and a covector ξ at p is the linear operator on the space $S_2(M_p)$ of 2-covariant symmetric tensors h at p given by

$$\sigma_\xi \cdot h = \frac{1}{2}(\xi \otimes i_\xi h + i_\xi h \otimes \xi - \xi \otimes \xi \mathrm{tr} h - \langle \xi, \xi \rangle h)$$

where $i_\xi h$ is the covector at p obtained by contracting h on the left by the vector corresponding to ξ. We see that for any given ξ and any other covector X, $h = \xi \otimes X + X \otimes \xi$ belongs to the null space $N(\sigma_\xi)$ of the symbol at ξ. Nevertheless, the E-V equations are seen to be hyperbolic when account is taken of the geometric equivalence of metrics related by a diffeomorphism. Since any 1-parameter group of diffeomorphisms is generated by a vector field X and the infinitesimal action of the group on the space of metrics is the Lie derivative

$$(\mathcal{L}_X \mathbf{g})_{\mu\nu} = \mathbf{D}_\mu X_\nu + \mathbf{D}_\nu X_\mu,$$

the symbol of which is $\xi \otimes X + X \otimes \xi$, we consider the quotient Q_ξ of $S_2(M_p)$ by the following equivalence relation: $h_1 \sim h_2$ if and only if $h_1 - h_2 = \xi \otimes X$

$+ X \otimes \xi$ for some covector X at p. The symbol σ_ξ when reduced to Q_ξ is seen to have zero null space whenever ξ is not null. Moreover, when ξ corresponds to a nonzero null vector, choosing a null conjugate $\underline{\xi}$ to ξ, that is, another null vector in the same component of the null cone at p such that $\langle \xi, \underline{\xi} \rangle = -2$, we can identify Q_ξ with the space of all $h \in S_2(M_p)$ such that $i_{\underline{\xi}} h = 0$. Then $N(\sigma_\xi)$ is found to be $\hat{S}_2(\Pi)$, that is, the space of trace-free symmetric 2-covariant tensors on the 2-plane Π, defined to be the orthogonal complement of the plane spanned by ξ and $\underline{\xi}$. Therefore the 2-dimensional space $\hat{S}_2(\Pi)$ is the space of *dynamical degrees of freedom of the gravitational field at a point.*

This discussion indicates that it must be possible to choose coordinates relative to which the E-V equations can be written as a system of nonlinear wave equations. One well-known such choice is that of "wave coordinates,"[4] namely, coordinates x^μ which satisfy the wave equation

$$\mathbf{g}^{\alpha\beta}\mathbf{D}_\alpha \mathbf{D}_\beta x^\mu = 0.$$

In this case the principal part of $\mathbf{R}_{\mu\nu}$ becomes $-\frac{1}{2}\mathbf{g}^{\alpha\beta}\partial_\alpha\partial_\beta \mathbf{g}_{\mu\nu}$. The well-posedness of the local Cauchy problem was proved by Y. Choquet-Bruhat (see [Br1]) in wave coordinates, yet as she pointed out later, these coordinates are unstable (see [Br2]) in the large. The problem of finding globally stable, well-posed coordinate conditions is the first major difficulty one has to overcome in the construction of global solutions to the Einstein equations.

To emphasize the dynamic character of the E-V equations, it is helpful to express them in terms of the parameters ϕ, g, k of an arbitrary t-foliation. Thus, assuming that the space-time $(\mathbf{M}, \mathbf{g})$ can be foliated by the level surfaces of a time function t, and writing $\mathbf{g}$ in its canonical form 1.0.1, the E-V equations are equivalent to the following:

Constraint Equations for E-V

$$\nabla^j k_{ji} - \nabla_i trk = 0 \tag{1.0.5a}$$
$$R - |k|^2 + (trk)^2 = 0. \tag{1.0.5b}$$

Evolution Equations for E-V

$$\partial_t g_{ij} = -2\phi k_{ij} \tag{1.0.6a}$$
$$\partial_t k_{ij} = -\nabla_i \nabla_j \phi + \phi(R_{ij} + trk k_{ij} - 2k_{ia}k^a_j). \tag{1.0.6b}$$

Indeed, the equivalence of the equations 1.0.5a, 1.0.5b, 1.0.6a, and 1.0.6b with the E-V is an immediate consequence of 1.0.3d, 1.0.4a, and 1.0.4b.

[4] These are inappropriately called harmonic coordinates in the literature.

Also, 1.0.4c becomes

$$\partial_t trk = -\triangle \phi + \phi(R + (trk)^2). \tag{1.0.7}$$

Given a t-foliation, we denote by E, H the electric-magnetic decomposition of the curvature tensor $\mathbf{R}$ of an E-V manifold with respect to T, the future oriented unit normal to the time foliation. Clearly E, H are symmetric, traceless 2-tensors tangent to the foliation. In view of these definitions the equations 1.0.3b and 1.0.3c become

$$\nabla_i k_{jm} - \nabla_j k_{im} = \in_{ij}{}^{l} H_{lm} \tag{1.0.8a}$$

$$R_{ij} - k_{ia}k^a{}_j + k_{ij}trk = E_{ij}. \tag{1.0.8b}$$

Remark that the total number of unknowns in the evolution equations 1.0.6a and 1.0.6b is 13 while the total number of equations is only 12. This discrepancy corresponds to the remaining freedom of choosing the time function t, which defines the foliation. To emphasize the crucial importance of making an appropriate choice of time function, we note that the natural choice $\phi = 1$, corresponding to the temporal distance function from an initial hypersurface, leads to finite time breakdown. This can be seen from equation 1.0.7, which becomes, in view of 1.0.5b,

$$\partial_t trk = |k|^2 \geq \frac{1}{3}(\mathrm{tr}k)^2.$$

We also remark that, in view of the twice-contracted Bianchi identities, if g, k satisfy the evolution equations, then the constraint equations 1.0.5a and 1.0.5b are automatically satisfied on any Σ_t provided they are satisfied on a given initial slice Σ_{t_0}. Therefore they can be regarded as constraints on given initial conditions for g and k. According to this an *initial data set* for E-V is defined to be a triplet (Σ, g, k) consisting of a 3-dimensional manifold Σ together with a Riemannian metric g and covariant symmetric 2-tensor k, which satisfy the constraint equations 1.0.5a and 1.0.5b on Σ.

A development of an initial data set consists of an Einstein-Vacuum spacetime $(\mathbf{M}, \mathbf{g})$ together with an embedding $i : \Sigma \longrightarrow \mathbf{M}$ such that g and k are the induced first and second fundamental forms of Σ in $\mathbf{M}$. The central problem in the mathematical theory of E-V equations is the study of the evolution of general initial data sets.

The simplest solution of E-V equations is the Minkowski space-time $\mathbf{R}^{3+1}$, that is, the space $\mathbf{R}^4$ together with a given Einstein metric $\langle , \rangle$ and a canonical coordinate system (x^0, x^1, x^2, x^3) such that

$$\langle \partial_\alpha, \partial_\beta \rangle = \eta_{\alpha\beta}; \qquad \alpha, \beta = 0, 1, 2, 3.$$

The issue we want to address in our work is that of the global nonlinear stability of the Minkowski space-time. More precisely, we want to investigate whether

Cauchy developments of initial data sets that are close, in an appropriate sense, to the trivial data set lead to global, smooth, geodesically complete solutions of the Einstein-Vacuum equations that remain close, in an appropriate, global sense, to the Minkowski space-time. We stress the fact that at the present time it is not even known whether there are, apart from the Minkowski space-time, *any smooth, geodesically complete solution, which becomes flat at infinity on any given spacelike direction*. Any attempt to simplify the problem significantly by looking for solutions with additional symmetries fails as a consequence of the well-known results of Lichnerowicz for static solutions[5] and Birkhoff for spherically symmetric solutions. According to Lichnerowicz, a static solution that is geodesically complete and flat at infinity on any spacelike hypersurface must be flat. The Birkhoff theorem asserts that all spherically symmetric solutions of the E-V equations are static. Thus, disregarding the Schwarzschild solution, which is not geodesically complete, the only such solution that becomes flat at spacelike infinity is the Minkowski space-time.

The problem of stability of the Minkowski space-time is closely related to that of characterizing the space-time solutions of the Einstein-Vacuum equations, which are *globally asymptotically flat*—as defined in the physics literature, space-times that become flat as we approach infinity in any direction. Despite the central importance that such space-times have in General Relativity as corresponding to isolated physical systems, it is not at all settled how to define them correctly, consistent with the field equations. Attempts to develop such a notion, however, have been made in the last 25 years (see [Ne-To] for a survey) beginning with the work of Bondi ([Bo-Bu-Me], [Bo]; see also [Sa]), who introduced the idea of analyzing solutions of the field equations along null hypersurfaces. The present state of understanding was set by Penrose ([Pe1], [Pe2]), who formalized the idea of asymptotic flatness by adding a boundary at infinity attached through a smooth *conformal compactification*. However, it remains questionable whether there exists any nontrivial[6] solution of the field equations that satisfies the Penrose requirements. Indeed, his regularity assumptions translate into fall-off conditions of the curvature that may be too stringent and thus may fail to be satisfied by any solution that would allow gravitational waves. Moreover, the picture given by the conformal compactification fails to address the crucial issue of the relationship between conditions in the past and behavior in the future.

We believe that a real understanding of asymptotically flat spaces can only be accomplished by constructing them from initial data and studying their asymptotic behavior. In addition, only such a construction can address the

[5] A space-time is said to be stationary if there exists a one-parameter group of isometries whose orbits are timelike curves It is said to be static if, in addition, the orbits of the group are orthogonal to a spacelike hypersurface.

[6] Namely, a nonstationary solution

crucial issue of the relationship between conditions in the past and behavior in the future, an issue that the conformal compactification leaves entirely open. This is precisely the objective we set out to achieve.

To bring our discussion more to the point, we have to introduce the notion of an asymptotically flat initial data set. By this we mean an initial data set (Σ, g, k) with the property that the complement of a finite set in Σ is diffeomorphic to the complement of a ball in R^3 (i.e., Σ is diffeomorphic to R^3 at *infinity*), and the notion of energy, linear, and angular momentum are well defined and finite. These can be unambiguously defined for the following class of initial data sets, which we will refer to as *strongly asymptotically flat*.

We say that an initial data set (Σ, g, k) satisfies the S.A.F. condition if g and k are sufficiently smooth and there exists a coordinate system (x^1, x^2, x^3) defined in a neighborhood of infinity such that, as $r = \left[\sum_{i=1}^{3}(x^i)^2\right]^{1/2} \to \infty$,

S.A.F. Initial Data Sets[7]

$$g_{ij} = (1 + 2M/r)\delta_{ij} + o_4(r^{-3/2}) \tag{1.0.9a}$$

$$k_{ij} = o_3(r^{-5/2}). \tag{1.0.9b}$$

We shall call the leading term, $(1 + 2M/r)\delta_{ij}$, in the expansion 1.0.9a the *Schwarzschild Part* of the metric g.[8]

Given such a data set, the ADM (Arnowitt, Deser, and Misner) definitions of energy E, linear momentum P, and angular momentum J are given by

$$E = \frac{1}{16\pi}\lim_{r\to\infty}\int_{S_r}\sum_{i,j}(\partial_i g_{ij} - \partial_j g_{ii})N^j dA$$

$$P_i = \frac{1}{8\pi}\lim_{r\to\infty}\int_{S_r}(k_{ij} - tr k g_{ij})N^j dA; \qquad i = 1, 2, 3$$

$$J_i = \frac{1}{8\pi}\lim_{r\to\infty}\int_{S_r}\in_{iab}\, x^a(k^{bj} - g^{bj} tr k)N_j dA; \qquad i = 1, 2, 3$$

where S_r is the coordinate sphere of radius r, N is the exterior unit normal to it, and dA is its area element. Clearly the limits on the right-hand side of the formulas defining E and P exist and are finite. To check that J is well defined, one has to remark that the difference between the integrals on two different spheres r_1, r_2, can be written as a volume integral of an expression

[7] A function f is said to be $o_m(r^{-k})$, respectively $O_m(r^{-k})$, as $r \to \infty$ if $\partial^l f(x) = o(r^{-k-l})$, respectively $O(r^{-l-k})$, for any $l = 0, 1, \ldots m$, where ∂^l denote all the partial derivatives of order l relative to the coordinates x^1, x^2, x^3.

[8] It is the same as that of a spacelike hypersurface, orthogonal to the Killing vectorfield of a Schwarzschild space-time

which involves $\partial_j k_b{}^j - \partial_b(trk) = \nabla_j k_b{}^j - \nabla_b(trk) + o(r^{-9/2})$. The assertion follows then, with the help of the constraint equation 1.0.5a.

Moreover, due to our conditions 1.0.9a and 1.0.9b, we have

$$E = M \qquad P = 0.$$

Thus, the S.A.F. condition implies that the initial data set is in a center-of-mass frame. In view of the positive mass theorem ([Sc-Ya], [Wi]), M must be a positive number vanishing only if the initial data set is flat.

The definitions of the energy momentum (E, P_1, P_2, P_3) and the angular momentum (J_1, J_2, J_3) are independent of the particular choice of the coordinates x^1, x^2, x^3 in the definition of S.A.F. initial data sets.[9] Moreover, they are preserved by the evolution equations 1.0.6a and 1.0.6b of a normally foliated E-V space-time. This can be easily checked by taking the time derivatives of the expressions defining E, P, and J.

We believe that the question we are investigating here, namely, the stability of the Minkowski space-time, requires initial data sets with finite energy and linear and angular momentum.

In its least precise version our main result asserts the following:

Theorem 1.0.1 (First Version of the Main Theorem) Any strongly asymptotically flat[10] initial data set that satisfies, in addition, a global smallness assumption, leads to a unique, globally hyperbolic, smooth, and geodesically complete solution of the Einstein-Vacuum equations. Moreover, this development is globally asymptotically flat, by which we mean that its Riemann curvature tensor approaches zero on any causal or spacelike geodesic, as the corresponding affine parameter tends to infinity.

The main difficulties one encounters in the proof of our result are the following:[11]

1. The problem of coordinates.
2. The strongly nonlinear hyperbolic features of the Einstein equations.

(1) The problem of coordinates is the first major difficulty one has to overcome when trying to solve the Cauchy problem for the Einstein equations. In short,

[9] Indeed, first remark that the definitions are invariant under rigid transformations of the coordinates x^1, x^2, x^3. It thus suffices to show that the variations of the integrals defining E, P, J, with respect to one-parameter groups of diffeomorphisms generated by vectorfields $\xi = O_3(1)$ as $r \to \infty$ vanish in the limit.

[10] We note that the precise fall-off conditions of the initial data set are given later in L^2-norms, and thus differ slightly from those introduced in 1.0.9a and 1.0.9b.

[11] Besides the three major difficulties listed, one may add as a fourth the strong interdependence of these due to the highly nonlinear features of the equations.

one is faced with the following dilemma: To write the equations in a meaningful way, one seems forced to introduce coordinates. Such coordinates seem to be necessary even to allow the formulation of well-posed Cauchy problems and a proof of a local in time existence result. Nevertheless, as the particular case of wave coordinates illustrates, the coordinates may lead, in the large, to problems of their own. Our strategy to overcome this problem is based on two ideas. First, we describe our space-time by specifying, instead of full coordinate conditions, only a time function whose level hypersurfaces are *maximal*.[12] More precisely, we impose, in addition to the equations 1.0.5a and 1.0.5b, the constraint

$$trk = 0. \tag{1.0.10}$$

With this choice we remove the indeterminacy of the evolution equations 1.0.6a and 1.0.6b and obtain the following *determined* system of equations for the maximal foliation of an E-V space-time:

Constraint Equations of a Maximal Foliation

$$trk = 0 \tag{1.0.11a}$$

$$\nabla^j k_{ji} = 0 \tag{1.0.11b}$$

$$R = |k|^2. \tag{1.0.11c}$$

Evolution Equations of a Maximal Foliation

$$\partial_t g_{ij} = -2\phi k_{ij} \tag{1.0.12a}$$

$$\partial_t k_{ij} = -\nabla_i \nabla_j \phi + \phi(R_{ij} - 2k_{ia}k^a{}_j). \tag{1.0.12b}$$

Lapse Equation of a Maximal Foliation

$$\triangle \phi = |k|^2 \phi. \tag{1.0.13}$$

Remark that the time function t is defined by specifying the level sets only up to a transformation of the form $t \longrightarrow f \circ t$ with f being any orientation-preserving diffeomorphism of the real line. However, we can specify a unique t by requiring that, as it is regarded as a parameter on an integral curve Γ_x of T that passes through a point x of Σ_0, it converges to the arc length on Γ_x as x tends to infinity on Σ_0. This is equivalent to the condition that ϕ tends to 1 at infinity on each Σ_t, which is precisely the normal foliation condition introduced previously. Indeed, with the exception of the Minkowski space-time, this definition specifies a unique time function. This is due to the fact

[12] In Einstein geometry a maximal hypersurface is one that is spacelike and maximizes the volume among all possible compact perturbations of it.

that, when the A.D.M. energy E is nonzero, there is a unique maximal foliation with respect to which the linear momentum P vanishes. In physical terms, this foliation constitutes the center-of-mass frame of the corresponding isolated system.

The second idea is to make use in a fundamental way of the Bianchi identities of the space-time and the Bel-Robinson tensor introduced here. The basic observation is that, once we have good estimates for the curvature tensor $\mathbf{R}$, all the parameters of the foliation, that is g, k, ϕ, are determined purely by solving the elliptic system

$$R_{ij} - k_{ia}k^a{}_j = E_{ij} \tag{1.0.14a}$$

$$\text{curl } k_{ij} = H_{ij} \tag{1.0.14b}$$

$$\nabla^j k_{ji} = 0 \tag{1.0.14c}$$

$$trk = 0 \tag{1.0.14d}$$

together with the lapse equation 1.0.13. The equations 1.0.14a and 1.0.14b are immediate consequences of, respectively, 1.0.8b and 1.0.8a with $curl\ k_{ij} = \in_i{}^{ab} k_{jab}$. Thus, all the evolution features of the Einstein equations are contained in the Bianchi identities, which have the great advantage of being covariant.

(2) The other major obstacle in the study of the Einstein equations consists in their hyperbolic and strongly nonlinear character. The only powerful analytic tool we have in the study of nonlinear hyperbolic equations in the physical space-time dimension are the energy estimates. Yet the classical energy estimates are limited to proving results that are local in time. The difficulty has to do with the fact that, in order to control the higher energy norms of the solutions, one has to control the integral in time of their bounds in uniform norm. In recent years, however, new techniques were developed, based on modified energy estimates and the invariance property of the corresponding linear equations, which were applied to prove global or long-term existence results for nonlinear wave equations (see [Kl1], [Kl3]). More precisely, one uses the Killing and conformal Killing vectorfields generated by the conformal group of the Minkowski space-time to define a global energy norm that is invariant relative to the linear evolution. The precise asymptotic behavior, including the uniform bounds previously mentioned, are then an immediate consequence of a global version of the Sobolev inequalities (see [Ho], [Kl2], [Kl3]).

The relevant linearized equations for the E-V field equations are the Bianchi equations (see page 3) in Minkowski space-time. As a first preliminary step in our program, we have analyzed the complete asymptotic properties of the Bianchi equations[13] in Minkowski space-time by using only energy estimates

[13] In [Ch-Kl] they were called Spin-2 equations.

and the conformal invariance properties of the equations in the spirit of the ideas just outlined (see [Ch-Kl]).

To derive a global existence result, however, one also needs to investigate the structure of the nonlinear terms.[14] It is well known that arbitrary quadratic nonlinear perturbations of the scalar wave equation, even when derivable from a Lagrangean, could lead to formation of singularities unless a certain structural condition, which we have called the *null condition*, is satisfied (see [Ch], [Kl1]). It turns out that the appropriate, tensorial version of this structural condition is satisfied by the Einstein equations. Roughly speaking, one could say that the troublesome nonlinear terms, which could have led to formation of singularities, are in fact excluded due to the covariance and algebraic properties of the Einstein equations. An important illustration of this is the following algebraic identity satisfied by 2-covariant symmetric trace-free tensors in two dimensions. We recall that the space of dynamical degrees of freedom of the gravitational field at a point can be identified with $\hat{S}_2(\Pi)$, the space of trace-free symmetric 2-covariant tensors on the 2-plane Π. Given two such tensors u and v, one has, relative to an orthonormal basis for Π,

$$u_{AC}v_{CB} + u_{BC}v_{CA} - u \cdot v\delta_{AB} = 0,$$

which implies that if u, v, and w are any three such tensors, the trilinear invariant $\mathrm{tr}(uvw)$ vanishes identically.

These basic algebraic properties of the Einstein equations, which allow us to prove the global existence result, are in sharp contrast with the nonlinear hyperbolic equations of classical continuum mechanics. Indeed, the equation of nonlinear elasticity [John] and of compressible fluids [Si], in four space and time dimensions, form singularities even for arbitrary small initial conditions.

(3) In implementing the strategy outlined in (1) and (2), one encounters a very serious technical difficulty. The *mass term* that appears in the Schwarzschild part of an (S.A.F.) initial data set, 1.0.9a, has the long-range effect of changing the asymptotic position of the null geodesic cones relative to the maximal foliation. They are expected to diverge logarithmically from their corresponding position in flat space-time. In addition to this, their asymptotic shear[15] differs drastically from that in the Minkowski space-time. This difference reflects the presence of gravitational radiation in any nontrivial perturbation of the Minkowski space-time. To take this effect into account, one has to appropriately modify the Killing and conformal Killing vectorfields used in the definition of the basic energy norm. We achieve this by an elaborate construction of an *optical function* whose

[14] These terms are generated each time we commute the Bianchi identities with one of the vectorfields used in the definition of the global energy norm.

[15] The traceless part of their null second fundamental form.

level surfaces are outgoing null hypersurfaces related by a translation at infinity. By an optical function we mean a solution u of

Eikonal Equation

$$\mathbf{g}^{\alpha\beta}\frac{\partial u}{\partial x^\alpha}\frac{\partial u}{\partial x^\beta} = 0.$$

The construction of the optical function and the approximate Killing and conformal Killing vectorfields related to it requires more than half of our work. The most demanding part in the construction is taken by the angular momentum vectorfields.[16] These are particularly important to our construction, as they are crucial in circumventing the problem of slow decay at infinity of the initial data set. Thus, we do not estimate $\mathbf{R}$ directly from the Bianchi identities but only its Lie derivatives with respect to these vectorfields. This allows us to consider higher-weighted norms than will be possible for $\mathbf{R}$. Yet, as it turns out, the latter can be easily estimated in terms of the former.[17] Similarly, we use the approximate Killing vectorfield T, the unit normal to the foliation, to allow higher-weighted norms for the Lie derivatives of the curvature tensor with respect to T.[18]

Our construction requires initial data sets that satisfy, in addition to the constraint equations, the maximal condition $trk = 0$. We will refer to them as maximal in what follows.

To make the statement of our main theorem precise, we need also to define what we mean by the global smallness assumption. Before stating this condition, we assume the metric g to be complete, and we introduce the following quantity:

$$Q(x_{(0)}, b) = \sup_{\Sigma}\{b^{-2}(d_0^2 + b^2)^3|Ric|^2\}$$
$$+ b^{-3}\left\{\int_\Sigma \sum_{l=0}^{3}(d_0^2 + b^2)^{l+1}|\nabla^l k|^2 + \int_\Sigma \sum_{l=0}^{1}(d_0^2 + b^2)^{l+3}|\nabla^l B|^2\right\}$$

where $d_0(x) = d(x_{(0)}, x)$ is the Riemannian geodesic distance between the point x and a given point $x_{(0)}$ on Σ, b is a positive constant, $|Ric|^2 = R^{ij}R_{ij}$, ∇^l denotes the l-covariant derivatives, and B is the symmetric, traceless 2-tensor tensor[19]

$$B_{ij} = \epsilon_j{}^{ab}\nabla_a(R_{ib} - \frac{1}{4}g_{ib}R).$$

[16] The vectorfields that can be viewed as a deformation of $\Omega_{ij} = x_i\partial_j - x_j\partial_i$, for $i, j = 1, 2, 3$, of Minkowski space-time.

[17] This fact seems entirely plausible in view of the Birkhoff theorem.

[18] In view of the Lichnerowicz theorem, this procedure allows us to obtain information about $\mathbf{R}$ itself.

[19] Remark that B, called the Bach tensor, is dual to the tensor $R_{ijk} = \nabla_k R_{ij} - \nabla_j R_{ik} + 1/4(g_{ik}\nabla_j R - g_{ij}\nabla_k R)$ whose vanishing characterizes locally conformally flat 3-dimensional manifolds (see [Eisen]). Thus, up to lower-order terms, the Schwarzschild part of g does not affect it.

The symmetry and tracelessness of B follow immediately from the twice-contracted Bianchi identities $\nabla^j R_{ij} - \frac{1}{2}\nabla_i R = 0$. In fact we can write

$$B_{ij} = \frac{1}{2}(\in_i{}^{ab}\, \nabla_a \hat{R}_{jb} + \in_j{}^{ab}\, \nabla_a \hat{R}_{ib})$$

where $\hat{R}_{ij}$ is the traceless part of R_{ij}, $R_{ij} = \hat{R}_{ij} + \frac{1}{3}Rg_{ij}$.

Theorem 1.0.2 (The Global Smallness Assumption) We say that an S.A.F. initial data set, (Σ, g, k), satisfies the global smallness assumption if the metric g is complete and there exists a sufficiently small positive ϵ such that

$$\inf_{x_{(0)}\in\Sigma, b\geq 0} Q(x_{(0)}, b) < \varepsilon. \tag{1.0.15}$$

Theorem 1.0.3 (Second Version of the Main Theorem)[20] Any strongly asymptotically flat, maximal, initial data set that satisfies the global smallness assumption 1.0.15 leads to a unique, globally hyperbolic, smooth, and geodesically complete solution of the Einstein-Vacuum equations foliated by a normal, maximal time foliation. Moreover, this development is globally asymptotically flat.[21]

Remark 1 *In view of the scale invariance property of the Einstein-Vacuum equations, any initial data set Σ, g, k for which $Q(x_0, b) < \varepsilon$ can be rescaled to the new initial data set Σ, g', k' with $g' = b^{-2}g$, $k' = b^{-1}k$ for which $Q(x_0, 1) < \varepsilon$. The global existence for the new set is equivalent to the global existence for the original set. This is due to the fact that the developments* $\mathbf{g}, \mathbf{g}'$ *of the two sets are related by* $\mathbf{g}' = b^{-2}\mathbf{g}$. *It thus suffices to prove the theorem under the global smallness assumption*

$$\inf_{x_{(0)}\in\Sigma} Q(x_{(0)}, 1) < \varepsilon.$$

We next indicate how to construct maximal initial data sets that are asymptotically flat and satisfy 1.0.15. This is based on the observation that the constraint equations 1.0.11a and 1.0.11b are conformal invariant. More precisely, they are invariant with respect to the transformations $g_{ij} \longrightarrow \Phi^4 g_{ij}$ and $k_{ij} \longrightarrow \Phi^{-2}k_{ij}$. Thus, given arbitrary solutions $\tilde{g}, \tilde{k}$ to the equations

$$tr_{\tilde{g}}\tilde{k} = 0 \tag{1.0.16a}$$

$$\tilde{\nabla}^j \tilde{k}_{ji} = 0 \tag{1.0.16b}$$

[20] The first version of the theorem is not an immediate consequence of the second. It can be proved, however, by first developing the initial data set locally in time and then by embedding in it a maximal hypersurface. Embedding results of the type one needs were obtained by Bartnik (see [Ba1]).

[21] A precise statement of the asymptotic behavior for the curvature tensor $\mathbf{R}$ and also for the lapse function ϕ and second fundamental form k of the foliation will be given in the third version of the theorem.

where $\tilde{\nabla}$ denotes the covariant differentiation with respect to the metric $\tilde{g}$, we infer that $g_{ij} = \Phi^4\tilde{g}_{ij}$ and $k_{ij} = \Phi^{-2}\tilde{k}_{ij}$ are solutions to the same equations for arbitrary function Φ. To satisfy also the equation 1.0.11c, we have to subject Φ to the Lichnerowicz equation

$$\tilde{\Delta}\Phi - \frac{1}{8}\tilde{R}\Phi + |\tilde{k}|^2_{\tilde{g}}\Phi^{-7} = 0. \tag{1.0.17a}$$

In practice one does not solve the Lichnerowicz equation directly. The standard approach is to look for Φ of the form $\Phi = \Omega\Psi$ where Ω and Ψ are the conformal factors corresponding to transformations that take, first, an arbitrary solution of the equations 1.0.16a and 1.0.16b to a solution $\bar{g}, \bar{k}$ of the same equations and then take $\bar{g}, \bar{k}$ to the desired solution g, k. The first conformal factor Ω is chosen so that the Ricci curvature $\bar{R}$ of $\bar{g}$ vanishes identically. Thus, Ω has to be a solution of the linear equation

$$\tilde{\Delta}\Omega - \frac{1}{8}\tilde{R}\Omega = 0. \tag{1.0.17b}$$

The second conformal factor Ψ is chosen such that the transformed variables g, k satisfy $R = |k|^2$. For this to happen Ψ has to be a solution to the equation

$$\bar{\Delta}\Psi + \frac{1}{8}|\bar{k}|^2_{\bar{g}}\Psi^{-7} = 0. \tag{1.0.17c}$$

Remark that, by virtue of the maximal principle, the equation 1.0.17c always has a smooth solution, $\Psi \geq 1$, with $\Psi \longrightarrow 1$ as $x \longrightarrow \infty$ on Σ. On the other hand, a condition which suffices to insure that the equation 1.0.17b has a positive solution with the same property is that the $L^{3/2}$ norm of the negative part of $\tilde{R}$ be sufficiently small. Therefore, (Σ, g, k) is an initial data set satisfying the S.A.F. conditions 1.0.9a and 1.0.9b provided that the corresponding solutions $\tilde{g}$ and $\tilde{k}$ of 1.0.16a and 1.0.16b verify

$$\tilde{g}_{ij} = \delta_{ij} + o_4(r^{-3/2})$$
$$\tilde{k}_{ij} = o_3(r^{-5/2})$$

and that the negative part of $\tilde{R}$ satisfies the smallness condition. Moreover, g and k satisfy the global smallness assumption of the theorem provided that the metric $\tilde{g}$ is complete and that there exists a small positive ε such that

$$\inf_{x_{(0)}\in\Sigma, a\geq 0}\{\sup_{\Sigma}(a^2 + \tilde{d}_0^2)^3|\widetilde{Ric}|^2 + \int_\Sigma \sum_{l=0}^{2}(a^2 + \tilde{d}_0^2)^{l+2}|\tilde{\nabla}\widetilde{Ric}|^2$$
$$+ \int_\Sigma \sum_{l=0}^{3}(a^2 + \tilde{d}_0^2)^{l+1}|\tilde{\nabla}^l\tilde{k}|^2\}$$
$$< \varepsilon$$

where $\tilde{d}_0(x)$ denotes the Riemannian geodesic distance relative to $\tilde{g}$ between the point x and a given point $x_{(0)}$ on Σ.

It remains to discuss whether the equations 1.0.16a and 1.0.16b have solutions verifying these properties. This can be done using the orthogonal York decomposition of any symmetric, traceless 2-covariant tensor h, on a 3-dimensional Riemannian manifold $(\Sigma, \tilde{g})$ into a divergence-free part $\tilde{k}$ and the traceless part of the deformation tensor of a vectorfield X

$$h_{ij} = \tilde{k}_{ij} + \widehat{\mathcal{L}_X \tilde{g}_{ij}}.$$

The vectorfield X has to be a solution of the York equation,

$$L_{\tilde{g}} X = \tilde{\mathrm{div}}\, h, \tag{1.0.18}$$

where

$$L_{\tilde{g}} X_j = \tilde{\nabla}^i (\tilde{\nabla}_i X_j + \tilde{\nabla}_j X_i - \frac{2}{3} \tilde{g}_{ij} \tilde{\nabla}^l X_l).$$

One can show that $L_{\tilde{g}}$ is injective on spaces of vectorfields X with appropriate decay at infinity. It is also onto for appropriate spaces of vectorfields. Thus, for a given $\tilde{g} = \delta_{ij} + o_4(r^{-3/2})$, we select an appropriate $\tilde{k}$ by decomposing any symmetric traceless tensor $h = o_3(r^{-5/2})$ according to the previous definition, where X is a solution to the York equation. For details of how to achieve this we refer to [Ch-Mu]. We also remark that the corresponding tensor $\tilde{k}$ is of order $o_3(r^{-5/2})$ if the principal term in the expansion of solutions X at infinity, namely, the term of order r^{-1}, vanishes. This is so if and only if the linear momentum P_i of the corresponding initial data set is zero.

The proof of the main theorem hinges on an elaborate comparison argument with the Minkowski space-time at the level of the three geometric structures with which this is equipped.

- *The canonical spacelike foliation* of Minkowski space-time is given by any choice of a one-parameter family of parallel spacelike hyperplanes, the level sets of the time function $t = x^0 = const.$
- *The null structure* of the Minkowski space-time is specified by one family of future null cones and another of past null cones with vertices on a timelike geodesic orthogonal to the canonical spacelike foliation. These families are the level sets of the *optical* functions $u = r - t$ and, respectively, $v = r + t$, where $r = (\sum_{i=1}^{3} |x^i|^2)^{1/2}$. The null vectors $e_+ = \partial_t + \partial_r$ and $e_- = \partial_t - \partial_r$ are parallel to their respective gradients and span all the asymptotic null directions.
- *The conformal group structure* is given by the 15-parameter group of translations, Lorentz rotations, scaling, and inverted translations. The corresponding infinitesimal generators of the group are

1. The four generators of translations,

$$T_\mu = \partial_\mu; \qquad \mu = 0, 1, 2, 3,$$

2. The six generators of the Lorentz group,

$$\Omega_{\mu\nu} = x_\mu \partial_\nu - x_\nu \partial_\mu; \qquad \mu, \nu = 0, 1, 2, 3$$

   where $x_\mu = \eta_{\mu\nu} x^\nu$,

3. The scaling vectorfield,

$$S = x^\mu \partial_\mu,$$

4. The four inverted translation vectorfields,

$$K_\mu = -2x_\mu S + \langle x, x \rangle \partial_\mu; \qquad \mu = 0, 1, 2, 3.$$

We recall that the vectorfields in the first two groups are Killing while all the others are conformal Killing.[22] In particular, the deformation tensors of S and K_0 are given by

$$^{(S)}\pi = 2\eta, \qquad ^{(K_0)}\pi = 4t\eta.$$

As small perturbations of the Minkowski space-time, the solutions of the E-V that we want to construct will mirror the structures outlined previously. In other words, we construct them together with the following:

- A maximal spacelike foliation, of the type described previously, given by the level surfaces of a time function t.
- An appropriately defined optical function u whose level surfaces describe the structure of future null infinity.
- A family of almost Killing and conformal Killing[23] vectorfields tied to the definition of t and u.

The intersection between a t-slice Σ_t and a u-null hypersurface C_u is a 2-surface $S_{t,u}$. Thus, the (t, u) foliations of the space-time define a codimension 2-foliation by 2-surfaces. This foliation is crucial in our work; the asymptotic behavior of the curvature tensor **R** and the Hessians of t and u can only be fully described by decomposing them to components tangent to $S_{t,u}$. We achieve this by introducing null pairs consisting of two future-directed null vectors e_4 and e_3 orthogonal to $S_{t,u}$, with e_4 tangent to C_u and

$$\langle e_4, e_3 \rangle = -2.$$

[22] A vectorfield S in a space-time (M, g) is called Killing, respectively conformal Killing, if its deformation tensor $^{(X)}\pi = \mathcal{L}_X g$ is zero, respectively proportional to g.

[23] By almost conformal Killing we mean vectorfields whose traceless part of their deformation tensors are small in an appropriate fashion.

The null frame can be standardized by picking e_4 such that $\langle e_4, T\rangle = -1$. Another possible choice is to take $e_4 = l$, $e_3 = \underline{l}$ with

$$l^\mu = -\mathbf{g}^{\mu\nu}\frac{\partial u}{\partial x^\nu}.$$

We call this the l-null pair of t, u. A null pair together with an orthonormal frame e_1, e_2 on $S_{t,u}$ forms a null frame. The null decomposition of a tensor relative to a null frame e_4, e_3, e_2, e_1 is obtained by taking contractions with the vector fields e_4, e_3. For example, the null decomposition of the Riemann curvature tensor of an Einstein-Vacuum space-time consists of two S-tangent[24] symmetric traceless[25] tensors $\underline{\alpha}, \alpha$; two S-tangent 1-forms $\underline{\beta}, \beta$; and two scalars ρ, σ. They are defined by

$$\begin{aligned} \mathbf{R}_{A3B3} &= \underline{\alpha}_{AB} & \mathbf{R}_{A4B4} &= \alpha_{AB} \\ \mathbf{R}_{A334} &= 2\underline{\beta}_A & \mathbf{R}_{A434} &= 2\beta_A \\ \mathbf{R}_{3434} &= 4\rho & {}^*\mathbf{R}_{3434} &= 4\sigma. \end{aligned} \tag{1.0.19}$$

As part of our main theorem we deduce the following asymptotic properties for the null components of the curvature tensor:

$$\sup_{\Sigma_t} \tau_+ \tau_-^{5/2} |\underline{\alpha}| \le c \tag{1.0.20a}$$

$$\sup_{\Sigma_t} \tau_+^2 \tau_-^{3/2} |\underline{\beta}| \le c \tag{1.0.20b}$$

$$\sup_{\Sigma_t} \tau_+^3 |\rho| \le c \tag{1.0.20c}$$

$$\sup_{\Sigma_t} \tau_+^3 \tau_-^{1/2} |\sigma| \le c \tag{1.0.20d}$$

$$\sup_{\Sigma_t} \tau_+^{7/2} |\beta| \le c \tag{1.0.20e}$$

$$\sup_{\Sigma_t} \tau_+^{7/2} |\alpha| \le c \tag{1.0.20f}$$

where $\tau_- = (1+u^2)^{1/2}$, $\tau_+ = (1+v^2)^{1/2}$ with $v = 2r - u$, and $r = r(t,u)$ defined by requiring that the area of a surface $S_{t,u}$ is equal to $4\pi r^2$. Note that the "peeling property," connected to the smoothness of the conformal compactification, fails to be satisfied for the components α and β. Nevertheless, our results confirm the picture given by the conformal compactification in its main outlines.

[24] A space-time tensorfield is called S-tangent if, at every point, it is tangent to the $S_{t,u}$ 2-surface passing through that point.

[25] This is an immediate consequence of the trace zero condition of the Einstein equations.

The proof of the main theorem relies on a continuity argument. Using an adequate version of the local existence theorem, we assume our space-time to be maximally extended up to a value t_* of the t-function. Starting with the "last slice" Σ_{t_*}, we then define an appropriate optical function u. We use u and t to define a time-translation vectorfield T, which is simply the future-directed unit normal to the t-foliation, an inverted time-translation[26] vectorfield K_0, a scaling vectorfield S, and angular momentum operators ${}^{(a)}\Omega$. These vector fields are used, in conjunction with the Bel-Robinson tensor associated with $\mathbf{R}$, to define the basic quantities $\mathcal{Q}_1(t)$, $\mathcal{Q}_2(t)$

$$\mathcal{Q}_1(t) = \int_{\Sigma_t} Q(\hat{\mathcal{L}}_{\mathrm{O}}\mathbf{R})(\bar{K},\bar{K},T,T) + \int_{\Sigma_t} Q(\hat{\mathcal{L}}_T\mathbf{R})(\bar{K},\bar{K},\bar{K},T) \tag{1.0.21}$$

$$\begin{aligned}\mathcal{Q}_2(t) = &\int_{\Sigma} Q(\hat{\mathcal{L}}_{\mathrm{O}}^2\mathbf{R})(\bar{K},\bar{K},T,T) + \int_{\Sigma} Q(\hat{\mathcal{L}}_{\mathrm{O}}\hat{\mathcal{L}}_T\mathbf{R})(\bar{K},\bar{K},\bar{K},T)\\ &+ \int_{\Sigma} Q(\hat{\mathcal{L}}_S\hat{\mathcal{L}}_T\mathbf{R})(\bar{K},\bar{K},\bar{K},T) + \int_{\Sigma} Q(\hat{\mathcal{L}}_T^2\mathbf{R})(\bar{K},\bar{K},\bar{K},T)\end{aligned} \tag{1.0.22}$$

where $Q(\hat{\mathcal{L}}_{\mathrm{O}}\mathbf{R}) = \sum_{a=1}^{3} Q(\hat{\mathcal{L}}_{{}^{(a)}\Omega}\mathbf{R})$.

In linear theory[27] the time derivatives of the corresponding quantities would be zero in flat space-time; in our case they give rise to cubic error terms that depend linearly on the deformation tensors of $K_0, T, S, {}^{(a)}\Omega$, and quadratically in $\mathbf{R}$ and its covariant and Lie derivatives in the direction of $T, S, {}^{(a)}\Omega$. The crucial point of our overall strategy is to control the time integral of these error terms. This depends on the asymptotic behavior of all components of $\mathbf{R}$ and its covariant derivatives, on the asymptotic behavior of the deformation tensors of the vectorfields, and finally, because of the general covariance of the equations, on the cancellations of the "worst possible" cubic terms. The asymptotic behavior of $\mathbf{R}$ and its covariant derivatives can be traced, due to global Sobolev and Poincaré inequalities and the Bianchi identities, back to the basic quantities $\mathcal{Q}_1$, $\mathcal{Q}_2$.[28] The same is true for the deformation tensors of $K_0, T, S, {}^{(a)}\Omega$; we show this by elaborate estimates for the lapse function ϕ, the second fundamental form k_{ij} of the t-foliation, and the components of the Hessian of the optical function u. We are thus able to control the time integral of the error terms and show that $\mathcal{Q}_1 and \mathcal{Q}_2$ cannot exceed a constant multiple of their values at $t = 0$. This fact allows us to continue our space-time beyond t_*.

The book is divided into three parts. In what follows we give a short summary of the contents of each:

[26] This is the analogue of the vectorfield $K_0 = (t^2 + |x|^2)\partial_t + 2tx^i\partial_i$ in Minkowski space-time.

[27] By this we mean the replacement of the curvature tensor of the space-time with an arbitrary Weyl tensor verifying the Bianchi equations.

[28] The "flat" version of this fact was discussed in [Ch-Kl].

First Part: In the first part we discuss preliminary results concerning 2- and 3-dimensional Riemannian geometry. In Chapter 2 we discuss L^2 and L^p estimates for Hodge Systems on 2-D compact Riemannian manifolds of positive Gauss curvature. The L^2 theory for 1-forms verifying div – curl equations is standard; however, we need to develop a similar theory for traceless symmetric 2-tensors. For technical reasons, which become apparent in Part 3 when we discuss the estimates for the optical function, we also need L^p estimates. We do this with the help of the uniformization theorem and the Calderon-Zygmund theory for the corresponding Hodge systems on the standard sphere. To obtain good bounds, we need a sharp version of the uniformization theorem. We discuss this at the end of Chapter 2.

The geometry of each slice Σ_t of the space-time we construct is that of a 3-dimensional Riemannian manifold, diffeomorphic to $\Re^3$ and Euclidean at infinity. Moreover, these slices come equipped with an additional "radial" structure induced on them by the level hypersurfaces of the optical function u. In Chapter 3 we discuss general properties of such Riemannian manifolds Σ. Under certain simple assumptions we discuss global Sobolev and Poincaré inequalities. Similar global Sobolev inequalities were derived in [Ho] and [K12] based on scaling arguments. The argument we give here, appropriate for curved spaces, is based on the 2-D form of the isoperimetric inequality. In the last paragraph of the chapter we make sufficient assumptions of asymptotic flatness of Σ that allow us to attach to it, smoothly, a sphere at infinity. We then use this sphere to define an action of $SO(3)$ on Σ and corresponding rotation vectorfields. The main result of the chapter is a form of the Poincaré inequality for Lie derivatives with respect to the rotation vectorfields of 1-forms and 2-covariant, symmetric, traceless tensorfields. Together with the global Sobolev inequalities these will allow us later to derive, from the basic energy inequalities[29] and Bianchi identities, the full asymptotic behavior of all components of the curvature tensor.

In Chapter 4 we derive global estimates for solutions of the Poisson equation on a 3-D Riemannian manifold that verifies the properties discussed in Chapter 3. Together with the estimates for Hodge systems of Chapter 2, these are our main "elliptic" tools. The estimates for the Poisson equations are important in the study of the elliptic system 1.0.14b–1.0.14d as well as the lapse equation 1.0.13. Unlike the usual situation of elliptic theory, where the estimates are uniform in all directions, here we are forced to consider estimates that differentiate between the angular and normal components, relative to the radial foliation.[30] We call such estimates degenerate, in contrast to the usual type that we refer to as nondegenerate. Throughout the chapter we are careful to derive

[29] More precisely, from the form of our basic quantities $\mathcal{Q}_1$, $\mathcal{Q}_2$ defined in Part II.

[30] Due to the overall hyperbolic character of the Einstein equations.

our estimates while making minimal assumptions on the geometry of the space Σ and foliation u.[31] We achieve this by relying heavily on geometric and commutation methods. In the last paragraph of Chapter 4 we describe estimates for symmetric Hodge systems in 3-D. These estimates are uniform in all directions and are to be applied in the interior regions of the space-time.[32]

In Chapter 5 we interpret the global smallness assumption of an initial data set in terms of a radial decomposition of the Ricci curvature of the set.

In Chapter 6 we describe a theorem of local deformation of 2-surfaces in a 3-D Riemannian manifold. This theorem is a preliminary step for the construction of the optical function on the last slice. It involves the study of an O.D.E. in the infinite dimensional space of 2-surfaces embedded in the given Riemannian manifold.

Second Part: The second part is entirely devoted to the study of the Bianchi equations in a 4-D space-time. We assume that the space-time comes equipped with an optical function u and a time function t that satisfy certain hypotheses.

In Chapter 7 we show how to define the vectorfields $T, S, \bar{K}$ and compute their deformation tensors in terms of u and t. Assuming also the existence of angular momentum vectorfields ${}^{(a)}\Omega$, we define the basic quantities $\mathcal{Q}_1, \mathcal{Q}_2$. The main result of this chapter is the comparison theorem which allows us to reinterpret $\mathcal{Q}_1, \mathcal{Q}_2$ in terms of weighted norms of the components of the Weyl tensorfield verifying the Bianchi equations.

In Chapter 8 we find sufficient conditions that imply the boundedness of the basic quantities. We prove the boundedness theorem, which plays a major role in the proof of the main theorem.

The two theorems of the second part lie at the heart of our entire approach. They both rely on lengthy calculations. Most of the calculations can, however, be circumvented with the help of the principle of conservation of signature, introduced in Chapter 7, which allows us to take into account the structure of the terms that can occur in any given space-time formula without performing the calculation. This principle is closely connected to the "null condition," to which we have alluded before, and plays a fundamental role throughout our work.

Third Part: The last part contains the construction of the optical function, the estimates of the lapse ϕ and second fundamental form k, and the estimates for the deformation tensors of ${}^{(a)}\Omega$. The aim is to derive estimates consistent with the assumptions made in Part II. This allows us to close the proof of the main theorem and describe our main conclusions.

[31] In particular, we allow the level surfaces of the u-foliation to be far from umbilical. This is essential to applications in the study of time slices of radiative space-times

[32] We divide the space-time into two different regions. The one we call the exterior region contains the crucial "wave zone" in which all components of the curvature tensor behave differently. In the interior region, which is far away from the wave zone, the behavior is uniform.

In Chapter 9 we describe the construction of the optical function and derive sup-norm estimates for the components of its Hessian. In fact, our global optical function consists in the matching of two proper optical functions,[33] the first, defined throughout the exterior region, is fundamental as it describes the structure of null infinity. It is constructed by solving the Eikonal equation with initial conditions on the last slice. To avoid the caustics in the interior that this solution might form, we are obliged to match it with another optical function chosen by prescribing initial conditions on a "central line" given by an integral curve of the vectorfield T. The discussion of the initial choice, on the last slice, for the exterior optical function is postponed until Chapter 14.

In Chapter 10 we introduce the basic norms involving the lapse function ϕ, the second fundamental form k, and the Hessian of the optical function. We state the third and definitive version of our main theorem, including a complete account of the asymptotic behavior of the geometry of our space-times. We then give a detailed description of the strategy of the proof and refer to the relevant chapters of Parts II and III where each step is proved. We also include the statement of the local existence theorem and give a summary of its proof.

In Chapter 11 we discuss the estimates for the second fundamental form k. This is one of the most technical parts of our work. We find an elaborate decomposition of the components of k, which can be estimated with the help of the elliptic estimates derived in Part I. A major difficulty lies in the directionally dependent character of the behavior of each tensor in the wave zone. Here, however, the difficulty is compounded by the nonlocal character of the equations satisfied by k. The same remark applies for the estimates of the lapse function ϕ, discussed in Chapter 12.

In Chapter 13, one of the key chapters, we discuss the estimates for the higher derivatives of the optical function. The main ideas of our estimates are so basic that we feel obliged to give a rough account of them here.

Relative to a null frame $e_4 = l$, $e_3 = \underline{l}, e_2, e_1$, the hessian of the optical function decomposes in the components, tangent to the 2-surfaces $S_{t,u}$, $\chi_{AB} = \langle \mathbf{D}_{e_A} l, e_B \rangle$, $\zeta_A = \frac{1}{2}\langle \mathbf{D}_{e_A} l, e_3 \rangle$, and $\omega = \frac{1}{4}\langle \mathbf{D}_{e_3} l, e_3 \rangle$, which satisfy the equations

$$\frac{d\chi_{AB}}{ds} = -\chi_{AC}\chi_{CB} + \alpha_{AB} \tag{1.0.23a}$$

$$\frac{d\zeta_A}{ds} = -\chi_{AB}\zeta_B + \chi_{AB}\underline{\zeta}_B - \beta_A \tag{1.0.23b}$$

$$\frac{d\omega}{ds} = 2\underline{\zeta} \cdot \zeta - |\zeta|^2 - \rho \tag{1.0.23c}$$

[33] As a matter of fact, our global optical function u fails to be a proper solution of the Eikonal equation in the matching region.

where $\underline{\alpha}, \alpha, \underline{\beta}, \beta, \rho, \sigma$ are the components of the null decomposition of $\mathbf{R}$ as defined by 1.0.19. At first glance it looks as if χ, ζ, ω have the same order of differentiability as the components of the curvature. Nevertheless α is traceless, thus $\mathrm{tr}\chi$ satisfies an equation with no curvature term,

$$\frac{d\mathrm{tr}\chi}{ds} = -|\chi|^2 = -\frac{1}{2}(\mathrm{tr}\chi)^2 + |\hat{\chi}|^2, \tag{1.0.24a}$$

where $\hat{\chi}$ is the traceless part of χ. On the other hand, the null-Codazzi equations relate the angular derivatives of $\hat{\chi}$ to the angular derivatives of $\mathrm{tr}\chi$,

$$\nabla\!\!\!\!/_B \hat{\chi}_{AB} = -\zeta_B \hat{\chi}_{AB} + \frac{1}{2}(\nabla\!\!\!\!/_A \mathrm{tr}\chi + \mathrm{tr}\chi \zeta_A) - \beta_A, \tag{1.0.24b}$$

where $\nabla\!\!\!\!/$ is the induced covariant differentiation on $S_{t,u}$. The divergence equation 1.0.24b is one of the Hodge systems we study in chapter 2; its elliptic character allows us to conclude, by a bootstrap argument that combines the equations 1.0.24a and 1.0.24b, that χ is one degree of differentiability smoother than the curvature.

To estimate $\nabla\!\!\!\!/\zeta$ we introduce the *mass aspect function*

$$\mu = -\mathrm{d}\!\!\!/\mathrm{iv}\,\zeta + \frac{1}{2}\hat{\chi} \cdot \hat{\underline{\chi}} - \rho,$$

where $\underline{\chi}$ is the null second fundamental form $\underline{\chi}_{AB} = \langle \mathbf{D}_{e_A} \underline{l}, e_B \rangle$, and we derive this propagation equation for it,

$$\begin{aligned}\frac{d}{ds}\mu + tr\chi\mu &= \hat{\chi} \cdot (\nabla\!\!\!\!/ \widehat{\otimes} \zeta) + (\zeta - \underline{\zeta}) \cdot k\!\!\!/ - \frac{1}{4}\mathrm{tr}\underline{\chi}|\hat{\chi}|^2 \\ &+ \frac{1}{2}\mathrm{tr}\chi\left(- \mathrm{d}\!\!\!/\mathrm{iv}\,\underline{\zeta} - \frac{1}{2}\hat{\chi} \cdot \hat{\underline{\chi}} + \rho - |\underline{\zeta}|^2\right) + 2\zeta \cdot \hat{\chi} \cdot \underline{\zeta} - 2\zeta \cdot \beta,\end{aligned} \tag{1.0.25a}$$

with $\underline{\zeta}_A = \frac{1}{2}\langle \mathbf{D}_l \underline{l}, e_A \rangle$. We use this propagation equation in conjunction with the Hodge system on $S_{t,u}$,

$$\mathrm{d}\!\!\!/\mathrm{iv}\,\zeta = -\mu + \frac{1}{2}\hat{\chi} \cdot \hat{\underline{\chi}} - \rho \tag{1.0.25b}$$

$$\mathrm{c}\!\!\!/\mathrm{url}\,\zeta = \sigma - \frac{1}{2}\hat{\chi} \wedge \hat{\underline{\chi}}, \tag{1.0.25c}$$

to conclude that ζ is also one degree of differentiability smoother than the curvature.

To show that ω is also smoother, we derive a propagation equation for the scalar quantity $\psi\!\!\!/ = \Delta\!\!\!\!/\,\omega - \mathrm{d}\!\!\!/\mathrm{iv}\,\underline{\beta}$ where $\Delta\!\!\!\!/$ is the Laplacean on $S_{t,u}$. This combines the equation obtained by applying $\not\Delta$ to 1.0.23c with the divergence of the equation[34]

$$\frac{d\underline{\beta}_A}{ds} + tr\chi\underline{\beta}_A = -\nabla\!\!\!/_A\rho + \in_{AB} \nabla\!\!\!/_B\sigma + 2\hat{\underline{\chi}}_{AB}\beta_B - 3\rho\underline{\zeta}_A + 3\sigma^*\underline{\zeta}_A$$

to cancel the term containing the second derivative of ρ. We derive an equation of the form

$$\frac{d\psi\!\!\!/}{ds} + tr\chi\psi\!\!\!/ = -2\hat{\chi}_{AB}\nabla\!\!\!/_A\nabla\!\!\!/_B\omega + \frac{1}{2}tr\chi \operatorname{d\!\!\!/iv}\underline{\beta} + \text{l.o.t.} \tag{1.0.26a}$$

Once again we conclude that ω gains tangential derivatives by combining the propagation equation for $\psi\!\!\!/$ with the Laplace equation on $S_{t,u}$:

$$\not\Delta\omega = \psi\!\!\!/ + \operatorname{d\!\!\!/iv}\underline{\beta} \tag{1.0.26b}$$

This treatment, which combines propagation equations of the type 1.0.24a, 1.0.25a, and 1.0.26a with elliptic Hodge systems on the 2-surfaces $S_{t,u}$, is also important in deriving the optimal asymptotic behavior of χ, ζ, ω and their covariant derivatives. The mass aspect function μ is fundamental for yet another reason. Its integral on the 2-surfaces $S_{t,u}$ is connected to the Hawking mass

$$m(t,u) = \frac{r}{2}\left(1 + \frac{1}{16\pi}\int_{S_{t,u}} \text{tr}\chi\text{tr}\underline{\chi}\right)$$

according to the formula

$$\int_{S_{t,u}} \mu = 8\pi r m.$$

It thus places a major role in the derivation of the Bondi mass loss formula.

In Chapter 14 we discuss the construction of the exterior optical function on the last slice.

In Chapter 15 we discuss the properties of the interior optical function as well as its matching with the exterior one.

In Chapter 16 we define the angular momentum vectorfields ${}^{(a)}\Omega$, and, using the estimates for the optical function, we derive estimates for their deformation tensors.

Finally, Chapter 17 includes certain simple but important consequences of the third version of our Main Theorem. These concern the behavior of various geometric quantities at null infinity. We show that the basic null structure equations as well as the equations of the maximal foliation have limits at null infinity. These limiting equations are nonlinear, a manifestation of the strong

[34] This equation is a consequence of the Bianchi equations expressed relative to the null frame.

nonlinear features of the Einstein equations, even at infinity. We give a rigorous derivation of the Bondi mass loss formula, a precise statement on the "logarithmic divergence" of the future null cones, as well as several other important new consequences.

For the reader who is impatient with the length of our work and who does not care for details, or for the one who prefers to get an overall picture of our approach before plunging into its excruciating technicalities, we suggest the following order of reading:

Undoubtedly the heart of our approach lies in the proof of the two theorems of Part II. They are formulated as general results concerning Weyl tensorfields that satisfy the Bianchi equations in a space-time verifying certain conditions. These conditions are then met in Part III. There we suggest that the reader first become acquainted with the construction of the exterior optical function described in Chapter 9 and then read carefully Chapter 10. We then recommend a reading of Chapters 13 and 14 followed immediately by Chapter 16. This will give the reader an overview of the main ideas of the book.

Part I

Preliminary Results in 2- and 3-Dimensional Riemannian Geometry

CHAPTER 2

Generalized Hodge Systems in 2-D

Throughout this chapter we assume that (S, γ) is a compact, 2-dimensional Riemannian manifold. We denote by $\nabla\!\!\!/$ the covariant differentiation on S and by K its Gauss curvature. Recall

Gauss-Bonnet Theorem

$$\int_S K d\mu_\gamma = 2\pi\chi(S)$$

where $d\mu_\gamma$ is the volume element of (S, γ) and $\chi(S)$ is the Euler characteristic of S. We also recall the following:

Uniformization Theorem

There exists a conformal transformation of the metric $\overset{\circ}{\gamma} = \Omega^2\gamma$ such that the Gauss curvature $\overset{\circ}{K}$ of the new metric is

$$\overset{\circ}{K} = \begin{cases} 1 \text{ if } \chi(S) = 2 \\ 0 \text{ if } \chi(S) = 0 \\ -1 \text{ if } \chi(S) \leq -2. \end{cases}$$

Though most of the results of this chapter hold for general compact surfaces, the case of interest to us is that of strictly positive curvature. We write

$$k_m > 0 \tag{2.0.1a}$$

where

$$k_m = \min_S r^2 K, \qquad k_M = \max_S r^2 K. \tag{2.0.1b}$$

Here r is the normalizing constant defined by

$$A(S) = 4\pi r^2, \tag{2.0.1c}$$

with $A(S)$ being the total area of S.

2.1. Isoperimetric Inequality and Radius of Injectivity

Let $I(S)$ denote the isoperimetric constant of S, namely,

$$I(S) = \sup_{\Gamma} \frac{\min\{A_1(\Gamma), A_2(\Gamma)\}}{L(\Gamma)^2}$$

where Γ is an arbitrary closed curve on S, with A_1, A_2 being the areas of the two components of $S \setminus \Gamma$ and $L(\Gamma)$ being the total length of Γ. We recall the following (see [Oss]),

The Isoperimetric-Sobolev Inequality

$$\int_S (\Phi - \overline{\Phi})^2 \leq I(S)\Big(\int_S |\nabla\Phi| d\mu_\gamma\Big)^2$$

where

$$\overline{\Phi} = \frac{1}{A(S)} \int_S \Phi d\mu_\gamma$$

is the average of Φ on the surface S.

For the sake of completeness we sketch below the proof of the Isoperimetric-Sobolev inequality on an arbitrary surface S.

Proof 2.1.1 *Let ϕ be a function whose mean value is equal to zero. By Sard's Theorem the set of singular values t such that $\nabla\phi = 0$ somewhere on $\phi = t$ has measure zero. Assume that $t = 0$ is not a singular value, otherwise an obvious modification of the argument works. Let S_+, S_- be the sets of points where $\phi > 0$, $\emptyset < 0$, respectively. Without loss of generality we can assume that the areas $A(S_+), A(S_-)$ of S_+, S_- are equal. If this is not the case we can just subtract from ϕ an appropriate constant c_ϕ. Clearly, if the inequality holds for $\phi - c_\phi$, it also holds for ϕ. Now let $A_+(t)$ be the area of the set of points where $\phi \geq t$ and let $L_+(t)$ be the perimeter of the set $\phi = t$. In view of the definition of $I(S)$, we have*

$$L_+(t) \geq \frac{1}{\sqrt{I(S)}} \sqrt{A_+(t)}.$$

On the other hand,

$$\int_{S_+} |\nabla\phi| d\mu_\gamma = \int_0^\infty L_+(t) dt$$

$$\int_{S_+} |\phi|^2 d\mu_\gamma = \int_0^\infty 2t A_+(t) dt,$$

and since $A_+(t)$ is a decreasing function of t,

$$\int_0^\infty 2t A_+(t) dt \leq \Big(\int_0^\infty \sqrt{A_+(t)} dt\Big)^2.$$

Therefore,

$$\int_{S_+} |\phi|^2 d\mu_\gamma \leq I(S)\Big(\int_{S_+} |\nabla\phi| d\mu_\gamma\Big)^2.$$

The inequality for S_- is proved in precisely the same manner.

We will also make use of the Poincaré inequality on S.

Poincaré Inequality

$$\int_S (\Phi - \bar{\Phi})^2 \leq \frac{1}{\lambda_1(S)} \int |\nabla\Phi|^2$$

where $\lambda_1(S)$ the first eigenvalue of of the negative Laplacian, $-\Delta$, on S.

Remark 1 *We have the following obvious bound on λ_1:*

$$\lambda_1 \geq A(S)^{-1/2} I(S).$$

Remark 2 *Let $c(S)$ be a positive number, strictly less then the radius of injectivity of S, such that any metric ball of radius less than c is diffeomorphic to a disk and, in geodesic normal coordinates, the metric γ satisfies $\frac{1}{2}\delta \leq \gamma \leq 2\gamma$, where δ is the Euclidean metric. Then if Γ is a closed curve on S with length $L(\Gamma) \leq c(S)$, it must lie in a such a ball. On such balls we can estimate the isoperimetric constant in terms of that in the plane. Therefore, we find that*

$$I(S) \leq \max\left(1/\pi, \frac{A(S)}{c(S)^2}\right).$$

In view of a theorem of Cheeger (see Theorem 5.8 in [Ch-Eb]), the radius of injectivity of a compact n-dimensional manifold depends only on lower bounds for its volume, a bound on the absolute value of the sectional curvature, and an upper bound for its diameter. Thus, for surfaces S, the dimensionless constant $c(S)/\sqrt{A(S)}$ can be estimated from below by a positive constant that depends only on

$$\sup_S r^2|K| \quad and \quad \frac{diam(S)}{r}.$$

Remark 3 *In view of the Bonnet formula, for surfaces with positive Gauss curvature the ratio $diam(S)/r$ can be estimated from above in terms of $k_m(S)^{-1}$. Therefore by virtue of Cheeger's result applied for such surfaces, the isoperimetric constant $I(S)$ can be estimated by a continuous increasing function depending only on our basic constants $k_m(S)^{-1}$ and $k_M(S)$. In fact, for even-dimensional manifolds with strictly positive sectional curvature, a much simpler* *argument, due to Klingenberg (see Theorem 5.9 in [Ch-Eb]), can be applied to estimate the radius of injectivity. For surfaces, his estimate gives*

$$i(S) \geq \pi / \sqrt{\sup_S K}.$$

Moreover, we can pick a $c(S)$ *somewhat smaller than the radius of injectivity* $i(S)$ *and still depending only on* k_M*. Therefore, for surfaces of strictly positive Gauss curvature, the isoperimetric constant can in fact be estimated from above by a constant that depends only on* k_M*.*

We next sketch the proof of the result of Klingenberg for the particular case of surfaces of strictly positive Gauss curvature. First, let P be an arbitrary point and $C(P)$ the cut locus of P.[1] We recall the following (see Lemma 5.6 in [Ch-Eb]):

Lemma 2.1.1 Let Q be the point on the cut locus of P of minimum distance C(P) to P. Then either there is a minimal geodesic from P to Q along which Q is conjugate to P or there are precisely two distinct minimal geodesics from P to Q that are tangent at Q.

Now let λ be the geodesic distance function from P. In the region $\lambda < C(P)$ we can introduce geodesic polar coordinates corresponding to which metric γ takes the form

$$ds^2 = d\lambda^2 + r^2(\lambda, \theta) d\theta^2 \tag{2.1.1a}$$

where, for $\lambda = 0$,

$$r = 0, \qquad \partial r / \partial \lambda = 1$$

and $d\sigma^2 = r^2(\lambda, \theta) d\theta^2$ represents the element of arc length for the "great circles" $\lambda = \text{const}$. Let $h = h(\lambda, \theta)$ be the geodesic curvature of the great circles:

$$\begin{aligned} h &= \frac{\langle \nabla_{\frac{\partial}{\partial\theta}} \partial_\lambda, \frac{\partial}{\partial\theta} \rangle}{\langle \frac{\partial}{\partial\theta}, \frac{\partial}{\partial\theta} \rangle} \\ &= \frac{1}{r} \frac{\partial r}{\partial \lambda}. \end{aligned} \tag{2.1.1b}$$

Differentiating twice, tangential to the great circles, the equation

$$\gamma^{AB} \partial_A \lambda \partial_B \lambda = 1,$$

[1] We recall that the cut point of a geodesic starting from P is the furthest point along it on which the corresponding geodesic segment is minimizing. The cut locus of P is the set of cut points of P relative to all geodesics starting from P. On a compact manifold every geodesic starting at P has a cut point.

commuting covariant derivatives, and using the expression of the Riemann curvature tensor of S in terms of the Gauss curvature, we find, along any radial geodesic from P, the second variation formula

$$\frac{dh}{d\lambda} = -h^2 - K \tag{2.1.1c}$$

or, in view of 2.1.1b,

$$\frac{d^2}{d\lambda^2} r + Kr = 0. \tag{2.1.1d}$$

Comparing 2.1.1d with the equation

$$\frac{d^2}{d\lambda^2} y + (\sup_S K) y = 0 \tag{2.1.1e}$$

for the same initial conditions at P, we find that

$$r'/r \geq y'/y \tag{2.1.1f}$$

as long as r, y are positive, or, since r'/r tends to 1 as λ tends to 0, we infer that $r \geq \frac{\sin(\lambda\sqrt{\sup_S K})}{\sqrt{\sup_S K}} > 0$ as long as

$$\lambda < \pi / \sqrt{\sup_S K}. \tag{2.1.1g}$$

Therefore, there can be no conjugate points within a radius of less than $\pi/\sqrt{\sup_S K}$ on any geodesic starting at any point on S. Now assume that the radius of injectivity $i(S)$ is strictly smaller than this number and let P, Q be such that the distance between them is equal to the radius of injectivity $i(S)$. Clearly $Q \in \mathcal{C}(P)$, $P \in \mathcal{C}(Q)$, and, in view of the previous lemma, there must be a smooth closed geodesic Γ_0 passing through P and Q of total length equal to $2i(S)$. Let Γ_ϵ be the closed curves obtained by traveling a distance ϵ in the normal direction from Γ_ϵ. Using a variational formula as in 2.1.1c, one can easily show that the second derivative with respect to ϵ of the total length of Γ_ϵ is strictly negative at $\epsilon = 0$. Thus, for small ϵ, the curves Γ_ϵ are strictly shorter then Γ_0. Let P_ϵ be the point on Γ_ϵ at distance ϵ from P and let Q_ϵ be the point on the same curve furthest away from P_ϵ. Since the distance between P_ϵ and Q_ϵ is strictly smaller than the radius of injectivity, we infer that the two points cannot be conjugate to each other. Let Σ_ϵ be the unique minimal geodesic segment between them. In view of the definition of Q_ϵ, it must be orthogonal to Γ_ϵ at Q_ϵ. Now, by continuity, the points Q_ϵ converge to Q, and the unit tangent vectors to Σ_ϵ have a subsequence that converge to a unit vector v at P_ϵ. The geodesic through P in the direction of v must be a minimal geodesic connecting P to Q and orthogonal to Γ_0 at Q. In view of the lemma, this contradicts the assumption that P and Q are not conjugate to each other and ends the proof of Klingenberg's estimate.

2.2. The L^2 Theory for Hodge Systems

Given a vectorfield ξ in S we define its Hodge dual ${}^*\xi$ by

$$ {}^*\xi_A = \in_{AB}\xi^B $$

where $\in_{AB}$ are the components of the area element relative to an arbitrary frame e_A, $A = 1, 2$.[2] If ξ is a symmetric, traceless 2-tensor, we define the following left, ${}^*\xi$, and right, ξ^*, Hodge duals:

$$ {}^*\xi_{AB} = \in_{AC}\xi^C_B, \qquad \xi^*_{AB} = \xi_A{}^C \in_{CB} $$

Observe that the tensors ${}^*\xi, \xi^*$ are also symmetric and traceless, and that ${}^*\xi = -\xi^*$. The main objective of this section is to present estimates for solutions of the following two types of Hodge systems.

H$_1$. Consider a vector ξ on S that verifies

$$ \begin{aligned} \mathrm{d\!\!/iv}\,\xi &= f \\ \mathrm{cu\!\!/rl}\,\xi &= f_* \end{aligned} $$

where f, f_* are given scalar functions on S and the operators $\mathrm{d\!\!/iv}$, $\mathrm{cu\!\!/rl}$ are defined according to

$$ \mathrm{d\!\!/iv}\,\xi = \nabla\!\!\!/^A\xi_A, \qquad \mathrm{cu\!\!/rl}\,\xi = \in^{AB}\nabla\!\!\!/_A\xi_B. $$

H$_2$. Consider ξ to be a symmetric, traceless 2-tensor on S that verifies

$$ \mathrm{d\!\!/iv}\,\xi = f $$

where f is now a given vector and

$$ \mathrm{d\!\!/iv}\,\xi_A = \nabla\!\!\!/^B\xi_{AB}. $$

As a particular case of **H**$_1$, we shall also consider the scalar Poisson equation on S.

H$_0$. Consider a scalar function ϕ on S that verifies the equation

$$ \Delta\!\!\!/\,\phi = f $$

where f is an arbitrary scalar function on S.

[2] Unless otherwise specified, we take the frame to be orthonormal.

Remark 4 *Both Hodge systems* $\mathbf{H}_1$, $\mathbf{H}_2$ *and the Poisson equation* $\mathbf{H}_0$ *are conformally invariant. Indeed, if*

$$\tilde{\gamma} = \Omega^2 \gamma \tag{2.2.1a}$$

and $\Gamma, \tilde{\Gamma}$ *denote the connections of* γ *and* $\tilde{\gamma}$, *then, relative to any local system of coordinates,*

$$\tilde{\Gamma}^C_{AB} - \Gamma^C_{AB} = \Omega^{-1}(\delta^C_A \nabla\!\!\!\!/_B \Omega + \delta^C_B \nabla\!\!\!\!/_A \Omega - \gamma^{CD}\gamma_{AB} \nabla\!\!\!\!/_D \Omega). \tag{2.2.1b}$$

Also

$$d\mu_{\tilde{\gamma}} = \Omega^2 d\mu_\gamma \tag{2.2.1c}$$

or, in components,

$$\tilde{\in}_{AB} = \Omega^2 \in_{AB}. \tag{2.2.1d}$$

Thus, if ξ *is either a 1-form or a a 2-covariant symmetric, traceless tensor and* ϕ *is a scalar, we have*

$$\widetilde{d\!\!/iv}\,\xi = \Omega^{-2} d\!\!/iv\,\xi \tag{2.2.1e}$$

$$\widetilde{cu\!\!/rl}\,\xi = \Omega^{-2} cu\!\!/rl\,\xi \tag{2.2.1f}$$

$$\tilde{\Delta\!\!\!/}\phi = \Omega^{-2} \Delta\!\!\!/ \phi. \tag{2.2.1g}$$

We present here a straightforward derivation of the the L^2 theory for solutions of the Hodge systems $\mathbf{H}_1$, $\mathbf{H}_2$. Our first result is the following:

Proposition 2.2.1 *Consider* (S, γ) *an arbitrary compact Riemannian manifold.*

(i.) Assume that the vectorfield ξ *is a solution of* $\mathbf{H}_1$. *Then*

$$\int_S \{|\nabla\!\!\!\!/\xi|^2 + K|\xi|^2\} = \int_S \{|f|^2 + |f_*|^2\}.$$

(ii.) Assume that the symmetric, traceless 2-tensor ξ *is a solution of* $\mathbf{H}_2$. *Then*

$$\int_S \{|\nabla\!\!\!\!/\xi|^2 + 2K|\xi|^2\} = 2\int_S |f|^2.$$

As a particular case of the first part of the proposition we have the following:

Corollary 2.2.1.1 *If* ϕ *is a solution of* $\mathbf{H}_0$, *then:*

$$\int_S \{|\nabla\!\!\!\!/^2\phi|^2 + K|\nabla\!\!\!\!/\phi|^2\} = \int_S |f|^2.$$

Another important corollary of the proposition is the following:

Corollary 2.2.1.2 *Let $\mathcal{D}_1$ be the operator that takes any 1-form ξ on S into the pairs of functions $(\text{div}\,\xi, \text{curl}\,\xi)$. Similarly, let $\mathcal{D}_2$ be the operator that takes a 2-covariant, symmetric, traceless tensor ξ into the 1-form $\text{div}\,\xi$.*

Assuming only that S is diffeomorphic to the standard sphere S^2, the kernels of both operators $\mathcal{D}_1$ and $\mathcal{D}_2$ in L^2 are trivial.

The proof of Corollary 2.2.1.2 is an immediate consequence of the proposition, when S is the standard metric, in view of the positivity of the curvature. The general case follows from the conformal covariance of $\mathcal{D}_1$, $\mathcal{D}_2$, and the uniformization theorem.

Remark 5 *The L^2-adjoint of $\mathcal{D}_1$ is the operator $\mathcal{D}_1^*$, which takes pairs of functions (f, f_*) into 1-forms on S given by*

$$\mathcal{D}_1^*(f, f_*) = -\nabla_A f + \in_{AB} \nabla^B f_*.$$

The equation

$$\mathcal{D}_1^*(f, f_*) = 0$$

is precisely the Cauchy-Riemann equation and is satisfied only by constant functions f, f_*. Thus, since the kernel of $\mathcal{D}_1$ is trivial, we infer that its range consists of all L^2-integrable pairs of functions on S with vanishing mean values.

Similarly, the L^2-adjoint of $\mathcal{D}_2$ is the operator $\mathcal{D}_2^*$, which takes 1-forms f into 2-covariant, symmetric, traceless tensors given by

$$\mathcal{D}_2^* f = -\frac{1}{2}\widehat{\mathcal{L}_f\gamma}$$

where $\widehat{\mathcal{L}_f\gamma}$ is the traceless part of the Lie derivative of γ with respect to the vectorfield corresponding to f, that is, $\widehat{\mathcal{L}_f\gamma}_{AB} = \nabla_B f_A + \nabla_A f_B - (\text{div}\, f)\gamma_{AB}$. Therefore, the kernel of $\mathcal{D}_2^*$ can be identified with the set of all conformal Killing vectorfields on S. We thus infer that the range of $\mathcal{D}_2$ consists of all L^2 integrable 1-forms on S that are L^2 orthogonal to the Lie algebra of the conformal group of S.

For future reference we also record here the formulas

$$\mathcal{D}_2^*\mathcal{D}_2 = -\frac{1}{2}\Delta + K \tag{2.2.2a}$$

$$\mathcal{D}_1^*\mathcal{D}_1 = -\Delta + K \tag{2.2.2b}$$

$$\mathcal{D}_2\mathcal{D}_2^* = -\frac{1}{2}\Delta - \frac{1}{2}K \tag{2.2.2c}$$

$$\mathcal{D}_1\mathcal{D}_1^* = -\Delta \tag{2.2.2d}$$

$$\not{\mathcal{D}}_2^* \not{\mathcal{D}}_1^*(f, f_*) = \left(\widehat{\not\nabla^2} f, {}^*\widehat{\not\nabla^2} f_*\right) \tag{2.2.2e}$$

$$\not{\mathcal{D}}_1 \not{\mathcal{D}}_2 \xi = (\not{d}\mathrm{iv}^2 \xi, \not{d}\mathrm{iv}^{2\,*}\xi) \tag{2.2.2f}$$

where $\widehat{\not\nabla^2}$ is the traceless part of the hessian operator while $\not{d}\mathrm{iv}^2$ is the double divergence operator applied to symmetric tensors. The proof of 2.2.2a is a consequence of the following important lemma:

Lemma 2.2.1 *Consider an arbitrary 4-covariant tensor u_{ABCD} that is symmetric and traceless relative to both groups of indices AB and CD. Then*

$$u_{ACCB} + u_{BCCA} = u_{CDCD}\delta_{AB}.$$

In particular, if ξ, η are 2-covariant, symmetric, traceless tensors, then

$$\xi_{AC}\eta_{CB} + \xi_{BC}\eta_{CA} = (\xi \cdot \eta)\delta_{AB}. \tag{2.2.3}$$

More generally, if ξ, η are arbitrary 2-covariant tensors, and we introduce $s(\xi)_{AB} = \frac{1}{2}(\xi_{AB} + \xi_{BA})$, $\hat{s}(\xi)_{AB} = \frac{1}{2}(\xi_{AB} + \xi_{BA} - \mathrm{tr}(\xi)\delta_{AB})$, and $a(\xi) = \in^{AB}\xi_{AB}$, we have

$$\xi_{AC}\eta_{CB} + \xi_{BC}\eta_{CA} = (\xi \cdot \eta)\delta_{AB} \tag{2.2.4}$$

$$+ \mathrm{tr}(\xi)\widehat{s}(\xi)_{AB} + \mathrm{tr}(\eta)\widehat{s}(\xi)_{AB}$$

$$+ 2a(\xi)^*\widehat{s}(\eta)_{AB} - 2a(\eta)^*\widehat{s}(\xi)_{AB}. \tag{2.2.5}$$

The proof of the lemma is straightforward and will be omitted.

The results of Proposition 2.2.1 can be generalized to $k + 1$ symmetric, traceless tensors[3] ξ that satisfy the following system:

$\mathbf{H}_{k+1}$.

$$\not{d}\mathrm{iv}\,\xi = f$$

$$\mathrm{c}\not{u}\mathrm{rl}\,\xi = g$$

where

$$\not{d}\mathrm{iv}\,\xi_{A_1\ldots A_k} = \not\nabla^B \xi_{A_1\ldots A_k B}$$

$$\mathrm{c}\not{u}\mathrm{rl}\,\xi_{A_1\ldots A_k} = \in^{BC} \not\nabla_B \xi_{A_1\ldots A_k C}.$$

Here f and g are given k covariant, symmetric tensors.

More generally, we have

[3] These are traceless relative to any pair of indices.

Lemma 2.2.2 *Let ξ an arbitrary, k+1 covariant, totally symmetric tensor that satisfies the following generalized Hodge system:*

$\mathbf{H'}_{k+1}$.

$$\begin{aligned} \not{d}iv\xi &= f \\ c\not{u}rl\xi &= g \\ tr\xi &= h \end{aligned}$$

with $\not{d}iv\xi$, $c\not{u}rl\xi$ defined as above,

$$tr\xi_{A_1\ldots A_{k-1}} = \gamma^{BC}\xi_{A_1\ldots A_{k-1}BC},$$

f and g are given k covariant, symmetric tensors and h is a covariant symmetric tensor of rank k-1. Also, if k = 0, we take $tr\xi = 0$.

Then

$$\int_S \{|\not\nabla\xi|^2 + (k+1)K|\xi|^2\} = \int_S \{|f|^2 + |g|^2 + kK|h|^2\}.$$

Remark 6 *Any k + 1 covariant, symmetric tensor ξ can be decomposed into a totally traceless part, $\hat{\xi}$, and a tensor that depends only on the trace of ξ. Thus, if ξ is a 2-tensor,*

$$\xi_{AB} = \hat{\xi}_{AB} + \frac{1}{2}h\gamma_{AB},$$

while if ξ is a symmetric 3-tensor,

$$\xi_{ABC} = \hat{\xi}_{ABC} + \frac{1}{4}(h_A\gamma_{BC} + h_B\gamma_{CA} + h_C\gamma_{AB}).$$

In general, the pure trace part of a k + 1 tensor can be computed by subtracting from ξ the symmetrized tensor product of h with the metric γ. Symmetric, totally traceless tensors have the advantage that their Hodge duals have the same property. More precisely if we denote by ξ^* the tensor $\xi^*_{A_1\ldots A_k B} = \xi^*_{A_1\ldots A_k C} \in^C{}_B$, remark that ξ^* is symmetric if and only if ξ is totally traceless. Also remark that the system $\mathbf{H}_{k+1}$ is conformally invariant while $\mathbf{H'}_{k+1}$ is not.

Proof 2.2.1 *According to the curl equation, we have*

$$\not\nabla_C\xi_{A_1\ldots A_k B} = \not\nabla_B\xi_{A_1\ldots A_k C} - \in_{BC} g_{A_1\ldots A_k}.$$

Differentiating and commuting covariant derivatives, we find

$$\begin{aligned} \not\nabla_D\not\nabla_C\xi_{A_1\ldots A_k B} &= \not\nabla_D\not\nabla_B\xi_{A_1\ldots A_k C} - \in_{BC} \not\nabla_D g_{A_1\ldots A_k B} \qquad (2.2.6a) \\ &= \not\nabla_B\not\nabla_D\xi_{A_1\ldots A_k C} + \sum_{i=1}^{k} \underline{R}_{A_i MBD}\xi^M{}_{A_1\ldots\hat{A}_i\ldots A_k C} \\ &\quad + \underline{R}_{CMBD}\xi^M{}_{A_1\ldots A_k} - \in_{BC} \not\nabla_D g_{A_1\ldots A_k} \end{aligned}$$

where $\nabla\!\!\!/_{AB} = \nabla\!\!\!/_A \nabla\!\!\!/_B$, *and*

$$\underline{R}_{ABCD} = (\gamma_{AC}\gamma_{BD} - \gamma_{AD}\gamma_{BC})K \tag{2.2.6b}$$

is the Riemannian curvature tensor of S. Remark that, by taking the traces of 2.2.6b, we also have

$$\underline{R}_{AC} = \gamma_{AC}K, \qquad \underline{R} = 2K. \tag{2.2.6c}$$

Taking the trace of the equation 2.2.6a relative to C, D, and making use of 2.2.6b, 2.2.6c together with the equation $\mathbf{H}_{k+1}$, *we deduce*

$$\begin{aligned}\Delta\!\!\!/\xi_{A_1\dots A_k B} &= \nabla\!\!\!/_B f_{A_1\dots A_k} - \in_{BC} \nabla\!\!\!/^C g_{A_1\dots A_k} \\ &\quad + (k+1)K\xi_{A_1\dots A_k B} - \sum_{i=1}^{k} \gamma_{A_i B} K h_{A_1\dots \hat{A}_i \dots A_k}.\end{aligned} \tag{2.2.6d}$$

Finally, multiplying 2.2.6d by $\xi_{A_1\dots A_k B}$, *integrating on S, and integrating by parts on the left-hand side, we derive the desired result.*

The proof of the first part of Proposition 2.2.1 is an immediate corollary of the lemma. To prove the second part, we remark that any solution of $\mathbf{H}_1$ *verifies*

$$c\!\!\!/url\,\xi = {}^*f$$

where ${}^*f$ *is the Hodge dual of f. Therefore* ξ *satisfies a generalized Hodge system with* $g = {}^*f$ *and* $h = 0$.

To estimate higher derivatives of solutions ξ to $\mathbf{H}_1$ and $\mathbf{H}_2$, we need the following:

Lemma 2.2.3 *Let* ξ *be a k+1 covariant, symmetric tensor and introduce the following notation for its symmetrized covariant derivative:*

$$\begin{aligned}\tilde{D}\xi_{A_1\dots A_{k+1}B} &= \nabla\!\!\!/_{(B}\xi_{A_1\dots A_{k+1})} \\ &= \frac{1}{k+2}\Big(\nabla\!\!\!/_B\xi_{A_1\dots A_{k+1}} + \sum_{i=1}^{k+1}\nabla\!\!\!/_{A_i}\xi_{A_1\dots B\dots A_{k+1}}\Big).\end{aligned}$$

We also define the symmetrized right dual of ξ *according to*

$$\xi^\star_{A_1\dots A_{k+1}} = \xi_{(A_1\dots A_k B} \in_{BA_{k+1})}.$$

Assume that ξ *verifies the generalized Hodge system of Lemma 2.2.2 for any given f,g, and h. Then* $\tilde{\xi} = \tilde{D}\xi$ *is a solution of a similar system,*

$$\begin{aligned} d\!\!\!/iv\,\tilde{\xi} &= \tilde{f} \\ c\!\!\!/url\,\tilde{\xi} &= \tilde{g} \\ tr\,\tilde{\xi} &= \tilde{h},\end{aligned}$$

where

$$\tilde{f} = \tilde{D}f + \frac{1}{k+2}(\tilde{D}g)^* + (k+1)K\xi + \frac{k}{k+2}Kh\tilde{\otimes}\gamma$$
$$\tilde{g} = \tilde{D}g - (k+1)K\xi^*$$
$$\tilde{h} = \frac{k}{k+2}\tilde{D}h + \frac{2}{k+2}f,$$

where $h\tilde{\otimes}\gamma$ is the symmetrized tensor product between h and γ.

Lemma 2.2.3 can now be used repeatedly, together with Lemma 2.2.2, to prove the following:

Proposition 2.2.2 *Assume that the condition 2.0.1a is satisfied. Let ξ be a solution of either $\mathbf{H}_1$ or $\mathbf{H}_2$.*

(i.) First Derivatives Estimates.
There exists a constant c_1 that depends only on ${k_m}^{-1}$, k_M such that

$$\int_S \{|\nabla\xi|^2 + r^{-2}|\xi|^2\} \leq c_1 \int_S |f|^2.$$

(ii.) Second Derivatives Estimates.
There exists a constant c_2 that depends only on ${k_m}^{-1}$, k_M such that

$$\int_S |\nabla^2\xi|^2 \leq c_2 \int_S \{|\nabla f|^2 + r^{-2}|f|^2\}.$$

(iii.) Third Derivatives Estimates.

Assume that in addition to the condition 2.0.1a the curvature K satisfies

$$k_1 = \left(\int_S |\nabla K|^2\right)^{1/2} \text{ is finite.}$$

There exists a constant c_3, depending on ${k_m}^{-1}$, k_M, and k_1 such that

$$\int_S |\nabla^3\xi|^2 \leq c_3 \int_S \{|\nabla^2 f|^2 + r^{-2}|\nabla f|^2 + r^{-4}|f|^2\}$$

where we define, for the particular case $\mathbf{H}_1$, $f = (f, f_*)$ and $|f|^2 = |f|^2 + |f_*|^2$.

To prove the lemma, we remark that the L^2 integral of $\tilde{D}\xi$ can be expressed in terms of the L^2 integrals of $\nabla\xi$ and g only. Also, in the estimates of the third derivatives of ξ the only terms that contain the derivatives of K are of the form

$\int_S |\nabla K|^2 |\xi|^2$, which can be estimated by $k_1^2 \sup_S |\xi|^2$. On the other hand, in view of the classical Sobolev inequality on S,

$$\sup_S |\xi| \leq c |\nabla^2 \xi|_{L^2}^{1/2} |\xi|_{L^2}^{1/2},$$

which is an immediate consequence of the isoperimetric inequality (see page 32).

Corollary 2.2.2.1 *Assume that the radial foliation given by u verifies the assumption 2.0.1a and let ϕ be a solution to the Poisson equation on S, $\mathbf{H}_0$. Then there exists a constant c that depends only on ${k_m}^{-1}$, k_M such that*

$$|\nabla^2 \phi|_{L^2} + r^{-1} |\nabla \phi|_{L^2} + r^{-2} |\phi - \bar{\phi}|_{L^2} \leq c |f|_{L^2}$$
$$|\nabla^3 \phi|_{L^2} \leq c(|\nabla f|_{L^2} + r^{-1} |f|_{L^2}).$$

Moreover, if in addition the first derivative of the Gauss curvature verifies the condition $k_1 < \infty$,

$$|\nabla^4 \phi|_{L^2} \leq c(|\nabla^2 f|_{L^2} + r^{-1} |\nabla f|_{L^2} + r^{-2} |f|_{L^2})$$

where c depends only on ${k_m}^{-1}$, k_M, and k_1.

2.3. The L^p Theory

We need in our work not only the L^2 estimates presented but also certain L^p estimates. These can be derived from the classical Calderon-Zygmund inequalities with the help of a sharp form of the uniformization Theorem, which we discuss below. Indeed , if the metric γ is conformal to the standard metric of the unit sphere S^2, which we denote $\overset{\circ}{\gamma}$,

$$\gamma = \Omega^{-2} \overset{\circ}{\gamma},$$

and if ξ is a solution of $\mathbf{H}_{k+1}$, then

$$\overset{\circ}{\mathrm{d\!/\!iv}}\ \xi = \overset{\circ}{f} \qquad (2.3.1a)$$
$$\overset{\circ}{\mathrm{c\!/\!url}}\ \xi = \overset{\circ}{g}$$

where $\overset{\circ}{f} = \Omega^{-2} f$, $\overset{\circ}{g} = \Omega^{-2} g$. According to the Calderon Zygmund L^p inequalities on $S = S^2$, we have the following:

Lemma 2.3.1 *Let ξ be a solution of the Hodge System 2.3.1a, with either $k = 0$ or $k = 1$ on the standard unit sphere $(S^2, \overset{\circ}{\gamma})$. For every $1 < p < \infty$, there exists a constant $c = c_p$ such that* *(i.) First Derivatives Estimates.*

$$\int_S \left\{ |\overset{\circ}{\nabla} \xi|^p_{\overset{\circ}{\gamma}} + |\xi|^p_{\overset{\circ}{\gamma}} \right\} d\mu_{\overset{\circ}{\gamma}} \le c_p \int_S \left\{ |\overset{\circ}{f}|^p_{\overset{\circ}{\gamma}} + |\overset{\circ}{g}|^p_{\overset{\circ}{\gamma}} \right\} d\mu_{\overset{\circ}{\gamma}}.$$

(ii.) Second Derivatives Estimates.

$$\int_S |\overset{\circ}{\nabla}{}^2 \xi|^p_{\overset{\circ}{\gamma}} d\mu_{\overset{\circ}{\gamma}} \le c_p \int_S \left\{ |\overset{\circ}{\nabla}\overset{\circ}{f}|^p_{\overset{\circ}{\gamma}} + |\overset{\circ}{f}|^p_{\overset{\circ}{\gamma}} + |\overset{\circ}{\nabla}\overset{\circ}{g}|^p_{\overset{\circ}{\gamma}} + |\overset{\circ}{g}|^p_{\overset{\circ}{\gamma}} \right\} d\mu_{\overset{\circ}{\gamma}}.$$

Now assume that ξ is a symmetric tensor of rank k on the surface (S, γ). Remark that

$$|\xi|_\gamma = \Omega^k |\xi|_{\overset{\circ}{\gamma}}. \tag{2.3.1b}$$

On the other hand, formally,

$$\overset{\circ}{\nabla} \xi = \nabla \xi + (\overset{\circ}{\Gamma} - \Gamma)\xi$$

and

$$\overset{\circ}{\nabla}{}^2 \xi = \nabla^2 \xi + 2(\overset{\circ}{\Gamma} - \Gamma)\nabla\xi + \left(\nabla(\overset{\circ}{\Gamma} - \Gamma) + (\overset{\circ}{\Gamma} - \Gamma)^2 \right) \xi$$

with $(\overset{\circ}{\Gamma} - \Gamma)$ given by the formula 2.2.1b. Thus, formally,

$$(\overset{\circ}{\Gamma} - \Gamma) = \Omega^{-1} \nabla \Omega$$
$$\nabla(\overset{\circ}{\Gamma} - \Gamma) = \Omega^{-1} \nabla^2 \Omega + (\Omega^{-1} \nabla \Omega)^2.$$

Introduce the following constants:

$$\Omega_m = \inf_S r\Omega, \qquad \Omega_M = \sup_S r\Omega \tag{2.3.1c}$$

$$\Omega_1 = \sup_S \Omega^{-2} |\nabla \Omega| \tag{2.3.1d}$$

$$\Omega_{2,p} = \left(\int_S \Omega^{-3p+2} |\nabla^2 \Omega|^p \right)^{1/p}. \tag{2.3.1e}$$

Then

$$\int_S |\nabla \xi|^p \le c(\Omega_m^{-1}, \Omega_M, \Omega_1) r^{-(k+1)p+2} \int_{S^2} (|\overset{\circ}{\nabla} \xi|^p_{\overset{\circ}{\gamma}} + |\xi|^p_{\overset{\circ}{\gamma}}) d\mu_{\overset{\circ}{\gamma}} \tag{2.3.1f}$$

where $c(\Omega_m^{-1}, \Omega_M, \Omega_1)$ is a constant that depends only on the specified quantities. Also, taking into account the classical Sobolev inequality on the standard sphere, we infer that

$$\int_S |\nabla^2 \xi|^p \leq c(\Omega_m^{-1}, \Omega_M, \Omega_1) r^{-(k+2)p+2} \int_{S^2} \Big(|\overset{\circ}{\nabla}{}^2 \xi|^p_{\overset{\circ}{\gamma}} + |\overset{\circ}{\nabla} \xi|^p_{\overset{\circ}{\gamma}} + |\xi|^p_{\overset{\circ}{\gamma}} \Big)$$
$$+ c r^{-(k+2)p+2} (\Omega_m^{-1}, \Omega_M, \Omega_1) \Big(\Omega_{2,p} \sup_S |\xi|_{\overset{\circ}{\gamma}} \Big)^p \tag{2.3.1g}$$
$$\leq c(\Omega_m^{-1}, \Omega_M, \Omega_1, \Omega_{2,p}) r^{-(k+2)p+2} \int_{S^2} \Big(|\overset{\circ}{\nabla}{}^2 \xi|^p_{\overset{\circ}{\gamma}} + |\overset{\circ}{\nabla} \xi|^p_{\overset{\circ}{\gamma}} + |\xi|^p_{\overset{\circ}{\gamma}} \Big).$$

Therefore, if ξ is a solution of the Hodge system $\mathbf{H}_{k+1}$ for $k = 0, 1$ and with $h = 0$, we infer from the first-order estimates of Lemma 2.3.1, 2.3.1b, and 2.3.1f that

$$\int_S (|\nabla \xi|^p + r^{-p} |\xi|^p) \leq c(\Omega_m^{-1}, \Omega_M, \Omega_1, p) \int_S (|f|^p + |g|^p). \tag{2.3.1h}$$

Similarly, using the second-order estimates of Lemma 2.3.1,

$$\int_S |\nabla^2 \xi|^p \leq c(\Omega_m^{-1}, \Omega_M, \Omega_1, \Omega_{2,p}, p) \int_S (|\nabla f|^p + r^{-p}|f|^p + |\nabla g|^p + r^{-p}|g|^p). \tag{2.3.1i}$$

The estimates 2.3.1h, 2.3.1i provide us with the desired estimates, as long as we can control the constants $\Omega_m^{-1}, \Omega_M, \Omega_1, \Omega_{2,p}$. This can be done in view of the following form of the uniformization theorem:

Lemma 2.3.2 *Let S be a compact 2-dimensional surface with $\chi(S) = 2$. There exists a conformal transformation of the metric $\overset{\circ}{\gamma} = \Omega^2 \gamma$ such that that $\overset{\circ}{K} = 1$. Moreover the conformal factor Ω can be chosen such that the quantities Ω_m^{-1}, Ω_M, Ω_1, and $\Omega_{2,p}$ depend only on*

$$\sup r^2 |K|, \qquad \frac{diam(S)}{r}$$

or, if the condition 2.0.1a is valid, only on k_m^{-1} and k_M.

As a consequence, we have the following:

Proposition 2.3.1 *Assume that S is a compact 2-surface verifying the condition 2.0.1a. Let ξ be a solution of either* $\mathbf{H}_1$ *or* $\mathbf{H}_2$:

(i.) First Derivatives Estimates in L^p. There exists a constant c that depends only on ${k_m}^{-1}$, k_M and p such that, for all $2 \leq p < \infty$,

$$\int_S (|\nabla \xi|^p + r^{-p} |\xi|^p) \leq c \int_S (|f|^p + |g|^p).$$

(ii.) Second Derivatives Estimates in L^p. There exists a constant c that depends only on ${k_m}^{-1}$, k_M and p such that, for all $2 \leq p < \infty$

$$\int_S |\nabla^2 \xi|^p \leq c \int_S (|\nabla f|^p + r^{-p}|f|^p + |\nabla g|^p + r^{-p}|g|^p).$$

As an immediate corollary we obtain the following:

Corollary 2.3.1.1 *Assume that the Gauss curvature of the surface S satisfies 2.0.1a and let ϕ be a solution to the Poisson equation on S, namely* $\mathbf{H_0}$. *Then there exists a constant c that depends only on ${k_m}^{-1}$, k_M, p such that*

$$|\nabla^2 \phi|_{L^p} + r^{-1}|\nabla \phi|_{L^p} + r^{-2}|\phi - \bar{\phi}|_{L^p} \leq c|f|_{L^p}$$
$$|\nabla^3 \phi|_{L^p} \leq c(|\nabla f|_{L^p} + r^{-1}|f|_{L^p}).$$

One important consequence of this Corollary, which we will make use of later in the study of the regularity properties of the optical function, concerns the space of distributions $W^{-2,p}(S)$. These are defined to be linear bounded functionals G on the space of test functions $C^\infty(S)$ with the property that there exists a constant c, such that the following inequality holds, independent of $\phi \in C^\infty(S)$:

$$|\langle G, \phi \rangle| \leq c \left(|\nabla^2 \phi|_{L^q(S)} + r^{-1}|\nabla \phi|_{L^q(S)} + r^{-2}|\phi|_{L^q(S)} \right) \tag{2.3.2}$$

where q is the number conjugate to p, $\frac{1}{p} + \frac{1}{q} = 1$. We define $|G|_{W^{-2,p}(S)}$ to be the smallest constant c with this property.

Corollary 2.3.1.2 *We consider solutions, in the sense of distributions, of the equation*

$$\Delta u = G. \tag{2.3.3}$$

In other words, $\langle u, \Delta \psi \rangle = \langle G, \psi \rangle$ for every $\psi \in C^\infty(S)$. Then, if $G \in W^{-2,p}(S)$ with $\int_S G = 0$, the solution u is in fact in $L^p(S)$ and, if we denote by $\bar{u}$ its mean,

$$|u - \bar{u}|_{L^p} \leq c|G|_{W^{-2,p}}$$

with c depending only on ${k_m}^{-1}$ and k_M.

To prove the corollary we need to show that

$$|\langle u, \phi \rangle| \leq |G|_{W^{-2,p}} |\phi|_{L^q}$$

for any test function ϕ with vanishing mean. Let ψ be the solution with vanishing mean of the equation $\Delta \psi = \phi$. According to the previous corollary, 2.3.1.1,

$$|\nabla^2 \psi|_{L^q} + r^{-1}|\nabla \psi|_{L^q} + r^{-2}|\psi|_{L^q} \leq c|\phi|_{L^q}.$$

Therefore $\langle u, \phi\rangle = \langle u, \Delta\psi\rangle = \langle G, \psi\rangle$ and

$$|\langle u, \phi\rangle| \leq |G|_{W^{-2,p}(S)} \left(|\nabla^2\psi|_{L^q(S)} + r^{-1}|\nabla\psi|_{L^q(S)} + r^{-2}|\psi|_{L^q(S)}\right)$$
$$\leq c|G|_{W^{-2,p}(S)}|\phi|_{L^q(S)},$$

which proves the corollary.

2.4. Proof of the Uniformization Theorem

In this paragraph we will provide a proof of the statement of the uniformization theorem given in Lemma 2.3.2. By a simple change of scale we can assume that the volume of S is equal to 4π, in other words $r = 1$. We start with the remark that if $\tilde{K}$ is the Gauss curvature of the transformed metric $\tilde{\gamma} = \Omega^2\gamma$, then, setting $\Omega = e^u$, we derive the equation

$$\tilde{K}e^{2u} = K - \Delta u \tag{2.4.1}$$

where K is the Gauss curvature of the original metric γ and Δ is the corresponding Laplacian. Therefore we aim to solve the following equation:

$$\Delta u + e^{2u} = K. \tag{2.4.2a}$$

Remark 7 *The first difficulty one encounters in trying to construct an appropriate conformal factor u, whose size is under control, is the lack of uniqueness for the equation 2.4.2a. This is a reflection of the richness of the conformal group of the standard sphere. Indeed, if* $(S^2, \overset{\circ}{\gamma})$ *is the standard sphere, with Gauss curvature* $\overset{\circ}{K} = 1$, *there exists a 6-parameter family of conformal diffeomorphisms,* $\Phi : S^2 \longrightarrow S^2$, *with* $\Phi_*\gamma = e^{2u}\ \overset{\circ}{\gamma}$ *and* u *a solution of the equation*

$$\Delta u + e^{2u} = 1 \tag{2.4.2b}$$

on $(S^2\ \overset{\circ}{\gamma})$. *These conformal diffeomorphisms can be described as follows: Consider the stereographic projection from the North Pole* $N(0,0,1)$ *of the sphere* $x^2 + y^2 + z^2 = 1$ *to the plane* $z = -1$, *given in terms of the polar angles* θ, ϕ *and distance r, from the foot of the projection to the origin of the 2-plane, by the formula*

$$(\theta, \phi) \longrightarrow (r = 2\cot(\theta/2),\ \phi).$$

If we denote the Euclidean metric of the 2-plane by $\mathbf{e} = dx^2 + dy^2 = |dz|^2$, *where* $z = x + iy$, *we have* $\mathbf{e} = e^{2w}\ \overset{\circ}{\gamma}$ *and* $\overset{\circ}{\gamma} = e^{2v}\mathbf{e}$, *where* $w = -2\log(\sin\theta/2)$, $v = -\log(1 + r^2/4)$. *Thus, the standard metric on* $S^2 \setminus \{N\}$ *is conformally equivalent to the Euclidean metric of the plane*

$$\overset{\circ}{\gamma} = \left(1 + \frac{1}{4}|z|^2\right)^{-2} |dz|^2.$$

Now the conformal transformation of the plane, regular at infinity, is given by the group $SL(2, \mathbb{C})$, that is, the group of the holomorphic transformations

$$f(z) = \frac{az + b}{cz + d}$$

for any complex numbers a, b, c, d with $ad - bc = 1$. Thus the stereographic projection allows us to identify the group of conformal diffeomorphisms of the sphere $(S^2, \overset{\circ}{\gamma})$ with the group $SL(2, \mathbb{C})$.

The lack of uniqueness of equation 2.4.2b, and thus of 2.4.2a, shows that we cannot have, in general, a priori estimates for solutions to 2.4.2b.[4] To circumvent this difficulty we construct a solution as follows:

Step 1: We consider two points, O and P, that maximize the geodesic distance on S; in other words, $d_\gamma(O, P) = \text{diam}(S)$. We then construct a solution to the equation

$$\Delta_\gamma w - K = -4\pi\delta_P \tag{2.4.3a}$$

where δ_P is the Dirac measure on S of pole P. Given such a solution, we remark that the metric $\tilde{\gamma} = e^{2w}\gamma$, defined on $S \setminus P$, is flat; indeed, in view of 2.4.1, its Gauss curvature is identically zero.

Step 2: We define $\tilde{d}$ to be the distance function on S, relative to the metric $\tilde{\gamma}$, as measured from the point O and let $\tilde{u} = -\log(1 + \frac{\tilde{d}^2}{4})$. Using the function w constructed in the previous step, we deduce that the metric $e^{2\tilde{u}}\tilde{\gamma}$, defined on $S \setminus \{P\}$, has Gauss curvature equal to 1. We then set

$$u = w + \tilde{u} \tag{2.4.3b}$$

and check that u extends, as a bounded function, to the whole surface S. Moreover, there exist constants c_1, c_2 depending only on ${k_m}^{-1}$ and k_M such that the conformal factor $\Omega = e^{-u}$ verifies the estimate

$$c_1 \leq \Omega \leq c_2 \tag{2.4.3c}$$

or, with the notation introduced in 2.3.1d, Ω_m^{-1}, Ω_M are bounded.

[4] Indeed, consider the family of solutions u_a on the standard sphere, which corresponds to the scaling transformations $f(z) = az$, and remark that their values, at a given point, blow up logarithmically with a.

Step 3: We enhance the regularity of the conformal factor u by using the L^p elliptic theory for the equation 2.4.2a, transformed to the standard sphere. More precisely, we consider the equation

$$\Delta_{\overset{\circ}{\gamma}} u = -1 + Ke^{-2u}. \tag{2.4.3d}$$

In view of the boundedness of u we remark that the right-hand side of the equation belongs to L^p for every $1 < p < \infty$. We are thus in a position to apply the Calderon Zygmund L^p estimates of Lemma 2.3.1 and deduce

$$\int_{S^2} \left(|\overset{\circ}{\nabla}\overset{\circ}{\nabla} u|^p_{\overset{\circ}{\gamma}} + |\overset{\circ}{\nabla} u|^p_{\overset{\circ}{\gamma}} + |u|^p_{\overset{\circ}{\gamma}} \right) \leq c(\, k_M \,, \Omega_m).$$

Transforming back to the metric γ, we derive the desired estimates on the conformal factor Ω stated in Lemma 2.2.6d.

The main idea in the construction of the solution w to 2.4.3a is to use the logarithm of the geodesic distance function, as measured from the point P relative to the given metric γ, as a local fundamental solution for the Laplacian on S. Denoting the distance function by λ, we recall (see 2.1.1a) that in a neighborhood of P the metric γ takes the form

$$ds^2 = d\lambda^2 + r^2(\lambda, \theta)d\theta^2$$

where $r^2(\lambda, \theta)d\theta^2$ represents the element of arc length for the great circles $\lambda =$ constant. By a straightforward computation we find that, for any $\lambda \neq 0$,

$$\Delta(\log \lambda) = \frac{h'}{\lambda} \tag{2.4.4a}$$

$$h' = h - \frac{1}{\lambda}$$

with $h = \frac{1}{r}\frac{\partial r}{\partial \lambda}$ the geodesic curvature of the great circles (see 2.1.1b). Remark that the term $1/\lambda$, which appears in the definition of h', corresponds to the value of h in Euclidean geometry. Now, recall that at $\lambda = 0$, $r(0) = 0$, $\partial r/\partial\lambda = 1$, and by virtue of the equation 2.1.1d, $\partial^2 r/\partial\lambda^2 = 0$. Therefore,

$$r(\lambda) = O(\lambda) \tag{2.4.4b}$$

$$\frac{r}{\lambda} = 1 + O(\lambda^2) \tag{2.4.4c}$$

$$\partial r/\partial\lambda = 1 + O(\lambda^2) \tag{2.4.4d}$$

as λ tends to 0.

Therefore, we infer that

$$\begin{aligned} h &= \tfrac{1}{\lambda} + O(\lambda) \qquad \text{as } \lambda \to 0 \\ \not\Delta(\log \lambda) &= O(1) \qquad\quad \text{as } \lambda \to 0,\ \lambda \neq 0. \end{aligned} \tag{2.4.4e}$$

On the other hand, let ϕ be a smooth cut-off function defined as follows,

$$\begin{cases} \phi = 1 & \text{on } D_\epsilon(P) \\ \phi = 0 & \text{on } S \setminus D_{2\epsilon}(P) \end{cases}$$

where $D_\epsilon(P)$ denotes the geodesic ball centered at P. Choosing ϵ sufficiently small, such that 2ϵ is not greater than the radius of injectivity $i(S)$, and $\eta < \epsilon$, we calculate, taking into account 2.4.4c,

$$\begin{aligned} \int_{S \setminus D_\eta} \not\Delta_\gamma(\phi \log \lambda) d\mu_\gamma &= -\int_0^{2\pi} \frac{d}{d\lambda}(\log \lambda) r d\theta \Big|_{\lambda=\eta} \\ &= -\int_0^{2\pi} \frac{r}{\lambda} d\theta \\ &= -2\pi + O(\eta^2) \quad \text{as } \eta \to 0. \end{aligned}$$

Thus, letting $\eta \to 0$, we infer that the total mass of $\not\Delta(\phi \log \lambda)$ is equal to -2π.

$$\int_S \not\Delta(\phi \log \lambda) = -2\pi \tag{2.4.5a}$$

We now set

$$f = \begin{cases} K + 2\not\Delta_\gamma(\phi \log \lambda) & \text{on } S \setminus P \\ = 0 & \text{at } P \end{cases}$$

and solve the equation

$$\not\Delta_\gamma v = f \tag{2.4.5b}$$

with the normalization condition $v(P) = 0$. This can be done since, in view of 2.4.4e, 2.4.5a, and the Gauss-Bonnet formula, we have $f \in L^\infty(S)$ and $\int_S f = 0$. We then define

$$w = v - 2\phi \log \lambda \tag{2.4.5c}$$

and remark that w is the desired solution of 2.4.3a.

To complete step 2, it remains to show that the function $u = w + \tilde{u}$, as we defined it, can be extended as an L^∞ function to all of S. To achieve this, we observe that, in a neighborhood of P, $e^u = e^w e^{\tilde{u}} = O((\tilde{d}\lambda)^{-2})$. It thus suffices to show that there exist constants c_1, c_2 such that

$$c_1 \leq \lambda\tilde{d} \leq c_2 \tag{2.4.5d}$$

in any neighborhood of P. To prove the upper bound for $\lambda\tilde{d}$, we consider Q to be an arbitrary point, different from P, in the interior of the geodesic disk $D_\epsilon(P)$ and let $\tilde{\Gamma}$ be the geodesic line, relative to the flat metric $\tilde{\gamma}$, that minimizes the distance between the points O and Q. Let Q_1 be the point where $\tilde{\Gamma}$ first meets the boundary of the disk $D_\epsilon(P)$ and let Q_2 be the point where the geodesic ray PQ meets the same disk.[5] Let $\tilde{d}(Q) = \tilde{d}(O, Q)$. Clearly, $\tilde{d}(Q) = \tilde{d}(O, Q_1) + \tilde{d}(Q_1, Q)$ and

$$\tilde{d}(O, Q_1) \leq \operatorname{diam}(S) \exp\left(\sup_{S\setminus D_\epsilon(P)} w\right) \leq c\frac{\operatorname{diam}(S)}{\epsilon^2}. \tag{2.4.5e}$$

Thus, choosing ϵ to be half the radius of injectivity, we infer that $\tilde{d}(O, Q_1)$ is bounded by a constant c, which depends only on the diameter of S, k_m, and ${k_m}^{-1}$. It remains to estimate $\tilde{d}(Q_1, Q)$. We accomplish this as follows:

$$\begin{aligned}\tilde{d}(Q_1, Q) &\leq \tilde{d}(Q, Q_2) + \tilde{d}(Q_2, Q_1)\\ &\leq \widetilde{\operatorname{Ray}}(Q, Q_2) + \widetilde{\operatorname{Arc}}(Q_2, Q_1)\end{aligned}$$

where $\widetilde{\operatorname{Ray}}(Q, Q_2)$ is the distance along the ray initiating at P and $\widetilde{\operatorname{Arc}}(Q_2, Q_1)$ is the length of the segment of the circle $\lambda = \epsilon$ measured relative to the metric $\tilde{\gamma}$. In view of its definition the metric $\tilde{\gamma}$ can be expressed relative to λ, θ by

$$\widetilde{ds}^2 = e^{2w(\lambda,\theta)}\left(d\lambda^2 + r^2(\lambda, \theta)d\theta^2\right).$$

Therefore,

$$\begin{aligned}\widetilde{\operatorname{Arc}}(Q_2, Q_1) &\leq \int_0^{2\pi} e^{w(\epsilon,\theta)} r(\epsilon, \theta)d\theta \leq c\\ \widetilde{\operatorname{Ray}}(Q, Q_2) &\leq \int_{\lambda(Q)}^{\epsilon} e^{w(\lambda',\theta)}d\lambda'\\ &\leq c\int_{\lambda(Q)}^{\epsilon} (\lambda')^{-2}d\lambda' \leq c\lambda(Q)^{-1}.\end{aligned}$$

[5] The ray is geodesic relative to the original metric γ

In other words,

$$\begin{aligned}\tilde{d}(Q) = \tilde{d}(O, Q) &\leq \tilde{d}(Q_1) + c\lambda(Q)^{-1} \\ &\leq c_1\lambda(Q)^{-1},\end{aligned}$$

which proves the required upper bound. The lower bound is easier to obtain. Indeed, parameterizing the segment of $\tilde{\Gamma}$ between Q_1 and Q by $\theta = \theta(\lambda)$, we write

$$\begin{aligned}\tilde{d}(Q_1, Q) &= \int_{\lambda(Q)}^{\epsilon} e^{w}\sqrt{1 + r^2(\frac{d\theta}{d\lambda})^2 d\lambda'} \\ &\geq c\int_{\lambda(Q)}^{\epsilon} \frac{1}{\lambda'^2} \geq c_2\frac{1}{\lambda(Q)}.\end{aligned}$$

Therefore, $\tilde{d}(Q)\lambda(Q) \geq c_2$ for all Q in $D_\epsilon(P)$, as desired.

CHAPTER 3

General Results in 3-D Geometry

3.1. Preliminaries

Throughout this chapter we assume (Σ, g) to be a 3-dimensional Riemannian manifold diffeomorphic to R^3 on which there exists a generalized *radial* function. By this we understand a differentiable, real, function u defined on all points of Σ, outside a center point O, which takes values onto the interval (u_0, ∞) and verifies the following assumptions:

1. u has no critical points.
2. The level surfaces of u, to be denoted by S_u, are diffeomorphic to the 2-dimensional spheres S^2.

   Also, denoting by $\text{Int}S_u$ the component of $\Sigma \setminus S_u$ that contains O, and by $\text{Ext}S_u$ the other one, we require that
3. $\bigcap_{u\in(u_0,\infty)}\text{Int}S_u = \{O\}$
4. $\bigcup_{u\in(u_0,\infty)}\text{Int}S_u = \Sigma$

The canonical form of the metric g, relative to the foliation induced by the level surfaces of u, is given by the formula

$$ds^2 = a^2du^2 + \sum_{A,B=1,2} \gamma_{AB}d\phi^A d\phi^B \tag{3.1.1a}$$

where $a = |\nabla u|^{-1}$, measures the normal separation of the surfaces S_u, and will be called the lapse of the foliation, while γ is the metric induced by g on the spheres S_u.

Let $\mathrm{N} = a^{-1}\partial_u$ be the unit exterior normal to the foliation and define $\Pi = \Pi(u)$ to be the projection operator from the tangent space of Σ to the tangent space of S_u. More precisely, $\Pi(u)$ is a 2-tensor, expressed relative to arbitrary coordinates on Σ, by the formula

$$\Pi^i_j = \delta^i_j - N^i N_j \tag{3.1.1b}$$

where δ^i_j is the Kronecker unit tensor.

Remark that Π satisfies $\Pi_i^m \Pi_m^j = \Pi_i^j$ and the induced metric γ is the restriction of $\Pi_i^m \Pi_j^n g_{mn}$ to the space of vectors tangent to S_u. The projection operator can be extended to arbitrary covariant tensors $U_{i_1 \ldots i_M}$ on Σ by forming the contractions

$$\Pi_{j_1}^{i_1} \ldots \Pi_{j_M}^{i_M} U_{i_1 \ldots i_M}.$$

The new tensor is tangent to S_u. In what follows we will consider only tensors, defined on Σ, that are tangent, at any point p, to the sphere S_u passing through p.

We denote by $\nabla\!\!\!\!/$ the intrinsic covariant differentiation on S_u and by $K = K(u)$ its Gauss curvature. Given a covariant tensorfield U, of rank M, defined on all of Σ and tangent to S_u at any point in S_u, $\nabla\!\!\!\!/ U$ can be regarded as an $M+1$ covariant tensorfield defined on all of Σ, obtained by projecting to S_u the Σ-covariant derivative of U. More precisely,

$$\nabla\!\!\!\!/_{i_{M+1}} U_{i_1 \ldots i_M} = \Pi_{i_1}^{j_1} \ldots \Pi_{i_{M+1}}^{j_{M+1}} \nabla_{j_{M+1}} U_{j_1 \ldots j_M}.$$

Also, we denote by $\nabla\!\!\!\!/_N U$ the projection of $\nabla_N U$ to S_u:

$$\nabla\!\!\!\!/_N U_{i_1 \ldots i_M} = \Pi_{i_1}^{j_1} \ldots \Pi_{i_M}^{j_M} \nabla_N U_{j_1 \ldots j_M}.$$

The second fundamental form to the foliation is given by the formula

$$\theta_{ij} = \Pi_i^m \Pi_j^n \nabla_m N_n = \nabla_i N_j - N_i \nabla_N N_j$$

or

$$\theta_{ij} - a^{-1} N_i \nabla\!\!\!\!/_j a = \nabla_i N_j \tag{3.1.1c}$$

and, in particular,

$$\mathrm{tr}\theta = \mathrm{div} N. \tag{3.1.1d}$$

If $N, \{e_A\}_{A=1,2}$ is an orthonormal frame in Σ, we have

$$\begin{aligned} \nabla_N e_A &= \nabla\!\!\!\!/_N e_A + a^{-1}(\nabla\!\!\!\!/_A a) N \\ \nabla_A N &= \theta_{AB} e_B \\ \nabla_B e_A &= \nabla\!\!\!\!/_B e_A - \theta_{AB} N \\ \nabla_N N &= -a^{-1} \nabla\!\!\!\!/_A a e_A. \end{aligned} \tag{3.1.1e}$$

A special local orthonormal frame on Σ [1] can be obtained by propagating a local orthonormal frame $(e_A)_{A=1,2}$ on a given sphere S_u such that $\nabla_N e_A = f_A N$ and $\langle e_A, N \rangle = 0$. In this case $\nabla\!\!\!\!/_N e_A = 0$ and $f_A = a^{-1}(\nabla\!\!\!\!/_A a)$.

[1] This frame is sometimes called Fermi propagated.

Relative to arbitrary coordinates on S_u, the second fundamental form is given by

$$\theta_{AB} = (2a)^{-1}\partial_u\gamma_{AB}.$$

Therefore, $\mathrm{tr}\theta = (2a)^{-1}\gamma^{AB}\partial_u\gamma_{AB} = \frac{1}{2a}\partial_u \log\det\gamma$ where $\det\gamma$ is the determinant of the metric γ relative to the given coordinates.
Consequently, if we denote by $A(u)$ the total area of S_u, we have

$$\frac{d}{du}A(u) = \int_{S_u} a\mathrm{tr}\theta d\mu_\gamma = 4\pi r^2\overline{a\mathrm{tr}\theta} \tag{3.1.1f}$$

where $d\mu_\gamma$ is the area element of S_u. Whenever f is a scalar function, $\overline{f}(u)$ denotes the average of f on S_u.

More general, for an arbitrary scalar function f,

$$\frac{d}{du}\int_{S_u} f d\mu_\gamma = \int_{S_u} a(\nabla_N f + \mathrm{tr}\theta f)d\mu_\gamma. \tag{3.1.1g}$$

In what follows we denote r the function of u defined according to

$$A = 4\pi r^2 \tag{3.1.1h}$$

and denote by λ the derivative of r relative to u. In view of 3.1.1f

$$\lambda := \frac{dr}{du} = \frac{r}{2}\overline{a\mathrm{tr}\theta} \tag{3.1.1i}$$

$$\nabla_N r = \frac{r}{2a}\overline{a\mathrm{tr}\theta} = a^{-1}\lambda. \tag{3.1.1j}$$

We also record the following consequence of 3.1.1g

$$\frac{d}{du}\bar{f} = \overline{a\nabla_N f} + r^{-1}\overline{a\kappa(f-\bar{f})} \tag{3.1.1k}$$

where

$$\kappa = ra^{-1}(a\mathrm{tr}\theta - \overline{a\mathrm{tr}\theta}). \tag{3.1.1l}$$

Given a continuous function f on Σ, we have this well-known formula:

Coarea Formula

$$\int_\Sigma f d\mu_g = \int_{u_0}^{\infty}\left(\int_{S_u} a f d\mu_\gamma\right)du$$

($d\mu_g$ is the volume element on Σ).

The second fundamental form θ, the lapse function a, and the Gauss curvature K of the sphere S_u are connected to the Ricci curvature R_{ij} of the ambient space Σ according to these formulas:

The Structure Equations of the u-Foliation

$$R_{NN} = N^i N^j R_{ij} = -a^{-1} \Delta a - \nabla_N \mathrm{tr}\theta - |\theta|^2 \tag{3.1.2a}$$

$$\Pi_i^m N^j R_{mj} = \nabla^l \theta_{il} - \nabla_i \mathrm{tr}\theta \tag{3.1.2b}$$

$$\Pi_i^m \Pi_j^n R_{mn} = -a^{-1} \nabla_i \nabla_j a - \nabla_N \theta_{ij} - \theta_i^m \theta_{mj} - K\gamma_{ij}. \tag{3.1.2c}$$

Taking the trace of 3.1.2c we derive, in view of 3.1.2a,

$$R - 2R_{NN} = 2K - (\mathrm{tr}\theta)^2 + |\theta|^2. \tag{3.1.2d}$$

We introduce the following fundamental constants associated to the u-foliation:

The Fundamental Constants of the u-Foliation

$$\begin{aligned} a_m &= \inf_u a_m(u) \ ; \ a_M = \sup_u a_M(u) \\ h_m &= \inf_u h_m(u) \ ; \ h_M = \sup_u h_M(u) \\ k_m &= \inf_u k_m(u) \ ; \ k_M = \sup_u k_M(u) \\ h &= \sup_u h(u) \ ; \ \varsigma = \sup_u \varsigma(u) \\ \kappa &= \sup_u \kappa(u) \end{aligned} \tag{3.1.3}$$

where

$$\begin{aligned} a_m(u) &= \inf_{S_u} a \ ; \ a_M(u) = \sup_{S_u} a \\ h_m(u) &= \sup_{S_u} r\mathrm{tr}\theta \ ; \ h_M(u) = \sup_{S_u} r\mathrm{tr}\theta \\ k_m(u) &= \inf_{S_u} r^2 K \ ; \ k_M(u) = \sup_{S_u} r^2 K \\ h(u) &= \sup_{S_u} |r\mathrm{tr}\theta| \ ; \ \varsigma(u) = \sup_{S_u} r|\hat{\theta}| \\ \kappa(u) &= \sup_{S_u} |\kappa| \end{aligned}$$

where $\hat{\theta}$ is the traceless part of θ. We defined κ in 3.1.11.

3.2. Sobolev and Poincaré Inequalities

In this section we consider the constants

$$I = \sup_u I(u), \qquad \Lambda = \sup_u \Lambda(u) \tag{3.2.1}$$

where $I(u)$ is the isoperimetric constant of S_u (see definition on page 32) and $\Lambda(u)$ is the dimensionless number

$$\Lambda(u) = \frac{1}{r^2 \lambda_1(u)} \tag{3.2.2}$$

with $\lambda_1(u)$ as the first eigenvalue of $-\Delta$ on S_u.

We recall that, in general, the isoperimetric constant $I(u)$ and the dimensionless number $\Lambda(u)$ can be estimated in terms of a continuous increasing function of the variables

$$\sup_{S_u} r^2|K| \quad \text{and} \quad \frac{\text{diam}(S_u)}{r}$$

and, if $k_m(u) > 0$, they can in fact be estimated by a continuous increasing function depending only on the fundamental constants $k_m(u)^{-1}$ and $k_M(u)$. On each surface S_u we have

The Isoperimetric Inequality

$$\int_{S_u} (\Phi - \overline{\Phi})^2 \leq I(u)\Big(\int |\nabla\!\!\!/ \Phi| d\mu_\gamma\Big)^2$$

where

$$\overline{\Phi} = \frac{1}{A(u)} \int_{S_u} \Phi d\mu_\gamma$$

is the average of Φ on the sphere S_u.

Poincaré Inequality

$$\int_{S_u} (\Phi - \bar{\Phi})^2 \leq \frac{1}{\lambda(u)} \int |\nabla\!\!\!/ \Phi|^2.$$

This holds for any sufficiently smooth scalar function ϕ.

Proposition 3.2.1 *Assume that the foliation given by the radial function u has positive mean curvature, that is,*

$$\text{tr}\theta \geq 0,$$

and that F is an arbitrary tensor on Σ, tangent to S_u at every point. Then

(i.) There exists a constant $c_{\text{Sob}} = c(I, \frac{a_M}{a_m})$ depending only on I and the ratio $\frac{a_M}{a_m}$ such that

$$\Big(\int_\Sigma r^6 |F|^6\Big)^{1/6} \leq c_{\text{Sob}} \Big(\int_\Sigma |F|^2 + r^2 |\nabla F|^2\Big)^{1/2}$$

provided that the right-hand side is finite.

(ii.) Let $\tau_- = (1+u^2)^{1/2}$. *There exists a constant* $\underline{c}_{\mathbf{Sob}} = c(I, a_M, a_m)$ *such that*

$$\left(\int_\Sigma r^4\tau_-^2|F|^6\right)^{1/6} \leq \underline{c}_{\mathbf{Sob}}\left(\int_\Sigma |F|^2 + r^2|\nabla\!\!\!/ F|^2 + \tau_-^2|\nabla\!\!\!/_N F|^2\right)^{1/2}$$

provided that the right-hand side is finite.

Remark 1 *We can replace* Σ *in both inequalities of the proposition by either* Int S_{u_1} *or* Ext S_{u_1}, *where* $u_1 \in (u_0, \infty)$. *Moreover, both inequalities remain true on* Ext S_{u_1} *under the weaker assumption that* $\mathrm{tr}\theta \geq 0$ *is satisfied only there. We also remark that if* u_1 *is positive, the constant* $\underline{c}_{\mathbf{Sob}}$ *depends only on* I *and on the ratio* $\frac{a_M}{a_m}$. *Finally, we remark that the condition* $\mathrm{tr}\theta \geq 0$ *can be completely dropped if we allow the constants* $c_{\mathbf{Sob}}$, $\underline{c}_{\mathbf{Sob}}$ *to depend also on* h.

Remark 2 *Remark that in the region of* Σ *where* $r \geq 2r_0$ *we have* $\tau_- \geq \frac{r}{a_M h}$. *Indeed, in view of 3.1.1i, this follows immediately from the formula*

$$r(u) - r_0 = \int_0^u \frac{dr}{du}du \leq \frac{1}{2}a_M hu.$$

As a consequence of this fact we infer that, everywhere on Σ,

$$\frac{r}{\tau_-} \leq c(h, a_M)r_0. \tag{3.2.3}$$

Proof 3.2.1 *To prove the first part of the inequality, we take* $\Phi = |F|^3$ *in the isoperimetric inequality applied to any sphere* S_u. *Thus, using the Hölder inequalities repeatedly, we obtain*

$$\int_{S_u} |F|^6 \leq c(I)\left(r^{-2}\int_{S_u} |F|^4\right)\left(\int_{S_u} |F|^2 + r^2|\nabla\!\!\!/ F|^2\right). \tag{3.2.4a}$$

Multiplying the inequality by $r^6 a_M(u)$, *and then integrating with respect to* u, *we infer, in view of the coarea formula,*

$$\int_\Sigma r^6|F|^6 \leq \frac{a_M}{a_m}c(I)\sup_u\left(r^4\int_{S_u} |F|^4\right)\left(\int_\Sigma |F|^2 + r^2|\nabla\!\!\!/ F|^2\right). \tag{3.2.4b}$$

On the other hand,

$$\begin{aligned}\int_{S_u} r^4|F|^4 &= -\int_{ExtS_u} div(r^4|F|^4 N)\\ &= -\int_{ExtS_u} \left[(divN)r^4|F|^4 + 4r^3(\nabla_N r|F|^4 + r|F|^2 F\cdot\nabla\!\!\!/_N F)\right].\end{aligned}$$

Therefore, in view of the equations 3.1.1d and 3.1.1j, we infer that

$$\int_{S_u} r^4|F|^4 = -\int_{ExtS_u} r^4|F|^4\left(\mathrm{tr}\theta + \frac{4}{ar}\lambda\right) - 4\int_{ExtS_u} r^4|F|^2F\cdot\nabla_N F.$$

Making use of the assumption $\mathrm{tr}\theta \geq 0$, *we obtain,*

$$\begin{aligned}\int_{S_u} r^4|F|^4 &\leq \int_{ExtS_u} r^4|F|^2F\cdot\nabla_N F\\ &\leq 4\left(\int_{ExtS_u} r^6|F|^6\right)^{1/2}\left(\int_{ExtS_u} r^2|\nabla_N F|^2\right)^{1/2}.\end{aligned} \tag{3.2.4c}$$

Finally, substituting this last inequality in 3.2.4b, we derive

$$\int_{S_u} r^6|F|^6 \leq c\left(I,\frac{a_M}{a_m}\right)\left(\int_\Sigma r^2|\nabla_N F|^2\right)\left(\int_\Sigma |F|^2+r^2|\nabla F|^2\right)^2, \tag{3.2.4d}$$

which implies the desired result.

To prove the second, degenerate, Sobolev inequality, we start by multiplying 3.2.4a by $r^4\tau_-^2 a_M(u)$. *Hence,*

$$\int_\Sigma r^4\tau_-^2|F|^6 \leq \frac{a_M}{a_m}c(I)\sup_u\left(\int_{S_u} r^2\tau_-^2|F|^4\right)\left(\int_\Sigma |F|^2+r^2|\nabla F|^2\right). \tag{3.2.5}$$

On the other hand we have

$$\begin{aligned}\int_{S_u} r^2\tau_-^2|F|^4 &= -\int_{ExtS_u} div(r^2\tau_-^2|F|^4N)\\ &= -\int_{ExtS_u}\left(\mathrm{tr}\theta + \frac{2}{ar}\lambda\right)r^2\tau_-^2|F|^4\\ &\quad -2\int_{ExtS_u} a^{-1}ur^2|F|^4 - 4\int_{ExtS_u} r^2\tau_-^2|F|^2F\cdot\nabla_N F.\end{aligned}$$

Since $\mathrm{tr}\theta \geq 0$, *we conclude that*

$$\int_{S_u} r^2\tau_-^2|F|^4 \leq -2\int_{ExtS_u} a^{-1}ur^2|F|^4 - 4\int_{ExtS_u} r^2\tau_-^2|F|^2F\cdot\nabla_N F. \tag{3.2.6}$$

Hence, in view of the fact that

$$\int_{ExtS_u} a^{-1}ur^2|F|^4 \leq c(a_m)\left(\int_{ExtS_u}|F|^2\right)^{1/2}\left(\int_{ExtS_u} r^4\tau_-^2|F|^6\right)^{1/2}$$

and

$$\int_{ExtS_u} r^2\tau_-^2|F|^2 F\nabla\!\!\!/_N F \le \left(\int_{ExtS_u} r^4\tau_-^2|F|^6\right)^{1/2}\left(\int_{ExtS_u} \tau_-^2|\nabla\!\!\!/_N F|^2\right)^{1/2}$$

we infer that

$$\int_{S_u} r^2\tau_-^2|F|^4 \le 4\left(\int_{ExtS_u} r^4\tau_-^2|F|^6\right)^{1/2}\left[\int_{ExtS_u} |F|^2+\tau_-^2|\nabla\!\!\!/_N F|^2\right]^{1/2}. \tag{3.2.7}$$

Thus, taking into account 3.2.5, we derive

$$\int_\Sigma r^4\tau_-^2|F|^6 \le c(I,a_m,a_M)\left(\int |F|^2+\tau_-^2|\nabla\!\!\!/_N F|^2\right)\left(\int |F|^2+r^2|\nabla\!\!\!/ F|^2\right)^2, \tag{3.2.8}$$

which implies the desired inequality.

In particular, we have also proved the following:

Corollary 3.2.1.1 *Assume that the foliation given by the radial function u has positive mean curvature and that F is an arbitrary tensor on Σ, tangent to S_u at every point. Then*

(i.) With the same constant $c_{\mathbf{Sob}}$ as in the first part of Proposition 3.2.1,

$$\sup_{u\in\Sigma}\left(\int_{S_u} r^4|F|^4\right)^{1/4} \le c_{\mathbf{Sob}}\left(\int_\Sigma |F|^2+r^2|\nabla F|^2\right)^{1/2}.$$

(ii.) With the same constant $\underline{c}_{\mathbf{Sob}}$ as in the second part of Proposition 3.2.1,

$$\sup_{u\in\Sigma}\left(\int_{S_u} r^2\tau_-^2|F|^4\right)^{1/4} \le \underline{c}_{\mathbf{Sob}}\left(\int_\Sigma |F|^2+r^2|\nabla\!\!\!/ F|^2+\tau_-^2|\nabla\!\!\!/_N F|^2\right)^{1/2}.$$

We remark that if instead of multiplying 3.2.4a by $r^6 a_M(u)$, we multiply by $a_M(u)$, we derive

$$\int_\Sigma |F|^6 \le \frac{a_M}{a_m}c(I)\sup_u\left(\int_{S_u}|F|^4\right)\left(\int_\Sigma \frac{1}{r^2}|F|^2+|\nabla\!\!\!/ F|^2\right), \tag{3.2.9a}$$

and proceeding precisely as in the proof of 3.2.4c, we infer that

$$\int_{S_u}|F|^4 \le 4\left(\int_{\mathrm{Ext}S_u}|F|^6\right)^{1/2}\left(\int_{\mathrm{Ext}S_u}|\nabla\!\!\!/ F|^2\right)^{1/2}. \tag{3.2.9b}$$

Therefore,

$$\int_\Sigma |F|^6 \le c\left(I, \frac{a_M}{a_m}\right)\left(\int_\Sigma |\nabla_N F|^2\right)\left(\int_\Sigma \frac{1}{r^2}|F|^2 + |\nabla F|^2\right)^2. \quad (3.2.9c)$$

The classical form of the L^6 Sobolev inequality follows from 3.2.9c together with a nondegenerate, global version of the Poincaré inequality on Σ, which we discuss in Proposition 3.2.2.

Corollary 3.2.1.2 *Under the same assumptions as in the proposition, assume that the foliation given by the radial function u has strictly positive mean curvature, that is,* $h_m > 0$. *There exists a constant* $c_{\mathbf{sob}} = c_{\mathbf{sob}}(I, \frac{a_M}{a_m}, h_m^{-1}, h_M)$, *such that*

$$\left(\int_\Sigma |F|^6\right)^{1/6} \le c_{\mathbf{sob}}\left(\int_\Sigma |\nabla F|^2\right)^{1/2}.$$

Also

$$\sup_{u\in\Sigma}\left(\int_{S_u} |F|^4\right)^{1/4} \le c_{\mathbf{sob}}\left(\int_\Sigma |\nabla F|^2\right)^{1/2}.$$

The global version of the Poincaré inequality on Σ has the following form:

Proposition 3.2.2 *Let u be an arbitrary radial function on* Σ *and introduce* $s = \int_{u_0}^u \bar{a} du$.

(i.) There exists a constant $c_{\mathbf{P}} = c_{\mathbf{P}}(a_M, a_m, \kappa, \Lambda, \sup_u(\frac{s}{r})^2)$ [2] *such that, for every* C^∞ *tensorfield F with compact support on* Σ,[3] *we have the inequality*

$$\int_\Sigma \frac{|F|^2}{r^2} \le c_{\mathbf{P}} \int_\Sigma |\nabla F|^2.$$

(ii.) Assume that the radius r_0 *corresponding to* $u = 0$ *is greater or equal to 1. There exists a constant* $\underline{c}_{\mathbf{P}} = \underline{c}_{\mathbf{P}}(a_m, a_M, \Lambda, h, \sup_u(\frac{s}{r})^2)$ *such that for every* C^∞ *tensorfield F with compact support on* Σ

$$\int_\Sigma \frac{|F|^2}{r_-^2} \le \underline{c}_{\mathbf{P}} r_0 \int_\Sigma |\nabla F|^2.$$

The proof of Corollary 3.2.1.2 follows as an immediate consequence of 3.2.9c and Proposition 3.2.2i together with the remark that by virtue of

$$\frac{dr}{ds} = \frac{dr}{du}\frac{du}{ds} = \frac{r}{2}\frac{\overline{a\mathrm{tr}\theta}}{\bar{a}} \ge \frac{h_m}{2}\frac{a_m}{a_M}$$

we can estimate $\sup_u(\frac{s}{r})^2$ in terms of the constants $(\frac{a_M}{a_m}, h_m^{-1}, h_M)$.

[2] In view of 3.1.11 we define $\kappa = \sup_\Sigma r a^{-1}|a\mathrm{tr}\theta - \overline{a\mathrm{tr}\theta}|$.
[3] For which the right-hand side of the inequality is finite.

Proof 3.2.2 *First remark that it suffices to prove the assertion for scalar functions F. In view of the Poincaré inequality on S_u we derive*

$$\int_\Sigma \frac{|F-\bar{F}|^2}{r^2} \leq c\frac{a_M}{a_m}\Lambda \int_\Sigma |\nabla\!\!\!/ F|^2 \tag{3.2.10a}$$

where $\bar{F}(u)$ is the average of F on S_u. It thus suffices to prove the following:

$$\int_\Sigma \frac{|\bar{F}|^2}{r^2} \leq c\left(a_M, a_m, \kappa, \sup_u \left(\frac{s}{r}\right)^2\right)\int_\Sigma |\nabla F|^2 \tag{3.2.10b}$$

In view of the definition of s we have

$$\begin{aligned}\int_\Sigma \frac{|\bar{F}|^2}{r^2} &= 4\pi\int_{u_0}^\infty \bar{a}\bar{F}^2 du \\ &= 4\pi\int_0^\infty \bar{F}^2 ds \leq 16\pi \int_0^\infty \left(\frac{d\bar{F}}{ds}\right)^2 s^2 ds \\ &= 16\pi \int_{u_0}^\infty \frac{s^2}{\bar{a}}\left(\frac{d\bar{F}}{du}\right)^2 du.\end{aligned}$$

On the other hand, with the help of 3.1.1g, we write

$$\frac{d}{du}\bar{F} = \overline{a\nabla_N F} + r^{-1}\overline{a\kappa(F-\bar{F})},$$

and using the fact that $(\bar{f})^2 \leq \overline{f^2}$, we infer that

$$\left(\frac{d}{du}\bar{F}\right)^2 \leq \frac{1}{4\pi r^2}\left\{\int_{S_u} a^2|\nabla\!\!\!/_N F|^2 + r^{-2}\int_{S_u} a^2\kappa^2(F-\bar{F})^2\right\},$$

and, taking 3.2.10a into account once again, the inequality 3.2.10b follows in a straightforward fashion.

To prove the degenerate form of the global Poincaré inequality, Proposition 3.2.2ii, we start by splitting the integral on the left-hand side into an integral on $\mathrm{Int}S_{2r_0}$ and one on $\mathrm{Ext}S_{2r_0}$. Now, since in $\mathrm{Ext}S_{2r_0}$[4] *we have $\tau_- \geq \frac{r}{a_M h}$, we derive, with the help of the first part of the proposition,*

$$\int_{\mathrm{Ext}S_{2r_0}} \frac{|F|^2}{\tau_-^2} \leq c_P a_m^2 h^2 \int_\Sigma |\nabla F|^2.$$

On the other hand, in view of the classical Sobolev inequality of Corollary 3.2.1.2, and using the coarea formula, we have

[4] See Remark 2 on page 58.

$$\int_{IntS_{2r_0}} \frac{|F|^2}{\tau_-^2} = \int_0^{u_1} \tau_-^{-2} du \left(\int_{S_u} a|F|^2 \right)$$
$$\leq a_M \sup_u \left(\int_{S_u} r^{-1}|F|^2 \right) \int_0^{u_1} r\tau_-^{-2} du$$
$$\leq a_M 2\pi^{1/2} c_{\mathbf{sob}}^2 \left(\int_\Sigma |\nabla F|^2 \right) \int_0^{u_1} r\tau_-^{-2} du$$

where $u_1 = u(2r_0)$.
The desired inequality is now an immediate consequence of

$$\int_0^{u_1} r\tau_-^{-2} du \leq 2r_0 \int_0^{\infty} (1+u^2)^{-1} du \leq 4\pi r_0.$$

Another important application of Proposition 3.2.1 is the following form of the global Sobolev inequalities:

Proposition 3.2.3 *Assume that the foliation given by the radial function u has positive mean curvature. Let* P_i *denote the* $\Pi_i^j R_{jk} N^k$ *component of the Ricci curvature of* Σ *and introduce the following numbers:*

$$a_1 = \sup_\Sigma (ra^{-1}|\nabla a|), \qquad r_N = \left(\int_\Sigma r^5 |P|^4 \right)^{1/4}.$$

(i.) There exists a constant c depending only on $k_m, k_M, a_M, a_m, h_m, h_M, \varsigma, a_1, r_N$ *such that*

$$\sup_\Sigma (r^{3/2}|F|) \leq \left(\int_\Sigma c\{|F|^2 + r^2|\nabla F|^2 + r^2|\nabla_N F|^2 + r^4|\nabla^2 F|^2 + r^4|\nabla\nabla_N F|^2\} \right)^{1/2}.$$

(ii.) There exists a constant $\underline{c}$ *depending only on* k_m, k_M, a_M, a_m, h_m, h_M, ς, a_1, r_N

$$\sup_\Sigma (r\tau_-^{1/2}|F|) \leq \underline{c} \left(\int_\Sigma \{|F|^2 + r^2|\nabla F|^2 + \tau_-^2|\nabla_N F|^2 + r^4|\nabla^2 F|^2 + r^2\tau_-^2|\nabla\nabla_N F|^2\} \right)^{1/2}.$$

Proof 3.2.3 *The proof of the proposition is based on Corollary 3.2.1.2, the classical Sobolev inequality on S_u,*

$$\sup_{S_u} |F| \leq c r^{-1/2} \left(\int_{S_u} |F|^4 + r^4 |\nabla\!\!\!/ F|^4 \right)^{1/4}$$

where the constant c depends only on k_m, k_M, and the following commutation lemma:

Lemma 3.2.1 *Given an arbitrary k-tensor U on Σ, tangent to S_u at any point, we have the following commutation formula:*

$$\begin{aligned}([\nabla\!\!\!/_N, \nabla\!\!\!/]U)_{i_1\ldots i_k j} &= a^{-1}\nabla\!\!\!/_j a \nabla\!\!\!/_N U_{i_1\ldots i_k} - \theta_{jm}\nabla\!\!\!/^m U_{i_1\ldots i_k} \\ &+ \sum_{q=1}^{k}(\gamma_j{}^m P_{i_q} - \gamma_{j i_q} P^m) U_{i_1\ldots m\ldots i_k} \\ &+ a^{-1}\sum_{q=1}^{k}(\theta_{i_q j}\nabla\!\!\!/^m a - \theta_j{}^m \nabla\!\!\!/_{i_q} a) U_{i_1\ \ .m\ldots i_k}.\end{aligned}$$

The simplest way to prove the lemma is to do the computations relative to an orthonormal frame $N, \{e_A\}_{A=1,2}$ and to use the frame equation 3.1.1e.

As a particular case of the lemma we have the following:

Corollary 3.2.3.1 *If ϕ is a scalar function on Σ, we have*

$$([\nabla\!\!\!/_N, \nabla\!\!\!/]\phi)_j = a^{-1}\nabla\!\!\!/_j a \nabla\!\!\!/_N \phi - \theta_{jm}\nabla\!\!\!/^m \phi.$$

The following corollary to Lemma 3.2.1 will be useful in the next chapter:

Corollary 3.2.3.2 *If ϕ is a scalar function on Σ and $\triangle\!\!\!/$ is the Laplace-Beltrami operator on S_u, we have,*

$$\begin{aligned}[\nabla_N, \triangle\!\!\!/]\phi &= -\mathrm{tr}\theta \triangle\!\!\!/ \phi - 2\hat{\theta}\cdot\nabla\!\!\!/^2\phi + 2a^{-1}\nabla\!\!\!/ a \cdot \nabla\!\!\!/ \nabla_N \phi \\ &+ a^{-1}\triangle\!\!\!/ a \nabla_N \phi - 2P\cdot\nabla\!\!\!/\phi - \nabla\!\!\!/ \mathrm{tr}\theta\cdot\nabla\!\!\!/\phi - 2a^{-1}\hat{\theta}\,\nabla\!\!\!/ a\cdot\nabla\!\!\!/\phi.\end{aligned}$$

To prove the corollary, we write

$$[\nabla_N, \triangle\!\!\!/] = \gamma^{ij}\left\{[\nabla\!\!\!/_N, \nabla\!\!\!/_i]\nabla\!\!\!/_j \phi + \nabla\!\!\!/_i[\nabla\!\!\!/_N, \nabla\!\!\!/_j]\phi\right\}.$$

In view of Lemma 3.2.1,

$$\gamma^{ij}[\nabla\!\!\!/_N, \nabla\!\!\!/_i]\nabla\!\!\!/_j\phi = a^{-1}\nabla\!\!\!/ a \cdot \nabla\!\!\!/_N \nabla\!\!\!/ \phi - \theta \cdot \nabla\!\!\!/^2\phi$$
$$- P \cdot \nabla\!\!\!/ \phi + a^{-1}(\mathrm{tr}\theta \nabla\!\!\!/ a \cdot \nabla\!\!\!/ \phi - \theta \nabla\!\!\!/ a \cdot \nabla\!\!\!/ \phi)$$

and, using Corollary 3.2.3.1 together with the Codazzi equations (see 3.1.2b), we find

$$\gamma^{ij}\nabla\!\!\!/_i[\nabla\!\!\!/_N, \nabla\!\!\!/_j]\phi = a^{-1}\nabla\!\!\!/ a \cdot \nabla\!\!\!/ \nabla\!\!\!/_N\phi - \theta \cdot \nabla\!\!\!/^2\phi$$
$$- \nabla\!\!\!/ \mathrm{tr}\theta \cdot \nabla\!\!\!/ \phi - P \cdot \nabla\!\!\!/ \phi + a^{-1} \Delta\!\!\!/ a \nabla_N \phi$$
$$- a^{-2}\nabla\!\!\!/ a \cdot \nabla\!\!\!/ a \nabla_N \phi.$$

Finally, the desired identity follows by combining the last two formulas and applying once again Corollary 3.2.3.1.

For future reference we also introduce here the truncation function

$$f(x) = h(r(x)/r_0) \tag{3.2.11a}$$

where x is an arbitrary point on Σ, $r(x)$ is the radius of the sphere u = constant passing through it, and h is the real function

$$h(s) = \begin{cases} 1 & : \ s \leq 1/2 \\ 0 & : \ s \geq 3/4. \end{cases} \tag{3.2.11b}$$

We define r_0 to be the radius corresponding to $u = 0$ and assume that $r_0 \geq 1$. Also we denote by E the region $r \geq \frac{r_0}{2}$, which we call the exterior region of Σ.

Lemma 3.2.2 *(i.) The truncation function f is supported in the set Int($S_{3r_0/4}$) while its gradient, ∇f, is supported in $E \cap$ Int($S_{3r_0/4}$). Moreover, there exists a numerical, positive constant c_0 that depends only on the fundamental constants such that*

$$\tau_- \geq c_0 r_0$$

on the support of f and

$$\sup_\Sigma |\nabla f| \leq c_0 r_0^{-1}.$$

(ii.) If in addition

$$r|\nabla a| \leq C$$
$$r|\nabla \lambda| \leq C$$

for all points in the region $Int(S_{3r_0/4})$, *then*

$$\sup_{\Sigma} |\nabla^2 f| \leq c_1 r_0^{-2}$$

with c_1 *being a constant depending only on the fundamental constants and on the bound* C.

Proof 3.2.4 *We first remark that, in view of the definition of* $\lambda = dr/du = r/2 \cdot a\mathrm{tr}\theta$ *and the bootstrap assumptions, we have* $1/2 \cdot a_m h_m \leq \inf \lambda \leq 2a_M h_M$. *Therefore,*

$$r_0 - r = \int_u^0 \lambda du \leq \frac{1}{2} a_M h_M |u|,$$

and therefore $|u| \geq 2(r_0 - r)/a_M h_M$, *or, in the interior of the 2-surface* $S_{3r_0/4}$, $|u| \geq (r_0/2)a_M h_M$. *That is,* $\tau_- \geq (r_0/2)a_M h_M$, *which proves our first assertion.*

To prove the other two assertions, we calculate

$$\begin{aligned}
\nabla_i f &= r_0^{-1} h' \lambda a^{-1} N_i \\
\nabla_i \nabla_j f &= r_0^{-2} h'' \lambda a^{-1} N_i N_j + r_0^{-1} h' (\nabla_N \lambda a^{-1} - a^{-1} \lambda \nabla_N a) N_i N_j \\
&\quad + r_0^{-1} h' \lambda a^{-1} (\theta_{ij} - a^{-1} \nabla\!\!\!/_j a N_i).
\end{aligned}$$

Thus immediately, in view of our assumptions:

$$\begin{aligned}
|\nabla f| &\leq c r_0^{-1} \\
|\nabla^2 f| &\leq c r_0^{-2},
\end{aligned}$$

which ends the proof of the lemma.

3.3. The Action of SO(3) on (Σ, g)

In this section we define the sphere at infinity of (Σ, g) and use it to define an action of SO(3) on Σ. In order to do it we require that the fundamental constants 3.1.3 are finite and in addition the integrals

$$\begin{aligned}
&\int_u^\infty r^{-1} \kappa(\tau) d\tau \\
&\int_u^\infty r^{-1} \varsigma(\tau) d\tau
\end{aligned} \tag{3.3.1}$$

exist and are uniformly bounded for all values of u. Also, we require that the integrals

$$\begin{aligned}
&\int_u^\infty \kappa(\tau)\sup_{S_\tau} |\nabla a|_{\gamma_\tau} d\tau \\
&\int_u^\infty \varsigma(\tau)\sup_{S_\tau} |\nabla a|_{\gamma_\tau} d\tau \\
&\int_u^\infty \sup_{S_\tau} |\nabla \kappa|_{\gamma_\tau} d\tau \\
&\int_u^\infty r \sup_{S_\tau} |\nabla \hat{\theta}|_{\gamma_\tau} d\tau
\end{aligned} \tag{3.3.2}$$

are finite as $u \to \infty$ and are uniformly bounded for all values of u. Finally we also require that

$$r^2 K(u) \to 1 \tag{3.3.3}$$

as $u \to \infty$, where $K(u)$ is the Gauss curvature of S_u.

These conditions allow us to attach a sphere at infinity of Σ as follows: Consider the family of diffeomorphisms $\varphi_{u_1,u_2} : S_{u_1} \longrightarrow S_{u_2}$ given by the gradient flow of u. Since each S_u is diffeomorphic to S^2, we can define a family of diffeomorphisms $\varphi_u : S^2 \longrightarrow S_u$ such that $\phi_{u_2} \circ \phi_{u_1}^{-1} = \varphi_{u_1,u_2}$. Let m_u be the normalized family of metrics on S^2 given by $r(u)^{-2}\varphi_u^* \gamma_u$ and let m_∞ be the standard metric of S^2. Remark that the Gauss curvatures of the metrics m_u are $r^2 K(u)$, and therefore they converge to 1 as a consequence of 3.3.3. We claim that the conditions 3.3.1, 3.3.2, and 3.3.3 assure us the convergence of the metrics m_u to the standard metric m_∞, as measured relative to m_∞.

In order to compare m_u to m_∞, we chose an orthonormal basis e_1, e_2 on (S^2, m_∞) such that the matrix $m_u(e_A, e_B)$ is diagonal with smallest eigenvalue λ and largest eigenvalue Λ. Remark that if F is a (p,q)-tensor[5] on S^2, we have,

$$\Lambda^{-p}\lambda^q |F|^2_{m_\infty} \le |F|^2_{m_u} \le \lambda^{-p}\Lambda^q |F|^2_{m_\infty}. \tag{3.3.4}$$

Let $\mu = \sqrt{\det_{m_\infty} m_u} = \sqrt{\lambda\Lambda}$ and $\nu = \frac{1}{\mu} sup_{|\xi|_{m_\infty}=1} m(\xi,\xi) = \sqrt{\frac{\Lambda}{\lambda}}$. Since $\partial_u \gamma = 2a\theta$, we find

$$\frac{1}{a}\partial_u m_{AB} = r^{-1}\kappa m_{AB} + 2r^{-2}\hat{\theta}_{AB} \tag{3.3.5a}$$

and

$$\partial_u \mu = \frac{1}{2}\mu m^{AB}\partial_u m_{AB} = a r^{-1}\kappa\mu. \tag{3.3.5b}$$

Therefore, taking into account 3.3.1, we integrate and find for all values of u,

$$\mu(u) = \exp\left(- \int_u^\infty a r^{-1}\kappa \right). \tag{3.3.5c}$$

[5] That is, it is p-covariant, q-contravariant.

Consequently, there exists a positive constant c such that everywhere on Σ $c^{-1} \leq \mu \leq c$.

Similarly, setting $m' = \frac{1}{\mu} m$, we find

$$\frac{1}{a} \partial_u m'_{AB} = 2r^{-2} \frac{1}{\mu} \hat{\theta}_{AB},$$

and thus, for all vectors ξ of unit length relative to m_∞,

$$\frac{1}{a} \frac{d}{du} m'(\xi, \xi) = 2r^{-2} \mu^{-1} \hat{\theta}(\xi, \xi). \tag{3.3.5d}$$

Now, in view of 3.3.4,

$$|\hat{\theta}|_{\gamma_u} = r^{-2} |\hat{\theta}|_{m_u} \geq r^{-2} \Lambda^{-1} |\hat{\theta}|_{m_\infty}. \tag{3.3.5e}$$

Hence

$$|\hat{\theta}(\xi, \xi)| \leq |\hat{\theta}|_{m_\infty} |\xi|_{m_\infty} \leq r^2 \Lambda |\hat{\theta}|_{\gamma_u}.$$

In view of 3.3.1.2, we can integrate 3.3.5d to ∞ and infer that

$$|m'(\xi, \xi)| \leq 1 + 2 \int_u^\infty (a |\hat{\theta}|_\gamma) \nu du,$$

hence,

$$\nu \leq 1 + 2 \int_u^\infty (a |\hat{\theta}|_\gamma) \nu du \tag{3.3.5f}$$

and thus deduce by Gronwall's inequality that $\nu \longrightarrow 1$ as $u \longrightarrow \infty$ and ν is bounded uniformly on Σ. We have thus concluded the proof of the first part of the following:

Lemma 3.3.1 *Assume that Σ is a 3-dimensional Riemannian manifold verifying the assumptions 3.1.3 and 3.3.1. Then the eigenvalues $\lambda(u) \leq \Lambda(u)$ of the metric m_u relative to m_∞ converge to 1 as $u \longrightarrow \infty$ and there exists a constant c such that*

$$c^{-1} \leq \lambda(u) \leq \Lambda(u) \leq c$$

uniformly for all $u > u_0$.

Moreover, let $\Gamma(u)$ be the connection of the metric m_u, let $\Gamma(\infty)$ be the connection of m_∞, and assume that the assumptions 3.3.2 and 3.3.3 are also verified. Then there exists a constant C such that, for all values of u, $|\Gamma(u) - \Gamma(\infty)|_{m_\infty} \longrightarrow 0$ as $u \longrightarrow \infty$ and, for all values of u,

$$|\Gamma(u) - \Gamma(\infty)|_{m_\infty} \leq C.$$

To prove the second part of the lemma, we let $\Delta(u)$ be the derivative of Γ relative to u. We have

$$\Gamma(u_1) - \Gamma(u_2) = -\int_{u_1}^{u_2} \Delta(u)du.$$

Therefore,

$$|\Gamma(u_1) - \Gamma(u_2)|_{m_\infty} \leq \int_{u_1}^{u_2} |\Delta(u)|_{m_\infty}. \tag{3.3.5g}$$

On the other hand, computing relative to our special orthonormal frame $(e_A)_{A=1,2}$ on S^2, we find that

$$|\Delta|_{m_\infty} \leq \frac{\Lambda}{\lambda^{1/2}}|\Delta|_m \tag{3.3.5h}$$

or, since $|\Delta|_m^2 = r^2|\Delta|_\gamma^2$, we deduce that

$$|\Gamma(u_1) - \Gamma(u_2)|_{m_\infty} \leq \int_{u_1}^{u_2} r|\Delta(u)|_\gamma. \tag{3.3.5i}$$

We now calculate

$$\begin{aligned}
\Delta_{AB}^C &= \frac{1}{2}m^{CD}\left[\nabla\!\!\!/_A(\partial_u m_{BD}) + \nabla\!\!\!/_B(\partial_u m_{AD}) - \nabla\!\!\!/_D(\partial_u m_{AB})\right] \\
&= \frac{1}{2}r^{-1}\kappa\left[\delta_B^C \nabla\!\!\!/_A a + \delta_A^C \nabla\!\!\!/_B a - \gamma_{AB}\gamma^{CD}\nabla\!\!\!/_D a\right] \\
&\quad + \frac{1}{2}ar^{-2}\left[\delta_B^C \nabla\!\!\!/_A \kappa + \delta_A^C \nabla\!\!\!/_B \kappa - \gamma_{AB}\gamma^{CD}\nabla\!\!\!/_D \kappa\right] \qquad (3.3.5j)\\
&\quad + \gamma^{CD}(\hat\theta_{BD}\nabla\!\!\!/_A a + \hat\theta_{AD}\nabla\!\!\!/_B a - \hat\theta_{AB}\nabla\!\!\!/_D a) \\
&\quad + a\gamma^{CD}(\nabla\!\!\!/_A \hat\theta_{BD} + \nabla\!\!\!/_B \hat\theta_{AD} - \nabla\!\!\!/_D \hat\theta_{AB}) \qquad (3.3.5k)
\end{aligned}$$

where $\nabla\!\!\!/$ denotes the covariant differentiation with respect to γ. Consequently,

$$|\Delta|_\gamma \leq r^{-1}\kappa|\nabla\!\!\!/ a|_\gamma + r^{-1}a|\nabla\!\!\!/ \kappa|_\gamma + |\hat\theta|_\gamma|\nabla\!\!\!/ a|_\gamma + 3a|\nabla\!\!\!/ \hat\theta|_\gamma. \tag{3.3.5l}$$

Therefore, in view of the assumptions 3.3.2, $\int_u^\infty r|\Delta|_\gamma$ exists, is uniformly bounded for all values of u and converges to 0 as $u \to \infty$. It follows, therefore, from 3.3.5i that $\Gamma(u)$ converges to $\Gamma(\infty)$ relative to the metric m_∞ and

$$|\Gamma(u) - \Gamma(\infty)|_{m_\infty} \leq \int_u^\infty r|\Delta(u')|_\gamma du', \tag{3.3.6}$$

and the assertions of the second part of Lemma 3.3.1 are now immediate.

We next define an action $SO(3) \times \Sigma \longrightarrow \Sigma$ of the isometry group $SO(3)$ of the standard sphere S^2 on our manifold Σ according to the formula

$$(O, x \in S_u) \longrightarrow \varphi_u(O\varphi_u^{-1}(x)).$$

Let ${}^{(a)}\Omega_{(\infty)}$, $a = 1, 2, 3$, be a basis for the Lie algebra of $SO(3)$ such that

$$[{}^a\Omega_{(\infty)}, {}^b\Omega_{(\infty)}] = \in_{abc}\, {}^c\Omega_{(\infty)}$$

and extend them to any S_u by the formula

$$ {}^{(a)}\Omega = (\varphi_u)_*\, {}^a\Omega_{(\infty)}. \tag{3.3.7}$$

This defines ${}^{(a)}\Omega$ as vectorfields on all Σ verifying the following properties:

$$[{}^{(a)}\Omega, \partial_u] = 0 \tag{3.3.8a}$$
$$\langle {}^{(a)}\Omega, N\rangle_g = 0 \tag{3.3.8b}$$

Moreover,

$$[{}^{(a)}\Omega, {}^{(b)}\Omega] = \in_{abc}\, {}^{(c)}\Omega. \tag{3.3.8c}$$

Moreover, if f is an arbitrary p-covariant tensor on Σ tangent to S_u, we have

$$(\varphi_u)^* \mathcal{L}_{{}^{(a)}\Omega} f_u = \mathcal{L}_{{}^a\Omega_{(\infty)}}(\varphi_u^* f_u) \tag{3.3.8d}$$

where f_u is the restriction of f to S_u.

As an immediate consequence of the definition of ${}^{(a)}\Omega$ we have, taking into account Lemma 3.3.1,

$$|{}^{(a)}\Omega|_{m_u} \leq \Lambda^{1/2} |{}^{(a)}\Omega|_{m_\infty} \leq 2\Lambda^{1/2},$$

therefore,

$$|r^{-1}\, {}^{(a)}\Omega|_{\gamma_u} \leq 2\Lambda^{1/2}. \tag{3.3.8e}$$

Similarly,

$$\begin{aligned} |\nabla\!\!\!/_{m_u}^{a} \Omega_{(\infty)}|_{m_\infty} &\leq |\nabla\!\!\!/_{m_\infty}\, {}^a\Omega_{(\infty)}|_{m_\infty} + |\Gamma(u) - \Gamma(\infty)|_{m_\infty} |{}^a\Omega_{(\infty)}|_{m_\infty} \\ &\leq 2\left(1 + c\int_u^\infty r|\Delta(u')|_\gamma du'\right). \end{aligned}$$

Therefore,

$$|\nabla\!\!\!/\, {}^{(a)}\Omega(u)|_\gamma \leq c\left(1 + \int_u^\infty r|\Delta(u')|_\gamma du'\right). \tag{3.3.8f}$$

Also, in view of the formulas 3.1.1e, we find

$$\begin{aligned}\nabla_N {}^{(a)}\Omega_N &= a^{-1}\nabla\!\!\!\!/_A a^{(a)}\Omega_A \\ \nabla_N {}^{(a)}\Omega_A &= \nabla\!\!\!\!/_N {}^{(a)}\Omega_A = \theta_{AB} {}^{(a)}\Omega_B \\ \nabla_A {}^{(a)}\Omega_N &= -\theta_{AB} {}^{(a)}\Omega_B.\end{aligned} \tag{3.3.8g}$$

Let ${}^{(a)}\pi$ denote the deformation tensor of ${}^{(a)}\Omega$. Then

$$\begin{aligned}{}^{(a)}\pi_{NN} &= 2a^{-1}\nabla\!\!\!\!/_A a^{(a)}\Omega_A \\ {}^{(a)}\pi_{AN} &= 0.\end{aligned} \tag{3.3.8h}$$

The angular components of ${}^{(a)}\pi$ can be calculated as follows:

Let ${}^{(a)}H$ be the tensor tangent to S_u defined by ${}^{(a)}H_{AB} = \frac{1}{2}{}^{(a)}\pi_{AB}$. Then, on the sphere at infinity, we set

$$ {}^{(a)}h := r^{-2}(\varphi_u)^{*(a)}H = \frac{1}{2}\mathcal{L}_{{}^a\Omega_{(\infty)}} m_u. $$

Now,

$$ (\mathcal{L}_{{}^a\Omega_{(\infty)}} m_u - \mathcal{L}_{{}^a\Omega_{(\infty)}} m_\infty)_{AB} = -2(\Gamma(u)^C_{AB} - \Gamma(\infty)^C_{AB})^a\Omega_{(\infty)C} $$

or, since according to the definition of ${}^a\Omega_{(\infty)}$ we have $\mathcal{L}_{{}^a\Omega_{(\infty)}} m_\infty = 0$,

$$ {}^{(a)}h_{AB} = -(\Gamma(u)^C_{AB} - \Gamma(\infty)^C_{AB})^a\Omega_{(\infty)C}. \tag{3.3.8i}$$

Hence, taking into account 3.3.6,

$$ |{}^{(a)}h|_{m_\infty} \leq 2|\Gamma(u) - \Gamma(\infty)|_{m_\infty} \leq 2\int_u^\infty r|\Delta(u)|_\gamma, \tag{3.3.8j}$$

and therefore, going back to the spheres S_u,

$$ |{}^{(a)}H|_{\gamma_u} \leq c\int_u^\infty r|\Delta(u')|_\gamma du'. \tag{3.3.8k}$$

We have thus proved the following:

Proposition 3.3.1 *The rotation vectorfields ${}^{(a)}\Omega$ verify the equation*

$$ \nabla\!\!\!\!/_N {}^{(a)}\Omega_A = \theta_{AB} {}^{(a)}\Omega_A $$

and the bounds

$$\begin{aligned} |r^{-1(a)}\Omega|_{\gamma_u} &\leq 2\Lambda^{1/2} \\ |\nabla\!\!\!\!/ {}^{(a)}\Omega|_{\gamma_u} &\leq c\left(1 + \int_u^\infty r|\Delta(u')|_\gamma\right)du' \\ &\leq c\int_u^\infty (1 + |\kappa\nabla\!\!\!\!/ a|_\gamma + |a\nabla\!\!\!\!/\kappa|_\gamma)du' \\ &\quad + c\int_u^\infty r(|\hat{\theta}|_\gamma|\nabla\!\!\!\!/ a|_\gamma + |a\nabla\!\!\!\!/\hat{\theta}|_\gamma)du'.\end{aligned}$$

Moreover, the components ${}^{(a)}\pi_{NN}, {}^{(a)}\pi_{AN}$ *of deformation tensor* ${}^{(a)}\pi$ *of* ${}^{(a)}\Omega$ *are given by the formulas 3.3.8h, and* ${}^{(a)}H_{AB} = \frac{1}{2}{}^{(a)}\pi_{AB}$ *verifies the bound*

$$
\begin{aligned}
|{}^{(a)}H|_{\gamma_u} &\le c\int_u^\infty r|\Delta(u')|_\gamma du' \\
&\le c\int_u^\infty (|\kappa \nabla a|_\gamma + |a\nabla\kappa|_\gamma)du' \\
&\quad + c\int_u^\infty r(|\hat{\theta}|_\gamma|\nabla a|_\gamma + |a\nabla\hat{\theta}|_\gamma)du'
\end{aligned}
$$

where c is a constant that depends only on λ, Λ.

To estimate higher angular derivatives of ${}^{(a)}H$, we proceed as follows:

First, setting $\delta(u) = \Gamma(u) - \Gamma(\infty)$, we differentiate the equation 3.3.8i relative to the metric m_∞ and write symbolically

$$
\nabla_\infty {}^{(a)}h = -\nabla_{m_\infty}\delta(u)\cdot{}^a\Omega_{(\infty)} - \delta(u)\cdot\nabla^a_{m_\infty}\Omega_{(\infty)}
$$

where ∇_{m_∞} is the covariant differentiation with respect to m_∞. Hence

$$
|\nabla_{m_\infty}{}^{(a)}h|_{m_\infty} \le 2(|\nabla_{m_\infty}\delta(u)|_{m_\infty} + |\delta(u)|_{m_\infty}). \tag{3.3.9a}
$$

Now,

$$
\begin{aligned}
|\nabla_{m_u}{}^{(a)}h| &\le |\nabla_{m_\infty}{}^{(a)}h|_{m_\infty} + |(\nabla_{m_u} - \nabla_{m_\infty}){}^{(a)}h|_{m_\infty} \\
&\le 2(|\nabla_{m_\infty}\delta(u)|_{m_\infty} + |\delta(u)|_{m_\infty}) + |\delta(u)|_{m_\infty}|{}^{(a)}h|_{m_\infty}.
\end{aligned}
$$

Thus, since according to Proposition 3.3.1 and the assumptions 3.3.1, 3.3.2, and 3.3.3 $|{}^{(a)}h|_{m_\infty} \le c|{}^{(a)}H|_{\gamma_u} \le c$, we infer that,

$$
|\nabla_{m_u}{}^{(a)}h|_{m_\infty} \le c(|\nabla_{m_\infty}\delta(u)|_{m_\infty} + |\delta(u)|_{m_\infty}) \tag{3.3.9b}
$$

with c depending only on $|{}^{(a)}h|_{m_\infty}$.

Given an arbitrary k-covariant tensorfield v on the standard sphere (S^2, m_∞) we denote by $|v|_{p,S,\infty}$ the L^p norm

$$
|v|_{p,S,\infty} = \left(\int_{S^2}|v|^p_{m_\infty}d\mu_{m_\infty}\right)^{1/p}.
$$

Similarly, for tensorfield V defined on the spheres S_u we introduce

$$
|V|_{p,S}(u) = \left(\int_{S_u}|V|^p_\gamma d\mu_{\gamma_u}\right)^{1/p}.
$$

Clearly, in view of the relations between the metrics m_∞, m_u, and γ,

$$c(\Lambda,\lambda)^{-1}|r^{k-2/p}V|_{p,S}(u) \leq |(\phi_u)^* V|_{p,S,\infty} \leq c(\Lambda,\lambda)|r^{k-2/p}V|_{p,S}(u). \tag{3.3.9c}$$

Thus, from 3.3.9b,

$$|\nabla\!\!\!/_{m_u}{}^{(a)}h|_{p,S,\infty} \leq c(|\nabla\!\!\!/_{m_\infty}\delta(u)|_{p,S,\infty} + |\delta(u)|_{p,S,\infty}), \tag{3.3.9d}$$

and consequently, by 3.3.9c and the definition of ${}^{(a)}h$,

$$|r^{1-2/p}\nabla\!\!\!/^{(a)}H|_{p,S}(u) \leq c(|\nabla\!\!\!/_{m_\infty}\delta(u)|_{p,S,\infty} + |\delta(u)|_{p,S,\infty}). \tag{3.3.9e}$$

Now, recalling 3.3.5g,

$$|\delta(u)|_{m_\infty} \leq \int_u^\infty |\Delta(u')|_{m_\infty} du'.$$

Hence,

$$\begin{aligned} |\delta(u)|_{p,S,\infty} &\leq \int_u^\infty |\Delta(u')|_{p,S,\infty} du' \\ &\leq c(\Lambda,\lambda)\int_u^\infty |r^{1-2/p}\Delta|_{p,S}(u')du'. \end{aligned} \tag{3.3.9f}$$

Similarly,

$$|\nabla\!\!\!/_{m_\infty}\delta(u)|_{m_\infty} \leq \int_u^\infty |\nabla\!\!\!/_{m_\infty}\Delta(u)|_{m_\infty},$$

hence,

$$|\nabla\!\!\!/_{m_\infty}\delta(u)|_{p,S,\infty} \leq \int_u^\infty |\nabla\!\!\!/_{m_\infty}\Delta(u')|_{p,S,\infty} du'. \tag{3.3.9g}$$

On the other hand,

$$\begin{aligned} |\nabla\!\!\!/_{m_\infty}\Delta(u')|_{m_\infty} &\leq |\nabla\!\!\!/_{m_u}\Delta(u')|_{m_\infty} + |\delta(u')|_{m_\infty}|\Delta(u')|_{m_\infty} \\ &\leq |\nabla\!\!\!/_{m_u}\Delta(u')|_{m_\infty} + c|\Delta(u')|_{m_\infty}. \end{aligned}$$

We thus derive

$$|\nabla\!\!\!/_{m_{\inf}}\Delta(u')|_{p,S,\infty} \leq c\big(|r^{2-2/p}\nabla\!\!\!/_{\gamma_{u'}}\Delta|_{p,S}(u') + |r^{1-2/p}\Delta|_{p,S}(u')\big),$$

and, back to 3.3.9g, we infer that

$$|\nabla\!\!\!/_{m_{\inf}}\delta(u)|_{p,S,\infty} \leq c\int_u^\infty \Big(|r^{2-2/p}\nabla\!\!\!/_{\gamma_{u'}}\Delta|_{p,S}(u') + |r^{1-2/p}\Delta|_{p,S}(u')\Big)du'. \tag{3.3.9h}$$

Using this in conjunction with 3.3.9e and 3.3.9f, we conclude with

$$|r^{1-2/p}\nabla\!\!\!/^{(a)}H|_{p,S} \le c\int_u^\infty \Big(|r^{2-2/p}\nabla\!\!\!/\Delta|_{p,S}(u') + |r^{1-2/p}\Delta|_{p,S}(u')\Big)du'. \tag{3.3.9i}$$

This proves the following:

Proposition 3.3.2 *The angular derivatives of ${}^{(a)}H$ can be estimated according to*

$$|r^{1-2/p}\nabla\!\!\!/^{(a)}H|_{p,S} \le c\int_u^\infty \Big(|r^{2-2/p}\nabla\!\!\!/\Delta|_{p,S}(u') + |r^{1-2/p}\Delta|_{p,S}(u')\Big)du'$$

where c is a constant depending only on 3.3.1, 3.3.2, and 3.3.3, and δ is given by the formula 3.3.5k.

We next recall the following:

Lemma 3.3.2 *Consider the standard sphere S^2 embedded in the Euclidean space $\mathbf{R}^3$ and consider ${}^{(a)}\Omega$ the rotation vectorfields ${}^{(a)}\Omega = \in_{abc}x_b\partial_c$ relative to the canonical system of coordinates.*

(i.) If U is a 1-form tangent to S^2, then

$$\sum_{a=1,2,3} |\mathcal{L}_{{}^{(a)}\Omega}U|^2 = |\nabla\!\!\!/ U|^2 + |U|^2.$$

(ii.) If U is a symmetric 2-covariant tensor tangent to S^2,

$$\sum_{a=1,2,3} |\mathcal{L}_{{}^{(a)}\Omega}U|^2 = |\nabla\!\!\!/ U|^2 + 4|U|^2 - 2(\mathrm{tr}U)^2.$$

The proof of the lemma follows by a straightforward calculation. We apply the lemma to the sphere (S^2, m_∞). Thus for any given 1-form F on (S^2, m_∞) we have

$$\sum_{a=1,2,3} |\mathcal{L}_{{}^a\Omega_{(\infty)}}F|^2_{m_\infty} = |\nabla\!\!\!/_{m_\infty}F|^2_{m_\infty} + |F|^2_{m_\infty}.$$

Now let f be a 1-form on Σ tangent to S_u at every point and let $F_u = \varphi_u^* f$. In view of 3.3.8d we have

$$\begin{aligned}\sum_{a=1,2,3} |\mathcal{L}_{{}^{(a)}\Omega}f|^2_\gamma &= r^{-2}\sum_{a=1,2,3} |\mathcal{L}_{{}^a\Omega_{(\infty)}}F|^2_{m_u}\\ &\ge \Lambda^{-1}r^{-2}\sum_{a=1,2,3} |\mathcal{L}_{{}^a\Omega_{(\infty)}}F|^2_{m_\infty}\\ &= \Lambda^{-1}r^{-2}(|\nabla\!\!\!/_{m_\infty}F|^2_{m_\infty} + |F|^2_{m_\infty}).\end{aligned} \tag{3.3.9j}$$

Therefore, $\Lambda^{-1}r^{-2}|F|^2_{m_\infty} \leq \sum_{a=1,2,3} |\mathcal{L}_{{}^{(a)}\Omega} f|^2_\gamma$, and, taking into account 3.3.4, we derive

$$|f|^2_\gamma \leq \frac{\Lambda}{\lambda} \sum_{a=1,2,3} |\mathcal{L}_{{}^{(a)}\Omega} f|^2_\gamma. \tag{3.3.9k}$$

On the other hand,

$$\begin{aligned} |\nabla_{m_\infty} F|^2_{m_\infty} + |F|^2_{m_\infty} &\geq \frac{1}{2}|\nabla_{m_\infty} F|^2_{m_\infty} + |F|^2_{m_\infty} \\ &\quad - |(\nabla_m - \nabla_{m_\infty})F|^2_{m_\infty} \\ &\geq \frac{1}{2}\lambda^2 r^2 |\nabla f|^2_{\gamma_u} + \lambda r^2 |f|^2_{\gamma_u} \\ &\quad - \lambda r^2 |f|^2_{\gamma_u} |\Gamma(u) - \Gamma(\infty)|^2_{m_\infty}. \end{aligned} \tag{3.3.9l}$$

Consequently, using 3.3.9j,

$$\sum_{a=1,2,3} |\mathcal{L}_{{}^{(a)}\Omega} f|^2_{\gamma_u} \geq \frac{\lambda^2}{2\Lambda} |\nabla f|^2_{\gamma_u} + \frac{\lambda}{\Lambda} |f|^2_{\gamma_u} (1 - |\Gamma(u) - \Gamma(\infty)|^2_{m_\infty}).$$

Therefore, taking into account 3.3.9k,

$$|\nabla f|^2_{\gamma_u} \leq \frac{2\Lambda}{\lambda^2}(1 + |\Gamma(u) - \Gamma(\infty)|^2_{m_\infty}) \sum_{a=1,2,3} |\mathcal{L}_{{}^{(a)}\Omega} f|^2_{\gamma_u}. \tag{3.3.9m}$$

Taking into account Lemma 3.3.1, the estimates 3.3.9k and 3.3.9m imply the following:

Proposition 3.3.3 *Assume that the assumption 3.3.1 is satisfied. Let f be a 1-form on Σ tangent to S_u at every point. There exists a constant C_1 that depends only on the fundamental constants 3.1.3 and the integrals 3.3.1 such that*

$$|f|^2_\gamma \leq C_1 \sum_{a=1,2,3} |\mathcal{L}_{{}^{(a)}\Omega} f|^2_\gamma.$$

Moreover, if the assumptions 3.3.2 are also verified, there exists another constant C_2, depending only on 3.1.3 and the integrals 3.3.1 and 3.3.2, such that

$$|\nabla f|^2_{\gamma_u} \leq C_2 \sum_{a=1,2,3} |\mathcal{L}_{{}^{(a)}\Omega} f|^2_{\gamma_u}.$$

We next consider a 2-covariant traceless tensor on Σ tangent to S_u at every point. Let $F_u = \varphi_u^* f$ and apply to it Lemma 3.3.2.

$$\sum_{a=1,2,3} |\mathcal{L}_{{}^{a}\Omega_{(\infty)}} F|^2_{m_\infty} = |\nabla_{m_\infty} F|^2_{m_\infty} + 4|F|^2_{m_\infty} - 2(\mathrm{tr}_{m_\infty} F)^2$$

Since F is traceless relative to m_u, we have $m_u^{AB}F_{AB} = \lambda^{-1}F_{11} + \Lambda^{-1}F_{22} = 0$. Consequently,

$$4|F|^2_{m_\infty} - 2(\mathrm{tr}_{m_\infty}F)^2 = 8F_{12}^2 + 2(F_{11} - F_{22})^2$$
$$= 8F_{12}^2 + 2\Big(1 + \frac{\Lambda}{\lambda}\Big)^2 F_{11}^2. \qquad (3.3.10a)$$

On the other hand,

$$|F|^2_{m_u} = 2\lambda^{-2}F_{11}^2 + 2\lambda^{-1}\Lambda^{-1}F_{12}^2$$

and therefore

$$4|F|^2_{m_\infty} - 2(\mathrm{tr}_{m_\infty}F)^2 \geq 4\lambda\Lambda|F|^2_{m_u}. \qquad (3.3.10b)$$

Now,

$$\sum_{a=1,2,3} |\mathcal{L}_{{}^{(a)}\Omega} f|^2_\gamma = r^{-4} \sum_{a=1,2,3} |\mathcal{L}_{{}^a\Omega_{(\infty)}} F|^2_{m_u}$$
$$\geq \Lambda^{-2}r^{-4} \sum_{a=1,2,3} |\mathcal{L}_{{}^a\Omega_{(\infty)}} F|^2_{m_\infty}$$
$$= \Lambda^{-2}r^{-4}\left(|\nabla\!\!\!\!/_{m_\infty}F|^2_{m_\infty} + 4|F|^2_{m_\infty} - 2(\mathrm{tr}_{m_\infty}F)^2\right)$$
$$\geq \Lambda^{-2}r^{-4}|\nabla\!\!\!\!/_{m_\infty}F|^2_{m_\infty} + 4r^{-4}\frac{\lambda}{\Lambda}|F|^2_{m_u}. \qquad (3.3.10c)$$

In particular, we infer that

$$|f|^2_\gamma \leq \frac{\Lambda}{\lambda} \sum_{a=1,2,3} |\mathcal{L}_{{}^{(a)}\Omega} f|^2_\gamma. \qquad (3.3.10d)$$

On the other hand,

$$|\nabla\!\!\!\!/_{m_\infty}F|^2_{m_\infty} \geq \frac{1}{2}|\nabla\!\!\!\!/_{m_\infty}F|^2_{m_\infty} - |(\nabla\!\!\!\!/_m - \nabla\!\!\!\!/_{m_\infty})F|^2_{m_\infty} \qquad (3.3.10e)$$
$$\geq \frac{1}{2}\lambda^3r^4|\nabla\!\!\!\!/ f|^2_{\gamma_u} - \lambda^2r^4|f|^2_{\gamma_u}|\Gamma(u) - \Gamma(\infty)|^2_{m_\infty}.$$

Thus, taking 3.3.10d into account, we derive from 3.3.10c

$$|\nabla\!\!\!\!/ f|^2_{\gamma_u} \leq \left(\frac{2\Lambda^2}{\lambda^3} + \frac{2\Lambda}{\lambda^2}|\Gamma(u) - \Gamma(\infty)|^2_{m_\infty}\right) \sum_{a=1,2,3} |\mathcal{L}_{{}^{(a)}\Omega} f|^2_{\gamma_u}. \qquad (3.3.10f)$$

Finally, taking into account Lemma 3.3.1, the estimates 3.3.10d and 3.3.10f imply the following:

Proposition 3.3.4 *Assume that the assumption 3.3.1 is satisfied. Let f be a 2-covariant tensor on Σ tangent to S_u at every point. There exists a constant C_1' that depends only on the fundamental constants 3.1.3 and the integrals 3.3.1 such that*

$$|f|_\gamma^2 \leq C_1' \sum_{a=1,2,3} |\mathcal{L}_{{}^{(a)}\Omega} f|_\gamma^2.$$

Moreover, if the assumptions 3.3.2 are also verified, there exists another constant C_2' depending only on all the above mentioned constants such that

$$|\nabla f|_{\gamma_u}^2 \leq C_2' \sum_{a=1,2,3} |\mathcal{L}_{{}^{(a)}\Omega} f|_{\gamma_u}^2.$$

CHAPTER 4

The Poisson Equation in 3-D

4.1. Preliminaries

Throughout this chapter we assume (Σ, g) to be a 3-dimensional Riemannian manifold diffeomorphic to R^3 on which there exists a generalized radial function u with second fundamental form $\theta$and Gaussian curvature K. We require that the radial function u is *quasiconvex*, by which we mean that

$$\mathrm{tr}\theta > 0, \qquad K > 0.$$

This is in fact implied by the stronger assumptions we make, namely, that the fundamental constants $k_m, k_M, a_M, a_m, h_m, h_M, \varsigma$ (see 3.1.3) verify that

$$k_m^{-1}, k_M, a_m{}^{-1}, a_M, h_m^{-1}, h_M, \varsigma \text{ are finite.} \tag{4.1.1a}$$

Also, we make the assumption that

$$r_0 = r(0) \geq 1 \tag{4.1.1b}$$

where r is the function of u defined by 3.1.1h. Other assumptions that will be required at various points in this chapter are that

$$a_1 = \sup_\Sigma(ra^{-1}|\nabla a|) \text{ is finite} \tag{4.1.2a}$$

$$\tilde{a}_1 = \sup_\Sigma(r^{3/2}\tau_-^{-1/2}a^{-1}|\nabla a|) \text{ is finite} \tag{4.1.2b}$$

$$h_1 = \left(\int_\Sigma r^5|\nabla \mathrm{tr}\theta|^4\right)^{1/4} \text{ is finite} \tag{4.1.2c}$$

$$a_2 = \left(\int_\Sigma r^5|a^{-1}\nabla^2 a|^4\right)^{1/4} \text{ is finite} \tag{4.1.2d}$$

$$r_N = \left(\int_\Sigma r^5|P|^4\right)^{1/4} \text{ is finite.} \tag{4.1.2e}$$

Throughout the chapter we denote by $\|U\|_p$ the L^p norm in Σ of an arbitrary tensor U. We will use $\|U\|$ for the particular case when $p = 2$. We also recall the definition of the weighted norms,

$$\|U\|^2_{H_{k,\delta}} = \sum_{0 \le i \le k} \int_\Sigma (1 + r^2)^{\delta+i} |\nabla^i U|^2 d\mu_g$$

and the corresponding weighted Sobolev spaces $H_{k,\delta}$.

Our object of study here is the Poisson equation.

Poisson Equation

$$\triangle \phi = f$$

where f is a given scalar function on Σ. We estimate the solutions subject to the Dirichlet boundary condition at infinity:

$$\sup_{S_u} |\phi| \to 0 \quad \text{as} \quad u \to \infty.$$

It is well known (see [Ba2], [Ch-Br]) that, under mild assumptions concerning the Euclidean structure of Σ at infinity, which are implied by our assumptions 4.1.1a and 4.1.2a, the Laplace-Beltrami operator $\triangle$ defined from $H_{2,\delta}$ to $H_{0,\delta+2}$ is a Fredholm operator for all values of δ that differ from $-3/2 + k$, with k an arbitrary integer. Moreover, for all values $-3/2 < \delta$ we have $\ker \triangle = 0$, for δ restricted to $-3/2 < \delta < -1/2$, $\triangle$ is also onto, while for $-1/2 < \delta < 1/2$ its range has codimension 1 and is given by all functions f in $H_{0,\delta+2}$ with $\int_\Sigma f = 0$.[1] The cases of interest for us here are $\delta = -1$ and $\delta = 0$.

The major part of the chapter consists in providing nondegenerate and degenerate estimates for this equation. The nondegenerate estimates are similar to the standard weighted L^2 estimates of solutions to linear elliptic equations in the Euclidean space. They are distinguished by the fact that the weights are uniform in all directions; they depend only on the order of differentiation. In contrast, the degenerate estimates distinguish between the angular derivatives, intrinsic to the leaves of the radial foliation, and the derivatives normal to the foliation. They appear naturally here due to the hyperbolic character of the Einstein-Vacuum equation. Thus, though the constraint equations, to which the estimates of this chapter are to be applied, are elliptic, their source terms remember the evolution character of the overall system and thus have degenerate behavior in the wave zone.

[1] Since the range of $\triangle$ is closed, it suffices to show that the kernel of the dual map, as defined from $H_{0,-\delta-2}$, has dimension 0 for $\delta < -1/2$ and consists only of constants for $\delta < 1/2$.

Parts i and ii of Proposition 4.2.1 are our main tools in the treatment of the constraint equations, which we address in Part III. In the spirit of the whole book, our derivation is entirely geometric; we use only commutation arguments and integration by parts. In this respect the position vectorfield of Σ is central to our approach. We recall that, in the Euclidean space $\Re^3$, this is defined to be the conformal Killing vectorfield that generates the scaling transformations. A good analogue in curved space would be a vectorfield whose deformation tensor has small traceless part. We define such a vectorfield by the following equation:

Position Vectorfield

$$Z = rN. \tag{4.1.3a}$$

Its deformation tensor $\pi = {}^{(Z)}\pi$ can be expressed as follows:

$$\hat{\pi}_{ij} = 2r\hat{\theta}_{ij} + \frac{1}{3}\kappa(g_{ij} - 3N_i N_j) - ra^{-1}(N_i \nabla\!\!\!/_j a + N_j \nabla\!\!\!/_i a)$$

$$\mathrm{tr}\pi = ra^{-1}(2a\mathrm{tr}\theta + \overline{a\mathrm{tr}\theta}) \tag{4.1.3b}$$

$$\pi_{ij} = \hat{\pi}_{ij} + \frac{1}{3}\mathrm{tr}\pi g_{ij} \tag{4.1.3c}$$

where

$$\kappa = ra^{-1}(a\mathrm{tr}\theta - \overline{a\mathrm{tr}\theta}). \tag{4.1.3d}$$

We also write

$$\nabla_i Z_j = \frac{1}{2}\aleph_{ij} + \frac{1}{3}(r\mathrm{tr}\theta + \lambda a^{-1})g_{ij} \tag{4.1.3e}$$

where, recall from 3.1.1i, $\lambda = \frac{dr}{du} = r/2\overline{a\mathrm{tr}\theta}$ and

$$\aleph_{ij} = \hat{\pi}_{ij} - ra^{-1}(N_i \nabla\!\!\!/_j a - N_j \nabla\!\!\!/_i a). \tag{4.1.3f}$$

Now if the assumptions 4.1.1a and 4.1.2a are satisfied then,

$$\sup_{\Sigma}(|\hat{\pi}|, |\mathrm{tr}\pi|) \leq c. \tag{4.1.3g}$$

In the following paragraph we provide estimates for the principal error term in the asymptotic expansion of the solution to the Poisson equation at infinity. We thus assume that f belongs to the space $H_{0,2}$ outside a sufficiently large compact set of Σ and estimate the difference between ϕ and the integral

$$P[\phi] = -\frac{1}{4\pi}\int_{\Sigma} f. \tag{4.1.4}$$

It is easy to see that the error term belongs to the space $L^2 = H_{0,0}$, outside a sufficiently large set. Indeed, choose χ a smooth function equal to $1/r$ in a neighborhood of infinity. In view of this definition $\int_\Sigma \Delta\chi = -4\pi$ and $\int_\Sigma (f + \Delta\chi P) = 0$. Consequently, the equation $\Delta\phi' = f + \Delta\chi P$ has a solution in the space $H_{0,0}$. Moreover, $\phi = \phi' + P\chi$ is the desired solution to $\Delta\phi = f$, which proves our assertion. We refer to $P[\phi]$ as the charge of ϕ.

4.2. Degenerate and Nondegenerate L^2 Estimates

Throughout this section we assume that $f \in H_{0,1}$ and therefore that the Poisson equation has a solution $\phi \in H_{2,-1}$. We start with the following:

Proposition 4.2.1 *Let ϕ be a solution of the Poisson equation.*

(i.) Assume that $\int_\Sigma r^2|f|^2$ is finite. There exists a constant $\tilde{c}_1$ such that

$$\int_\Sigma |\nabla\phi|^2 \leq \tilde{c}_1 \int_\Sigma r^2|f|^2.$$

(ii.) Assume that $\int_\Sigma \tau_-^2|f|^2$ is finite. There exists a constant $\tilde{c}_2$ such that

$$\int_\Sigma |\nabla\phi|^2 \leq \tilde{c}_2 r_0 \int_\Sigma \tau_-^2|f|^2.$$

Proof 4.2.1 *To prove the first part of the proposition we start by multiplying the Poisson equation with ϕ and then integrate by parts on Σ. We thus derive*

$$\begin{aligned}\int_\Sigma |\nabla\phi|^2 &= -\int_\Sigma \phi f\\ &\leq \left(\int_\Sigma r^{-2}\phi^2\right)^{1/2}\left(\int_\Sigma r^2|f|^2\right)^{1/2},\end{aligned}$$

and the desired statement is now an immediate consequence of the nondegenerate version of the global Poincaré inequality (see Proposition 3.2.2i).

The second part of the proposition is proved precisely in the same manner by making use of the degenerate version of the global Poincaré inequality (see Proposition 3.2.2 ii).

Remark 1 *The constants $\tilde{c}_1$, $\tilde{c}_2$ depend only on the constants that appear in the corresponding Poincaré inequalities.*

Proposition 4.2.2 *Assume that in addition to the condition 4.1.1a the radial function u verifies 4.1.2a, and consider ϕ a solution of the Poisson equation.*

(i.) Under the assumption that $\int_\Sigma r^2|f|^2$ is finite there exists a constant $\tilde{c}_3$ such that

$$\int_\Sigma r^2|\nabla^2\phi|^2 \leq \tilde{c}_3 \int_\Sigma r^2|f|^2.$$

(ii.) Under the assumption that $\int_\Sigma \{\tau_-^2|f|^2 + r^2\tau_-^2|\nabla\!\!\!/ f|^2\}$ is finite there exists a constant $\tilde{c}_4$ such that

$$\int_\Sigma \{r^2|\nabla\!\!\!/^2\phi|^2 + r^2|\nabla\!\!\!/\nabla\!\!\!/_N\phi|^2 + w_1^2|\nabla\!\!\!/_N^2\phi|^2\} \leq \tilde{c}_4 r_0 \int_\Sigma \{\tau_-^2|f|^2 + r^2\tau_-^2|\nabla\!\!\!/ f|^2\}$$

where w_1 is defined according to

$$w_p = \min\{\tau_-^p r_0^{1/2}, r^p\}. \tag{4.2.1}$$

Remark that an equivalent definition of w_p is given by

$$w_p^{-1} = \tau_-^{-p} r_0^{-1/2} + r^{-p}. \tag{4.2.2}$$

Proof 4.2.2 *The main idea is to use the commutation property of the position vectorfield, with the Laplacian on Σ. For a general vectorfield X this reads*

$$\begin{aligned}[\triangle, \nabla_X]\phi = {}& {}^{(X)}\hat{\pi}^{ij}\nabla_i\nabla_j\phi + \frac{1}{3}\mathrm{tr}\,{}^{(X)}\pi\triangle\phi \\ & + \nabla_j{}^{(X)}\hat{\pi}^{ij}\nabla_i\phi - \frac{1}{6}\nabla^i\mathrm{tr}\,{}^{(X)}\pi\nabla_i\phi\end{aligned} \tag{4.2.3}$$

where ${}^{(X)}\pi$ is the deformation tensor of X and ${}^{(X)}\hat{\pi}$ is its traceless part.

Applying the commutation formula to our Poisson equation, we have

$$\begin{aligned}\triangle(\nabla_Z\phi) = {}& \nabla_Z f + \hat{\pi}^{ij}\nabla_i\nabla_j\phi + \frac{1}{3}\mathrm{tr}\pi f \\ & + \nabla_j\hat{\pi}^{ij}\nabla_i\phi - \frac{1}{6}\nabla^i\mathrm{tr}\pi\nabla_i\phi.\end{aligned} \tag{4.2.4a}$$

After multiplying 4.2.4a by $\nabla_Z\phi$ and integrating by parts, we derive:

$$\begin{aligned}\int_\Sigma |\nabla\nabla_Z\phi|^2 = {}& \overbrace{\int_\Sigma f\nabla_Z^2\phi}^{I_1} + \overbrace{\int_\Sigma \hat{\pi}^{ij}\nabla_i\phi\nabla_j\nabla_Z\phi}^{I_2} \\ & - \frac{1}{6}\overbrace{\int_\Sigma \mathrm{tr}\pi\nabla^i\phi\nabla_i\nabla_Z\phi}^{I_3}.\end{aligned} \tag{4.2.4b}$$

Since $Z = rN$ we estimate I_1 in a straightforward fashion:

$$I_1 \leq \left(\int_\Sigma r^2 |f|^2 \right)^{1/2} \left(\int_\Sigma |\nabla_N \nabla_Z \phi|^2 \right)^{1/2}.$$

For the second integral I_2 we write

$$I_2 \leq \sup_\Sigma |\hat{\pi}| \|\nabla \phi\| \|\nabla \nabla_Z \phi\|.$$

Finally we have

$$I_3 \leq \frac{1}{6} \sup_\Sigma |\mathrm{tr}\pi| \|\nabla \phi\| \|\nabla \nabla_Z \phi\|,$$

and, in view of 4.1.3g and Proposition 4.2.1 we infer that

$$\int_\Sigma r^2 \{ |\nabla_N \nabla_N \phi|^2 + |\nabla\!\!\!/ \nabla_N \phi|^2 \} \leq c \int_\Sigma r^2 |f|^2. \tag{4.2.4c}$$

It thus remains to estimate only the second angular derivatives, $\nabla\!\!\!/^2 \phi$. They can be estimated directly from the Poisson equation by using the polar decomposition of the Laplacian on Σ:

$$\triangle = \nabla_N^2 + \mathrm{tr}\theta \nabla_N + \triangle\!\!\!/ + a^{-1} \nabla\!\!\!/ a \cdot \nabla\!\!\!/. \tag{4.2.5}$$

Therefore, ϕ can be viewed as a solution to the following equation on each S_u:

$$\begin{aligned} \triangle\!\!\!/ \phi &= f' \\ &= f - \nabla_N^2 \phi - \mathrm{tr}\theta \nabla_N \phi - a^{-1} \nabla\!\!\!/ a \cdot \nabla\!\!\!/ \phi. \end{aligned} \tag{4.2.6a}$$

Therefore, making use of the corollary to Proposition 2.2.1, we have

$$\int_{S_u} \{ |\nabla\!\!\!/^2 \phi|^2 + K |\nabla\!\!\!/ \phi|^2 \} = \int_{S_u} |f'|^2, \tag{4.2.6b}$$

with K being the Gauss curvature of S_u.

Multiplying 4.2.6b by ar^2 and integrating with respect to u, we derive, with the help of the first-order estimates of Proposition 4.2.1,

$$\begin{aligned} \int_\Sigma r^2 |\nabla\!\!\!/^2 \phi|^2 &\leq c' \left(\int_\Sigma r^2 |f'|^2 + \int_\Sigma r^2 |f|^2 \right) \\ \int_\Sigma r^2 |f'|^2 &\leq c \int_\Sigma r^2 |f|^2, \end{aligned} \tag{4.2.6c}$$

which ends the proof of the first part of the proposition.

To prove the second part, we introduce the following 1-form on Σ, tangent to S_u at any point:

$$v_i = r\nabla\!\!\!/_i\phi. \tag{4.2.7a}$$

$$\nabla\!\!\!/_N v = [\nabla\!\!\!/_N, r\nabla\!\!\!/]\phi + r\nabla\!\!\!/\nabla_N\phi$$
$$\nabla\!\!\!/^2_N v = \nabla\!\!\!/_N([\nabla\!\!\!/_N, r\nabla\!\!\!/]\phi) + [\nabla\!\!\!/_N, r\nabla\!\!\!/]\nabla_N\phi + r\nabla\!\!\!/\nabla^2_N\phi$$
$$[\Delta\!\!\!/, \nabla\!\!\!/]\phi = K\nabla\!\!\!/\phi$$

together with the expansion 4.2.5 and the formula[2]

$$\begin{aligned}\Pi_i^j(\Delta v_j) = {} & \Delta\!\!\!/ v_i + \nabla\!\!\!/^2_N v_i - \theta_i^k\theta_{kj}v^j \\ & + \mathrm{tr}\theta\nabla\!\!\!/_N v_i + a^{-1}\nabla\!\!\!/^j a\nabla\!\!\!/_j v_i - a^{-2}\nabla\!\!\!/_i a\nabla\!\!\!/^j a v_j,\end{aligned} \tag{4.2.7b}$$

we derive

$$\begin{aligned}\Pi_i^j(\Delta v_j) = {} & r\nabla\!\!\!/_i f + \nabla\!\!\!/_N([\nabla\!\!\!/_N, r\nabla\!\!\!/]_i\phi) + \mathrm{tr}\theta[\nabla\!\!\!/_N, r\nabla\!\!\!/]_i\phi \\ & + [\nabla\!\!\!/_N, r\nabla\!\!\!/]_i\nabla_N\phi - r(\nabla\!\!\!/_i tr\theta)\nabla_N\phi - \nabla\!\!\!/_i(a^{-1}\nabla\!\!\!/^j a)v_j \\ & + Kv_i - \theta_i^k\theta_{kj}v^j - a^{-2}\nabla\!\!\!/_i a\nabla\!\!\!/^j a v_j.\end{aligned} \tag{4.2.7c}$$

Contracting 4.2.7c with v_i, integrating on Σ, and making use of the integration-by-parts formula,[3]

$$\int_\Sigma (\nabla_N F + \mathrm{tr}\theta F)d\mu_g = 0.$$

For scalar function F, we derive

$$\begin{aligned}\int_\Sigma\{|\nabla v|^2 + K|v|^2\} = {} & \overbrace{\int_\Sigma|\theta\cdot v|^2}^{J_1} - \overbrace{\int_\Sigma rv\cdot\nabla\!\!\!/ f}^{J_2} \\ & + \overbrace{\int_\Sigma \nabla\!\!\!/_N v\cdot[\nabla\!\!\!/_N, r\nabla\!\!\!/]\phi}^{J_3} - \overbrace{\int_\Sigma (v\cdot[\nabla\!\!\!/_N, r\nabla\!\!\!/])\nabla_N\phi}^{J_4} \\ & + \overbrace{\int_\Sigma r(v\cdot\nabla\!\!\!/\mathrm{tr}\theta)\nabla_N\phi}^{J_5} + \overbrace{\int_\Sigma a^{-1}v\cdot(\nabla\!\!\!/^2 a)\cdot v}^{J_6}.\end{aligned} \tag{4.2.7d}$$

[2] This formula holds for arbitrary 1-form v, tangent to S_u. To check the formula, we can consider a local orthonormal frame $N, (e_A)_{A=1,2}$ on Σ and compute $\Delta v_A = \nabla^2_N v_A + \delta^{BC}\nabla_B\nabla_C v_A$ using the formulas 3.1.1e.

[3] This follows immediately from integrating 3.1.1g with respect to u and using the coarea formula.

The first two integrals can be estimated with no difficulties in view of Proposition 4.2.1. and the degenerate form of the global Poincaré inequality (see Proposition 3.2.2):

$$J_1 \leq c \int_\Sigma |\nabla\!\!\!/ \phi|^2 \leq c r_0 \int_\Sigma \tau_-^2 |f|^2$$

$$J_2 \leq c r_0 \Big(\int_\Sigma \tau_-^2 r^2 |\nabla\!\!\!/ f|^2 \Big)^{1/2} \Big(\int_\Sigma |\nabla v|^2 \Big)^{1/2}.$$

Remark that J_2 cannot be estimated if we first integrate by parts, as one would naturally be tempted to do in order to avoid the loss of derivatives on f. Indeed if we do that, we will be stuck with the term $\int_\Sigma \frac{r^4}{\tau_-^2} |\nabla\!\!\!/^2 \phi|^2$, which is bad in the wave zone where τ_- is small in comparison to r.

To estimate J_3 and J_4, we need to apply the commutation formula (see the Lemma 3.2.1 on page 64)

$$[\nabla\!\!\!/_N, r\nabla\!\!\!/]_i \phi = -\frac{1}{2}\kappa \nabla\!\!\!/_i \phi - r\hat{\theta}_i^j \nabla\!\!\!/_j \phi + r a^{-1} \nabla\!\!\!/_i a \nabla_N \phi \qquad (4.2.7e)$$

where κ is given by 4.1.3d. Therefore, in view of Proposition 4.2.1,

$$\|[\nabla\!\!\!/_N, r\nabla\!\!\!/]_i \phi\| \leq \|\nabla \phi\| \leq c r_0^{1/2} \Big(\int_\Sigma \tau_-^2 |f|^2 \Big)^{1/2}. \qquad (4.2.7f)$$

Hence,

$$J_3 \leq c r_0^{1/2} \Big(\int_\Sigma \tau_-^2 |f|^2 \Big)^{1/2} \Big(\int_\Sigma |\nabla v|^2 \Big)^{1/2}.$$

Now, in view of 4.2.7e, we write

$$[\nabla\!\!\!/_N, r\nabla\!\!\!/] \nabla_N \phi = -\frac{1}{2}\kappa \nabla\!\!\!/ \nabla_N \phi - r\hat{\theta} \cdot \nabla\!\!\!/ \nabla_N \phi + r a^{-1} \nabla\!\!\!/ a \nabla_N^2 \phi$$

or, substituting ∇_N^2 from 4.2.5 and using the definition of v,

$$\begin{aligned}
[\nabla\!\!\!/_N, r\nabla\!\!\!/] \nabla_N \phi = & -\frac{1}{2}\kappa r^{-1} \nabla_N v - \hat{\theta} \cdot \nabla\!\!\!/_N v + \frac{1}{2} r^{-1} \kappa [\nabla\!\!\!/_N, r\nabla\!\!\!/] \phi \\
& + \hat{\theta} \cdot [\nabla\!\!\!/_N, r\nabla\!\!\!/] \phi \\
& + r a^{-1} \nabla\!\!\!/ a (f - \mathrm{tr}\theta \nabla_N \phi - r^{-1} d\!\!\!/iv\, v - r^{-1} a^{-1} \nabla\!\!\!/ a \cdot v).
\end{aligned}$$

Therefore, proceeding as in the estimate for J_3, we find

$$\begin{aligned}
J_4 \leq & \, c \|r^{-1} v\| (\|\nabla v\| + \|r^{-1} v\| + \|[\nabla\!\!\!/_N, r\nabla\!\!\!/]\phi\| + \|\nabla_N \phi\|) \\
& + c \|\tau_-^{-1} v\| \|\tau_- f\| \\
\leq & \, r_0^{1/2} \|\tau_- f\| (\|\nabla v\| + r_0^{1/2} \|\tau_- f\|).
\end{aligned}$$

To estimate J_5 we integrate by parts using the following

Remark 2 *Let v be a vectorfield on Σ, tangent to S_u at any point, and g a scalar function. Then*

$$\int_\Sigma g\, \not{d}\mathrm{iv}\, v = -\int_\Sigma a^{-1}\not\nabla(ag)\cdot v$$
$$\int_\Sigma \not\nabla g\cdot v = -\int_\Sigma g\, \not{d}\mathrm{iv}\, v - \int_\Sigma g(a^{-1}\not\nabla a)\cdot v.$$

With the help of this remark we write

$$J_5 = -\int_\Sigma r\mathrm{tr}\theta(\not{d}\mathrm{iv}\, v + a^{-1}\not\nabla a\cdot v)\nabla_N\phi - \int_\Sigma \mathrm{tr}\theta v(\not\nabla_N v - [\not\nabla_N, r\not\nabla]\phi).$$

Then, as before,

$$J_5 \le c r_0^{1/2}\|\tau_- f\|(\|\nabla v\| + r_0^{1/2}\|\tau_- f\|).$$

Similarly, after integrating by parts,

$$J_6 \le c r_0^{1/2}\|\tau_- f\|\|\nabla v\|.$$

Finally, collecting all the inequalities for $J_1 \dots J_6$, we infer from 4.2.7d

$$\int_\Sigma |\nabla v|^2 \le c r_0 \int_\Sigma \{\tau_-^2|f|^2 + r^2\tau_-^2|\not\nabla f|^2\}$$

or,

$$\int_\Sigma \{r^2|\not\nabla^2\phi|^2 + r^2|\not\nabla\not\nabla_N\phi|^2 \le c r_0 \int_\Sigma \{\tau_-^2|f|^2 + r^2\tau_-^2|\not\nabla f|^2\}. \qquad (4.2.8)$$

It remains only to estimate the second normal derivatives of ϕ. According to 4.2.5

$$\nabla_N^2\phi = f - \not\Delta\phi - \mathrm{tr}\theta\nabla_N\phi - a^{-1}\not\nabla a\not\nabla\phi. \qquad (4.2.9)$$

Multiplying this equation by $w_1 = \min\{\tau_- r_0^{1/2}, r\}$, integrating, and using the previous estimates, we infer that

$$\int_\Sigma w_1^2|\not\nabla_N\phi|^2 \le c r_0 \int_\Sigma |\tau_- f|^2 + \int_\Sigma r^2|(\not\Delta\phi + \mathrm{tr}\theta\nabla_N\phi + a^{-1}\not\nabla a\cdot\not\nabla\phi)|^2$$
$$\le c r_0 \int_\Sigma \{\tau_-^2|f|^2 + r^2\tau_-^2|\not\nabla f|^2\}. \qquad (4.2.10)$$

which, together with 4.2.8, ends the proof of the second part of the proposition.

Proposition 4.2.3 *Assume that in addition to the conditions 4.1.1a and 4.1.2a required in Proposition 4.2.2 the assumption 4.1.2e is also satisfied and let ϕ be a solution of the Poisson equation.*

(i.) If $\int_\Sigma \{r^2|f|^2 + r^4|\nabla\!\!\!/ f|^2\}$ is finite there exists a constant $\tilde{c}_5$, depending only on the quantities in 4.1.1a, 4.1.2a, 4.1.2c, 4.1.2d, and 4.1.2e, such that

$$\int_\Sigma r^4\{|\nabla\!\!\!/^3\phi|^2 + |\nabla\!\!\!/^2\nabla_N\phi|^2 + |\nabla\!\!\!/\nabla_N^2\phi|^2\} \leq \tilde{c}_5 \int_\Sigma \{r^2|f|^2 + r^4|\nabla\!\!\!/ f|^2\}.$$

(ii.) Assume in addition that the assumption 4.1.2b is satisfied and that the integral $\int_\Sigma \{\tau_-^2|f|^2 + r^2\tau_-^2|\nabla\!\!\!/ f|^2 + r^4\tau_-^2|\nabla\!\!\!/^2 f|^2\}$ is finite. There exists a constant $\tilde{c}_6$, depending only on the quantities in 4.1.1a, 4.1.2a, 4.1.2c, 4.1.2d, 4.1.2e, and 4.1.2b, such that

$$\int_\Sigma r^4\{|\nabla\!\!\!/^3\phi|^2 + |\nabla\!\!\!/^2\nabla_N\phi|^2\} + \int_\Sigma r^2 w_1^2|\nabla\!\!\!/\nabla_N^2\phi|^2$$
$$\leq \tilde{c}_6 r_0 \int_\Sigma \{\tau_-^2|f|^2 + r^2\tau_-^2|\nabla\!\!\!/ f|^2 + r^4\tau_-^2|\nabla\!\!\!/^2 f|^2\}.$$

Proof 4.2.3 *Consider the function $\chi = r^2 \Delta\!\!\!\!/\,\phi$. We have*

$$\triangle\chi = r^2\Delta\!\!\!\!/\, f + [\triangle, r^2\Delta\!\!\!\!/\,]\phi. \tag{4.2.11a}$$

Therefore,

$$\int_\Sigma |\nabla\chi|^2 = -\int_\Sigma r^2\chi\Delta\!\!\!\!/\, f - \int_\Sigma \chi[\triangle, r^2\Delta\!\!\!\!/\,]\phi. \tag{4.2.11b}$$

The first term in 4.2.11b can be easily estimated, after an integration by parts (see the remark on page 86), as follows:

$$\int_\Sigma r^2\chi\Delta\!\!\!\!/\, f = -\int_\Sigma r^2\nabla\!\!\!/\chi\cdot\nabla\!\!\!/ f + \int_\Sigma r^2\chi a^{-1}\nabla\!\!\!/ a\cdot\nabla\!\!\!/ f \tag{4.2.11c}$$
$$\leq c(\|\nabla\!\!\!/\chi\| + \|r^{-1}\chi\|)\|r^2\nabla\!\!\!/ f\|.$$

To estimate the commutator term in 4.2.11b,[4]*, we write, with the help of 4.2.3,*

$$[\triangle, r^2\Delta\!\!\!\!/\,]\phi = (\nabla_N A + \mathrm{tr}\theta A) + B + C \tag{4.2.11d}$$

where

$$A = [\nabla_N, r^2\Delta\!\!\!\!/\,]\phi \tag{4.2.11e}$$
$$B = [\nabla_N, r^2\Delta\!\!\!\!/\,]\nabla_N\phi \tag{4.2.11f}$$
$$C = -2r^2\nabla\!\!\!/\mathrm{tr}\theta\cdot\nabla\!\!\!/\nabla_N\phi - r^2\Delta\!\!\!\!/\,\mathrm{tr}\theta\nabla_N\phi + [a^{-1}\nabla\!\!\!/ a\cdot\nabla\!\!\!/, r^2\Delta\!\!\!\!/\,]\phi. \tag{4.2.11g}$$

[4] Remark that the value of this commutator in Euclidean space is zero

In view of Corollary 3.2.3.2 we have

$$A = -\kappa r^{-1} r^2 \not\Delta \phi - 2r^2 \hat{\theta} \cdot \not\nabla^2 \phi + 2r^2 a^{-1} \not\nabla a \cdot \not\nabla \nabla_N \phi \qquad (4.2.11h)$$
$$- r^2 \not\nabla \mathrm{tr}\theta \cdot \not\nabla \phi - 2r^2 P \cdot \not\nabla \phi - 2r^2 a^{-1} \not\nabla a \cdot \hat{\theta} \cdot \not\nabla \phi + r^2 a^{-1} \not\Delta a \nabla_N \phi$$

and similarly for B. Now,

$$\int_\Sigma \chi[\Delta, r^2 \not\Delta]\phi = \int_\Sigma (\nabla_N A + \mathrm{tr}\theta A)\chi + \int_\Sigma (B + C)\chi. \qquad (4.2.11i)$$

The first term on the right can be estimated, after integration by parts, as follows:

$$\int_\Sigma (\nabla_N A + \mathrm{tr}\theta A)\chi = -\int_\Sigma A \nabla_N \chi \qquad (4.2.11j)$$
$$\leq \|A\| \|\nabla \chi\|$$
$$\leq c \|r f\| \|\nabla \chi\|.$$

The last part of 4.2.11j is quite straightforward. Indeed, a typical term to estimate is $\|r^2 \not\nabla \mathrm{tr}\theta \cdot \not\nabla \phi\|$:

$$\|r^2 \not\nabla \mathrm{tr}\theta \cdot \not\nabla \phi\| \leq \|r^{5/4} \not\nabla \mathrm{tr}\theta\|_4 \|r^{3/4} \not\nabla \phi\|_4.$$

On the other hand, in view of the global Sobolev inequalities of Proposition 3.2.1i, for a given tensor F, tangent on any $S_{t,u}$,

$$\|r^{3/4} F\|_4 \leq \|F\|^{1/2} (\|F\| + \|r \nabla F\|)^{1/2}. \qquad (4.2.11k)$$

Therefore, with the help of Proposition 4.2.1; and Proposition 4.2.2i, respectively, we deduce that

$$\|r^{3/4} \not\nabla \phi\|_4 \leq c \|r f\|.$$

To estimate $\int_\Sigma B\chi$ *we write, after an integration by parts*[5] *and application of the Codazzi equations 3.1.1e,* $\not{d}\mathrm{iv}\, \hat{\theta} = \frac{1}{2} \not\nabla \mathrm{tr}\theta + P$,

$$\int_\Sigma B\chi = \int_\Sigma r\kappa \not\nabla \chi \cdot \not\nabla \nabla_N \phi + \int_\Sigma r^2 \chi \not\nabla \mathrm{tr}\theta \cdot \not\nabla \nabla_N \phi \qquad (4.2.11l)$$
$$+ 2 \int_\Sigma r^2 \not\nabla \chi \hat{\theta} \cdot \not\nabla \nabla_N \phi + \int_\Sigma r^2 \chi \mathrm{tr}\theta a^{-1} \not\nabla a \cdot \not\nabla \nabla_N \phi$$
$$- 2 \int_\Sigma r^2 a^{-1} \not\nabla a \cdot \not\nabla \chi \nabla_N^2 \phi - \int_\Sigma r^2 a^{-1} \chi \not\Delta a \nabla_N^2 \phi.$$

[5] See the remark on page 86.

Now proceeding precisely in the same manner as in the estimates involving A, we find

$$\int_\Sigma B\chi \leq c\|rf\|(\|rf\| + \|\nabla\chi\|) \tag{4.2.11m}$$

$$\int_\Sigma C\chi \leq c\|rf\|(\|rf\| + \|\nabla\chi\|). \tag{4.2.11n}$$

Finally, combining 4.2.11m and 4.2.11n with 4.2.11j and 4.2.11c, we infer that

$$\|\nabla\chi\|^2 \leq c(\|rf\| + \|r^2\nabla\!\!\!/ f\|)(\|\nabla\chi\| + \|rf\|)$$

or

$$\|r^4\nabla\Delta\!\!\!/\phi\| \leq c(\|rf\| + \|r^2\nabla\!\!\!/ f\|). \tag{4.2.11o}$$

On the other hand, since $\Delta\!\!\!/\phi = r^{-2}\chi$, we can use the 2-dimensional elliptic estimates (see Proposition 2.2.2i a pplied to $\xi = \nabla\!\!\!/\phi$) on each $S_u$to obtain, after an integration relative to u,

$$\int_\Sigma r^4|\nabla\!\!\!/^3\phi|^2 \leq c\int_\Sigma \{r^2|f|^2 + r^4|\nabla\!\!\!/ f|^2\}. \tag{4.2.11p}$$

To estimate $\nabla\!\!\!/^2\nabla_N\phi$, we write

$$\Delta\!\!\!/\nabla_N\phi = r^{-2}(\nabla_N\chi - A)$$

where A is the commutator 4.2.11e. Thus using the 2-dimensional elliptic theory and the expression 4.2.11h for A, we find

$$\begin{aligned}\int_\Sigma r^4|\nabla\!\!\!/^2\nabla_N\phi|^2 &\leq c\int_\Sigma \{|\nabla_N\chi|^2 + |A|^2\} \\ &\leq c\int_\Sigma \{r^2|f|^2 + r^4|\nabla\!\!\!/ f|^2\}.\end{aligned} \tag{4.2.11q}$$

Finally, we estimate $\int_\Sigma r^4|\nabla\!\!\!/\nabla_N^2\phi|$ by taking tangential derivatives of the equation 4.2.9 and estimating the right-hand side in a straightforward fashion. This ends the proof of the first part of this proposition.

To prove the second part of the proposition, we start in the same manner, but instead of integrating by parts, in the first integral of the right-hand side of 4.2.11b, we write with the help of the degenerate form of the global Poincaré inequality, Proposition 3.2.2,

$$\begin{aligned}\int_\Sigma r^2\chi\Delta\!\!\!/ f &\leq \|\tau_-^{-1}\chi\|\|\tau_- r^2\Delta\!\!\!/ f\| \\ &\leq cr_0^{1/2}\|\nabla\chi\|\|\tau_- r^2\nabla\!\!\!/^2 f\|.\end{aligned} \tag{4.2.12a}$$

The commutator term in 4.2.11b is estimated as before with the help of the formulas 4.2.11d–4.2.11i. Then, proceeding as in 4.2.11j, we estimate $\|A\|$ as follows:

$$\begin{aligned}\|A\| \leq c\{r_0^{1/2}(\|\tau_- f\| + \|\tau_- r \nabla\!\!\!/ f\|) \\ + \|r^{3/4}\nabla\!\!\!/ \phi\|_4 + \|r^{3/4}\nabla_N \phi\|_4\}.\end{aligned} \tag{4.2.12b}$$

In view of the global Sobolev inequality 4.2.11k and the second derivative estimates for ϕ,

$$\|r^{3/4}\nabla\!\!\!/ \phi\|_4 \leq c r_0^{1/2}\left(\int_\Sigma \tau_-^2\{|f|^2 + r^2|\nabla\!\!\!/ f|^2\}\right)^{1/2}. \tag{4.2.12c}$$

On the other hand, to estimate $\|r^{3/4}\nabla_N \phi\|_4$, we need a degenerate form of 4.2.11k:

Lemma 4.2.1 *Let F be an arbitrary tensor on Σ, tangent to $S_u$at every point, and assume, as in Proposition 3.2.1, that the foliation given by the radial function u has positive mean curvature. There exists a constant $c = c(I, a_M, a_m)$ such that*

$$\int_\Sigma r^3|F|^4 \leq c\left(\int_\Sigma w_1^2|\nabla\!\!\!/_N F|^2\right)\left(\int_\Sigma \{|F|^2 + r^2|\nabla\!\!\!/ F|^2\}\right).$$

Therefore,

$$\begin{aligned}\int_\Sigma |r^{3/4}\nabla_N \phi|^4 &\leq \left(\int_\Sigma w_1^2|\nabla\!\!\!/_N \nabla\!\!\!/_N \phi|^2\right)\left(\int_\Sigma \{|\nabla\phi|^2 + r^2|\nabla\!\!\!/ \nabla_N \phi|^2\}\right) \\ &\leq c r_0^2\left(\int_\Sigma \tau_-^2\{|f|^2 + r^2|\nabla\!\!\!/ f|^2\}\right)^2\end{aligned} \tag{4.2.12d}$$

and consequently

$$\|A\| \leq r_0^{1/2}(\|\tau_- f\| + \|r\tau_- \nabla\!\!\!/ f\|), \tag{4.2.12e}$$

which implies the estimate

$$\int_\Sigma (\nabla_N A + \mathrm{tr}\theta A)\chi \leq r_0^{1/2}(\|\tau_- f\| + \|r\tau_- \nabla\!\!\!/ f\|)\|\nabla\chi\|. \tag{4.2.12f}$$

To estimate $\int_\Sigma B\chi$, we recall formula 4.2.11l. The only terms whose estimation requires some subtlety are the last two terms on the right, which involve $\nabla_N^2\phi$. To do this, we write, using 4.2.5,

$$\begin{aligned}\nabla_N^2\phi &= f - g \\ g &= r^{-2}\chi + \mathrm{tr}\theta\nabla_N\phi + a^{-1}\nabla\!\!\!/ a \cdot \nabla\!\!\!/ \phi.\end{aligned} \tag{4.2.12g}$$

In view of the results of the second part of Proposition 4.2.3, we deduce

$$\|rg\| \leq r_0^{1/2}(\|\tau_- f\| + \|r\tau_- \nabla f\|). \tag{4.2.12h}$$

If we denote by I_1, I_2 the contribution of g to the last two terms in 4.2.11l, we can thus write

$$\begin{aligned} r_0^{1/2} I_1 &\leq c I_2 \|r \nabla \chi\| \|rg\| && (4.2.12\text{i}) \\ &\leq c r_0^{1/2} \|\nabla \chi\| (\|\tau_- f\| + \|r\tau_- \nabla f\|). \\ I_2 &\leq c \|r^{5/4} a^{-1} \Delta a\|_4 \cdot \|r^{-1/4} \chi\|_4 \cdot \|rg\| && (4.2.12\text{j}) \\ &\leq c r_0^{1/2} \|\nabla \chi\| (\|\tau_- f\| + \|r\tau_- \nabla f\|) + r_0 (\|\tau_- f\| + \|r\tau_- \nabla f\|)^2. \end{aligned}$$

To estimate the contribution of f to the last two terms in 4.2.11l, we first integrate by parts in the last integral. The resulting sum can then be estimated as follows, with the help of the degenerate form of the global Poincaré inequality:[6]

$$\begin{aligned} \int_\Sigma r^2 a^{-1} \nabla a \cdot (f \nabla \chi - \chi \nabla f) &\leq (\sup_\Sigma r\tau_-^{-1})^{1/2} \sup_\Sigma r^{3/2} \tau_-^{-1/2} a^{-1} |\nabla a| \\ &\quad \cdot \left\{ \|\tau_- f\| \|\nabla \chi\| + \|r^{-1} \chi\| \|r\tau_- \nabla f\| \right\} \\ &\leq r_0^{1/2} \tilde{a}_1 \|\nabla \chi\| (\|\tau_- f\| + \|r\tau_- \nabla f\|). \end{aligned}$$

Therefore,

$$\int_\Sigma B\chi \leq c r_0^{1/2} \|\nabla \chi\| (\|\tau_- f\| + \|r\tau_- \nabla f\|). \tag{4.2.12k}$$

Finally, the last term, $\int_\Sigma C\chi$, can be estimated without further difficulties, and, together with 4.2.11b, 4.2.11d, and 4.2.12a–4.2.12k, we conclude that

$$\|\nabla \Delta \phi\| \leq c r_0 (\|\tau_- f\| + \|r\tau_- \nabla f\| + \|\tau_- r^2 \nabla^2 f\|). \tag{4.2.12l}$$

Thus it remains to estimate only $\int_\Sigma r^2 w_1^2 |\nabla \nabla_N^2 \phi|^2$. This is now straightforward, by taking the angular derivatives in 4.2.9 and proceeding as in 4.2.10.

Proof 4.2.4 *We first apply the isoperimetric inequality (see page 32) to the scalar $|F|^2$, on each S_u. Hence,*

$$\int_{S_u} |F|^4 \leq c(I) \left(r^{-2} \int_{S_u} |F|^2 \right) \left(\int_{S_u} \{ |F|^2 + r^2 |\nabla F|^2 \} \right).$$

[6] Recall, from Remark 2 on page 58, that $\frac{r}{\tau_-} \leq c r_0$.

Multiplying this by r^3 and integrating in u, we derive

$$\int_\Sigma r^3|F|^4 \le c(I) \sup_u \left(r \int_{S_u} |F|^2 \right) \cdot \left(\int_\Sigma \{|F|^2 + r^2|\nabla F|^2\} \right).$$

It thus suffices to show that

$$\sup_u \left(r \int_{S_u} |F|^2 \right) \le \int_\Sigma w_1^2 |\nabla_N F|^2. \tag{4.2.13}$$

We proceed as follows:

$$\begin{aligned}
\int_{S_u} |F|^2 &= -\int_{ExtS_u} div(N|F|^2) \\
&= -\int_{Ext} \mathrm{tr}\theta |F|^2 - 2\int_{ExtS_u} F \cdot \nabla_N F \\
&\le 2a_M \int_u^\infty \left(\int_{S_{u'}} |F|^2 \right)^{1/2} \left(\int_{S_{u'}} |\nabla_N F|^2 \right)^{1/2} du'.
\end{aligned}$$

Therefore,

$$\begin{aligned}
\int_{S_u} |F|^2 &\le a_M \left[\int_u^\infty \left(\int_{S_{u'}} |\nabla_N F|^2 \right)^{1/2} du' \right]^2 \\
&\le \frac{a_M}{a_m} \left(\int_u^\infty w_1^{-2} du' \right) \left(\int_\Sigma w_1^2 |\nabla_N F|^2 \right).
\end{aligned}$$

But

$$\int_u^\infty w_1^{-2} du' \le 2 \int_u^\infty (\tau_-^{-2} r_0^{-1} + r^{-2}) du \le c\frac{1}{r},$$

which ends the proof of 4.2.13 and the lemma.

In the end of this paragraph we discuss sup-norm estimates for solutions of the Poisson equation. For the purpose of this book we derive them from the L^2-estimates with the help of the sup-norm version of the global Sobolev inequalities (see Proposition 3.2.3). This is straightforward in the case when we use only the nondegenerate estimates included in the first parts of Propositions 4.2.1–4.2.3; however, the application of the global Sobolev inequalities to the degenerate L^2 estimates of the second parts of Propositions 4.2.1–4.2.3 encounters the difficulty that the terms involving the highest normal derivatives on the left-hand sides of Proposition 4.2.2 and Proposition 4.2.3, respectively, are weighted by the factor w_1, which does not appear in the global Sobolev inequalities. We circumvent this difficulty by using instead of sup-norm estimates on Σ the sup-norm in u of L^3 norms on the spheres S_u. More precisely, we prove the following:

Proposition 4.2.4 *Let ϕ be a solution of the Poisson equation.*

- *Under the same assumptions as in Proposition 4.2.2 there exists a constant c such that*

$$\sup_u \left[r^{5/6} \left(\int_{S_u} |\nabla\!\!\!/ \phi|^3 + |\nabla_N \phi|^3 \right)^{1/3} \right] \leq c r_0^{1/2} \left(\int_\Sigma \{ \tau_-^2 |f|^2 + r^2 \tau_-^2 |\nabla\!\!\!/ f|^2 \} \right)^{1/2}.$$

- *Under the same assumptions as in Proposition 4.2.3 there exists a constant c such that*

$$\sup_u \left[r^{11/6} \left(\int_{S_u} |\nabla\!\!\!/^2 \phi|^3 + |\nabla\!\!\!/ \nabla_N \phi|^3 \right)^{1/3} \right] \leq c r_0^{1/2} \left(\int_\Sigma \{ \tau_-^2 |f|^2 + r^2 \tau_-^2 |\nabla\!\!\!/ f|^2 + r^4 \tau_-^2 |\nabla\!\!\!/^2 f|^2 \} \right)^{1/2}.$$

Proof 4.2.5 *It suffices to prove the desired inequalities for the terms on the left that involve normal derivatives. Indeed, the other terms can be estimated by remarking that, for a given tensor u on S_u, $(\int_S |u|^3)^{1/3} \leq r^{1/6} (\int_S |u|^4)^{1/4}$ and then by appealing to the nondegenerate form of Corollary 3.2.11.*

Thus, letting $\psi = \nabla_N \phi$, we have to show that

$$\sup_u r^{5/6} \left(\int_{S_u} |\psi|^3 \right)^{1/3} \leq c r_0^{1/2} Q \tag{4.2.14a}$$

where $Q = \left(\int_\Sigma \{ \tau_-^2 |f|^2 + r^2 \tau_-^2 |\nabla\!\!\!/ f|^2 \} \right)^{1/2}$. Now, since in view of 4.2.12g,

$$\nabla_N \psi = \nabla_N^2 \phi = f - g, \tag{4.2.14b}$$

we have

$$\begin{aligned} \int_{S_u} |\psi|^3 &= - \int_{Ext(S_u)} div(N|\psi|^3) \\ &\leq 3 \int_{Ext(S_u)} |\psi|^2 |\nabla_N \psi| \qquad (4.2.14c) \\ &\leq 3 \int_{Ext(S_u)} |\psi|^2 (|f| + |g|). \qquad (4.2.14d) \end{aligned}$$

On the other hand, according to 4.2.12h, Lemma 4.2.1, and Proposition 4.2.2, we estimate

$$\begin{aligned}\int_{Ext(S_u)} |\psi|^2|g| &\leq c r_0^{1/2} r^{-5/2} Q \Big(\int_\Sigma r^3 |\psi|^4 \Big)^{1/2} \\ &\leq c r_0^{1/2} r^{-5/2} Q \int_\Sigma \{ |\nabla\!\!\!/_N \phi|^2 + r^2 |\nabla\!\!\!/ \nabla_N \phi|^2 + w_1^2 |\nabla_N^2 \phi|^2 \} \\ &\leq c r_0 r^{-5/2} Q^3. \end{aligned} \tag{4.2.14e}$$

With the help of the isoperimetric inequality on each S_u we have $(\int_{S_u} |f|^3)^{1/3} \leq r^{-1/3} (\int_{S_{u'}} |f|^2 + r^2 |\nabla\!\!\!/ f|^2)^{1/2}$ and therefore

$$\int_{Ext(S_u)} |\psi|^2 |f| \leq c \int_u^\infty du' \Big(\int_{S_{u'}} |\psi|^3 \Big)^{2/3} \Big(\int_{S_{u'}} |f|^3 \Big)^{1/3} \tag{4.2.14f}$$

$$\leq c \int_u^\infty du' r^{-1/3} \Big(\int_{S_{u'}} |f|^2 + r^2 |\nabla\!\!\!/ f|^2 \Big)^{1/2} \Big(\int_{S_{u'}} |\psi|^3 \Big)^{2/3}.$$

Therefore, we derive the integral equation in $\int_{S_u} |\psi|^3$,

$$\begin{aligned}\int_{S_u} |\psi|^3 &\leq c r_0 r^{-5/2} Q^2 \\ &+ \int_u^\infty du' r^{-1/3} \Big(\int_{S_{u'}} |f|^2 + r^2 |\nabla\!\!\!/ f|^2 \Big)^{1/2} \Big(\int_{S_{u'}} |\psi|^3 \Big)^{2/3}.\end{aligned}$$

Now apply the following:

Lemma 4.2.2 *Consider A and B and X three real, positive, continuous functions of u with B, X vanishing sufficiently fast as $u \to \infty$. We assume also that A is nonincreasing and that X satisfies the integral inequality*

$$X(u) \leq A(u) + \int_u^\infty B(s) X(s)^{1-1/p}$$

for some $p > 1$. Then,

$$X(u) \leq 2^p A(u) + \Big(\frac{4}{p} \int_u^\infty B(s) ds \Big)^p.$$

In view of this we infer

$$\int_{S_u} |\psi|^3 \leq c r_0 r^{-5/2} Q^3 + c I^3 \tag{4.2.14g}$$

with I being the integral

$$I = \int_u^\infty du' r^{-1/3} \left(\int_{S_{u'}} \{|f|^2 + r^2 |\nabla f|^2\} \right)^{1/2} \tag{4.2.14h}$$
$$\leq \int_u^\infty du' \tau_-^2 \left(\int_{S_{u'}} \{|f|^2 + r^2 |\nabla f|^2\} \right)^{1/2} \left(\int_u^\infty du' r^{-2/3} \tau_-^{-2} \right)^{1/2}$$
$$\leq cQ r^{-1/3} \tau_-^{-1/2} \leq cQ r_0^{1/2} r^{-5/6}$$

and consequently

$$\int_{S_u} |\psi|^3 \leq c r_0^{3/2} r^{-5/2} Q^3, \tag{4.2.14i}$$

which proves the desired inequality.

The proof of the second part of the proposition is proved in precisely the same manner.

To prove Lemma 4.2.2, we first remark that if A, B, and X are three real, continuous, and positive functions of t in some interval $[0, T]$ and X satisfies the inequality

$$X(t) \leq A(t) + \int_0^t B(s) X(s)^{1-1/p} ds$$

for some $p > 1$, and if, in addition, A is nondecreasing,

$$X(t) \leq 2^p A(t) + \left(\frac{4}{p} \int_0^t B(s) ds \right)^p.$$

To show this, consider $Y(t) = \int_0^t B(s) X(s)^{1-1/p}$ and remark that $\frac{d}{dt} Y(t) \leq B(t)(Y(t) + A(t))^{1-1/p}$. If at an arbitrary point t we have $Y \leq A$, then $X(t) \leq 2A(t)$ and we are done. If, however, the opposite is true, then let t_0 be the largest value in $[0, t]$ for which $Y(t_0) \leq A(t_0)$. This value exists since $Y(0) = 0$. Then in the interval $[t_0, t]$ we have $\frac{d}{ds} Y(s) \leq 2^{1-1/p} B(s) Y(s)^{1-1/p}$. Integrating, we derive

$$Y(t) \leq 2^p Y(t_0) + 2^{2p-1} \left(\int_{t_0}^t B(s) ds \right)^p$$
$$\leq 2^p A(t_0) + \left(4 \int_{t_0}^t B(s) ds \right)^p$$
$$\leq 2^p A(t) + \left(4 \int_{t_0}^t B(s) ds \right)^p,$$

which proves our assertion.

Finally, to prove the lemma, set $t = 1/u$, $\overline{A}(\sigma) = A(1/\sigma)$, $\overline{B}(\sigma) = B(1/\sigma)\frac{1}{\sigma^2}$, $\overline{X}(\sigma) = X(1/\sigma)$. Then

$$\overline{X}(t) \leq \overline{A}(t) + \int_0^t \overline{B}(s)\overline{X}(s)^{1-1/p} ds,$$

and we are in a position to apply the result.

4.3. L^2 Estimates for the Principal Error Term

In this paragraph we replace the lapse function a by introducing the normalized lapse function

$$b = a\lambda^{-1}. \tag{4.3.1}$$

In view of this we find that

$$\overline{b\mathrm{tr}\theta} = \frac{2}{r}.$$

Our aim here is to provide estimates for the error term $\phi - r^{-1}P$ as $r \to \infty$, where $P = P[\phi]$ is the charge integral $-\frac{1}{4\pi}\int_\Sigma f$ (see 4.1.4). Therefore we assume that both ϕ and f are supported in the region $r \geq R$, for some large value of R, and that in the same region the normalized lapse function b and the second fundamental form of the foliation θ verify the following assumptions:

$$\begin{aligned}\sup_{r\geq R} r^2\left(b\mathrm{tr}\theta - \frac{2}{r}\right) &\leq c_0 \\ \sup_{r\geq R} r^2|\hat{\theta}| &\leq c_0 \\ \sup_{r\geq R} r^2|\nabla b| &\leq c_0.\end{aligned} \tag{4.3.2a}$$

We also assume that

$$\begin{aligned}\sup_{r\geq R}\left(\int_{S_r} r^7|\nabla^2 b|^3\right)^{1/3} &\leq c_1 \\ \sup_{r\geq R}\left(\int_{S_r} r^7|\nabla\!\!\!/\mathrm{tr}\theta|^3\right)^{1/3} &\leq c_1 \\ \sup_{r\geq R}\left(\int_{S_r} r^7|\mathrm{Ric}|^3\right)^{1/3} &\leq c_1.\end{aligned} \tag{4.3.2b}$$

Remark 3 *We remark that both assumptions 4.3.2a and 4.3.2b can be significantly weakened. In fact the weight r^2 in 4.3.1 can be replaced by r^q for any*

$q > 3/2$ while the weight r^7 in 4.3.2b can be replaced by r^p for any $p > 11/2$. The exact assumptions can be easily derived from the definition of the quantities $\mu_1 - \mu_6$ in 4.3.8c and 4.3.8d. We simply need them to be bounded and tend to zero as $R \longrightarrow \infty$.

To justify what follows, we first consider the Poisson equation $\triangle\phi = f$ in the Euclidean space $\mathcal{R}^3$ and perform the inversion $\overline{x} = \frac{x}{r^2}$ where $r = |x|$. Also let $\overline{r} = |\overline{x}|$, we have $\overline{r} = r^{-1}$. Relative to the new coordinate system $\overline{x}$, the Poisson equation takes the form $\overline{\triangle}(\overline{\phi}) = \overline{f}$ where $\overline{\phi} = r\phi$ and $\overline{f} = r^5\phi$. The basic L^2 identity for the second derivatives of $\overline{\phi}$, relative to $\overline{x}$, is

$$\int |\overline{\nabla}^2(\overline{\phi})|^2 d\overline{x}^3 = \int |\overline{f}|^2 d\overline{x}^3$$
$$= \int r^4 |f|^2 dx^3.$$

On the other hand, changing back to the x coordinates, we have $\overline{\nabla}_i\overline{\phi} = \frac{\partial(r\phi)}{\partial\overline{x}^i} = \frac{\partial(r\phi)}{\partial x^j}\frac{\partial x^j}{\partial\overline{x}^i} = rw_i$ where $w_i = rh_i^j\nabla_j(r\phi)$ and $h_i^j = \frac{1}{r^2}\frac{\partial x^j}{\partial\overline{x}^i} = \delta_i^j - 2\frac{x_i x^j}{r^2}$. Thus also, $\overline{\nabla}_k\overline{\nabla}_i\overline{\phi} = r^3 h_k^l \nabla_l w_i - r x_k w_i$. Hence, taking into account that $h_k^i h_j^k = \delta_j^i$ and $h_j^i x^j = -x^i$, we infer that $|\overline{\nabla}^2(\overline{\phi})|^2 = r^6|\nabla w|^2 + r^4 x^j \nabla_j(|w|^2) + r^4|w|^2$, and, after an integration by parts, we find

$$\int |\overline{\nabla}^2(\overline{\phi})|^2 d\overline{x}^3 = \int |\nabla w|^2 d^3x.$$

We therefore deduce the identity

$$\int |\nabla w|^2 d^3x = \int r^4 |f|^2 dx^3,$$

which we want to generalize in curved space.

This motivates us to introduce, on our manifold Σ, the equation

$$w_i = rh_i{}^j\nabla_j(r\phi) \tag{4.3.3a}$$

where h is the Einstein metric

$$h_i{}^j = \delta_i{}^j - 2N_i N^j. \tag{4.3.3b}$$

Remark that

$$h_i{}^j h_j^k = \delta_i^k. \tag{4.3.3c}$$

We calculate Δw and find the following:

Lemma 4.3.1 *Let ϕ be a solution of the Poisson equation and let w be defined as above. Then*

$$\triangle w_i = (r^2 \delta_i{}^j - 2Z_i Z^j)\nabla_j f - \left(5b^{-1} + \frac{4}{3}\kappa\right) Z_i f \tag{4.3.4a}$$
$$+ A_i{}^{jk}\nabla_j \nabla_k \phi + B_i{}^j \nabla_j \phi + C_i \phi$$

where

$$A_i{}^{jk} = -\frac{4}{3}\kappa\delta_i{}^k Z^j - 2(\aleph^k{}_i Z^j + \aleph^{kj} Z_i) \tag{4.3.4b}$$

$$B_i{}^j = \left(-\frac{4}{3}\kappa\left(b^{-1} + \frac{1}{3}\kappa\right) - 2rb^{-2}\nabla_N b\right)\delta_i{}^j \tag{4.3.4c}$$
$$- 2\left(b^{-1} + \frac{1}{3}\kappa\right)(\aleph^j{}_i + \aleph_i{}^j) - \aleph_{mi}\aleph^{mj} - b^{-1}\aleph^j{}_i$$
$$+ b^{-2} Z_i \nabla^j b - 2(Z_i \triangle Z^j + Z^j \triangle Z_i) + (r^2 R_i{}^j - 2Z_i Z^m R_m^j)$$

$$C_i = (b^{-2}\triangle b - 2b^{-3}|\nabla b|^2)Z_i + 2\left(b^{-1} + \frac{1}{3}\kappa\right)b^{-2}\nabla_i b \tag{4.3.4d}$$
$$+ b^{-2}\aleph_{ij}\nabla^j b - b^{-1}\triangle Z$$

with

$$\triangle Z_j = \frac{1}{2}\nabla^i \aleph_{ij} + \nabla_j\left(b^{-1} + \frac{1}{3}\kappa\right).$$

Now, multiplying the equation 4.3.4a by w and integrating on Σ, we find, for the first term on the right-hand side,

$$-\int_\Sigma \left(r^2 w \cdot \nabla f - 2Z(f) w \cdot Z\right) =$$
$$\int_\Sigma \left(r^2 \mathrm{div} w - 2\nabla_Z(w \cdot Z)\right) f + \text{ error terms}$$

and

$$r^2 \mathrm{div} w - 2\nabla_Z(w \cdot Z) = r^2 \nabla^i (h_i{}^j w_j) + \ldots$$
$$= r^4 f + r^2(b^{-1} - 2\kappa)\nabla_Z \phi$$
$$+ r^2 b^{-1}(b^{-1} - \kappa - b^{-1}\nabla_Z b)\phi.$$

On the other hand, we write

$$\nabla_i \phi = r^{-2} h_i{}^j w_j - r^{-2} b^{-1} r\phi N_i \tag{4.3.5a}$$
$$\nabla_j \nabla_k \phi = r^{-2} h_k{}^m \nabla_j w_m + r^{-3} T_{jk}{}^m w_m + r^{-3} S_{jk} \tag{4.3.5b}$$

where

$$T_{jk}{}^{m} = b^{-1}\left(-2N_j\delta_k{}^m - N_k\delta_j{}^m + 10N_jN_kN^m\right) \tag{4.3.5c}$$
$$- 2\left(b^{-1} + \frac{1}{3}\kappa\right)(g_{jk}N^m + N_k\delta_j{}^m) - \aleph_{jk}N^m - N_k\aleph_j{}^m$$
$$S_{jk} = b^{-1}\left(3N_jN_k - \left(b^{-1} + \frac{1}{3}\kappa\right) - \frac{1}{2}\aleph_{jk} + rb^{-1}\nabla_j bN_k\right). \tag{4.3.5d}$$

We are now ready to derive the following crucial identity:

Lemma 4.3.2 *Let ϕ be a solution of the Poisson equation and let w be defined by the formula 4.3.3a. Then the following integral identity holds true:*

$$\int_\Sigma |\nabla w|^2 = \int_\Sigma r^4|f|^2 \tag{4.3.6a}$$
$$+ \int_\Sigma \frac{4}{3}r\kappa N\cdot wf + \int_\Sigma rb^{-1}(\kappa - rb^{-1}\nabla_N b)r\phi f$$
$$- \int_\Sigma w^i\hat{A}_i{}^{jm}\nabla_j w_m - \int_\Sigma w^i\hat{B}_i{}^m w_m - \int_\Sigma w^i\hat{C}_i r\phi$$

where

$$\hat{A}_i{}^{jm} = r^{-2}A_i{}^{jn}h_n{}^m$$
$$\hat{B}_i{}^m = r^3A_i{}^{jn}T_{jn}{}^m + r^{-2}B_i{}^jh_j{}^m \tag{4.3.6b}$$
$$\hat{C}_i = r^{-1}C_i - r^{-2}b^{-1}B_i{}^jN_j - r^{-3}A_i{}^{jm}S_{jm}.$$

We now proceed to estimate the error terms, which are present in the general case. First remark that, in view of the local Sobolev inequality and the global Poincaré inequality (see Corollary 3.2.1.2 and Proposition 3.2.2), and with the help of the identity $|\nabla(r\phi)|^2 = r^{-2}|w|^2$, we infer that

$$\|r\phi - P\|_6 \le c_{\mathrm{sob}}\|\nabla(r\phi)\|_2$$
$$\le c_{\mathrm{sob}}c_{\mathrm{P}}\|\nabla w\|_2. \tag{4.3.7}$$

Thus writing $w\cdot\hat{C}r\phi = Pw\cdot\hat{C} + (r\phi - P)w\cdot\hat{C}$, we find

$$\int_\Sigma w\cdot\hat{C}r\phi \le c_{\mathrm{sob}}|P|\|\hat{C}\|_{6/5,R}\|\nabla w\|_2$$
$$+ c_{\mathrm{sob}}^2c_{\mathrm{P}}\|\hat{C}\|_{3/2,R}\|\nabla w\|_2^2$$
$$\int_\Sigma w^i\hat{B}_i{}^m w_m \le c_{\mathrm{sob}}^2\|\hat{B}\|_{3/2,R}\|\nabla w\|_2^2$$
$$\int_\Sigma w^i\hat{A}_i{}^{jm}\nabla_j w_m \le c_{\mathrm{sob}}\|\hat{A}\|_{3,R}\|\nabla w\|_2^2,$$

also

$$\int_\Sigma \frac{4}{3} r\kappa N \cdot w f \leq c_P \sup_{r\geq R} |\kappa| \|r^2 f\|_2 \|\nabla w\|_2$$

$$\int_\Sigma rb^{-1}(\kappa - rb^{-1}\nabla_N b) r\phi f \leq |P| \|r^{-1}(\kappa - rb^{-1}\nabla_N b)\|_{2,R} \|r^2 f\|_2 + c_P c_{sob} \|r^{-1}(\kappa - rb^{-1}\nabla_N b)\|_{3,R} \|\nabla w\|_2 \|r^2 f\|_2$$

where $\| \ \|_{p,R}$ denotes the L^p in the exterior region $r \geq R$. Therefore,

Corollary 4.3.1.1 *As a consequence of Lemma 4.3.2 we derive the following estimate:*

$$\|\nabla w\|^2 \leq \|r^2 f\|^2 + \nu_0 \|\nabla w\|^2 + (\nu_1 \|r^2 f\| + \nu_2 |P|) \|\nabla w\| \tag{4.3.8a}$$

where

$$\begin{aligned} \nu_0 &= c_{sob} \left(\|\hat{A}\|_{3,R} + c_{sob}(\|\hat{B}\|_{3/2,R} + c_P \|\hat{C}\|_{3/2,R}) \right) \\ \nu_1 &= c_P \left(\sup_{r\geq R} |\kappa| + c_{sob} \|r^{-1}(\kappa - rb^{-1}\nabla_N b)\|_{3,R} \right) \\ \nu_2 &= \|r^{-1}(\kappa - rb^{-1}\nabla_N b)\|_{2,R} + c_{sob} \|\hat{C}\|_{6/5,R} \end{aligned} \tag{4.3.8b}$$

where c_{sob}, c_P are the constants appearing in the local Sobolev inequality (see Corollary 3.2.1.2), and the nondegenerate global Poincaré inequality (Proposition 3.2.2), respectively.

We now introduce the following:

$$\begin{aligned} \mu_o &= \sup_{r\geq R} r|b\text{tr}\theta - \frac{2}{r}| \\ \mu_1 &= \|b\text{tr}\theta - \frac{2}{r}\|_{2,R} + \|\nabla b\|_{2,R} + \|\hat{\theta}\|_{2,R} \\ \mu_2 &= \|b\text{tr}\theta - \frac{2}{r}\|_{3,R} + \|\nabla b\|_{3,R} + \|\hat{\theta}\|_{3,R} \\ \mu_3 &= \|r^{-1}(b\text{tr}\theta - \frac{2}{r})\|_{6/5,R} + \|r^{-1}\nabla b\|_{6/5,R} + \|r^{-1}\hat{\theta}\|_{6/5,R} \\ \mu_4 &= \|r^{-1}(b\text{tr}\theta - \frac{2}{r})\|_{3/2,R} + \|r^{-1}\nabla b\|_{3/2,R} + \|r^{-1}\hat{\theta}\|_{3/2,R} \end{aligned} \tag{4.3.8c}$$

and,

$$\begin{aligned} \mu_5 &= \|\nabla^2 b\|_{6/5,R} + \|\nabla\!\!\!/ \text{tr}\theta\|_{6/5,R} + \|\text{Ric}\|_{6/5,R} \\ \mu_6 &= \|\nabla^2 b\|_{3/2,R} + \|\nabla\!\!\!/ \text{tr}\theta\|_{3/2,R} + \|\text{Ric}\|_{3/2,R} \end{aligned} \tag{4.3.8d}$$

and check that

$$\nu_0 \leq C c_P (\mu_0 + c_{sob}\mu_2)$$
$$\nu_1 \leq C \left(\mu_2 + c_{sob}^2(1 + c_P)(\mu_4 + \mu^2 + \mu_6)\right)$$
$$\nu_2 \leq C (\mu_1 + c_{sob}(\mu_3 + \mu_1\mu_2 + \mu_5)) \tag{4.3.8e}$$

where C is a purely numerical constant. Now, in view of the assumptions 4.3.2a, we infer that

$$\mu_1, \ \mu_3 \leq c c_0 R^{-1/2} \tag{4.3.9}$$
$$\mu_2, \ \mu_4 \leq c c_0 R^{-1} \tag{4.3.10}$$

where c depends only on c_{sob}, c_P, and c_0, the constant in 4.3.2a. Also, taking into account the assumptions 4.3.2b,

$$\mu_5 \leq c c_1 R^{-1/2} \tag{4.3.11}$$
$$\mu_6 \leq c c_1 R^{-1}.$$

Therefore, by virtue of 4.3.8a, choosing R sufficiently large,

$$\|\nabla w\|^2 \leq c\|r^2 f\|^2,$$

and, by the global Poincaré inequality,

$$\|\nabla(r\phi)\|_2 = \|r^{-1}w\|_2 \leq \|r^2 f\|_2.$$

On the other hand, since

$$r^2|\nabla(r\phi)| \leq c(|\nabla w|^2 + r^{-2}|w|^2),$$
$$\|r^2\nabla^2(r\phi)\| \leq c\|r^2 f\|^2,$$

and, using the global Poincaré inequality again,

$$\|\phi - \frac{1}{r}P\|_2 \leq c\|r^2 f\|^2.$$

We summarize our results in the following:

Proposition 4.3.1 *Let ϕ be a solution of the Poisson equation, supported in the exterior region $r \geq R$. Assume that the normalized lapse function b and the second fundamental form of the u-foliation on Σ verify, in that region, the assumptions 4.3.2a and 4.3.2b. Then, choosing R sufficiently large, depending only on c_0, c_1, and c_{sob}, c_P, there exists a constant c, which depends on the same quantities, such that*

$$\|\phi - \frac{1}{r}P\|_2 + \|\nabla(r\phi)\|_2 + \|r\nabla^2(r\phi)\| \leq c\|r^2 f\|_2. \tag{4.3.12a}$$

Moreover, under the same assumptions,

$$\|r\nabla^3(r\phi)\| \leq c(\|r^2 f\|_2 + \|r^3\nabla f\|_2). \tag{4.3.12b}$$

To prove the last statement of the proposition, we proceed as in the proof of the nondegenerate estimates of Proposition 4.2.3. Instead of multiplying 4.2.11a by χ, however, we multiply by $r^2\chi$, integrate by parts, and use the estimates 4.3.12a.

4.4. Symmetric Hodge Systems in 3-D

Lemma 4.4.1 *Let U be a smooth, compactly supported $p+1$ symmetric tensor on Σ. Consider the following expressions:*

$$\begin{aligned} A(U)_{a_1\ldots a_p bc} &= \nabla_c U_{a_1\ldots a_p b} - \nabla_b U_{a_1\ldots a_p c} \\ D(U)_{a_1\ldots a_p} &= \nabla^m U_{a_1\ldots a_p m} \\ T(U)_{a_1\ldots a_{p-1}} &= g^{bc} U_{a_1\ldots a_{p-1} bc}. \end{aligned}$$

The following integral identity is valid:

$$\begin{aligned} \int_\Sigma |\nabla U|^2 = &- \int_\Sigma \left((2p+1)R_{mn}U^{\mathbf{I}m}U_{\mathbf{I}}{}^n - \frac{p}{2}R|U|^2\right) \\ &+ \int_\Sigma \left(\frac{1}{2}|A(U)|^2 + |D(U)|^2\right) \\ &+ \int_\Sigma \left(2pT(U)\cdot Ric\cdot U - \frac{p}{2}R|T(U)|^2\right) \end{aligned}$$

where $\mathbf{I}$, here, stands for the sequence of indices $a_1 \ldots a_p$, R is the scalar curvature of Σ, and $T(U)\cdot Ric\cdot U = T(U)_{a_1\ldots a_{p-1}} R_{mn} U^{a_1\ldots a_{p-1}mn}$.

Proof 4.4.1 *We first integrate $A(U)$ on Σ,*

$$\int_\Sigma |A(U)|^2 = 2\int_\Sigma |\nabla U|^2 - 2\int_\Sigma \nabla_c U_{\mathbf{I}b}\nabla^b U^{\mathbf{I}c}. \tag{4.4.1}$$

Integrating by parts and commuting derivatives, we find

$$\begin{aligned} \int_\Sigma \nabla_c U_{\mathbf{I}b}\nabla^b U^{\mathbf{I}c} &= -\int_\Sigma U^{\mathbf{I}a}\nabla^b\nabla_a U_{\mathbf{I}b} \qquad (4.4.2) \\ &= -\int_\Sigma U^{\mathbf{I}a}\left(\nabla_a D(U) + \sum_{i=1}^{p} R_{a_i}{}^{sc}{}_a U_{a_1\ldots s\ldots c} + R^s{}_a U_{\mathbf{I}s}\right). \end{aligned}$$

In view of the 3-dimensional formula expressing the Riemann curvature tensor in terms of the Ricci tensor Ric and scalar curvature R, (see page 5), we find, for each $i = 1 \ldots p$,

$$\sum_{i=1}^{p} R_{a_i scb} U_{a_1 \ldots a_p}^{\ b} U^{a_1 \ldots s \ldots c} = 2R_{mn} U^{lm} U_l^{\ n} - \frac{p}{2} R|U|^2$$
$$- 2pT(U) \cdot Ric \cdot U + \frac{p}{2} RT(U)^2.$$

Taking this formula into account, in 4.4.2 we immediately derive the desired result.

We will apply the preceding lemma to the following Hodge type system, for a symmetric tensor of rank 2. Given such a tensor V, we define its divergence and symmetrized curl according to

$$(\text{div} V)_i = \nabla^j V_{ij} \tag{4.4.3}$$
$$(\text{curl} V)_{ij} = \frac{1}{2}(\in_i^{\ lm} \nabla_l V_{mj} + \in_j^{\ lm} \nabla_l V_{mi}). \tag{4.4.4}$$

Remark that $\text{curl} V$ is also a symmetric traceless tensor.

We consider the following system:

Rank-2 Symmetric Hodge System on Σ

$$\begin{aligned} \text{div} V &= \rho \\ \text{curl} V &= \sigma \\ \text{tr} V &= 0 \end{aligned} \tag{4.4.5}$$

where ρ is a given 1-form and σ is a given symmetric traceless 2-covariant tensor.

Remark that, with the notation introduced in Lemma 4.4.1,

$$A(V)_{iab} = \left(\sigma_{im} + \frac{1}{2}\rho_n \in^n_{\ im}\right) \in^m_{\ ab}$$

and

$$|A(V)|^2 = 2\left(|\sigma|^2 + \frac{1}{2}|\rho|^2\right), \tag{4.4.6}$$

therefore we derive the following corollary of Lemma 4.4.1:

Proposition 4.4.1 *Let V be a smooth, compactly supported 2-symmetric tensor on Σ, which verifies the previous Hodge system. Then*

$$\int_\Sigma \left(|\nabla V|^2 + 3R_{mn}V^{im}V_i{}^n - \frac{1}{2}R|V|^2 \right) = \int_\Sigma \left(|\sigma|^2 + \frac{1}{2}|\rho|^2 \right).$$

Using Lemma 4.4.1, we can also derive estimates for the higher derivatives of the solutions V to 4.4.5. More precisely, we have

Proposition 4.4.2 *Under the same assumptions as in the previous proposition there exists a constant c such that*

$$\int_\Sigma |\nabla^2 V|^2 \le c\left(\int_\Sigma (|\nabla\sigma|^2 + |\nabla\rho|^2) + \int_\Sigma (|Ric||\nabla V|^2 + |Ric|^2|V|^2) \right).$$

Remark 4 *Lemma 4.4.1 can also be applied to derive estimates for solutions to the scalar Laplace equation:*

$$\Delta\phi = f.$$

Indeed, taking $V = \nabla\phi$, we derive, for compactly supported solutions ϕ,

$$\int_\Sigma |\nabla^2\phi|^2 \le \int_\Sigma |f|^2 + \int_\Sigma |Ric||\nabla\phi|^2$$
$$\int_\Sigma |\nabla^3\phi|^2 \le c\left(\int_\Sigma |\nabla f|^2 + \int_\Sigma (|Ric||\nabla^2\phi|^2 + |Ric|^2|\nabla\phi|^2) \right).$$

The same technique can be applied to the following system of coupled rank-2 symmetric Hodge systems:

$$\begin{aligned} \text{div} E &= \rho_E \\ \text{curl} E - \nabla_X H &= \sigma_H \\ \text{div} H &= \rho_H \\ \text{curl} H + \nabla_X E &= \sigma_E \end{aligned} \tag{4.4.7a}$$

where E and H are symmetric traceless tensors and X is an arbitrary vectorfield that verifies the assumption

$$\mu(X) = \inf_\Sigma (1 - |X|_g^2) > 0. \tag{4.4.7b}$$

Let $h_{ij} = g_{ij} - X_i X_j$. Clearly, in view of 4.4.7b, for any vector u,

$$\mu(X)|u|^2 \leq h^{ij} u_i u_j \leq |u|^2. \tag{4.4.7c}$$

According to Proposition 4.4.1 applied to each Hodge system separately we have

$$\int_\Sigma \left(|\nabla E|^2 - |\nabla_X H|^2 + 3R_{mn} E^{im} E_i{}^n - \frac{1}{2} R |E|^2 \right)$$
$$= \int_\Sigma \left(|\sigma_E|^2 + 2\sigma_H \cdot \nabla_X H + \frac{1}{2} |\rho_E|^2 \right)$$
$$\int_\Sigma \left(|\nabla H|^2 - |\nabla_X E|^2 + 3R_{mn} H^{im} H_i{}^n - \frac{1}{2} R |H|^2 \right)$$
$$= \int_\Sigma \left(|\sigma_H|^2 - 2\sigma_E \cdot \nabla_X E + \frac{1}{2} |\rho_H|^2 \right).$$

Hence we add the following:

$$\int_\Sigma (h^{ij} \nabla_i E^m \nabla_j E_m + h^{ij} \nabla_i H^m \nabla_j H_m)$$
$$+ \int_\Sigma \left(3R_{mn} E^{im} E_i{}^n - \frac{1}{2} R |E|^2 + 3R_{mn} H^{im} H_i{}^n - \frac{1}{2} R |H|^2 \right)$$
$$= \int_\Sigma \left(|\sigma_E|^2 + 2\sigma_H \cdot \nabla_X H + \frac{1}{2} |\rho_E|^2 + |\sigma_H|^2 - 2\sigma_E \cdot \nabla_X E + \frac{1}{2} |\rho_H|^2 \right).$$

Thus, estimating in a straightforward manner with the help of 4.4.7c, we infer the following:

Corollary 4.4.2.1 *Let E and H be symmetric traceless tensors verifying the coupled Hodge system 4.4.7a for an arbitrary vector X of length strictly less then unity. Then there exists a constant c, which depends only on the constant* $\mu(X)$ *such that*

$$\int_\Sigma (|\nabla E|^2 + |\nabla H|^2) \leq c \left(\int_\Sigma (|\rho_E|^2 + |\rho_H|^2 + |\sigma_E|^2 + |\sigma_H|^2) \right.$$
$$\left. + \int_\Sigma |Ric| (|E|^2 + |H|^2) \right).$$

We next record here, for future reference, the form of the Hodge system 4.4.5 decomposed relative to the spheres S_u. We decompose the symmetric traceless tensor V into the scalar δ, the S-tangent 1-form ϵ, and the S-tangent symmetric 2-tensor η as follows:

$$V_{NN} = \delta$$

$$
\begin{aligned}
V_{AN} &= \epsilon_A \\
V_{AB} &= \eta_{AB}.
\end{aligned}
\tag{4.4.8}
$$

Let $\hat{\eta}$ be the traceless part of η, that is, $\eta_{AB} = \hat{\eta}_{AB} - \frac{1}{2}\delta\gamma_{AB}$. We also decompose the 1-form ρ into the scalar ρ_N and the S-tangent 1-form $\not{\rho}_A = \rho_A$. Similarly, we decompose σ into the scalar σ_{NN}, the S-tangent 1-form $(\sigma_{\not{N}})_A = \sigma_{AN}$, and the symmetric S-tangent 2-form $\not{\sigma}_{AB} = \sigma_{AB}$.

Proposition 4.4.3 *Relative to the previous decomposition, the Hodge system 4.4.5 can be written in the following form:*

$$\not{d}iv\,\epsilon = \rho_N - \nabla_N\delta - \frac{3}{2}\mathrm{tr}\theta\delta + \hat{\eta}\cdot\hat{\theta} - 2(\not{\nabla}\tilde{a})\cdot\epsilon \tag{4.4.9}$$

$$c\not{u}rl\,\epsilon = \sigma_{NN} + \hat{\theta}\wedge\hat{\eta} \tag{4.4.10}$$

$$
\begin{aligned}
\not{\nabla}_N\epsilon + \mathrm{tr}\theta\epsilon &= \left(\frac{1}{2}\not{\rho} + {}^*\sigma_{\not{N}}\right) + \not{\nabla}\delta - 2\hat{\theta}\cdot\epsilon \\
&\quad + \frac{3}{2}(\not{\nabla}\tilde{a})\cdot\delta - \hat{\eta}\cdot(\not{\nabla}\tilde{a})
\end{aligned}
\tag{4.4.11}
$$

$$\not{d}iv\,\hat{\eta} = \left(\frac{1}{2}\not{\rho} - {}^*\sigma_{\not{N}}\right) - \frac{1}{2}\not{\nabla}\delta + \hat{\theta}\cdot\epsilon - \frac{1}{2}\mathrm{tr}\theta\epsilon \tag{4.4.12}$$

$$\not{\nabla}_N\hat{\eta} + \frac{1}{2}\mathrm{tr}\theta\hat{\eta} = \widehat{{}^*(\not{\sigma})} + \frac{1}{2}\not{\nabla}\widehat{\otimes}\epsilon + \frac{3}{2}\delta\hat{\theta} - (\not{\nabla}\tilde{a})\widehat{\otimes}\epsilon \tag{4.4.13}$$

where $\tilde{a} := \log(a)$, *and, given two S-tangent 1-forms u and v, we denote by $u\widehat{\otimes}v$ the symmetric traceless 2-tensor*

$$(u\widehat{\otimes}v)_{AB} = u_A v_B + u_B v_A - (u\cdot v)\gamma_{AB}.$$

Given two symmetric traceless tensors u,v, $u\wedge v$ denotes the scalar

$$u\wedge v = \in^{AB} u_{AC}v_{CB}.$$

Also, given a S-tangent 1-form u, the expression $\not{\nabla}\widehat{\otimes}u$ is defined by

$$(\not{\nabla}\widehat{\otimes}u)_{AB} = \not{\nabla}_A u_B + \not{\nabla}_B u_A - (\not{d}iv\,u)\gamma_{AB}.$$

Proof 4.4.2 *We first calculate, with the help of 3.1.1e,*

$$
\begin{aligned}
\nabla_N V_{NN} &= \nabla_N\delta + 2(\not{\nabla}\tilde{a})\cdot\epsilon \\
\nabla_B V_{NA} &= \not{\nabla}_B\epsilon_A + \frac{3}{4}\delta\mathrm{tr}\theta\gamma_{AB} - \frac{1}{2}\mathrm{tr}\theta\hat{\eta}_{AB} \\
&\quad - \hat{\eta}_{AC}\hat{\theta}_{CB} + \frac{3}{2}\delta\hat{\theta}_{AB}
\end{aligned}
$$

$$
\begin{aligned}
\gamma^{AB}\nabla_B V_{NA} &= \not{d}iv\,\epsilon + \frac{3}{2}\delta \mathrm{tr}\theta - \hat{\eta}\cdot\hat{\theta} \\
\in^{BA}\nabla_B V_{NA} &= c\not{u}rl\,\epsilon + \hat{\eta}\wedge\hat{\theta} \\
\nabla_C V_{AB} &= \not{\nabla}_C \eta_{AB} + \theta_{AC}\epsilon_B + \theta_{BC}\epsilon_A \\
\gamma^{BC}\nabla_C V_{AB} &= (\not{d}iv\,\eta)_A + \frac{3}{2}\mathrm{tr}\theta\epsilon_A + \hat{\theta}_{AB}\epsilon_B \\
\nabla_A V_{NN} &= \nabla_A\delta - 2\theta_{AB}\epsilon_B \\
\nabla_N V_{AN} &= \not{\nabla}_N \epsilon_A - \frac{3}{2}\delta\not{\nabla}_A\tilde{a} + (\not{\nabla}_B\tilde{a})\hat{\eta}_{AB} \\
\nabla_N V_{AB} &= \not{\nabla}_N V_{AB} - (\not{\nabla}_A\tilde{a})\epsilon_B - (\not{\nabla}_B\tilde{a})\epsilon_A.
\end{aligned}
$$

From the normal component of the divergence equation we infer

$$\rho_N = \not{d}iv\,\epsilon + \nabla_N\delta + \frac{3}{2}\mathrm{tr}\theta\delta - \hat{\eta}\cdot\hat{\theta} + 2(\not{\nabla}\tilde{a})\cdot\epsilon.$$

From the angular component of the divergence equation we infer

$$
\begin{aligned}
\not{\rho}_A = (\not{d}iv\,\hat{\eta})_A - \frac{1}{2}\nabla_A\delta + \not{\nabla}_N\epsilon_A + \frac{3}{2}\mathrm{tr}\theta\epsilon_A \\
+ \hat{\theta}_{AB}\epsilon_B - \frac{3}{2}(\nabla_A\tilde{a})\delta + \hat{\eta}_{AB}(\nabla_B\tilde{a}).
\end{aligned}
\qquad (4.4.14)
$$

From the normal-normal component of the curl equation for V we infer

$$\sigma_{NN} = c\not{u}rl\,\epsilon + \hat{\eta}\wedge\hat{\theta}.$$

From the normal-angular component of the curl equation we infer;

$$
\begin{aligned}
2{}^*\sigma_{N\not{A}} = -(\not{d}iv\,\hat{\eta})_A - \frac{3}{2}\nabla_A\delta + \left(\not{\nabla}_N\epsilon_A + \frac{1}{2}\mathrm{tr}\theta\epsilon_A\right) \\
+ 3\hat{\theta}_{AB} - \frac{3}{2}\delta\nabla_A\tilde{a} + \hat{\eta}_{AB}(\nabla_B\tilde{a}).
\end{aligned}
\qquad (4.4.15)
$$

From the angular-angular component of the curl equation we infer

$$
\begin{aligned}
{}^*\widehat{(\not{\sigma})}_{AB} = &-\frac{1}{2}\left(\not{\nabla}_A\epsilon_B + \not{\nabla}_B\epsilon_A - (\not{d}iv\,\epsilon)\gamma_{AB}\right) \\
&+ \not{\nabla}_N\hat{\eta}_{AB} + \frac{1}{2}\mathrm{tr}\theta\hat{\eta}_{AB} - \frac{3}{2}\delta\hat{\theta}_{AB} \\
&+ \left(\nabla_A\tilde{a}\epsilon_B + \nabla_A\tilde{a}\epsilon_B - (\not{\nabla}\tilde{a}\cdot\epsilon_B)\gamma_{AB}\right).
\end{aligned}
$$

To check the last equation, note that if u is an S-tangent 1-form and v is the traceless symmetric S-tangent tensor

$$v_{AB} = \in_{AC}\not{\nabla}_C u_B + \in_{BC}\not{\nabla}_C u_A - (\in^{MN}\not{\nabla}_N u_M)\gamma_{AB},$$

then

$$^{\star}v_{AB} = \nabla\!\!\!/_A u_B + \nabla\!\!\!/_B u_A - (\mathrm{d}\!\!\!/\mathrm{iv}\, u)\gamma_{AB}.$$

In what follows we record an integral identity concerning divergence equations of the type

$$\mathrm{div} E = J \tag{4.4.16a}$$

for a symmetric traceless tensor E. We consider Z the position vectorfield introduced in 4.1.3a, page 80, with $\hat{\pi}$ the traceless part of its deformation tensor. Multiplying the equation 4.4.16a by Z we derive

$$\begin{aligned}\mathrm{div}(i_Z E) &= \frac{1}{2}\hat{\pi}\cdot E + Z\cdot J \\ &= \frac{1}{2}\hat{\pi}\cdot E + Z\cdot J.\end{aligned} \tag{4.4.16b}$$

In view of 4.1.3b we have, relative to an orthonormal frame e_1, e_2, N,

$$\begin{aligned}\hat{\pi}_{AB} &= 2r\hat{\theta}_{AB} + \frac{1}{3}\kappa\delta_{AB} \\ \hat{\pi}_{AN} &= -ra^{-1}\nabla\!\!\!/_A a \\ \hat{\pi}_{NN} &= -\frac{2}{3}\kappa.\end{aligned} \tag{4.4.16c}$$

Letting $\rho = E_{NN}$ be the normal-normal component of E, we write

$$\hat{\pi}\cdot E = 2r\hat{\theta}_{AB}E_{AB} - 2ra^{-1}\nabla\!\!\!/_A a E_{AN} - \kappa\rho.$$

Now, integrating the equation 4.4.16b inside the ball B_r of radius r and applying the divergence theorem, we find

$$4\pi r^3\overline{\rho} = \int_{B_r}\Big(\frac{1}{2}\hat{\pi}\cdot E + Z\cdot J\Big). \tag{4.4.16d}$$

In view of the coarea formula and the definition of κ, we have

$$\begin{aligned}\int_{B_r}\kappa\rho &= \int_{u_0}^{u} du\Big(\int_{S_{u'}} a\kappa\rho\Big) \\ &= \int_{u_0}^{u} du\Big(\int_{S_{u'}} a\kappa(\rho - \overline{\rho})\Big).\end{aligned} \tag{4.4.16e}$$

Hence,

$$4\pi r^3\overline{\rho} = \int_{B_r}\Big(-\frac{1}{2}\kappa(\rho - \overline{\rho}) + r\hat{\theta}_{AB}E_{AB} - ra^{-1}(\nabla\!\!\!/_A a)E_{AN} + rJ_N\Big). \tag{4.4.16f}$$

A similar formula can be derived for the exterior of the ball B_r. We summarize the result in the following:

Proposition 4.4.4 *Assume that the symmetric traceless tensor E satisfies the equation 4.4.16a. Then, for any value of r, the mean value of $\rho = E_{NN}$ on the surface S_r is given by the formula*

$$4\pi r^3\overline{\rho}(r) = \int_{B_r}\left(-\frac{1}{2}\kappa(\rho-\overline{\rho}) + r\hat{\theta}\cdot \not{E} - ra^{-1}\not{\nabla}a\cdot \not{E}_N + rJ_N\right) \quad (4.4.16g)$$

where $\not{E}$ is the purely tangential part of E, and $\not{E}_N$ is its tangential-normal part.

Finally, in the end of this section we record a simple identity concerning products between symmetric traceless tensors of rank 2 in Σ.

Lemma 4.4.2 *Consider two symmetric traceless tensors of rank 2, A and B, and introduce the following operation between them:*

$$(A\times B)_{ij} = \in_i{}^{ab}\in_j{}^{cd}A_{ac}B_{bd} + \frac{1}{3}(A\cdot B)g_{ij}. \quad (4.4.17)$$

Remark that the operation is commutative and takes symmetric traceless tensors of rank 2 into tensors of the same rank. We have the following identity:

$$(A\times B)_{ij} = A_i{}^m B_{mj} + A_j{}^m B_{mi} - \frac{2}{3}(A\cdot B)g_{ij}.$$

Proof 4.4.3 *We consider the totally symmetric trilinear expression*

$$\{A, B, C\} = (A\times B)\cdot C$$

where A, B, and C are symmetric traceless tensors. To prove the lemma, it suffices to show that

$$\{A, B, C\} = 2\mathrm{tr}_g(ABC).$$

To prove this, we first remark that, for $A = B = C$,

$$\{A, A, A\} = 6\det A = 2\mathrm{tr}(A\cdot A\cdot A)$$

with $\det A$ *and* $\mathrm{tr} A$ *defined relative to the metric g. The result is then an immediate consequence of the polarization formula*

$$24\{A, B, C\} = \{A+B+C\}^3 + \{A-B-C\}^3 - \{A+B-C\}^3 - \{A-B+C\}^3$$

where, for a given X, $\{X\}^3$ denotes the trilinear expression $\{X, X, X\}$.

CHAPTER 5

Curvature of an Initial Data Set

In this chapter we use the results of the previous chapter to analyze the global smallness assumption of our main theorem (see 1.0.15).

In view of the remark following the statement of the second version of the main theorem in the introduction we can assume that, given an initial data set Σ, g, k verifying the constraint equations and the global smallness assumption, there exists a point $x_0 \in \Sigma$ such that

$$Q(x_{(0)}, 1) \leq \varepsilon \tag{5.0.1a}$$

where

$$Q(x_{(0)}, b) = \sup_{\Sigma}\{(d_0^2 + 1)^3 |Ric|^2\} \\ + \left\{ \int_\Sigma \sum_{l=0}^{3} (d_0^2 + 1)^{l+1} |\nabla^l k|^2 + \int_\Sigma \sum_{l=0}^{1} (d_0^2 + 1)^{l+3} |\nabla^l B|^2 \right\}$$

where d is the distance function from x_0 on Σ and B is the Bach tensor; $B_{ij} = (\mathrm{curl}\, \hat{R})_{ij}$ as defined in the introduction.

The traceless symmetric tensor $\hat{\mathrm{Ric}}$ verifies the following equations:

$$\mathrm{curl}(\hat{\mathrm{Ric}}) = B \tag{5.0.1b}$$

$$\mathrm{div}(\hat{\mathrm{Ric}}) = \frac{1}{6}\nabla_i R \tag{5.0.1c}$$

$$\mathrm{tr}(\hat{\mathrm{Ric}}) = 0. \tag{5.0.1d}$$

In view of the constraint equations the scalar curvature R is given by

$$R = |k|^2. \tag{5.0.1e}$$

We are going to prove the following:

Theorem 5.0.1 *Let Σ, g, k be an initial data set verifying the constraint equations and the smallness assumption 5.0.1a.*

1. *There exists a point O on Σ such that the distance function centered at O defines a global radial foliation on Σ, with lapse function $a = 1$ and bounded fundamental constants h_m, h_M, k_m, k_M, $h(u)$, $\varsigma(u)$, and $\kappa(u)$.*
2. *Let N be the exterior unit normal to the foliation and define the decomposition of $\widehat{Ric}$ relative to the foliation*

$$Q = \widehat{Ric}_{NN}, \;\; P_A = \widehat{Ric}_{AN}, \;\; S_{AB} = \widehat{Ric}_{AB}$$

where Q is a scalar, P a 1-form tangent to the level surfaces of the foliation, S a 2-covariant symmetric tensor tangent to the foliation. Also let $\hat{S}$ be the traceless part of S, that is $S_{AB} = \hat{S}_{AB} - \frac{1}{2}Q\gamma_{AB}$ with γ the induced metric on the level surfaces of the foliation. Let also $Q_N = \nabla_N Q + \frac{3}{2}\mathrm{tr}\theta Q$ and let $\bar{Q}$ be the average of Q on the level surfaces of the foliation. Finally let $\mathcal{D}_0$ be the norm

$$\begin{aligned}
\mathcal{D}_0^2 = {} & |\sigma_0^2(Q-\bar{Q})\|_2^2 + \|\sigma_0^3 Q_N\|_2^2 + \|\sigma_0^3 \nabla\!\!\!/ Q\|_2^2 \\
& + \|\sigma_0^4 \nabla_N Q_N\|_2^2 + \|\sigma_0^4 \nabla\!\!\!/ Q_N\|_2^2 + \|\sigma_0^4 \nabla\!\!\!/^2 Q\|_2^2 \\
& + \|\sigma_0^2 P\|_2^2 + \|\sigma_0^3 \nabla\!\!\!/ P\|_2^2 + \|\sigma_0^3 \nabla\!\!\!/_N P\|_2^2 \\
& + \|\sigma_0^4 \nabla\!\!\!/ \nabla\!\!\!/_N P\|_2^2 + \|\sigma_0^4 \nabla\!\!\!/_N P\|_2^2 \\
& + \|\sigma_0^2 \hat{S}\|_2^2 + \|\sigma_0^3 \nabla\!\!\!/ \hat{S}\|_2^2 + \|\sigma_0^3 \nabla\!\!\!/_N \hat{S}\|_2^2 \\
& + \|\sigma_0^4 \nabla\!\!\!/ \nabla\!\!\!/_N \hat{S}\|_2^2 + \|\sigma_0^4 \nabla\!\!\!/_N^2 \hat{S}\|_2^2 + \left(\sup_\Sigma \sigma_0^3 \bar{Q}\right)^2
\end{aligned}$$

where $\sigma_0 = (1 + d_0^2)^{1/2}$.

There exists a constant c, independent of ε, such that

$$\mathcal{D}_0 \le c\varepsilon. \tag{5.0.2}$$

To prove the theorem it suffices to prove the following more general result:

Proposition 5.0.1 *Let Σ, g be a Riemannian 3-dimensional manifold, diffeomorphic to $\Re^3$ and assume there exists a sufficient small ε_0 such that the following estimates hold true:*

$$\begin{aligned}
\sup_\Sigma (1+d^2)^{3/2}|Ric| &\le \varepsilon_0 \\
\left(\int_\Sigma (1+d^2)^2|R|^2\right)^{1/2} &\le \varepsilon_0 \\
\left(\int_\Sigma (1+d^2)^3(|B|^2+|\nabla R|^2)\right)^{1/2} &\le \varepsilon_0 \\
\left(\int_\Sigma (1+d^2)^4(|\nabla B|^2+|\nabla^2 R|^2)\right)^{1/2} &\le \varepsilon_0
\end{aligned} \tag{5.0.3}$$

where d is the geodesic distance function from a point O on Σ, Ric is the Ricci curvature of g, R is its scalar curvature, and B is the Bach tensor $B_{ij} = (curl\hat{R})_{ij}$.

Then

1. *The geodesic distance function defines a radial function $d = u$, with lapse $a = 1$, verifying the following conditions:*

$$\sup_{S_u} \frac{(1+u)^3 u[1+\log(1+u)}{|}\hat{\theta}| \leq c\varepsilon_0 \tag{5.0.4a}$$

$$\sup_{S_u} \frac{(1+u)^3}{u[1+\log(1+u)]}|(\mathrm{tr}\theta - \frac{2}{u})| \leq c\varepsilon_0 \tag{5.0.4b}$$

The Gauss curvature $K(u)$ of the surfaces S_u verifies the inequality

$$|K(u) - \frac{1}{u^2}| \leq c\varepsilon_0 \frac{(1+\log(1+u))}{(1+u)^3}, \tag{5.0.4c}$$

and the function $r(u)$, which corresponds to the area of the surfaces S_u that is, Area(S_u) $= 4\pi r(u)^2$, differs from u according to the formula

$$|r(u) - u| \leq c\varepsilon_0 \frac{u^3(1+\log(1+u))}{(1+u)^3}. \tag{5.0.4d}$$

2. *The following weighted L^2 estimates for $\widehat{Ric}$ hold true, for any $p < 5/2$:*

$$\int_\Sigma (1+u^2)^p |\nabla \widehat{Ric}|^2 \leq c\varepsilon_0$$
$$\int_\Sigma (1+u^2)^{p+1} |\nabla^2 \widehat{Ric}|^2 \leq c\varepsilon_0. \tag{5.0.4e}$$

Also, for any $q < 7/2$,

$$\left(\int_{S_u} (1+u^2)^{2q} |\nabla \widehat{Ric}|^4\right)^{1/4} \leq c\varepsilon_0. \tag{5.0.4f}$$

3. *The first angular derivatives of θ verify, for every $0 \leq q < 5/2$, the estimates*

$$\left(\int_{S_u} (1+u^2)^{2q} |\nabla\!\!\!/ \theta|^4\right)^{1/4} \leq c\varepsilon_0. \tag{5.0.4g}$$

4. *Let $Q = \widehat{Ric}_{NN}$, $P_A = \widehat{Ric}_{AN}$, $S_{AB} = \widehat{Ric}_{AB}$ be the decomposition of $\widehat{Ric}$ relative to the u foliation. Let $\hat{S}$ be the traceless part of S, that is, $S_{AB} = \hat{S}_{AB} - \frac{1}{2}Q\gamma_{AB}$. Also $Q_N = \nabla_N Q + \frac{3}{2}\mathrm{tr}\theta Q$. Then*

$$\|\sigma_0^2(Q - \bar{Q})\|_2 + \|\sigma_0^3 Q_N\|_2 + \|\sigma_0^3 \nabla\!\!\!/ Q\|_2$$

$$+ \|\sigma_0^4 \nabla_N Q_N\|_2 + \|\sigma_0^4 \nabla\!\!\!/ Q_N\|_2 + \|\sigma_0^4 \nabla\!\!\!/^2 Q\|_2 \leq c\varepsilon_0 \quad (5.0.4h)$$

$$\|\sigma_0^2 P\|_2 + \|\sigma_0^3 \nabla\!\!\!/ P\|_2 + \|\sigma_0^3 \nabla\!\!\!/_N P\|_2$$
$$+ \|\sigma_0^4 \nabla\!\!\!/ \nabla\!\!\!/_N P\|_2 + \|\sigma_0^4 \nabla\!\!\!/_N P\|_2 \leq c\varepsilon_0 \quad (5.0.4i)$$

$$\|\sigma_0^2 \hat{S}\|_2 + \|\sigma_0^3 \nabla\!\!\!/ \hat{S}\|_2 + \|\sigma_0^3 \nabla\!\!\!/_N \hat{S}\|_2$$
$$+ \|\sigma_0^4 \nabla\!\!\!/ \nabla\!\!\!/_N \hat{S}\|_2 + \|\sigma_0^4 \nabla\!\!\!/_N \hat{S}\|_2 \leq c\varepsilon_0. \quad (5.0.4j)$$

Proof 5.0.1 *We start with the proof of the first part of the proposition. This requires an argument similar to but much simpler than those we develop at length in Part III.*

First we remark that the structure equations (see 3.1.2a, 3.1.2b, 3.1.2c, and 3.1.2d) of the foliation given by $u = d$ take the form

$$\frac{d}{du}\mathrm{tr}\theta = -\frac{1}{2}(\mathrm{tr}\theta)^2 - |\hat{\theta}|^2 - R_{NN} \quad (5.0.5a)$$
$$\frac{d}{du}\hat{\theta}_{AB} = -\mathrm{tr}\theta\hat{\theta}_{AB} - \hat{S}_{AB} \quad (5.0.5b)$$
$$\nabla\!\!\!/_B \theta_{AB} = -\nabla\!\!\!/_A \mathrm{tr}\theta + P_A \quad (5.0.5c)$$
$$2K - (\mathrm{tr}\theta)^2 + |\theta|^2 = R - 2R_{NN} \quad (5.0.5d)$$

where

$$P_A = R_{AN}, \qquad S_{AB} = R_{AB}, \qquad \hat{S}_{AB} = S_{AB} - \frac{1}{2}(R - R_{NN})\gamma_{AB}.$$

Setting $h = \frac{2}{\mathrm{tr}\theta}$, we derive

$$\frac{d}{du}h = 1 + \frac{1}{2}h^2(|\hat{\theta}|^2 + R_{NN}). \quad (5.0.5e)$$

Therefore,

$$\frac{d}{du}h = 1 + \frac{1}{2}h^2(|\hat{\theta}|^2 + R_{NN}) \quad (5.0.5f)$$
$$\frac{d}{du}|\hat{\theta}| + 2h^{-1}|\hat{\theta}| \leq |\hat{S}|. \quad (5.0.5g)$$

Clearly,[1] *as $u \longrightarrow 0$*

$$|\mathrm{tr}\theta - \frac{2}{u}| \leq \frac{1}{2} \quad (5.0.5h)$$
$$|\hat{\theta}| \leq \frac{1}{2}. \quad (5.0.5i)$$

[1] One can also see this as a bootstrap condition that we will recover at the end of the argument.

Hence also, as $u \longrightarrow 0$,

$$|h - u| < u^2. \tag{5.0.5j}$$

Now let Γ be the property

$$|h - u| < \frac{u^2}{(1+u)^{3/2}}. \tag{5.0.5k}$$

In view of 5.0.5j the property Γ is valid for sufficiently small u. We shall show that, under the assumption

$$|Ric(u)| \leq B(1+u)^{-3} \tag{5.0.5l}$$

with B sufficiently small, the property Γ holds globally in u. Indeed, assume that Γ holds true for u in the interval $[0, u_1)$. Then, in $[0, u_1]$,

$$|h - u| \leq \frac{u^2}{(1+u)^{3/2}}.$$

We first multiply the equation 5.0.5g by u^2 and derive

$$\frac{d}{du}u^2|\hat{\theta}| + 2(h^{-1} - u^{-1})u^2|\hat{\theta}| \leq u^2|\hat{S}|.$$

In view of 5.0.5k we have, for some constant c,

$$|h| \leq cu$$
$$|h^{-1} - u^{-1}| \leq c(1+u)^{-3/2}$$

for all $u \in [0, u_1]$. Therefore, taking into account 5.0.5l, we find a constant c such that

$$u^2|\hat{\theta}(u)| \leq c\int_0^u u'^2|\hat{S}(u')|du'$$
$$\leq cB\frac{u^3[1+log(1+u)]}{(1+u)^3}.$$

Hence,

$$|\hat{\theta}(u)| \leq cB\frac{u[1+\log(1+u)]}{(1+u)^3} \tag{5.0.5m}$$

for all $u \in [0, u_1]$.

Using the above inequalities for $\mathrm{tr}\theta, \hat{\theta}$, and the assumption 5.0.5l, we can now estimate all terms on the right-hand side of 5.0.5f and integrate to derive

$$|h(u) - u| \leq cB\frac{u^3[1+\log(1+u)]}{(1+u)^3}. \tag{5.0.5n}$$

Finally, let k be the maximum of the function $\frac{u[1+\log(1+u)]}{(1+u)^{3/2}}$ and choose B such that $ckB \leq 1$. Then, from 5.0.5m and 5.0.5n, we infer that the property Γ holds in the whole interval $[0, u_1]$ and therefore also holds globally for all u. Therefore,

$$|h(u) - u| \leq cB\frac{u^3[1 + \log(1 + u)]}{(1 + u)^3} \tag{5.0.5o}$$

$$|\hat{\theta}(u)| \leq cB\frac{u[1 + \log(1 + u)]}{(1 + u)^3} \tag{5.0.5p}$$

for all values of u. This proves the formulas 5.0.4a and 5.0.4b.

To check 5.0.4c, we consider the Gauss equation 5.0.5d, which we write in the form

$$K = \frac{1}{4}(\mathrm{tr}\theta)^2 - \frac{1}{2}|\hat{\theta}|^2 + \frac{1}{2}R - R_{NN}.$$

Using the estimates 5.0.4a, 5.0.4b, and 5.0.5l, we infer that;

$$|K - \frac{1}{u^2}| \leq cB\frac{[1 + \log(1 + u)]}{1 + u^3}, \tag{5.0.6}$$

which proves 5.0.4c. Integrating 5.0.6 on the surfaces S_u and taking into account the Gauss-Bonnet theorem

$$\overline{K} = \frac{1}{r^2},$$

we derive

$$|r - u| \leq cB\frac{u^3[1 + \log(1 + u)]}{1 + u^3},$$

which proves 5.0.4d.

To end the proof of the first part of the proposition, it only remains to show that the level surfaces of the distance function from O provide a foliation of Σ. If this was not the case, then the cut-locus **C** *of O will be nonempty. This is the set of points P on $T_O(\Sigma)$ for which the image of the segment $[O, P]$ through the exponential map* $\exp_O$ *ceases to be minimizing in Σ. Let* **D** *be the star-shaped open domain in $T_O(\Sigma)$ with the boundary* **C** *and* $\overline{\mathbf{D}}$ *its closure. Since Σ is a complete Riemannian manifold, the restriction of* $\exp_O$ *to* $\overline{\mathbf{D}}$ *is onto. The estimates we have established show, in particular,*[2] *that* $\exp_O$ *is a local diffeomorphism at every point of* $\overline{\mathbf{D}}$. *We can thus extend* $\exp_O$ *to a covering map from an open, star-shaped domain $\Omega \supset \overline{\mathbf{D}}$ onto Σ. Since Σ is diffeomorphic to R^3, we infer that* $\exp_O : \Omega \longrightarrow \Sigma$ *must be a diffeomorphism.*

[2] This argument requires only that $\mathrm{tr}\theta$ be bounded from below.

This implies in particular that **C** *is empty and that, in fact, the exponential map is a diffeomorphism from* $T_O(\Sigma)$ *to* Σ.

To prove the second part of the proposition, we define V *to be the symmetric 2-tensor* $V_{ij} = f_p \widehat{Ric}_{ij}$ *where* $f_p = (1+u^2)^{p/2}$. *We recall that* $\widehat{Ric}$ *verifies the rank-2 symmetric Hodge system on* Σ:

$$curl(\hat{Ric})_{ij} = B_{ij} \tag{5.0.7a}$$

$$div(\hat{Ric}) = \frac{1}{6}\nabla_i R \tag{5.0.7b}$$

$$\mathrm{tr}(\hat{Ric}) = 0. \tag{5.0.7c}$$

Hence also

$$\begin{aligned} curlV &= f_p B + \nabla f_p \wedge \widehat{Ric} \\ divV &= f_p\left(\frac{1}{6}\nabla R\right) + \nabla f_p \cdot \widehat{Ric} \\ \mathrm{tr}V &= 0. \end{aligned}$$

Note that, in view of the first part of the proposition,

$$\begin{aligned} |\nabla f_p| &\le c(1+u^2)^{(p-1)/2} \\ |\nabla^2 f_p| &\le c(1+u^2)^{(p-2)/2}. \end{aligned}$$

The estimates 5.0.4e follow now immediately as an application of Propositions 4.4.1, and 4.4.2 to this system. The estimate 5.0.4f is then a consequence of 5.0.4e and the nondegenerate form of the global Sobolev inequalities, see Corollary 3.2.1.1.

To estimate the angular derivatives of θ, *we first differentiate the equation 5.0.5b. With the help of the commutation lemma (3.2.1), we derive*

$$\frac{d}{du}\nabla_A\hat{\theta}_{BC} + \frac{3}{2}\mathrm{tr}\theta\nabla_A\hat{\theta}_{BC} = F_{ABC} \tag{5.0.8a}$$

where

$$\begin{aligned} F_{ABC} = &-\hat{\theta}_{AD}\nabla_D\hat{\theta}_{BC} - \nabla\mathrm{tr}\theta\hat{\theta}_{BC} - \nabla_A\hat{S}_{BC} + (P_B\hat{\theta}_{AC} + P_C\hat{\theta}_{AB}) \\ &+ P^D(-\delta_{AB}\hat{\theta}_{DC} - \delta_{AC}\hat{\theta}_{BC}). \end{aligned} \tag{5.0.8b}$$

We next state and use the following lemma:[3]

[3] This is an obvious analogue of Lemma 13.1.1, which will be used repeatedly in Part III.

Lemma 5.0.1 *Let $U_{A_1\ldots A_k}$ and $F_{A_1\ldots A_k}$ be two k-covariant tensorfields tangent to the spheres S_u at every point on Σ satisfying the equation*

$$\frac{dU_{A_1\ldots A_k}}{du}+\lambda_0 tr\theta U_{A_1\ldots A_k}=F_{A_1\ldots A_k}$$

where λ_0 is a nonnegative real number. We let $\lambda_1=2(\lambda_0-\frac{1}{p})$ and define the norms

$$|r^\lambda U|_{p,S}=\left(\int_{S_u} r^\lambda |U|^p d\mu_\gamma\right)^{1/p}.$$

Then

$$|r^{\lambda_1}U|_{p,S}(u)\le c\left(|r^{\lambda_1}U|_{p,S}(0)+\int_0^u |r^{\lambda_1}F|_{p,S}(u')du'\right)$$

where c is a constant independent of u.

Applying the lemma to 5.0.8a, for p = 4, we derive

$$|r^{5/2}\nabla\!\!\!/\,\hat\theta|_{4,S}(u)\le\int_0^u |r^{5/2}\nabla\!\!\!/\,F|_{4,S}(u')du.$$

Taking into account the estimates 5.0.4a–5.0.4d as well as the equation 5.0.5c[4] *and the assumption on the sup-norm of Ric, we derive*

$$\begin{aligned}|r^{5/2}\nabla\!\!\!/\,F|_{4,S}&\le c\varepsilon_0\frac{u[1+\log(1+u)]}{(1+u)^3}|r^{5/2}\nabla\!\!\!/\,\hat\theta|_{4,S}+|r^{5/2}\nabla Ric|_{4,S}\\&\quad+c\varepsilon_0 r(u)^3\frac{u[1+log(1+u)]}{(1+u)^6}.\end{aligned}$$

Taking into account 5.0.4f, we infer that, for any $q<7/2$,

$$|r^{5/2}\nabla\!\!\!/\,\hat\theta|_{4,S}(u)\le c\varepsilon_0\int_0^u r(u')^{5/2}(1+r(u')^2)^{-q/2}du',$$

from which 5.0.4g follows.

To derive the estimates of the last part of the proposition we first decompose the Hodge system 5.0.7a–5.0.7c according to Proposition 4.4.3. We thus derive

$$\mathrm{d\!\!\!/iv}\, P=\nabla_N R-\nabla_N Q-\frac{3}{2}\mathrm{tr}\theta Q+\hat S\cdot\hat\theta \tag{5.0.9a}$$

$$\mathrm{cu\!\!\!/rl}\, P=B_{NN}+\hat\theta\wedge\hat S \tag{5.0.9b}$$

$$\nabla\!\!\!/_N P+\mathrm{tr}\theta P=\left(\frac{1}{12}\nabla\!\!\!/ R+{}^*B\!\!\!/_N\right)+\nabla\!\!\!/ Q-2\hat\theta\cdot P \tag{5.0.9c}$$

[4] This allows us to estimate $\nabla\!\!\!/\mathrm{tr}\theta$ relative to $\nabla\!\!\!/\hat\theta$.

$$\begin{aligned} \mathrm{d}\!\!/\mathrm{iv}\, \hat{S} &= \left(\frac{1}{12}\nabla\!\!\!/ R - {}^*\!B\!\!\!/_N\right) - \frac{1}{2}\nabla\!\!\!/ Q \\ &\quad + \hat{\theta}\cdot P - \frac{1}{2}\mathrm{tr}\theta P \end{aligned} \tag{5.0.9d}$$

$$\nabla\!\!\!/_N \hat{S} + \frac{1}{2}\mathrm{tr}\theta\hat{S} = {}^*\widehat{B\!\!\!/} + \frac{1}{2}\nabla\!\!\!/\widehat{\otimes}P + \frac{3}{2}\hat{\theta}Q \tag{5.0.9e}$$

where $B\!\!\!/_N$ is the S-tangent 1-form $(B\!\!\!/_N)_A = B_{AN}$ and $B\!\!\!/$ is the S-tangent symmetric 2-tensor $B\!\!\!/_{AB} = B_{AB}$.

We remark that both P and S can be completely determined in terms of the first derivatives of Q. To estimate Q, we first derive a Poisson equation for it. To achieve this, we consider the position vectorfield $Z = rN$ first introduced in 4.1.3a (page 80), with π its deformation tensor and $\hat{\pi}$ the traceless part of π. We have

$$\begin{aligned} \hat{\pi}_{ij} &= 2r\hat{\theta}_{ij} + \frac{1}{3}\kappa(g_{ij} - 3N_iN_j) \\ \mathrm{tr}\pi &= r(2\mathrm{tr}\theta + \overline{\mathrm{tr}\theta}) \\ \pi_{ij} &= \hat{\pi}_{ij} + \frac{1}{3}\mathrm{tr}\pi g_{ij} \end{aligned} \tag{5.0.10a}$$

where

$$\kappa = r(\mathrm{tr}\theta - \overline{\mathrm{tr}\theta}). \tag{5.0.10b}$$

Moreover,

$$\nabla_i Z_j = \frac{1}{2}\hat{\pi}_{ij} + \frac{1}{6}\mathrm{tr}\pi g_{ij} \tag{5.0.10c}$$

and

$$\triangle Z + \mathrm{Ric}\cdot Z = \mathrm{div}\hat{\pi} - \frac{1}{6}\nabla(\mathrm{tr}\pi). \tag{5.0.10d}$$

Relative to an orthonormal frame N, e_1, e_2,

$$\hat{\pi}_{AN} = 0 \tag{5.0.10e}$$

$$\hat{\pi}_{NN} = -\frac{2}{3}\kappa \tag{5.0.10f}$$

$$\hat{\pi}_{AB} = 2r\hat{\theta}_{AB} + \frac{1}{3}\kappa\delta_{AB}. \tag{5.0.10g}$$

Let $i_Z\widehat{\mathrm{Ric}}$ and remark, in view of equations 5.0.7a and 5.0.7b, that

$$\mathrm{curl}(i_Z\widehat{\mathrm{Ric}}) = F + G \tag{5.0.11a}$$

$$\mathrm{div}(i_Z\widehat{\mathrm{Ric}}) = \frac{1}{6}\nabla_Z R + \frac{1}{2}\hat{\pi}\cdot\widehat{\mathrm{Ric}} := I \tag{5.0.11b}$$

with

$$F = i_Z B \tag{5.0.11c}$$

$$G = \frac{1}{2}\hat{\pi} \wedge \widehat{\text{Ric}} \tag{5.0.11d}$$

where, given 2-covariant symmetric traceless tensors A, B, we denote by $A \wedge B$ the 1-form $(A \wedge B)_i = \in_i{}^{mn} A_m{}^l B_{ln}$. Now, we take the curl of the equation 5.0.11a with the help of the following formula, which holds for an arbitrary 1-form v:

$$\text{curl}(\text{curl}\, v)_i = R_i{}^j v_j - \Delta v_i + \nabla_i(\text{div}\, v) \tag{5.0.11e}$$

Thus

$$\Delta(i_Z\widehat{\text{Ric}}) = -\text{curl}\, F - \text{curl}\, G + \nabla I + \text{Ric} \cdot (i_Z\widehat{\text{Ric}}). \tag{5.0.11f}$$

Contracting the last equation once more with Z and setting $q = r^2 Q$, we infer that

$$\Delta q = Z \cdot \big(- \text{curl}\, F - \text{curl}\, G + \nabla I + \text{Ric} \cdot (i_Z\widehat{\text{Ric}})\big)$$
$$+ \hat{\pi}_{ij} \nabla_i (i_Z\widehat{\text{Ric}})_j + \Delta Z \cdot (i_Z\widehat{\text{Ric}}) + \frac{1}{3}\text{tr}\pi I.$$

We now appeal to the formula 5.0.10d to express ΔZ. Remark also that for any 1-form A on Σ we have

$$Z \cdot \text{curl}\, A = r \text{c}\!\!/\text{url}\, A\!\!\!/$$

where $A\!\!\!/ = \Pi_i{}^j A_j$. Consequently,

$$\Delta q = f \tag{5.0.11g}$$

$$f = -r \text{c}\!\!/\text{url}\, F\!\!\!/ - r \text{c}\!\!/\text{url}\, G\!\!\!/ + r\nabla_N I$$
$$+ \hat{\pi} \cdot \nabla(i_Z\widehat{\text{Ric}}) + \frac{1}{3}\text{tr}\pi I$$
$$+ \left(\text{div}\hat{\pi} - \frac{1}{6}\nabla(\text{tr}\pi)\right) (i_Z\widehat{\text{Ric}}) \tag{5.0.11h}$$

where $F\!\!\!/, G\!\!\!/$ are the projections of F, G to the spheres S_u. Clearly,

$$F\!\!\!/ = rB_{AN} \tag{5.0.11i}$$

$$G\!\!\!/ = \frac{1}{2} \in_{AB} (\pi_{BC} P_C - \pi_{AN} P_B). \tag{5.0.11j}$$

We are now in a position to apply Proposition 4.3.1. Indeed, we easily check that $\int_\Sigma |r^4| f|^2$ is finite. Moreover, in view of the remark following the assumptions 4.3.2a and 4.3.2b, the estimates 5.0.4a, 5.0.4b, and 5.0.4g suffice to apply that proposition. Therefore,

$$\|q - \frac{1}{r}[q]\|_2 + \|\nabla(rq)\|_2 + \|r\nabla^2(rq)\| \leq c\|r^2 f\|_2 \leq c\varepsilon_0 \tag{5.0.12}$$

where $[q]$ is the charge of q.

Since $q = r^2 Q$ and

$$r^3 Q_N = r^3 \left(\nabla_N Q + \frac{3}{2} \mathrm{tr}\theta Q \right) = \nabla_N (rq) + \frac{3}{2} (\mathrm{tr}\theta - \overline{\mathrm{tr}\theta}) Q,$$

we infer that

$$\|r^3 Q_N\|_2 + \|r^3 \nabla\!\!\!/ Q\|_2 + \|r^4 \nabla_N Q_N\|_2 \\ + \|r^4 \nabla\!\!\!/ Q_N\|_2 + \|r^4 \nabla\!\!\!/^2 Q\|_2 \leq c\varepsilon_0.$$

Finally, taking into account the Poincaré inequality on the surfaces S_u, we conclude that

$$\|r^2 (Q - \bar{Q})\|_2 \leq c\varepsilon_0,$$

which proves 5.0.4h:

$$\|r^2 (Q - \bar{Q})\|_2 + \|r^3 Q_N\|_2 + \|r^3 \nabla\!\!\!/ Q\|_2 \\ + \|r^4 \nabla_N Q_N\|_2 + \|r^4 \nabla\!\!\!/ Q_N\|_2 + \|r^4 \nabla\!\!\!/^2 Q\|_2 \leq c\varepsilon_0.$$

We next estimate P from the Hodge system on the spheres S_u :

$$\begin{aligned} \mathrm{d}\!\!\!/\mathrm{iv}\, P &= \nabla_N R - Q_N + \hat{S} \cdot \hat{\theta} \\ \mathrm{c}\!\!\!/\mathrm{url}\, P &= B_{NN} + \hat{\theta} \wedge \hat{S}. \end{aligned}$$

This is a system of the type $\mathbf{H_1}$, discussed in Chapter 2. Therefore, according to Proposition 2.2.2 and the previous estimates, we infer that

$$\|r^2 P\|_2 + \|r^3 \nabla\!\!\!/ P\|_2 + \|r^4 \nabla\!\!\!/^2 P\|_2 \leq c\varepsilon_0. \tag{5.0.13a}$$

Taking into account the equation 5.0.9c, we also infer that

$$\|r^3 \nabla\!\!\!/_N P\|_2 + \|r^4 \nabla\!\!\!/ \nabla\!\!\!/_N P\|_2 \leq c\varepsilon_0. \tag{5.0.13b}$$

Differentiating 5.0.9c relative to $\nabla\!\!\!/_N$, we also derive

$$\|r^4 \nabla\!\!\!/_N^2 P\|_2 \leq c\varepsilon_0. \tag{5.0.13c}$$

This proves 5.0.4i.

Finally, to estimate $\hat{S}$, we proceed in the same manner. We use the equation 5.0.9d as Hodge system of type $\mathbf{H_2}$ on S_u to estimate the tangential derivatives and the equation 5.0.9e to estimate the normal derivatives. Therefore,

$$\begin{aligned} \|r^2 \hat{S}\|_2 + \|r^3 \nabla\!\!\!/ \hat{S}\|_2 + \|r^4 \nabla\!\!\!/^2 \hat{S}\|_2 &\leq c\varepsilon_0 \\ \|r^3 \nabla\!\!\!/_N \hat{S}\|_2 + \|r^4 \nabla\!\!\!/ \nabla\!\!\!/_N \hat{S}\|_2 &\leq c\varepsilon_0 \\ \|r^4 \nabla\!\!\!/_N^2 \hat{S}\|_2 &\leq c\varepsilon_0, \end{aligned}$$

which ends the proof of the proposition and the theorem.

CHAPTER 6

Deformation of 2-Surfaces in 3-D

The aim of this chapter is to present a method of producing foliations of a 3-dimensional Riemannian manifold.

Definition 6.0.1 A surface S in Σ, diffeomorphic to S^2, is an equivalent class of embeddings $i : S^2 \longrightarrow \Sigma$. Two embeddings i_1, i_2 are said to be equivalent if there exists a diffeomorphism $h : S^2 \longrightarrow S^2$ s.t. $i_2 = i_1 \circ h$.

Definition 6.0.2 A homotopy of embeddings of S^2 into Σ is a differentiable map $f : [0,1] \times S^2 \longrightarrow \Sigma$ such that, for each $t \in [0,1]$, the map $f_t : S^2 \longrightarrow \Sigma$, defined by $f_t(\omega) = f(t, \omega)$, is an embedding.

Two homotopies f_1, f_2 are said to be equivalent if there exists a differentiable map $h : [0,1] \times S^2 \longrightarrow S^2$ such that, for each fixed $t \in [0,1]$, the map $h_t : S^2 \longrightarrow S^2$ is a diffeomorphism and

$$f_2(t, \omega) = f_1(t, h(t, \omega)).$$

The homotopy is called normal if, for each fixed $\omega \in S^2$, the curves C_ω, given by the image of $[0,1] \times \{\omega\}$ through the map f, are orthogonal to the surfaces $S_t = f_t(S^2)$.

Let ω^A denote local coordinates in S^2 and let x^i denote local coordinates in Σ. Given a homotopy of embeddings f, we can write

$$\frac{\partial f^i}{\partial t} = aN^i + \beta^i \tag{6.0.1a}$$

where N denotes the normal to the surfaces S_t and the vector β, called the shift of the homotopy, is tangent to S_t. The scalar a, called the lapse function of the homotopy, is independent of the class of equivalence of f. It measures the normal separation between the surfaces S_t. We call it the lapse function of the homotopy.

The vectors $\frac{\partial f^i}{\partial \omega^A}$ form a basis of the tangent space to S_t. Let β^A be the coordinates of β relative to this basis and let $\gamma_{AB} = g_{ij} \frac{\partial f^i}{\partial \omega^A} \frac{\partial f^j}{\partial \omega^B}$ the corresponding coefficients of the metric induced on S_t. Thus, setting, $\beta_A = \gamma_{AB} \beta^B$, we have

$$\beta_A = g_{ij} \frac{\partial f^i}{\partial t} \frac{\partial f^j}{\partial \omega^A}. \tag{6.0.1b}$$

Now let $\omega = \omega(t, \tilde{\omega})$ be the solution to the equation on $[0, 1] \times S^2$,

$$\frac{d\omega^A}{dt} = -\beta^A(t, \omega), \tag{6.0.1c}$$

subject to the initial condition $\omega(0, \tilde{\omega}) = \tilde{\omega}$, and let

$$\tilde{f}(t, \tilde{\omega}) = f(t, \omega(t, \tilde{\omega})).$$

One easily checks that the shift vector of $\tilde{f}$ vanishes. Therefore, any homotopy of embeddings is equivalent to a normal one.

Definition 6.0.3 A homotopy of surfaces in Σ, diffeomorphic to S^2, is a class of equivalence of homotopies of embeddings of S^2 into Σ. The lapse function of such a homotopy is simply the lapse function of any representative homotopy of embeddings.

The structure equations of a homotopy of surfaces is given, in terms of a Fermi-propagated local orthonormal basis $(e_A)_{A=1,2}$,[1] by

$$\begin{aligned} \frac{\partial \mathrm{tr}\theta}{\partial t} &= -a\left(\frac{1}{2}(\mathrm{tr}\theta)^2 + |\hat{\theta}|^2 + R_{NN}\right) - \Delta\!\!\!/ a \\ \frac{\partial \hat{\theta}_{AB}}{\partial t} &= -a(\mathrm{tr}\theta\hat{\theta}_{AB} + \hat{S}_{AB}) - \nabla\!\!\!/_A\nabla\!\!\!/_B a + \frac{1}{2}\delta_{AB}\Delta\!\!\!/ a \\ \mathrm{d\!\!\!/iv}\,\hat{\theta}_{AB} &= \frac{1}{2}\nabla\!\!\!/_A \mathrm{tr}\theta + P_A \\ 2K &= R - 2R_{NN} + \frac{1}{2}(\mathrm{tr}\theta)^2 - |\hat{\theta}|^2 \end{aligned} \tag{6.0.1d}$$

where S_{AB} is the restriction of R_{ij} to S_t and $P_A = R_{AN}$

Given an homotopy of surfaces, represented by the normal homotopy of embeddings $f : [0, 1] \times S^2 \longrightarrow \Sigma$ of lapse function a, we define its arc length $\mathcal{L}_k^p$ to be

$$\mathcal{L}_k^p = \int_0^1 \|a\|_{W_k^p(S_t)} dt$$

where, for a given function u on S,

$$\|u\|_{W_k^p(S)} = \left(\sum_{l=0}^{k} A(S)^{\frac{pl}{2}-1} \int_S |\nabla\!\!\!/^l u|^p d\mu_\gamma\right)^{1/p},$$

In what follows we consider the Sobolev spaces $W_k^p(S)$ with $2 < p \leq 4$ and $k \geq 3$.

[1] that is, $\nabla_N e_A = f_A N$ and $\langle N, e_A \rangle_g = 0$.

Definition 6.0.4 Let S_0 be a C^2 surface in Σ, diffeomorphic to S^2, such that, its second fundamental form θ_{AB} belongs to $W^p_{k-2}(S_0)$. We define $\mathcal{S}^p_k(S_0)$ to be the space of all C^2 surfaces in Σ that can be joined to S_0 by a C^2 homotopy of finite arc length $\mathcal{L}^p_k$. We also make the supplementary assumption that the Ricci curvature of Σ satisfies the regularity assumption, Ric $\in^{(\mathrm{loc})} H_{k-1}(\Sigma)$. We define a distance function on $\mathcal{S}^p_k$ by

$$d(S_0, S_1) = \inf \mathcal{L}^p_k$$

with the infinium taken over all homotopies of surfaces joining S_0 to S_1.

Remark 1 *In view of the structure equations 6.0.1d we remark that, for all $S \in \mathcal{S}^p_k(S_0)$, we have $\theta_{AB} \in W^p_{k-2}(S)$. Also, for all $S \in \mathcal{S}^p_k(S_0)$ we have $K \in W^p_{k-2}(S)$.*

All surfaces in $\mathcal{S}^p_k$ that belong to a ball $B_r(S)$, centered at S of radius r, are graphs over S provided that r is sufficiently small and $k \geq 3$. This allows us to define a linear structure in $B_r(S)$ that, together with the metric previously defined, makes $\mathcal{S}^p_k$ a pre-Banach manifold. In what follows we provide a detailed description of such balls in terms of normal geodesic charts based on a given S in $\mathcal{S}^p_k$.

Let S_0 be an arbitrary surface in $\mathcal{S}^p_k$. For each $p \in S_0$ there is a maximal $\epsilon > 0$ such that for any $d < \epsilon$ the geodesics passing through any two distinct points of S_0, normal to S_0, do not intersect within distance d from S_0. In other words ϵ is the radius of injectivity of the normal exponential map based on S_0. We denote by $\Omega_\epsilon(S_0)$ the ϵ geodesic neighborhood, in other words the set $\{q \in \Sigma / \ d(q, S_0) < \epsilon\}$. Given any function ϕ on S_0 such that

$$\sup_{S_0} |\phi| < \epsilon,$$

we can construct its graph over S_0 as the surface

$$S = \{\exp_p(\phi N_0) / \ p \in S_0\} \tag{6.0.2a}$$

where N_0 is the unit normal to S_0; in other words, the locus of points obtained by moving along the normal geodesic through $p \in S_0$ a distance equal to $\phi(p)$. We extend N_0 to $\Omega_\epsilon(S_0)$ by setting it equal to the geodesic vectorfield. Relative to geodesic normal coordinates, based on S_0, the metric g takes the form

$$ds^2 = du^2 + \overset{\circ}{\gamma}_{AB} (u, \omega) d\omega^A d\omega^B \tag{6.0.2b}$$

where $\overset{\circ}{\gamma} (u, \omega)$ is the metric induced on each level surface of u. We denote by $\overset{\circ}{\nabla}$ the covariant differentiation with respect to the metric $\overset{\circ}{\gamma} (u, \omega)$, by $\overset{\circ}{K}$ the

Gauss curvature, and by $\overset{\circ}{\theta}$ the second fundamental form of the level surfaces of u.

$$\overset{\circ}{\theta}_{AB} = \frac{1}{2}\frac{\partial \overset{\circ}{\gamma}_{AB}}{\partial u}.$$

Relative to these coordinates, the surface S introduced in 6.0.2a can be expressed by

$$u = \phi(\omega).$$

The unit normal N to S is given by,

$$N = a\left(\frac{\partial}{\partial u} - \overset{\circ}{\gamma}^{AB}\frac{\partial \phi}{\partial \omega_B}\frac{\partial}{\partial \omega_A}\right) \tag{6.0.2c}$$

where

$$a = \frac{1}{\sqrt{1 + \overset{\circ}{\gamma}^{AB}\frac{\partial \phi}{\partial \omega_A}\frac{\partial \phi}{\partial \omega_B}}}. \tag{6.0.2d}$$

The induced metric on S is given by

$$\gamma_{AB}(\omega)d\omega^A d\omega^B$$

where

$$\gamma_{AB}(\omega) = \frac{\partial \phi}{\partial \omega_A}\frac{\partial \phi}{\partial \omega_B} + \overset{\circ}{\gamma}_{AB}\,(\phi(\omega), \omega). \tag{6.0.2e}$$

The second fundamental form of S is given by the formula,

$$\theta_{AB} = a\Big[-\overset{\circ}{\nabla}_A\overset{\circ}{\nabla}_B\,\phi + \overset{\circ}{\theta}_{AB} \tag{6.0.2f}$$

$$+\overset{\circ}{\gamma}^{CD}\frac{\partial \phi}{\partial \omega^C}\left(\overset{\circ}{\theta}_{AD}\frac{\partial \phi}{\partial \omega^B} + \overset{\circ}{\theta}_{BD}\frac{\partial \phi}{\partial \omega^A}\right)\Big] \tag{6.0.2g}$$

and the Gauss curvature K is given by the Gauss equation

$$2K = R - 2R_{NN} + \frac{1}{2}(\mathrm{tr}\theta)^2 - |\hat{\theta}|^2. \tag{6.0.2h}$$

Therefore, if $\phi \in W_3^{p}(S_0)$, the surface S given by the graph of ϕ according to 6.0.2a has second fundamental form θ and Gauss curvature K, which, expressed as functions on S_0, belong to $W_1^{p}(S_0)$ relative to the metric $\overset{\circ}{\gamma}$. If we want to change the covariant derivatives with respect to $\overset{\circ}{\gamma}$ into covariant derivatives relative to the induced metric on S_0, we need one more degree of differentiation of the second fundamental form $\overset{\circ}{\theta}$.[2] If $\overset{\circ}{\theta}(0) \in W_2^p(S_0)$, then, assuming that Ric $\in^{(loc)} W_2^p(\Sigma)$, we have $\overset{\circ}{\theta}(u), \overset{\circ}{K}(u) \in W_1^p(S_0)$. Therefore, if $\phi \in W_3^p(S_0)$ and $\overset{\circ}{\theta}(0) \in W_2^p(S_0)$, then $\theta, K \in W_1^p(S_0)$.

Now suppose we have a differentiable function $\phi : [0,1] \times S_0 \longrightarrow \mathcal{R}$. The surfaces $S_t = S_t[\phi] = \{\exp_p(\phi_t N_0)/\ p \in S_0\}$ define a homotopy of surfaces of lapse function

$$a = \frac{\partial \phi}{\partial t}\langle N_0, N\rangle \tag{6.0.3a}$$

or, since $N_0 = \frac{\partial}{\partial u}$, we deduce in view of 6.0.2c

$$-\frac{1}{a^2}(\frac{\partial \phi}{\partial t})^2 + \overset{\circ}{\gamma}{}^{AB}(\phi(t,\omega),\omega)\frac{\partial \phi}{\partial \omega_A}\frac{\partial \phi}{\partial \omega_B} = -1. \tag{6.0.3b}$$

We consider $\mathcal{A}$ to be a nonlinear operator that takes functions on S_0 to functions on S_0, more precisely, $\mathcal{A} : W_3^p(S_0) \longrightarrow W_3^p(S_0)$, and we ask the following:

Inverse Lapse Problem: Given a nonlinear operator $\mathcal{A}$, find a one-parameter family of functions on S_0, $\phi_t : S_0 \longrightarrow \mathcal{R}$, $t \in I = [0,\epsilon]$, such that the lapse function corresponding to the homotopy of surfaces induced by ϕ_t, that is, $S_t = \{\exp_p(\phi_t N_0)/\ p \in S_0\}$, viewed as a one-parameter family of functions $a_t : S_0 \longrightarrow \mathcal{R}$, is identical to $\mathcal{A}[\phi_t]$. In other words, in view of the equation 6.0.3b, we want $\phi : I \times S_0 \longrightarrow \mathcal{R}$ such that

$$-\frac{1}{\mathcal{A}[\phi_t]^2(\omega)}(\frac{\partial \phi}{\partial t})^2 + \overset{\circ}{\gamma}{}^{AB}(\phi(t,\omega),\omega)\frac{\partial \phi}{\partial \omega_A}\frac{\partial \phi}{\partial \omega_B} = -1 \tag{6.0.4a}$$

and

$$\phi(0,\omega) = 0. \tag{6.0.4b}$$

We solve the problem by a contraction argument. Thus let $\phi_t(\omega) = \phi(t,\omega)$ be a given one-parameter family of functions on S_0 and introduce

$$\tilde{a}(t,\omega) = \mathcal{A}[\phi_t](\omega) \tag{6.0.4c}$$

$$\tilde{\gamma}_{AB}(t.\omega) = \overset{\circ}{\gamma}_{AB}(\phi(t,\omega),\omega). \tag{6.0.4d}$$

Let $\check{\phi}(t,\omega)$ be the solution to the equation

$$-\frac{1}{\tilde{a}^2}\left(\frac{\partial \check{\phi}}{\partial t}\right)^2 + \tilde{\gamma}^{AB}(t,\omega)\frac{\partial \check{\phi}}{\partial \omega_A}\frac{\partial \check{\phi}}{\partial \omega_B} = 1 \tag{6.0.4e}$$

[2] Indeed, $\overset{\circ}{\Gamma}{}^C_{AB}(u) - \overset{\circ}{\Gamma}{}^C_{AB}(0) = \int_0^u \frac{\partial}{\partial u}\overset{\circ}{\Gamma}{}^C_{AB}(u)du$ and $\frac{\partial}{\partial u}\overset{\circ}{\Gamma}{}^C_{AB}(u) = \overset{\circ}{\gamma}{}^{CD}(\overset{\circ}{\nabla}_A\overset{\circ}{\theta}_{BD} + \overset{\circ}{\nabla}_B\overset{\circ}{\theta}_{AD} - \overset{\circ}{\nabla}_D\overset{\circ}{\theta}_{AB})$, where $\overset{\circ}{\Gamma}(u)$ denotes the Christoffel symbols relative to $\overset{\circ}{\gamma}(u)$.

subject to the initial condition

$$\check{\phi}(0,\omega)=0. \tag{6.0.4f}$$

To solve 6.0.4a, we have to establish a fixed point of the map

$$\phi\longrightarrow\check{\phi}.$$

We now introduce $\tilde{g}$ to be the following Einstein metric on $I\times S_0$:

$$d\tilde{s}^2=\tilde{g}_{\mu\nu}dx^\mu dx^\nu=-\tilde{a}^2dt^2+\tilde{\gamma}_{AB}d\omega^A d\omega^B \tag{6.0.5a}$$

In view of this definition the equation 6.0.4e takes the form

$$\tilde{g}^{\mu\nu}\partial_\mu\check{\phi}\partial_\nu\check{\phi}=-1. \tag{6.0.5b}$$

Therefore the desired solution of 6.0.4e, 6.0.4f can be interpreted as the temporal distance function, as measured from $t=0$, on the cylinder $I\times S_0$.

Denote by $\tilde{D}$ the covariant derivative and by $\tilde{R}$ the curvature tensor with respect to $\tilde{g}$. Also, let $\tilde{T}=\tilde{a}^{-1}\partial/\partial t$ be the unit normal and $\tilde{\theta}$ and $\tilde{K}$ be respectively the second fundamental form and Gauss curvature of the level surfaces of t. All these geometric quantities can be expressed in terms of $\overset{\circ}{\gamma}$, $\overset{\circ}{\theta}$, ϕ, and the operator $\mathcal{A}$. Indeed, we have

$$\tilde{\theta}_{AB}=-\frac{1}{2\tilde{a}}\partial_t\tilde{\gamma}_{AB}=-\frac{1}{\tilde{a}}\overset{\circ}{\theta}_{AB}\,\partial_t\phi \tag{6.0.5c}$$

$$\partial_t\tilde{\theta}_{AB}=-\frac{1}{\tilde{a}}\partial_u\overset{\circ}{\theta}_{AB}\,(\partial_t\phi)^2+\frac{1}{\tilde{a}^2}\overset{\circ}{\theta}_{AB}\,\partial_t\tilde{a}\partial_t\phi-\frac{1}{\tilde{a}}\overset{\circ}{\theta}_{AB}\,\partial_t^2\phi \tag{6.0.5d}$$

$$\partial_t\tilde{a}=\mathcal{A}'[\phi]\partial_t\phi \tag{6.0.5e}$$

where $\mathcal{A}'$ is the Fréchet derivative of $\mathcal{A}$. On the other hand, $\tilde{R}_{\alpha\beta}$ depends only on

$$\tilde{\nabla}\tilde{a},\ \tilde{\nabla}^2\tilde{a},\ \tilde{\nabla}\tilde{\theta},\ \partial_t\tilde{\theta},\ \tilde{K}$$

where we denote by $\tilde{\nabla}$ the covariant derivative on the level surfaces of t with respect to the induced metric $\tilde{\gamma}$. In order to be able to control these quantities, we assume that the nonlinear operator $\mathcal{A}$ is a C^1 operator from a neighborhood of the origin in the Sobolev space $W_3^p(S_0)$ to $W_3^p(S_0)$ and that $\inf_{S_0}\mathcal{A}[\phi]$ is positive for every ϕ in that neighborhood. More precisely, for some $\delta_0>0$,

Assumptions for $\mathcal{A}$

$$\sup_{\|\phi\|_{W_3^p} \leq \delta_0} (\inf_{S_0} \mathcal{A}[\phi])^{-1} \leq c_{\mathcal{A}}$$

$$\sup_{\|\phi\|_{W_3^p} \leq \delta_0} (\|\mathcal{A}[\phi]\|_{W_3^p} + |||\mathcal{A}'[\phi]|||_{W_2^p}) \leq c_{\mathcal{A}} \tag{6.0.5f}$$

where $|||\ |||_{W_2^p}$ denotes the norm for linear operators on W_2^p and $c_{\mathcal{A}}$ denotes constants depending only on the operator $\mathcal{A}$ and the surface S_0.

Let $\phi \in C^0(I, W_3^p(S_0)) \bigcap C^1(I, W_2^p(S_0)) \bigcap C^2(I, W_1^p(S_0))$ verify

$$\phi(0, \omega) = 0, \qquad \partial_t\phi(0, \omega) = \mathcal{A}[0](\omega) \tag{6.0.5g}$$

and let

$$\sup_{t\in I} \|\phi(t)\|^2_{W_3^p} \leq \delta \leq \delta_0,$$

$$\sup_{t\in I} \|\partial_t\phi(t)\|^2_{W_2^p} \leq r, \qquad \sup_{t\in I} \|\partial_t^2\phi(t)\|^2_{W_1^p} \leq s. \tag{6.0.5h}$$

In view of 6.0.4c, 6.0.4d, and these assumptions, for every $t \in I$ we then have

$$\begin{aligned}
(\inf_{S_0} \tilde{a}(t))^{-1} &\leq c_{\mathcal{A}} \\
\|\tilde{a}(t)\|_{W_3^p(S_0)} &\leq c_{\mathcal{A}} \\
\|\partial_t\tilde{a}(t)\|_{W_2^p(S_0)} &\leq c_{\mathcal{A}} r \\
\|\tilde{\theta}(t)\|_{W_2^p(S_0)} &\leq c_{\mathcal{A}} r \\
\|\partial_t\tilde{\theta}(t)\|_{W_1^p(S_0)} &\leq c_{\mathcal{A}}(r^2 + s) \\
\|\tilde{K}\|_{W_1^p(S_0)} &\leq c_{\mathcal{A}} \\
\|\tilde{R}(t)\|_{W_1^p(S_0)} &\leq c_{\mathcal{A}}(1 + r^2 + s).
\end{aligned} \tag{6.0.5i}$$

Let $\check{T}^\mu = -\tilde{g}^{\mu\nu}\partial_\nu\check{\phi}$ be the unit normal of the level surfaces of $\check{\phi}$. We decompose $\check{T}$ into its components orthogonal and parallel to $\tilde{T}$:

$$\check{T} = \cosh v\tilde{T} + \tilde{Y}, \qquad |\tilde{Y}| = \sinh v. \tag{6.0.6a}$$

Denoting by τ the arc length along the timelike geodesics, integral curves of $\check{T}$, and by $d/d\tau$ covariant differentiation along $\check{T}$, we have

$$\frac{d}{d\tau}\cosh v = \tilde{\theta}_{\tilde{Y}\tilde{Y}} - \cosh v\tilde{a}^{-1}\tilde{\nabla}_{\tilde{Y}}\tilde{a}. \tag{6.0.6b}$$

Taking into account the fact that

$$\frac{dt}{d\tau} = \tilde{a}^{-1} \cosh v, \tag{6.0.6c}$$

we then find

$$|\frac{dv}{dt}| \leq \tilde{a}|\tilde{\theta}| + |\tilde{\nabla}\tilde{a}|. \tag{6.0.6d}$$

Integrating this equation on I under the initial condition

$$v|_{t=0} = 0$$

and taking into account the estimates 6.0.5i, we infer that

$$\sup_{I \times S_0} |v| \leq \epsilon \mathcal{A}(1 + r). \tag{6.0.6e}$$

Now let $\check{\theta}$ be the second fundamental form of the level surfaces of $\check{\phi}$. We remark that, by conditions 6.0.5g, the metric induced by $\tilde{g}$ on the level surface $t = 0$ coincides with the original metric on S_0 and

$$\check{\theta}(0, \omega) = \tilde{\theta}(0, \omega) = -\overset{\circ}{\theta}(0, \omega). \tag{6.0.7a}$$

Differentiating 6.0.5b, we deduce

$$\frac{d\check{\theta}_{\alpha\beta}}{d\tau} = \check{\theta}_{\alpha\mu}\check{\theta}^{\mu}{}_{\beta} + \tilde{R}_{\alpha\check{T}\beta\check{T}}. \tag{6.0.7b}$$

Let

$$\tilde{\Pi}^{\mu}_{\nu} = \delta^{\mu}_{\nu} + \tilde{T}^{\mu}\tilde{T}_{\nu} \tag{6.0.7c}$$

be the operator of projection to the level surfaces of t. We define $\bar{\theta}$ to be the projection of $\check{\theta}$ to the level surfaces of t :

$$\bar{\theta}_{\mu\nu} = \tilde{\Pi}^{\alpha}_{\mu}\tilde{\Pi}^{\beta}_{\nu}\check{\theta}_{\alpha\beta}. \tag{6.0.7d}$$

We can then express

$$\bar{\nabla}_{\mu}\bar{\nabla}_{\nu}\check{\phi} = \bar{\theta}_{\mu\nu} - \cosh v\tilde{\theta}_{\mu\nu}. \tag{6.0.7e}$$

Using the fact that

$$\frac{d}{d\tau}\tilde{\Pi}^{\mu}_{\nu} = \tilde{T}^{\mu}\tilde{b}_{\nu} + \tilde{T}_{\nu}\tilde{b}^{\mu}$$

where

$$\tilde{b}^{\mu} = \frac{d}{d\tau}\tilde{T}^{\mu} = \cosh v\tilde{a}^{-1}\tilde{\nabla}^{\mu}\tilde{a} + \tilde{\theta}^{\mu}{}_{\nu}\tilde{Y}^{\nu}.$$

we find from 6.0.7b

$$\tilde{\Pi}^{\mu}_{\alpha}\tilde{\Pi}^{\nu}_{\beta}\frac{d\bar{\theta}_{\mu\nu}}{d\tau} = \bar{\theta}_{\alpha\mu}\bar{\theta}^{\mu}{}_{\beta} - \tilde{\eta}_{\alpha}\tilde{\eta}_{\beta} + \tilde{b}_{\alpha}\tilde{\eta}_{\beta} + \tilde{b}_{\beta}\tilde{\eta}_{\alpha} + F_{\alpha\beta} \tag{6.0.7f}$$

where

$$\tilde{\eta}_{\alpha} = \tilde{\Pi}^{\mu}_{\alpha}\check{\theta}_{\mu\nu}\tilde{T}^{\nu} = -(\cosh v)^{-1}\bar{\theta}_{\alpha\nu}\tilde{Y}^{\nu}$$

and

$$F_{\alpha\beta} = \tilde{\Pi}^{\mu}_{\alpha}\tilde{\Pi}^{\nu}_{\beta}\tilde{R}_{\mu\tilde{T}\nu\tilde{T}}.$$

The initial condition for $\bar{\theta}$ is

$$\bar{\theta}(0,\omega) = \check{\theta}(0,\omega) = -\overset{\circ}{\theta}(0,\omega). \tag{6.0.7g}$$

From 6.0.7f and 6.0.7g, using the estimate 6.0.6e as well as the estimates 6.0.5i, we establish that for sufficiently small ϵ and all $t \in I = [0,\epsilon]$

$$\|\bar{\theta}(t) - \bar{\theta}(0)\|_{W_1^p(S_0)} \leq c_{\mathcal{A}}\epsilon(1 + r^2 + s). \tag{6.0.8a}$$

Thus, in view of 6.0.7e,

$$\sup_{t\in I}\|\check{\phi}(t)\|_{W_3^p} \leq c_{\mathcal{A}}\epsilon(1 + r^2 + s), \tag{6.0.8b}$$

and from equation 6.0.4e,

$$\sup_{t\in I}\|\partial_t\check{\phi}(t)\|_{W_2^p} \leq c_{\mathcal{A}}, \qquad \sup_{t\in I}\|\partial_t^2\check{\phi}(t)\|_{W_1^p} \leq c_{\mathcal{A}}r. \tag{6.0.8c}$$

Let $\mathcal{K}_{\delta,r,s}$ be the closed set in the space

$$\mathcal{E}_3(S_0, I) = \mathcal{C}^0(I, W_3^p(S_0))\bigcap\mathcal{C}^1(I, W_2^p(S_0))\bigcap\mathcal{C}^2(I, W_1^p(S_0))$$

defined by the initial conditions

$$\phi(0,\omega) = 0, \qquad \partial_t\phi(0,\omega) = \mathcal{A}[0](\omega)$$

and the inequalities:

$$\sup_{t\in I}\|\phi(t)\|_{W_3^p} \leq \delta$$

$$\sup_{t\in I}\|\partial_t\phi(t)\|_{W_2^p} \leq r$$

$$\sup_{t\in I}\|\partial_t^2\phi(t)\|_{W_1^p} \leq s.$$

By virtue of the estimates 6.0.8b and 6.0.8c, choosing first $r \geq c_{\mathcal{A}}$, then $s \geq c_{\mathcal{A}} r$, and finally ϵ small enough so that

$$\epsilon c_{\mathcal{A}}(1 + r^2 + s) \leq \delta,$$

we conclude that the mapping $\phi \longrightarrow \check{\phi}$ maps $\mathcal{K}_{\delta,r,s}$ into itself. Moreover, one finds that this mapping is in fact a contraction on $\mathcal{K}_{\delta,r,s}$ relative to the $C^0(I, W_2^p)$ norm provided that a further smallness condition on ϵ relative to δ, r, s is satisfied. This proves the following:

Proposition 6.0.1 *Let (Σ, g) be a 3-dimensional Riemannian manifold with Ric $\in^{(loc)} W_2^p(\Sigma)$. Let S_0 be a C^2 surface embedded in (Σ, g) with induced metric γ_0 and second fundamental form $\theta_0 \in W_2^p(S_0)$. Let $\mathcal{A}$ be a nonlinear operator from the closed ball of radius δ_0 in $W_3^p(S_0)$ to $W_3^p(S_0)$ satisfying the assumptions 6.0.5f. Then there exist positive constants ϵ, δ, r, s, such that the equation 6.0.4a, subject to the initial condition 6.0.4b, has a unique solution in $I = [0, \epsilon]$ belonging to the closed set $\mathcal{K}_{\delta,r,s} \subset \mathcal{E}_3(S_0, I)$.*

For applications we need a slightly stronger version of the proposition, which requires only that $\theta_0 \in W_1^p(S_0)$ and Ric $\in^{(\text{loc})} H_2(\Sigma)$. In fact the assumptions $\theta_0 \in W_2^p(S_0)$, Ric $\in^{(\text{loc})} W_2^p(\Sigma)$ were made only in order to ensure that the geodesic normal coordinates based on S_0 are sufficiently regular. We relax these assumptions by replacing the geodesic distance function u with a new function u^* defined as the solution to the Laplace equation

$$\Delta u^* = 0$$

subject to the boundary conditions $u^*|_{S_\eta} = 1$ and $u^*|_{S_{-\eta}} = -1$, where $S_\eta, S_{-\eta}$ are the surfaces obtained by traveling distances η, respectively, $-\eta$ from S_0. In view of the interior regularity properties of the Laplace equation we infer that u^* is a function in $H_5(\Omega_\eta(S_0))$, where $\Omega_\eta(S_0)$ is the domain bounded by the surfaces S_η and $S_{-\eta}$. We then use the integral curves of the gradient of u^*, which is nonvanishing as a consequence of the maximal principle, to parametrize $\Omega_\eta(S_0)$ by coordinates u^*, ω such that

$$ds^2 = \overset{\circ}{a}{}^2 (du^*)^2 + \overset{\circ}{\gamma}_{AB} (u^*, \omega) d\omega^A d\omega^B$$

where $\overset{\circ}{a} = \frac{1}{|\nabla u^*|}$ is the lapse function of the u^* foliation. We now consider all surfaces in $\Omega_\eta(S_0)$ as graphs over the surface S_0^* defined by $u^* = 0$ and reinterpret the inverse lapse problem as a problem in the function space $W_3^p(S_0^*)$. We thus obtain the following:

Proposition 6.0.2 *Let (Σ, g) be a 3-dimensional Riemannian manifold with Ric $\in^{(loc)} H_2(\Sigma)$. Let S_0 be a C^2 surface embedded in (Σ, g) with induced metric γ_0 and second fundamental form $\theta_0 \in W_1^p(S_0)$. Then S_0 is the graph of a function $\phi_0 \in W_3^p(S_0^*)$. Let $\mathcal{A}$ be a C^1 nonlinear operator from a ball centered at ϕ_0 in $W_3^p(S_0^*)$ to $W_3^p(S_0^*)$ such that* $\inf_{S_0^*} \mathcal{A}[\phi]$ *is positive for every ϕ in that ball. Then there exists an $\epsilon > 0$ such that the inverse lapse problem*

$$-\frac{1}{\mathcal{A}[\phi_t]^2(\omega)}\left(\frac{\partial \phi}{\partial t}\right)^2 + \overset{\circ}{a}{}^2 \overset{\circ}{\gamma}{}^{AB}(\phi(t,\omega),\omega)\frac{\partial \phi}{\partial \omega_A}\frac{\partial \phi}{\partial \omega_B} = -1$$

subject to the initial condition

$$\phi(0,\omega) = \phi_0$$

has a unique solution in $I = [0, \epsilon]$ belonging to $\mathcal{E}_3(S_0^*, I)$. Furthermore, the solution has a C^1 dependence on the initial data.

Part II

Bianchi Equations in Space-Time

CHAPTER 7

The Comparison Theorem

This part provides the main ideas of our treatment of the Bianchi identities of an Einstein-Vacuum space-time, in order to control its curvature tensor. These ideas are at the heart of this book.

7.1. PreliminaryResults

Given an Einstein space-time $(\mathbf{M}, \mathbf{g})$, we consider Weyl tensors W, which are four tensors verifying all the symmetry properties of the Riemann curvature tensor, that is,

$$W_{\alpha\beta\gamma\delta} = -W_{\beta\alpha\gamma\delta} = -W_{\alpha\beta\delta\gamma} \tag{7.1.1a}$$

$$W_{\alpha\beta\gamma\delta} + W_{\alpha\gamma\delta\beta} + W_{\alpha\delta\beta\gamma} = 0 \tag{7.1.1b}$$

$$W_{\alpha\beta\gamma\delta} = W_{\gamma\delta\alpha\beta} \tag{7.1.1c}$$

(7.1.1d)

plus the trace condition

$$W_{\beta\delta} = W^{\alpha}{}_{\beta\alpha\delta} = 0. \tag{7.1.1e}$$

We recall the well-known fact that 7.1.1c is in fact a consequence of 7.1.1a and 7.1.1b. Thus, $W'_{\alpha\beta\gamma\delta} = W'_{\gamma\delta\alpha\beta}$.

The left and right duals of W have been defined by

$$^{\star}W_{\alpha\beta\gamma\delta} = \frac{1}{2} \in_{\alpha\beta\mu\nu} W^{\mu\nu}{}_{\gamma\delta} \tag{7.1.1f}$$

$$W^{\star}_{\alpha\beta\gamma\delta} = W_{\alpha\beta}{}^{\mu\nu} \frac{1}{2} \in_{\mu\nu\gamma\delta}.$$

It is easy to check that, if W is a Weyl tensor, both duals are equal and define another Weyl field. Moreover, $^{\star}(^{\star}W) = -W$. For future reference we recall here the multiplication properties of the coefficients $\in$ of the volume element of $\mathbf{M}$.

$$\in^{\alpha_1\alpha_2\alpha_3\alpha_4} \in_{\beta_1\beta_2\beta_3\beta_4} = -\det(\delta^{\alpha_i}_{\beta_j})_{i,j=1,\dots 4}$$

also, by contraction,

$$\begin{aligned}
\in^{\alpha_1\alpha_2\alpha_3\alpha_4}\in_{\alpha_1\beta_2\beta_3\beta_4} &= -\det(\delta^{\alpha_i}_{\beta_j})_{i,j=2,3,4} \\
\in^{\alpha_1\alpha_2\alpha_3\alpha_4}\in_{\alpha_1\alpha_2\beta_3\beta_4} &= -2\det(\delta^{\alpha_i}_{\beta_j})_{i,j=3,4} \\
\in^{\alpha_1\alpha_2\alpha_3\alpha_4}\in_{\alpha_1\alpha_2\alpha_3\beta_4} &= -6\delta^{\alpha_4}_{\beta_4} \\
\in^{\alpha_1\alpha_2\alpha_3\alpha_4}\in_{\alpha_1\alpha_2\alpha_3\alpha_4} &= -24.
\end{aligned}$$

Our main object of study in this chapter is the Bianchi equations (see page 3):

$$\mathbf{D}_{[\sigma}W_{\gamma\delta]\alpha\beta} = 0. \tag{7.1.2}$$

However, since the commutation of 7.1.2 with Lie derivatives leads to error terms, it is convenient to discuss the more general inhomogeneous equations

$$\mathbf{D}^{\alpha}W_{\alpha\beta\gamma\delta} = J_{\beta\gamma\delta} \tag{7.1.3a}$$

$$\mathbf{D}^{\alpha}{}^{\star}W_{\alpha\beta\gamma\delta} = J^{\star}_{\beta\gamma\delta} \tag{7.1.3b}$$

where J, $J^{\star} = \frac{1}{2}J_{\beta\mu\nu}\in^{\mu\nu}{}_{\gamma\delta}$ satisfy the obvious symmetries required by the equations. In view of the fact that ${}^{\star}W = W^{\star}$, the equations 7.1.3a and 7.1.3b are equivalent. We also remark that the following equations are an immediate consequence of 7.1.3a and 7.1.3b:

$$\mathbf{D}_{[\sigma}W_{\gamma\delta]\alpha\beta} = \tilde{J}_{\sigma\gamma\delta\alpha\beta} = \in_{\mu\sigma\gamma\delta}J^{\star\mu}{}_{\alpha\beta} \tag{7.1.3c}$$

$$\mathbf{D}_{[\sigma}{}^{\star}W_{\gamma\delta]\alpha\beta} = \tilde{J}^{\star}{}_{\sigma\gamma\delta\alpha\beta} = -\in_{\mu\sigma\gamma\delta}J^{\mu}{}_{\alpha\beta}. \tag{7.1.3d}$$

7.1.1. *Basic Properties of the Bianchi Equations*

In this subsection we recall the definition of the Bell-Robinson tensor and its properties. We also investigate the commutation properties of the Bianchi equations with the Lie derivative of an arbitrary vectorfield.

Let $Q(W)$ be the Bell-Robinson tensor associated to W:

$$Q(W)_{\alpha\beta\gamma\delta} = W_{\alpha\rho\gamma\sigma}W_{\beta}{}^{\rho}{}_{\delta}{}^{\sigma} + {}^{\star}W_{\alpha\rho\gamma\sigma}{}^{\star}W_{\beta}{}^{\rho}{}_{\delta}{}^{\sigma}. \tag{7.1.4}$$

We recall the following remarkable fact:

Lemma 7.1.1 *Given an arbitrary Weyl tensor W,*

1. *$Q(W)$ is symmetric and traceless in all pairs of indices.*
2. *$Q(W)(X_1, X_2, X_3, X_4)$ is positive for any nonspacelike future directed vectorfields X_1, X_2, X_3, X_4, whenever two of them at most are distinct.*[1]

[1] The assertion holds true, in fact, even when all vectorfields are distinct.

Proof 7.1.1 *We indicate only the main steps in the proof of the symmetry of* Q.[2] *First, we remark that the symmetry* $Q_{\alpha\beta\gamma\delta} = Q_{\gamma\delta\alpha\beta}$ *is immediate from the definition of* Q. *Next, to check the symmetry between the first and second indices, we write*

$$\begin{aligned}
Q_{\alpha\beta\gamma\delta} - Q_{\gamma\beta\alpha\delta} &= (W_{\alpha\rho\gamma\sigma} - W_{\gamma\rho\alpha\sigma})W_{\beta\ \delta}^{\ \rho\ \sigma} + ({}^*W_{\alpha\rho\gamma\sigma} - {}^*W_{\gamma\rho\alpha\sigma}){}^*W_{\beta\ \delta}^{\ \rho\ \sigma} \\
&= W_{\rho\sigma\alpha\gamma}W_{\beta\ \delta}^{\ \rho\ \sigma} + {}^*W_{\rho\sigma\alpha\gamma}{}^*W_{\beta\ \delta}^{\ \rho\ \sigma} \\
&= \frac{1}{2}(W^{\rho\sigma}_{\ \ \alpha\gamma}W_{\rho\sigma\beta\delta} + {}^*W^{\rho\sigma}_{\ \ \alpha\gamma}{}^*W_{\rho\sigma\beta\delta}) \\
&= 0.
\end{aligned}$$

Finally, the symmetry relative to the first two indices is an immediate consequence of the formula

$$Q_{\alpha\beta\gamma\delta} = W_{\alpha\rho\gamma\sigma}W_{\beta\ \delta}^{\ \rho\ \sigma} + W_{\alpha\rho\delta\sigma}W_{\beta\ \gamma}^{\ \rho\ \sigma} - \frac{1}{8}g_{\alpha\beta}g_{\gamma\delta}W_{\rho\sigma\mu\nu}W^{\rho\sigma\mu\nu}. \quad (7.1.5a)$$

The formula is a consequence of the following identities:

$$\begin{aligned}
{}^*W_{\alpha\rho\gamma\sigma}{}^*W_{\beta\ \delta}^{\ \rho\ \sigma} &= W_{\alpha\rho\delta\sigma}W_{\beta\ \gamma}^{\ \rho\ \sigma} - \frac{1}{2}g_{\gamma\delta}W_{\alpha\sigma\mu\nu}W_{\beta}^{\ \sigma\mu\nu} && (7.1.5b) \\
W_{\alpha\sigma\mu\nu}W_{\beta}^{\ \sigma\mu\nu} &= \frac{1}{4}g_{\alpha\beta}W_{\rho\sigma\mu\nu}W^{\rho\sigma\mu\nu}. && (7.1.5c)
\end{aligned}$$

The first identity follows easily from the product formulas for the $\in$*'s. To prove the last identity, we contract the indices* α, β *in the previous one and derive*

$${}^*W_{\gamma\sigma\alpha\rho}{}^*W_{\delta}^{\ \sigma\alpha\rho} = W_{\gamma\sigma\alpha\rho}W_{\delta}^{\ \sigma\alpha\rho} - \frac{1}{2}g_{\gamma\delta}W_{\alpha\sigma\mu\nu}W^{\alpha\sigma\mu\nu}.$$

On the other hand, by contracting the indices γ, δ *in the formula* ${}^*W_{\alpha\gamma}^{\ \ \ \rho\sigma}{}^*W_{\beta\delta\rho\sigma} = -W_{\alpha\gamma}^{\ \ \ \rho\sigma}W_{\beta\delta\rho\sigma}$, *we obtain* ${}^*W_{\alpha}^{\ \gamma\rho\sigma}{}^*W_{\beta\gamma\rho\sigma} = -W_{\alpha}^{\ \gamma\rho\sigma}W_{\beta\gamma\rho\sigma}$. *Substituting this on the left of 7.1.5b, we derive 7.1.5c and thus complete the proof for 7.1.5a.*

We next prove the following:

Proposition 7.1.1 *Let* W *be a Weyl field verifying the equations 7.1.3a and 7.1.3b. Then*

$$\begin{aligned}
\mathbf{D}^{\alpha}Q_{\alpha\beta\gamma\delta} &= W_{\beta\ \delta}^{\ \mu\ \nu}J_{\mu\gamma\nu} + W_{\beta\ \gamma}^{\ \mu\ \nu}J_{\mu\delta\nu} \\
&\quad + {}^*W_{\beta\ \delta}^{\ \mu\ \nu}J^{\star}_{\mu\gamma\nu} + {}^*W_{\beta\ \gamma}^{\ \mu\ \nu}J^{\star}_{\mu\delta\nu}.
\end{aligned}$$

[2] For a complete proof of the lemma we refer to [Ch-Kl].

Proof 7.1.2 *To prove the proposition, we first remark that*

$$
\begin{aligned}
W^{\alpha}{}_{\mu\gamma\nu}\mathbf{D}_\alpha W_\beta{}^\mu{}_\delta{}^\nu &= \frac{1}{2}W^{\alpha\mu}{}_\gamma{}^\nu(\mathbf{D}_\alpha W_{\beta\mu\delta\nu} - \mathbf{D}_\mu W_{\beta\alpha\delta\nu})\\
&= \frac{1}{2}W^{\alpha\mu}{}_\gamma{}^\nu(\mathbf{D}_\alpha W_{\beta\mu\delta\nu} + \mathbf{D}_\mu W_{\alpha\beta\delta\nu})\\
&= \frac{1}{2}W^{\alpha\mu}{}_\gamma{}^\nu(\mathbf{D}_{[\alpha} W_{\beta\mu]\delta\nu} + \mathbf{D}_\beta W_{\alpha\mu\delta\nu}) \qquad (7.1.6)
\end{aligned}
$$

where

$$\mathbf{D}_{[\alpha} W_{\beta\mu]\delta\nu} = \mathbf{D}_\alpha W_{\beta\mu\delta\nu} + \mathbf{D}_\beta W_{\mu\alpha\delta\nu} + \mathbf{D}_\mu W_{\alpha\beta\delta\nu}.$$

Therefore,

$$
\begin{aligned}
\mathbf{D}^\alpha Q_{\alpha\beta\gamma\delta} = {} & W_\beta{}^\mu{}_\delta{}^\nu J_{\mu\gamma\nu} + {}^*W_\beta{}^\mu{}_\delta{}^\nu J^*_{\mu\gamma\nu}\\
& + \frac{1}{2}(W^{\alpha\mu}{}_\gamma{}^\nu \mathbf{D}_\beta W_{\alpha\mu\delta\nu} + {}^*W^{\alpha\mu}{}_\gamma{}^\nu \mathbf{D}_\beta {}^*W_{\alpha\mu\delta\nu})\\
& + \frac{1}{2}(W^{\alpha\mu}{}_\gamma{}^\nu \mathbf{D}_{[\alpha} W_{\beta\mu]\delta\nu} + {}^*W^{\alpha\mu}{}_\gamma{}^\nu \mathbf{D}_{[\alpha} {}^*W_{\beta\mu]\delta\nu}).
\end{aligned}
$$

Now, the proposition is an immediate consequence of the identity

$$W^{\alpha\mu}{}_\gamma{}^\nu \mathbf{D}_\beta W_{\alpha\mu\delta\nu} + {}^*W^{\alpha\mu}{}_\gamma{}^\nu \mathbf{D}_\beta {}^*W_{\alpha\mu\delta\nu} = 0$$

and equations 7.1.3c and 7.1.3d.

Proposition 7.1.1 can be used to derive approximate conservation laws as follows:

Assume that the space-time $(\mathbf{M}, \mathbf{g})$ is foliated by a time function t and let T be the unit normal to the foliation. Consider three arbitrary vectorfields X, Y, Z and define $P_\alpha = Q(W)_{\alpha\beta\gamma\delta}X^\beta Y^\gamma Z^\delta$. Then

$$
\begin{aligned}
\mathbf{Div}P = {} & X^\beta Y^\gamma Z^\delta (\mathbf{Div}Q)_{\beta\gamma\delta}\\
& + \frac{1}{2}Q^{\alpha\beta\gamma\delta}({}^{(X)}\pi_{\alpha\beta}Y^\gamma Z^\delta + {}^{(Y)}\pi_{\alpha\beta}Z^\gamma X^\delta + {}^{(Z)}\pi_{\alpha\beta}X^\gamma Y^\delta)
\end{aligned}
$$

where ${}^{(X)}\pi$ is the deformation tensor of X. Integrating on a slab $\bigcup_{t'\in[t_0,t]}\Sigma_{t'}$, we derive the following:

Corollary 7.1.1.1 *Let $Q(W)$ be the Bell-Robinson of a solution of the inhomogeneous Bianchi equations 7.1.3a and 7.1.3b in a space-time foliated by a time-function t. Then for any vectorfields X, Y, Z,*[3]

$$\int_{\Sigma_t} Q(W)(X, Y, Z, T)d\mu_g = \int_{\Sigma_{t_0}} Q(W)(X, Y, Z, T)d\mu_g$$

[3] We also make suitable assumptions at spacelike infinity on each slice $\Sigma_{t'}$.

$$+\int_{t_0}^{t} dt' \left[\int_{\Sigma_{t'}} (\mathbf{Div}Q)_{\beta\gamma\delta} X^\beta Y^\gamma Z^\delta \phi d\mu_g \right.$$
$$+\frac{1}{2}\int_{\Sigma_{t'}} Q^{\alpha\beta\gamma\delta} \Big({}^{(X)}\pi_{\alpha\beta} Y_\gamma Z_\delta$$
$$\left. + {}^{(Y)}\pi_{\alpha\beta} Z_\gamma X_\delta + {}^{(Z)}\pi_{\alpha\beta} X_\gamma Y_\delta \Big) \phi d\mu_g \right]$$

with ϕ being the lapse function and g being the induced metric of the time foliation (see the definition on page 5). The divergence term $\mathbf{Div}Q$ *is given by Proposition 7.1.1.*

Given a Weyl field W and a vectorfield X, the Lie derivative of W with respect to X is not, in general, a Weyl field. Indeed, while the conditions 7.1.1a, 7.1.1b, and 7.1.1c are satisfied, the trace condition 7.1.1e is not. In fact,

$$\mathbf{g}^{\alpha\gamma}(\mathcal{L}_X W_{\alpha\beta\gamma\delta}) = \pi^{\alpha\gamma} W_{\alpha\beta\gamma\delta}$$

where $\pi = {}^{(X)}\pi$ is the deformation tensor of X.[4] To compensate, we define the following modified Lie derivative:

$$\hat{\mathcal{L}}_X W := \mathcal{L}_X W - \tfrac{1}{2}{}^{(X)}[W] + \tfrac{3}{8}\mathrm{tr}\,{}^{(X)}\pi W$$

where,

$$ {}^{(X)}[W]_{\alpha\beta\gamma\delta} := \pi^\mu{}_\alpha W_{\mu\beta\gamma\delta} + \pi^\mu{}_\beta W_{\alpha\mu\gamma\delta}$$
$$\pi^\mu{}_\gamma W_{\alpha\beta\mu\delta} + \pi^\mu{}_\delta W_{\alpha\beta\gamma\mu} \tag{7.1.7}$$

Decomposing π between its traceless part $\hat{\pi}$ and trace $\mathrm{tr}\pi$, we also write

$$\hat{\mathcal{L}}_X W_{\alpha\beta\gamma\delta} = \mathcal{L}_X W_{\alpha\beta\gamma\delta} - \frac{1}{8}\mathrm{tr}\,{}^{(X)}\pi$$
$$-\frac{1}{2}\Big(\hat{\pi}^\mu{}_\alpha W_{\mu\beta\gamma\delta} + \hat{\pi}^\mu{}_\beta W_{\alpha\mu\gamma\delta}$$
$$+ \hat{\pi}^\mu{}_\gamma W_{\alpha\beta\mu\delta} + \hat{\pi}^\mu{}_\delta W_{\alpha\beta\gamma\mu} \Big).$$

Lemma 7.1.2 *Let W be a Weyl field, and let X be an arbitrary vectorfield. Then,*

1. *$\hat{\mathcal{L}}_X W$ is a Weyl field.*

[4] We have ${}^{(X)}\pi_{\alpha\beta} = (\mathcal{L}_X \mathbf{g})_{\alpha\beta}$ and $(\mathcal{L}_X \mathbf{g})^{\alpha\beta} = -{}^{(X)}\pi^{\alpha\beta}$.

2. *the modified Lie derivative commutes with the Hodge dual of W, that is,*

$$\hat{\mathcal{L}}_X {}^*W = {}^*\hat{\mathcal{L}}_X W.$$

To check that $\hat{\mathcal{L}}_X W$ is a Weyl field, it suffices to remark that ${}^{(X)}[W]$ verifies conditions 7.1.1a and 7.1.1b and therefore also 7.1.1c. Condition 7.1.1e is automatically satisfied. The second part of the lemma is an immediate consequence of the following formulas:

$$\frac{1}{2} \in_{\alpha\beta}{}^{\sigma\tau} \mathcal{L}_X W_{\sigma\tau\gamma\delta} = \mathcal{L}_X {}^*W_{\alpha\beta\gamma\delta} + \frac{1}{2}\mathrm{tr}\pi\, {}^*W_{\alpha\beta\gamma\delta} - \pi_\alpha{}^\mu\, {}^*W_{\mu\beta\gamma\delta} - \pi_\beta{}^\mu\, {}^*W_{\alpha\mu\gamma\delta} \tag{7.1.8a}$$

$$\frac{1}{2} \in_{\alpha\beta}{}^{\sigma\tau}\, {}^{(X)}[W]_{\sigma\tau\gamma\delta} = \mathrm{tr}\pi\, {}^*W_{\alpha\beta\gamma\delta} - (\pi_\alpha{}^\mu\, {}^*W_{\mu\beta\gamma\delta} + \pi_\beta{}^\mu\, {}^*W_{\alpha\mu\gamma\delta}) + \pi^\mu{}_\gamma W_{\alpha\beta\mu\delta} + \pi^\mu{}_\delta W_{\alpha\beta\gamma\mu}. \tag{7.1.8b}$$

To check formula 7.1.8a, we recall that $(\mathcal{L}_X \in)_{\alpha\beta\gamma\delta} = \frac{1}{2}(\mathrm{tr}\,{}^{(X)}\pi) \in_{\alpha\beta\gamma\delta}$. With the help of the multiplication properties of $\in$, we then derive

$$\frac{1}{2} \in_{\alpha\beta}{}^{\sigma\tau} \mathcal{L}_X W_{\sigma\tau\gamma\delta} = \mathcal{L}_X\left(\frac{1}{2} \in_{\alpha\beta}{}^{\sigma\tau} W_{\sigma\tau\gamma\delta}\right) - \frac{1}{2}(\mathcal{L}_X \in)_{\alpha\beta}{}^{\sigma\tau} W_{\sigma\tau\gamma\delta}$$
$$= \mathcal{L}_X {}^*W_{\alpha\beta\gamma\delta} - \frac{1}{2}\mathrm{tr}\pi\, {}^*W_{\alpha\beta\gamma\delta} + \pi^{\mu\nu} \in_{\alpha\beta\mu}{}^\lambda W_{\nu\lambda\gamma\delta}$$

and 7.1.8a follows, in view of

$$\pi^{\mu\nu} \in_{\alpha\beta\mu}{}^\lambda W_{\nu\lambda\gamma\delta} = -\frac{1}{2} \in_{\alpha\beta\mu}{}^\lambda \in^{\xi\zeta}{}_{\nu\lambda}\, \pi^{\mu\nu}\, {}^*W_{\xi\zeta\gamma\delta}$$
$$= \mathrm{tr}\pi\, {}^*W_{\alpha\beta\gamma\delta} - (\pi_\alpha{}^\mu\, {}^*W_{\mu\beta\gamma\delta} + \pi_\beta{}^\mu\, {}^*W_{\alpha\mu\gamma\delta}).$$

To verify 7.1.8b, we write

$$\frac{1}{2} \in_{\alpha\beta}{}^{\sigma\tau}\, {}^{(X)}[W]_{\sigma\tau\gamma\delta} = -\frac{1}{4} \in_{\alpha\beta}{}^{\sigma\tau} \big(\in_{\mu\tau}{}^{\xi\zeta}\, {}^*W_{\xi\zeta\gamma\delta}\pi^\mu{}_\sigma + \in_{\sigma\mu}{}^{\xi\zeta}\, {}^*W_{\xi\zeta\gamma\delta}\pi^\mu{}_\tau$$
$$+ \in_{\sigma\tau}{}^{\xi\zeta}\, {}^*W_{\xi\zeta\mu\delta}\pi^\mu{}_\gamma + \in_{\sigma\tau}{}^{\xi\zeta}\, {}^*W_{\xi\zeta\gamma\mu}\pi^\mu{}_\delta \big)$$

and check that

$$-\frac{1}{4} \in_{\alpha\beta}{}^{\sigma\tau} \in_{\mu\tau}{}^{\xi\zeta}\, {}^*W_{\xi\zeta\gamma\delta}\pi^\mu{}_\sigma = \frac{1}{2}\mathrm{tr}\pi\, {}^*W_{\alpha\beta\gamma\delta} - \frac{1}{2}(\pi_\alpha{}^\mu\, {}^*W_{\mu\beta\gamma\delta} + \pi_\beta{}^\mu\, {}^*W_{\alpha\mu\gamma\delta})$$
$$-\frac{1}{4} \in_{\alpha\beta}{}^{\sigma\tau} \in_{\sigma\mu}{}^{\xi\zeta}\, {}^*W_{\xi\zeta\gamma\delta}\pi^\mu{}_\tau = \frac{1}{2}\mathrm{tr}\pi\, {}^*W_{\alpha\beta\gamma\delta} - \frac{1}{2}(\pi_\alpha{}^\mu\, {}^*W_{\mu\beta\gamma\delta} + \pi_\beta{}^\mu\, {}^*W_{\alpha\mu\gamma\delta})$$

$$-\frac{1}{4}\in_{\alpha\beta}{}^{\sigma\tau}\in_{\sigma\tau}{}^{\xi\zeta}\,{}^\star W_{\xi\zeta\mu\delta}\pi^\mu{}_\gamma=\pi_\gamma{}^{\mu\star}W_{\alpha\beta\mu\delta}$$
$$-\frac{1}{4}\in_{\alpha\beta}{}^{\sigma\tau}\in_{\sigma\tau}{}^{\xi\zeta}\,{}^\star W_{\xi\zeta\gamma\mu}\pi^\mu{}_\delta=\pi_\delta{}^{\mu\star}W_{\alpha\beta\gamma\mu}.$$

We next prove the following:

Proposition 7.1.2 *Let W be a Weyl field with*

$$\mathbf{D}^\alpha W_{\alpha\beta\gamma\delta}=J_{\beta\gamma\delta}$$

and let X be an arbitrary vectorfield with deformation tensor π. Then,

$$\begin{aligned}\mathbf{D}^\alpha(\hat{\mathcal{L}}_X W_{\alpha\beta\gamma\delta})=\hat{\mathcal{L}}_X J_{\beta\gamma\delta}&+\frac{1}{2}\hat{\pi}^{\mu\nu}\mathbf{D}_\nu W_{\mu\beta\gamma\delta}+\frac{1}{2}\mathbf{D}^\alpha\hat{\pi}_{\alpha\lambda}W^\lambda{}_{\beta\gamma\delta}\\&+\frac{1}{2}\{(\mathbf{D}_\beta\hat{\pi}_{\alpha\lambda}-\mathbf{D}_\lambda\hat{\pi}_{\alpha\beta})W^{\alpha\lambda}{}_{\gamma\delta}\\&+(\mathbf{D}_\gamma\hat{\pi}_{\alpha\lambda}-\mathbf{D}_\lambda\hat{\pi}_{\alpha\gamma})W^\alpha{}_\beta{}^\lambda{}_\delta\\&+(\mathbf{D}_\delta\hat{\pi}_{\alpha\lambda}-\mathbf{D}_\lambda\hat{\pi}_{\alpha\delta})W^\alpha{}_{\beta\gamma}{}^\lambda\}\end{aligned}$$

where

$$\hat{\mathcal{L}}_X J_{\beta\gamma\delta}:=\mathcal{L}_X J_{\beta\gamma\delta}-\frac{1}{2}(\hat{\pi}_\beta{}^\mu J_{\mu\gamma\delta}+\hat{\pi}_\gamma{}^\mu J_{\beta\mu\delta}+\hat{\pi}_\delta{}^\mu J_{\beta\gamma\mu})+\frac{1}{8}\mathrm{tr}\pi J_{\beta\gamma\delta}.$$

To prove Proposition 7.1.2 we need to appeal to the following:

Lemma 7.1.3 *Let V be an arbitrary k-covariant tensorfield and let X be a vectorfield with deformation tensor π. The following commutation formula holds:*

$$\mathbf{D}_\sigma(\mathcal{L}_X V_{\alpha_1\ldots\alpha_k})-\mathcal{L}_X(\mathbf{D}_\sigma V_{\alpha_1\ldots\alpha_k})=\sum_{i=1}^{k}{}^{(X)}\Gamma_{\alpha_i\sigma\lambda}V_{\alpha_1\ldots}{}^\lambda{}_{\ldots\alpha_k}$$

where

$${}^{(X)}\Gamma_{\alpha\beta\gamma}=\frac{1}{2}(\mathbf{D}_\beta^{(X)}\pi_{\alpha\gamma}+\mathbf{D}_\alpha^{(X)}\pi_{\beta\gamma}-\mathbf{D}_\gamma^{(X)}\pi_{\alpha\beta}).$$

Proof 7.1.3 *Proof of Proposition 7.1.2 In view of the definition of $\hat{\mathcal{L}}$, we have*

$$\begin{aligned}\mathbf{D}^\alpha(\hat{\mathcal{L}}_X W_{\alpha\beta\gamma\delta})=\mathbf{D}^\alpha(\mathcal{L}_X W_{\alpha\beta\gamma\delta})&-\frac{1}{2}\mathbf{D}^\alpha({}^{(X)}[W]_{\alpha\beta\gamma\delta})\\&+\frac{3}{8}\mathrm{tr}\pi J_{\beta\gamma\delta}+\frac{3}{8}\mathbf{D}^\lambda\mathrm{tr}\pi W_{\lambda\beta\gamma\delta}.\end{aligned}$$

Now, using Commutation Lemma 7.1.3,

$$\begin{aligned}\mathbf{D}^\alpha(\mathcal{L}_X W_{\alpha\beta\gamma\delta})=\mathcal{L}_X J_{\beta\gamma\delta}&+\pi^{\mu\nu}\mathbf{D}_\nu W_{\mu\beta\gamma\delta}+{}^{(X)}\Gamma_\lambda W^\lambda{}_{\beta\gamma\delta}\\&+{}^{(X)}\Gamma_{\beta\alpha\lambda}W^{\alpha\lambda}{}_{\gamma\delta}+{}^{(X)}\Gamma_{\gamma\alpha\lambda}W^\alpha{}_\beta{}^\lambda{}_\delta\\&+{}^{(X)}\Gamma_{\delta\alpha\lambda}W^\alpha{}_{\beta\gamma}{}^\lambda\end{aligned}\qquad(7.1.9)$$

where ${}^{(X)}\Gamma_\lambda = \mathbf{g}^{\alpha\sigma(X)}\Gamma_{\alpha\sigma\lambda}$. *We also find, by a straightforward computation,*

$$\begin{aligned}\mathbf{D}^\alpha({}^{(X)}[W]_{\alpha\beta\gamma\delta}) &= \pi_\beta{}^\mu J_{\mu\gamma\delta} + \pi_\gamma{}^\mu J_{\beta\mu\delta} + \pi_\delta{}^\mu J_{\beta\gamma\mu}\\ &\quad + \pi^{\mu\nu}\mathbf{D}^\nu W_{\mu\beta\gamma\delta} + W^\lambda{}_{\beta\gamma\delta}\mathbf{D}^\alpha\pi_{\alpha\lambda}\\ &\quad + W^{\alpha\lambda}{}_{\gamma\delta}\mathbf{D}_\alpha\pi_{\lambda\beta} + W^\alpha{}_\beta{}^\lambda{}_\delta\mathbf{D}_\alpha\pi_{\lambda\gamma} + W^\alpha{}_{\beta\gamma}{}^\lambda\mathbf{D}_\alpha\pi_{\lambda\delta}.\end{aligned}$$

Therefore,

$$\begin{aligned}\mathbf{D}^\alpha(\hat{\mathcal{L}}_X W_{\alpha\beta\gamma\delta}) &= \mathcal{L}_X J_{\beta\gamma\delta} - \frac{1}{2}(\pi_\beta{}^\mu J_{\mu\gamma\delta} + \pi_\gamma{}^\mu J_{\beta\mu\delta} + \pi_\delta{}^\mu J_{\beta\gamma\mu})\\ &\quad + \frac{3}{8}\mathrm{tr}\pi J_{\beta\gamma\delta} + \frac{1}{2}\pi^{\mu\nu}\mathbf{D}_\nu W_{\mu\beta\gamma\delta}\\ &\quad + \left({}^{(X)}\Gamma_\lambda - \frac{1}{2}\mathbf{D}^\alpha\pi_{\lambda\alpha}\right)W^\lambda{}_{\beta\gamma\delta} + \frac{3}{8}\mathbf{D}^\lambda\mathrm{tr}\pi W_{\lambda\beta\gamma\delta}\\ &\quad + \left({}^{(X)}\Gamma_{\beta\alpha\lambda} - \frac{1}{2}\mathbf{D}_\alpha\pi_{\lambda\beta}\right)W^{\alpha\lambda}{}_{\gamma\delta}\\ &\quad + \left({}^{(X)}\Gamma_{\gamma\alpha\lambda} - \frac{1}{2}\mathbf{D}_\alpha\pi_{\lambda\gamma}\right)W^\alpha{}_\beta{}^\lambda{}_\delta\\ &\quad + \left({}^{(X)}\Gamma_{\delta\alpha\lambda} - \frac{1}{2}\mathbf{D}_\alpha\pi_{\lambda\delta}\right)W^\alpha{}_{\beta\gamma}{}^\lambda.\end{aligned}$$

In view of the definition of ${}^{(X)}\Gamma$ *we find*

$$\left({}^{(X)}\Gamma_{\beta\alpha\lambda} - \frac{1}{2}\mathbf{D}_\alpha\pi_{\lambda\beta}\right)W^{\alpha\lambda}{}_{\gamma\delta} = \frac{1}{2}(\mathbf{D}_\beta\pi_{\alpha\lambda} - \mathbf{D}_\lambda\pi_{\alpha\beta})$$

and remark that the expression

$$\begin{aligned}\Big\{(\mathbf{D}_\beta\pi_{\alpha\lambda} - \mathbf{D}_\lambda\pi_{\alpha\beta})W^{\alpha\lambda}{}_{\gamma\delta} + (\mathbf{D}_\gamma\pi_{\alpha\lambda} - \mathbf{D}_\lambda\pi_{\alpha\gamma})W^\alpha{}_\beta{}^\lambda{}_\delta&\\ + (\mathbf{D}_\delta\pi_{\alpha\lambda} - \mathbf{D}_\lambda\pi_{\alpha\delta})W^\alpha{}_{\beta\gamma}{}^\lambda\Big\}&\end{aligned}$$

depends only on the traceless part of π. *Thus, the statement of the proposition is now immediate, by calculating*

$$\left({}^{(X)}\Gamma_\lambda - \frac{1}{2}\mathbf{D}^\alpha\pi_{\lambda\alpha}\right)W^\lambda{}_{\beta\gamma\delta} + \frac{3}{8}\mathbf{D}^\lambda\mathrm{tr}\pi W_{\lambda\beta\gamma\delta} = \frac{1}{2}\mathbf{D}^\alpha\hat{\pi}_{\lambda\alpha}W^\lambda{}_{\beta\gamma\delta}.$$

Proof 7.1.4 (of Lemma 7.1.3) *For simplicity we take* V *to be a 1-form and write*

$$\begin{aligned}\mathcal{L}_X V_\alpha &= X^\lambda\mathbf{D}_\lambda V_\alpha + (\mathbf{D}_\alpha X^\lambda)V_\lambda\\ \mathbf{D}_\sigma(\mathcal{L}_X V_\alpha) &= (\mathbf{D}_\sigma X^\lambda)\mathbf{D}_\lambda V_\alpha + X^\lambda\mathbf{D}_\sigma\mathbf{D}_\lambda V_\alpha + (\mathbf{D}_\sigma\mathbf{D}_\alpha X^\lambda)V_\lambda + (\mathbf{D}_\alpha X^\lambda)\mathbf{D}_\sigma V_\lambda\\ \mathcal{L}_X(\mathbf{D}_\sigma V_\alpha) &= X^\lambda\mathbf{D}_\lambda\mathbf{D}_\sigma V_\alpha + (\mathbf{D}_\alpha X^\lambda)\mathbf{D}_\sigma V_\lambda + (\mathbf{D}_\sigma X^\lambda)\mathbf{D}_\lambda V_\alpha.\end{aligned}$$

Thus, subtracting the last two equalities and making use of the following consequence of the Ricci identity for X,

$$\mathbf{D}_\sigma \mathbf{D}_\alpha X_\lambda = \mathbf{R}_{\lambda\alpha\sigma\mu} X^\mu + {}^{(X)}\Gamma_{\alpha\sigma\lambda},$$

together with the Ricci identity for V,

$$\mathbf{D}_\sigma \mathbf{D}_\lambda V_\alpha - \mathbf{D}_\lambda \mathbf{D}_\sigma V_\alpha = \mathbf{R}_{\alpha\mu\sigma\lambda} V^\mu,$$

we derive the desired result.

7.2. The Electric-Magnetic Decomposition

In this paragraph we assume that our space-time $(\mathbf{M}, \mathbf{g})$ is foliated by a time function t with lapse ϕ and second fundamental form k.[5] We assume that the time foliation is maximal; that is, on each slice Σ_t of the foliation $\mathrm{tr}k = 0$. Given an arbitrary vectorfield X we recall that the electric-magnetic decomposition of a Weyl field W consists of the pairs of symmetric traceless 2-tensors orthogonal to X, $ii_X(W)_{\alpha\beta} = W_{\mu\alpha\nu\beta} X^\mu X^\nu$, $ii_X({}^\star W)_{\alpha\beta} = {}^\star W_{\mu\alpha\nu\beta} X^\mu X^\nu$. They completely determine W whenever X is not null. This is an immediate consequence of the more general formula

$$\begin{aligned}
\langle X, X\rangle \langle Y, Y\rangle W_{\alpha\beta\gamma\delta} = &- \in_{\gamma\delta}{}^{\mu\nu} \in_{\alpha\beta}{}^{\rho\sigma} (i_{(X,Y)} W)_{\rho\mu} X_\sigma Y_\nu \\
&+ \in_{\gamma\delta}{}^{\mu\nu} \Big[(i_{(X,Y)} {}^\star W)_{\alpha\mu} X_\beta \\
&- (i_{(X,Y)} {}^\star W)_{\beta\mu} X_\alpha \Big] Y_\nu \\
&+ \in_{\alpha\beta}{}^{\mu\nu} \Big[(i_{(X,Y)} {}^\star W)_{\mu\gamma} Y_\delta \\
&- (i_{(X,Y)} {}^\star W)_{\mu\delta} Y_\gamma \Big] X_\nu \\
&+ (i_{(X,Y)} W)_{\alpha\gamma} X_\beta Y_\delta - (i_{(X,Y)} W)_{\alpha\delta} X_\beta Y_\gamma \\
&- (i_{(X,Y)} W)_{\beta\gamma} X_\alpha Y_\delta + (i_{(X,Y)} W)_{\beta\delta} X_\alpha Y_\gamma
\end{aligned}$$

where $i_{(X,Y)}$ is the contraction of W with X and Y on its first and, third components respectively.

For the special case of the vectorfield T, the unit normal to the foliation, we write $E = ii_T(W)$, $H = ii_T({}^\star W)$. Thus E and H are t-tangent traceless symmetric 2-tensors.[6]

[5] See formulas 1.0.1 and 1.0.2.

[6] For convenience we call t-tangent any space-time tensor that is tangent to the leaves of the t-foliation.

Given an arbitrary frame e_1, e_2, e_3 tangent to the foliation and setting $e_0 = T$ we have[7]

$$
\begin{aligned}
W_{abc0} &= -\in_{ab}{}^{s} H_{sc} & {}^*W_{abc0} &= \in_{ab}{}^{s} E_{sc} \\
W_{abcd} &= -\in_{abs}\in_{cdt} E^{st} & {}^*W_{abcd} &= -\in_{abs}\in_{cdt} H^{st}.
\end{aligned}
\tag{7.2.1}
$$

In what follows we write the homogeneous Bianchi equations 7.1.3a and 7.1.3b relative to the electric-magnetic decomposition E, H. For simplicity we calculate relative to a Fermi-propagated orthonormal frame e_1, e_2, e_3. We recall from the introduction the following formulas:

$$
\begin{aligned}
\mathbf{D}_i e_j &= \nabla_i e_j - k_{ij} T \\
\mathbf{D}_i T &= -k_{ij} e_j \\
\mathbf{D}_T e_i &= (\phi^{-1}\nabla_i \phi) T \\
\mathbf{D}_T T &= (\phi^{-1}\nabla_i \phi) e_i.
\end{aligned}
$$

First remark that

$$
\mathbf{D}_c W_{a0b0} = \nabla_c E_{ab} - (\in_{al}{}^{m} H_{mb} - \in_{bl}{}^{m} H_{ma}) k^l{}_c \tag{7.2.2a}
$$

$$
\mathbf{D}_c {}^*W_{a0b0} = \nabla_c H_{ab} + (\in_{al}{}^{m} E_{mb} + \in_{bl}{}^{m} E_{ma}) k^l{}_c. \tag{7.2.2b}
$$

Therefore taking the trace and using the equations 7.1.3a, we infer that

$$
(\mathrm{div} E)_a = -(H \wedge k)_a + J_{0a0} \tag{7.2.2c}
$$

$$
(\mathrm{div} H)_a = (E \wedge k)_a + J^*_{0a0} \tag{7.2.2d}
$$

where, given t-tangent symmetric traceless tensors A, B, we denote by $A \wedge B$ the 1-form

$$
(A \wedge B)_a = \in_a{}^{mn} A_m^l B_{ln}; \tag{7.2.2e}
$$

that is, $A \wedge B = {}^*[A, B]$.

On the other hand,

$$
\in_{acs} \mathbf{D}_c W_{a0b0} = \in_{acs} \nabla_c E_{ab} + H_{bc} k_{cs} - \in_{sac} \in_{bml} H_{am} k_{lc}.
$$

and, using the notation introduced in Lemma 4.4.2 on page 109, we derive

$$
\in_{acs} \mathbf{D}_c W_{a0b0} = \in_{acs} \nabla_c E_{ab} + H_{bc} k_{cs} - (H \times k)_{sb} + \frac{1}{3}(H \cdot k) g_{sb}.
$$

[7] Note that $\in_{abc} = \in_{0abc}$ for $a, b, c = 1, 2, 3$ are the coefficients of the induced volume element on the leaves of the foliation.

Therefore symmetrizing in s, b and using the identity of Lemma 4.4.2, we obtain

$$\in_{acs} \mathbf{D}_c W_{a0b0} + \in_{acb} \mathbf{D}_c W_{a0s0} = 2(\mathrm{curl} E)_{sb} - (H \times k)_{sb} + \frac{4}{3}(H \cdot k) g_{sb} \tag{7.2.2f}$$

where we recall that for symmetric 2-tensors the symmetrized curl is defined by the formula curl $E_{ab} = \frac{1}{2}(\in_a{}^{st} \nabla_t E_{sb} + \in_b{}^{st} \nabla_t E_{sa})$.

Now using 7.1.3c, we write

$$\mathbf{D}_c W_{a0b0} - \mathbf{D}_a W_{0c0b} = -\mathbf{D}_0 W_{cab0} + \in_{sac} J_{s0b}.^{\star} \tag{7.2.2g}$$

On the other hand,

$$\mathbf{D}_0 W_{cab0} = \phi^{-1} (- \in_{cas} \partial_t H_{sb} + \nabla_c \phi E_{ab} - \nabla_a \phi E_{bc} - W_{cabs} \nabla_s \phi). \tag{7.2.2h}$$

Hence,

$$\begin{aligned} \mathbf{D}_c W_{a0b0} - \mathbf{D}_a W_{0c0b} = \phi^{-1} (&\in_{cas} \partial_t H_{sb} - \nabla_c \phi E_{ab} \phi + \nabla_a \phi E_{bc} \\ &- \nabla_a \phi E_{bc} - W_{cabs} \nabla_s \phi) + \in_{sac} J_{s0b} \end{aligned}$$

or

$$\begin{aligned} 2 \in_{acs} \mathbf{D}_c W_{a0b0} &= 2\phi^{-1} \left(\partial_t H_{sb} + \in_{acs} \nabla_c \phi E_{ab} - {}^{\star}W_{bls0} \nabla_l \phi\right) + 2J^{\star}_{s0b} \\ &= 2\phi^{-1} (\partial_t H_{sb} + \in_{sac} \nabla_c \phi E_{ab} + \in_{bac} \nabla_c \phi E_{as}) + 2J^{\star}_{s0b}. \end{aligned}$$

Introducing the notation for the product between a 1-form v and a symmetric traceless 2-form A tangent to the foliation,

$$(v \wedge A)_{ij} = \in_i{}^{mn} v_m A_{nj} + \in_j{}^{mn} v_m A_{in},$$

we infer that

$$\in_{acs} \mathbf{D}_c W_{a0b0} = \phi^{-1} (\partial_t H_{sb} + (\nabla \phi \wedge E)_{sb}) + J^{\star}_{s0b}. \tag{7.2.2i}$$

We now combine 7.2.2f with 7.2.2i to derive

$$\begin{aligned} -\phi^{-1} \partial_t H_{ab} + (\mathrm{curl} E)_{ab} = -\phi^{-1} (\nabla \phi \wedge E)_{ab} - \frac{1}{2}(k \times H)_{ab} \\ + \frac{2}{3}(k \cdot H) g_{ab} - J^{\star}_{a0b}. \end{aligned} \tag{7.2.2j}$$

Similarly, by duality,

$$\begin{aligned} \phi^{-1} \partial_t E_{ab} + (\mathrm{curl} H)_{ab} = -\phi^{-1} (\nabla \phi \wedge H)_{ab} + \frac{1}{2}(k \times E)_{ab} \\ - \frac{2}{3}(k \cdot E) g_{ab} - J_{a0b}. \end{aligned} \tag{7.2.2k}$$

Now, since $\phi^{-1}\partial_t H_{ij} = \mathcal{L}_T H_{ij}$ and $\mathcal{L}_T H_{ij} = \hat{\mathcal{L}}_T H_{ij} - \frac{2}{3}(k \cdot H)g_{ij}$, we rewrite:

$$-\hat{\mathcal{L}}_T H + \mathrm{curl} E = -\phi^{-1}\nabla\phi \wedge E - \frac{1}{2}k \times H - J^\star \tag{7.2.2l}$$

$$\hat{\mathcal{L}}_T E - \mathrm{curl} H = -\phi^{-1}\nabla\phi \wedge H + \frac{1}{2}k \times E - J. \tag{7.2.2m}$$

We summarize the results of this section in the following:

Proposition 7.2.1 *Let E, H be the electric-magnetic decomposition, relative to the maximal t-foliation, of a Weyl field W that verifies the inhomogeneous Bianchi equations 7.1.3a and 7.1.3b. Then,*

$$div E = k \wedge H + J$$
$$div H = -k \wedge E + J^\star$$

$$-\hat{\mathcal{L}}_T H + curl E = -\phi^{-1}\nabla\phi \wedge E - \frac{1}{2}k \times H - J^\star$$
$$\hat{\mathcal{L}}_T E + curl H = -\phi^{-1}\nabla\phi \wedge H + \frac{1}{2}k \times E - J.$$

7.3. Null Decomposition of a Weyl Field

In this section we assume that our space-time $(\mathbf{M}, \mathbf{g})$ is foliated by a time function t with lapse function ϕ and second fundamental form k. We also assume a sufficiently smooth space-time function u that induces a radial foliation on any slice t = constant. We denote by $S_{t,u}$ the 2-surfaces obtained by intersecting the level surfaces of t with those of u. Let T be the future-oriented unit normal to the level surfaces of t and let N be the outward unit normal of $S_{t,u}$ with respect to the corresponding level surface of t. The case of interest to us is when u is an optical function, in other words, a solution to

Eikonal Equation

$$\mathbf{g}^{\alpha\beta}\frac{\partial u}{\partial x^\alpha}\frac{\partial u}{\partial x^\beta} = 0.$$

To control the properties of W along null directions, we decompose it relative to a fixed null pair.

Definition 7.3.1 We say that two vectors $e_4 = e_+, e_3 = e_-$, defined in some domain of the space-time, form a null pair compatible with t and u if $\langle e_+, e_- \rangle = -2$ and, at every given point, they are normal to the sphere $S_{t,u}$ passing through that point. Clearly, the pair $e_+ = T + N$, $e_- = T - N$ is such a pair; we call it the standard null pair. If, in particular, u is an optical function whose level surfaces are outgoing null hypersurfaces, we can also define the important null pair $l^\mu = -\mathbf{g}^{\mu\nu}\frac{\partial u}{\partial x^\nu} = a^{-1}(T + N)$ and $\underline{l} = a(T - N)$. We call $(l, \underline{l})$ the l-null pair of t, u.

Given a null pair, we define $\Pi(t, u)$ to be the tensor of projection from the tangent space of $\mathbf{M}$ to the tangent space of $S_{t,u}$. In other words,

$$\Pi^{\mu\nu} = \mathbf{g}^{\mu\nu} + \frac{1}{2}(e_-^\nu e_+^\mu + e_+^\nu e_-^\mu). \tag{7.3.1a}$$

Clearly, $\Pi_\alpha^\mu \Pi_\mu^\beta = \Pi_\alpha^\beta$, and the induced metric γ on $S_{t,u}$ is the restriction of $\Pi_\alpha^\mu \Pi_\beta^{nu} \mathbf{g}_{\mu\nu}$ to the vectors tangent to $S_{t,u}$. We extend it to arbitrary tensors $U_{\alpha_1 \ldots \alpha_m}$ on $\mathbf{M}$ by forming the contractions $\Pi_{\alpha_1}^{\beta_1} \ldots \Pi_{\alpha_m}^{\beta_m} U_{\beta_1 \ldots \beta_m}$.

Together with an orthonormal frame $e_A, A = 1, 2$ on $S_{t,u}$, a given null pair defines a null frame for the space-time $\mathbf{M}$.

The Ricci rotation coefficients of a null frame e_1, e_2, e_3, e_4 are defined by the formulas

$$\begin{aligned}
\langle \mathbf{D}_A e_3, e_B \rangle &= \underline{H}_{AB} & \langle \mathbf{D}_A e_4, e_B \rangle &= H_{AB} \\
\langle \mathbf{D}_3 e_3, e_A \rangle &= 2\underline{Y}_A & \langle \mathbf{D}_4 e_4, e_A \rangle &= 2Y_A \\
\langle \mathbf{D}_4 e_3, e_A \rangle &= 2\underline{Z}_A & \langle \mathbf{D}_3 e_4, e_A \rangle &= 2Z_A \\
\langle \mathbf{D}_3 e_3, e_4 \rangle &= 4\underline{\Omega} & \langle \mathbf{D}_4 e_4, e_3 \rangle &= 4\Omega \\
\langle \mathbf{D}_A e_4, e_3 \rangle &= 2V_A.
\end{aligned} \tag{7.3.1b}$$

Thus,

$$\begin{aligned}
\mathbf{D}_A e_3 &= \underline{H}_{AB} e_B + V_A e_3 & \mathbf{D}_A e_4 &= H_{AB} e_B - V_A e_4 \\
\mathbf{D}_3 e_3 &= 2\underline{Y}_A e_A - 2\underline{\Omega} e_3 & \mathbf{D}_3 e_4 &= 2Z_A e_A + 2\underline{\Omega} e_4 \\
\mathbf{D}_4 e_3 &= 2\underline{Z}_A e_A + 2\Omega e_3 & \mathbf{D}_4 e_4 &= 2Y_A e_A - 2\Omega e_4.
\end{aligned} \tag{7.3.1c}$$

Also,

$$\begin{aligned}
\mathbf{D}_B e_A &= \nabla\!\!\!/_B e_A + \frac{1}{2} H_{AB} e_3 + \frac{1}{2} \underline{H}_{AB} e_4 \\
\mathbf{D}_3 e_A &= \mathbf{D}\!\!\!/_3 e_A + Z_A e_3 + \underline{Y}_A e_4 \\
\mathbf{D}_4 e_A &= \mathbf{D}\!\!\!/_4 e_A + Y_A e_3 + \underline{Z}_A e_4.
\end{aligned} \tag{7.3.1d}$$

Given any covariant tensor U at a point of the space-time, we define a null component of it to be any tensor tangent to the sphere $S_{t,u}$ at that point,[8] which is derived from U through contractions with either e_3 or e_4 and projections to $S_{t,u}$. To any such component we assign a signature that is defined as the difference between the total numbers of contractions with e_4 and the total number of contractions with e_3. We are now ready to state the following heuristic principle.

Principle of Conservation of Signature: *Consider an arbitrary covariant tensor U that can be expressed as a multilinear form in an arbitrary number of covariant tensors $U_1 \dots U_p$, with coefficients depending only on the space-time metric and its volume form. Then the signature of any null term of U, expressed in terms of the null components of $U_1 \dots U_p$, is equal to the sum of the signatures of each constituent in the decomposition.*

This principle, which is essentially self-evident,[9] is extremely useful in assessing the structure of terms that appear in the formulas throughout this chapter without actually performing the long calculations required in their derivation.

In addition, for any null component ψ of signature s and a covariant space-time tensorfield U, we denote its covariant derivative intrinsic to $S_{t,u}$ by $\nabla\!\!\!/\psi$ and assign to it the same signature s. We also define $\mathbf{D}\!\!\!\!/_3\psi$, $\mathbf{D}\!\!\!\!/_4\psi$ as the projections to $S_{t,u}$ of $\mathbf{D}_3\psi$, and $\mathbf{D}_4\psi$, respectively, and assign to them the signatures $s-1$ and $s+1$, respectively. Now $\mathbf{D}\!\!\!\!/_3\psi$, $\nabla\!\!\!/\psi$ and $\mathbf{D}_4\psi$ are connected with the null components of the tensor $\mathbf{D}U$ through expressions that involve the the frame coefficients 7.3.1b. For consistency we assign the following signatures to the Ricci rotation coefficients:

$$\begin{aligned} s(\underline{Y}) &= -2 \qquad & s(Y) &= 2 \\ s(\underline{H},\underline{\Omega}) &= -1 \qquad & s(H,\Omega) &= 1 \\ s(\underline{Z},Z,V) &= 0. \end{aligned} \tag{7.3.2}$$

Then the equations expressing the null components of $\mathbf{D}U$ in terms of the $\mathbf{D}\!\!\!\!/_3, \nabla\!\!\!/, \mathbf{D}\!\!\!\!/_4$-derivatives of the null components of U and the Ricci coefficients satisfy the principle of conservation of signature. We will give an illustration of this in the discussion of the expression of the Bianchi equations relative to a null frame (see Proposition 7.3.2).

We now define the following null components of W by

$$\underline{\alpha}_{\mu\nu}(W) = \Pi_\mu{}^\rho \Pi_\nu{}^\sigma W_{\rho\gamma\sigma\delta} e_3^\gamma e_3^\delta \tag{7.3.3a}$$

[8] We will refer to such tensors as S-tangent.

[9] It is due to the fact that the only nonvanishing null components of the metric and the volume form have signature zero

$$\underline{\beta}_\mu(W) = \frac{1}{2}\Pi_\mu{}^\rho W_{\rho\sigma\gamma\delta} e_3^\sigma e_3^\gamma e_4^\delta$$
$$\rho(W) = \frac{1}{4} W_{\alpha\beta\gamma\delta} e_3^\alpha e_4^\beta e_3^\gamma e_4^\delta$$
$$\sigma(W) = \frac{1}{4} {}^\star W_{\alpha\beta\gamma\delta} e_3^\alpha e_4^\beta e_3^\gamma e_4^\delta$$
$$\beta_\mu(W) = \frac{1}{2}\Pi_\mu{}^\rho W_{\rho\sigma\gamma\delta} e_4^\sigma e_3^\gamma e_4^\delta$$
$$\alpha_{\mu\nu}(W) = \Pi_\mu{}^\rho \Pi_\nu{}^\sigma W_{\rho\gamma\sigma\delta} e_4^\gamma e_4^\delta.$$

In view of the previous discussion we assign the following signatures:

$$\begin{aligned} s(\underline{\alpha}) &= -2 & s(\alpha) &= 2 \\ s(\underline{\beta}) &= -1 & s(\beta) &= 1 \\ s(\rho) &= 0 & s(\sigma) &= 0. \end{aligned} \tag{7.3.3b}$$

Clearly $\underline{\alpha}, \alpha$ are symmetric traceless 2-tensors tangent to $S_{t,u}$, while $\underline{\beta}, \beta$ are 1-forms tangent to $S_{t,u}$. We denote by ${}^\star\underline{\alpha}, {}^\star\alpha, {}^\star\underline{\beta}, {}^\star\beta$ their left duals relative to $S_{t,u}$. Given an orthonormal frame $\{e_A : A = 1, 2\}$ on $S_{t,u}$, we easily check the following:

$$\begin{aligned} W_{A3B3} &= \underline{\alpha}_{AB} & W_{A4B4} &= \alpha_{AB} \\ W_{A334} &= 2\underline{\beta}_A & W_{A434} &= 2\beta_A \\ W_{3434} &= 4\rho & W_{AB34} &= 2\sigma \in_{AB} \\ W_{ABC3} &= \in_{AB} {}^\star\underline{\beta}_C & W_{ABC4} &= -\in_{AB} {}^\star\beta_C \\ W_{A3B4} &= -\rho\delta_{AB} + \sigma \in_{AB} & W_{ABCD} &= -\in_{AB}\in_{CD} \rho \end{aligned} \tag{7.3.3c}$$

while for the space-time dual of W we have

$$
\begin{aligned}
{}^*W_{A3B3} &= {}^*\underline{\alpha}_{AB} & {}^*W_{A4B4} &= -{}^*\alpha_{AB} \\
{}^*W_{A334} &= 2{}^*\underline{\beta}_A & {}^*W_{A434} &= -2{}^*\beta_A \\
{}^*W_{3434} &= 4\sigma & {}^*W_{AB34} &= -2\rho \in_{AB} \\
{}^*W_{ABC3} &= -\in_{AB} \underline{\beta}_C & {}^*W_{ABC4} &= -\in_{AB} \beta_C \\
{}^*W_{A3B4} &= -\sigma\delta_{AB} - \rho \in_{AB} & {}^*W_{ABCD} &= \in_{AB}\in_{CD} \sigma.
\end{aligned}
\tag{7.3.3d}
$$

In particular, we remark that the null components 7.3.3a completely determine both W and ${}^*W$. We refer to them as the null decomposition of W.

We also record here the following formulas, which relate the E-H decomposition to the null decomposition of a Weyl field W :

$$
\begin{aligned}
E_{AB} &= \tfrac{1}{4}\alpha_{AB} + \tfrac{1}{4}\underline{\alpha}_{AB} - \tfrac{1}{2}\rho\delta_{AB} & H_{AB} &= -\tfrac{1}{4}{}^*\alpha_{AB} + \tfrac{1}{4}{}^*\underline{\alpha}_{AB} - \tfrac{1}{2}\sigma\delta_{AB} \\
E_{AN} &= \tfrac{1}{2}\underline{\beta}_A + \tfrac{1}{2}\beta_A & H_{AN} &= \tfrac{1}{2}{}^*\underline{\beta}_A - \tfrac{1}{2}{}^*\beta_A \\
E_{NN} &= \rho & H_{NN} &= \sigma.
\end{aligned}
\tag{7.3.3e}
$$

Given any Weyl tensor W with $Q(W)$ the corresponding Bell-Robison tensor, we have the following:

Lemma 7.3.1 *Relative to any null pair e_3, e_4, we have*

$$
\begin{aligned}
Q(W)(e_3, e_3, e_3, e_3) &= 2|\underline{\alpha}|^2 & Q(W)(e_4, e_4, e_4, e_4) &= 2|\alpha|^2 \\
Q(W)(e_3, e_3, e_3, e_4) &= 4|\underline{\beta}|^2 & Q(W)(e_3, e_4, e_4, e_4) &= 4|\beta|^2 \\
Q(W)(e_3, e_3, e_4, e_4) &= 4(\rho^2 + \sigma^2)
\end{aligned}
$$

As a corollary we infer that given the vectorfields $T = \frac{1}{2}(e_+ + e_-)$ and Y, an arbitrary vector field of the form $Y = \frac{1}{2}(\tau_+^2 e_+ + \tau_-^2 e_-)$, then

$$
\begin{aligned}
Q(W)(Y, Y, T, T) = {} & \frac{1}{8}\tau_-^4 |\underline{\alpha}|^2 + \frac{1}{8}\tau_+^4 |\alpha|^2 \\
& + \frac{1}{2}\tau_-^2\left(\tau_-^2 + \frac{1}{2}\tau_+^2\right)|\underline{\beta}|^2 + \frac{1}{2}\tau_+^2\left(\frac{1}{2}\tau_-^2 + \tau_+^2\right)|\beta|^2 \\
& + \left(\frac{1}{4}(\tau_-^2 + \tau_+^2)^2 + \frac{1}{2}(\tau_-^2\tau_+^2)\right)(\rho^2 + \sigma^2)
\end{aligned}
\tag{7.3.4a}
$$

$$Q(W)(Y,Y,Y,T) = \frac{1}{8}\tau_-^6|\underline{\alpha}|^2 + \frac{1}{8}\tau_+^6|\alpha|^2$$
$$+ \frac{1}{4}\tau_-^4(\tau_-^2 + 3\tau_+^2)|\underline{\beta}|^2 + \frac{1}{4}\tau_+^4(3\tau_-^2 + \tau_+^2)|\beta|^2$$
$$+ \frac{3}{4}(\tau_-^2 + \tau_+^2)\tau_-^2\tau_+^2(\rho^2 + \sigma^2). \qquad (7.3.4b)$$

The following lemma will also be useful in the following chapter:

Lemma 7.3.2 *Relative to any null pair e_3, e_4,*

$$Q_{A444} = 4\alpha_{AB}\beta_B$$
$$Q_{A344} = 4\rho\beta_A - 4\sigma^\star\beta_A$$
$$Q_{A433} = -4\rho\underline{\beta}_A - 4\sigma^\star\underline{\beta}_A$$
$$Q_{AB44} = 2\delta_{AB}|\beta|^2 + 2\rho\alpha_{AB} - 2\sigma^\star\alpha_{AB}$$
$$Q_{AB34} = 2(\delta_{AB}\beta \cdot \underline{\beta} - \beta_A\underline{\beta}_B - \beta_B\underline{\beta}_A)$$
$$+ 2\delta_{AB}(\rho^2 + \sigma^2)$$
$$Q_{AB33} = 2\delta_{AB}|\underline{\beta}|^2 + 2\rho\underline{\alpha}_{AB} + 2\sigma^\star\underline{\alpha}_{AB}$$
$$Q_{A333} = -4\underline{\alpha}_{AB}\underline{\beta}_B.$$

Remark 1 *We note that the exact form of the numerical coefficients in front of the quadratic terms on the right-hand side of these equations is not important in this book. In fact, it suffices to take into account only the structure of the error terms, without performing the precise calculations, by appealing to the principle of conservation of signature. Using this principle, we can write*

$$Q_{AB34} = Qr\left[\beta\,;\,\underline{\beta}\right] + Qr\left[\rho,\sigma\,;\,\rho,\sigma\right] + \boxed{\quad Qr[\alpha\,;\,\underline{\alpha}]\quad}$$

where we denote by $Qr[\;;\;]$ any quadratic form with coefficients that depend only on the induced metric and area form of $S_{t,u}$. The cancellation of the terms, which could appear in view of signature considerations alone, such as the underlined term $\boxed{\quad Qr[\alpha\,;\,\underline{\alpha}]\quad}$ *in the formula for Q_{AB34}, can be easily explained in view of the following remarks:*[10]

Remark 2 *For any two vectors u and v tangent to $S_{t,u}$ we have*

$$u_Av_B + u_Bv_A + {}^\star u_A{}^\star v_B + {}^\star u_B{}^\star v_A = 2\delta_{AB}u \cdot v. \qquad (7.3.5a)$$

[10] See Lemma 2.2.1.

Remark 3 *Given any two traceless symmetric forms u, v tangent to $S_{t,u}$ we have*

$$u_{AC}v_{CB} + u_{BC}v_{CA} = \delta_{AB} u \cdot v. \tag{7.3.5b}$$

Thus the cancellation of the term involving the product between α and $\underline{\alpha}$ in the expression for Q_{AB34} can be easily seen from Remark 2 and from the fact that the 2-tensor $Q_{AB34} - 2\delta_{AB}(\rho^2 + \sigma^2)$, viewed as a tensor on $S_{t,u}$, is traceless. On the other hand, the cancellation of the terms involving the product between α and $\underline{\beta}$ in the expression for Q_{A344} or the product between $\underline{\alpha}$ and β in the expression for Q_{A433} can be explained by remarking that

$$\alpha \cdot \underline{\beta} - {}^*\alpha \cdot {}^*\underline{\beta} = 0$$
$$\underline{\alpha} \cdot \beta - {}^*\underline{\alpha} \cdot {}^*\beta = 0.$$

In fact, $\alpha \cdot \underline{\alpha} - {}^*\alpha \cdot {}^*\underline{\alpha} = 0$, which gives another explanation for the cancellation of the terms involving the product between $\alpha, \underline{\alpha}$ in the expression for Q_{AB34}.

We next prove the following basic commutation formulas between Lie derivatives of W and its null decomposition.

Proposition 7.3.1 *Let X be an arbitrary vectorfield and consider the commutators of X with the null frame e_1, e_2, e_3, e_4:*

$$\begin{aligned}
[X, e_3] &= {}^{(X)}\underline{P}_A e_A + {}^{(X)}\underline{M} e_3 + {}^{(X)}\underline{N} e_4 \\
[X, e_4] &= {}^{(X)}P_A e_A + {}^{(X)}N e_3 + {}^{(X)}M e_4 \\
[X, e_A] &= \Pi\,[X, e_A] + \frac{1}{2}{}^{(X)}Q_A e_3 + \frac{1}{2}{}^{(X)}\underline{Q}_A e_4
\end{aligned} \tag{7.3.6a}$$

where

$$\begin{aligned}
{}^{(X)}\underline{P}_A &= \mathbf{g}(\mathbf{D}_X e_3, e_A) - \mathbf{D}_3 X_A & {}^{(X)}P_A &= \mathbf{g}(\mathbf{D}_X e_4, e_A) - \mathbf{D}_4 X_A \\
{}^{(X)}\underline{M} &= -\tfrac{1}{2}\mathbf{g}(\mathbf{D}_X e_3, e_4) + \tfrac{1}{2}\mathbf{D}_3 X_4 & {}^{(X)}M &= -\tfrac{1}{2}\mathbf{g}(\mathbf{D}_X e_4, e_3) + \tfrac{1}{2}\mathbf{D}_4 X_3 \\
{}^{(X)}\underline{N} &= \tfrac{1}{2}\mathbf{D}_3 X_3 & {}^{(X)}N &= \tfrac{1}{2}\mathbf{D}_4 X_4 \\
{}^{(X)}\underline{Q}_A &= \mathbf{g}(\mathbf{D}_X e_3, e_A) + \mathbf{D}_A X_3 & {}^{(X)}Q_A &= \mathbf{g}(\mathbf{D}_X e_4, e_A) + \mathbf{D}_A X_4
\end{aligned} \tag{7.3.6b}$$

are the Lie coefficients of X relative to the null frame. Remark that

$$4^{(X)}N = {}^{(X)}\hat{\pi}_{44} \qquad 4^{(X)}\underline{N} = {}^{(X)}\hat{\pi}_{33}$$

$${}^{(X)}Q_A - {}^{(X)}P_A = {}^{(X)}\hat{\pi}_{4A} \qquad {}^{(X)}\underline{Q}_A - {}^{(X)}\underline{P}_A = {}^{(X)}\hat{\pi}_{3A}$$

$$2({}^{(X)}M + {}^{(X)}\underline{M}) = {}^{(X)}\pi_{34} = {}^{(X)}\hat{\pi}_{34} - \tfrac{1}{2}\mathrm{tr}^{(X)}\pi. \tag{7.3.6c}$$

Let W be an arbitrary Weyl tensor and consider $\alpha(W), \ldots, \underline{\alpha}(W)$ *as well as* $\alpha(\hat{\mathcal{L}}_X W), \ldots, \underline{\alpha}(\hat{\mathcal{L}}_X W)$ *the null decompositions of W and* $\hat{\mathcal{L}}_X W$. *Let* $\not{\mathcal{L}}_X\alpha, \ldots, \not{\mathcal{L}}_X\underline{\alpha}$ *denote the projection on* $S_{t,u}$ *of* $\mathcal{L}_X\alpha, \ldots, \mathcal{L}_X\underline{\alpha}$ *and let* $\hat{\not{\mathcal{L}}}_X\alpha$, $\hat{\not{\mathcal{L}}}_X\underline{\alpha}$ *be the traceless parts of the 2-tensors* $\not{\mathcal{L}}_X\alpha, \not{\mathcal{L}}_X\underline{\alpha}$. *Then,*

$$\begin{aligned}\alpha(\hat{\mathcal{L}}_X W)_{AB} = {} & \hat{\not{\mathcal{L}}}_X\alpha_{AB} + \Big(-2^{(X)}M - \frac{1}{8}\mathrm{tr}^{(X)}\pi\Big)\alpha_{AB} \\ & - ({}^{(X)}P_A + {}^{(X)}Q_A)\beta_B - ({}^{(X)}P_B + {}^{(X)}Q_B)\beta_A \\ & + \delta_{AB}({}^{(X)}P + {}^{(X)}Q)\cdot\beta\end{aligned} \tag{7.3.6d}$$

$$\begin{aligned}\beta(\hat{\mathcal{L}}_X W)_A = {} & \not{\mathcal{L}}_X\beta_A - \frac{1}{2}{}^{(X)}\hat{\pi}_{AB}\beta_B + \Big(-{}^{(X)}M - \frac{1}{8}\mathrm{tr}^{(X)}\pi\Big)\beta_A \\ & - \frac{3}{4}({}^{(X)}P_A + {}^{(X)}Q_A)\rho - \frac{3}{4}\in_{AB}({}^{(X)}P_B + {}^{(X)}Q_B)\sigma \\ & - \frac{1}{4}({}^{(X)}\underline{P}_B + {}^{(X)}\underline{Q}_B)\alpha_{AB}\end{aligned} \tag{7.3.6e}$$

$$\begin{aligned}\rho(\hat{\mathcal{L}}_X W) = {} & \not{\mathcal{L}}_X\rho - \frac{1}{8}\mathrm{tr}^{(X)}\pi\rho \\ & - \frac{1}{2}({}^{(X)}\underline{P}_A + {}^{(X)}\underline{Q}_A)\beta_A + \frac{1}{2}({}^{(X)}P_A + {}^{(X)}Q_A)\underline{\beta}_A\end{aligned} \tag{7.3.6f}$$

$$\begin{aligned}\sigma(\hat{\mathcal{L}}_X W) = {} & \not{\mathcal{L}}_X\sigma - \frac{1}{8}\mathrm{tr}^{(X)}\pi\sigma \\ & + \frac{1}{2}({}^{(X)}\underline{P}_A + {}^{(X)}\underline{Q}_A)^*\beta_A + \frac{1}{2}({}^{(X)}P_A + {}^{(X)}Q_A)^*\underline{\beta}_A\end{aligned} \tag{7.3.6g}$$

$$\begin{aligned}\underline{\beta}(\hat{\mathcal{L}}_X W)_A = {} & \not{\mathcal{L}}_X\underline{\beta}_A - \frac{1}{2}{}^{(X)}\hat{\pi}_{AB}\underline{\beta}_B + \Big(-{}^{(X)}\underline{M} - \frac{1}{8}\mathrm{tr}^{(X)}\pi\Big)\underline{\beta}_A \\ & + \frac{3}{4}({}^{(X)}\underline{P}_A + {}^{(X)}\underline{Q}_A)\rho - \frac{3}{4}\in_{AB}({}^{(X)}\underline{P}_B + {}^{(X)}\underline{Q}_B)\sigma \\ & + \frac{1}{4}({}^{(X)}P_B + {}^{(X)}Q_B)\underline{\alpha}_{AB}\end{aligned} \tag{7.3.6h}$$

$$\underline{\alpha}(\hat{\mathcal{L}}_X W)_{AB} = \hat{\not{\mathcal{L}}}_X \underline{\alpha}_{AB} + \left(-2^{(X)}\underline{M} - \frac{1}{8}\mathrm{tr}^{(X)}\pi \right)\underline{\alpha}_{AB} \tag{7.3.6i}$$
$$+ ({}^{(X)}\underline{P}_A + {}^{(X)}\underline{Q}_A)\underline{\beta}_B + ({}^{(X)}\underline{P}_B + {}^{(X)}\underline{Q}_B)\underline{\beta}_A$$
$$- \delta_{AB}({}^{(X)}\underline{P} + {}^{(X)}\underline{Q}) \cdot \underline{\beta}.$$

To prove the proposition, we first calculate

$$\mathcal{L}_X W_{4A4B} = \not{\mathcal{L}}_X \alpha_{AB} - 2^{(X)}M\alpha_{AB} + 2^{(X)}N\rho\delta_{AB}$$
$$-{}^{(X)}Q_A\beta_B - {}^{(X)}Q_B\beta_A$$
$$-{}^{(X)}P_A\beta_B - {}^{(X)}P_B\beta_A + 2\delta_{AB}^{(X)}P \cdot \beta$$

$$\frac{1}{2}\mathcal{L}_X W_{A434} = \not{\mathcal{L}}_X \beta_A + (-{}^{(X)}\underline{M} + 2^{(X)}M)\beta_A - {}^{(X)}N\underline{\beta}_B$$
$$- \left(\frac{1}{2}{}^{(X)}P_A + {}^{(X)}Q_A\right)\rho - \frac{3}{2}\in_{AB}^{(X)} P_B\sigma - \frac{1}{2}{}^{(X)}\underline{P}_B\alpha_{AB}$$

$$\frac{1}{4}\mathcal{L}_X W_{3434} = \not{\mathcal{L}}_X \rho - 2\rho({}^{(X)}M + {}^{(X)}\underline{M})$$
$$-{}^{(X)}\underline{P}_A\beta_A + {}^{(X)}P_A\underline{\beta}_A.$$

On the other hand, recalling 7.1.7, we calculate,

$${}^{(X)}[W]_{A4B4} = -2\delta_{AB}\beta_C\,{}^{(X)}\hat{\pi}_{4C} + {}^{(X)}\hat{\pi}_{AC}\alpha_{CB} + {}^{(X)}\hat{\pi}_{BC}\alpha_{CA}$$
$$-{}^{(X)}\hat{\pi}_{34}\alpha_{AB} + \rho\delta_{AB}\,{}^{(X)}\hat{\pi}_{44} + \mathrm{tr}^{(X)}\pi\alpha_{AB}$$

$$\frac{1}{2}{}^{(X)}[W]_{A434} = \left({}^{(X)}\hat{\pi}_{AB} - \frac{3}{2}{}^{(X)}\hat{\pi}_{34}\delta_{AB}\right)\beta_B + \frac{1}{2}{}^{(X)}\hat{\pi}_{B3}\alpha_{AB}$$
$$- \frac{1}{2}{}^{(X)}\hat{\pi}_{4A}\rho + \frac{3}{2}\in_{AB}^{(X)}\hat{\pi}_{4B}\sigma - \frac{1}{2}{}^{(X)}\hat{\pi}_{44}\underline{\beta}_A + \mathrm{tr}^{(X)}\pi\beta_A$$

$$\frac{1}{4}{}^{(X)}[W]_{3434} = \beta_A^{(X)}\hat{\pi}_{A3} - \underline{\beta}_A^{(X)}\hat{\pi}_{A4} - 2^{(X)}\hat{\pi}_{34}\rho + \mathrm{tr}^{(X)}\pi\rho.$$

The formulas 7.3.6d, 7.3.6e, and 7.3.6f of the proposition follow by combining the above formulas together with Remarks 1 and 2 on page 151. Finally, the formulas 7.3.6g, 7.3.6h, and 7.3.6i can be obtained in the same manner.[11]

[11] Observe that 7.3.6g can be derived by dualizing 7.3 6f relative to the * operator while 7.3.6h and 7 3.6i are duals to 7.3.6e and 7.3 6f relative to interchanging e_3 and e_4.

Remark 4 *As in the remark following Lemma 7.3.2, we make the observation that the exact form of the numerical coefficients in front of the error terms in formulas 7.3.6d–7.3.6i is irrelevant in this book. We can in fact assess the structure of the error terms, without performing the precise calculations, by appealing to the principle of conservation of signature. Thus, if we assign to the Lie coefficients 7.3.6b signatures*

$$s({}^{(X)}\underline{N}) = -2 \qquad s({}^{(X)}N) = 2$$
$$s({}^{(X)}\underline{Q}, {}^{(X)}\underline{P}) = -1 \qquad s({}^{(X)}Q, {}^{(X)}\underline{P}) = 1,$$
$$s({}^{(X)}\underline{M}, {}^{(X)}M) = 0$$

we remark that each error term is a product between a null component of W and either a component of the Lie coefficients or a null component of the deformation tensor of X,[12] *such that the sum of their assigned signatures equals the signature of the corresponding null component of $\hat{\mathcal{L}}_X W$ on the left. In view of this we could simply write*

$$\alpha(\hat{\mathcal{L}}_X W) = \hat{\mathcal{L}}_X \alpha + Qr\left[({}^{(X)}M, {}^{(X)}\underline{M}, \underline{\mathrm{tr}{}^{(X)}\pi}) \,;\, \alpha\right] + Qr\left[({}^{(X)}P, {}^{(X)}Q) \,;\, \beta\right] + \boxed{Qr\left[{}^{(X)}N \,;\, (\rho, \sigma)\right]} .$$

Note that the underlined term, which involves only scalars, cannot appear, since the left-hand side is a traceless symmetric 2-tensor. The other term involving ${}^{(X)}\hat{\pi}_{AB}$ and α, which could appear by signature considerations, is excluded in view of Remark 2 on page 152. Similar formulas can be derived for all the other null components of $\hat{\mathcal{L}}_X W$. In all cases, simple considerations show that the terms involving the ${}^{(X)}N$ and ${}^{(X)}\underline{N}$ do not appear.

We now express Bianchi equations 7.1.3a and 7.1.3b relative to a null frame. First we write, from 7.1.3d,

$$\mathbf{D}_4 W_{A3B3} = -\frac{1}{2}(\mathbf{D}_B W_{A334} + \mathbf{D}_A W_{B334}) + \frac{1}{2}\mathbf{D}_3(W_{A3B4} + W_{B3A4}) + \frac{1}{2}(\tilde{J}_{A3B34} + \tilde{J}_{B3A34}) \tag{7.3.7a}$$

$$\mathbf{D}_3 W_{A4B4} = \frac{1}{2}(\mathbf{D}_B W_{A434} + \mathbf{D}_A W_{B434}) + \frac{1}{2}\mathbf{D}_4(W_{A3B4} + W_{B3A4}) + \frac{1}{2}(\tilde{J}_{A4B43} + \tilde{J}_{B4A43}). \tag{7.3.7b}$$

[12] Each null component of ${}^{(X)}\pi$ is assigned a signature according to the general principle described on page 148.

Also, from 7.1.3a,

$$\mathbf{D}_4 W_{A434} = 2\mathbf{D}^C W_{A4C4} - 2J_{4A4} \tag{7.3.7c}$$
$$\mathbf{D}_3 W_{A434} = -2\mathbf{D}^C W_{C3A4} + 2J_{3A4} \tag{7.3.7d}$$
$$\mathbf{D}_4 W_{A334} = 2\mathbf{D}^C W_{A3C4} - 2J_{4A3} \tag{7.3.7e}$$
$$\mathbf{D}_3 W_{A334} = -2\mathbf{D}^C W_{A3C3} + 2J_{3A3} \tag{7.3.7f}$$

$$\mathbf{D}_4 W_{3434} = 2\mathbf{D}^C W_{C434} - 2J_{434} \tag{7.3.7g}$$
$$\mathbf{D}_3 W_{3434} = -2\mathbf{D}^C W_{C334} + 2J_{334}. \tag{7.3.7h}$$

Now, we compute with the help of 7.3.1b, 7.3.1c, and 7.3.1d, and taking into account 7.3.5a,

$$\begin{aligned}\mathbf{D}_3 W_{A4B4} = {} & \not{D}_3\alpha_{AB} - 4\underline{\Omega}\alpha_{AB} \\ & - 4(Z_A\beta_B + Z_B\beta_A - Z\cdot\beta\delta_{AB})\end{aligned} \tag{7.3.8a}$$
$$\begin{aligned}\mathbf{D}_4 W_{A3B3} = {} & \not{D}_4\underline{\alpha}_{AB} - 4\Omega\underline{\alpha}_{AB} \\ & + 4(\underline{Z}_A\underline{\beta}_B + \underline{Z}_B\underline{\beta}_A - \underline{Z}\cdot\underline{\beta})\end{aligned} \tag{7.3.8b}$$

$$\begin{aligned}\mathbf{D}_4 W_{A434} = {} & 2\not{D}_4\beta_A + 4\Omega\beta_A - 2\underline{Z}_B\alpha_{AB} \\ & - 6Y_A\rho - 6{}^\star Y_A\sigma\end{aligned} \tag{7.3.8c}$$
$$\begin{aligned}\mathbf{D}_3 W_{A434} = {} & 2\not{D}_3\beta_A - 4\underline{\Omega}\beta_A - 2\underline{Y}_B\alpha_{AB} \\ & - 6Z_A\rho - 6{}^\star Z_A\sigma\end{aligned} \tag{7.3.8d}$$
$$\begin{aligned}\mathbf{D}_4 W_{A334} = {} & 2\not{D}_4\underline{\beta}_A - 4\Omega\underline{\beta}_A + 2Y_B\underline{\alpha}_{AB} \\ & + 6\underline{Z}_A\rho - 6{}^\star\underline{Z}\sigma\end{aligned} \tag{7.3.8e}$$
$$\begin{aligned}\mathbf{D}_3 W_{A334} = {} & 2\not{D}_3\underline{\beta}_A - 4\underline{\Omega}\underline{\beta}_A - 2Z_B\underline{\alpha}_{AB} \\ & + 6\underline{Y}_A\rho - 6{}^\star\underline{Y}\sigma\end{aligned} \tag{7.3.8f}$$

$$\begin{aligned}\frac{1}{2}\mathbf{D}_3(W_{A3B4} + W_{B3A4}) = {} & -\mathbf{D}_3\rho\delta_{AB} \\ & + (\underline{Y}_A\beta_B + \underline{Y}_B\beta_A) + ({}^\star\underline{Y}_A{}^\star\beta_B + {}^\star\underline{Y}_B{}^\star\beta_A) \\ & - (Z_A\underline{\beta}_B + Z_B\underline{\beta}_A) - ({}^\star Z_A{}^\star\underline{\beta}_B + {}^\star Z_B{}^\star\underline{\beta}_A) \\ = {} & (-\mathbf{D}_3\rho + 2\underline{Y}\cdot\beta - 2Z\cdot\underline{\beta})\delta_{AB}\end{aligned} \tag{7.3.8g}$$

$$\begin{aligned}\frac{1}{2}\mathbf{D}_4(W_{A3B4} + W_{B3A4}) = {} & -\mathbf{D}_4\rho\delta_{AB} \\ & + (\underline{Z}_A\beta_B + \underline{Z}_B\beta_A) + ({}^\star\underline{Z}_A{}^\star\beta_B + {}^\star\underline{Z}_B{}^\star\beta_A) \\ & - (Y_A\underline{\beta}_B + Y_B\underline{\beta}_A) - ({}^\star Y_A{}^\star\underline{\beta}_B + {}^\star Y_B{}^\star\underline{\beta}_A) \\ = {} & (-\mathbf{D}_4\rho - 2Y\cdot\underline{\beta} + 2\underline{Z}\cdot\beta)\delta_{AB}\end{aligned} \tag{7.3.8h}$$

$$\mathbf{D}_4 W_{3434} = 4\mathbf{D}_4\rho - 8\underline{Z}_A\beta_A + 8Y_A\underline{\beta}_A \tag{7.3.8i}$$

$$\mathbf{D}_3 W_{3434} = 4\mathbf{D}_3\rho + 8Z_A\underline{\beta}_A - 8\underline{Y}_A\beta_A. \tag{7.3.8j}$$

On the other hand,[13]

$$\mathbf{D}^B W_{A4B4} = \nabla\!\!\!/^B \alpha_{AB} - 2\mathrm{tr}H\beta_A + 2V_B\alpha_{AB} \tag{7.3.9a}$$

$$\mathbf{D}^B W_{A3B3} = \nabla\!\!\!/^B \underline{\alpha}_{AB} + 2\mathrm{tr}\underline{H}\underline{\beta}_A - 2V_B\underline{\alpha}_{AB}, \tag{7.3.9b}$$

and, taking into account 7.3.5b,

$$\begin{aligned}\frac{1}{2}(\mathbf{D}_B W_{A434} + \mathbf{D}_A W_{B434}) = {} & \nabla\!\!\!/_A\beta_B + \nabla\!\!\!/_B\beta_A - \frac{1}{2}\mathrm{tr}\underline{H}\alpha_{AB} \\ & - 3H_{AB}\rho - 3{}^\star\hat{H}_{AB}\sigma - \frac{1}{2}(\alpha\cdot\underline{\hat{H}})\delta_{AB} \\ & + (V_A\beta_B + V_B\beta_A)\end{aligned} \tag{7.3.9c}$$

$$\begin{aligned}\frac{1}{2}(\mathbf{D}_B W_{A334} + \mathbf{D}_A W_{B334}) = {} & \nabla\!\!\!/_A\underline{\beta}_B + \nabla\!\!\!/_B\underline{\beta}_A + \frac{1}{2}\mathrm{tr}H\underline{\alpha}_{AB} \\ & + 3\underline{H}_{AB}\rho - 3{}^\star\underline{\hat{H}}_{AB}\sigma + \frac{1}{2}(\underline{\alpha}\cdot\hat{H})\delta_{AB} \\ & - (V_A\underline{\beta}_B + V_B\underline{\beta}_A)\end{aligned} \tag{7.3.9d}$$

$$\begin{aligned}\mathbf{D}^B W_{A3B4} = {} & -\nabla\!\!\!/_A\rho + \in_{AB}\nabla\!\!\!/_B\sigma \\ & - \mathrm{tr}H\underline{\beta}_A + 2\underline{\hat{H}}_{AB}\beta_B\end{aligned} \tag{7.3.9e}$$

$$\begin{aligned}\mathbf{D}^B W_{B3A4} = {} & -\nabla\!\!\!/_A\rho + \in_{BA}\nabla\!\!\!/_B\sigma \\ & + \mathrm{tr}\underline{H}\beta_A - 2\hat{H}_{AB}\underline{\beta}_B.\end{aligned} \tag{7.3.9f}$$

(7.3.9g)

By contracting 7.3.9c and 7.3.9d, we infer

$$\begin{aligned}\mathbf{D}^A W_{A434} = {} & 2\mathrm{d}\!\!\!/\mathrm{iv}\,\beta - 3\mathrm{tr}H\rho \\ & - \alpha\cdot\underline{H} + 2V\cdot\beta\end{aligned} \tag{7.3.9h}$$

$$\begin{aligned}\mathbf{D}^A W_{A334} = {} & 2\mathrm{d}\!\!\!/\mathrm{iv}\,\underline{\beta} + 3\mathrm{tr}\underline{H}\rho \\ & + H\cdot\underline{\alpha} - 2V\cdot\underline{\beta}.\end{aligned} \tag{7.3.9i}$$

Combining 7.3.7g and 7.3.8i with 7.3.9h on the one hand, and combining 7.3.7h and 7.3.8j with 7.3.9i on the other, we infer

$$\mathbf{D}_4\rho = \mathrm{d}\!\!\!/\mathrm{iv}\,\beta - \frac{3}{2}\mathrm{tr}H\rho - \frac{1}{2}\underline{\hat{H}}\cdot\alpha \tag{7.3.10a}$$

[13] Remark that, for any given S-tangent symmetric traceless 2-tensor u and 1-form v we have ${}^\star u_{AB}{}^\star v_B = {}^\star u^\star_{AB} v_B = u_{AB} v_B$.

$$+V\cdot\beta+2(\underline{Z}\cdot\beta-Y\cdot\underline{\beta})-\frac{1}{2}J_{434}$$

$$\mathbf{D}_3\rho=-\not{\mathrm{div}}\,\underline{\beta}-\frac{3}{2}\mathrm{tr}\underline{H}\rho-\frac{1}{2}\hat{H}\cdot\underline{\alpha} \tag{7.3.10b}$$

$$+V\cdot\underline{\beta}+2(\underline{Y}\cdot\beta-Z\cdot\underline{\beta})+\frac{1}{2}J_{334}.$$

and, proceeding in the same fashion for the dual of W,

$$\mathbf{D}_4\sigma=-\not{\mathrm{div}}\,{}^*\beta-\frac{3}{2}\mathrm{tr}H\sigma+\frac{1}{2}\underline{\hat{H}}\cdot{}^*\alpha \tag{7.3.10c}$$

$$-V\cdot{}^*\beta-2(\underline{Z}\cdot{}^*\beta+2Y\cdot{}^*\underline{\beta})-\frac{1}{2}J^*_{434}$$

$$\mathbf{D}_3\sigma=-\not{\mathrm{div}}\,{}^*\underline{\beta}-\frac{3}{2}\mathrm{tr}\underline{H}\sigma+\frac{1}{2}\hat{H}\cdot{}^*\underline{\alpha} \tag{7.3.10d}$$

$$-V\cdot{}^*\underline{\beta}-2(\underline{Y}\cdot{}^*\beta+Z\cdot{}^*\underline{\beta})+\frac{1}{2}J^*_{334}.$$

Similarly, combining 7.3.7c and 7.3.8c with 7.3.9a, then combining 7.3.7d and 7.3.8d with 7.3.9f, then combining 7.3.7e and 7.3.8e with 7.3.9e, and finally combining 7.3.7f and 7.3.8b with 7.3.9b, we derive

$$\mathbf{D}_4\beta_A=\not\nabla^B\alpha_{AB}-2\mathrm{tr}H\beta_A-2\Omega\beta_A \tag{7.3.10e}$$

$$+(2V_B+\underline{Z}_B)\alpha_{AB}+3(Y_A\rho+{}^*Y_A\sigma)-J_{4A4}$$

$$\mathbf{D}_3\beta_A=\not\nabla_A\rho+\in_{AB}\not\nabla_B\sigma-\mathrm{tr}\underline{H}\beta_A \tag{7.3.10f}$$

$$+2\hat{H}_{AB}\underline{\beta}_{AB}+2\underline{\Omega}\beta_A+\underline{Y}_B\alpha_{AB}+3(Z_A\rho+{}^*Z_A\sigma)+J_{3A4}$$

$$\mathbf{D}_4\underline{\beta}_A=-\not\nabla_A\rho+\in_{AB}\not\nabla_B\sigma-\mathrm{tr}H\underline{\beta}_A \tag{7.3.10g}$$

$$+2\underline{\hat{H}}_{AB}\beta_B+2\Omega\underline{\beta}_A-Y_B\underline{\alpha}_{AB}-3(\underline{Z}_A\rho-{}^*\underline{Z}_A\sigma)-J_{4A3}$$

$$\mathbf{D}_3\underline{\beta}_A=-\not\nabla^B\underline{\alpha}_{AB}-2\mathrm{tr}\underline{H}\underline{\beta}_A-2\underline{\Omega}\underline{\beta}_A \tag{7.3.10h}$$

$$-(-2V_B+Z_B)\underline{\alpha}_{AB}+3(-\underline{Y}_A\rho+{}^*\underline{Y}_A\sigma)+J_{3A3}.$$

Finally, combining 7.3.7a, 7.3.8b, 7.3.8g, and 7.3.9d with 7.3.10b, and combining 7.3.7b, 7.3.8a, 7.3.8g, and 7.3.9c with 7.3.9i, we deduce

$$\mathbf{D}_4\underline{\alpha}_{AB}=-(\not\nabla_A\underline{\beta}_B+\not\nabla_B\underline{\beta}_A-\not{\mathrm{div}}\,\underline{\beta}\delta_{AB})-\frac{1}{2}\mathrm{tr}H\underline{\alpha}_{AB} \tag{7.3.10i}$$

$$+4\Omega\underline{\alpha}_{AB}-3(\underline{\hat{H}}_{AB}\rho-{}^*\underline{\hat{H}}_{AB}\sigma)$$

$$+(V_A-4\underline{Z}_A)\underline{\beta}_B+(V_B-4\underline{Z}_B)\underline{\beta}_A-(V-4\underline{Z})\cdot\underline{\beta}\delta_{AB}$$

$$+\frac{1}{2}(\tilde{J}_{A3B34}+\tilde{J}_{B3A34}-J_{334}\delta_{AB})$$

$$\not{D}_3\alpha_{AB} = (\nabla\!\!\!/_A\beta_B + \nabla\!\!\!/_B\beta_A - \not{d}\mathrm{iv}\,\beta\delta_{AB}) - \frac{1}{2}\mathrm{tr}\underline{H}\alpha_{AB} \tag{7.3.10j}$$

$$+4\underline{\Omega}\alpha_{AB} - 3(\hat{H}_{AB}\rho + {}^*\hat{H}_{AB}\sigma)$$
$$+(V_A + 4Z_A)\beta_B + (V_B + 4Z_B)\beta_A - (V + 4Z)\cdot\beta\delta_{AB}$$
$$+\frac{1}{2}(\tilde{J}_{A4B43} + \tilde{J}_{B4A43} + J_{434}\delta_{AB}). \tag{7.3.10k}$$

Similar formulas can be derived for the second derivatives of the null components of W relative to e_3, e_4. To do this, we need the following commutation lemma:

Lemma 7.3.3 *Let $U_{A_1\ldots A_k}$ be an S-tangent k-covariant tensor on* $(\mathbf{M}, \mathbf{g})$*. Then*

$$\not{D}_4\nabla\!\!\!/_B U_{A_1\ldots A_k} - \nabla\!\!\!/_B\not{D}_4 U_{A_1\ldots A_k} + H_{BC}\nabla\!\!\!/_C U_{A_1\ldots A_k} = F_{4BA_1\ldots A_k}$$
$$\not{D}_3\nabla\!\!\!/_B U_{A_1\ldots A_k} - \nabla\!\!\!/_B\not{D}_3 U_{A_1\ldots A_k} + \underline{H}_{BC}\nabla\!\!\!/_C U_{A_1\ldots A_k} = F_{3BA_1\ldots A_k}$$
$$\not{D}_3\not{D}_4 U_{A_1\ldots A_k} - \not{D}_4\not{D}_3 U_{A_1\ldots A_k} = F_{34A_1\ldots A_k}$$

where

$$F_{4BA_1\ldots A_k} = Y_B\not{D}_3 U_{A_1\ldots A_k} + (\underline{Z}_B + V_B)\not{D}_4 U_{A_1\ldots A_k}$$
$$= \sum_{i=1}^{k}\big(\underline{H}_{A_iB}Y_C - \underline{H}_{BC}Y_{A_i}$$
$$+ H_{A_iB}\underline{Z}_C - H_{BC}\underline{Z}_{A_i} + \in_{A_iC}{}^*\beta(\mathbf{R})_B\big)\,U_{A_1\ldots\overset{C}{\check{A}_i}\ldots A_k}$$

$$F_{3BA_1\ldots A_k} = \underline{Y}_B\not{D}_4 U_{A_1\ldots A_k} + (Z_B - V_B)\not{D}_3 U_{A_1\ldots A_k}$$
$$+\sum_{i=1}^{k}\big(H_{A_iB}\underline{Y}_C - H_{BC}\underline{Y}_{A_i}$$
$$+ \underline{H}_{A_iB}Z_C - \underline{H}_{BC}Z_{A_i} - \in_{A_iC}{}^*\underline{\beta}(\mathbf{R})_B\big)\,U_{A_1\ldots\overset{C}{\check{A}_i}\ldots A_k}$$

$$F_{34A_1\ldots A_k} = 2\Omega\not{D}_3 U_{A_1\ldots A_k} + 2\underline{\Omega}\not{D}_4 U_{A_1\ldots A_k} + (Z_B - \underline{Z}_B)\nabla\!\!\!/_B U_{A_1\ldots A_k}$$
$$+2\sum_{i=1}^{k}\big(Y_{A_i}\underline{Y}_C - Y_C\underline{Y}_{A_i}$$
$$+ \underline{Z}_{A_i}Z_C - \underline{Z}_{A_i}Z_C + \in_{A_iC}\sigma(\mathbf{R})\big)\,U_{A_1\ldots\overset{C}{\check{A}_i}\ldots A_k}.$$

Here $\beta(\mathbf{R})_A = \frac{1}{2}\mathbf{R}_{A434}$, $\underline{\beta}(\mathbf{R})_A = \frac{1}{2}\mathbf{R}_{A334}$, *and* $\in_{AC}\sigma(\mathbf{R}) = \frac{1}{4}\mathbf{R}_{AC34}$ *are null components of the Riemann curvature tensor of the space-time.*

In what follows we reexpress the equations 7.3.10a–7.3.10i and those involving the second derivatives of the null frame with the help of the following notation:

Definition 7.3.2 Given an arbitrary Weyl tensor W, and u any of its null components, we introduce the S-tangent tensors

$$u_3 = \not{D}_3 u + \frac{3-s}{2}\mathrm{tr}(\underline{H})u$$
$$u_4 = \not{D}_4 u + \frac{3+s}{2}\mathrm{tr}(H)u$$

where $s = s(u)$ is the signature of that particular component. In addition, we define

$$u_{33} = \not{D}_3 u_3 + \frac{3-(s-1)+1}{2}\mathrm{tr}(\underline{H})u_3$$
$$u_{34} = \not{D}_4 u_3 + \frac{3+(s-1)+1}{2}\mathrm{tr}(H)u_3$$
$$u_{43} = \not{D}_3 u_4 + \frac{3-(s+1)+1}{2}\mathrm{tr}(\underline{H})u_4$$
$$u_{44} = \not{D}_4 u_4 + \frac{3+(s+1)+1}{2}\mathrm{tr}(H)u_4$$

or, setting the signature $s(u_3) = s(u) - 1$, $s(u_4) = s(u) + 1$, and introducing the rank $k(u) = 0$, $k(u_3) = k(u_4) = 1$, we can combine the two definitions in

$$v_3 = \not{D}_3 v + \frac{3-s+k}{2}\mathrm{tr}(\underline{H})v$$
$$v_4 = \not{D}_4 v + \frac{3+s+k}{2}\mathrm{tr}(H)v$$

where v is any S-tangent tensor of signature s and rank k.

We also recall the following operators introduced in Corollary 2.2.1.2 and the remark following it.

1. *$\not{\mathcal{D}}_1$ takes any S-tangent 1-form ξ into the pairs of functions $(\not{d}\mathrm{iv}\,\xi,\ \not{c}\mathrm{url}\,\xi)$.*
2. *$\not{\mathcal{D}}_1^*$, the L^2-adjoint of $\not{\mathcal{D}}_1$, takes any pairs of scalars ρ, σ into the S-tangent 1-form $-\not{\nabla}_A\rho + \in_{AB}\not{\nabla}^B\sigma$.*
3. *$\not{\mathcal{D}}_2$ takes any S-tangent 2-covariant symmetric traceless tensors ξ into the S-tangent 1-form $\not{d}\mathrm{iv}\,\xi$.*
4. *$\not{\mathcal{D}}_2^*$, the L^2-adjoint of $\not{\mathcal{D}}_2$ takes any S-tangent 1-form ξ into the 2-form $-\frac{1}{2}\widehat{\mathcal{L}_\gamma\xi} = -\frac{1}{2}(\not{\nabla}_B\xi_A + \not{\nabla}_A\xi_B - (\not{d}\mathrm{iv}\,\xi)\gamma_{AB})$.*

Also, given two covariant S-tangent tensors u, v we denote by $u\widehat{\otimes}v$ the 2-covariant symmetric traceless tensor $u_A v_B + u_B v_A - (u\cdot v)\delta_{AB}$.

Finally, we are ready to state the following:

Proposition 7.3.2 *Relative to a null frame, compatible to the t, u foliations, the homogeneous Weyl equations 7.1.3a and 7.1.3b take the following form:*

$$
\begin{aligned}
\alpha_3 &= -2\not{\mathcal{D}}_2^*\beta + E_3(\alpha) \\
E_3(\alpha) &= 4\underline{\Omega}\alpha - 3(\hat{H}\rho + {}^*\hat{H}\sigma) + (V + 4Z)\hat{\otimes}\beta
\end{aligned}
\tag{7.3.11a}
$$

$$
\begin{aligned}
\beta_4 &= \not{\mathrm{div}}\,\alpha + E_4(\beta) \\
E_4(\beta) &= -2\Omega\beta + (2V + \underline{Z})\cdot\alpha + 3(Y\rho + {}^*Y\sigma)
\end{aligned}
\tag{7.3.11b}
$$

$$
\begin{aligned}
\beta_3 &= \not{\mathcal{D}}_1^*(-\rho, \sigma) + E_3(\beta) \\
E_3(\beta) &= 2\hat{H}\cdot\underline{\beta} + 2\underline{\Omega}\beta + \underline{Y}\cdot\alpha + 3(Z\rho + {}^*Z\sigma)
\end{aligned}
\tag{7.3.11c}
$$

$$
\begin{aligned}
\rho_4 &= \not{\mathrm{div}}\,\beta + E_4(\rho) \\
E_4(\rho) &= -\frac{1}{2}\underline{\hat{H}}\cdot\alpha + V\cdot\beta + 2(\underline{Z}\cdot\beta - Y\cdot\underline{\beta})
\end{aligned}
\tag{7.3.11d}
$$

$$
\begin{aligned}
\rho_3 &= -\not{\mathrm{div}}\,\underline{\beta} + E_3(\rho) \\
E_3(\rho) &= -\frac{1}{2}\hat{H}\cdot\underline{\alpha} + V\cdot\underline{\beta} + 2(\underline{Y}\cdot\beta - Z\cdot\underline{\beta})
\end{aligned}
\tag{7.3.11e}
$$

$$
\begin{aligned}
\sigma_4 &= -\not{\mathrm{curl}}\,\beta + E_4(\sigma) \\
E_4(\sigma) &= \frac{1}{2}\underline{\hat{H}}\cdot{}^*\alpha - V\cdot{}^*\beta - 2(\underline{Z}\cdot{}^*\beta + 2Y\cdot{}^*\underline{\beta})
\end{aligned}
\tag{7.3.11f}
$$

$$
\begin{aligned}
\sigma_3 &= -\not{\mathrm{curl}}\,\underline{\beta} + E_3(\sigma) \\
E_3(\sigma) &= -\frac{1}{2}\hat{H}\cdot{}^*\underline{\alpha} + V\cdot{}^*\underline{\beta} - 2(\underline{Y}\cdot{}^*\beta + Z\cdot{}^*\underline{\beta})
\end{aligned}
\tag{7.3.11g}
$$

$$
\begin{aligned}
\underline{\beta}_4 &= \not{\mathcal{D}}_1^*(\rho, \sigma) + E_4(\underline{\beta}) \\
E_4(\underline{\beta}) &= 2\underline{\hat{H}}\cdot\beta + 2\Omega\underline{\beta} - Y\cdot\underline{\alpha} - 3(\underline{Z}\rho - {}^*\underline{Z}\sigma)
\end{aligned}
\tag{7.3.11h}
$$

$$
\begin{aligned}
\underline{\beta}_3 &= -\not{\mathrm{div}}\,\underline{\alpha} + E_3(\underline{\beta}) \\
E_3(\underline{\beta}) &= -2\underline{\Omega}\underline{\beta} - (-2V + Z)\cdot\underline{\alpha} + 3(-\underline{Y}\rho + {}^*\underline{Y}\sigma)
\end{aligned}
\tag{7.3.11i}
$$

$$
\begin{aligned}
\underline{\alpha}_4 &= 2\not{\mathcal{D}}_2^*\underline{\beta} + E_4(\underline{\alpha}) \\
E_4(\underline{\alpha}) &= 4\Omega\underline{\alpha} - 3(\underline{\hat{H}}\rho - {}^*\underline{\hat{H}}\sigma) + (V - 4\underline{Z})\hat{\otimes}\underline{\beta}.
\end{aligned}
\tag{7.3.11j}
$$

Also, for the second derivatives,

$$\begin{aligned}\alpha_{33} = &-2\not{\mathcal{D}}_2^{\star}\beta_3 \\ &- \underline{\hat{H}}\not{d}iv\,\beta + {}^{\star}\underline{\hat{H}}c\not{u}rl\,\beta - (\not{\nabla}\mathrm{tr}\underline{H})\widehat{\otimes}\beta \\ &+ \not{\mathbf{D}}_3E_3(\alpha) + \frac{3}{2}\mathrm{tr}\underline{H}E_3(\alpha) + 2\hat{s}(F_3(\beta))\end{aligned} \tag{7.3.12a}$$

$$\begin{aligned}\alpha_{34} = &-2\not{\mathcal{D}}_2^{\star}\beta_4 \\ &- 2(\not{\nabla}\mathrm{tr}H)\widehat{\otimes}\beta - \hat{H}\not{d}iv\,\beta + {}^{\star}\hat{H}c\not{u}rl\,\beta \\ &+ \not{\mathbf{D}}_4E_3(\alpha) + \frac{5}{2}\mathrm{tr}HE_3(\alpha) + 2\hat{s}(F_4(\beta))\end{aligned} \tag{7.3.12b}$$

$$\begin{aligned}\beta_{44} = &\not{d}iv\,\alpha_4 \\ &- \frac{5}{2}\not{\nabla}\mathrm{tr}H\cdot\alpha - \hat{H}\cdot\not{\nabla}\alpha \\ &+ \not{\mathbf{D}}_4E_4(\beta) + 3\mathrm{tr}HE_4(\beta) + \mathrm{tr}F_4(\alpha)\end{aligned} \tag{7.3.12c}$$

$$\begin{aligned}\beta_{34} = &\not{\nabla}\rho_4 + {}^{\star}\not{\nabla}\sigma_4 \\ &- \frac{3}{2}\not{\nabla}\mathrm{tr}H\rho - \frac{3}{2}{}^{\star}\not{\nabla}\mathrm{tr}H\sigma - \hat{H}\cdot\not{\nabla}\rho - {}^{\star}\hat{H}\cdot\not{\nabla}\sigma \\ &+ \not{\mathbf{D}}_4E_3(\beta) + 2\mathrm{tr}HE_3(\beta) + F_4(\rho) + {}^{\star}F_4(\sigma)\end{aligned} \tag{7.3.12d}$$

$$\beta_{43} = \beta_{34} + \beta\cdot(2\not{\mathbf{D}}_3\mathrm{tr}H - \not{\mathbf{D}}_4\mathrm{tr}\underline{H}) + F_{34}(\beta) \tag{7.3.12e}$$

$$\begin{aligned}\beta_{33} = &\not{\nabla}\rho_3 + {}^{\star}\not{\nabla}\sigma_3 \\ &- \frac{3}{2}\rho\not{\nabla}(\mathrm{tr}\underline{H}) - \frac{3}{2}\sigma{}^{\star}\not{\nabla}(\mathrm{tr}\underline{H}) - \underline{\hat{H}}\cdot\not{\nabla}\rho - {}^{\star}\underline{\hat{H}}\cdot\not{\nabla}\sigma \\ &+ \not{\mathbf{D}}_3E_3(\beta) + 2\mathrm{tr}\underline{H}E_3(\beta) + F_3(\rho) + {}^{\star}F_3(\sigma)\end{aligned} \tag{7.3.12f}$$

$$\begin{aligned}\rho_{44} = &\not{d}iv\,\beta_4 \\ &- 2\beta\cdot\not{\nabla}\mathrm{tr}H - \hat{H}\cdot\not{\nabla}\beta \\ &+ \mathbf{D}_4E_4(\rho) + \frac{5}{2}\mathrm{tr}HE_4(\rho) + \mathrm{tr}F_4(\beta)\end{aligned} \tag{7.3.12g}$$

$$\begin{aligned}\rho_{43} = &\not{d}iv\,\beta_3 \\ &- \beta\cdot\not{\nabla}\mathrm{tr}\underline{H} - \underline{\hat{H}}\cdot\not{\nabla}\beta \\ &+ \mathbf{D}_3E_4(\rho) + \frac{3}{2}\mathrm{tr}\underline{H}E_4(\rho) + \mathrm{tr}F_3(\beta)\end{aligned} \tag{7.3.12h}$$

$$\rho_{34} = \rho_{43} + \frac{3}{2}(\mathbf{D}_4\mathrm{tr}\underline{H} - \mathbf{D}_3\mathrm{tr}H)\rho + F_{34}(\rho) \tag{7.3.12i}$$

$$\begin{aligned}\rho_{33} = {} & \not{d}iv\,\underline{\beta}_3 \\ & + 2\underline{\beta}\cdot\not\nabla\mathrm{tr}\underline{H} + \underline{\hat{H}}\cdot\not\nabla\underline{\beta} \\ & + \mathbf{D}_3E_3(\rho) + \frac{5}{2}\mathrm{tr}\underline{H}E_3(\rho) - \mathrm{tr}F_3(\underline{\beta})\end{aligned} \tag{7.3.12j}$$

$$\begin{aligned}\sigma_{44} = {} & -c\not{u}rl\,\beta_4 \\ & + 2{}^*\beta\cdot\not\nabla\mathrm{tr}H + \hat{H}\cdot\not\nabla^*\beta \\ & + \mathbf{D}_4E_4(\sigma) + \frac{5}{2}(\mathrm{tr}H)E_4(\sigma) - \mathrm{tr}F_4({}^*\beta)\end{aligned} \tag{7.3.12k}$$

$$\begin{aligned}\sigma_{43} = {} & -c\not{u}rl\,\beta_3 \\ & + {}^*\beta\cdot\not\nabla\mathrm{tr}\underline{H} + \underline{\hat{H}}\cdot\not\nabla^*\beta \\ & + \mathbf{D}_3E_4(\sigma) + \frac{3}{2}(\mathrm{tr}\underline{H})E_4(\sigma) - \mathrm{tr}F_3({}^*\beta)\end{aligned} \tag{7.3.12l}$$

$$\sigma_{34} = \sigma_{43} + \frac{3}{2}(\mathbf{D}_4\mathrm{tr}\underline{H} - \mathbf{D}_3\mathrm{tr}H)\sigma + F_{34}(\sigma) \tag{7.3.12m}$$

$$\begin{aligned}\sigma_{33} = {} & -c\not{u}rl\,\beta_3 \\ & - 2{}^*\underline{\beta}\cdot\not\nabla\mathrm{tr}\underline{H} - \underline{\hat{H}}\cdot\not\nabla^*\underline{\beta} \\ & + \mathbf{D}_3E_3(\sigma) + \frac{5}{2}\mathrm{tr}\underline{H}E_3(\sigma) - \mathrm{tr}F_3({}^*\underline{\beta})\end{aligned} \tag{7.3.12n}$$

$$\begin{aligned}\underline{\beta}_{44} = {} & -\not\nabla\rho_4 - {}^*\not\nabla\sigma_4 \\ & + \frac{3}{2}\rho\not\nabla(\mathrm{tr}H) - \frac{3}{2}\sigma{}^*\not\nabla(\mathrm{tr}H) + \hat{H}\cdot\not\nabla\rho - {}^*\hat{H}\cdot\not\nabla\sigma \\ & + \not{\mathbf{D}}_4E_4(\underline{\beta}) + 2\mathrm{tr}HE_4(\underline{\beta}) - F_4(\rho) - {}^*F_4(\sigma)\end{aligned} \tag{7.3.12o}$$

(7.3.12p)

$$\begin{aligned}\underline{\beta}_{43} = {} & -\not\nabla\rho_3 - {}^*\not\nabla\sigma_3 \\ & + \frac{3}{2}\not\nabla\mathrm{tr}\underline{H}\rho - \frac{3}{2}{}^*\not\nabla\mathrm{tr}\underline{H}\sigma + \underline{\hat{H}}\cdot\not\nabla\rho - {}^*\underline{\hat{H}}\cdot\not\nabla\sigma \\ & + \not{\mathbf{D}}_3E_4(\underline{\beta}) + 2\mathrm{tr}\underline{H}E_4(\underline{\beta}) - F_3(\rho) - {}^*F_3(\sigma)\end{aligned} \tag{7.3.12q}$$

$$\underline{\beta}_{34} = \underline{\beta}_{43} + \underline{\beta}\cdot(2\not{\mathbf{D}}_4\mathrm{tr}\underline{H} - \not{\mathbf{D}}_3\mathrm{tr}H) + F_{43}(\underline{\beta}) \tag{7.3.12r}$$

$$\begin{aligned}\underline{\beta}_{33} &= -\not{d}iv\,\underline{\alpha}_3 \\ &+ \frac{5}{2}\not{\nabla}\mathrm{tr}\underline{H}\cdot\underline{\alpha} + \underline{\hat{H}}\cdot\not{\nabla}\underline{\alpha} \\ &+ \not{D}_3 E_3(\underline{\beta}) + 3\mathrm{tr}\underline{H}E_3(\underline{\beta}) - \mathrm{tr}F_3(\underline{\alpha})\end{aligned} \tag{7.3.12s}$$

$$\begin{aligned}\underline{\alpha}_{44} &= +2\not{\mathcal{D}}_2^{\star}\underline{\beta}_4 \\ &+ \hat{H}\not{d}iv\,\underline{\beta} - {}^{\star}\hat{H}c\not{u}rl\,\underline{\beta} + (\not{\nabla}\mathrm{tr}H)\widehat{\otimes}\underline{\beta} \\ &+ \not{D}_4 E_4(\underline{\alpha}) + \frac{3}{2}\mathrm{tr}H E_4(\underline{\alpha}) - 2\hat{s}(F_4(\underline{\beta}))\end{aligned} \tag{7.3.12t}$$

$$\begin{aligned}\underline{\alpha}_{43} &= +2\not{\mathcal{D}}_2^{\star}\underline{\beta}_3 \\ &+ 2(\not{\nabla}\mathrm{tr}\underline{H})\widehat{\otimes}\underline{\beta} + \underline{\hat{H}}\not{d}iv\,\underline{\beta} - {}^{\star}\underline{\hat{H}}c\not{u}rl\,\underline{\beta} \\ &+ \not{D}_3 E_4(\underline{\alpha}) + \frac{5}{2}\mathrm{tr}\underline{H}E_4(\underline{\alpha}) - 2\hat{s}(F_3(\underline{\beta})).\end{aligned} \tag{7.3.12u}$$

Remark 5 *The terms $F_3(\alpha), F_4(\alpha), F_{34}(\alpha), \ldots,$ which appear in the null Bianchi formulas for the second derivatives, correspond to the commutation formulas of Lemma 7.3.3 applied to the null components of W. The symbols $\hat{s}(F_3(\beta)), \hat{s}(F_3(\beta))$ are shorthand notation for the traceless parts of the symmetrization of $\overline{F_3}(\beta)$ and $F_3(\underline{\beta})$.*

Remark 6 *As in Proposition 7.3.1 we note that, with the exception of the main terms that are present even in Minkowski space, in all the other terms, which we consider error terms, the exact form of the numerical constants is irrelevant to our considerations. We also note that the precise form of the angular operators $\not{d}iv, c\not{u}rl, \not{\nabla}, {}^{\star}\not{\nabla}, \not{\mathcal{D}}_1, \not{\mathcal{D}}_2, \not{\mathcal{D}}_1^{\star}$, and$\not{\mathcal{D}}_1^{\star}$ as well as that of the numerical constant in front of them is also not important for our considerations in this chapter. They all count as one-order angular derivatives in our estimates. Moreover, the rough structure of the null Bianchi formulas can be obtain by signature considerations, as in Remark 3 following Proposition 7.3.1, as well as the space-time dualities.[14] Certain terms, which could be present by signature considerations, cancel due to the identity in Remark 2. Finally, we note that among the terms we consider as error terms one has to pay special attention to the terms involving $\underline{\hat{H}}$. To get a feeling for the strength of each term in the equations, we suggest that the reader look at Assumptions 0, 1, and 2 for the Ricci coefficients on page 176 as well as at the results of Theorem 7.6.1 on page 181.*

[14] These are relative to the e_3-e_4 components, and Hodge duality, defined relative to the volume element of space-time.

7.4. The Null-Structure Equations of a Space-Time

In this section we write down the structure equations of a space-time $(\mathbf{M}, \mathbf{g})$ relative to the $S_{t,u}$ foliation discussed in the previous section. These equations will play a major role in Part III.

Let e_3, e_4 be a null pair compatible with t, u and consider the corresponding Ricci coefficients as defined in 7.3.1b.

Now,

$$\langle \mathbf{D}_3 \mathbf{D}_B e_3, e_A \rangle = \not{\mathbf{D}}_3 \underline{H}_{AB} + 2(V_B - Z_B)\underline{Y}_A - 2\underline{Z}_A \underline{Y}_B$$
$$\langle \mathbf{D}_B \mathbf{D}_3 e_3, e_A \rangle = 2\not{\nabla}_B \underline{Y}_A - 2\underline{\Omega}\underline{H}_{AB} - \underline{H}_{AC}\underline{H}_{CB} - 2V_B \underline{Y}_A$$
$$\langle \mathbf{D}_3 \mathbf{D}_B e_3, e_A \rangle = \langle \mathbf{D}_B \mathbf{D}_3 e_3, e_A \rangle + \mathbf{R}_{A33B}.$$

On the other hand,

$$\mathbf{R}_{A33B} = -\underline{\alpha}_{AB}.$$

Henceforth,

$$\not{\mathbf{D}}_3 \underline{H}_{AB} = 2\not{\nabla}_B \underline{Y}_A - 2\underline{\Omega}\underline{H}_{AB} - \underline{H}_{AC}\underline{H}_{CB} - 4V_B \underline{Y}_A + 2Z_B \underline{Y}_A + 2\underline{Z}_A \underline{Y}_A - \underline{\alpha}_{AB}. \qquad (7.4.1a)$$

Taking the trace and splitting $\underline{H}$ into its trace $\mathrm{tr}\underline{H}$ and traceless part $\hat{\underline{H}}$, we derive

$$\mathbf{D}_3(\mathrm{tr}\underline{H}) + \frac{1}{2}(\mathrm{tr}\underline{H})^2 = 2\not{\mathrm{d}}\mathrm{iv}\,\underline{Y} - 2\underline{\Omega}\mathrm{tr}\underline{H} + 2\underline{Y} \cdot (Z + \underline{Z} - 2V) - \hat{\underline{H}} \cdot \hat{\underline{H}}. \qquad (7.4.1b)$$

The antisymmetric part of 7.4.1a yields[15]

$$\not{\mathrm{c}}\mathrm{url}\,\underline{Y} - \underline{Y} \wedge (Z + \underline{Z} - 2V) = 0 \qquad (7.4.1c)$$

while the symmetric traceless part becomes

$$\not{\mathbf{D}}_3 \hat{\underline{H}}_{AB} + (\mathrm{tr}\underline{H})\hat{\underline{H}}_{AB} = (\not{\nabla}_A \underline{Y}_B + \not{\nabla}_B \underline{Y}_A - \not{\mathrm{d}}\mathrm{iv}\,\underline{Y}\gamma_{AB}) - 2\underline{\Omega}\hat{\underline{H}}_{AB} + ((Z + \underline{Z} - 2V)\widehat{\otimes}\underline{Y})_{AB} - \underline{\alpha}_{AB}. \qquad (7.4.1d)$$

Next we calculate

$$\langle \mathbf{D}_4 \mathbf{D}_B e_3, e_A \rangle = \not{\mathbf{D}}_4 \underline{H}_{AB} + 2\underline{Z}_A V_B - 2Y_B \underline{Y}_A - 2\underline{Z}_B \underline{Z}_A$$
$$\langle \mathbf{D}_B \mathbf{D}_4 e_3, e_A \rangle = 2\not{\nabla}_B \underline{Z}_A + 2V_B \underline{Z}_A + 2\omega \underline{H}_{AB} - H_{BC}\underline{H}_{BC}$$
$$\langle \mathbf{D}_4 \mathbf{D}_B e_3, e_A \rangle = \langle \mathbf{D}_4 \mathbf{D}_B e_3, e_A \rangle + \mathbf{R}_{A34B}$$

[15] Given two S-tangent 1-forms $\xi, \underline{\xi}$ we write $\xi \wedge \underline{\xi} = \in^{AB} \xi_A \underline{\xi}_B$. Similarly, given S-tangent symmetric, traceless 2-tensors u, v we write, $u \wedge v = \in^{AB} u_{AC} v_{CB}$.

and

$$\mathbf{R}_{A34B} = \rho\gamma_{AB} - \sigma \in_{AB} .$$

Henceforth,

$$\begin{aligned} \not{D}_4 \underline{H}_{AB} = 2\nabla\!\!\!/_B \underline{Z}_A + 2\Omega \underline{H}_{AB} - H_{BC}\underline{H}_{AC} \\ + 2(Y_B \underline{Y}_A + \underline{Z}_B \underline{Z}_A) + \rho\gamma_{AB} - \sigma\epsilon_{AB}. \end{aligned} \tag{7.4.1e}$$

Taking the trace, we derive

$$\begin{aligned} \not{D}_4 \mathrm{tr}\underline{H} + \frac{1}{2}(\mathrm{tr}H)\mathrm{tr}\underline{H} = 2\,\mathrm{d\!\!\!/iv}\,\underline{Z} + 2\Omega\mathrm{tr}\underline{H} - \hat{H}\cdot\hat{\underline{H}} \\ + 2(Y\cdot\underline{Y} + \underline{Z}\cdot\underline{Z}) + 2\rho. \end{aligned} \tag{7.4.1f}$$

The antisymmetric part of 7.4.1e yields

$$\mathrm{c\!\!\!/url}\,\underline{Z} = \frac{1}{2}\hat{H}\wedge\hat{\underline{H}} - Y\wedge\underline{Y} - \sigma. \tag{7.4.1g}$$

Next we take the symmetric traceless part of of 1.0.5b and derive

$$\begin{aligned} \not{D}_4 \hat{\underline{H}}_{AB} + \frac{1}{2}\mathrm{tr}H\hat{\underline{H}}_{AB} = (\nabla\!\!\!/_B \underline{Z}_A + \nabla\!\!\!/_A \underline{Z}_B - \mathrm{d\!\!\!/iv}\,\underline{Z}\gamma_{AB}) + 2\Omega\hat{\underline{H}}_{AB} \\ - \frac{1}{2}\mathrm{tr}\underline{H}\hat{H}_{AB} + (Y\widehat{\otimes}\underline{Y})_{AB} + (\underline{Z}\widehat{\otimes}\underline{Z})_{AB}. \end{aligned} \tag{7.4.1h}$$

Then we calculate

$$\begin{aligned} \langle \mathbf{D}_C\mathbf{D}_B e_3, e_A\rangle &= \nabla\!\!\!/_C \underline{H}_{AB} + V_B \underline{H}_{AC} - \underline{Z}_A \underline{H}_{BC} - \underline{Y}_A H_{BC} \\ \langle \mathbf{D}_B\mathbf{D}_C e_3, e_A\rangle &= \nabla\!\!\!/_B \underline{H}_{AC} + V_C \underline{H}_{AB} - \underline{Z}_A \underline{H}_{BC} - \underline{Y}_A H_{BC} \\ \langle \mathbf{D}_C\mathbf{D}_B e_3, e_A\rangle &= \langle \mathbf{D}_B\mathbf{D}_C e_3, e_A\rangle + \mathbf{R}_{A3CB}. \end{aligned}$$

On the other hand,

$$\mathbf{R}_{A3CB} = - \in_{CB} {}^*\underline{\beta}_A,$$

hence

$$\nabla\!\!\!/_C \underline{H}_{AB} + V_B \underline{H}_{AC} = \nabla\!\!\!/_B \underline{H}_{AC} + V_C \underline{H}_{AB} + \in_{CB} {}^*\underline{\beta}_A.$$

Taking the trace in C, A, we infer that

$$(\mathrm{d\!\!\!/iv}\,\underline{H})_A - V_B \underline{H}_{AB} = \nabla\!\!\!/_A \mathrm{tr}\underline{H} - V_A \mathrm{tr}\underline{H} + \underline{\beta}_A. \tag{7.4.1i}$$

Now let u be an arbitrary Σ-tangent vectorfield. We calculate

$$\begin{aligned} \langle \mathbf{D}_C\mathbf{D}_B u, e_A\rangle = \nabla\!\!\!/_C \nabla\!\!\!/_B u_A + \frac{1}{2}(H_{AC}\underline{H}_{BD} + \underline{H}_{AC} H_{BD})u_D \\ + \text{terms symmetric in } B,\, C \end{aligned}$$

$$\langle \mathbf{D}_B\mathbf{D}_C u, e_A\rangle = \nabla\!\!\!/_B \nabla\!\!\!/_C u_A + \frac{1}{2}(H_{AB}\underline{H}_{CD} + \underline{H}_{AB}H_{CD})u_D$$
$$+ \text{ terms symmetric in } B,\ C$$
$$\langle \mathbf{D}_C\mathbf{D}_B u, e_A\rangle = \langle \mathbf{D}_B\mathbf{D}_C u, e_A\rangle + \mathbf{R}_{ADCB}u_D$$
$$\langle \nabla\!\!\!/_C \nabla\!\!\!/_B u, e_A\rangle = \langle \nabla\!\!\!/_B \nabla\!\!\!/_C u, e_A\rangle + R\!\!\!/_{ADCB}u_D$$

where $R\!\!\!/$ is the intrinsic Riemann curvature tensor of the 2-surfaces $S_{t,u}$. Hence

$$\mathbf{R}_{ADCB} = R\!\!\!/_{ADCB} + \frac{1}{2}(H_{AC}\underline{H}_{BD} - H_{AB}\underline{H}_{CD})$$
$$+ \frac{1}{2}(\underline{H}_{AC}H_{BD} - \underline{H}_{AB}H_{CD})$$

and by contraction, in view of the fact that $R\!\!\!/_{ADCB} = (\gamma_{AC}\gamma_{DB} - \gamma_{AB}\gamma_{CD})K$, where K is the Gauss curvature of $S_{t,u}$,

$$\gamma^{AC}\gamma^{BD}\mathbf{R}_{ADCB} = 2K + \frac{1}{2}\text{tr}H\text{tr}\underline{H} - \hat{H}\cdot\hat{\underline{H}}.$$

On the other hand,

$$\gamma^{AC}\gamma^{BD}\mathbf{R}_{ADCB} = -2\rho,$$

hence

$$K = -\frac{1}{4}\text{tr}H\text{tr}\underline{H} + \frac{1}{2}\hat{H}\cdot\hat{\underline{H}} - \rho. \tag{7.4.1j}$$

Next we calculate

$$\langle \mathbf{D}_3\mathbf{D}_A e_3, e_4\rangle = -2D\!\!\!/_3 V_A + 4\underline{\Omega}(V_A - Z_A) + 4\Omega\underline{Y}_A - 2\underline{H}_{AB}Z_B$$
$$\langle \mathbf{D}_A\mathbf{D}_3 e_3, e_4\rangle = 4\nabla_A\underline{\Omega} - 2H_{AB}\underline{Y}_B + 2\underline{H}_{AB}V_B$$
$$\langle \mathbf{D}_3\mathbf{D}_A e_3, e_4\rangle = \langle \mathbf{D}_A\mathbf{D}_3 e_3, e_4\rangle + \mathbf{R}_{A334}$$

and $\mathbf{R}_{A334} = 2\underline{\beta}_A$; hence

$$D\!\!\!/_3 V_A = -2\nabla\!\!\!/\underline{\Omega} - \underline{H}_{AB}(V_B + Z_B) + 2\underline{\Omega}(V_A - Z_A)$$
$$+ H_{AB}\underline{Y}_B + 2\Omega\underline{Y}_A - \underline{\beta}_A. \tag{7.4.1k}$$

Next,

$$\langle \mathbf{D}_3\mathbf{D}_4 e_3, e_A\rangle = 2D\!\!\!/_3\underline{Z}_A + 4\Omega\underline{Y}_A - 2\underline{H}_{AB}Z_B - 4\underline{\Omega}\underline{Z}_A$$
$$\langle \mathbf{D}_4\mathbf{D}_3 e_3, e_A\rangle = 2D\!\!\!/_4\underline{Y}_A - 4\underline{\Omega}\underline{Z}_A - 2\underline{H}_{AB}\underline{Z}_B - 4\Omega\underline{Y}_A$$
$$\langle \mathbf{D}_3\mathbf{D}_4 e_3, e_A\rangle = \langle \mathbf{D}_4\mathbf{D}_3 e_3, e_A\rangle + \mathbf{R}_{A334};$$

hence,

$$\not{D}_4\underline{Y}_A - \not{D}_3\underline{Z}_A = 4\Omega\underline{Y}_A + \underline{H}_{AB}(\underline{Z}_B - Z_B) - \underline{\beta}_A. \tag{7.4.1l}$$

Next,

$$\begin{aligned}\langle \mathbf{D}_4\mathbf{D}_3 e_3, e_4\rangle &= 4(\mathbf{D}_4\underline{\Omega} - Y\cdot\underline{Y} + V\cdot\underline{Z})\\ \langle \mathbf{D}_3\mathbf{D}_4 e_3, e_4\rangle &= -4(\mathbf{D}_3\Omega + Z\cdot\underline{Z} - V\cdot Z - 4\Omega\underline{\Omega})\\ \langle \mathbf{D}_4\mathbf{D}_3 e_3, e_4\rangle &= \langle \mathbf{D}_3\mathbf{D}_4 e_3, e_4\rangle + \mathbf{R}_{3434}\end{aligned}$$

and $\mathbf{R}_{3434} = 4\rho$; hence,

$$\mathbf{D}_4\underline{\Omega} + \mathbf{D}_3\Omega = Y\cdot\underline{Y} + V\cdot(Z - \underline{Z}) - Z\cdot\underline{Z} + 4\Omega\underline{\Omega} + \rho. \tag{7.4.1m}$$

We combine the results in the following:

Proposition 7.4.1 *Given an arbitrary Σ foliation of a space-time, the structure equations of the foliation are given by*

$$\begin{aligned}\not{D}_3\underline{\hat{H}}_{AB} + (\mathrm{tr}\underline{H})\underline{\hat{H}}_{AB} = (&\not{\nabla}_A\underline{Y}_B + \not{\nabla}_B\underline{Y}_A - \not{d}iv\,\underline{Y}\gamma_{AB})\\ &- 2\underline{\Omega}\underline{\hat{H}}_{AB} &&(7.4.2a)\\ &+ ((Z + \underline{Z} - 2V)\widehat{\otimes}\underline{Y})_{AB} - \underline{\alpha}_{AB} &&(7.4.2b)\end{aligned}$$

$$c\not{u}rl\,\underline{Y} = \underline{Y}\wedge(Z + \underline{Z} - 2V) \tag{7.4.2c}$$

$$\begin{aligned}\mathbf{D}_3(\mathrm{tr}\underline{H}) + \frac{1}{2}(\mathrm{tr}\underline{H})^2 = {}&2\not{d}iv\,\underline{Y} - 2\underline{\Omega}\mathrm{tr}\underline{H}\\ &+ 2\underline{Y}\cdot(Z + \underline{Z} - 2V) - \underline{\hat{H}}\cdot\underline{\hat{H}} &&(7.4.2d)\end{aligned}$$

$$\begin{aligned}\not{D}_4\underline{\hat{H}}_{AB} + \frac{1}{2}\mathrm{tr}H\underline{\hat{H}}_{AB} = {}&(\not{\nabla}_B\underline{Z}_A + \not{\nabla}_A\underline{Z}_B - \not{d}iv\,\underline{Z}\gamma_{AB}) + 2\Omega\underline{\hat{H}}_{AB}\\ &- \frac{1}{2}\mathrm{tr}\underline{H}\hat{H}_{AB} + (Y\widehat{\otimes}\underline{Y})_{AB} + (\underline{Z}\widehat{\otimes}\underline{Z})_{AB} &&(7.4.2e)\end{aligned}$$

$$c\not{u}rl\,\underline{Z} = \frac{1}{2}\hat{H}\wedge\underline{\hat{H}} - Y\wedge\underline{Y} - \sigma \tag{7.4.2f}$$

$$\begin{aligned}\not{D}_4\mathrm{tr}\underline{H} + \frac{1}{2}(\mathrm{tr}H)\mathrm{tr}\underline{H} = {}&2\not{d}iv\,\underline{Z} + 2\Omega\mathrm{tr}\underline{H} - \hat{H}\cdot\underline{\hat{H}}\\ &+ 2(Y\cdot\underline{Y} + \underline{Z}\cdot\underline{Z}) + 2\rho &&(7.4.2g)\end{aligned}$$

$$(\not{d}iv\, \underline{H})_A - V_B \underline{H}_{AB} = \nabla\!\!\!/_A \mathrm{tr}\underline{H} - V_A \mathrm{tr}\underline{H} + \underline{\beta}_A \tag{7.4.2h}$$

$$(\not{d}iv\, H)_A + V_B H_{AB} = \nabla\!\!\!/_A \mathrm{tr}H + V_A \mathrm{tr}H - \beta_A \tag{7.4.2i}$$

$$K = -\frac{1}{4}\mathrm{tr}H\mathrm{tr}\underline{H} + \frac{1}{2}\hat{H}\cdot\hat{\underline{H}} - \rho \tag{7.4.2j}$$

$$\begin{aligned}\not{D}_3 V_A = &-2\nabla\!\!\!/_A \underline{\Omega} - \underline{H}_{AB}(V_B + Z_B) + 2\underline{\Omega}(V_A - Z_A)\\ &+ H_{AB}\underline{Y}_B + 2\Omega\underline{Y}_A - \underline{\beta}_A\end{aligned} \tag{7.4.2k}$$

$$\begin{aligned}\not{D}_4 V_A = &\, 2\nabla\!\!\!/_A \Omega + H_{AB}(-V_B + \underline{Z}_B) + 2\Omega(V_A + \underline{Z}_A)\\ &- \underline{H}_{AB}Y_B - 2\underline{\Omega}Y_A - \beta_A .\end{aligned} \tag{7.4.2l}$$

Also,

$$\begin{aligned}\not{D}_3 \hat{H}_{AB} + \frac{1}{2}\mathrm{tr}\underline{H}\hat{H}_{AB} = &(\nabla\!\!\!/_B Z_A + \nabla\!\!\!/_A Z_B - \not{d}iv\, Z\gamma_{AB}) + 2\underline{\Omega}\hat{H}_{AB}\\ &- \frac{1}{2}\mathrm{tr}H\hat{\underline{H}}_{AB} + (\underline{Y}\widehat{\otimes}Y)_{AB} + (Z\widehat{\otimes}Z)_{AB}\end{aligned} \tag{7.4.2m}$$

$$c\not{u}rl\, Z = -\frac{1}{2}\hat{H}\wedge\hat{\underline{H}} + Y\wedge\underline{Y} + \sigma \tag{7.4.2n}$$

$$\begin{aligned}\not{D}_3 \mathrm{tr}H + \frac{1}{2}(\mathrm{tr}\underline{H})\mathrm{tr}H = &\, 2\not{d}iv\, Z + 2\underline{\Omega}\mathrm{tr}H - \hat{H}\cdot\hat{\underline{H}}\\ &+ 2(\underline{Y}\cdot Y + Z\cdot Z) + 2\rho\end{aligned} \tag{7.4.2o}$$

$$\begin{aligned}\mathbf{D}_4 \hat{H}_{AB} + (\mathrm{tr}H)\hat{H}_{AB} = &(\nabla\!\!\!/_A Y_B + \nabla\!\!\!/_B Y_A - \not{d}iv\, Y\gamma_{AB}) - 2\Omega\hat{H}_{AB}\\ &+ ((Z + \underline{Z} + 2V)\widehat{\otimes}Y)_{AB} - \alpha_{AB}\end{aligned} \tag{7.4.2p}$$

$$c\not{u}rl\, Y = Y\wedge(Z + \underline{Z} + 2V) \tag{7.4.2q}$$

$$\begin{aligned}\mathbf{D}_4(\mathrm{tr}H) + \frac{1}{2}(\mathrm{tr}H)^2 = &\, 2\not{d}iv\, Y - 2\Omega\mathrm{tr}H\\ &+ 2Y\cdot(Z + \underline{Z} + 2V) - \hat{H}\cdot\hat{H}.\end{aligned} \tag{7.4.2r}$$

Also,

$$\not{D}_4\underline{Y}_A - \not{D}_3\underline{Z}_A = 4\Omega\underline{Y}_A + \underline{H}_{AB}(\underline{Z}_B - Z_B) - \underline{\beta}_A \qquad (7.4.2s)$$

$$\not{D}_3 Y_A - \not{D}_4 Z_A = 4\underline{\Omega} Y_A + H_{AB}(Z_B - \underline{Z}_B) + \beta_A \qquad (7.4.2t)$$

$$\mathbf{D}_4\underline{\Omega} + \mathbf{D}_3\Omega = Y \cdot \underline{Y} + V \cdot (Z - \underline{Z}) - Z \cdot \underline{Z} + 4\Omega\underline{\Omega} + \rho. \qquad (7.4.2u)$$

7.5. Ricci Coefficients and the Vectorfields $\bar{K}$, S, T, and $\mathbf{O}$

Throughout this section we assume that our space-time $(\mathbf{M}, \mathbf{g})$ is foliated by a maximal t-function with lapse ϕ and second fundamental form k and a radial function u. We denote by $a = \frac{1}{|\nabla u|}$ the lapse function of the foliation induced by u on each Σ_t and by θ the second fundamental form of the surfaces $S_{t,u}$ relative to Σ_t. Given an orthonormal frame N, e_1, e_2 on Σ_t, we decompose k by introducing the following:

$$\begin{aligned} \eta_{AB} &= k_{AB} \\ \epsilon_A &= k_{AN} \\ \delta &= k_{NN} \end{aligned} \qquad (7.5.1)$$

We recall $r = r(u)$ was defined such that $4\pi r^2$ is the area of the surface $S_{t,u}$. Also recall that r_0 is the value of r corresponding to $u = 0$. We denote by Σ_t^e the exterior region of Σ_t corresponding to points for which

$$r \geq \frac{r_0}{2}$$

and by $\Sigma_t^i = I$ the interior region corresponding to points for which $r \leq \frac{r_0}{2}$. We assume that $r_0 = r_0(t) \geq 1$ for all $t \geq 0$, and, moreover, we make the basic assumption that u is optical in the exterior region of the space-time. In other words, u is a solution to the Eikonal equation at all points for which $r \geq \frac{r_0}{2}$. We will also denote by $\hat{I}$ the extended interior region that consists of the set of points with $r \leq \frac{3r_0}{4}$.

We next calculate the Ricci coefficients of a standard null frame:

Proposition 7.5.1 *Let* $e_3 = T - N, e_4 = T + N$ *be the standard null pair corresponding to the t,u foliations, and let* $e_1, e_2, e_3 = e_-, e_4 = e_+$ *be a corre-*

sponding null frame. The Ricci rotation coefficients are given by the formulas 7.3.1b with,

$$
\begin{aligned}
H &= \chi & \underline{H} &= \underline{\chi} \\
Z &= \zeta & \underline{Z} &= \underline{\zeta} \\
Y &= 0 & \underline{Y} &= \underline{\xi} \\
\Omega &= \tfrac{1}{2}\nu & \underline{\Omega} &= \tfrac{1}{2}\underline{\nu} \\
V &= \epsilon
\end{aligned}
\tag{7.5.2a}
$$

where

$$
\begin{aligned}
\chi_{AB} &= \theta_{AB} - k_{AB} \\
\underline{\chi}_{AB} &= -\theta_{AB} - k_{AB} \\
\underline{\xi}_A &= \phi^{-1}\nabla\!\!\!\!/_A\phi - a^{-1}\nabla\!\!\!\!/_A a \\
\underline{\zeta}_A &= \phi^{-1}\nabla\!\!\!\!/_A\phi - \epsilon_A \\
\zeta_A &= a^{-1}\nabla\!\!\!\!/_A a + \epsilon_A \\
\nu &= -\phi^{-1}\nabla_N\phi + \delta \\
\underline{\nu} &= \phi^{-1}\nabla_N\phi + \delta.
\end{aligned}
\tag{7.5.2b}
$$

Proof 7.5.1 *To check the proposition we recall the following formulas:*[16]

$$
\begin{aligned}
\mathbf{D}_i e_j &= \nabla_i e_j - k_{ij}T \\
\mathbf{D}_i T &= -k_{ij}e_j \\
\mathbf{D}_T e_i &= Proj[\mathbf{D}_T e_i] + (\phi^{-1}\nabla_i\phi)T \\
\mathbf{D}_T T &= (\phi^{-1}\nabla_i\phi)e_i \\
\nabla_N e_A &= \nabla\!\!\!\!/_N e_A + a^{-1}(\nabla\!\!\!\!/_A a)N \\
\nabla_A N &= \theta_{AB}e_B \\
\nabla_B e_A &= \nabla\!\!\!\!/_B e_A - \theta_{AB}N \\
\nabla_N N &= -a^{-1}\nabla\!\!\!\!/_A a e_A.
\end{aligned}
$$

The formulas for $\chi, \underline{\chi}$ are immediate. To calculate ζ, we write $e_3 = e_4 - 2N$ and substitute in the definition

$$
\begin{aligned}
\zeta_A &= \frac{1}{2}\langle \mathbf{D}_3 e_4, e_A\rangle = \frac{1}{2}\langle \mathbf{D}_4 e_4, e_A\rangle - \langle \mathbf{D}_N e_4, e_A\rangle \\
&= a^{-1}\nabla_A a + \epsilon_A.
\end{aligned}
$$

[16] See pages 5 and 54.

To calculate $\underline{\zeta}$, we write $e_3 = 2T - e_4$ and substitute,

$$\underline{\zeta}_A = \frac{1}{2}\langle \mathbf{D}_4 e_3, e_A \rangle = \langle \mathbf{D}_4 T, e_A \rangle = \phi^{-1}\nabla_A \phi - \epsilon_A.$$

In the same manner,

$$\begin{aligned} \underline{\xi}_A &= \frac{1}{2}\langle \mathbf{D}_3 e_3, e_A \rangle = \langle \mathbf{D}_3 T, e_A \rangle - \frac{1}{2}\langle \mathbf{D}_3 e_4, e_A \rangle \\ &= \phi^{-1}\nabla_A \phi + \epsilon_A - \zeta_A = \phi^{-1}\nabla_A \phi - a^{-1}\nabla_A a. \end{aligned}$$

All other formulas can be checked in the same way.

We now set $\tau^2 = 1 + u^2$, $v = 2r - u$ and define the following vectorfields:

$$T = \phi^{-1}\partial_t = \frac{1}{2}(e_+ + e_-) \tag{7.5.3a}$$

$$S = \frac{1}{2}(ve_+ - ue_-) \tag{7.5.3b}$$

$$K = \frac{1}{2}(v^2 e_+ + u^2 e_-). \tag{7.5.3c}$$

Also,

$$\bar{K} = \frac{1}{2}(\tau_+^2 e_+ + \tau_-^2 e_-) \tag{7.5.3d}$$

where

$$\begin{aligned} \tau_-^2 &= 1 + u^2 \\ \tau_+^2 &= 1 + v^2. \end{aligned} \tag{7.5.3e}$$

Calculating the deformation tensor of T, we find

$${}^{(T)}\pi_{\alpha\beta} = -2k_{\alpha\beta} - \phi^{-1}(T_\alpha \nabla_\beta \phi + T_\beta \nabla_\alpha \phi) \tag{7.5.4a}$$

or, in components relative to the null frame,

$$\begin{aligned} {}^{(T)}\pi_{44} &= -2\nu \\ {}^{(T)}\pi_{34} &= 2\delta \\ {}^{(T)}\pi_{33} &= -2\underline{\nu} \\ {}^{(T)}\pi_{AB} &= -2\eta_{AB}. \\ {}^{(T)}\pi_{A3} &= 2\epsilon_A + \phi^{-1}\nabla\!\!\!/_A \phi := 2\underline{m}_A \\ {}^{(T)}\pi_{A4} &= -2\epsilon_A + \phi^{-1}\nabla\!\!\!/_A \phi := 2m_A. \end{aligned} \tag{7.5.4b}$$

Also, for the Lie coefficients of T,

$$
\begin{aligned}
{}^{(T)}\underline{P}_A &= \underline{\zeta}_A - \zeta_A & {}^{(T)}P_A &= \zeta_A - \underline{\zeta}_A \\
{}^{(T)}\underline{M} &= \tfrac{1}{2}\nu & {}^{(T)}M &= \tfrac{1}{2}\underline{\nu} \\
{}^{(T)}\underline{N} &= \tfrac{1}{4}{}^{(T)}\pi_{33} & {}^{(T)}N &= \tfrac{1}{4}{}^{(T)}\pi_{44} \\
{}^{(T)}\underline{Q}_A &= \underline{\xi}_A + \underline{\zeta}_A + \epsilon_A & {}^{(T)}Q_A &= \zeta_A - \epsilon_A .
\end{aligned}
\tag{7.5.4c}
$$

To calculate the deformation tensor of S in the exterior region, we first remark that, along the level surfaces of u,

$$\frac{\partial r}{\partial t} = \frac{r}{2}\overline{\phi \mathrm{tr}\chi}. \tag{7.5.5a}$$

This follows by computing the rate of change of the area $A(t,u) = 4\pi r^2$ of the spheres $S_{t,u}$ along the hypersurfaces given by the level sets of u. To do this, we pick s to be an affine parameter for the null geodesics generated by the vectorfield l, that is, $\frac{d}{ds} = \mathbf{D}_l$ along the geodesic. We can parametrize the surfaces $S_{t,u}$, for fixed a u by picking arbitrary coordinates ϕ^1, ϕ^2 at some fixed value of $t = t_0$ and following the trajectories $x^\mu = x^\mu(s(t,\phi^1,\phi^2),\phi^1,\phi^2)$, with $x^\mu(s,\phi^1,\phi^2)$ being the null geodesic starting at a given point on $S_{t_0,u}$ and $s(t,\phi^1,\phi^2)$ being the corresponding value of s for which this geodesic intersects $S_{t,u}$. Relative to these coordinates, we calculate $\frac{d}{dt}\gamma$ where γ is the determinant of metric γ_{AB} on $S_{t,u}$:

$$
\begin{aligned}
\frac{d}{dt}\gamma &= \frac{ds}{dt}\frac{d}{ds}\gamma \\
\frac{d}{ds}\gamma &= \left(\gamma^{AB}\frac{d}{ds}\gamma_{AB}\right)\gamma = 2a^{-1}\gamma^{AB}\chi_{AB}\gamma \\
&= 2a^{-1}\mathrm{tr}\chi\gamma
\end{aligned}
$$

or, since $\partial_\mu t = -\phi^{-1}T_\mu$, we calculate $\frac{dt}{ds} = a^{-1}\phi^{-1}$, $\frac{ds}{dt} = a\phi$, and therefore $\frac{d}{dt}\gamma = 2\phi\mathrm{tr}\chi\gamma$. Thus, we conclude that $\frac{\partial A}{\partial t} = \int_S \phi\mathrm{tr}\chi$, which proves 7.5.5a.

With this remark we introduce[17]

$$\Lambda = \mathbf{D}_4 r = \phi^{-1}\frac{r}{2}\overline{\phi\mathrm{tr}\chi} \tag{7.5.5b}$$

$$
\begin{aligned}
\underline{\Lambda} = \mathbf{D}_3 r &= \mathbf{D}_4 r - 2\nabla_N r \\
&= \alpha^{-1}\frac{r}{2}\overline{a\mathrm{tr}\underline{\chi}} + \frac{r}{2}(\phi^{-1}\overline{\phi\mathrm{tr}\chi} - a^{-1}\overline{a\mathrm{tr}\chi})
\end{aligned}
\tag{7.5.5c}
$$

[17] Recall also the formula $\nabla_N r = \frac{r}{2a}\overline{a\mathrm{tr}\theta}$ on page 55.

and observe that

$$\mathbf{D}_3 u = -2a^{-1} \tag{7.5.5d}$$
$$\mathbf{D}_4 u = 0 \tag{7.5.5e}$$
$$\mathbf{D}_3 v = 2\underline{\Lambda} + 2a^{-1} \tag{7.5.5f}$$
$$\mathbf{D}_4 v = 2\Lambda. \tag{7.5.5g}$$

The deformation tensor of S is given by the formula

$$\begin{aligned}
{}^{(S)}\pi_{44} &= 2u\underline{\nu} \\
{}^{(S)}\pi_{34} &= -2a^{-1} - 2\Lambda - u\underline{\nu} + v\nu \\
{}^{(S)}\pi_{33} &= -4\underline{\Lambda} - 4a^{-1} - 2v\underline{\nu} \\
{}^{(S)}\pi_{AB} &= v\chi_{AB} - u\underline{\chi}_{AB} \\
{}^{(S)}\pi_{A3} &= -u\underline{\xi}_A + v(\epsilon_A + \zeta_A) \\
{}^{(S)}\pi_{A4} &= u(\epsilon_A - \underline{\zeta}_A).
\end{aligned} \tag{7.5.6a}$$

Or, since

$$\mathrm{tr}\,{}^{(S)}\pi = 2a^{-1} + 2\Lambda + u(\underline{\nu} - \mathrm{tr}\underline{\chi}) + v(-\nu + \mathrm{tr}\chi), \tag{7.5.6b}$$

we have

$$\begin{aligned}
{}^{(S)}\hat{\pi}_{44} &= 2u\nu \\
{}^{(S)}\hat{\pi}_{34} &= -a^{-1} - \Lambda - \frac{u}{2}(\underline{\nu} + \mathrm{tr}\underline{\chi}) + \frac{v}{2}(\nu + \mathrm{tr}\chi) \\
{}^{(S)}\hat{\pi}_{33} &= -4\underline{\Lambda} - 4a^{-1} - 2v\underline{\nu} \\
{}^{(S)}\hat{\pi}_{AB} &= v\hat{\chi}_{AB} - u\underline{\hat{\chi}}_{AB} + \frac{1}{2}\delta_{AB}{}^{(S)}\hat{\pi}_{34} \\
{}^{(S)}\hat{\pi}_{A3} &= -u\underline{\xi}_A + v(\epsilon_A + \zeta_A) \\
{}^{(S)}\hat{\pi}_{A4} &= u(\epsilon_A - \underline{\zeta}_A).
\end{aligned} \tag{7.5.6c}$$

Remark that ${}^{(S)}\hat{\pi}_{34}$ can also be expressed in the form

$$\begin{aligned}
{}^{(S)}\hat{\pi}_{34} = &-(a^{-1} - 1) - (\Lambda - 1) - \frac{u}{2}\left(\underline{\nu} + \mathrm{tr}\underline{\chi} + \frac{2}{r}\right) \\
&+ \frac{v}{2}\left(\nu + \mathrm{tr}\chi - \frac{2}{r}\right).
\end{aligned} \tag{7.5.6d}$$

We also calculate the Lie coefficients of S :

$$
\begin{aligned}
{}^{(S)}\underline{P}_A &= v(\underline{\zeta}_A - \zeta_A) & {}^{(S)}P_A &= u(\underline{\zeta}_A - \zeta_A) \\
{}^{(S)}\underline{M} &= \tfrac{1}{2}v\nu - a^{-1} & {}^{(S)}M &= -\tfrac{1}{2}u\underline{\nu} - \Lambda \\
{}^{(S)}\underline{N} &= \tfrac{1}{4}{}^{(S)}\pi_{33} & {}^{(S)}N &= \tfrac{1}{4}{}^{(S)}\pi_{44} \\
{}^{(S)}\underline{Q}_A &= \tfrac{1}{2}v(\epsilon_A + \underline{\zeta}_A) - \tfrac{1}{2}u\underline{\xi}_A & {}^{(S)}Q_A &= \tfrac{1}{2}(u\epsilon_A + v\zeta_A)
\end{aligned}
\tag{7.5.6e}
$$

Finally we calculate the null components of the traceless part of the deformation tensor of K :

$$
\begin{aligned}
{}^{(K)}\hat{\pi}_{44} &= -2u^2\nu \\
{}^{(K)}\hat{\pi}_{4A} &= u^2(\underline{\zeta}_A - \epsilon_A) \\
{}^{(K)}\hat{\pi}_{34} &= 2u(a^{-1} - 1) - 2v(\Lambda - 1) \\
&\quad + \frac{1}{2}u^2\Big(\mathrm{tr}\underline{\chi} + \frac{2}{r}\Big) + \frac{1}{2}v^2\Big(\mathrm{tr}\chi - \frac{2}{r}\Big) \\
&\quad + \frac{1}{2}u^2\underline{\nu} + \frac{1}{2}v^2\nu \\
{}^{(K)}\hat{\pi}_{33} &= -8v(\underline{\Lambda} + a^{-1}) - 2v^2\underline{\nu} \\
{}^{(K)}\hat{\pi}_{3A} &= u^2\underline{\xi}_A + v^2(\zeta_A + \epsilon_A) \\
{}^{(K)}\hat{\pi}_{AB} &= u^2\hat{\underline{\chi}}_{AB} + v^2\hat{\chi}_{AB} + \frac{1}{2}{}^{(K)}\hat{\pi}_{34}\delta_{AB}.
\end{aligned}
\tag{7.5.7}
$$

To check these calculations, note that

$$
\begin{aligned}
\mathrm{tr}^{(K)}\pi &= {}^{(K)}\pi_A^A - {}^{(K)}\pi_{34} \\
{}^{(K)}\pi_{AB} &= u^2\underline{\chi}_{AB} + v^2\chi_{AB} \\
{}^{(K)}\pi_{34} &= 4(a^{-1}u - v\Lambda) + u^2\underline{\nu} + v^2\nu.
\end{aligned}
$$

In what follows we make certain assumptions concerning the basic parameters of the (t, u) foliations and the Ricci coefficients of standard null frame. These assumptions will then be used in the derivation of the comparison theorem, which we present below.

Assumption 0: We assume that at each t, r as a function of u satisfies the following: $\frac{2}{3} \le \frac{dr}{du} \le \frac{3}{2}$. We also assume that the fundamental constants $k_m, k_M, a_m, a_M, a_1, h_m, h_M, \varsigma$, and ϕ_m of each slice Σ_t are uniformly bounded for all values of t. We recall that k_m, k_M, a_m, a_M, h_m, and h_M, ς

were defined on page 56. We define $\phi_m = (\inf_{\Sigma_t} \phi)^{-1}$.[18] *We also recall that* $a_1 = \sup_{\Sigma_t} r|a^{-1}\nabla\!\!\!/ a|$.

Assumption 1: At all points in the exterior region of the space-time,

$$\begin{aligned}\sup_{\Sigma_t^e} r^2|(\underline{\xi}, \underline{\zeta}, \zeta, \nu, \underline{\nu})| &\leq \varepsilon_0 \\ \sup_{\Sigma_t^e} r(|r\mathrm{tr}\chi - 2|, |r\mathrm{tr}\underline{\chi} - 2|) &\leq \varepsilon_0 \\ \sup_{\Sigma_t^e} r^2|\hat{\chi}| &\leq \varepsilon_0 \\ \sup_{\Sigma_t^e} r\tau_-|\underline{\hat{\chi}}| &\leq \varepsilon_0.\end{aligned} \tag{7.5.8a}$$

At all points in the extended interior region of the space-time $\hat{I}$,[19] *we have*

$$\|\phi^{-1}\nabla\phi\|_{3,\hat{I}} + \|k\|_{3,\hat{I}} \leq \varepsilon_0 \tag{7.5.8b}$$

$$\|r\hat{\theta}\|_{2,\hat{I}} + \|ra^{-1}\nabla\!\!\!/ a\|_{2,\hat{I}} \leq \varepsilon_0 r_0^{3/2} \tag{7.5.8c}$$

where $\| \ \|_{p,\hat{I}}$ *are the usual* L^p *norms defined in the region* $\hat{I}$ *of* Σ_t.

Assumption 2: We assume that, at all points in the exterior region of the space-time,

$$\begin{aligned}|||r^{5/2}\nabla\!\!\!/(\underline{\xi}, \underline{\zeta}, \zeta, \nu, \underline{\nu})|||_{4,e} &\leq c \\ |||r^{5/2}\nabla\!\!\!/(\mathrm{tr}\chi, \mathrm{tr}\underline{\chi})|||_{4,e} &\leq c \\ |||r^{5/2}\nabla\!\!\!/\hat{\chi}|||_{4,e} &\leq c \\ |||r^{3/2}\tau_-\nabla\!\!\!/\underline{\hat{\chi}}|||_{4,e} &\leq c\end{aligned} \tag{7.5.9a}$$

$$\begin{aligned}|||r^{3/2}\tau_-\mathbf{D}\!\!\!/_3(\underline{\xi}, \underline{\zeta}, \zeta)|||_{4,e} &\leq c \\ |||r^{3/2}\tau_-\mathbf{D}\!\!\!/_3\hat{\chi}|||_{4,e} &\leq c \\ |||r^{1/2}\tau_-^2\mathbf{D}\!\!\!/_3\underline{\hat{\chi}}|||_{4,e} &\leq c\end{aligned} \tag{7.5.9b}$$

$$\begin{aligned}|||r^{5/2}\mathbf{D}\!\!\!/_4(\underline{\xi}, \underline{\zeta}, \zeta)|||_{4,e} &\leq c \\ |||r^{5/2}\mathbf{D}\!\!\!/_4\hat{\chi}|||_{4,e} &\leq c\end{aligned} \tag{7.5.9c}$$

[18] Note that $\phi_M = \sup_{\Sigma_t} \phi = 1$ in view of the maximum principle for the lapse equation.

[19] Note that these assumptions are a consequence of the exterior assumptions for $\frac{r_0}{2} \leq r \leq \frac{3r_0}{2}$, thus they are really interior assumptions in I.

$$|||r^{3/2}\tau_-\not{\mathbf{D}}_4\hat{\underline{\chi}}|||_{4,e} \leq c$$

$$|||r^{5/2}(\mathbf{D}_4 \mathrm{tr}\chi + \frac{1}{2}(\mathrm{tr}\chi)^2)|||_{4,e} \leq c \tag{7.5.9d}$$

$$|||r^{3/2}\tau_-(\mathbf{D}_3 \mathrm{tr}(\underline{\chi}) + \frac{1}{2}(\mathrm{tr}\underline{\chi})^2)|||_{4,e} \leq c$$

$$|||r^{1/2}\tau_-\mathbf{D}_3(\nu, \underline{\nu})|||_{4,e} \leq cr_0^{-1} \tag{7.5.9e}$$

$$|||r^{3/2}\mathbf{D}_4(\nu, \underline{\nu})|||_{4,e} \leq cr_0^{-1}, \tag{7.5.9f}$$

and in the far exterior region $r \geq 2r_0$,

$$|||r^{5/2}\nabla_N(\nu, \underline{\nu})|||_{4,fe} \leq c \tag{7.5.9g}$$

where for any S-tangent tensor U the norms $|||U|||_{p,e}$ *and* $|||U|||_{p,fe}$ *are defined on* Σ_t *by the formulas*

$$|||U|||_{p,e} = sup_{r \geq \frac{r_0}{2}} \left(\int_{S_r} |U|^p d\mu_\gamma \right)^{1/p}$$

$$|||U|||_{p,fe} = sup_{r \geq 2r_0} \left(\int_{S_r} |U|^p d\mu_\gamma \right)^{1/p}.$$

On the other hand, in the interior region $\hat{I}$ *of the space-time, we assume that*

$$\|(\mathbf{D}\phi, k)\|_{\infty, I} \leq cr_0^{-1} \tag{7.5.9h}$$

$$\|(\mathbf{D}^2\phi, \mathbf{D}k)\|_{3, I} \leq cr_0^{-1}. \tag{7.5.9i}$$

Auxiliary Curvature Assumption *We also assume that the null components* $\beta(\mathbf{R})$, $\sigma(\mathbf{R})$, *and* $\underline{\beta}(\mathbf{R})$ *of the curvature tensor* $\mathbf{R}$ *of the space-time verify the following estimates in the exterior region:*

$$|||r^{5/2}\beta(\mathbf{R})|||_{4,e} \leq c$$

$$|||r^2\tau_-^{1/2}\sigma(\mathbf{R})|||_{4,e} \leq c \tag{7.5.10a}$$

$$|||r\tau_-^{3/2}\underline{\beta}(\mathbf{R})|||_{4,e} \leq c;$$

and in the interior region

$$\|Ric\|_{3/2, I} \leq \varepsilon_0 \tag{7.5.10b}$$

where Ric is the Ricci curvature tensor of the t-foliation.

Using these assumptions, we easily deduce the following:

Proposition 7.5.2 *The null components of ${}^{(T)}\hat{\pi}$ and the Lie coefficients of T satisfy the following:*

$$\sup_{\Sigma_t^e} r|({}^{(T)}M,\ {}^{(T)}\underline{M})| \leq c\varepsilon_0$$
$$\sup_{\Sigma_t^e} r|({}^{(T)}P,\ {}^{(T)}\underline{P})| \leq c\varepsilon_0 \tag{7.5.11a}$$
$$\sup_{\Sigma_t^e} r|({}^{(T)}Q,\ {}^{(T)}\underline{Q})| \leq c\varepsilon_0.$$

Also,

$$|||r^2 \nabla\!\!\!/ ({}^{(T)}M,\ {}^{(T)}\underline{M})|||_{4,e} \leq c$$
$$|||r^2 \nabla\!\!\!/ ({}^{(T)}P,\ {}^{(T)}\underline{P})|||_{4,e} \leq c \tag{7.5.11b}$$
$$|||r^2 \nabla\!\!\!/ ({}^{(T)}Q, {}^{(T)}\underline{Q})|||_{4,e} \leq c$$

$$|||r \not{\mathcal{L}}_S ({}^{(T)}M,\ {}^{(T)}\underline{M})|||_{4,e} \leq c$$
$$|||r \not{\mathcal{L}}_S ({}^{(T)}P,\ {}^{(T)}\underline{P})|||_{4,e} \leq c \tag{7.5.11c}$$
$$|||r \not{\mathcal{L}}_S ({}^{(T)}Q,\ {}^{(T)}\underline{Q})|||_{4,e} \leq c$$

$$|||r \not{\mathcal{L}}_T ({}^{(T)}M,\ {}^{(T)}\underline{M})|||_{4,e} \leq c$$
$$|||r \not{\mathcal{L}}_T ({}^{(T)}P,\ {}^{(T)}\underline{P})|||_{4,e} \leq c \tag{7.5.11d}$$
$$|||r \not{\mathcal{L}}_T ({}^{(T)}Q,\ {}^{(T)}\underline{Q})|||_{4,e} \leq c.$$

The Lie coefficients of S satisfy the following:

$$\sup_{\Sigma_t^e} |\mathrm{tr}^{(S)}\pi - 4| \leq c$$
$$\sup_{\Sigma_t^e} |\left(({}^{(S)}M + 1), ({}^{(S)}\underline{M} + 1)\right)| \leq c$$
$$\sup_{\Sigma_t^e} |({}^{(S)}Q,\ {}^{(S)}\underline{Q})| \leq c \tag{7.5.11e}$$
$$\sup_{\Sigma_t^e} |({}^{(S)}P,\ {}^{(S)}\underline{P})| \leq c.$$

Besides the vectorfields T, S, and $\bar{K}$, we assume that we can define rotation vectorfields ${}^{(i)}\Omega$, $i = 1, 2, 3$. A precise construction of these vectorfields will be given in the succeeding chapters. Here we assume that there exist vectorfields satisfying the requirements of the following proposition.

Proposition 7.5.3 *There exist rotation vectorfields ${}^{(i)}\Omega$ that span the tangent space of $S_{t,u}$ at every point in the space-time and that verify the following properties:*

Property 0: Given any S-tangent covariant tensor f on $\mathbf{M}$ *there exists a constant* c_O *such that*

$$c_O^{-1}\int_{S_{t,u}} r^2|\nabla f|^2 d\mu_\gamma \leq \int_{S_{t,u}} |\mathcal{L}_O f|^2 d\mu_\gamma \leq c_O \int_{S_{t,u}} (|f|^2 + r^2|\nabla f|^2) d\mu_\gamma \tag{7.5.12a}$$

where γ *is the metric induced on* $S_{t,u}$*;* $d\mu_\gamma$ *is the corresponding area element on* $S_{t,u}$*; and* $|\mathcal{L}_O f|^2 = \sum_{i=1,2,3} |\mathcal{L}_{{}^{(i)}\Omega} f|^2$*. Moreover, if f is either a 1-form or a symmetric 2-covariant traceless tensor tangent to the surfaces* $S_{t,u}$*,*

$$c_O^{-1}\int_{S_{t,u}} |f|^2 d\mu_\gamma \leq \int_{S_{t,u}} |\mathcal{L}_O f|^2 d\mu_\gamma. \tag{7.5.12b}$$

Property 1: The Lie coefficients ${}^{(O)}N$, ${}^{(O)}P$, ${}^{(O)}\underline{Q}$, ${}^{(O)}Q$ *of* ${}^{(O)}\Omega$ *are identically zero everywhere in the exterior region. The Lie coefficients* ${}^{(O)}\underline{M}$ *are given by the formula*

$${}^{({}^{(i)}\Omega)}\underline{M} = -a^{-1}\,{}^{(i)}\Omega(a)$$

in the exterior region.

Also,

$${}^{(O)}\pi_{44} = {}^{(O)}\pi_{A4} = 0; \qquad A = 1,2, \tag{7.5.13a}$$

and, for a given, small, positive number h,

$$\sup_{\Sigma_t^e} r^{1/2+h}|({}^{(O)}\pi_{33}, {}^{(O)}\pi_{34})| \leq \varepsilon_0 \tag{7.5.13b}$$

$$\sup_{\Sigma_t^e} r^{1/2+h}\left(\sum_{A,B} |{}^{(O)}\pi_{AB}|^2\right)^{1/2} \leq \varepsilon_0$$

in the exterior region.

Property 2:

$$\begin{gathered} |||r^{1+h}\nabla({}^{(O)}\pi_{33}, {}^{(O)}\pi_{34})|||_{4,e} \leq c \\ \sum_{A,B} |||r^{1+h}\nabla{}^{(O)}\pi_{AB}|||_{4,e} \leq c \end{gathered} \tag{7.5.13c}$$

everywhere in the exterior region.

7.6. The Statement of the Comparison Theorem

We introduce the following notation, on each slice Σ_t,

$$\mathcal{R}_q(W) = \max({}^e\mathcal{R}_q(W), {}^i\mathcal{R}_q(W)). \tag{7.6.1a}$$

The norms ${}^e\mathcal{R}_q(W), {}^i\mathcal{R}_q(W)$ are, respectively, the exterior and interior L^2 norms of the curvature defined as follows:

$$ {}^i\mathcal{R}_q(W) = r_0^{2+q}\|\mathbf{D}^q W\|_{2,I}. \tag{7.6.1b}$$

On the other hand, the exterior norms ${}^e\mathcal{R}_q(W)$ are defined as follows:[20]

$$\begin{aligned} {}^e\mathcal{R}_0(W)^2 = &\int_{\Sigma_t^e} \tau_-^4|\underline{\alpha}|^2 + \int_{\Sigma_t^e} \tau_-^2 r^2|\underline{\beta}|^2 + \int_{\Sigma_t^e} r^4|\rho - \bar{\rho}|^2 \\ &+ \int_{\Sigma_t^e} r^4|\sigma - \bar{\sigma}|^2 + \int_{\Sigma_t^e} r^4|\beta|^2 + \int_{\Sigma} r^4|\alpha|^2 \end{aligned} \tag{7.6.1c}$$

$$ {}^e\mathcal{R}_1(W)^2 = \int_{\Sigma_t^e} r^2\tau_-^4|\nabla\!\!\!\!/\,\underline{\alpha}|^2 + \int_{\Sigma_t^e} r^4\tau_-^2|\nabla\!\!\!\!/\,\underline{\beta}|^2 + \int_{\Sigma_t^e} r^6|\nabla\!\!\!\!/\,\rho|^2 \tag{7.6.1d}$$

$$\begin{aligned} &+ \int_{\Sigma_t^e} r^6|\nabla\!\!\!\!/\,\sigma|^2 + \int_{\Sigma_t^e} r^6|\nabla\!\!\!\!/\,\beta|^2 + \int_{\Sigma_t^e} r^6|\nabla\!\!\!\!/\,\alpha|^2 \\ &+ \int_{\Sigma_t^e} \tau_-^6|\underline{\alpha}_3|^2 + \int_{\Sigma_t^e} \tau_-^2 r^4|\underline{\alpha}_4|^2 + \int_{\Sigma_t^e} \tau_-^4 r^2|\underline{\beta}_3|^2 + \int_{\Sigma_t^e} r^6|\underline{\beta}_4|^2 \\ &+ \int_{\Sigma} \tau_-^2 r^4|\rho_3|^2 + \int_{\Sigma_t^e} r^6|\rho_4|^2 + \int_{\Sigma_t^e} \tau_-^2 r^4|\sigma_3|^2 + \int_{\Sigma_t^e} r^6|\sigma_4|^2 \\ &+ \int_{\Sigma_t^e} r^6|\beta_3|^2 + \int_{\Sigma_t^e} r^6|\beta_4|^2 + \int_{\Sigma_t^e} r^6|\alpha_3|^2 + \int_{\Sigma_t^e} r^6|\alpha_4|^2 \end{aligned} \tag{7.6.1e}$$

$$ {}^e\mathcal{R}_2(W)^2 = \int_{\Sigma_t^e} r^4\tau_-^4|\nabla\!\!\!\!/^2\underline{\alpha}|^2 + \int_{\Sigma_t^e} r^6\tau_-^2|\nabla\!\!\!\!/^2\underline{\beta}|^2 + \int_{\Sigma_t^e} r^8|\nabla\!\!\!\!/^2\rho|^2 \tag{7.6.1f}$$

$$\begin{aligned} &+ \int_{\Sigma_t^e} r^8|\nabla\!\!\!\!/^2\sigma|^2 + \int_{\Sigma_t^e} r^8|\nabla\!\!\!\!/^2\beta|^2 + \int_{\Sigma_t^e} r^8|\nabla\!\!\!\!/^2\alpha|^2 \\ &+ \int_{\Sigma_t^e} r^2\tau_-^6|\nabla\!\!\!\!/\,\underline{\alpha}_3|^2 + \int_{\Sigma_t^e} \tau_-^2 r^6|\nabla\!\!\!\!/\,\underline{\alpha}_4|^2 + \int_{\Sigma_t^e} \tau_-^4 r^4|\nabla\!\!\!\!/\,\underline{\beta}_3|^2 \\ &+ \int_{\Sigma_t^e} r^8|\nabla\!\!\!\!/\,\underline{\beta}_4|^2 + \int_{\Sigma_t^e} \tau_-^2 r^6|\nabla\!\!\!\!/\,\rho_3|^2 + \int_{\Sigma_t^e} r^8|\nabla\!\!\!\!/\,\rho_4|^2 \end{aligned}$$

[20] Note that $\bar{\rho}, \bar{\sigma}$ are the mean values of ρ, σ on $S_{t,u}$.

$$
\begin{aligned}
&+\int_{\Sigma_t^e}\tau_-^2r^6|\nabla\!\!\!/\sigma_3|^2+\int_{\Sigma_t^e}r^8|\nabla\!\!\!/\sigma_4|^2+\int_{\Sigma_t^e}r^8|\nabla\!\!\!/\beta_3|^2\\
&+\int_{\Sigma_t^e}r^8|\nabla\!\!\!/\beta_4|^2+\int_{\Sigma_t^e}r^8|\nabla\!\!\!/\alpha_3|^2+\int_{\Sigma_t^e}r^8|\nabla\!\!\!/\alpha_4|^2\\
&+\int_{\Sigma_t^e}\tau_-^8|\underline{\alpha}_{33}|^2+\int_{\Sigma_t^e}\tau_-^4r^4|\underline{\alpha}_{34}|^2+\int_{\Sigma_t^e}\tau_-^4r^4|\underline{\alpha}_{43}|^2\\
&+\int_{\Sigma_t^e}r^6\tau_-^2|\underline{\alpha}_{44}|^2+\int_{\Sigma_t^e}\tau_-^6r^2|\underline{\beta}_{33}|^2+\int_{\Sigma_t^e}\tau_-^2r^6|\underline{\beta}_{34}|^2\\
&+\int_{\Sigma_t^e}\tau_-^2r^6|\underline{\beta}_{43}|^2+\int_{\Sigma_t^e}r^8|\underline{\beta}_{44}|^2+\int_{\Sigma_t^e}\tau_-^4r^4|\rho_{33}|^2\\
&+\int_{\Sigma_t^e}\tau_-r^7|\rho_{34}|^2+\int_{\Sigma_t^e}\tau_-r^7|\rho_{43}|^2+\int_{\Sigma_t^e}r^8|\rho_{44}|^2\\
&+\int_{\Sigma_t^e}\tau_-^4r^4|\sigma_{33}|^2+\int_{\Sigma}\tau_-r^7|\sigma_{34}|^2+\int_{\Sigma_t^e}\tau_-r^7|\sigma_{43}|^2+\int_{\Sigma_t^e}r^8|\sigma_{44}|^2\\
&+\int_{\Sigma_t^e}\tau_-^2r^6|\beta_{33}|^2+\int_{\Sigma}r^8|\beta_{34}|^2+\int_{\Sigma_t^e}r^8|\beta_{43}|^2+\int_{\Sigma}r^8|\beta_{44}|^2\\
&+\int_{\Sigma_t^e}r^8|\alpha_{33}|^2+\int_{\Sigma}r^8|\alpha_{34}|^2+\int_{\Sigma_t^e}r^8|\alpha_{43}|^2+\int_{\Sigma_t^e}r^8|\underline{\alpha}_{44}|^2.
\end{aligned}
$$

We also introduce the following notation involving the Bell-Robinson tensor Q of a Weyl tensor W and the vectorfields T, $\bar{K}$, S, and $\mathbf{O}$.

$$
\mathcal{Q}_1(W)=\int_\Sigma Q(\hat{\mathcal{L}}_{\mathbf{O}}W)(\bar{K},\bar{K},T,T)+\int_\Sigma Q(\hat{\mathcal{L}}_TW)(\bar{K},\bar{K},\bar{K},T) \tag{7.6.1g}
$$

$$
\begin{aligned}
\mathcal{Q}_2(W)=&\int_\Sigma Q(\hat{\mathcal{L}}_{\mathbf{O}}^2W)(\bar{K},\bar{K},T,T)+\int_\Sigma Q(\hat{\mathcal{L}}_{\mathbf{O}}\hat{\mathcal{L}}_TW)(\bar{K},\bar{K},\bar{K},T)\\
&+\int_\Sigma Q(\hat{\mathcal{L}}_S\hat{\mathcal{L}}_TW)(\bar{K},\bar{K},\bar{K},T)+\int_\Sigma Q(\hat{\mathcal{L}}_T^2W)(\bar{K},\bar{K},\bar{K},T).
\end{aligned} \tag{7.6.1h}
$$

The aim of this section is to compare the norms $\mathcal{R}_1(W),\mathcal{R}_2(W)$, and $\mathcal{Q}_1^{1/2},\mathcal{Q}_2^{1/2}$. This is the content of the following.

Theorem 7.6.1 *We assume given a space-time* $(\mathbf{M},\mathbf{g})$ *endowed with a (t, u)-foliation as described in the previous section and consider an arbitrary Weyl tensor W that satisfies the homogeneous Bianchi equations 7.1.2.*

(i.) Assume that the space-time verifies the Assumptions 0 and 1 on page 176 and that the auxiliary curvature assumption (7.5.10b) is satisfied in the interior region. Moreover, we assume that there exist angular momentum vectorfields ${}^{(i)}\Omega$, $i = 1, 2, 3$ that satisfy the assumptions 7.5.12a and 7.5.13b of Proposition 7.5.3.

Then, there exists a sufficiently small ε_0 and a constant c such that, if $\mathcal{Q}_1(\tilde{W})$ is finite,

$$\mathcal{R}_0(\tilde{W}) + \mathcal{R}_1(\tilde{W}) \leq c\mathcal{Q}_1(\tilde{W})^{1/2}$$

and

$$\sup_{r \geq r_0} r^3 \bar{\rho}, \bar{\sigma} \leq c\varepsilon_0 \mathcal{Q}_1(\tilde{W})^{1/2}$$

$$\sup_{r \leq r_0} \tau_- r^3 \bar{\rho}, \bar{\sigma} \leq c\varepsilon_0 \mathcal{Q}_1(\tilde{W})^{1/2}.$$

In particular, introducing the norm

$$\mathcal{R}_{[1]}(\tilde{W}) = \mathcal{R}_0(\tilde{W}) + \mathcal{R}_1(\tilde{W}) + \sup_{r \geq r_0} r^3 \bar{\rho}, \bar{\sigma} + \sup_{r \leq r_0} \tau_- r^3 \bar{\rho}, \bar{\sigma}$$

we have

$$\mathcal{R}_{[1]}(\tilde{W}) \leq c\mathcal{Q}_1(\tilde{W})^{1/2}.$$

(ii.) Moreover, if the space-time also verifies Assumption 2 and the auxiliary curvature assumption (7.5.10a), and the vectorfields ${}^{(i)}\Omega$ also verifies 7.5.13c of Proposition 7.5.3, then there exists a constant c such that, if $\mathcal{Q}_2(\tilde{W})$ is finite,

$$\mathcal{R}_2(\tilde{W}) \leq c(\mathcal{Q}_1(\tilde{W})^{1/2} + \mathcal{Q}_2(\tilde{W})^{1/2}).$$

(iii.) Under the same assumptions the results in i and ii hold when W is the Riemann curvature tensor **R** *of an Einstein metric. In that case we drop the auxiliary curvature assumption, which is in fact a consequence of the estimates for $\mathcal{R}_1(\mathbf{R}) + \mathcal{R}_2(\mathbf{R})$.*

7.7. Proof of the Comparison Theorem

We divide the theorem into four propositions, which we prove separately. We first state the following:

Proposition 7.7.1 *Assume that the space-time $(\mathbf{M}, \mathbf{g})$ is endowed with a (t, u)-foliation verifying Assumption 0 and the exterior Assumption 1 as described in the previous sections. Finally, assume that there exist angular momentum operators ${}^{(i)}\Omega$, $i = 1, 2, 3$ verifying the assumptions 7.5.12a and 7.5.13b.*

Let W be a Weyl tensor verifying the homogeneous Bianchi equations. Let ${}^e\mathcal{R}_1(W)$ be defined as in the previous section and divide it into the following parts:

$$
\begin{aligned}
{}^e\mathcal{R}_{1,1}{}^2 &= \int_{\Sigma_t^e} r^2\tau_-^4|\nabla\!\!\!/\,\underline{\alpha}|^2 + \int_{\Sigma_t^e} r^4\tau_-^2|\nabla\!\!\!/\,\underline{\beta}|^2 + \int_{\Sigma_t^e} r^6|\nabla\!\!\!/\,\rho|^2 \\
&\quad + \int_{\Sigma_t^e} r^6|\nabla\!\!\!/\,\sigma|^2 + \int_{\Sigma_t^e} r^6|\nabla\!\!\!/\,\beta|^2 + \int_{\Sigma_t^e} r^6|\nabla\!\!\!/\,\alpha|^2 \\
{}^e\mathcal{R}_{1,2}{}^2 &= \int_{\Sigma_t^e} \tau_-^2 r^4|\underline{\alpha}_4|^2 + \int_{\Sigma_t^e} \tau_-^4 r^2|\underline{\beta}_3|^2 + \int_{\Sigma_t^e} r^6|\underline{\beta}_4|^2 + \int_{\Sigma} \tau_-^2 r^4|\rho_3|^2 \\
&\quad + \int_{\Sigma_t^e} r^6|\rho_4|^2 + \int_{\Sigma_t^e} \tau_-^2 r^4|\sigma_3|^2 + \int_{\Sigma_t^e} r^6|\sigma_4|^2 + \int_{\Sigma_t^e} r^6|\beta_3|^2 \\
&\quad + \int_{\Sigma_t^e} r^6|\beta_4|^2 + \int_{\Sigma_t^e} r^6|\alpha_3|^2 \\
{}^e\mathcal{R}_{1,3}{}^2 &= \int_{\Sigma_t^e} \tau_-^6|\underline{\alpha}_3|^2 + \int_{\Sigma_t^e} r^6|\alpha_4|^2.
\end{aligned}
$$

There exists a constant c such that for sufficiently small ε_0

$$
{}^e\mathcal{R}_0{}^2 + {}^e\mathcal{R}_{1,1}{}^2 \le c\int_\Sigma Q(\hat{\mathcal{L}}_O W)(\bar{K},\bar{K},T,T) + \varepsilon_0^3 r_0^2 \int_I |W|^2. \tag{7.7.1a}
$$

Moreover,

$$
\sup_{\frac{r_0}{2}\le r\le r_0} \tau_- r^3|(\overline{\rho},\overline{\sigma})|,\ \sup_{r\ge r_0} r^3|(\bar{\rho},\bar{\sigma})| \le c\varepsilon_0\left(\int_\Sigma Q(\hat{\mathcal{L}}_O W)(\bar{K},\bar{K},T,T)\right)^{1/2} + c\varepsilon_0 r_0\left(\int_I |W|^2\right)^{1/2}. \tag{7.7.1b}
$$

Also,

$$
{}^e\mathcal{R}_{1,2}{}^2 \le c\int_\Sigma Q(\hat{\mathcal{L}}_O W)(\bar{K},\bar{K},T,T) + c\varepsilon_0^2 r_0^2 \int_I |W|^2. \tag{7.7.1c}
$$

Finally,

$$
{}^e\mathcal{R}_{1,3}{}^2 \le c\Big(\int_\Sigma Q(\hat{\mathcal{L}}_O W)(\bar{K},\bar{K},T,T) + \int_\Sigma Q(\hat{\mathcal{L}}_T W)(\bar{K},\bar{K},\bar{K},T) + c\varepsilon_0^3 r_0^2 \int_I |W|^2\Big). \tag{7.7.1d}
$$

Proof 7.7.1 *Indeed, in view of Lemma 7.3.1 and the definition of $\bar{K}$, we have*

$$Q(\hat{\mathcal{L}}_O W)(\bar{K},\bar{K},T,T) \geq \frac{1}{8}\big(\tau_-^4|\underline{\alpha}(\hat{\mathcal{L}}_O W)|^2 + \tau_-^2\tau_+^2|\underline{\beta}(\hat{\mathcal{L}}_O W)|^2 + \tau_+^4|\rho(\hat{\mathcal{L}}_O W)|^2 + \tau_+^4|\sigma(\hat{\mathcal{L}}_O W)|^2 + \tau_+^4|\beta(\hat{\mathcal{L}}_O W)|^2 + \tau_+^4|\alpha(\hat{\mathcal{L}}_O W)|^2\big).$$

Consequently, by virtue of the commutation formulas of Proposition 7.3.1 and assumptions 7.5.13b of Proposition 7.5.3, we deduce

$$\begin{aligned}\int_\Sigma Q(\hat{\mathcal{L}}_O W)(\bar{K},\bar{K},T,T) \geq c^{-1}\Big[&\int_{\Sigma_t^e}\tau_-^4|\hat{\mathcal{L}}_O\underline{\alpha}|^2 + \int_{\Sigma_t^e}\tau_-^2\tau_+^2|\mathcal{L}_O\underline{\beta}|^2 \\ &+ \int_{\Sigma_t^e}\tau_+^4|\mathcal{L}_O\rho|^2 + \int_{\Sigma_t^e}\tau_+^4|\mathcal{L}_O\sigma|^2 \\ &+ \int_{\Sigma_t^e}\tau_+^4|\mathcal{L}_O\beta|^2 + \int_{\Sigma_t^e}\tau_+^4|\hat{\mathcal{L}}_O\alpha|^2\Big] \\ &- c\varepsilon_0\int_{\Sigma_t^e}\tau_-^{3-2h}(|\underline{\alpha}|^2 + |\underline{\beta}|^2) \\ &+ r^{3-2h}(|\rho|^2 + |\sigma|^2 + |\beta|^2 + |\alpha|^2).\end{aligned}$$

To check this, we recall that the Lie coefficients ${}^{(O)}N$, ${}^{(O)}P$, ${}^{(O)}\underline{Q}$, and ${}^{(O)}Q$ of ${}^{(O)}\Omega$ are identically zero, while

$$\sup_{\Sigma_t^e} r^{1/2+h}|({}^{(O)}M,\ {}^{(O)}\underline{M})| \leq \varepsilon_0$$

$$\sup_{\Sigma_t^e} r^{1/2+h}|({}^{(O)}N,\ {}^{(O)}\underline{P})| \leq \varepsilon_0.$$

Thus, using Property 0 of Proposition 7.5.3 and the Poincaré inequality for ρ, σ,

$$\int_{\Sigma_t^e} r^4|\rho-\bar{\rho}|^2 \leq c\int_{\Sigma_t^e} r^6|\nabla\!\!\!/\rho|^2$$

$$\int_{\Sigma_t^e} r^4|\sigma-\bar{\sigma}|^2 \leq c\int_{\Sigma_t^e} r^6|\nabla\!\!\!/\sigma|^2$$

and choosing ε_0 sufficiently small, we infer that

$$\begin{aligned}\int_\Sigma Q(\hat{\mathcal{L}}_O W)(\bar{K},\bar{K},T,T) \geq c^{-1}\Big[&\int_{\Sigma_t^e}\tau_-^4|\underline{\alpha}|^2 + \int_{\Sigma_t^e}\tau_-^2 r^2|\underline{\beta}|^2 \\ &+ \int_{\Sigma_t^e} r^4|\rho-\bar{\rho}|^2 + \int_{\Sigma_t^e} r^4|\sigma-\bar{\sigma}|^2 + \int_{\Sigma_t^e} r^4|\beta|^2\end{aligned}$$

$$+\int_{\Sigma_t^e} r^4|\alpha|^2+\int_{\Sigma_t^e} r^2\tau_-^4|\nabla\underline{\alpha}|^2+\int_{\Sigma_t^e} r^4\tau_-^2|\nabla\underline{\beta}|^2$$
$$+\int_{\Sigma_t^e} r^6|\nabla\rho|^2+\int_{\Sigma_t^e} r^6|\nabla\sigma|^2+\int_{\Sigma_t^e} r^6|\nabla\beta|^2$$
$$+\int_{\Sigma_t^e} r^6|\nabla\alpha|^2\Big]-c\varepsilon_0\int_{\Sigma_t^e} r^{3-2h}(|\bar{\rho}|^2+|\bar{\sigma}|^2). \tag{7.7.2a}$$

Therefore, to derive 7.7.1a, it suffices to estimate $\bar{\rho}$, $\bar{\sigma}$. Before doing this, however, we first indicate how to derive the estimates 7.7.1c and 7.7.1d, assuming 7.7.1a and 7.7.1b.

To check 7.7.1c we use formulas 7.3.11a–7.3.11j of Proposition 7.3.2 together with the assumptions 7.5.8a for the frame coefficients. To obtain 7.7.1d, we first remark, according to Lemma 7.3.1, that

$$\int_\Sigma \tau_-^6\,|\underline{\alpha}(\hat{\mathcal{L}}_TW)|^2+\int_\Sigma \tau_+^6|\alpha(\hat{\mathcal{L}}_TW)|^2\le 8\int_\Sigma Q(\hat{\mathcal{L}}_TW)(\bar{K},\bar{K},\bar{K},T). \tag{7.7.2b}$$

Consequently, in view of Propositions 7.3.1 and 7.7.1a,

$$\begin{aligned}\int_\Sigma \tau_-^6\,|\hat{\not{\mathcal{L}}}_T\underline{\alpha}|^2+\int_\Sigma \tau_+^6|\hat{\not{\mathcal{L}}}_T\alpha|^2 &\le \int_\Sigma Q(\hat{\mathcal{L}}_TW)(\bar{K},\bar{K},\bar{K},T)\\ &\quad+c\varepsilon_0\int_\Sigma Q(\hat{\mathcal{L}}_{\mathbf{O}}W)(\bar{K},\bar{K},T,T)\\ &\quad+c\varepsilon_0^4r_0^2\int_I |W|^2\end{aligned} \tag{7.7.2c}$$

provided that the the Lie coefficients of T verify

$$\sup_{\Sigma_t^e} r|{}^{(T)}M,\ {}^{(T)}\underline{M}|\le c\varepsilon_0$$
$$\sup_{\Sigma_t^e} r|{}^{(T)}P,\ {}^{(T)}Q,\ {}^{(T)}\underline{P},\ {}^{(T)}\underline{Q})|\le c\varepsilon_0,$$

which is precisely the estimate 7.5.11a of Proposition 7.5.2.

Now, since

$$\begin{aligned}\not{\mathcal{L}}_T\alpha_{AB}&=\Pi_A^{\ \rho}\Pi_B^{\ \sigma}\mathcal{L}_T\alpha_{\rho,\sigma}\\ &=\not{\mathbf{D}}_T\alpha-\eta_{AC}\alpha_{CB}-\eta_{BC}\alpha_{AC}\\ &=\not{\mathbf{D}}_T\alpha-\alpha\cdot\hat{\eta}\delta_{AB}+\delta\alpha_{AB}\end{aligned}$$

$$\begin{aligned}\hat{\not{\mathcal{L}}}_T\alpha_{AB}&=\not{\mathbf{D}}_T\alpha+\delta\alpha_{AB}\\ \hat{\not{\mathcal{L}}}_T\underline{\alpha}_{AB}&=\not{\mathbf{D}}_T\underline{\alpha}+\delta\underline{\alpha}_{AB},\end{aligned}$$

we derive, taking 7.7.1a into account,

$$\int_{\Sigma_t^e} \tau_-^6 |\underline{\alpha}_3|^2 + \int_\Sigma r^6 |\alpha_4|^2 \le c \int_\Sigma Q(\hat{\mathcal{L}}_T W)(\bar{K}, \bar{K}, \bar{K}, T)$$
$$+ c \int_\Sigma Q(\hat{\mathcal{L}}_O W)(\bar{K}, \bar{K}, T, T) + c\varepsilon_0^3 r_0^3 \int_I |W|^2.$$

The estimates 7.5.11a follow immediately from the formulas 7.5.4b and 7.5.4c and the assumptions 7.5.8a.

It remains to estimate $\bar{\rho}, \bar{\sigma}$. In view of Proposition 7.2.1 the electric and magnetic parts of W satisfy the divergence equations

$$\operatorname{div} E = +k \wedge H$$
$$\operatorname{div} H = -k \wedge E.$$

We now appeal to the interior identity of Proposition 4.4.4 and infer that

$$4\pi r^3 \bar{\rho} = \int_{B_r} \Big(-\frac{1}{2}\kappa(\rho - \bar{\rho})$$
$$+ r\hat{\theta}_{AB} E_{AB} - r a^{-1} (\nabla\!\!\!/_A a) E_{AN} + r(k \wedge H)_N \Big)$$
$$4\pi r^3 \bar{\sigma} = \int_{B_r} \Big(-\frac{1}{2}\kappa(\sigma - \bar{\sigma})$$
$$+ r\hat{\theta}_{AB} H_{AB} - r a^{-1} (\nabla\!\!\!/_A a) H_{AN} - r(k \wedge E)_N \Big).$$

Taking into account 7.3.3e, we write

$$r\hat{\theta}_{AB} E_{AB} = \frac{1}{4} r\hat{\theta}_{AB} (\alpha_{AB} + \underline{\alpha}_{AB})$$
$$r(a^{-1} \nabla\!\!\!/_A a) E_{AN} = \frac{1}{2} r a^{-1} \nabla\!\!\!/_A a (\beta_A + \underline{\beta}_A)$$
$$r(k \wedge H)_N = \frac{1}{4} r\eta_{AB} (\alpha_{AB} - \underline{\alpha}_{AB}) + \frac{1}{2} r\epsilon_A (\beta_A - \underline{\beta}_A);$$

or, since $\chi_{AB} = \theta_{AB} - \eta_{AB}$, $\underline{\chi}_{AB} = -\theta_{AB} - \eta_{AB}$, we derive

$$4\pi r^3 \bar{\rho} = \int_{B_r} -\frac{1}{2}\kappa(\rho - \bar{\rho}) \tag{7.7.3a}$$
$$+ \int_{B_r} \frac{1}{4} r \big((\underline{\alpha} \cdot \hat{\chi} - \alpha \cdot \hat{\underline{\chi}})$$
$$+ 2\beta \cdot (-a^{-1} \nabla\!\!\!/ a + \epsilon) - 2\underline{\beta} \cdot (a^{-1} \nabla\!\!\!/ a + \epsilon) \big)$$

$$4\pi r^3\overline{\sigma} = \int_{B_r} -\frac{1}{2}\kappa(\sigma - \overline{\sigma}) \tag{7.7.3b}$$
$$+ \int_{B_r} \frac{1}{4}r\big(({}^*\underline{\alpha}\cdot\hat{\chi} + {}^*\alpha\cdot\hat{\underline{\chi}})$$
$$- 2{}^*\beta\cdot(-a^{-1}\nabla\!\!\!/ a + \epsilon) - 2{}^*\underline{\beta}\cdot(a^{-1}\nabla\!\!\!/ a + \epsilon)\big).$$

We then easily derive the following estimates:

$$\sup_{\frac{r_0}{2}\le r\le r_0} \tau_- r^3|\overline{\rho}|,\ \sup_{r\ge r_0} r^3|\overline{\rho}| \le c\varepsilon_0\Big[\int_{\Sigma_t^e}\tau_-^4|\underline{\alpha}|^2 + \int_{\Sigma_t^e}\tau_-^2 r^2|\underline{\beta}|^2$$
$$+ \int_{\Sigma_t^e} r^4|\rho - \bar{\rho}|^2 + \int_{\Sigma_t^e} r^4|\beta|^2$$
$$+ \int_{\Sigma_t^e} r^4|\alpha|^2\Big]^{1/2} + c\varepsilon_0\Big[r_0^2\int_I |W|^2\Big]^{1/2} \tag{7.7.3c}$$

$$\sup_{\frac{r_0}{2}\le r\le r_0} r^3\tau_-|\overline{\sigma}|,\ \sup_{r\ge r_0} r^3|\overline{\sigma}| \le c\varepsilon_0\Big[\int_{\Sigma_t^e}\tau_-^4|\underline{\alpha}|^2 + \int_{\Sigma_t^e}\tau_-^2 r^2|\underline{\beta}|^2$$
$$+ \int_{\Sigma_t^e} r^4|\sigma - \bar{\sigma}|^2 + \int_{\Sigma_t^e} r^4|\beta|^2$$
$$+ \int_{\Sigma} r^4|\alpha|^2\Big]^{1/2} + c\varepsilon_0\Big[r_0^2\int_I |W|^2\Big]^{1/2} \tag{7.7.3d}$$

if the following condition is verified in the interior:

$$\|rk\|_{2,\hat{I}} + \|r\hat{\theta}\|_{2,\hat{I}} + \|ra^{-1}\nabla\!\!\!/ a\|_{2,\hat{I}} \le c\varepsilon_0 r_0.$$

Substituting these estimates into 7.7.2a yields 7.7.1a, provided that ε_0 is sufficiently small. This ends the proof of the proposition.

We are now ready to consider the estimates in the interior region I given by the set of points on each Σ_t for which $r \le \frac{r_0}{2}$. We first prove the following:

Proposition 7.7.2 *We assume that in addition to the exterior assumptions of Proposition 7.7.1 the interior assumptions 7.5.8b and 7.5.8c on page 176 and the auxiliary curvature assumption (7.5.10b) are also satisfied in the extended interior region $\hat{I}$.*[21] *Then, for any Weyl field W that verifies the homogeneous*

[21] Recall that $\hat{I}$ is the set of points for which $r \le \frac{3r_0}{4}$.

Bianchi equations, there exists a constant c such that for ε_0 sufficiently small we have

$$\int_I |\mathbf{D}W|^2 \leq c r_0^{-6} \mathcal{Q}_1(W) \tag{7.7.4a}$$

$$\int_I |W|^2 \leq c r_0^{-4} \mathcal{Q}_1(W). \tag{7.7.4b}$$

Finally, we remark that if W is the Riemann curvature tensor **R** *of an Einstein metric and $\mathcal{Q}_1(\mathbf{R})$ is sufficiently small, the auxiliary curvature assumption (7.5.10b) can be dropped.*

To prove the propositions we first observe that,

$$\int_{\hat{I}} (|E(\hat{\mathcal{L}}_T W)|^2 + |H(\hat{\mathcal{L}}_T W)|^2) \leq c r_0^{-6} \int_\Sigma Q(\hat{\mathcal{L}}_T W)(\bar{K}, \bar{K}, \bar{K}, T). \tag{7.7.5}$$

Indeed, this is an immediate consequence of Lemma 7.3.1 and the definition of τ_+, τ_-. Now, by virtue of 7.1.7, the definition of E, H, and 7.5.4a, we have

$$E(\hat{\mathcal{L}}_T W) = \hat{\mathcal{L}}_T E + k \times E - 2\phi^{-1} \nabla\phi \wedge H$$
$$H(\hat{\mathcal{L}}_T W) = \hat{\mathcal{L}}_T H + k \times H + 2\phi^{-1} \nabla\phi \wedge E.$$

Therefore, we rewrite the homogeneous Bianchi equations expressed relative to E, H (see Proposition 7.2.1) in the form

$$\operatorname{div} E = +k \wedge H \tag{7.7.6a}$$
$$\operatorname{div} H = -k \wedge E \tag{7.7.6b}$$

$$\operatorname{curl} E = H(\hat{\mathcal{L}}_T W) - 3\phi^{-1} \nabla\phi \wedge E - \frac{3}{2} k \times H \tag{7.7.6c}$$

$$\operatorname{curl} H = -E(\hat{\mathcal{L}}_T W) - 3\phi^{-1} \nabla\phi \wedge H + \frac{3}{2} k \times E. \tag{7.7.6d}$$

Now consider the truncation function f introduced on page 66. Thus on a given slice Σ_t we define

$$\mathring{E} = fE \tag{7.7.6e}$$
$$\mathring{H} = fH. \tag{7.7.6f}$$

We form the elliptic Hodge systems

$$\operatorname{div} \mathring{E} = \rho_E = k \wedge \mathring{H} + \nabla f \cdot E \tag{7.7.6g}$$

$$\operatorname{curl} \overset{\circ}{E} = fH(\hat{\mathcal{L}}_T W) + \sigma_E \tag{7.7.6h}$$

$$\sigma_E = -3\phi^{-1}\nabla\phi \wedge \overset{\circ}{E} - \frac{3}{2}k \times \overset{\circ}{H} + \frac{1}{2}\nabla f \wedge E$$

$$\operatorname{div} \overset{\circ}{H} = \rho_H = -k \wedge \overset{\circ}{E} + \nabla f \cdot H \tag{7.7.6i}$$

$$\operatorname{curl} \overset{\circ}{H} = -fE(\hat{\mathcal{L}}_T W) + \sigma_H \tag{7.7.6j}$$

$$\sigma_H = -3\phi^{-1}\nabla\phi \wedge \overset{\circ}{H} + \frac{3}{2}k \times \overset{\circ}{E} + \frac{1}{2}\nabla f \wedge H$$

to which we apply Proposition 4.4.1. Therefore,

$$\begin{aligned}
\int_{\Sigma_t} |\nabla \overset{\circ}{E}|^2 + |\nabla \overset{\circ}{H}|^2 = {} & -3\int_{\Sigma_t} R_{mn}\Big(\overset{\circ}{E}{}^{im}\, \overset{\circ}{E}{}_i^{\,n} + \overset{\circ}{H}{}^{im}\, \overset{\circ}{H}{}_i^{\,n}\Big) \\
& + \frac{1}{2}\int_{\Sigma_t} R\Big(|\overset{\circ}{E}|^2 + |\overset{\circ}{H}|^2\Big) \\
& + \int_{\mathit{l}} (|E(\hat{\mathcal{L}}_T W)|^2 + |H(\hat{\mathcal{L}}_T W)|^2) \\
& + \int_{\Sigma_t} \Big(|\sigma_E|^2 + |\sigma_H|^2 + \frac{1}{2}|\rho_E|^2 + \frac{1}{2}|\rho_H|^2\Big) \\
\leq {} & cr_0^{-6}\int_{\Sigma_t} Q(\hat{\mathcal{L}}_T W)(\bar{K}, \bar{K}, \bar{K}, T) \\
& + c\|\mathrm{Ric}\|_{3/2,\mathit{l}}\Big(\|\overset{\circ}{E}\|_6^2 + \|\overset{\circ}{H}\|_6^2\Big) \\
& + \int_{\Sigma_t} \Big(|\sigma_E|^2 + |\sigma_H|^2 + \frac{1}{2}|\rho_E|^2 + \frac{1}{2}|\rho_H|^2\Big).
\end{aligned}$$

On the other hand,

$$\begin{aligned}
\int_{\Sigma_t} \Big(|\sigma_E|^2 + |\sigma_H|^2 + \frac{1}{2}|\rho_E|^2 + \frac{1}{2}|\rho_H|^2\Big) \leq {} & c(\|\phi^{-1}\nabla\phi\|_{3,\mathit{l}} \\
& + \|k\|_{3,\mathit{l}})\Big(\|\overset{\circ}{E}\|_6^2 + \|\overset{\circ}{H}\|_6^2\Big) \\
& + c\int_{\Sigma^e \cap \mathit{l}} |\nabla f|^2(|E|^2 + |H|^2).
\end{aligned} \tag{7.7.6k}$$

Hence:

$$\begin{aligned}
\int_{\Sigma_t} |\nabla \overset{\circ}{E}|^2 + |\nabla \overset{\circ}{H}|^2 \leq {} & cr_0^{-6}\int_{\Sigma_t} Q(\hat{\mathcal{L}}_T W)(\bar{K}, \bar{K}, \bar{K}, T) \\
& + c\varepsilon_0\Big(\int_{\Sigma_t^e} |\overset{\circ}{E}|^6 + |\overset{\circ}{H}|^6\Big)^{1/3}
\end{aligned}$$

$$+ c \int_{\Sigma_t^e \bigcap I} |\nabla f|^2 (|E|^2 + |H|^2), \quad (7.7.6l)$$

provided that

$$\|\mathrm{Ric}\|_{3/2,I} + \|\phi^{-1}\nabla\phi\|_{3,I} + \|k\|_{3,I} \leq c\varepsilon_0,$$

which holds in view of assumption 7.5.8b and the auxiliary assumption 7.5.10b. Now, in view of the properties of the truncation function f and the formulas 7.3.3e we can write

$$\begin{aligned}
\int_{\Sigma^e \bigcap I} |\nabla f|^2 (|E|^2 + |H|^2) &\leq c r_0^{-2} \int_{\Sigma^e \bigcap I} (|E|^2 + |H|^2) \\
&\leq c r_0^{-6} \int_{\Sigma^e \bigcap I} (r^4 |\alpha|^2 + r^4 |\beta|^2 \\
&\quad + r^2 \tau_-^2 |\underline{\beta}|^2 + \tau_-^4 |\underline{\alpha}|^2) \\
&\quad + c r_0^{-6} \int_{\Sigma^e \bigcap I} r^4 (|\rho - \bar{\rho}|^2 + r^4 |\sigma - \bar{\sigma}|^2) \\
&\quad + c r_0^{-2} \int_{\Sigma^e \bigcap I} (|\bar{\rho}|^2 + |\bar{\sigma}|^2).
\end{aligned}$$

Now, by the estimate 7.7.1b of Proposition 7.7.1,

$$\begin{aligned}
\int_{\Sigma^e \bigcap I} (|\bar{\rho}|^2 + |\bar{\sigma}|^2) &\leq c\varepsilon_0^2 r_0^{-5} \int_{\Sigma} Q(\hat{\mathcal{L}}_O W)(\bar{K}, \bar{K}, T, T) \\
&\quad + c\varepsilon_0^2 r_0^{-1} \int_I |W|^2. \quad (7.7.6m)
\end{aligned}$$

Henceforth, also using the estimate 7.7.1a of Proposition 7.7.1,

$$\begin{aligned}
\int_{\Sigma^e \bigcap I} |\nabla f|^2 (|E|^2 + |H|^2) &\leq c r_0^{-6} \int_{\Sigma} Q(\hat{\mathcal{L}}_O W)(\bar{K}, \bar{K}, T, T) \\
&\quad + c\varepsilon_0^2 r_0^{-3} \int_I |W|^2 \\
&\leq c r_0^{-6} \int_{\Sigma} Q(\hat{\mathcal{L}}_O W)(\bar{K}, \bar{K}, T, T) \\
&\quad + c\varepsilon_0^2 r_0^{-1} \left(\int_I |W|^6 \right)^{1/3}. \quad (7.7.6n)
\end{aligned}$$

Going back to 7.7.6l, we infer that

$$\int_{\Sigma_t} |\nabla \mathring{E}|^2 + |\nabla \mathring{H}|^2 \leq c r_0^{-6} \int_{\Sigma_t} Q(\hat{\mathcal{L}}_T W)(\bar{K}, \bar{K}, \bar{K}, T)$$
$$+ c r_0^{-6} \int_{\Sigma_t} Q(\hat{\mathcal{L}}_O W)(\bar{K}, \bar{K}, T, T)$$
$$+ c\varepsilon_0 \left(\int_{\Sigma_t} |\mathring{E}|^6 + |\mathring{H}|^6 \right)^{1/3}$$
$$+ c\varepsilon_0^2 r_0^{-1} \left(\int_{\Sigma_t^i} |E|^6 + |H|^6 \right)^{1/3}.$$

By the Sobolev inequality of Corollary 3.2.1.2 we have

$$\left(\int_I |E|^6 + |H|^6 \right)^{1/3} \leq \left(\int_{\Sigma_t} |\mathring{E}|^6 + |\mathring{H}|^6 \right)^{1/3} \leq c \int_{\Sigma_t} |\nabla \mathring{E}|^2 + |\nabla \mathring{H}|^2.$$

Therefore, if ε_0 is sufficiently small, we conclude:

$$\int_{\Sigma_t} |\nabla \mathring{E}|^2 + |\nabla \mathring{H}|^2 \leq c r_0^{-6} \int_{\Sigma_t} Q(\hat{\mathcal{L}}_T W)(\bar{K}, \bar{K}, \bar{K}, T)$$
$$+ c r_0^{-6} \int_{\Sigma_t} Q(\hat{\mathcal{L}}_O W)(\bar{K}, \bar{K}, T, T). \quad (7.7.6o)$$

Taking into account 7.7.5, we derive the inequality 7.7.4a. Then 7.7.4b follows since

$$\int_I (|E|^2 + |H|^2) \leq c r_0^2 \left(\int_I |E|^6 + |H|^6 \right)^{1/3}.$$

Next, we prove the following:

Proposition 7.7.3 *In addition to the hypotheses of Proposition 7.7.1 we assume that the space-time also verifies the exterior assumptions 2 and that the angular momentum operators ${}^{(i)}\Omega$ verify also the assumptions 7.5.13c of Proposition 7.5.3.*

Let W be a Weyl tensor verifying the homogeneous Bianchi equations. Let ${}^e\mathcal{R}_2(W)$ be defined as in the previous section and divide it into the following parts:

$${}^e\mathcal{R}_{2,1}{}^2 = \int_{\Sigma_t^e} r^4 \tau_-^4 |\not\nabla^2 \underline{\alpha}|^2 + \int_{\Sigma_t^e} r^6 \tau_-^2 |\not\nabla^2 \underline{\beta}|^2 + \int_{\Sigma_t^e} r^8 |\not\nabla^2 \rho|^2$$
$$+ \int_{\Sigma_t^e} r^8 |\not\nabla^2 \sigma|^2 + \int_{\Sigma_t^e} r^8 |\not\nabla^2 \beta|^2 + \int_{\Sigma_t^e} r^8 |\not\nabla^2 \alpha|$$

$$
\begin{aligned}
{}^e\mathcal{R}_{2,2}{}^2 = & \int_{\Sigma_t^e} \tau_-^2 r^6 |\nabla \underline{\alpha}_4|^2 + \int_{\Sigma_t^e} \tau_-^4 r^4 |\nabla \underline{\beta}_3|^2 + \int_{\Sigma_t^e} r^8 |\nabla \underline{\beta}_4|^2 \\
& + \int_{\Sigma_t^e} \tau_-^2 r^6 |\nabla \rho_3|^2 + \int_{\Sigma_t^e} r^8 |\nabla \rho_4|^2 + \int_{\Sigma_t^e} \tau_-^2 r^6 |\nabla \sigma_3|^2 \\
& + \int_{\Sigma_t^e} r^8 |\nabla \sigma_4|^2 + \int_{\Sigma_t^e} r^8 |\nabla \beta_3|^2 + \int_{\Sigma_t^e} r^8 |\nabla \beta_4|^2 + \int_{\Sigma_t^e} r^8 |\nabla \alpha_3|^2
\end{aligned}
$$

$$
{}^e\mathcal{R}_{2,3}{}^2 = \int_{\Sigma_t^e} (r^2 \tau_-^6 |\nabla \underline{\alpha}_3|^2 + r^8 |\nabla \alpha_4|^2)
$$

$$
\begin{aligned}
{}^e\mathcal{R}_{2,4}{}^2 = & \int_{\Sigma_t^e} \tau_-^4 r^4 |\underline{\alpha}_{34}|^2 + \int_{\Sigma_t^e} \tau_-^4 r^4 |\underline{\alpha}_{43}|^2 + \int_{\Sigma_t^e} r^6 \tau_-^2 |\underline{\alpha}_{44}|^2 \\
& + \int_{\Sigma_t^e} \tau_-^6 r^2 |\underline{\beta}_{33}|^2 + \int_{\Sigma_t^e} \tau_-^2 r^6 |\underline{\beta}_{34}|^2 + \int_{\Sigma_t^e} \tau_-^2 r^6 |\underline{\beta}_{43}|^2 + \int_{\Sigma_t^e} r^8 |\underline{\beta}_{44}|^2 \\
& + \int_{\Sigma_t^e} \tau_-^4 r^4 |\rho_{33}|^2 + \int_{\Sigma_t^e} \tau_- r^7 |\rho_{34}|^2 + \int_{\Sigma_t^e} \tau_- r^7 |\rho_{43}|^2 + \int_{\Sigma_t^e} r^8 |\rho_{44}|^2 \\
& + \int_{\Sigma_t^e} \tau_-^4 r^4 |\sigma_{33}|^2 + \int_{\Sigma} \tau_- r^7 |\sigma_{34}|^2 + \int_{\Sigma_t^e} \tau_- r^7 |\sigma_{43}|^2 + \int_{\Sigma_t^e} r^8 |\sigma_{44}|^2 \\
& + \int_{\Sigma_t^e} \tau_-^2 r^6 |\beta_{33}|^2 + \int_{\Sigma} r^8 |\beta_{34}|^2 + \int_{\Sigma_t^e} r^8 |\beta_{43}|^2 + \int_{\Sigma} r^8 |\beta_{44}|^2 \\
& + \int_{\Sigma_t^e} r^8 |\alpha_{33}|^2 + \int_{\Sigma} r^8 |\alpha_{34}|^2 + \int_{\Sigma_t^e} r^8 |\alpha_{43}|^2
\end{aligned}
$$

$$
{}^e\mathcal{R}_{2,5}{}^2 = \int_{\Sigma_t^e} \tau_-^8 |\underline{\alpha}_{33}|^2 + \int_{\Sigma_t^e} r^8 |\alpha_{44}|^2 .
$$

Under the assumption just mentioned we claim that there exists a constant c such that

$$
{}^e\mathcal{R}_{2,1}{}^2 \leq c \int_{\Sigma} Q(\hat{\mathcal{L}}_O^2 W)(\bar{K}, \bar{K}, T, T) + c\mathcal{Q}_1 \tag{7.7.7a}
$$

$$
{}^e\mathcal{R}_{2,2}{}^2 \leq c \int_{\Sigma} Q(\hat{\mathcal{L}}_O^2 W)(\bar{K}, \bar{K}, T, T) + c\mathcal{Q}_1 \tag{7.7.7b}
$$

while

$$
{}^e\mathcal{R}_{2,3}{}^2 \leq c \int_{\Sigma} Q(\hat{\mathcal{L}}_O \hat{\mathcal{L}}_T W)(\bar{K}, \bar{K}, \bar{K}, T) \tag{7.7.7c}
$$

$$+c\int_\Sigma Q(\hat{\mathcal{L}}_O^2 W)(\bar{K},\bar{K},T,T)$$
$$+c\mathcal{Q}_1. \tag{7.7.7d}$$

Also,

$${}^e\mathcal{R}_{2,4}{}^2 \leq c\int_\Sigma Q(\hat{\mathcal{L}}_O^2 W)(\bar{K},\bar{K},T,T) \tag{7.7.7e}$$
$$+c\int_\Sigma Q(\hat{\mathcal{L}}_O\hat{\mathcal{L}}_T W)(\bar{K},\bar{K},\bar{K},T)+c\mathcal{Q}_1.$$

Finally,

$${}^e\mathcal{R}_{2,5}{}^2 \leq c\int_\Sigma Q(\hat{\mathcal{L}}_S\hat{\mathcal{L}}_T W)(\bar{K},\bar{K},T,T)+c\int_\Sigma Q(\hat{\mathcal{L}}_T^2 W)(\bar{K},\bar{K},T,T)$$
$$+c\int_\Sigma Q(\hat{\mathcal{L}}_O^2 W)(\bar{K},\bar{K},T,T)+c\mathcal{Q}_1. \tag{7.7.7f}$$

Proof 7.7.2 *The inequality 7.7.7a is an immediate consequence of Propositions 7.5.3, 7.3.1, and 7.7.1, making use of the assumptions 7.5.13b and 7.5.13c of 7.5.3. Indeed, in view of Lemma 7.3.1 and the definition of $\bar{K}$, we have*

$$Q(\hat{\mathcal{L}}_O^2 W)(\bar{K},\bar{K},T,T) \geq \frac{1}{8}\Big(\tau_-^4|\underline{\alpha}(\hat{\mathcal{L}}_O^2 W)|^2+\tau_-^2\tau_+^2|\underline{\beta}(\hat{\mathcal{L}}_O^2 W)|^2$$
$$+\tau_+^4|\rho(\hat{\mathcal{L}}_O^2 W)|^2+\tau_+^4|\sigma(\hat{\mathcal{L}}_O^2 W)|^2$$
$$+\tau_+^4|\beta(\hat{\mathcal{L}}_O^2 W)|^2+\tau_+^4|\alpha(\hat{\mathcal{L}}_O^2 W)|^2\Big).$$

Consequently, by virtue of the commutation formulas of Proposition 7.3.1, the assumptions 7.5.13b of Proposition 7.5.3, and the first step in the proof of Proposition 7.7.1, we infer that

$$\int_{\Sigma_t^e}\tau_-^4|\hat{\mathcal{L}}_O\underline{\alpha}(\hat{\mathcal{L}}_O W)|^2+\int_{\Sigma_t^e}\tau_-^2\tau_+^2|\mathcal{L}_O\underline{\beta}(\hat{\mathcal{L}}_O W)|^2+\int_{\Sigma_t^e}\tau_+^4|\mathcal{L}_O\rho(\hat{\mathcal{L}}_O W)|^2$$
$$+\int_{\Sigma_t^e}\tau_+^4|\mathcal{L}_O\sigma(\hat{\mathcal{L}}_O W)|^2+\int_{\Sigma_t^e}\tau_+^4|\mathcal{L}_O\beta(\hat{\mathcal{L}}_O W)|^2+\int_{\Sigma_t^e}\tau_+^4|\hat{\mathcal{L}}_O\alpha(\hat{\mathcal{L}}_O W)|^2$$
$$\leq c\left(\int_\Sigma Q(\hat{\mathcal{L}}_O^2 W)(\bar{K},\bar{K},T,T)+\varepsilon_0\int_\Sigma Q(\hat{\mathcal{L}}_O W)(\bar{K},\bar{K},T,T)\right).$$

On the other hand, referring once more to the commutation formulas of Proposition 7.3.1 we write

$$\hat{\mathcal{L}}_O\alpha(\hat{\mathcal{L}}_O W)_{AB}=\hat{\mathcal{L}}_O^2\alpha_{AB}+\hat{\mathcal{L}}_O\left(\left(-2^{(O)}M-\frac{1}{8}\mathrm{tr}^{(O)}\pi\right)\alpha_{AB}\right)$$

$$\not{\mathcal{L}}_{\mathbf{O}}\beta(\hat{\mathcal{L}}_{\mathbf{O}}W)_A = \not{\mathcal{L}}^2_{\mathbf{O}}\beta_A - \not{\mathcal{L}}_{\mathbf{O}}\left(\frac{1}{2}{}^{(\mathbf{O})}\hat{\pi}_{AB}\beta_B + \left({}^{(\mathbf{O})}M + \frac{1}{8}\mathrm{tr}{}^{(\mathbf{O})}\pi\right)\beta_A + \frac{1}{4}{}^{(\mathbf{O})}\underline{P}_B\alpha_{AB}\right)$$

$$\not{\mathcal{L}}_{\mathbf{O}}\rho(\hat{\mathcal{L}}_{\mathbf{O}}W) = \not{\mathcal{L}}^2_{\mathbf{O}}\rho + \not{\mathcal{L}}_{\mathbf{O}}\left(-\frac{1}{8}\mathrm{tr}{}^{(\mathbf{O})}\pi\rho - \frac{1}{2}{}^{(\mathbf{O})}\underline{P}_A\beta_A\right)$$

and similarly for all other null components of $\hat{\mathcal{L}}_{\mathbf{O}}W$. Henceforth, in view of 7.5.13b,

$$\int_{\Sigma_t^e} \tau_+^4|\hat{\not{\mathcal{L}}}_{\mathbf{O}}\alpha(\hat{\mathcal{L}}_{\mathbf{O}}W)|^2 \geq \int_{\Sigma_t^e} \tau_+^4|\hat{\not{\mathcal{L}}}^2_{\mathbf{O}}\alpha|^2 - c\varepsilon_0\int_{\Sigma_t^e} \tau_+^4|\hat{\not{\mathcal{L}}}_{\mathbf{O}}\alpha|^2 - c\int_{\Sigma_t^e} \tau_+^4|\alpha|^2 r^2(|\nabla\!\!\!/{}^{(\mathbf{O})}M|^2 + |\nabla\!\!\!/\mathrm{tr}{}^{(\mathbf{O})}\pi|^2).$$

The last term can be estimated as follows:

$$\int_{\Sigma_t^e} \tau_+^4|\alpha|^2 r^2(|\nabla\!\!\!/{}^{(\mathbf{O})}M|^2 + |\nabla\!\!\!/\mathrm{tr}{}^{(\mathbf{O})}\pi|^2 \leq c\int_{\frac{r_0}{2}}^{\infty} \tau_+^4 r^2 dr\left(\int_{S_{r'}}|\alpha|^4\right)^{1/2} \cdot\left(\int_{S_{r'}}|\nabla\!\!\!/{}^{(\mathbf{O})}M|^4 + |\nabla\!\!\!/\mathrm{tr}{}^{(\mathbf{O})}\pi|^4\right)^{1/2}$$
$$\leq cr_0^{-1}|||r^3\alpha|||^2_{4,e}(|||r\nabla\!\!\!/{}^{(\mathbf{O})}M|||^2_{4,e} + |||r\nabla\!\!\!/\mathrm{tr}{}^{(\mathbf{O})}\pi|||^2_{4,e}).$$

Now, in view of the corollary to the nondegenerate form of the global Sobolev inequality of Corollary 3.2.1.1,

$$|||r^3\alpha|||^2_{4,e} \leq c\int_{\Sigma_t^e} r^4\left(|\alpha|^2 + r^2|\nabla\!\!\!/\alpha|^2 + r^2|\nabla\!\!\!/_N\alpha|^2\right)$$
$$\leq \int_{\Sigma_t^e} r^4\left(|\alpha|^2 + r^2|\nabla\!\!\!/\alpha|^2 + r^2|\mathbf{D}\!\!\!\!/_3\alpha|^2 + r^2|\mathbf{D}\!\!\!\!/_4\alpha|^2\right).$$

Therefore, using 7.5.13c of Proposition 7.5.3 as well as the results of the previous two propositions, we find

$$\int_{\Sigma_t^e} \tau_+^4|\alpha|^2 r^2(|\nabla\!\!\!/{}^{(\mathbf{O})}M|^2 + |\nabla\!\!\!/\mathrm{tr}{}^{(\mathbf{O})}\pi|^2) \leq cr_0^{-1}\mathcal{Q}_1.$$

Proceeding in the same fashion with all the other terms, we infer that

$$\int_{\Sigma_t^e} \tau_-^4|\hat{\not{\mathcal{L}}}^2_{\mathbf{O}}\underline{\alpha}|^2 + \int_{\Sigma_t^e} \tau_-^2\tau_+^2|\not{\mathcal{L}}^2_{\mathbf{O}}\underline{\beta}|^2 + \int_{\Sigma_t^e} \tau_+^4|\not{\mathcal{L}}^2_{\mathbf{O}}\rho|^2$$

$$+\int_{\Sigma_t^e} \tau_+^4 |\hat{\not{\mathcal{L}}}_{\mathbf{O}}^2 \sigma|^2 + \int_{\Sigma_t^e} \tau_+^4 |\hat{\not{\mathcal{L}}}_{\mathbf{O}}^2 \beta|^2 + \int_{\Sigma_t^e} \tau_+^4 |\hat{\not{\mathcal{L}}}_{\mathbf{O}}^2 \alpha|^2$$

$$\leq c \int_\Sigma Q(\hat{\mathcal{L}}_{\mathbf{O}}^2 W)(\bar{K}, \bar{K}, T, T) + c\mathcal{Q}_1 .$$

Now, in view of Proposition 7.5.3,

$$\int_{\Sigma_t^e} \tau_-^4 |\hat{\not{\mathcal{L}}}_{\mathbf{O}}^2 \underline{\alpha}|^2 \geq c \int_{\Sigma_t^e} r^2 \tau_-^4 |\nabla\!\!\!/ \hat{\not{\mathcal{L}}}_{\mathbf{O}} \underline{\alpha}|^2 ,$$

and using the 2-dimensional analogue of Lemma 7.1.3 to commute $\nabla\!\!\!/$ *with* $\hat{\not{\mathcal{L}}}_{\mathbf{O}}$, *we find*

$$\int_{\Sigma_t^e} \tau_-^4 |\nabla\!\!\!/ \hat{\not{\mathcal{L}}}_{\mathbf{O}} \underline{\alpha}|^2 \geq \int_{\Sigma_t^e} \tau_-^4 |\hat{\not{\mathcal{L}}}_{\mathbf{O}} \nabla\!\!\!/ \alpha|^2 - c \int_{\Sigma_t^e} |{}^{(\mathbf{O})}\Gamma|^2 |\alpha|^2$$

where ${}^{(\mathbf{O})}\Gamma_{ABC} = \frac{1}{2}(\nabla\!\!\!/_B {}^{(\mathbf{O})}\pi_{AC} + \nabla\!\!\!/_A {}^{(\mathbf{O})}\pi_{BC} - \nabla\!\!\!/_C {}^{(\mathbf{O})}\pi_{AB})$. *Therefore, using once more the Property 0 of Proposition 7.5.3 and estimating the error terms as before, we deduce that*

$$\int_{\Sigma_t^e} \tau_-^4 r^4 |\nabla\!\!\!/^2 \underline{\alpha}|^2 \leq c \int_\Sigma Q(\hat{\mathcal{L}}_{\mathbf{O}}^2 W)(\bar{K}, \bar{K}, T, T) + c\mathcal{Q}_1 .$$

Proceeding in precisely the same manner with all the other terms yields the estimate 7.7.7a.

The estimate 7.7.7b can now be derived by taking the angular derivatives of 7.3.11a–7.3.11j and using the assumptions 7.5.8a and 7.5.9a on the Ricci coefficients of the frame.

To prove 7.7.7c, we first write, with the help of Lemma 7.3.1,

$$\int_\Sigma Q(\hat{\mathcal{L}}_{\mathbf{O}} \hat{\mathcal{L}}_T W)(\bar{K}, \bar{K}, \bar{K}, T) \geq \int_{\Sigma_t^e} (\tau_-^6 |\underline{\alpha}(\hat{\mathcal{L}}_{\mathbf{O}} \hat{\mathcal{L}}_T W)|^2 + r^6 |\alpha(\hat{\mathcal{L}}_{\mathbf{O}} \hat{\mathcal{L}}_T W)|^2).$$

According to the commutation formulas of Proposition 7.3.1, the assumptions 7.5.13b on the Lie coefficients of the rotation vectorfields, together with the L^2 estimates for the null decomposition of $\hat{\mathcal{L}}_T W$, we have

$$\int_{\Sigma_t^e} \tau_-^6 |\underline{\alpha}(\hat{\mathcal{L}}_{\mathbf{O}} \hat{\mathcal{L}}_T W)|^2 \geq \int_{\Sigma_t^e} \tau_-^6 |\hat{\not{\mathcal{L}}}_{\mathbf{O}} \underline{\alpha}(\hat{\mathcal{L}}_T W)|^2$$

$$- c\varepsilon_0 \int_\Sigma Q(\hat{\mathcal{L}}_T W)(\bar{K}, \bar{K}, \bar{K}, T).$$

Now,

$$\hat{\mathcal{L}}_O \underline{\alpha}(\hat{\mathcal{L}}_T W)_{AB} = \hat{\mathcal{L}}_O \Big(\hat{\mathcal{L}}_T \underline{\alpha}_{AB} - 2\,{}^{(T)}\underline{M}\alpha_{AB} + ({}^{(T)}\underline{P}_A + {}^{(T)}\underline{Q}_A)\underline{\beta}_B + ({}^{(T)}\underline{P}_B + {}^{(T)}\underline{Q}_B)\underline{\beta}_A - \delta_{AB}({}^{(T)}\underline{P} + {}^{(T)}\underline{Q}) \cdot \underline{\beta} \Big).$$

Hence,

$$\begin{aligned} \int_{\Sigma_t^e} \tau_-^6 |\hat{\mathcal{L}}_O \hat{\mathcal{L}}_T \underline{\alpha}|^2 \le & \int_\Sigma Q(\hat{\mathcal{L}}_O \hat{\mathcal{L}}_T W)(\bar{K}, \bar{K}, \bar{K}, T) \\ & + c\varepsilon_0 \int_\Sigma Q(\hat{\mathcal{L}}_T W)(\bar{K}, \bar{K}, \bar{K}, T) \\ & + c \int_{\Sigma_t^e} \tau_-^6 |\hat{\mathcal{L}}_O {}^{(T)}\underline{M}|^2 |\underline{\alpha}|^2 + c \int_{\Sigma_t^e} \tau_-^6 |{}^{(T)}\underline{M}|^2 |\hat{\mathcal{L}}_O \underline{\alpha}|^2 \\ & + c \int_{\Sigma_t^e} \tau_-^6 |\hat{\mathcal{L}}_O ({}^{(T)}\underline{P}, {}^{(T)}\underline{Q})|^2 |\underline{\beta}|^2 \\ & + c \int_{\Sigma_t^e} \tau_-^6 |({}^{(T)}\underline{P}, {}^{(T)}\underline{Q})|^2 |\hat{\mathcal{L}}_O \underline{\beta}|^2. \end{aligned}$$

The terms on the right-hand side of this inequality, which involve the Lie derivatives of $\underline{\alpha}$, $\underline{\beta}$, can be easily estimated by appealing to the estimates 7.5.11a of Proposition 7.5.2 for the Lie coefficients of T together with the estimates for $\hat{\mathcal{L}}_O \underline{\alpha}$, $\hat{\mathcal{L}}_O \underline{\beta}$ of Proposition 7.7.1. The other terms can be estimated as follows:

$$\int_{\Sigma_t^e} \tau_-^6 |\hat{\mathcal{L}}_O {}^{(T)}\underline{M}|^2 |\underline{\alpha}|^2 \le c r_0^{-1} |||r \hat{\mathcal{L}}_O {}^{(T)}\underline{M}|||_{4,e}^2 |||r^{1/2} \tau_-^{5/2} \underline{\alpha}|||_{4,e}^2.$$

Now, as a consequence of the degenerate form of the global Sobolev inequality as it appears in Corollary 3.2.1.1, we have

$$|||r^{1/2} \tau_-^{5/2} \underline{\alpha}|||_{4,e}^2 \le \int_{\Sigma_t^e} \tau_-^4 (|\underline{\alpha}|^2 + r^2 |\nabla\!\!\!/ \underline{\alpha}|^2 + \tau_-^2 |\nabla\!\!\!/_N \underline{\alpha}|^2).$$

Moreover, since in view of 7.5.11b, (see Proposition 7.5.2 and Property 0 of Proposition 7.5.3) we have $|||r \hat{\mathcal{L}}_O {}^{(T)}\underline{M}|||_{4,e} \le c$, hence, according to the estimates of Proposition 7.7.1,

$$\int_{\Sigma_t^e} \tau_-^6 |\hat{\mathcal{L}}_O {}^{(T)}\underline{M}|^2 |\underline{\alpha}|^2 \le c\varepsilon_0 r_0^{-2} \int_\Sigma Q(\hat{\mathcal{L}}_O W)(\bar{K}, \bar{K}, T, T).$$

Therefore, proceeding in precisely the same manner for all other terms, we infer that

$$\int_{\Sigma_t^e} \tau_-^6 |\hat{\not{\mathcal{L}}}_O \hat{\not{\mathcal{L}}}_T \underline{\alpha}|^2 \leq \int_\Sigma Q(\hat{\mathcal{L}}_O \hat{\mathcal{L}}_T W)(\bar{K}, \bar{K}, \bar{K}, T) + c\mathcal{Q}_1.$$

Finally, recalling that $\hat{\not{\mathcal{L}}}_T \underline{\alpha} = \not{\mathbf{D}}_T \underline{\alpha} + \delta \underline{\alpha}$ and using once more the estimate of Proposition 7.5.3, we conclude that, $\int_{\Sigma_t^e} \tau_-^6 r^2 |\nabla \not{\mathbf{D}}_T \underline{\alpha}|^2$ can be also estimated in terms of the same right-hand side as in the previous inequality. Since $T = \frac{1}{2}(e_4 + e_3)$, and $\nabla \underline{\alpha}_4$ has already been estimated in 7.7.7b, we obtain the desired estimate for $\nabla \underline{\alpha}_3$. The estimate for $\nabla \alpha_4$ can then be derived in a similar manner.

The estimates 7.7.7e then follow from formulas 7.3.12a–7.3.12u and from Proposition 7.7.1 with the help of assumptions 7.5.9b, 7.5.9c, and 7.5.9d. As a typical term we consider $\underline{\alpha}_{44}$. According to formula 7.3.12t we have

$$\begin{aligned} \int_{\Sigma_t^e} r^6 \tau_-^2 |\underline{\alpha}_{44}|^2 \leq c \int_{\Sigma_t^e} r^6 \tau_-^2 |\nabla \underline{\beta}_4|^2 \\ + c \int_{\Sigma_t^e} r^6 \tau_-^2 (|\hat{\chi}|^2 |\nabla \underline{\beta}|^2 + |\nabla \mathrm{tr}\chi|^2 |\underline{\beta}|^2) \\ + c \int_{\Sigma_t^e} r^6 \tau_-^2 \left(|\not{\mathbf{D}}_4 E_4(\underline{\alpha}) + \frac{3}{2} \mathrm{tr}\chi E_4(\underline{\alpha})|^2 + |\hat{s}(F_4(\underline{\beta}))|^2 \right). \end{aligned} \tag{7.7.8}$$

The first term on the right-hand side has already been estimated in 7.7.7b. In fact we have the stronger estimate, with $r^6 \tau_-^2$ replaced by r^8. The second integral can easily be estimated with the help of the estimates of Proposition 7.7.1 and assumptions 7.5.8a and 7.5.9a. Once again the estimate allows the weight r^8 instead of $r^6 \tau_-^2$. Now consider the integral of $|\not{\mathbf{D}}_4 E_4(\underline{\alpha}) + \frac{3}{2} \mathrm{tr}\chi E_4(\underline{\alpha})|^2$. By virtue of formula 7.3.11j, expressed relative to the standard frame according to Proposition 7.5.1,

$$E_4(\underline{\alpha}) = 2\nu \underline{\alpha} - 3(\hat{\underline{\chi}} \rho - {}^*\hat{\underline{\chi}} \sigma) + (\epsilon - 4\underline{\zeta}) \widehat{\otimes} \underline{\beta}.$$

By far the worst contribution is due to $\nu \underline{\alpha}$. In fact all other terms can be estimated even with the weight r^8. We thus consider the integrals $\int_{\Sigma_t^e} r^6 \tau_-^2 |\underline{\alpha} \mathbf{D}_4 \nu|^2$, $\int_{\Sigma_t^e} r^6 \tau_-^2 |\not{\mathbf{D}}_4 \underline{\alpha} \nu|^2$, and $\int_{\Sigma_t^e} r^6 \tau_-^2 |\nu \mathrm{tr}\chi \underline{\alpha}|^2$. To estimate the first integral, we write, using the Sobolev inequality on the 2-surfaces $S_{t,u}$,

$$\begin{aligned} \int_{\Sigma_t^e} r^6 \tau_-^2 |\not{\mathbf{D}}_4 \nu \underline{\alpha}|^2 &\leq c |||r^{3/2} \mathbf{D}_4 \nu|||_{4,e}^2 \int_{\frac{r_0}{2}}^\infty r^2 \tau_-^2 dr \left(\int_{S_r} r^2 |\underline{\alpha}|^4 \right)^{1/2} \\ &\leq c |||r^{3/2} \mathbf{D}_4 \nu|||_{4,e}^2 \int_{\frac{r_0}{2}}^\infty r^2 \tau_-^2 dr \left(\int_{S_r} |\underline{\alpha}|^2 + r^2 |\nabla \underline{\alpha}|^2 \right) \end{aligned}$$

$$\leq c|||r^{3/2}\mathbf{D}_4\nu|||_{4,e}^2 \int_{\Sigma_t^e} \tau_-^2(r_0^2+\tau_-^2)(|\underline{\alpha}|^2 + r^2|\nabla\!\!\!/\,\underline{\alpha}|^2)$$

$$\leq cr_0^2|||r^{3/2}\mathbf{D}_4\nu|||_{4,e}^2 \int_{\Sigma_t^e} \tau_-^2(|\underline{\alpha}|^2 + r^2|\nabla\!\!\!/\,\underline{\alpha}|^2)$$

$$+ |||r^{3/2}\mathbf{D}_4\nu|||_{4,e}^2 \int_{\Sigma_t^e} \tau_-^4(|\underline{\alpha}|^2 + r^2|\nabla\!\!\!/\,\underline{\alpha}|^2).$$

Hence, by virtue of assumption 7.5.9f and the results of the previous propositions,

$$\int_{\Sigma_t^e} r^6\tau_-^2|\mathbf{D}\!\!\!\!/\,_4\nu\underline{\alpha}|^2 \leq c\varepsilon_0\mathcal{Q}_1.$$

The second integral is easier to estimate. To estimate the last term in 7.7.8, we appeal to the corresponding commutation formula of Lemma 7.3.3. Therefore, making use of the assumptions 7.5.8a, we write, with the weight r^8 replacing $r^6\tau_-^2$,

$$\int_{\Sigma_t^e} r^8|\hat{s}(F_4(\underline{\beta}))|^2 \leq c\int_{\Sigma_t^e} r^4|\mathbf{D}\!\!\!\!/\,_4\underline{\beta}|^2 + c\int_{\Sigma_t^e} r^2|\underline{\beta}|^2$$

$$+c\int_{\Sigma_t^e} r^8|\beta(\mathbf{R})|^2|\underline{\beta}|^2.$$

The term $\int_{\Sigma_t^e} r^8|\beta(\mathbf{R})|^2|\underline{\beta}|^2$ can be estimated with the help of the auxiliary curvature assumptions 7.5.10a and the L^4 form of the Sobolev inequality on $S_{t,u}$:

$$\int_{\Sigma_t^e} r^8|\beta(\mathbf{R})|^2|\underline{\beta}|^2 \leq c|||r^{5/2}\beta(\mathbf{R})|||_{4,e}^2 \int_{\Sigma_t^e}(r^2|\underline{\beta}|^2 + r^4|\nabla\!\!\!/\,\underline{\beta}|^2).$$

We thus conclude that $\int_{\Sigma_t^e} r^6\tau_-^2|\underline{\alpha}_{44}|^2$ can be estimated by the terms appearing on the right-hand side of 7.7.7e as desired. Moreover, we make the remark that if instead of the standard pair $e_4 = T+N, e_3 = T-N$ we use the l-pair $e_4 = l^\mu = -\mathbf{g}^{\mu\nu}\frac{\partial u}{\partial x^\nu} = a^{-1}(T+N)$, $e_3 = \underline{l} = a(T-N)$, the corresponding term $\underline{\alpha}_{44}$ can be estimated with the stronger weight r^8 replacing $r^6\tau_-^2$. This is due to the fact that the term $4\Omega\underline{\alpha}$ in 7.3.11j, which is equal to $2\nu\underline{\alpha}$ in the standard frame, is in fact zero in the l-frame.

All the other terms on the left-hand side of 7.7.7e can be treated in the same manner.

Remark 7 *Before continuing we do remark, however, that* ρ_{34}, ρ_{43}, σ_{34}, *and* σ_{43} *behave worse than expected.*[22] *Indeed, the components of a Weyl-Bianchi*

[22] The remark remains true even in the l-frame.

tensor in Minkowski space allow the weight r^8 in the corresponding expression instead of $r^7\tau_-$ as in 7.7.7e. This is due, in the case of ρ_{43}, to the "nonlinear" error terms $\mathbf{D}_3 E_4(\rho)$ in formula 7.3.12h. Indeed, according to 7.3.11d,

$$\mathbf{D}_3 E_4(\rho) = -\frac{1}{2}\not{\mathbf{D}}_3\hat{\underline{\chi}} \cdot \alpha + l.o.t.,$$

and, in view of the assumptions 7.5.9c, the integral $\int_{\Sigma_t^e} r^8 |\mathbf{D}_3\hat{\underline{\chi}}|^2|\alpha|^2$ diverges. Instead we have

$$\begin{aligned}\int_{\Sigma_t^e} r^7\tau_- |\mathbf{D}_3\hat{\underline{\chi}}|^2|\alpha|^2 &\leq c|||r^{1/2}\tau_- \not{\mathbf{D}}_3\hat{\underline{\chi}}|||_{4,e}^2 |||r^3\alpha|||_{4,e}^2 \\ &\leq c\epsilon_0^2 \mathcal{Q}_1.\end{aligned}$$

The other terms ρ_{34}, σ_{34}, and σ_{43} are estimated in precisely the same manner.

Finally, to prove 7.7.7f we proceed as follows:
First, according to Lemma 7.3.1, we have

$$\frac{1}{8}\left(\int_\Sigma \tau_-^6 |\underline{\alpha}(\hat{\mathcal{L}}_S\hat{\mathcal{L}}_T W)|^2 + \int_\Sigma \tau_+^6 |\alpha(\hat{\mathcal{L}}_S\hat{\mathcal{L}}_T W)|^2\right) \leq \int_\Sigma Q(\hat{\mathcal{L}}_S\hat{\mathcal{L}}_T W)(\bar{K},\bar{K},\bar{K},T).$$

Now, in view of Proposition 7.3.1,

$$\begin{aligned}\int_{\Sigma_t^e} \tau_-^6 |\underline{\alpha}(\hat{\mathcal{L}}_S\hat{\mathcal{L}}_T W)|^2 &\geq \int_{\Sigma_t^e} \tau_-^6 |\not{\mathcal{L}}_S\underline{\alpha}(\hat{\mathcal{L}}_T W)|^2 \\ &\quad - c\int_{\Sigma_t^e} \tau_-^6 (|\underline{\alpha}(\hat{\mathcal{L}}_T W)|^2 + |\underline{\beta}(\hat{\mathcal{L}}_T W)|^2),\end{aligned}$$

provided that

$$\sup_{\Sigma_t^e} |({}^{(S)}\underline{M}, {}^{(S)}\underline{P}, {}^{(S)}\underline{Q})| \leq c$$

$$\sup_{\Sigma_t^e} |\mathrm{tr}\,{}^{(S)}\pi| \leq c,$$

which are satisfied in view of formulas 7.5.11e of Proposition 7.5.2.
On the other hand, since

$$\int_{\Sigma_t^e} \tau_-^6 (|\underline{\alpha}(\hat{\mathcal{L}}_T W)|^2 + |\underline{\beta}(\hat{\mathcal{L}}_T W)|^2) \leq \int_\Sigma Q(\hat{\mathcal{L}}_T W)(\bar{K},\bar{K},\bar{K},T),$$

we infer that

$$\int_{\Sigma_t^e} \tau_-^6 |\hat{\not{\mathcal{L}}}_S \underline{\alpha}(\hat{\mathcal{L}}_T W)|^2 \le c \int_{\Sigma_t^e} \tau_-^6 |\underline{\alpha}(\hat{\mathcal{L}}_S \hat{\mathcal{L}}_T W)|^2 + c\varepsilon_0 \int_\Sigma Q(\hat{\mathcal{L}}_T W)(\bar{K}, \bar{K}, \bar{K}, T).$$

Now, in view of the commutation formulas of Proposition 7.3.1, taking into account that $\mathrm{tr}^{(T)}\pi = 0$,

$$\hat{\not{\mathcal{L}}}_S \underline{\alpha}(\hat{\mathcal{L}}_T W) = \hat{\not{\mathcal{L}}}_S \left(\hat{\not{\mathcal{L}}}_T \underline{\alpha} - 2^{(T)}\underline{M}\underline{\alpha} + ({}^{(T)}\underline{P} + {}^{(T)}\underline{Q})\widehat{\otimes}\underline{\beta} \right),$$

and hence, with the help of 7.5.11e again,

$$\int_{\Sigma_t^e} \tau_-^6 |\hat{\not{\mathcal{L}}}_S \hat{\not{\mathcal{L}}}_T \underline{\alpha}|^2 \le \int_{\Sigma_t^e} \tau_-^6 |\hat{\not{\mathcal{L}}}_S \underline{\alpha}(\hat{\mathcal{L}}_T W)|^2 + c \int_{\Sigma_t^e} \tau_-^4 (|\hat{\not{\mathcal{L}}}_S \underline{\alpha}|^2 + |\hat{\not{\mathcal{L}}}_S \underline{\beta}|^2) + c r_0^{-1} |||r(\not{\mathcal{L}}_S^{(T)}\underline{M}, \not{\mathcal{L}}_S^{(T)}\underline{P}, \not{\mathcal{L}}_S^{(T)}\underline{Q})|||_{4,e}^2 |||r^{1/2}\tau_-^{5/2}(\underline{\alpha}, \underline{\beta})|||_{4,e}^2.$$

Thus, taking into account 7.5.11c and in view of Corollary 3.2.1.1 to the degenerate form of the global Sobolev inequalities, we obtain

$$\int_{\Sigma_t^e} \tau_-^6 |\hat{\not{\mathcal{L}}}_S \hat{\not{\mathcal{L}}}_T \underline{\alpha}|^2 \le c \int_\Sigma Q(\hat{\mathcal{L}}_S \hat{\mathcal{L}}_T W)(\bar{K}, \bar{K}, \bar{K}, T) + c\mathcal{Q}_1.$$

Now given an arbitrary S-tangent 2-tensor w,

$$\begin{aligned} \not{\mathcal{L}}_S w_{AB} &= \not{\mathbf{D}}_S w_{AB} - \Pi_A^{\ \rho}\Pi_B^{\ \sigma}(\mathbf{D}_\rho S^\mu w_{\mu\sigma} + \mathbf{D}_\sigma S^\mu w_{\rho\mu}) \\ &= \not{\mathbf{D}}_S w_{AB} - \frac{1}{2}(v\chi_{AC} - u\underline{\chi}_{AC})w_{CB} - \frac{1}{2}(v\chi_{BC} - u\underline{\chi}_{BC})w_{AC}. \end{aligned}$$

Hence, if w is symmetric traceless, taking into account identity 7.3.5b, we infer that

$$\hat{\not{\mathcal{L}}}_S w_{AB} = \not{\mathbf{D}}_S w - \frac{1}{2}(v\mathrm{tr}\chi - u\mathrm{tr}\underline{\chi})w.$$

Also, from a previous calculation, $\hat{\not{\mathcal{L}}}_T \underline{\alpha} = \not{\mathbf{D}}_T \underline{\alpha} + \delta\underline{\alpha}$. Therefore we deduce that

$$\int_{\Sigma_t^e} \tau_-^6 |\not{\mathbf{D}}_S \not{\mathbf{D}}_T \underline{\alpha}|^2 \le c \int_\Sigma Q(\hat{\mathcal{L}}_S \hat{\mathcal{L}}_T W)(\bar{K}, \bar{K}, \bar{K}, T) + c\mathcal{Q}_1$$

and consequently

$$\int_{\Sigma_t^e} \tau_-^6 u^2 |\underline{\alpha}_{33}|^2 \leq c\mathcal{Q}_2 + c\mathcal{Q}_1.$$

On the other hand, making use of the integral $\int_\Sigma Q(\hat{\mathcal{L}}_T^2 W)(\bar{K}, \bar{K}, T, T)$, we can also estimate $\int_{\Sigma_t^e} \tau_-^6 |\underline{\alpha}_{33}|^2$. Thus,

$$\int_{\Sigma_t^e} \tau_-^8 |\underline{\alpha}_{33}|^2 \leq c(\mathcal{Q}_2 + \mathcal{Q}_1).$$

Proceeding in precisely the same manner for $\alpha(\hat{\mathcal{L}}_S \hat{\mathcal{L}}_T W)$, we find

$$\int_{\Sigma_t^e} \tau_+^6 |\mathbf{D}\!\!\!/_S \mathbf{D}\!\!\!/_T \alpha|^2 \leq c(\mathcal{Q}_2 + \mathcal{Q}_1)$$

and hence

$$\int_{\Sigma_t^e} \tau_+^6 v^2 |\mathbf{D}\!\!\!/_4^2 \alpha|^2 \leq c(\mathcal{Q}_2 + \mathcal{Q}_1),$$

which ends the proof of the proposition.

Proposition 7.7.4 *We assume that in addition to the assumptions of Propositions 7.7.1–7.7.2 the interior assumptions 7.5.9h are also verified. Then, there exists a constant c such that for sufficiently small* ε_0

$$\int_I |\mathbf{D}^2 W|^2 \leq c r_0^{-8} (\mathcal{Q}_2 + \mathcal{Q}_1). \tag{7.7.9}$$

The main idea in the proof of the proposition is to appeal to the scaling vectorfield S defined by 7.5.3b, which we write in the form $S = (r - u)T + Z$ with $Z = rN$ the position vectorfield of Σ_t. Now let $W' = \hat{\mathcal{L}}_T W$, with E', H' its decomposition, and consider the $E - H$ decomposition of $\hat{\mathcal{L}}_S W'$,

$$E(\hat{\mathcal{L}}_S W') = (r - u) E(\hat{\mathcal{L}}_T W') + E(\hat{\mathcal{L}}_Z W') + \mathrm{Qr}[\mathbf{D}(r - u); W'],$$

where $\mathrm{Qr}[\mathbf{D}(r - u); W']$ is a quadratic expression in $\mathbf{D}(r - u), (E', H')$.

Now, according to the definition of $\hat{\mathcal{L}}_X$ in 7.1.7,

$$E(\hat{\mathcal{L}}_T W') = \hat{\mathcal{L}}_T E' + k \times E' - 2\phi^{-1} \nabla\phi \wedge H'$$
$$E(\hat{\mathcal{L}}_Z W') = \hat{\mathcal{L}}_Z E' + \mathrm{Qr}[{}^{(Z)}\pi; W'] + \mathrm{Qr}[[Z, T]; W']$$

where $Qr[^{(Z)}\pi; W']$ is a quadratic expression in ${}^{(Z)}\pi, E', H'$. Here ${}^{(Z)}\pi$ is the deformation tensor of Z with respect to the space-time metric. Therefore,

$$
\begin{aligned}
(r-u)\hat{\mathcal{L}}_T E' &= E(\hat{\mathcal{L}}_S W') - \hat{\mathcal{L}}_Z E' + Qr[\mathbf{D}(r-u); W'] \\
&\quad + Qr[^{(Z)}\pi; W'] + Qr[[Z,T]; W'] \\
&\quad + (r-u) Qr[(k, \phi^{-1}\nabla\phi); W'] \\
&= E(\hat{\mathcal{L}}_S W') - \nabla_Z E' + Qr[\mathbf{D}(r-u); W'] \\
&\quad + Qr[(^{(Z)}\pi; W'] + Qr[[Z,T]; W'] + Qr[\nabla Z; E'] \\
&\quad + (r-u) Qr[(k, \phi^{-1}\nabla\phi); W'].
\end{aligned}
\tag{7.7.10a}
$$

On the other hand, according to Proposition 7.1.2, W' satisfies the Bianchi equations

$$\mathbf{D}^\alpha W'_{\alpha\beta\gamma\delta} = J_{\beta\gamma\delta}$$

with J of the form

$$J = Qr[^{(T)}\hat{\pi}; \mathbf{D}W] + Qr[\mathbf{D}^{(T)}\hat{\pi}; W].$$

Therefore, according to Proposition 7.2.1, E', H' verify the equations

$$
\begin{aligned}
\operatorname{div} E' &= +k \wedge H' + Qr \\
\operatorname{div} H' &= -k \wedge E' + Qr
\end{aligned}
$$

$$
\begin{aligned}
-\hat{\mathcal{L}}_T H' + \operatorname{curl} E' &= -\phi^{-1}\nabla\phi \wedge E' - \frac{1}{2} k \times H' + Qr \\
\hat{\mathcal{L}}_T E' + \operatorname{curl} H' &= -\phi^{-1}\nabla\phi \wedge H' + \frac{1}{2} k \times E' + Qr
\end{aligned}
$$

where the error terms Qr are of the form $Qr[^{(T)}\hat{\pi}; \mathbf{D}W] + Qr[\mathbf{D}^{(T)}\hat{\pi}; W]$. Henceforth, taking into account the formula 7.7.10a for $\hat{\mathcal{L}}_T H', \hat{\mathcal{L}}_T E'$, we write

$$
\begin{aligned}
\operatorname{div} E' &= \rho_{E'} \\
\operatorname{curl} E' - \nabla_X H' &= \frac{1}{r-u} H(\hat{\mathcal{L}}_S W') + \sigma_{E'}
\end{aligned}
\tag{7.7.10b}
$$

$$
\begin{aligned}
\operatorname{div} H' &= \rho_{H'} \\
\operatorname{curl} H' + \nabla_X E' &= -\frac{1}{r-u} E(\hat{\mathcal{L}}_S W') + \sigma_{H'}
\end{aligned}
\tag{7.7.10c}
$$

where X is the vectorfield $X = -\frac{1}{r-u} Z$ and

$$\rho_{E'}, \rho_{H'} = \mathrm{Qr}[k; W'] + \mathrm{Qr}[{}^{(T)}\hat{\pi}; \mathbf{D}W] + \mathrm{Qr}[\mathbf{D}{}^{(T)}\hat{\pi}; W] \tag{7.7.10d}$$

$$\begin{aligned} \sigma_{E'}, \sigma_{H'} = {} & \mathrm{Qr}[(k, \phi^{-1}\nabla\phi); W'] + \mathrm{Qr}[{}^{(T)}\hat{\pi}; \mathbf{D}W] + \mathrm{Qr}[\mathbf{D}{}^{(T)}\hat{\pi}; W] \\ & + \frac{1}{r-u}\mathrm{Qr}[({}^{(Z)}\pi, [Z,T], \nabla Z); W'] \\ & + \frac{1}{r-u}\mathrm{Qr}[\mathbf{D}(r-u); W']. \end{aligned} \tag{7.7.10e}$$

We now introduce, as before the truncations of E', H',

$$\overset{\circ}{E}{}' = fE' \tag{7.7.11a}$$

$$\overset{\circ}{H}{}' = fH'. \tag{7.7.11b}$$

Hence,

$$\begin{aligned} \mathrm{div}\ \overset{\circ}{E}{}' &= f\rho_{E'} + \mathrm{Qr}[\nabla f; W'] \\ \mathrm{curl}\ \overset{\circ}{E}{}' - \nabla_X \overset{\circ}{H}{}' &= \frac{1}{r-u} f H(\hat{\mathcal{L}}_S W') + f\sigma_{E'} \\ &\quad + \mathrm{Qr}[\nabla f; W'] + \mathrm{Qr}[\nabla_X f; W'] \end{aligned} \tag{7.7.11c}$$

$$\begin{aligned} \mathrm{div}\ \overset{\circ}{H}{}' &= f\rho_{H'} + \mathrm{Qr}[\nabla f; W'] \\ \mathrm{curl}\ \overset{\circ}{H}{}' + \nabla_X \overset{\circ}{E}{}' &= -\frac{1}{r-u} f E(\hat{\mathcal{L}}_S W') + f\sigma_{H'} \\ &\quad + \mathrm{Qr}[\nabla f; W'] + \mathrm{Qr}[\nabla_X f; W']. \end{aligned}$$

We now remark that in $\hat{I}$ the vectorfield X has length strictly less than 1, and thus we are in a position to apply the corollary of Proposition 4.4.1 to the system 7.7.11c and derive

$$\begin{aligned} \int_{St} \left(|\nabla \overset{\circ}{E}{}'|^2 + |\nabla \overset{\circ}{H}{}'|^2 \right) \leq {} & c r_0^{-8} \int_{\hat{I}} r^6 (|E(\hat{\mathcal{L}}_S W')|^2 + |H(\hat{\mathcal{L}}_S W')|^2) \\ & + c \int_{\hat{I}} (|\rho_{E'}|^2 + |\sigma_{E'}|^2 + |\rho_{H'}|^2 + |\sigma_{H'}|^2) \\ & + \int_{\Sigma^e \cap \hat{I}} |\nabla f| (|E'|^2 + |H'|^2) \\ & + c \int_{\Sigma_t} |\mathrm{Ric}| \left(|\overset{\circ}{E}{}'|^2 + |\overset{\circ}{H}{}'|^2 \right). \end{aligned} \tag{7.7.11d}$$

Now,

$$\int_{\hat{I}} |f|^2(|\rho_{E'}|^2 + |\sigma_{E'}|^2 + |\rho_{H'}|^2 + |\sigma_{H'}|^2) \leq cr_0^{-2}\|\mathbf{D}W\|^2_{2,\hat{I}} + cr_0^{-4}\|W\|^2_{2,\hat{I}} \tag{7.7.11e}$$

provided that we check that

$$\begin{aligned} \|{}^{(T)}\hat{\pi}\|_{\infty,\hat{I}} &\leq cr_0^{-1} \\ \|({}^{(Z)}\pi, \nabla Z, [Z,T], \mathbf{D}(r-u))\|_{\infty,\hat{I}} &\leq c \\ \|\mathbf{D}{}^{(T)}\hat{\pi}\|_{3,\hat{I}} &\leq cr_0^{-1}. \end{aligned}$$

Therefore, taking into account also that $\|\mathrm{Ric}\|_{3/2,\hat{I}} \leq \varepsilon_0$, we infer from 7.7.11c that

$$\begin{aligned} \int_{St}\left(|\nabla \overset{\circ}{E}'|^2 + |\nabla \overset{\circ}{H}'|^2\right) \leq\ & cr_0^{-8}\int_{\Sigma} Q(\hat{\mathcal{L}}_S\hat{\mathcal{L}}_T W)(\bar{K},\bar{K},\bar{K},T) \\ & + cr_0^{-2}\|\mathbf{D}W\|^2_{2,\hat{I}} + cr_0^{-4}\|W\|^2_{2,\hat{I}} \\ & + c\varepsilon_0\left(\int_{\Sigma_t} |\overset{\circ}{E}'|^6 + |\overset{\circ}{H}'|^6\right)^{1/3}. \end{aligned}$$

Or, applying the Sobolev inequality to the last term on the right, taking into account the results of 7.7.2 and choosing ε_0 sufficiently small, we infer that

$$\begin{aligned} \int_{St}\left(|\nabla \overset{\circ}{E}'|^2 + |\nabla \overset{\circ}{H}'|^2\right) \leq\ & cr_0^{-8}\int_{\Sigma} Q(\hat{\mathcal{L}}_S\hat{\mathcal{L}}_T W)(\bar{K},\bar{K},\bar{K},T) \\ & + cr_0^{-8}\mathcal{Q}_1. \end{aligned}$$

which proves the desired estimates for $\int_I |\nabla E(\hat{\mathcal{L}}_T W)|^2 + |\nabla H(\hat{\mathcal{L}}_T W)|^2$. To prove 7.7.9, it only remains to obtain estimates for $\nabla^2 E$, $\nabla^2 H$. We achieve this by going back to the equations 7.7.6g–7.7.6j for $\overset{\circ}{E}$, $\overset{\circ}{H}$ and applying to them the estimates of Proposition 4.4.2.

CHAPTER 8

The Error Estimates

In this chapter we provide the error estimates generated in the process of estimating the norms $\mathcal{Q}_1$ and $\mathcal{Q}_2$ (see the definition in 7.6.1g and 7.6.1h). We prove here the main result of Part II, which we call the *boundedness theorem.*

8.1. Preliminaries

We start with the following:

Definition 8.1.1 Given an arbitrary vectorfield X with deformation tensor ${}^{(X)}\pi$, we introduce the following notation:

$$
\begin{aligned}
{}^{(X)}p_\gamma &= (\mathbf{Div}^{(X)}\hat{\pi})_\gamma = \mathbf{D}^\alpha\, {}^{(X)}\hat{\pi}_{\alpha\gamma} && (8.1.1a)\\
{}^{(X)}q_{\alpha\beta\gamma} &= \mathbf{D}_\beta^{(X)}\hat{\pi}_{\gamma\alpha} - \mathbf{D}_\gamma^{(X)}\hat{\pi}_{\beta\alpha} \\
&\quad - \frac{1}{3}({}^{(X)}p_\gamma \mathbf{g}_{\alpha\beta} - {}^{(X)}p_\beta \mathbf{g}_{\alpha\gamma}) && (8.1.1b)
\end{aligned}
$$

where ${}^{(X)}\hat{\pi}$ is the traceless part of ${}^{(X)}\pi$.

Let W be a Weyl tensor satisfying the homogeneous Bianchi equations 7.1.2, and consider $\hat{\mathcal{L}}_X W$. Then, recalling Proposition 7.1.2, we write

$$\mathbf{D}^\alpha(\hat{\mathcal{L}}_X W)_{\alpha\beta\gamma\delta} = J(X;W)_{\beta\gamma\delta} \qquad (8.1.2a)$$

where

$$
\begin{aligned}
J(X;W) &= \frac{1}{2}(J^1(X;W) + J^2(X;W) + J^3(X;W)) \qquad (8.1.2b)\\
J^1(X;W)_{\beta\gamma\delta} &= {}^{(X)}\hat{\pi}^{\mu\nu}\mathbf{D}_\nu W_{\mu\beta\gamma\delta}\\
J^2(X;W)_{\beta\gamma\delta} &= {}^{(X)}p_\lambda W^\lambda{}_{\beta\gamma\delta}\\
J^3(X;W)_{\beta\gamma\delta} &= {}^{(X)}q_{\alpha\beta\lambda}W^{\alpha\lambda}{}_{\gamma\delta} + {}^{(X)}q_{\alpha\gamma\lambda}W^\alpha{}_\beta{}^\lambda{}_\delta + {}^{(X)}q_{\alpha\delta\lambda}W^\alpha{}_{\beta\gamma}{}^\lambda.
\end{aligned}
$$

Taking once more the Lie derivative relative to a second vectorfield Y, we write, according to Proposition 7.1.2,

$$\mathbf{D}^\alpha(\hat{\mathcal{L}}_Y\hat{\mathcal{L}}_X W)_{\alpha\beta\gamma\delta} = J(X,Y;W)_{\beta\gamma\delta} \qquad (8.1.2c)$$

where

$$J(X,Y;W) = J^0(X,Y;W) + \frac{1}{2}(J^1(X,Y;W) + J^2(X,Y;W) + J^3(X,Y;W)) \tag{8.1.2d}$$

$$\begin{aligned} J^0(X,Y;W) &= \hat{\mathcal{L}}_Y J(X;W) \\ J^1(X,Y;W) &= J^1(Y;\hat{\mathcal{L}}_X W) \\ J^2(X,Y;W) &= J^2(Y;\hat{\mathcal{L}}_X W) \\ J^3(X,Y;W) &= J^3(Y;\hat{\mathcal{L}}_X W) \end{aligned}$$

with

$$\begin{aligned} \hat{\mathcal{L}}_Y J(X;W)_{\beta\gamma\delta} = \mathcal{L}_Y J(X,W)_{\beta\gamma\delta} & - \frac{1}{2}\Big({}^{(Y)}\hat{\pi}_\beta{}^\mu J(X,W)_{\mu\gamma\delta} + {}^{(Y)}\hat{\pi}_\gamma{}^\mu J(X,W)_{\beta\mu\delta} \\ & + {}^{(Y)}\hat{\pi}_\delta{}^\mu J(X,W)_{\beta\gamma\mu}\Big) + \frac{1}{8}\mathrm{tr}\,{}^{(Y)}\hat{\pi} J(X,W)_{\beta\gamma\delta}. \end{aligned}$$

Now, in view of Corollary 7.1.1.1, we have

$$\begin{aligned} \int_{\Sigma_t} Q(\hat{\mathcal{L}}_X W)(\bar{K},\bar{K},T,T) = & \int_{\Sigma_0} Q(\hat{\mathcal{L}}_X W)(\bar{K},\bar{K},T,T) \\ & + \int_{V_t} \phi \mathbf{Div} Q(\hat{\mathcal{L}}_X W)_{\beta\gamma\delta}\bar{K}^\beta\bar{K}^\gamma T^\delta \\ & + \int_{V_t} \phi Q(\hat{\mathcal{L}}_X W)_{\alpha\beta\gamma\delta}({}^{(\bar{K})}\pi^{\alpha\beta}\bar{K}^\gamma T^\delta \\ & + \frac{1}{2}{}^{(T)}\pi^{\alpha\beta}\bar{K}_\gamma\bar{K}_\delta) \end{aligned} \tag{8.1.3a}$$

$$\begin{aligned} \int_{\Sigma_t} Q(\hat{\mathcal{L}}_X W)(\bar{K},\bar{K},\bar{K},T) = & \int_{\Sigma_0} Q(\hat{\mathcal{L}}_X W)(\bar{K},\bar{K},\bar{K},T) \\ & + \int_{V_t} \phi \mathbf{Div} Q(\hat{\mathcal{L}}_X W)_{\beta\gamma\delta}\bar{K}^\beta\bar{K}^\gamma \bar{K}^\delta \\ & + \frac{3}{2}\int_{V_t} \phi Q(\hat{\mathcal{L}}_X W)_{\alpha\beta\gamma\delta}{}^{(\bar{K})}\pi^{\alpha\beta}\bar{K}^\gamma\bar{K}^\delta \end{aligned} \tag{8.1.3b}$$

where, according to Proposition 7.1.1,

$$\mathbf{Div}Q(\hat{\mathcal{L}}_X W)_{\beta\gamma\delta} = (\hat{\mathcal{L}}_X W)_{\beta\ \delta}^{\ \mu\ \nu} J(X,W)_{\mu\gamma\nu} + (\hat{\mathcal{L}}_X W)_{\beta\ \gamma}^{\ \mu\ \nu} J(X,W)_{\mu\delta\nu}$$

$$+ {}^{\star}(\hat{\mathcal{L}}_X W)_{\beta\ \delta}^{\ \mu\ \nu} J(X,W)^{\star}_{\mu\gamma\nu} + {}^{\star}(\hat{\mathcal{L}}_X W)_{\beta\ \gamma}^{\ \mu\ \nu} J(X,W)^{\star}_{\mu\delta\nu} \tag{8.1.3c}$$

and V_t is the slab $\bigcup_{t'\in[0,t]} \Sigma_{t'}$.

Recalling the definition of $\mathcal{Q}_1$, which appears in 7.6.1g, we write

$$\mathcal{Q}_1(W,t) \leq \mathcal{Q}_1(W,0) + \mathcal{E}_1(W,t) \tag{8.1.4a}$$

where

$$\begin{aligned}
\mathcal{E}_1(W,t) = &\int_{V_t} |\phi \mathbf{Div}Q(\hat{\mathcal{L}}_{\mathbf{O}}W)_{\beta\gamma\delta}\bar{K}^\beta \bar{K}^\gamma T^\delta| \\
&+ \int_{V_t} |\phi \mathbf{Div}Q(\hat{\mathcal{L}}_T W)_{\beta\gamma\delta}\bar{K}^\beta \bar{K}^\gamma \bar{K}^\delta| \\
&+ \int_{V_t} |\phi Q(\hat{\mathcal{L}}_{\mathbf{O}}W)_{\alpha\beta\gamma\delta}{}^{(\bar{K})}\pi^{\alpha\beta}\bar{K}^\gamma T^\delta| \\
&+ \frac{1}{2}\int_{V_t} |\phi Q(\hat{\mathcal{L}}_{\mathbf{O}}W)_{\alpha\beta\gamma\delta}{}^{(T)}\pi_{\alpha\beta}\bar{K}_\gamma \bar{K}_\delta| \\
&+ \frac{3}{2}\int_{V_t} |\phi Q(\hat{\mathcal{L}}_T W)_{\alpha\beta\gamma\delta}{}^{(\bar{K})}\pi^{\alpha\beta}\bar{K}^\gamma \bar{K}^\delta|.
\end{aligned} \tag{8.1.4b}$$

Similarly, for the second derivatives we have, according to the definition of $\mathcal{Q}_2$ in 7.6.1h,

$$\mathcal{Q}_2(W,t) \leq \mathcal{Q}_2(W,0) + \mathcal{E}_2(W,t) \tag{8.1.4c}$$

where

$$\begin{aligned}
\mathcal{E}_2(W,t) = &\int_{V_t} |\phi \mathbf{Div}Q(\hat{\mathcal{L}}_{\mathbf{O}}^2 W)_{\beta\gamma\delta}\bar{K}^\beta \bar{K}^\gamma T^\delta| \\
&+ \int_{V_t} |\phi \mathbf{Div}Q(\hat{\mathcal{L}}_{\mathbf{O}}\hat{\mathcal{L}}_T W)_{\beta\gamma\delta}\bar{K}^\beta \bar{K}^\gamma \bar{K}^\delta| \\
&+ \int_{V_t} |\phi \mathbf{Div}Q(\hat{\mathcal{L}}_S\hat{\mathcal{L}}_T W)_{\beta\gamma\delta}\bar{K}^\beta \bar{K}^\gamma \bar{K}^\delta| \\
&+ \int_{V_t} |\phi \mathbf{Div}Q(\hat{\mathcal{L}}_T^2 W)_{\beta\gamma\delta}\bar{K}^\beta \bar{K}^\gamma \bar{K}^\delta| \\
&+ \int_{V_t} |\phi Q(\hat{\mathcal{L}}_{\mathbf{O}}^2 W)_{\alpha\beta\gamma\delta}{}^{(\bar{K})}\pi^{\alpha\beta}\bar{K}^\gamma T^\delta|
\end{aligned}$$

$$
\begin{aligned}
&+\frac{1}{2}\int_{V_t} |\phi Q(\hat{\mathcal{L}}_O^2 W)_{\alpha\beta\gamma\delta}{}^{(T)}\pi^{\alpha\beta}\bar{K}_\gamma \bar{K}_\delta| \\
&+\frac{3}{2}\int_{V_t} |\phi Q(\hat{\mathcal{L}}_O\hat{\mathcal{L}}_T W)_{\alpha\beta\gamma\delta}{}^{(\bar{K})}\pi^{\alpha\beta}\bar{K}^\gamma \bar{K}^\delta| \\
&+\frac{3}{2}\int_{V_t} |\phi Q(\hat{\mathcal{L}}_S\hat{\mathcal{L}}_T W)_{\alpha\beta\gamma\delta}{}^{(\bar{K})}\pi^{\alpha\beta}\bar{K}^\gamma \bar{K}^\delta| \\
&+\frac{3}{2}\int_{V_t} |\phi Q(\hat{\mathcal{L}}_T^2 W)_{\alpha\beta\gamma\delta}{}^{(\bar{K})}\pi^{\alpha\beta}\bar{K}^\gamma \bar{K}^\delta|. \qquad (8.1.4d)
\end{aligned}
$$

In the proof of the boundedness theorem we will also need, besides $\mathcal{Q}_1$ and $\mathcal{Q}_2$, the following quantities, defined along any null cone[1] C_u contained in the exterior region $r \geq \frac{r_0}{2}$ of the space-time:

$$
\begin{aligned}
\tilde{\mathcal{Q}}(W; u,t) = &\int_{C_u(t_0,t)} Q(\hat{\mathcal{L}}_O W)(\bar{K},\bar{K},T,e_4) \\
&+\int_{C_u(t_0,t)} Q(\hat{\mathcal{L}}_T W)(\bar{K},\bar{K},\bar{K},e_4) \qquad (8.1.4e)
\end{aligned}
$$

$$
\begin{aligned}
\tilde{\mathcal{Q}}_2(W; u,t) = &\int_{C_u(t_0,t)} Q(\hat{\mathcal{L}}_O^2 W)(\bar{K},\bar{K},T,e_4) \\
&+\int_{C_u(t_0,t)} Q(\hat{\mathcal{L}}_O\hat{\mathcal{L}}_T W)(\bar{K},\bar{K},\bar{K},e_4) \\
&+\int_{C_u(t_0,t)} Q(\hat{\mathcal{L}}_S\hat{\mathcal{L}}_T W)(\bar{K},\bar{K},\bar{K},e_4) \\
&+\int_{C_u(t_0,t)} Q(\hat{\mathcal{L}}_T^2 W)(\bar{K},\bar{K},\bar{K},e_4) \qquad (8.1.4f)
\end{aligned}
$$

where $0 \leq t_0 = t_0(u) \leq t$ is the value of t, for which the cone C_u intersects the boundary of the interior region and $C_u(t_0,t)$ is the region of C_u contained between Σ_{t_0} and Σ_t. For scalar functions f we define

$$
\int_{C_u(t_0,t_1)} f = \int_{t_0}^{t_1} dt \left(\int_{S_{u,t}} a\phi f d\mu_\gamma \right).
$$

Now, for a given fixed value t_* of t we introduce

$$
\mathcal{Q}_1* = \sup_{[0,t_*]} \mathcal{Q}_1(W,t) + \sup_{t\in[0,t_*]}\ \sup_{u\geq u_0(t)} \tilde{\mathcal{Q}}_1(W;u,t) \qquad (8.1.4g)
$$

$$
\mathcal{Q}_2* = \sup_{[0,t_*]} \mathcal{Q}_2(W,t) + \sup_{t\in[0,t_*]}\ \sup_{u\geq u_0(t)} \tilde{\mathcal{Q}}_2(W;u,t) \qquad (8.1.4h)
$$

[1] By null cone we mean a level hypersurface of the optical function u.

where, for a given t, $u_0(t)$ is the value of u that corresponds to the boundary of the interior region $r \leq \frac{r_0}{2}$. We claim the following;

Proposition 8.1.1 *With the notation introduced above we have*

$$\mathcal{Q}_1* \leq 2(\mathcal{Q}_1(0) + \mathcal{E}_1(t_*))$$
$$\mathcal{Q}_2* \leq 2(\mathcal{Q}_2(0) + \mathcal{E}_2(t_*)).$$

The proof of the proposition is based on formulas 8.1.4a and 8.1.4c, and the following lemma, which is an analogue of Corollary 7.1.1.1 for domains $D(u, t_0, t_1)$ of the space-time, bounded by a null hypersurface C_u given by a level hypersurface of u and the spacelike hypersurfaces Σ_{t_0} and Σ_{t_1}.

Lemma 8.1.1 *Let P^μ be a vector verifying the divergence equation*

$$\mathbf{D}^\mu P_\mu = F$$

on a domain $D(u, t_0, t_1)$ as above. Then

$$\int_{\Sigma_{t_1}} \langle P, T\rangle d\mu_g - \int_{C_u(t_0,t_1)} \langle P, e_4\rangle = \int_{\Sigma_{t_0}} \langle P, T\rangle d\mu_g - \int_D F$$

where

$$\int_{C_u(t_0,t_1)} \langle P, e_4\rangle = \int_{t_0}^{t_1} dt \left(\int_{S_{u,t}} \langle P, e_4\rangle a\phi d\mu_\gamma \right)$$

and

$$\int_D F = \int_{t_0}^{t_1} dt \left(\int_{\Sigma_t \bigcap D} \phi F d\mu_g \right).$$

To prove the proposition, we recall from the proof of Corollary 7.1.1.1 that if we denote, for arbitrary vectorfields X, Y, Z,

$$P_\alpha(W; X, Y, Z) = Q(W)_{\alpha\beta\gamma\delta} X^\beta Y^\gamma Z^\delta,$$

then

$$\begin{aligned}\mathbf{Div}P &= X^\beta Y^\gamma Z^\delta (\mathbf{Div}Q)_{\beta\gamma\delta} \\ &+ \frac{1}{2} Q^{\alpha\beta\gamma\delta} ({}^{(X)}\pi_{\alpha\beta} Y^\gamma Z^\delta + {}^{(Y)}\pi_{\alpha\beta} Z^\gamma X^\delta + {}^{(Z)}\pi_{\alpha\beta} X^\gamma Y^\delta).\end{aligned}$$

We apply the lemma first to W replaced by $\hat{\mathcal{L}}_O W$ and $X = \bar{K}, Y = \bar{K}, Z = T$, and then to W replaced by $\hat{\mathcal{L}}_T W$ and $X = \bar{K}, Y = \bar{K}, Z = \bar{K}$ in any domain

$D(u, t_1, t_2)$ bounded by a null hypersurface $C_{u(t_1,t_2)}$ that is contained in the exterior region. We can then estimate the integrals on the spacelike hypersurfaces by the quantities $\mathcal{Q}_1(t)$ and the volume integral on D by $\sup_{[0,t_*]} |\mathcal{E}_1(t)|$. Since clearly

$$\mathcal{Q}_1(t) \leq \mathcal{Q}_1(0) + \mathcal{E}_1(t_*)$$

we derive the first estimate of the proposition. The second estimate is proved in precisely the same manner.

Proof 8.1.1 (of Lemma 8.1.1) *To prove the lemma we apply the Stokes theorem to the differential form*

$$^\star P_{\alpha\beta\gamma} = \in_{\alpha\beta\gamma}{}^{\nu} P_\nu.$$

Then

$$^\star(d^\star P) = \mathbf{D}^\mu P_\mu = F,$$

or

$$d^\star P = -^\star F.$$

Integrating this last formula on the domain D we have, in view of the Stokes formula,

$$\begin{aligned} -\int_D {}^\star F &= \int_{\partial D} {}^\star P \qquad (8.1.5) \\ &= \int_{\Sigma_2} {}^\star P - \int_{\Sigma_1} {}^\star P - \int_{C_u} {}^\star P \end{aligned}$$

where C_u is the corresponding null hypersurface of the boundary of D. To estimate the integral of $^\star P$ on C_u we introduce coordinates on C by choosing angular coordinates θ^A, $A = 1, 2$ on Σ_{t_0}, and by following the curves $x^\mu(s)$, with s being the affine parameter of s, until they meet Σ_{t_1}. In other words, we have

$$x^\mu(\theta, t) = x^\mu(\theta, s(t, \theta))$$

where $\frac{d}{ds}x^\mu = l^\mu$; hence,

$$\frac{d}{dt}x^\mu = a\phi l^\mu.$$

Now,

$$\begin{aligned} \int_C {}^\star P &= \int_C {}^\star P_{\alpha\beta\gamma} dx^\alpha dx^\beta dx^\gamma \\ &= \int_C {}^\star P_{\alpha\beta\gamma} \frac{\partial x^\alpha}{\partial \theta^A} \frac{\partial x^\beta}{\partial \theta^B} a\phi l^\gamma d\theta^A d\theta^B dt \\ &= \int_C a\phi \langle P, l \rangle \in_{\alpha\beta} \frac{\partial x^\alpha}{\partial \theta^A} \frac{\partial x^\beta}{\partial \theta^B} d\theta^A d\theta^B dt \end{aligned}$$

where $\in_{\alpha\beta} = \frac{1}{2}\in_{\alpha\beta\gamma\delta}\underline{l}^{\gamma} l^{\delta}$ *is the area form of the 2-surfaces* $S_{t,u}$. *Therefore,*

$$\int_C {}^\star P = \int_{t_0}^{t_1} dt \left(\int_{S_{u,t}} \langle P, l \rangle a\phi d\mu_\gamma \right).$$

Proceeding in a similar manner for the other integrals on the left-hand side of 8.1.5, we derive the desired formula.

Next we note the following simple corollary of Lemma 7.3.1 on page 151.

Proposition 8.1.2 *Given an arbitrary Weyl tensor W, we have*

$$\begin{aligned} Q(W)(\bar{K}, \bar{K}, T, e_4) = & \frac{1}{4}\tau_+^4|\alpha|^2 + \frac{1}{2}(\tau_+^4 + 2\tau_+^2\tau_-^2)|\beta|^2 \\ & + \frac{1}{2}(\tau_-^4 + 2\tau_+^2\tau_-^2)(|\rho|^2 + |\sigma|^2) + \tau_-^4|\underline{\beta}|^2 \end{aligned}$$

$$\begin{aligned} Q(W)(\bar{K}, \bar{K}, \bar{K}, e_4) = & \frac{1}{4}\tau_+^6|\alpha|^2 + \frac{1}{2}\tau_+^4\tau_-^2|\beta|^2 \\ & + \frac{1}{2}\tau_+^2\tau_-^4(|\rho|^2 + |\sigma|^2) + \frac{1}{2}\tau_-^6|\underline{\beta}|^2. \end{aligned}$$

In the end of this section we prove a particular form of the coarea formula adapted to domains bounded by null hypersurfaces.

Lemma 8.1.2 *Let* $D(u_1, u_2, t_0, t_1)$ *be a domain of the space-time bounded by the null hypersurfaces* $u = u_1$, $u = u_2$ *and the spacelike hypersurfaces* $\Sigma_{t_1}, \Sigma_{t_2}$. *Then, given any scalar function f on D, we have*

$$\int_D f d\mu_{\mathbf{g}} = \int_{u_1}^{u_2} du \left(\int_{C_u(t_0,t_1)} f \right)$$

where

$$\int_{C_u(t_0,t_1)} f = \int_{t_0}^{t_1} dt \left(\int_{S_{u,t}} a\phi f d\mu_\gamma \right)$$

where $C_{u(t_0,t_1)}$ *is the region of the cone* C_u *between* t_0 *and* t_1.

8.1.1. *The Null Decomposition of a Weyl Current*

Definition 8.1.2 A tensorfield $J_{\beta\gamma\delta}$ is called a Weyl current if it satisfies the following:

$$\begin{aligned} J_{\beta\gamma\delta} + J_{\gamma\delta\beta} + J_{\delta\beta\gamma} &= 0 \\ J_{\beta\gamma\delta} + J_{\beta\delta\gamma} &= 0 \\ \mathrm{tr} J_\delta = \mathbf{g}^{\beta\gamma} J_{\beta\gamma\delta} &= 0. \end{aligned}$$

Given a Weyl current J we introduce the following null decomposition:

$$\begin{aligned}
\Lambda(J) &= \tfrac{1}{4}J_{434} &;\quad \underline{\Lambda}(J) &= \tfrac{1}{4}J_{343}\\
K(J) &= \tfrac{1}{4}\in^{AB} J_{4AB} &;\quad \underline{K}(J) &= \tfrac{1}{4}\in^{AB} J_{3AB}\\
\Xi(J)_A &= \tfrac{1}{2}J_{44A} &;\quad \underline{\Xi}(J)_A &= \tfrac{1}{2}J_{33A}\\
I(J)_A &= \tfrac{1}{2}J_{34A} &;\quad \underline{I}(J)_A &= \tfrac{1}{2}J_{43A}\\
\Theta(J)_{AB} &= J_{A4B} + J_{B4A} &;\quad \underline{\Theta}(J)_{AB} &= J_{A3B} + J_{B3A}\\
&- (\delta^{CD} J_{C4D})\delta_{AB} &;\quad &- (\delta^{CD} J_{C3D})\delta_{AB}
\end{aligned} \tag{8.1.6a}$$

Remark that

$$J_{A4B} = \Theta(J)_{AB} - \Lambda\delta_{AB} + K \in_{AB}$$
$$J_{A3B} = \underline{\Theta}(J)_{AB} - \underline{\Lambda}\delta_{AB} + \underline{K} \in_{AB}$$

and

$$J_{ABC} = \in_{BC}({}^{\star}I(J)_A + {}^{\star}\underline{I}(J)_A).$$

Also,

$$\begin{aligned}
\Lambda(J^\star) &= K(J) &;\quad \underline{\Lambda}(J^\star) &= -\underline{K}(J)\\
K(J^\star) &= -\Lambda(J) &;\quad \underline{K}(J^\star) &= \underline{\Lambda}(J)\\
\Xi(J^\star) &= -{}^\star\Xi(J) &;\quad \underline{\Xi}(J^\star) &= {}^\star\underline{\Xi}(J)\\
I(J^\star) &= -{}^\star I(J) &;\quad \underline{I}(J^\star) &= {}^\star\underline{I}(J)\\
\Theta(J^\star) &= -{}^\star\Theta(J) &;\quad \underline{\Theta}(J^\star) &= {}^\star\underline{\Theta}(J).
\end{aligned} \tag{8.1.6b}$$

The following proposition is an immediate consequence of formula 8.1.3c. We note that the precise numerical coefficients are not important.

Proposition 8.1.3 *Let W be an arbitrary Weyl tensor, let X be a given vectorfield, and set $D(X, W) = \mathbf{Div}Q(\hat{\mathcal{L}}_X W)$. The following formulas hold everywhere in the exterior region:*

$$D(X, W)(\bar{K}, \bar{K}, T) = \frac{1}{8}\tau_+^4(D(X, W)_{444} + D(X, W)_{344}) \tag{8.1.7a}$$

$$
\begin{aligned}
&+\frac{1}{4}\tau_+^2\tau_-^2(D(X,W)_{344}+D(X,W)_{334})\\
&+\frac{1}{8}\tau_-^4(D(X,W)_{334}+D(X,W)_{333})
\end{aligned}
$$

$$
\begin{aligned}
D(X,W)(\bar{K},\bar{K},\bar{K}) = &\frac{1}{8}\tau_+^6 D(X,W)_{444}\\
&+\frac{3}{8}\tau_+^4\tau_-^2 D(X,W)_{344}\\
&+\frac{3}{8}\tau_+^2\tau_-^4 D(X,W)_{334}\\
&+\frac{1}{8}\tau_-^6 D(X,W)_{333}
\end{aligned}
\tag{8.1.7b}
$$

where

$$D(X,W)_{333} = 4\underline{\alpha}(\hat{\mathcal{L}}_X W)\cdot\underline{\Theta}(X,W) + 8\underline{\beta}(\hat{\mathcal{L}}_X W)\cdot\underline{\Xi}(X,W)$$

$$
\begin{aligned}
D(X,W)_{334} = &\,8\rho(\hat{\mathcal{L}}_X W)\underline{\Lambda}(X,W) - 8\sigma(\hat{\mathcal{L}}_X W)\underline{K}(X,W)\\
&- 8\underline{\beta}(\hat{\mathcal{L}}_X W)\cdot\underline{I}(X,W)
\end{aligned}
\tag{8.1.7c}
$$

$$
\begin{aligned}
D(X,W)_{443} = &\,8\rho(\hat{\mathcal{L}}_X W)\Lambda(X,W) + 8\sigma(\hat{\mathcal{L}}_X W)K(X,W)\\
&+ 8\beta(\hat{\mathcal{L}}_X W)\cdot I(X,W)
\end{aligned}
$$

$$D(X,W)_{444} = 4\alpha(\hat{\mathcal{L}}_X W)\cdot\Theta(X,W) - 8\beta(\hat{\mathcal{L}}_X W)\cdot\Xi(X,W)$$

where $\Lambda(X,W), \underline{\Lambda}(X,W), K(X,W), \underline{K}(X,W), \ldots$ *are the null components of* $J(X, W)$ *as defined above.*

Proof 8.1.2 *The first formula can be easily checked by signature considerations. To check the second formula, note that* $D_{334} = D_{433}$ *and use 8.1.3c for the latter. The last two formulas can then be checked by using the duality relative to components* 3, *and* 4.

We next express the null decomposition of the Weyl current $J = J(X,W)$, which appears on the right-hand side of 8.1.2a, relative to the null decomposition of $\hat{\mathcal{L}}_X W$ and that of ${}^{(X)}\hat{\pi}$, ${}^{(X)}p$, ${}^{(X)}q$. In doing this, we remark that ${}^{(X)}q$, as defined by 8.1.1b, satisfies all the properties of a Weyl current. It can thus be decomposed into the null components $\Lambda({}^{(X)}q)$, $\underline{\Lambda}({}^{(X)}q)$, $K({}^{(X)}q)$, $\underline{K}({}^{(X)}q), \ldots$. We also introduce the null components of ${}^{(X)}\hat{\pi}$ and ${}^{(X)}p$ according to the following:

Definition 8.1.3 Given the deformation tensor ${}^{(X)}\pi$ of an arbitrary vectorfield X, we introduce the null decomposition of its traceless part ${}^{(X)}\hat{\pi}$ according to

$$
\begin{aligned}
{}^{(X)}\mathbf{i}_{AB} &= {}^{(X)}\hat{\pi}_{AB} \quad ; \quad {}^{(X)}\mathbf{j} = {}^{(X)}\hat{\pi}_{34} \\
{}^{(X)}\mathbf{m}_A &= {}^{(X)}\hat{\pi}_{4A} \quad ; \quad {}^{(X)}\underline{\mathbf{m}}_A = {}^{(X)}\hat{\pi}_{3A} \\
{}^{(X)}\mathbf{n} &= {}^{(X)}\hat{\pi}_{44} \quad ; \quad {}^{(X)}\underline{\mathbf{n}}_A = {}^{(X)}\hat{\pi}_{33}.
\end{aligned}
\tag{8.1.8}
$$

We decompose the vector ${}^{(X)}p$, defined according to 8.1.1a, into the two scalars ${}^{(X)}p_3$, ${}^{(X)}p_4$ and the S-tangent vector ${}^{(X)}\not{p}_A = {}^{(X)}p_A$.

We are now ready to state the following:

Proposition 8.1.4 *Let X be an arbitrary vectorfield with ${}^{(X)}\hat{\pi}$, ${}^{(X)}p$, ${}^{(X)}q$ defined as before, and let $J = J(X, W) = J^1 + J^2 + J^3$ be the Weyl current defined by 8.1.2a.*

(i.) The null decomposition of J^1 is given by the following formulas:[2]

$$
\begin{aligned}
\underline{\Xi}(J^1) = {} & Qr\left[{}^{(X)}\mathbf{i}\,;\,\not\nabla\underline{\alpha}\right] + Qr\left[{}^{(X)}\mathbf{m}\,;\,\underline{\alpha}_3\right] + Qr\left[{}^{(X)}\underline{\mathbf{m}}\,;\,\underline{\alpha}_4\right] \\
& + Qr\left[{}^{(X)}\underline{\mathbf{m}}\,;\,\not\nabla\underline{\beta}\right] + Qr\left[{}^{(X)}\mathbf{j}\,;\,\underline{\beta}_3\right] + Qr\left[{}^{(X)}\underline{\mathbf{n}}\,;\,\underline{\beta}_4\right] \\
& + \mathrm{tr}\underline{\chi}\left(Qr\left[{}^{(X)}\mathbf{m}\,;\,\underline{\alpha}\right] + Qr\left[({}^{(X)}\mathbf{i},\,{}^{(X)}\mathbf{j})\,;\,\underline{\beta}\right] + Qr\left[{}^{(X)}\underline{\mathbf{m}}\,;\,(\rho,\sigma)\right]\right) \\
& + \mathrm{tr}\chi\left(Qr\left[{}^{(X)}\underline{\mathbf{m}}\,;\,\underline{\alpha}\right] + Qr\left[{}^{(X)}\underline{\mathbf{n}}\,;\,\underline{\beta}\right]\right) + l.o.t.
\end{aligned}
$$

$$
\begin{aligned}
\underline{\Theta}(J^1) = {} & Qr\left[{}^{(X)}\mathbf{m}\,;\,\not\nabla\underline{\alpha}\right] + Qr\left[{}^{(X)}\mathbf{n}\,;\,\underline{\alpha}_3\right] + Qr\left[{}^{(X)}\mathbf{j}\,;\,\underline{\alpha}_4\right] \\
& + Qr\left[{}^{(X)}\mathbf{i}\,;\,\not\nabla\underline{\beta}\right] + Qr\left[{}^{(X)}\mathbf{m}\,;\,\underline{\beta}_3\right] + Qr\left[{}^{(X)}\underline{\mathbf{m}}\,;\,\underline{\beta}_4\right] \\
& + Qr\left[{}^{(X)}\underline{\mathbf{m}}\,;\,\not\nabla(\rho,\sigma)\right] + Qr\left[{}^{(X)}\mathbf{j}\,;\,(\rho_3,\sigma_3)\right] + Qr\left[{}^{(X)}\underline{\mathbf{n}}\,;\,(\rho_4,\sigma_4)\right] \\
& + \mathrm{tr}\underline{\chi}\Big(Qr\left[{}^{(X)}\mathbf{n}\,;\,\underline{\alpha}\right] + Qr\left[{}^{(X)}\mathbf{m}\,;\,\underline{\beta}\right] + Qr\left[({}^{(X)}\mathbf{i},\,{}^{(X)}\mathbf{j})\,;\,(\rho,\sigma)\right] \\
& + Qr\left[{}^{(X)}\underline{\mathbf{m}}\,;\,\beta\right]\Big) + \mathrm{tr}\chi\Big(\boxed{\;Qr\left[{}^{(X)}\mathbf{i}\,;\,\underline{\alpha}\right]\;} \\
& + Qr\left[{}^{(X)}\mathbf{j}\,;\,\underline{\alpha}\right] + Qr\left[{}^{(X)}\underline{\mathbf{m}}\,;\,\underline{\beta}\right] + Qr\left[{}^{(X)}\underline{\mathbf{n}}\,;\,(\rho,\sigma)\right]\Big) + l.o.t.
\end{aligned}
$$

[2] We recall from the remark on page 155 that Qr[;] is a generic notation for any quadratic form with coefficients that depend only on the induced metric and area form of $S_{t,u}$. We note also that the terms that are boxed are in fact vanishing; we include them to stress that fact

$$\begin{aligned}\underline{\Lambda}(J^1) = &\, Qr\left[{}^{(X)}\mathbf{i}\,;\, \nabla\!\!\!/\underline{\beta}\right] + Qr\left[{}^{(X)}\mathbf{m}\,;\, \underline{\beta}_3\right] + Qr\left[{}^{(X)}\underline{\mathbf{m}}\,;\, \underline{\beta}_4\right]\\
&+ Qr\left[{}^{(X)}\underline{\mathbf{m}}\,;\, \nabla\!\!\!/(\rho,\sigma)\right] + Qr\left[{}^{(X)}\mathbf{j}\,;\, (\rho_3,\sigma_3)\right] + Qr\left[{}^{(X)}\underline{\mathbf{n}}\,;\, (\rho_4,\sigma_4)\right]\\
&+ \mathrm{tr}\underline{\chi}\left(Qr\left[{}^{(X)}\mathbf{m}\,;\, \underline{\beta}\right] + Qr\left[({}^{(X)}\mathbf{i},\, {}^{(X)}\mathbf{j})\,;\, (\rho,\sigma)\right] + Qr\left[{}^{(X)}\underline{\mathbf{m}}\,;\, \beta\right]\right)\\
&+ \mathrm{tr}\chi\left(Qr\left[({}^{(X)}\mathbf{i},\, {}^{(X)}\mathbf{j})\,;\, \underline{\alpha}\right] + Qr\left[{}^{(X)}\underline{\mathbf{m}}\,;\, \underline{\beta}\right] + Qr\left[{}^{(X)}\underline{\mathbf{n}}\,;\, (\rho,\sigma)\right]\right)\\
&+ l.o.t.\end{aligned}$$

$$\begin{aligned}\underline{K}(J^1) = &\, Qr\left[{}^{(X)}\mathbf{i}\,;\, \nabla\!\!\!/\underline{\beta}\right] + Qr\left[{}^{(X)}\mathbf{m}\,;\, \underline{\beta}_3\right] + Qr\left[{}^{(X)}\underline{\mathbf{m}}\,;\, \underline{\beta}_4\right]\\
&+ Qr\left[{}^{(X)}\underline{\mathbf{m}}\,;\, \nabla\!\!\!/(\rho,\sigma)\right] + Qr\left[{}^{(X)}\mathbf{j}\,;\, (\rho_3,\sigma_3)\right] + Qr\left[{}^{(X)}\underline{\mathbf{n}}\,;\, (\rho_4,\sigma_4)\right]\\
&+ \mathrm{tr}\underline{\chi}\left(Qr\left[{}^{(X)}\mathbf{m}\,;\, \underline{\beta}\right] + Qr\left[({}^{(X)}\mathbf{i},\, {}^{(X)}\mathbf{j})\,;\, (\rho,\sigma)\right] + Qr\left[{}^{(X)}\underline{\mathbf{m}}\,;\, \beta\right]\right)\\
&+ \mathrm{tr}\chi\left(Qr\left[({}^{(X)}\mathbf{i},\, {}^{(X)}\mathbf{j})\,;\, \underline{\alpha}\right] + Qr\left[{}^{(X)}\underline{\mathbf{m}}\,;\, \underline{\beta}\right] + Qr\left[{}^{(X)}\underline{\mathbf{n}}\,;\, (\rho,\sigma)\right]\right)\\
&+ l.o.t.\end{aligned}$$

$$\begin{aligned}\underline{I}(J^1) = &\, Qr\left[{}^{(X)}\mathbf{m}\,;\, \nabla\!\!\!/\underline{\beta}\right] + Qr\left[{}^{(X)}\mathbf{n}\,;\, \underline{\beta}_3\right] + Qr\left[{}^{(X)}\mathbf{j}\,;\, \underline{\beta}_4\right]\\
&+ Qr\left[{}^{(X)}\mathbf{i}\,;\, \nabla\!\!\!/(\rho,\sigma)\right] + Qr\left[{}^{(X)}\mathbf{m}\,;\, (\rho_3,\sigma_3)\right] + Qr\left[{}^{(X)}\underline{\mathbf{m}}\,;\, (\rho_4,\sigma_4)\right]\\
&+ \mathrm{tr}\underline{\chi}\left(Qr\left[{}^{(X)}\mathbf{n}\,;\, \underline{\beta}\right] + Qr\left[{}^{(X)}\mathbf{m}\,;\, (\rho,\sigma)\right] + Qr\left[{}^{(X)}\mathbf{i}\,;\, \beta\right]\right)\\
&+ \mathrm{tr}\chi\left(Qr\left[{}^{(X)}\mathbf{m}\,;\, \underline{\alpha}\right] + Qr\left[({}^{(X)}\mathbf{i},\, {}^{(X)}\mathbf{j})\,;\, \underline{\beta}\right] + Qr\left[{}^{(X)}\underline{\mathbf{m}}\,;\, (\rho,\sigma)\right]\right)\\
&+ l.o.t.\end{aligned}$$

$$\begin{aligned}\Xi(J^1) = &\, Qr\left[{}^{(X)}\mathbf{i}\,;\, \nabla\!\!\!/\alpha\right] + Qr\left[{}^{(X)}\underline{\mathbf{m}}\,;\, \alpha_4\right] + Qr\left[{}^{(X)}\mathbf{m}\,;\, \alpha_3\right]\\
&+ Qr\left[{}^{(X)}\mathbf{m}\,;\, \nabla\!\!\!/\beta\right] + Qr\left[{}^{(X)}\mathbf{j}\,;\, \beta_4\right] + Qr\left[{}^{(X)}\mathbf{n}\,;\, \beta_3\right]\\
&+ \mathrm{tr}\chi\left(Qr\left[{}^{(X)}\underline{\mathbf{m}}\,;\, \alpha\right] + Qr\left[({}^{(X)}\mathbf{i},\, {}^{(X)}\mathbf{j})\,;\, \beta\right] + Qr\left[{}^{(X)}\mathbf{m}\,;\, (\rho,\sigma)\right]\right)\\
&+ \mathrm{tr}\underline{\chi}\left(Qr\left[{}^{(X)}\mathbf{m}\,;\, \alpha\right] + Qr\left[{}^{(X)}\mathbf{n}\,;\, \beta\right]\right) + l.o.t.\end{aligned}$$

$$\begin{aligned}\Theta(J^1) = &\, Qr\left[{}^{(X)}\underline{\mathbf{m}}\,;\, \nabla\!\!\!/\alpha\right] + Qr\left[{}^{(X)}\underline{\mathbf{n}}\,;\, \alpha_4\right] + Qr\left[{}^{(X)}\mathbf{j}\,;\, \alpha_3\right]\\
&+ Qr\left[{}^{(X)}\mathbf{i}\,;\, \nabla\!\!\!/\beta\right] + Qr\left[{}^{(X)}\underline{\mathbf{m}}\,;\, \beta_4\right] + Qr\left[{}^{(X)}\mathbf{m}\,;\, \beta_3\right]\\
&+ Qr\left[{}^{(X)}\mathbf{m}\,;\, \nabla\!\!\!/(\rho,\sigma)\right] + Qr\left[{}^{(X)}\mathbf{j}\,;\, (\rho_4,\sigma_4)\right] + Qr\left[{}^{(X)}\mathbf{n}\,;\, (\rho_3,\sigma_3)\right]\\
&+ \mathrm{tr}\chi\Big(Qr\left[{}^{(X)}\underline{\mathbf{n}}\,;\, \alpha\right] + Qr\left[{}^{(X)}\underline{\mathbf{m}}\,;\, \beta\right] + Qr\left[({}^{(X)}\mathbf{i},\, {}^{(X)}\mathbf{j})\,;\, (\rho,\sigma)\right]\\
&+ Qr\left[{}^{(X)}\mathbf{m}\,;\, \underline{\beta}\right]\Big) + \mathrm{tr}\underline{\chi}\Big(\boxed{\quad Qr\left[{}^{(X)}\mathbf{i}\,;\, \alpha\right]\quad}\\
&+ Qr\left[{}^{(X)}\mathbf{j}\,;\, \alpha\right] + Qr\left[{}^{(X)}\mathbf{m}\,;\, \beta\right] + Qr\left[{}^{(X)}\mathbf{n}\,;\, (\rho,\sigma)\right]\Big) + l.o.t.\end{aligned}$$

$$\begin{aligned}\Lambda(J^1) = &Qr\left[{}^{(X)}\mathbf{i}\,;\,\nabla\!\!\!/\beta\right] + Qr\left[{}^{(X)}\underline{\mathbf{m}}\,;\,\beta_4\right] + Qr\left[{}^{(X)}\mathbf{m}\,;\,\beta_3\right]\\ &+ Qr\left[{}^{(X)}\mathbf{m}\,;\,\nabla\!\!\!/(\rho,\sigma)\right] + Qr\left[{}^{(X)}\mathbf{j}\,;\,(\rho_4,\sigma_4)\right] + Qr\left[{}^{(X)}\mathbf{n}\,;\,(\rho_3,\sigma_3)\right]\\ &+ \mathrm{tr}\chi\left(Qr\left[{}^{(X)}\underline{\mathbf{m}}\,;\,\beta\right] + Qr\left[({}^{(X)}\mathbf{i},\,{}^{(X)}\mathbf{j})\,;\,(\rho,\sigma)\right] + Qr\left[{}^{(X)}\mathbf{m}\,;\,\underline{\beta}\right]\right)\\ &+ \mathrm{tr}\underline{\chi}\left(Qr\left[({}^{(X)}\mathbf{i},\,{}^{(X)}\mathbf{j})\,;\,\alpha\right] + Qr\left[{}^{(X)}\mathbf{m}\,;\,\beta\right] + Qr\left[{}^{(X)}\mathbf{n}\,;\,(\rho,\sigma)\right]\right)\\ &+ l.o.t.\end{aligned}$$

$$\begin{aligned}K(J^1) = &Qr\left[{}^{(X)}\mathbf{i}\,;\,\nabla\!\!\!/\beta\right] + Qr\left[{}^{(X)}\underline{\mathbf{m}}\,;\,\beta_4\right] + Qr\left[{}^{(X)}\mathbf{m}\,;\,\beta_3\right]\\ &+ Qr\left[{}^{(X)}\mathbf{m}\,;\,\nabla\!\!\!/(\rho,\sigma)\right] + Qr\left[{}^{(X)}\mathbf{j}\,;\,(\rho_4,\sigma_4)\right] + Qr\left[{}^{(X)}\mathbf{n}\,;\,(\rho_3,\sigma_3)\right]\\ &+ \mathrm{tr}\chi\left(Qr\left[{}^{(X)}\underline{\mathbf{m}}\,;\,\beta\right] + Qr\left[({}^{(X)}\mathbf{i},\,{}^{(X)}\mathbf{j})\,;\,(\rho,\sigma)\right] + Qr\left[{}^{(X)}\mathbf{m}\,;\,\underline{\beta}\right]\right)\\ &+ \mathrm{tr}\underline{\chi}\left(Qr\left[({}^{(X)}\mathbf{i},\,{}^{(X)}\mathbf{j})\,;\,\alpha\right] + Qr\left[{}^{(X)}\mathbf{m}\,;\,\beta\right] + Qr\left[{}^{(X)}\mathbf{n}\,;\,(\rho,\sigma)\right]\right)\\ &+ l.o.t.\end{aligned}$$

$$\begin{aligned}I(J^1) = &Qr\left[{}^{(X)}\underline{\mathbf{m}}\,;\,\nabla\!\!\!/\beta\right] + Qr\left[{}^{(X)}\underline{\mathbf{n}}\,;\,\beta_4\right] + Qr\left[{}^{(X)}\mathbf{j}\,;\,\beta_3\right]\\ &+ Qr\left[{}^{(X)}\mathbf{i}\,;\,\nabla\!\!\!/(\rho,\sigma)\right] + Qr\left[{}^{(X)}\underline{\mathbf{m}}\,;\,(\rho_4,\sigma_4)\right] + Qr\left[{}^{(X)}\mathbf{m}\,;\,(\rho_3,\sigma_3)\right]\\ &+ \mathrm{tr}\chi\left(Qr\left[{}^{(X)}\underline{\mathbf{n}}\,;\,\beta\right] + Qr\left[{}^{(X)}\underline{\mathbf{m}}\,;\,(\rho,\sigma)\right] + Qr\left[{}^{(X)}\mathbf{i}\,;\,\underline{\beta}\right]\right)\\ &+ \mathrm{tr}\underline{\chi}\left(Qr\left[{}^{(X)}\underline{\mathbf{m}}\,;\,\alpha\right] + Qr\left[({}^{(X)}\mathbf{i},\,{}^{(X)}\mathbf{j})\,;\,\beta\right] + Qr\left[{}^{(X)}\mathbf{m}\,;\,(\rho,\sigma)\right]\right)\\ &+ l.o.t.\end{aligned}$$

Here the terms that we denote by l.o.t. are cubic with respect to ${}^{(X)}\hat{\pi}$, W *and the Ricci coefficients* $\underline{\xi}$, ζ, $\underline{\zeta}$, ν, $\underline{\nu}$, $\hat{\chi}$, *and* $\hat{\underline{\chi}}$ *and linear with regard to each of them separately. They are manifestly lower order by comparison to all other terms both in regard to their asymptotic behavior in the wave zone and the order of differentiability relative to* W. *Their precise form can be tracked down from the formulas 8.1.9a and 8.1.9b.*[3]

(ii.) The null decomposition of J^2 *is given by the following formulas:*

$$\begin{aligned}\underline{\Xi}(J^2) &= Qr\left[{}^{(X)}\not{p}\,;\,\underline{\alpha}\right] + Qr\left[{}^{(X)}p_3\,;\,\underline{\beta}\right]\\ \underline{\Theta}(J^2) &= Qr\left[{}^{(X)}p_4\,;\,\underline{\alpha}\right] + Qr\left[{}^{(X)}\not{p}\,;\,\underline{\beta}\right] + Qr\left[{}^{(X)}p_3\,;\,(\rho,\sigma)\right]\\ \underline{\Lambda}(J^2) &= Qr\left[{}^{(X)}\not{p}\,;\,\underline{\beta}\right] + Qr\left[{}^{(X)}p_3\,;\,(\rho,\sigma)\right]\\ \underline{K}(J^2) &= Qr\left[{}^{(X)}\not{p}\,;\,\underline{\beta}\right] + Qr\left[{}^{(X)}p_3\,;\,(\rho,\sigma)\right]\\ \underline{I}(J^2) &= Qr\left[{}^{(X)}p_4\,;\,\underline{\beta}\right] + Qr\left[{}^{(X)}\not{p}\,;\,(\rho,\sigma)\right]\\ I(J^2) &= Qr\left[{}^{(X)}p_3\,;\,\beta\right] + Qr\left[{}^{(X)}\not{p}\,;\,(\rho,\sigma)\right]\end{aligned}$$

[3] See also the remark following those formulas.

$$K(J^2) = Qr\left[{}^{(X)}\not{p};\,\beta\right] + Qr\left[{}^{(X)}p_4\,;\,(\rho,\sigma)\right]$$
$$\Lambda(J^2) = Qr\left[{}^{(X)}\not{p};\,\beta\right] + Qr\left[{}^{(X)}p_4\,;\,(\rho,\sigma)\right]$$
$$\Theta(J^2) = Qr\left[{}^{(X)}p_3\,;\,\alpha\right] + Qr\left[{}^{(X)}\not{p};\,\beta\right] + Qr\left[{}^{(X)}p_4\,;\,(\rho,\sigma)\right]$$
$$\Xi(J^2) = Qr\left[{}^{(X)}\not{p};\,\alpha\right] + Qr\left[{}^{(X)}p_4\,;\,\beta\right].$$

(iii.) The null decomposition of J^3 is given by the following formulas:

$$\begin{aligned}\underline{\Xi}(J^3) = {} & Qr\left[\underline{\alpha}\,;\,(I,\underline{I})({}^{(X)}q)\right]\\ & + Qr\left[\underline{\beta}\,;\,(\underline{K},\underline{\Lambda},\underline{\Theta})({}^{(X)}q)\right]\\ & + Qr\left[(\rho,\sigma)\,;\,\underline{\Xi}({}^{(X)}q)\right]\end{aligned}$$

$$\begin{aligned}\underline{\Theta}(J^3) = {} & Qr\left[\underline{\alpha}\,;\,K({}^{(X)}q)\right] + Qr\left[\underline{\alpha}\,;\,\Lambda({}^{(X)}q)\right]\\ & + \boxed{Qr\left[\underline{\alpha}\,;\,\Theta({}^{(X)}q)\right]}\\ & + Qr\left[\underline{\beta}\,;\,(I,\underline{I})({}^{(X)}q)\right] + Qr\left[(\rho,\sigma)\,;\,\underline{\Theta}({}^{(X)}q)\right]\\ & + \boxed{Qr\left[\beta\,;\,\underline{\Xi}({}^{(X)}q)\right]}\end{aligned}$$

$$\begin{aligned}\underline{\Lambda}(J^3) = {} & Qr\left[\underline{\alpha}\,;\,\Theta({}^{(X)}q)\right]\\ & + \boxed{Qr\left[\underline{\beta}\,;\,(I,\underline{I})({}^{(X)}q)\right]}\\ & + Qr\left[(\rho,\sigma)\,;\,(\underline{K},\underline{\Lambda})({}^{(X)}q)\right]\\ & + Qr\left[\beta\,;\,\underline{\Xi}({}^{(X)}q)\right]\end{aligned}$$

$$\begin{aligned}\underline{K}(J^3) = {} & Qr\left[\underline{\alpha}\,;\,\Theta({}^{(X)}q)\right]\\ & + \boxed{Qr\left[\underline{\beta}\,;\,(I,\underline{I})({}^{(X)}q)\right]}\\ & + Qr\left[(\rho,\sigma)\,;\,(\underline{K},\underline{\Lambda})({}^{(X)}q)\right]\\ & + Qr\left[\beta\,;\,\underline{\Xi}({}^{(X)}q)\right]\end{aligned}$$

$$\begin{aligned}\underline{I}(J^3) = {} & Qr\left[\underline{\alpha}\,;\,\Xi({}^{(X)}q)\right]\\ & + Qr\left[\underline{\beta}\,;\,(K,\Lambda,\Theta)({}^{(X)}q)\right]\end{aligned}$$

$$
\begin{aligned}
&+ Qr\left[(\rho, {}^{(X)}\sigma)\,;\,(I, \underline{I})({}^{(X)}q)\right] \\
&+ Qr\left[\beta\,;\,(\underline{K}, \underline{\Lambda}, \underline{\Theta})({}^{(X)}q)\right] \\
&+ \boxed{Qr\left[\alpha\,;\,\underline{\Xi}({}^{(X)}q)\right]}
\end{aligned}
$$

$$
\begin{aligned}
I(J^3) = &\boxed{Qr\left[\underline{\alpha}\,;\,\Xi({}^{(X)}q)\right]} \\
&+ Qr\left[\underline{\beta}\,;\,(K, \Lambda, \Theta)({}^{(X)}q)\right] \\
&+ Qr\left[(\rho, {}^{(X)}\sigma)\,;\,(I, \underline{I})({}^{(X)}q)\right] \\
&+ Qr\left[\beta\,;\,(\underline{K}, \underline{\Lambda}, \underline{\Theta})({}^{(X)}q)\right] \\
&+ Qr\left[\alpha\,;\,\underline{\Xi}({}^{(X)}q)\right]
\end{aligned}
$$

$$
\begin{aligned}
K(J^3) = &\; Qr\left[\alpha\,;\,\underline{\Theta}({}^{(X)}q)\right] \\
&+ \boxed{Qr\left[\beta\,;\,(I, \underline{I})({}^{(X)}q)\right]} \\
&+ Qr\left[(\rho, \sigma)\,;\,(K, \Lambda)({}^{(X)}q)\right] \\
&+ Qr\left[\underline{\beta}\,;\,\Xi({}^{(X)}q)\right]
\end{aligned}
$$

$$
\begin{aligned}
\Lambda(J^3) = &\; Qr\left[\alpha\,;\,\underline{\Theta}({}^{(X)}q)\right] \\
&+ \boxed{Qr\left[\beta\,;\,(I, \underline{I})({}^{(X)}q)\right]} \\
&+ Qr\left[(\rho, \sigma)\,;\,(K, \Lambda)({}^{(X)}q)\right] \\
&+ Qr\left[\underline{\beta}\,;\,\Xi({}^{(X)}q)\right]
\end{aligned}
$$

$$
\begin{aligned}
\Theta(J^3) = &\; Qr\left[\alpha\,;\,\underline{K}({}^{(X)}q)\right] + Qr\left[\alpha\,;\,\underline{\Lambda}({}^{(X)}q)\right] \\
&+ \boxed{Qr\left[\alpha\,;\,\underline{\Theta}({}^{(X)}q)\right]} \\
&+ Qr\left[\beta\,;\,(I, \underline{I})({}^{(X)}q)\right] + Qr\left[(\rho, \sigma)\,;\,\Theta({}^{(X)}q)\right] \\
&+ \boxed{Qr\left[\underline{\beta}\,;\,\Xi({}^{(X)}q)\right]}
\end{aligned}
$$

$$
\begin{aligned}
\Xi(J^3) = &\; Qr\left[\alpha\,;\,(I, \underline{I})({}^{(X)}q)\right] \\
&+ Qr\left[\beta\,;\,(K, \Lambda, \Theta)({}^{(X)}q)\right] \\
&+ Qr\left[(\rho, \sigma)\,;\,\Xi({}^{(X)}q)\right].
\end{aligned}
$$

Proof 8.1.3 *To prove the first part of the proposition, we write, in view of formulas 7.3.8a and 7.3.8j,*

$$
\begin{aligned}
\mathbf{D}_4 W_{A4B4} &= \not{\mathbf{D}}_4 \alpha_{AB} + l.o.t. \\
\mathbf{D}_3 W_{A4B4} &= \not{\mathbf{D}}_3 \alpha_{AB} + l.o.t. \\
\mathbf{D}_4 W_{A3B3} &= \not{\mathbf{D}}_4 \underline{\alpha}_{AB} + l.o.t. \\
\mathbf{D}_3 W_{A3B3} &= \not{\mathbf{D}}_3 \underline{\alpha}_{AB} + l.o.t. \\
\mathbf{D}_4 W_{A434} &= 2\not{\mathbf{D}}_4 \beta_A + l.o.t. \\
\mathbf{D}_3 W_{A434} &= 2\not{\mathbf{D}}_3 \beta_A + l.o.t. \\
\mathbf{D}_4 W_{A334} &= 2\not{\mathbf{D}}_4 \underline{\beta}_A + l.o.t. \\
\mathbf{D}_3 W_{A334} &= 2\not{\mathbf{D}}_3 \underline{\beta}_A + l.o.t. \\
\mathbf{D}_4 W_{3434} &= 4\mathbf{D}_4 \rho + l.o.t. \\
\mathbf{D}_3 W_{3434} &= 4\mathbf{D}_3 \rho + l.o.t. \\
\mathbf{D}_3 W_{ABC3} &= \in_{AB} \in_{CD} \mathbf{D}_3 \underline{\beta}_D + l.o.t. \\
\mathbf{D}_4 W_{ABC3} &= \in_{AB} \in_{CD} \mathbf{D}_4 \underline{\beta}_D + l.o.t. \\
\mathbf{D}_3 W_{AB34} &= 2 \in_{AB} \mathbf{D}_3 \sigma + l.o.t. \\
\mathbf{D}_4 W_{AB34} &= 2 \in_{AB} \mathbf{D}_4 \sigma + l.o.t. \\
\mathbf{D}_3 W_{A3B4} &= -\mathbf{D}_3 \rho \delta_{AB} + \mathbf{D}_3 \sigma \in_{AB} + l.o.t. \\
\mathbf{D}_4 W_{A3B4} &= -\mathbf{D}_4 \rho \delta_{AB} + \mathbf{D}_4 \sigma \in_{AB} + l.o.t.
\end{aligned}
\tag{8.1.9a}
$$

On the other hand,

$$
\begin{aligned}
\mathbf{D}_C W_{A4B4} &= \not\nabla_C \alpha_{AB} - \mathrm{tr}\chi(\delta_{AC}\beta_B + \delta_{BC}\beta_A - \delta_{AB}\beta_C) + l.o.t. \\
\mathbf{D}_C W_{A3B3} &= \not\nabla_C \underline{\alpha}_{AB} + \mathrm{tr}\underline{\chi}(\delta_{AC}\underline{\beta}_B + \delta_{BC}\underline{\beta}_A - \delta_{AB}\underline{\beta}_C) + l.o.t. \\
\mathbf{D}_B W_{A434} &= 2\not\nabla_B \beta_A - \mathrm{tr}\underline{\chi}\alpha_{AB} - 3\mathrm{tr}\chi(\rho\delta_{AB} + \sigma \in_{AB}) + l.o.t. \\
\mathbf{D}_B W_{A334} &= 2\not\nabla_B \underline{\beta}_A + \mathrm{tr}\chi\alpha_{AB} + 3\mathrm{tr}\underline{\chi}(\rho\delta_{AB} - \sigma \in_{AB}) + l.o.t. \\
\mathbf{D}_A W_{3434} &= 4\not\nabla_A \rho - 2\mathrm{tr}\underline{\chi}\beta_A + 2\mathrm{tr}\chi\underline{\beta}_A + l.o.t.
\end{aligned}
$$

$$
\begin{aligned}
\mathbf{D}_A W_{BCD4} = {}& -\delta_{BD}\not\nabla\beta_C + \delta_{CD}\not\nabla\beta_B + \frac{1}{4}\mathrm{tr}\underline{\chi}(\delta_{AB}\alpha_{CD} - \delta_{AC}\alpha_{BD}) \\
& + \frac{3}{4}\mathrm{tr}\chi\rho(\delta_{AC}\delta_{BD} - \delta_{AB}\delta_{CD}) \\
& + \frac{1}{4}\mathrm{tr}\chi\sigma(\delta_{AB} \in_{CD} - \delta_{AC} \in_{BD} - 2\delta_{AB} \in_{BC}) \\
& + l.o.t. \\
\mathbf{D}_A W_{BCD3} = {}& +\delta_{BD}\not\nabla\underline{\beta}_C - \delta_{CD}\not\nabla\underline{\beta}_B + \frac{1}{4}\mathrm{tr}\chi(\delta_{AB}\underline{\alpha}_{CD} - \delta_{AC}\underline{\alpha}_{BD}) \\
& + \frac{3}{4}\mathrm{tr}\underline{\chi}\rho(\delta_{AC}\delta_{BD} - \delta_{AB}\delta_{CD})
\end{aligned}
$$

$$- \frac{1}{4}\text{tr}\underline{\chi}\sigma(\delta_{AB} \in_{CD} - \delta_{AC} \in_{BD} - 2\delta_{AB} \in_{BC}) + l.o.t.$$

$$\mathbf{D}_A W_{CD34} = (2\nabla\!\!\!/_A \sigma + \text{tr}\underline{\chi}^\star \beta_A + \text{tr}\chi^\star \underline{\beta}_A) \in_{CD} + l.o.t.$$

$$\mathbf{D}_C W_{A3B4} = -\nabla\!\!\!/_C \rho \delta_{AB} + \nabla\!\!\!/_C \sigma \in_{AB} + \frac{1}{2}\text{tr}\underline{\chi}(\delta_{AB}\beta_C - \delta_{BC}\beta_A + \delta_{AB}\beta_C) + \frac{1}{2}\text{tr}\chi(\delta_{AC}\underline{\beta}_B - \delta_{BC}\underline{\beta}_A - \delta_{AB}\underline{\beta}_C) + l.o.t. \tag{8.1.9b}$$

Remark 1 *The terms that we denote by l.o.t. in 8.1.9a are terms involving products between a null component of W and one of the Ricci coefficients*

$$\underline{\xi}, \zeta, \underline{\zeta}, \nu, \underline{\nu}$$

introduced in Proposition 7.5.1. They can be easily accounted for, by the cautious reader, either by direct calculations or roughly, with the help of the principle of conservation of signature described in the previous section. In any case they are lower order, by comparison with all other terms that we took into consideration, both with regard to the order of differentiability relative in W as well as relative to their asymptotic behavior. This can be seen by recalling that the asymptotic behavior of the Ricci coefficients was postulated in Assumptions 0, 1, and 2 on page 176, while the asymptotic behavior of the null components of W was described by the comparison theorem on page 181. We can thus assess the behavior of each of the terms we denote by l.o.t. Thus, for example, by signature considerations we can write

$$\mathbf{D}_3 W_{A4B4} = \mathbf{D}\!\!\!\!/_3 \alpha_{AB} + Qr[\underline{\nu}; \alpha] + Qr\left[(\zeta, \underline{\zeta}); \beta\right] + Qr[\nu; (\rho, \sigma)];$$

hence the terms denoted by l.o.t. in the first formula of 8.1.9a have the form

$$l.o.t. = Qr[\underline{\nu}; \alpha] + Qr\left[(\zeta, \underline{\zeta}); \beta\right] + Qr[\nu; (\rho, \sigma)].$$

In view of the comparison theorem we have

$$\int_{\Sigma_t^e} r^6 |\mathbf{D}\!\!\!\!/_3 \alpha|^2 \leq c\mathcal{Q}_1.$$

Also,

$$\int_{\Sigma_t^e} r^4 |\alpha|^2 \leq c\mathcal{Q}_1$$

$$\int_{\Sigma_t^e} r^4 |\beta|^2 \leq c\mathcal{Q}_1,$$

and, for any small $h > 0$,

$$\int_{\Sigma_t^e} r^{3-2h}(|\rho|^2 + |\sigma|^2) \leq c\mathcal{Q}_1.$$

Therefore, taking into account Assumption 1 regarding the Ricci coefficients ζ, $\underline{\zeta}$, ν, and $\underline{\nu}$, we find that

$$\int_{\Sigma_t^e} r^{7-2h}|l.o.t.|^2 \leq c\varepsilon_0\mathcal{Q}_1,$$

which substantiates our claim that l.o.t. contains only lower-order terms by comparison to those we take into account.

The terms that we denote by l.o.t. in 8.1.9b contain, in addition to $\underline{\xi}$, ζ, $\underline{\zeta}$, ν, and $\underline{\nu}$, the Ricci coefficients $\hat{\chi}$ and $\hat{\underline{\chi}}$. We can take them into account by signature considerations. For example we can write

$$\begin{aligned} \mathbf{D}_C W_{A3B4} = &-\nabla\!\!\!/_C \rho\delta_{AB} + \nabla\!\!\!/_C\sigma \in_{AB} \\ &+ \mathrm{Qr}\left[\underline{\xi}\,;\,\alpha\right] + \mathrm{Qr}\left[(\hat{\underline{\chi}},\underline{\nu})\,;\,\beta\right] \\ &+ \mathrm{Qr}\left[(\zeta,\underline{\zeta})\,;\,(\rho,\sigma)\right] + \mathrm{Qr}\left[(\hat{\chi},\nu)\,;\,\underline{\beta}\right]. \end{aligned}$$

Thus, in the last formula of 8.1.9b, we keep the angular derivatives of ρ, σ and the quadratic terms[4] containing $\mathrm{tr}\chi$ and $\mathrm{tr}\underline{\chi}$ and treat the remaining terms as lower order. Thus,

$$\begin{aligned} \text{l.o.t.} = &\;\mathrm{Qr}\left[\underline{\xi}\,;\,\alpha\right] + \mathrm{Qr}\left[(\hat{\underline{\chi}},\underline{\nu})\,;\,\beta\right] \\ &+ \mathrm{Qr}\left[(\zeta,\underline{\zeta})\,;\,(\rho,\sigma)\right] + \mathrm{Qr}\left[(\hat{\chi},\nu)\,;\,\underline{\beta}\right]. \end{aligned}$$

One can then make precise the meaning of "lower order" in the same manner.

The proof of the formulas on part i of the proposition are an immediate consequence of equations 8.1.9a and 8.1.9b and the formula for J^1 given in 8.1.2a.

The proof for the formulas corresponding to J^2 follows in a straightforward manner from its expression in formula 8.1.2a. Finally, to prove part iii of the proposition, we rely on the formula for J^3 in 8.1.2a and the null decomposition of ${}^{(X)}q$.

One can easily assess the structure of the terms that appear in the above formulas with the help of the principle of conservation of signature. In order to

[4] The precise form of these terms is immaterial in our considerations.

do this, we associate to the null components of both J^3 and ${}^{(X)}q$ the following signatures:

$$
\begin{aligned}
s(\underline{\Xi}) &= -2 & s(\Xi) &= 2 \\
s(\underline{K}, \underline{\Lambda}, \underline{\Theta}) &= -1 & s(K, \Lambda, \Theta) &= +1 \\
s(I, \underline{I}) &= 0
\end{aligned}
\tag{8.1.10}
$$

The terms that cannot be eliminated by signature arguments or other straightforward considerations and that are nevertheless boxed in our formulas in order to stress their vanishing can easily be computed by direct calculations.

8.2. Statement of the Boundedness Theorem

In order to control the error terms, we need to make the following assumptions on the traceless parts of the deformation tensors for the vectorfields $T, \bar{K}, S, \mathbf{O}$. These assumptions are expressed relative to the exterior and interior norms $\| \, \|_{p,e}$, $\| \, \|_{p,\imath}$, for $p = 2, \infty$, as well as the local norms $||| \, |||_{p,e}$ for $p = 2, 4$. These norms were introduced in the previous section.[5] We also introduce the following local L^2 norm:

$$
{}^{(\mathrm{loc})}\|V\|_{2,e}(t) = \sup_{\lambda \geq u_0(t)} \left(\int_{D(\lambda,t)} |V|^2 \right)^{1/2}
$$

where $D(\lambda, t)$ is the annulus $\lambda \leq u \leq \lambda + 1$ on Σ_t.

8.2.0.1. Interior Assumptions

IA 0:

$$
\|{}^{(T)}\hat{\pi}\|_{\infty,\imath} \leq \varepsilon_0 r_0^{-3/2} \tag{8.2.1a}
$$

$$
\|{}^{(\mathbf{O})}\hat{\pi}\|_{\infty,\imath} \leq \varepsilon_0 r_0^{-1/2} \tag{8.2.1b}
$$

$$
\|{}^{(\bar{K})}\hat{\pi}\|_{\infty,\imath} \leq \varepsilon_0 r_0^{1/2}. \tag{8.2.1c}
$$

IA 1:

$$
\|\mathbf{D}{}^{(T)}\hat{\pi}\|_{2,\imath} \leq \varepsilon_0 r_0^{-1} \tag{8.2.1d}
$$

$$
\|\mathbf{D}{}^{(\mathbf{O})}\hat{\pi}\|_{2,\imath} \leq \varepsilon_0. \tag{8.2.1e}
$$

[5] We note that the interior norm $\| \, \|_{p,\imath}$ used here is the same as $\| \, \|_{p,I}$ used in the previous section.

IA 2:

$$\|\mathbf{D}^{2\,(T)}\hat{\pi}\|_{2,\imath} \leq \varepsilon_0 r_0^{-2} \tag{8.2.1f}$$
$$\|r\nabla\!\!\!/\mathbf{D}^{\,(\mathbf{O})}\hat{\pi}\|_{2,\imath} \leq \varepsilon_0 \tag{8.2.1g}$$

Also,

$$\|r^{1/2}\mathbf{D}^{(\mathbf{O})}\hat{\pi}\|_{4,\imath} \leq \varepsilon_0 r_0^{-1/2} \tag{8.2.1h}$$

$$\|^{(S)}\hat{\pi}\|_{\infty,\imath} \leq \varepsilon_0 r_0^{-1/2} \tag{8.2.1i}$$
$$\|\mathbf{D}^{(S)}\hat{\pi}\|_{2,\imath} \leq \varepsilon_0. \tag{8.2.1j}$$

8.2.0.2. Exterior Assumptions

EA 0:

$$\begin{aligned} \|r\tau_-\,{}^{(T)}\mathbf{i}\|_{\infty,e} &\leq \varepsilon_0 \\ \|r^{2\,(T)}\mathbf{j}\|_{\infty,e} &\leq \varepsilon_0 \\ \|r^2({}^{(T)}\mathbf{m},\ {}^{(T)}\underline{\mathbf{m}},\ {}^{(T)}\mathbf{n},\ {}^{(T)}\underline{\mathbf{n}})\|_{\infty,e} &\leq \varepsilon_0 \end{aligned} \tag{8.2.2a}$$

$$\begin{aligned} {}^{(\mathbf{O})}\mathbf{m},\ {}^{(\mathbf{O})}\mathbf{n} &= 0 \\ \|r({}^{(\mathbf{O})}\mathbf{i},\ {}^{(\mathbf{O})}\mathbf{j},\ {}^{(\mathbf{O})}\underline{\mathbf{m}},\ {}^{(\mathbf{O})}\underline{\mathbf{n}})\|_{\infty,e} &\leq \varepsilon_0 \end{aligned} \tag{8.2.2b}$$

$$\begin{aligned} \|({}^{(\bar{K})}\mathbf{i},\ {}^{(\bar{K})}\mathbf{j},\ {}^{(\bar{K})}\underline{\mathbf{m}},\ {}^{(\bar{K})}\underline{\mathbf{n}})\|_{\infty,e} &\leq \varepsilon_0 \\ \|r^2\tau_-^{-2}({}^{(\bar{K})}\mathbf{m},\ {}^{(\bar{K})}\mathbf{n})\|_{\infty,e} &\leq \varepsilon_0. \end{aligned} \tag{8.2.2c}$$

EA 1:

$$\begin{aligned} {}^{(\mathrm{loc})}\|r\nabla\!\!\!/({}^{(\mathbf{O})}\mathbf{i},\ {}^{(\mathbf{O})}\mathbf{j},\ {}^{(\mathbf{O})}\underline{\mathbf{m}},\ {}^{(\mathbf{O})}\underline{\mathbf{n}})\|_{2,e} &\leq \varepsilon_0 \\ {}^{(\mathrm{loc})}\|\tau_-\mathbf{D}\!\!\!/_3({}^{(\mathbf{O})}\mathbf{i},\ {}^{(\mathbf{O})}\mathbf{j},\ {}^{(\mathbf{O})}\underline{\mathbf{m}},\ {}^{(\mathbf{O})}\underline{\mathbf{n}})\|_{2,e} &\leq \varepsilon_0 \\ {}^{(\mathrm{loc})}\|r\mathbf{D}\!\!\!/_4({}^{(\mathbf{O})}\mathbf{i},\ {}^{(\mathbf{O})}\mathbf{j},\ {}^{(\mathbf{O})}\underline{\mathbf{m}},\ {}^{(\mathbf{O})}\underline{\mathbf{n}})\|_{2,e} &\leq \varepsilon_0. \end{aligned} \tag{8.2.2d}$$

Moreover,

$$|||r^{3/2}\mathbf{D}\!\!\!/_4({}^{(\mathbf{O})}\mathbf{i},{}^{(\mathbf{O})}\mathbf{j})|||_{4,e} \leq \varepsilon_0 \tag{8.2.2e}$$

$$\begin{aligned} \|r^{-1/2}\tau_-^{1/2}\nabla\!\!\!/^{(T)}\mathbf{i}\|_{2,e} &\leq \varepsilon_0(1+t)^{-3/2} \\ \|r^{-3/2}\tau_-^{3/2}\mathbf{D}\!\!\!/_3^{(T)}\mathbf{i}\|_{2,e} &\leq \varepsilon_0(1+t)^{-3/2} \end{aligned}$$

$$\begin{aligned}
\|\not{D}_4^{(T)}\mathbf{i} + \frac{1}{2}\mathrm{tr}\chi^{(T)}\mathbf{i}\|_{2,e} &\le \varepsilon_0(1+t)^{-3/2}\\
\|\not\nabla({}^{(T)}\mathbf{j},\ {}^{(T)}\mathbf{m},\ {}^{(T)}\underline{\mathbf{m}},\ {}^{(T)}\mathbf{n},\ {}^{(T)}\underline{\mathbf{n}})\|_{2,e} &\le \varepsilon_0(1+t)^{-3/2}\\
\|\not{D}_4({}^{(T)}\mathbf{j},\ {}^{(T)}\mathbf{m},\ {}^{(T)}\underline{\mathbf{m}},\ {}^{(T)}\mathbf{n},\ {}^{(T)}\underline{\mathbf{n}})\|_{2,e} &\le \varepsilon_0(1+t)^{-3/2}\\
\|r^{-1/2}\tau_-^{1/2}\not{D}_3({}^{(T)}\mathbf{j},\ {}^{(T)}\mathbf{m},\ {}^{(T)}\underline{\mathbf{m}},\ {}^{(T)}\underline{\mathbf{n}})\|_{2,e} &\le \varepsilon_0(1+t)^{-3/2}\\
\|\not{D}_3^{(T)}\mathbf{n}\|_{2,e} &\le \varepsilon_0(1+t)^{-3/2}.
\end{aligned} \tag{8.2.2f}$$

EA 2:

$$\begin{aligned}
\|r({}^{(S)}\mathbf{i},\ {}^{(S)}\mathbf{j},\ {}^{(S)}\underline{\mathbf{m}},\ {}^{(S)}\underline{\mathbf{n}})\|_{\infty,e} &\le \varepsilon_0\\
\|r^2\tau_-({}^{(S)}\mathbf{m},\ {}^{(S)}\mathbf{n})\|_{\infty,e} &\le \varepsilon_0
\end{aligned} \tag{8.2.2g}$$

$$\begin{aligned}
{}^{(loc)}\|\not\nabla({}^{(S)}\mathbf{i},\ {}^{(S)}\mathbf{j},\ {}^{(S)}\underline{\mathbf{m}},\ {}^{(S)}\underline{\mathbf{n}})\|_{2,e} &\le \varepsilon_0(1+t)^{-1}\\
{}^{(loc)}\|r\tau_-^{-1}\not\nabla({}^{(S)}\mathbf{m},\ {}^{(S)}\mathbf{n})\|_{2,e} &\le \varepsilon_0(1+t)^{-1}\\
{}^{(loc)}\|\not{D}_4({}^{(S)}\mathbf{i},\ {}^{(S)}\mathbf{j},\ {}^{(S)}\underline{\mathbf{m}},\ {}^{(S)}\underline{\mathbf{n}})\|_{2,e} &\le \varepsilon_0(1+t)^{-1}\\
{}^{(loc)}\|r\tau_-^{-1}\not{D}_4({}^{(S)}\mathbf{m},\ {}^{(S)}\mathbf{n})\|_{2,e} &\le \varepsilon_0(1+t)^{-1}\\
{}^{(loc)}\|r^{-1}\tau_-\not{D}_3({}^{(S)}\mathbf{i},\ {}^{(S)}\mathbf{j},\ {}^{(S)}\underline{\mathbf{m}},\ {}^{(S)}\underline{\mathbf{n}})\|_{2,e} &\le \varepsilon_0(1+t)^{-1}\\
{}^{(loc)}\|\not{D}_3({}^{(S)}\mathbf{m},\ {}^{(S)}\mathbf{n})\|_{2,e} &\le \varepsilon_0(1+t)^{-1}
\end{aligned} \tag{8.2.2h}$$

$$\begin{aligned}
|||r^{3/2}\not\nabla({}^{(\mathrm{O})}\mathbf{i},{}^{(\mathrm{O})}\mathbf{j},{}^{(\mathrm{O})}\underline{\mathbf{m}},{}^{(\mathrm{O})}\underline{\mathbf{n}})|||_{4,e} &\le \varepsilon_0\\
|||r^{1/2}\tau_-\not{D}_3({}^{(\mathrm{O})}\mathbf{i},{}^{(\mathrm{O})}\mathbf{j},{}^{(\mathrm{O})}\underline{\mathbf{m}},{}^{(\mathrm{O})}\underline{\mathbf{n}})|||_{4,e} &\le \varepsilon_0\\
|||r^{3/2}\not{D}_4({}^{(\mathrm{O})}\mathbf{i},{}^{(\mathrm{O})}\mathbf{j},{}^{(\mathrm{O})}\underline{\mathbf{m}},{}^{(\mathrm{O})}\underline{\mathbf{n}})|||_{4,e} &\le \varepsilon_0
\end{aligned} \tag{8.2.2i}$$

$$\begin{aligned}
{}^{(loc)}\|r^2\not\nabla^2({}^{(\mathrm{O})}\mathbf{i},\ {}^{(\mathrm{O})}\mathbf{j},\ {}^{(\mathrm{O})}\underline{\mathbf{m}},\ {}^{(\mathrm{O})}\underline{\mathbf{n}})\|_{2,e} &\le \varepsilon_0\\
{}^{(loc)}\|r\tau_-\not\nabla\not{D}_3({}^{(\mathrm{O})}\mathbf{i},\ {}^{(\mathrm{O})}\mathbf{j},\ {}^{(\mathrm{O})}\underline{\mathbf{m}},\ {}^{(\mathrm{O})}\underline{\mathbf{n}})\|_{2,e} &\le \varepsilon_0\\
{}^{(loc)}\|r^2\not\nabla\not{D}_4({}^{(\mathrm{O})}\mathbf{i},\ {}^{(\mathrm{O})}\mathbf{j},\ {}^{(\mathrm{O})}\underline{\mathbf{m}},\ {}^{(\mathrm{O})}\underline{\mathbf{n}})\|_{2,e} &\le \varepsilon_0.
\end{aligned} \tag{8.2.2j}$$

Moreover,

$$\begin{aligned}
|||r^{5/2}\not\nabla\not{D}_4({}^{(\mathrm{O})}\mathbf{i},\ {}^{(\mathrm{O})}\mathbf{j})|||_{4,e} &\le \varepsilon_0\\
{}^{(loc)}\|r^3\not\nabla^2\not{D}_4({}^{(\mathrm{O})}\mathbf{i},\ {}^{(\mathrm{O})}\mathbf{j})\|_{2,e} &\le \varepsilon_0
\end{aligned} \tag{8.2.2k}$$

$$\|r^{1/2}\tau_-^{1/2}\nabla\!\!\!\!/^2\,{}^{(T)}\mathbf{i}\|_{2,e} \le \varepsilon_0(1+t)^{-3/2}$$
$$\|r^{-1/2}\tau_-^{3/2}\nabla\!\!\!\!/\mathbf{D}\!\!\!\!/_3\,{}^{(T)}\mathbf{i}\|_{2,e} \le \varepsilon_0(1+t)^{-3/2}$$
$$\|r\nabla\!\!\!\!/(\mathbf{D}\!\!\!\!/_4^{(T)}\mathbf{i}+\frac{1}{2}\mathrm{tr}\chi^{(T)}\mathbf{i})\|_{2,e} \le \varepsilon_0(1+t)^{-3/2}$$
$$\|r\nabla\!\!\!\!/^2({}^{(T)}\mathbf{j},\ {}^{(T)}\mathbf{m},\ {}^{(T)}\underline{\mathbf{m}},\ {}^{(T)}\mathbf{n},\ {}^{(T)}\underline{\mathbf{n}})\|_{2,e} \le \varepsilon_0(1+t)^{-3/2}$$
$$\|r\nabla\!\!\!\!/\mathbf{D}\!\!\!\!/_4({}^{(T)}\mathbf{j},\ {}^{(T)}\mathbf{m},\ {}^{(T)}\underline{\mathbf{m}},\ {}^{(T)}\mathbf{n},\ {}^{(T)}\underline{\mathbf{n}})\|_{2,e} \le \varepsilon_0(1+t)^{-3/2}$$
$$\|r^{1/2}\tau_-^{1/2}\nabla\!\!\!\!/\mathbf{D}\!\!\!\!/_3({}^{(T)}\mathbf{j},\ {}^{(T)}\mathbf{m},\ {}^{(T)}\underline{\mathbf{m}},\ {}^{(T)}\underline{\mathbf{n}})\|_{2,e} \le \varepsilon_0(1+t)^{-3/2}$$
$$\|r\nabla\!\!\!\!/\mathbf{D}\!\!\!\!/_3^{(T)}\mathbf{n}\|_{2,e} \le \varepsilon_0(1+t)^{-3/2}$$

$$\|r^{-1/2}\tau_-^{1/2}\mathbf{D}\!\!\!\!/_S\nabla\!\!\!\!/\,{}^{(T)}\mathbf{i}\|_{2,e} \le \varepsilon_0(1+t)^{-3/2}$$
$$\|r^{-3/2}\tau_-^{3/2}\mathbf{D}\!\!\!\!/_S\mathbf{D}\!\!\!\!/_3\,{}^{(T)}\mathbf{i}\|_{2,e} \le \varepsilon_0(1+t)^{-3/2}$$
$$\|\mathbf{D}\!\!\!\!/_S(\mathbf{D}\!\!\!\!/_4^{(T)}\mathbf{i}+\frac{1}{2}\mathrm{tr}\chi^{(T)}\mathbf{i})\|_{2,e} \le \varepsilon_0(1+t)^{-3/2}$$
$$\|\mathbf{D}\!\!\!\!/_S\nabla\!\!\!\!/({}^{(T)}\mathbf{j},\ {}^{(T)}\mathbf{m},\ {}^{(T)}\underline{\mathbf{m}},\ {}^{(T)}\mathbf{n},\ {}^{(T)}\underline{\mathbf{n}})\|_{2,e} \le \varepsilon_0(1+t)^{-3/2}$$
$$\|\mathbf{D}\!\!\!\!/_S\mathbf{D}\!\!\!\!/_4({}^{(T)}\mathbf{j},\ {}^{(T)}\mathbf{m},\ {}^{(T)}\underline{\mathbf{m}},\ {}^{(T)}\mathbf{n},\ {}^{(T)}\underline{\mathbf{n}})\|_{2,e} \le \varepsilon_0(1+t)^{-3/2}$$
$$\|r^{-1/2}\tau_-^{1/2}\mathbf{D}\!\!\!\!/_S\mathbf{D}\!\!\!\!/_3({}^{(T)}\mathbf{j},\ {}^{(T)}\mathbf{m},\ {}^{(T)}\underline{\mathbf{m}},\ {}^{(T)}\underline{\mathbf{n}})\|_{2,e} \le \varepsilon_0(1+t)^{-3/2}$$
$$\|\mathbf{D}\!\!\!\!/_S\mathbf{D}\!\!\!\!/_3^{(T)}\mathbf{n}\|_{2,e} \le \varepsilon_0(1+t)^{-3/2}. \tag{8.2.21}$$

Proposition 8.2.1 *Let ${}^{(\mathbf{O})}p,\ {}^{(T)}p,\ {}^{(S)}p$ and ${}^{(\mathbf{O})}q,\ {}^{(T)}q,\ {}^{(S)}q$, associated with the vectorfields $\mathbf{O}, T, S$, be defined according to 8.1.2a and 8.1.2b. The Interior and Exterior Assumptions 0, 1, and 2 imply the following:*

8.2.0.3. Interior Estimates for ${}^{(\mathbf{O})}q,\ {}^{(\mathbf{O})}p$

The Interior Assumptions 8.2.1e and 8.2.1gimply the following:

$$\|{}^{(\mathbf{O})}p,\ {}^{(\mathbf{O})}q\|_{2,\imath} \le \varepsilon_0 \tag{8.2.3a}$$
$$\|\mathbf{D}({}^{(\mathbf{O})}p,\ {}^{(\mathbf{O})}q)\|_{2,\imath} \le \varepsilon_0 r_0^{-1}. \tag{8.2.3b}$$

8.2.0.4. Exterior Estimates for ${}^{(\mathbf{O})}q$, ${}^{(\mathbf{O})}p$

1. The assumptions 8.2.2d for ${}^{(\mathbf{O})}\hat{\pi}$ imply

$$
\begin{aligned}
{}^{(loc)}\|r\,{}^{(\mathbf{O})}p_4\|_{2,e} &\le c & {}^{(loc)}\|\tau_-^{(\mathbf{O})}p_3\|_{2,e} &\le c\varepsilon_0\\
{}^{(loc)}\|r\,{}^{(\mathbf{O})}\not{p}\|_{2,e} &\le c\varepsilon_0 & &\\
{}^{(loc)}\|r\Lambda({}^{(\mathbf{O})}q)\|_{2,e} &\le c\varepsilon_0 & {}^{(loc)}\|\tau_-\underline{\Lambda}({}^{(\mathbf{O})}q)\|_{2,e} &\le c\varepsilon_0\\
{}^{(loc)}\|rK({}^{(\mathbf{O})}q)\|_{2,e} &\le c\varepsilon_0 & {}^{(loc)}\|r\underline{K}({}^{(\mathbf{O})}q)\|_{2,e} &\le c\varepsilon_0\\
{}^{(loc)}\|rI({}^{(\mathbf{O})}q)\|_{2,e} &\le c\varepsilon_0 & {}^{(loc)}\|r\underline{I}({}^{(\mathbf{O})}q)\|_{2,e} &\le c\varepsilon_0\\
{}^{(loc)}\|r\Theta({}^{(\mathbf{O})}q)\|_{2,e} &\le c\varepsilon_0 & {}^{(loc)}\|\tau_-\underline{\Theta}({}^{(\mathbf{O})}q)\|_{2,e} &\le c\varepsilon_0\\
\Xi({}^{(\mathbf{O})}q) &= 0 & {}^{(loc)}\|\tau_-\underline{\Xi}({}^{(\mathbf{O})}q)\|_{2,e} &\le c\varepsilon_0.
\end{aligned}
\tag{8.2.3c}
$$

Moreover, if the assumption 8.2.2e is also verified, then

$$
\begin{aligned}
|||r^{3/2}\,{}^{(\mathbf{O})}p_4|||_{4,e} &\le c\varepsilon_0\\
|||r^{3/2}\,K({}^{(\mathbf{O})}q)|||_{4,e} &\le c\varepsilon_0\\
|||r^{3/2}\,\Lambda({}^{(\mathbf{O})}q)|||_{4,e} &\le c\varepsilon_0\\
|||r^{3/2}\,\Theta({}^{(\mathbf{O})}q)|||_{4,e} &\le c\varepsilon_0.
\end{aligned}
\tag{8.2.3d}
$$

2. The assumptions 8.2.2j imply

$$
\begin{aligned}
&{}^{(loc)}\|r^2\nabla\!\!\!/\,{}^{(\mathbf{O})}p_4\|_{2,e} \leq c\varepsilon_0 && {}^{(loc)}\|r\tau_-\nabla\!\!\!/\,{}^{(\mathbf{O})}p_3\|_{2,e} \leq c\varepsilon_0\\
&{}^{(loc)}\|r^2\nabla\!\!\!/\,{}^{(\mathbf{O})}p\!\!\!/\|_{2,e} \leq c\varepsilon_0
\end{aligned}
$$

$$
\begin{aligned}
&{}^{(loc)}\|r^2\nabla\!\!\!/\,\Lambda({}^{(\mathbf{O})}q)\|_{2,e} \leq c\varepsilon_0 && {}^{(loc)}\|r\tau_-\nabla\!\!\!/\,\underline{\Lambda}({}^{(\mathbf{O})}q)\|_{2,e} \leq c\varepsilon_0\\
&{}^{(loc)}\|r^2\nabla\!\!\!/\,K({}^{(\mathbf{O})}q)\|_{2,e} \leq c\varepsilon_0 && {}^{(loc)}\|r^2\nabla\!\!\!/\,\underline{K}({}^{(\mathbf{O})}q)\|_{2,e} \leq c\varepsilon_0\\
&{}^{(loc)}\|r^2\nabla\!\!\!/\,I({}^{(\mathbf{O})}q)\|_{2,e} \leq c\varepsilon_0 && {}^{(loc)}\|r^2\nabla\!\!\!/\,\underline{I}({}^{(\mathbf{O})}q)\|_{2,e} \leq c\varepsilon_0\\
&{}^{(loc)}\|r^2\nabla\!\!\!/\,\Theta({}^{(\mathbf{O})}q)\|_{2,e} \leq c\varepsilon_0 && {}^{(loc)}\|r\tau_-\nabla\!\!\!/\,\underline{\Theta}({}^{(\mathbf{O})}q)\|_{2,e} \leq c\varepsilon_0\\
& && {}^{(loc)}\|r\tau_-\nabla\!\!\!/\,\Xi({}^{(\mathbf{O})}q)\|_{2,e} \leq c\varepsilon_0.
\end{aligned}
\tag{8.2.3e}
$$

Moreover, if the auxiliary assumption 8.2.2k is verified, then

$$
\begin{aligned}
{}^{(loc)}\|r^3\nabla\!\!\!/^{2}\,{}^{(\mathbf{O})}p_4\|_{2,e} &\leq c\varepsilon_0\\
{}^{(loc)}\|r^3\nabla\!\!\!/^2 K({}^{(\mathbf{O})}q)\|_{2,e} &\leq c\varepsilon_0\\
{}^{(loc)}\|r^3\nabla\!\!\!/^2 \Lambda({}^{(\mathbf{O})}q)\|_{2,e} &\leq c\varepsilon_0\\
{}^{(loc)}\|r^3\nabla\!\!\!/^2 \Theta({}^{(\mathbf{O})}q)\|_{2,e} &\leq c\varepsilon_0.
\end{aligned}
\tag{8.2.3f}
$$

8.2.0.5. Interior Estimates for ${}^{(T)}q$, ${}^{(T)}p$

$$\|{}^{(T)}p,\ {}^{(T)}q\|_{2,\imath} \leq cr_0^{-1} \tag{8.2.4a}$$
$$\|\mathbf{D}({}^{(T)}p,\ {}^{(T)}q)\|_{2,\imath} \leq \varepsilon_0 c r_0^{-2}. \tag{8.2.4b}$$

8.2.0.6. Exterior Estimates for ${}^{(T)}q$, ${}^{(T)}p$

1. *The assumptions 8.2.2f imply*

$$
\begin{aligned}
\|{}^{(T)}p_4\|_{2,e} &\leq c\varepsilon_0(1+t)^{-3/2}\\
\|r^{-1/2}\tau_-^{1/2}\,{}^{(T)}p_3\|_{2,e} &\leq c\varepsilon_0(1+t)^{-3/2}\\
\|r^{-1/2}\tau_-^{1/2}\,{}^{(T)}p\!\!\!/\|_{2,e} &\leq c\varepsilon_0(1+t)^{-3/2}
\end{aligned}
$$

$$
\begin{aligned}
\|\Lambda({}^{(T)}q)\|_{2,e} &\le c\varepsilon_0(1+t)^{-3/2}\\
\|r^{-1/2}\tau_-^{1/2}\ \underline{\Lambda}({}^{(T)}q)\|_{2,e} &\le c\varepsilon_0(1+t)^{-3/2}\\
\|K({}^{(T)}q)\|_{2,e} &\le c\varepsilon_0(1+t)^{-3/2}\\
\|r^{-1/2}\tau_-^{1/2}\ \underline{K}({}^{(T)}q)\|_{2,e} &\le c\varepsilon_0(1+t)^{-3/2}\\
\|\Theta({}^{(T)}q)\|_{2,e} &\le c\varepsilon_0(1+t)^{-3/2}\\
\|r^{-3/2}\tau_-^{3/2}\ \underline{\Theta}({}^{(T)}q)\|_{2,e} &\le c\varepsilon_0(1+t)^{-3/2}\\
\|r^{-1/2}\tau_-^{1/2}\ I({}^{(T)}q)\|_{2,e} &\le c\varepsilon_0(1+t)^{-3/2}\\
\|r^{-1/2}\tau_-^{1/2}\ \underline{I}({}^{(T)}q)\|_{2,e} &\le c\varepsilon_0(1+t)^{-3/2}\\
\|\Xi({}^{(T)}q)\|_{2,e} &\le c\varepsilon_0(1+t)^{-3/2}\\
\|r^{-1/2}\tau_-^{1/2}\ \underline{\Xi}({}^{(T)}q)\|_{2,e} &\le c\varepsilon_0(1+t)^{-3/2}.
\end{aligned}
\tag{8.2.4c}
$$

2. *The assumptions 8.2.2l imply*

$$
\begin{aligned}
\|r\nabla\!\!\!/^{(T)}p_4\|_{2,e} &\le \varepsilon_0 c(1+t)^{-3/2}\\
\|r^{1/2}\tau_-^{1/2}\ \nabla\!\!\!/^{(T)}p_3\|_{2,e} &\le \varepsilon_0 c(1+t)^{-3/2}\\
\|r^{1/2}\tau_-^{1/2}\ \nabla\!\!\!/^{(T)}p\!\!\!/\|_{2,e} &\le \varepsilon_0 c(1+t)^{-3/2} \qquad (8.2.4d)\\
\|r\nabla\!\!\!/\Lambda({}^{(T)}q)\|_{2,e} &\le \varepsilon_0 c(1+t)^{-3/2}\\
\|r^{1/2}\tau_-^{1/2}\ \nabla\!\!\!/\underline{\Lambda}({}^{(T)}q)\|_{2,e} &\le \varepsilon_0 c(1+t)^{-3/2}\\
\|r\nabla\!\!\!/K({}^{(T)}q)\|_{2,e} &\le \varepsilon_0 c(1+t)^{-3/2}\\
\|r^{1/2}\tau_-^{1/2}\ \nabla\!\!\!/\underline{K}({}^{(T)}q)\|_{2,e} &\le \varepsilon_0 c(1+t)^{-3/2}\\
\|r\nabla\!\!\!/\Theta({}^{(T)}q)\|_{2,e} &\le \varepsilon_0 c(1+t)^{-3/2}\\
\|r^{-1/2}\tau_-^{3/2}\ \nabla\!\!\!/\underline{\Theta}({}^{(T)}q)\|_{2,e} &\le \varepsilon_0 c(1+t)^{-3/2}\\
\|r^{1/2}\tau_-^{1/2}\ \nabla\!\!\!/I({}^{(T)}q)\|_{2,e} &\le \varepsilon_0 c(1+t)^{-3/2}\\
\|r^{1/2}\tau_-^{1/2}\ \nabla\!\!\!/\underline{I}({}^{(T)}q)\|_{2,e} &\le \varepsilon_0 c(1+t)^{-3/2}\\
\|r\nabla\!\!\!/\Xi({}^{(T)}q)\|_{2,e} &\le \varepsilon_0 c(1+t)^{-3/2}\\
\|r^{1/2}\tau_-^{1/2}\ \nabla\!\!\!/\underline{\Xi}({}^{(T)}q)\|_{2,e} &\le \varepsilon_0 c(1+t)^{-3/2}. \qquad (8.2.4e)
\end{aligned}
$$

Also,

$$
\begin{aligned}
\|\mathbf{D}\!\!\!/_S^{(T)}p_4\|_{2,e} &\le \varepsilon_0 c(1+t)^{-3/2}\\
\|r\tau_-^{1/2}\ \mathbf{D}\!\!\!/_S^{(T)}p_3\|_{2,e} &\le \varepsilon_0 c(1+t)^{-3/2}\\
\|r^{-1/2}\tau_-^{1/2}\ \mathbf{D}\!\!\!/_S^{(T)}p\!\!\!/\|_{2,e} &\le \varepsilon_0 c(1+t)^{-3/2} \qquad (8.2.4f)\\
\|\mathbf{D}\!\!\!/_S\Lambda({}^{(T)}q)\|_{2,e} &\le \varepsilon_0 c(1+t)^{-3/2}
\end{aligned}
$$

$$
\begin{aligned}
\|r^{-1/2}\tau_-^{1/2}\, \not{D}_S \underline{\Lambda}({}^{(T)}q)\|_{2,e} &\leq \varepsilon_0 c(1+t)^{-3/2} \\
\|\not{D}_S K({}^{(T)}q)\|_{2,e} &\leq \varepsilon_0 c(1+t)^{-3/2} \\
\|r^{-1/2}\tau_-^{1/2}\, \not{D}_S \underline{K}({}^{(T)}q)\|_{2,e} &\leq \varepsilon_0 c(1+t)^{-3/2} \\
\|\not{D}_S \Theta({}^{(T)}q)\|_{2,e} &\leq \varepsilon_0 c(1+t)^{-3/2} \\
\|r^{-3/2}\tau_-^{3/2}\, \not{D}_S \underline{\Theta}({}^{(T)}q)\|_{2,e} &\leq \varepsilon_0 c(1+t)^{-3/2} \\
\|r^{-1/2}\tau_-^{1/2}\, \not{D}_S I({}^{(T)}q)\|_{2,e} &\leq \varepsilon_0 c(1+t)^{-3/2} \\
\|r^{-1/2}\tau_-^{1/2}\, \not{D}_S \underline{I}({}^{(T)}q)\|_{2,e} &\leq \varepsilon_0 c(1+t)^{-3/2} \\
\|\not{D}_S \Xi({}^{(T)}q)\|_{2,e} &\leq \varepsilon_0 c(1+t)^{-3/2} \\
\|r^{-1/2}\tau_-^{1/2}\, \not{D}_S \underline{\Xi}({}^{(T)}q)\|_{2,e} &\leq \varepsilon_0 c(1+t)^{-3/2}.
\end{aligned} \tag{8.2.4g}
$$

8.2.0.7. Interior Estimates for ${}^{(S)}q$, ${}^{(S)}p$

The Interior Assumptions 8.2.1j imply the following:

$$\|{}^{(S)}p,\ {}^{(S)}q\|_{2,\imath} \leq \varepsilon_0 c. \tag{8.2.5a}$$

(8.2.5b)

8.2.0.8. Exterior Estimates for ${}^{(S)}q$, ${}^{(S)}p$

$$
\begin{aligned}
(loc)\|{}^{(S)}p_4\|_{2,e} &\leq \varepsilon_0 c r(1+t)^{-1} & (loc)\|r^{-1}\tau_-^{(S)}p_3\|_{2,e} &\leq \varepsilon_0 c r(1+t)^{-1} \\
(loc)\|{}^{(S)}\not{p}\|_{2,e} &\leq \varepsilon_0 c r(1+t)^{-1}
\end{aligned}
$$

$$
\begin{aligned}
(loc)\|\Lambda({}^{(S)}q)\|_{2,e} &\leq \varepsilon_0 c r(1+t)^{-1} & (loc)\|r^{-1}\tau_-\underline{\Lambda}({}^{(S)}q)\|_{2,e} &\leq \varepsilon_0 c r(1+t)^{-1} \\
(loc)\|r\tau_-^{-1}K({}^{(S)}q)\|_{2,e} &\leq \varepsilon_0 c r(1+t)^{-1} & (loc)\|\underline{K}({}^{(S)}q)\|_{2,e} &\leq \varepsilon_0 c r(1+t)^{-1} \\
(loc)\|I({}^{(S)}q)\|_{2,e} &\leq \varepsilon_0 c r(1+t)^{-1} & (loc)\|\underline{I}({}^{(S)}q)\|_{2,e} &\leq \varepsilon_0 c r(1+t)^{-1} \\
(loc)\|\Theta({}^{(S)}q)\|_{2,e} &\leq \varepsilon_0 c r(1+t)^{-1} & (loc)\|r^{-1}\tau_-\underline{\Theta}({}^{(S)}q)\|_{2,e} &\leq \varepsilon_0 c r(1+t)^{-1} \\
(loc)\|r\tau_-^{-1}\Xi({}^{(S)}q)\|_{2,e} &\leq \varepsilon_0 c r(1+t)^{-1} & (loc)\|r^{-1}\tau_-\underline{\Xi}({}^{(S)}q)\|_{2,e} &\leq \varepsilon_0 c r(1+t)^{-1}
\end{aligned} \tag{8.2.5c}
$$

Proof 8.2.1 *The proof follows easily from the following calculations:*[6]

[6] Once again the precise calculations are not necessary; the structure of most terms in the formulas below can be deduced by signature considerations alone.

$$
\begin{aligned}
\mathbf{D}_3^{(X)}\hat{\pi}_{43} &= \mathbf{D}_3^{(X)}\mathbf{j} + Qr\left[\underline{\nu}\,;\,{}^{(X)}\mathbf{j}\right] + Qr\left[\zeta\,;\,{}^{(X)}\underline{\mathbf{m}}\right] + Qr\left[\underline{\xi}\,;\,{}^{(X)}\mathbf{m}\right]\\
\mathbf{D}_4^{(X)}\hat{\pi}_{33} &= \mathbf{D}_4^{(X)}\underline{\mathbf{n}} + Qr\left[\underline{\zeta}\,;\,{}^{(X)}\underline{\mathbf{m}}\right] + Qr\left[\nu\,;\,{}^{(X)}\underline{\mathbf{n}}\right]\\
\mathbf{D}_3^{(X)}\hat{\pi}_{33} &= \mathbf{D}_3^{(X)}\underline{\mathbf{n}} + Qr\left[\underline{\nu}\,;\,{}^{(X)}\underline{\mathbf{n}}\right] + Qr\left[\underline{\xi}\,;\,{}^{(X)}\underline{\mathbf{m}}\right]\\
\mathbf{D}_4^{(X)}\hat{\pi}_{44} &= \mathbf{D}_4^{(X)}\mathbf{n} + Qr\left[\nu\,;\,{}^{(X)}\mathbf{n}\right]\\
\mathbf{D}_3^{(X)}\hat{\pi}_{44} &= \mathbf{D}_3^{(X)}\mathbf{n} + Qr\left[\zeta\,;\,{}^{(X)}\mathbf{m}\right] + Qr\left[\underline{\nu}\,;\,{}^{(X)}\mathbf{n}\right]\\
\mathbf{D}_4^{(X)}\hat{\pi}_{43} &= \mathbf{D}_4^{(X)}\mathbf{j} + Qr\left[\nu\,;\,{}^{(X)}\mathbf{j}\right] + Qr\left[\underline{\zeta}\,;\,{}^{(X)}\mathbf{m}\right].
\end{aligned}
$$

Also,

$$
\begin{aligned}
\mathbf{D}_3^{(X)}\hat{\pi}_{3A} &= \not{\mathbf{D}}_3^{(X)}\underline{\mathbf{m}} + Qr\left[\underline{\nu}\,;\,{}^{(X)}\underline{\mathbf{m}}\right] + Qr\left[\zeta\,;\,{}^{(X)}\underline{\mathbf{n}}\right]\\
&\quad + Qr\left[\underline{\xi}\,;\,({}^{(X)}\mathbf{i},{}^{(X)}\mathbf{j})\right]\\
\mathbf{D}_4^{(X)}\hat{\pi}_{3A} &= \not{\mathbf{D}}_4^{(X)}\underline{\mathbf{m}} + Qr\left[\nu\,;\,{}^{(X)}\underline{\mathbf{m}}\right]\\
&\quad + Qr\left[\underline{\zeta}\,;\,({}^{(X)}\mathbf{i},{}^{(X)}\mathbf{j})\right]
\end{aligned}
$$

$$
\begin{aligned}
\mathbf{D}_4^{(X)}\hat{\pi}_{4A} &= \not{\mathbf{D}}_4^{(X)}\mathbf{m} + Qr\left[\nu\,;\,{}^{(X)}\mathbf{m}\right] + Qr\left[\underline{\zeta}\,;\,{}^{(X)}\mathbf{n}\right]\\
\mathbf{D}_3^{(X)}\hat{\pi}_{4A} &= \not{\mathbf{D}}_3^{(X)}\mathbf{m} + Qr\left[\nu\,;\,{}^{(X)}\underline{\mathbf{m}}\right] + Qr\left[\underline{\nu}\,;\,{}^{(X)}\mathbf{m}\right] + Qr\left[\zeta\,;\,({}^{(X)}\mathbf{i},{}^{(X)}\mathbf{j})\right]\\
&\quad + Qr\left[\underline{\xi}\,;\,{}^{(X)}\mathbf{n}\right]\\
\mathbf{D}_3^{(X)}\hat{\pi}_{AB} &= \not{\mathbf{D}}_3^{(X)}\mathbf{i}_{AB} + Qr\left[\zeta\,;\,{}^{(X)}\underline{\mathbf{m}}\right] + Qr\left[\underline{\xi}\,;\,{}^{(X)}\mathbf{m}\right]\\
\mathbf{D}_4^{(X)}\hat{\pi}_{AB} &= \not{\mathbf{D}}_4^{(X)}\mathbf{i}_{AB} + Qr\left[\underline{\zeta}\,;\,{}^{(X)}\mathbf{m}\right].
\end{aligned}
$$

Also,

$$
\begin{aligned}
\mathbf{D}_A^{(X)}\hat{\pi}_{33} &= \nabla\!\!\!\!/_A^{(X)}\underline{\mathbf{n}} - \mathrm{tr}\underline{\chi}\,{}^{(X)}\underline{\mathbf{m}}_A\\
&\quad + Qr\left[\underline{\hat{\chi}}\,;\,{}^{(X)}\underline{\mathbf{m}}\right] + Qr\left[\epsilon\,;\,{}^{(X)}\underline{\mathbf{n}}\right]\\
\mathbf{D}_A^{(X)}\hat{\pi}_{34} &= \nabla\!\!\!\!/_A^{(X)}\mathbf{j} - \frac{1}{2}(\mathrm{tr}\chi\,{}^{(X)}\underline{\mathbf{m}}_A + \mathrm{tr}\underline{\chi}\,{}^{(X)}\mathbf{m}_A)\\
&\quad + Qr\left[\hat{\chi}\,;\,{}^{(X)}\underline{\mathbf{m}}\right] + Qr\left[\underline{\hat{\chi}}\,;\,{}^{(X)}\mathbf{m}\right]\\
\mathbf{D}_A^{(X)}\hat{\pi}_{44} &= \nabla\!\!\!\!/_A^{(X)}\mathbf{n} - \mathrm{tr}\chi\,{}^{(X)}\mathbf{m}_A\\
&\quad + Qr\left[\hat{\chi}\,;\,{}^{(X)}\mathbf{m}\right] + Qr\left[\epsilon\,;\,{}^{(X)}\mathbf{n}\right]\\
\mathbf{D}_B^{(X)}\hat{\pi}_{3A} &= \nabla\!\!\!\!/_B^{(X)}\underline{\mathbf{m}}_A - \frac{1}{2}\mathrm{tr}\underline{\chi}\,{}^{(X)}\mathbf{i}_{AB} - \frac{1}{4}\delta_{AB}(\mathrm{tr}\underline{\chi}\,{}^{(X)}\mathbf{j} + \mathrm{tr}\chi\,{}^{(X)}\underline{\mathbf{n}})\\
&\quad + Qr\left[\hat{\chi}\,;\,{}^{(X)}\underline{\mathbf{n}}\right] + Qr\left[\underline{\hat{\chi}}\,;\,({}^{(X)}\mathbf{i},{}^{(X)}\mathbf{j})\right] + Qr\left[\epsilon\,;\,{}^{(X)}\underline{\mathbf{m}}\right]
\end{aligned}
$$

$$\mathbf{D}_B^{(X)}\hat{\pi}_{4A} = \nabla\!\!\!/_B^{(X)}\mathbf{m}_A - \frac{1}{2}\mathrm{tr}\chi^{(X)}\mathbf{i}_{AB} - \frac{1}{4}\delta_{AB}(\mathrm{tr}\chi^{(X)}\mathbf{j} + \mathrm{tr}\underline{\chi}^{(X)}\mathbf{n})$$
$$+ Qr\left[\hat{\underline{\chi}}\,;\,{}^{(X)}\mathbf{n}\right] + Qr\left[\hat{\chi}\,;\,({}^{(X)}\mathbf{i},{}^{(X)}\mathbf{j})\right] + Qr\left[\epsilon\,;\,{}^{(X)}\mathbf{m}\right]$$
$$\mathbf{D}_C^{(X)}\hat{\pi}_{AB} = \nabla\!\!\!/_C^{(X)}\mathbf{i}_{AB} - \frac{1}{4}\mathrm{tr}\underline{\chi}(\delta_{AC}^{(X)}\mathbf{m}_B + \delta_{BC}^{(X)}\mathbf{m}_A) - \frac{1}{4}\mathrm{tr}\chi(\delta_{AC}^{(X)}\underline{\mathbf{m}}_B + \delta_{BC}^{(X)}\underline{\mathbf{m}}_A)$$
$$+ Qr\left[\hat{\chi}\,;\,{}^{(X)}\underline{\mathbf{m}}\right] + Qr\left[\hat{\underline{\chi}}\,;\,{}^{(X)}\mathbf{m}\right].$$

We then calculate

$${}^{(X)}p_3 = d\!\!\!/iv\,{}^{(X)}\underline{\mathbf{m}} - \frac{1}{2}(\mathbf{D}_4^{(X)}\underline{\mathbf{n}} + \mathbf{D}_3^{(X)}\mathbf{j})$$
$$- \frac{1}{2}\mathrm{tr}\underline{\chi}({}^{(X)}\mathbf{j} + \mathrm{tr}^{(X)}\mathbf{i}) - \frac{1}{2}\mathrm{tr}\chi^{(X)}\underline{\mathbf{n}} + Qr\left[\hat{\chi}\,;\,{}^{(X)}\underline{\mathbf{n}}\right]$$
$$+ Qr\left[\hat{\underline{\chi}}\,;\,({}^{(X)}\mathbf{i},{}^{(X)}\mathbf{j})\right] + Qr\left[(\epsilon,\zeta,\underline{\zeta})\,;\,{}^{(X)}\underline{\mathbf{m}}\right] + Qr\left[\underline{\xi}\,;\,{}^{(X)}\mathbf{m}\right]$$
$${}^{(X)}p_4 = d\!\!\!/iv\,{}^{(X)}\mathbf{m} - \frac{1}{2}(\mathbf{D}_3^{(X)}\mathbf{n} + \mathbf{D}_4^{(X)}\mathbf{j})$$
$$- \frac{1}{2}\mathrm{tr}\chi({}^{(X)}\mathbf{j} + \mathrm{tr}^{(X)}\mathbf{i}) - \frac{1}{2}\mathrm{tr}\underline{\chi}^{(X)}\mathbf{n}$$
$$+ Qr\left[\hat{\underline{\chi}}\,;\,{}^{(X)}\mathbf{n}\right] + Qr\left[\hat{\chi}\,;\,({}^{(X)}\mathbf{i},{}^{(X)}\mathbf{j})\right] + Qr\left[\epsilon,\zeta,\underline{\zeta}\,;\,{}^{(X)}\mathbf{m}\right]$$
$${}^{(X)}p\!\!\!/_A = d\!\!\!/iv\,{}^{(X)}\mathbf{i} - \frac{1}{2}(\mathbf{D}_4^{(X)}\underline{\mathbf{m}} + \mathbf{D}_3^{(X)}\mathbf{m})$$
$$- \frac{3}{4}\mathrm{tr}\underline{\chi}^{(X)}\mathbf{m} - \frac{3}{4}\mathrm{tr}\chi^{(X)}\underline{\mathbf{m}} + Qr\left[\hat{\chi}\,;\,{}^{(X)}\underline{\mathbf{m}}\right] + Qr\left[\hat{\underline{\chi}}\,;\,{}^{(X)}\mathbf{m}\right]$$
$$+ Qr\left[\nu\,;\,{}^{(X)}\underline{\mathbf{m}}\right] + Qr\left[\underline{\nu}\,;\,{}^{(X)}\mathbf{m}\right] + Qr\left[\zeta,\underline{\zeta}\,;\,({}^{(X)}\mathbf{i},{}^{(X)}\mathbf{j})\right]$$
$$+ Qr\left[\underline{\xi}\,;\,{}^{(X)}\mathbf{n}\right].$$

Also,

$$\Xi_A({}^{(X)}q) = \frac{1}{2}(\mathbf{D}\!\!\!\!/_3^{(X)}\underline{\mathbf{m}}_A - \nabla\!\!\!/_A^{(X)}\underline{\mathbf{n}}) + \frac{1}{2}\mathrm{tr}\underline{\chi}^{(X)}\underline{\mathbf{m}}_A$$
$$+ Qr\left[\hat{\underline{\chi}}\,;\,{}^{(X)}\underline{\mathbf{m}}\right] + Qr\left[\underline{\nu}\,;\,{}^{(X)}\underline{\mathbf{m}}\right] + Qr\left[\zeta,\epsilon\,;\,{}^{(X)}\underline{\mathbf{n}}\right]$$
$$+ Qr\left[\underline{\xi}\,;\,({}^{(X)}\mathbf{i},{}^{(X)}\mathbf{j})\right]$$

$$\underline{\Lambda}({}^{(X)}q) = \mathbf{D}_4^{(X)}\underline{\mathbf{n}} - \mathbf{D}_3^{(X)}\mathbf{j} + \frac{2}{3}{}^{(X)}p_3$$
$$+ Qr\left[\zeta,\underline{\zeta}\,;\,{}^{(X)}\underline{\mathbf{m}}\right] + Qr\left[\underline{\nu}\,;\,{}^{(X)}\mathbf{j}\right]$$
$$\underline{K}({}^{(X)}q) = \frac{1}{2}cu\!\!\!/rl\,{}^{(X)}\underline{\mathbf{m}} + Qr\left[\hat{\chi}\,;\,{}^{(X)}\underline{\mathbf{n}}\right] + Qr\left[\hat{\underline{\chi}}\,;\,({}^{(X)}\mathbf{i},{}^{(X)}\mathbf{j})\right] + Qr\left[\epsilon\,;\,{}^{(X)}\underline{\mathbf{m}}\right]$$
$$\underline{\Theta}({}^{(X)}q)_{AB} = 2\mathbf{D}\!\!\!\!/_3^{(X)}\mathbf{i}_{AB} - (\nabla\!\!\!/_A^{(X)}\underline{\mathbf{m}}_B + \nabla\!\!\!/_B^{(X)}\underline{\mathbf{m}}_A - d\!\!\!/iv\,{}^{(X)}\underline{\mathbf{m}}\delta_{AB})$$

$$+\operatorname{tr}\underline{\chi}({}^{(X)}\mathbf{i}_{AB}-\frac{1}{2}\operatorname{tr}{}^{(X)}\mathbf{i}\delta_{AB})+Qr\left[\underline{\xi}\,;\,{}^{(X)}\mathbf{m}\right]+Qr\left[(\zeta,\epsilon)\,;\,{}^{(X)}\underline{\mathbf{m}}\right]$$
$$+Qr\left[\hat{\chi}\,;\,{}^{(X)}\underline{\mathbf{n}}\right]+Qr\left[\underline{\hat{\chi}}\,;\,({}^{(X)}\mathbf{i},{}^{(X)}\mathbf{j})\right]$$

$$\underline{I}_A({}^{(X)}q)=\frac{1}{2}\left(\mathbf{D}\!\!\!/_3{}^{(X)}\mathbf{m}_A-\nabla\!\!\!\!/_A{}^{(X)}\mathbf{j}+\frac{1}{2}(\operatorname{tr}\chi{}^{(X)}\underline{\mathbf{m}}_A+\operatorname{tr}\underline{\chi}{}^{(X)}\mathbf{m}_A)+\frac{2}{3}{}^{(X)}\!\!\not{p}_A\right)$$
$$+Qr\left[\hat{\chi}\,;\,{}^{(X)}\underline{\mathbf{m}}\right]+Qr\left[\underline{\hat{\chi}}\,;\,{}^{(X)}\mathbf{m}\right]+Qr\left[\nu\,;\,{}^{(X)}\underline{\mathbf{m}}\right]+Qr\left[\underline{\nu}\,;\,{}^{(X)}\mathbf{m}\right]$$
$$+Qr\left[\underline{\xi}\,;\,{}^{(X)}\mathbf{n}\right]+Qr\left[\zeta\,;\,({}^{(X)}\mathbf{i},{}^{(X)}\mathbf{j})\right]$$

$$I_A({}^{(X)}q)=\frac{1}{2}\left(\mathbf{D}\!\!\!/_4{}^{(X)}\underline{\mathbf{m}}_A-\nabla\!\!\!\!/_A{}^{(X)}\mathbf{j}+\frac{1}{2}(\operatorname{tr}\underline{\chi}{}^{(X)}\mathbf{m}_A+\operatorname{tr}\chi{}^{(X)}\underline{\mathbf{m}}_A)+\frac{2}{3}{}^{(X)}\!\!\not{p}_A\right)$$
$$+Qr\left[\underline{\hat{\chi}}\,;\,{}^{(X)}\mathbf{m}\right]+Qr\left[\hat{\chi}\,;\,{}^{(X)}\underline{\mathbf{m}}\right]+Qr\left[\underline{\nu}\,;\,{}^{(X)}\mathbf{m}\right]+Qr\left[\nu\,;\,{}^{(X)}\underline{\mathbf{m}}\right]$$
$$+Qr\left[\underline{\zeta}\,;\,({}^{(X)}\mathbf{i},{}^{(X)}\mathbf{j})\right]$$

$$\Theta({}^{(X)}q)_{AB}=2\mathbf{D}\!\!\!/_4{}^{(X)}\mathbf{i}_{AB}-(\nabla\!\!\!\!/_A{}^{(X)}\mathbf{m}_B+\nabla\!\!\!\!/_B{}^{(X)}\mathbf{m}_A-\operatorname{d\!\!/iv}{}^{(X)}\mathbf{m}\delta_{AB})$$
$$+\operatorname{tr}\chi\left({}^{(X)}\mathbf{i}_{AB}-\frac{1}{2}\operatorname{tr}{}^{(X)}\mathbf{i}\delta_{AB}\right)+Qr\left[(\underline{\zeta},\epsilon)\,;\,{}^{(X)}\mathbf{m}\right]$$
$$+Qr\left[\underline{\hat{\chi}}\,;\,{}^{(X)}\mathbf{n}\right]+Qr\left[\hat{\chi}\,;\,({}^{(X)}\mathbf{i},{}^{(X)}\mathbf{j})\right]$$

$$K({}^{(X)}q)=\frac{1}{2}\operatorname{c\!\!/url}{}^{(X)}\mathbf{m}+Qr\left[\underline{\hat{\chi}}\,;\,{}^{(X)}\mathbf{n}\right]+Qr\left[\hat{\chi}\,;\,({}^{(X)}\mathbf{i},{}^{(X)}\mathbf{j})\right]+Qr\left[\epsilon\,;\,{}^{(X)}\mathbf{m}\right]$$

$$\Lambda({}^{(X)}q)=\mathbf{D}_3{}^{(X)}\mathbf{n}-\mathbf{D}_4{}^{(X)}\mathbf{j}+\frac{2}{3}{}^{(X)}p_4$$
$$+Qr\left[\zeta,\underline{\zeta}\,;\,{}^{(X)}\mathbf{m}\right]+Qr\left[\nu\,;\,{}^{(X)}\mathbf{j}\right]$$

$$\Xi_A({}^{(X)}q)=\frac{1}{2}(\mathbf{D}\!\!\!/_4{}^{(X)}\mathbf{m}_A-\nabla\!\!\!\!/_A{}^{(X)}\mathbf{n})+\frac{1}{2}\operatorname{tr}\chi{}^{(X)}\mathbf{m}_A$$
$$+Qr\left[\hat{\chi}\,;\,{}^{(X)}\mathbf{m}\right]+Qr\left[\nu\,;\,{}^{(X)}\mathbf{m}\right]+Qr\left[\underline{\zeta},\epsilon\,;\,{}^{(X)}\mathbf{n}\right].$$

We are now ready to state our main theorem.

Theorem 8.2.1 *We assume a given space-time* $(\mathbf{M},\mathbf{g})$ *endowed with a (t, u)-foliation as described in the previous section and consider an arbitrary Weyl tensor W that satisfies the homogeneous Bianchi equations 7.1.2. Assume that the assumptions of the comparison theorem are verified in the space-time slab* $V_{t*}=\bigcup_{t\in[0,t_*]}\Sigma_t$, *and hence, as a consequence, there exists a constant* $c_{\mathbf{eq}}$ *such that*

$$\mathcal{R}_{[1]}\leq c_{\mathbf{eq}}\mathcal{Q}_1$$
$$\mathcal{R}_2\leq c_{\mathbf{eq}}(\mathcal{Q}_1+\mathcal{Q}_2)$$

in the space-time slab V_{t*}.

i. We require that the vectorfields $T,\mathbf{O}$ *verify Assumptions 0, and 1. Then there exists a constant* c_1 *such that, with the notation* $\mathcal{Q}_1*,\mathcal{Q}_2*$ *introduced in 8.1.4g* *and 8.1.4h,*

$$\mathcal{Q}_1* \leq \mathcal{Q}_1(0) + c_1\varepsilon_0(\mathcal{Q}_1* + \mathcal{Q}_2*). \tag{8.2.6a}$$

ii. *In addition we also require that Assumption 2 is also verified. Then*

$$\mathcal{Q}_2* \leq \mathcal{Q}_2(0) + c_2\varepsilon_0(\mathcal{Q}_1* + \mathcal{Q}_2*). \tag{8.2.6b}$$

Thus, if ε_0 is chosen sufficiently small, we infer that

$$\mathcal{Q}_1* + \mathcal{Q}_2* \leq 2(\mathcal{Q}_1(0) + \mathcal{Q}_2(0)). \tag{8.2.6c}$$

8.3. Proof of the Theorem

In view of Proposition 8.1.1, to prove the theorem, we have to estimate the error terms $\mathcal{E}_1(t_*)$, $\mathcal{E}_2(t_*)$, where $\mathcal{E}_1(t), \mathcal{E}_2(t)$ are defined by 8.1.4b and 8.1.4d, respectively.

(i.) **Estimates for $\mathcal{E}_1(t)$:**

There are five integrals appearing in the definition of $\mathcal{E}_1(t)$ that we split into the interior region V_t^i, for which $r \leq \frac{r_0}{2}$ and the exterior region V_t^e where $r \geq \frac{r_0}{2}$. The interior integrals are estimated in straightforward manner using the interior assumptions $IA\,0, IA\,1$, and the comparison theorem. Indeed, in the interior all components of the tensors W, $\hat{\mathcal{L}}_{\mathbf{O}}W$, $\hat{\mathcal{L}}_T W$, and $\mathbf{D}W$ and all components of the deformation tensors of $\bar{K}, \mathbf{O}, T$ behave in the same manner. The exterior estimates are far more complicated; they depend crucially on the structure of the nonlinear terms.

We start with

A: Estimates for $\text{Integral}_A = \int_{V_t} |\phi \mathbf{Div} Q(\hat{\mathcal{L}}_{\mathbf{O}}W)_{\beta\gamma\delta}\bar{K}^\beta\bar{K}^\gamma T^\delta|$

In view of formula 8.1.7a of Proposition 8.1.3 we write

$$\begin{aligned}\text{Integral}_A = \frac{1}{8}\int_{V_t} & \phi\tau_+^4(D(\mathbf{O},W)_{444} + D(\mathbf{O},W)_{344}) \qquad (8.3.1a)\\ + \int_{V_t} & \phi\tau_+^2\tau_-^2(D(\mathbf{O},W)_{344} + D(\mathbf{O},W)_{334})\\ + \int_{V_t} & \phi\frac{1}{8}\tau_-^4(D(\mathbf{O},W)_{334} + D(\mathbf{O},W)_{333}).\end{aligned}$$

The most sensitive term to estimate is $\int_{V_t^e}\phi\tau_+^4 D(\mathbf{O},W)_{344}$. In view of 8.1.7c we write

$$\int_{V_t^e}\phi\tau_+^4 D(\mathbf{O},W)_{344} = 8\int_{V_t^e}\phi\tau_+^4\rho(\hat{\mathcal{L}}_{\mathbf{O}}W)\Lambda(\mathbf{O},W)$$

$$+8\int_{V_t^e}\phi\tau_+^4\sigma(\hat{\mathcal{L}}_{\mathbf{O}}W)K(\mathbf{O},W)$$
$$+8\int_{V_t^e}\phi\tau_+^4\beta(\hat{\mathcal{L}}_{\mathbf{O}}W)\cdot I(\mathbf{O},W). \quad (8.3.1b)$$

In view of the coarea formula we write $\int_{V_t^e}=\int_0^t dt'\int_{\Sigma_t^e}\phi d\mu_g$. With the integrals on Σ_t^e we proceed as follows:

$$\int_{\Sigma_t^e}|\phi\tau_+^4\rho(\hat{\mathcal{L}}_{\mathbf{O}}W)\Lambda(\mathbf{O},W)|\leq c\|r^2\rho(\hat{\mathcal{L}}_{\mathbf{O}}W)\|_{2,e}\|r^2\Lambda(\mathbf{O},W)\|_{2,e}$$
$$\leq c\mathcal{Q}_1^{1/2}\|r^2\Lambda(\mathbf{O},W)\|_{2,e} \quad (8.3.1c)$$

$$\int_{\Sigma_t^e}|\phi\tau_+^4\sigma(\hat{\mathcal{L}}_{\mathbf{O}}W)K(\mathbf{O},W)|\leq c\|r^2\sigma(\hat{\mathcal{L}}_{\mathbf{O}}W)\|_{2,e}\|r^2K(\mathbf{O},W)\|_{2,e}$$
$$\leq c\mathcal{Q}_1^{1/2}\|r^2K(\mathbf{O},W)\|_{2,e} \quad (8.3.1d)$$

$$\int_{\Sigma_t^e}|\phi\tau_+^4\beta(\hat{\mathcal{L}}_{\mathbf{O}}W)\cdot I(\mathbf{O},W)|\leq c\|r^2\beta(\hat{\mathcal{L}}_{\mathbf{O}}W)\|_{2,e}\|r^2I(\mathbf{O},W)\|_{2,e}$$
$$\leq c\mathcal{Q}_1^{1/2}\|r^2I(\mathbf{O},W)\|_{2,e}. \quad (8.3.1e)$$

According to the decomposition of Proposition 8.1.4 we write

$$\Lambda(\mathbf{O},W)=\frac{1}{2}(\Lambda^1(\mathbf{O},W)+\Lambda^2(\mathbf{O},W)+\Lambda^3(\mathbf{O},W))$$
$$K(\mathbf{O},W)=\frac{1}{2}(K^1(\mathbf{O},W)+K^2(\mathbf{O},W)+K^3(\mathbf{O},W))$$
$$I(\mathbf{O},W)=\frac{1}{2}(I^1(\mathbf{O},W)+I^2(\mathbf{O},W)+I^3(\mathbf{O},W)).$$

On the other hand, according to Proposition 8.1.4,

$$\Lambda^1(\mathbf{O},W)=\mathrm{Qr}\left[{}^{(\mathbf{O})}\mathbf{i}\,;\,\nabla\!\!\!/\beta\right]+\mathrm{Qr}\left[{}^{(\mathbf{O})}\underline{\mathbf{m}}\,;\,\beta_4\right]+\boxed{\mathrm{Qr}\left[{}^{(\mathbf{O})}\mathbf{m}\,;\,\beta_3\right]}$$
$$+\boxed{\mathrm{Qr}\left[{}^{(\mathbf{O})}\mathbf{m}\,;\,\nabla\!\!\!/(\rho,\sigma)\right]}+\mathrm{Qr}\left[{}^{(\mathbf{O})}\mathbf{j}\,;\,(\rho_4,\sigma_4)\right]$$
$$+\boxed{\mathrm{Qr}\left[{}^{(\mathbf{O})}\mathbf{n}\,;\,(\rho_3,\sigma_3)\right]}$$
$$+\,\mathrm{tr}\chi\Big(\mathrm{Qr}\left[{}^{(\mathbf{O})}\underline{\mathbf{m}}\,;\,\beta\right]$$
$$+\,\mathrm{Qr}\left[({}^{(\mathbf{O})}\mathbf{i},{}^{(\mathbf{O})}\mathbf{j})\,;\,(\rho,\sigma)\right]+\boxed{\mathrm{Qr}\left[{}^{(\mathbf{O})}\mathbf{m}\,;\,\underline{\beta}\right]}\Big)$$

$$+\operatorname{tr}\underline{\chi}\Big(\mathrm{Qr}\left[({}^{(\mathbf{O})}\mathbf{i},{}^{(\mathbf{O})}\mathbf{j});\,\alpha\right]+\boxed{\mathrm{Qr}\left[{}^{(\mathbf{O})}\mathbf{m}\,;\,\beta\right]}$$

$$+\boxed{\mathrm{Qr}\left[{}^{(\mathbf{O})}\mathbf{n}\,;\,(\rho,\sigma)\right]}\Big)$$

$$\Lambda^2(\mathbf{O},W)=\mathrm{Qr}\left[{}^{(\mathbf{O})}\not{p}\,;\,\beta\right]+\mathrm{Qr}\left[{}^{(\mathbf{O})}p_4\,;\,(\rho,\sigma)\right]$$

$$\Lambda^3(\mathbf{O},W)=\mathrm{Qr}\left[\alpha\,;\,\underline{\Theta}({}^{(\mathbf{O})}q)\right]+\boxed{\mathrm{Qr}\left[\beta\,;\,(I,\underline{I})({}^{(\mathbf{O})}q)\right]}$$

$$+\,\mathrm{Qr}\left[(\rho,\sigma)\,;\,(K,\Lambda)({}^{(\mathbf{O})}q)\right]+\boxed{\mathrm{Qr}\left[\underline{\beta}\,;\,\Xi({}^{(\mathbf{O})}q)\right]}\,.$$

In view of Assumption 0 for ${}^{(\mathbf{O})}\hat{\pi}$ (see 8.2.2b), the comparison between $\mathcal{R}_{[1]}$ and $\mathcal{Q}_1$, the Assumption 0 on the Ricci coefficients, and the global Sobolev inequalities of Propositions 3.2.1 and 3.2.3, we easily deduce that

$$\begin{aligned}
\|r^2\Lambda^1(\mathbf{O},W)\|_{2,e} &\le c r_0^{-2}\|r\,{}^{(\mathbf{O})}\mathbf{i}\|_{\infty,e}\|r^3\nabla\!\!\!/\,\beta\|_{2,e}\\
&\quad+c r_0^{-2}\|r\,{}^{(\mathbf{O})}\underline{\mathbf{m}}\|_{\infty,e}\|r^3\beta_4\|_{2,e}\\
&\quad+c r_0^{-2}\|r\,{}^{(\mathbf{O})}\mathbf{j}\|_{\infty,e}\|r^3(\rho_4,\sigma_4)\|_{2,e}\\
&\quad+c r_0^{-2}\|r\,{}^{(\mathbf{O})}\underline{\mathbf{m}}\|_{\infty,e}\|r\chi\|_{\infty,e}\|r^2\beta\|_{2,e}\\
&\quad+c r_0^{-2}\|r({}^{(\mathbf{O})}\mathbf{i},{}^{(\mathbf{O})}\mathbf{j})\|_{\infty,e}\|r\underline{\chi}\|_{\infty,e}\|r^2\alpha\|_{2,e}\\
&\quad+c r_0^{-3/2}\|r({}^{(\mathbf{O})}\mathbf{i},{}^{(\mathbf{O})}\mathbf{j})\|_{\infty,e}\|r\chi\|_{\infty,e}\|r^3(\rho,\sigma)\|_{\infty,e}\\
&\le \varepsilon_0 r_0^{-3/2}(\mathcal{Q}_1+\mathcal{Q}_2)^{1/2}.
\end{aligned}\tag{8.3.1f}$$

Also,

$$\begin{aligned}
\|r^2\Lambda^2(\mathbf{O},W)\|_{2,e} &\le c r_0^{-3/2}\Big(\|r^{7/2}\beta\|_{\infty,e}\ {}^{(\mathrm{loc})}\|r\,{}^{(\mathbf{O})}\not{p}\|_{2,e}\\
&\quad+\|r^3(\rho,\sigma)\|^{(\mathrm{loc})}_{\infty,e}\|r\,{}^{(\mathbf{O})}p_4\|_{2,e}\Big)\\
&\le \varepsilon_0 c r_0^{-3/2}(\mathcal{Q}_1+\mathcal{Q}_2)^{1/2}
\end{aligned}\tag{8.3.1g}$$

provided that

$$\begin{aligned}
{}^{(\mathrm{loc})}\|r\,{}^{(\mathbf{O})}\not{p}\|_{2,e}&\le\varepsilon_0 c\\
{}^{(\mathrm{loc})}\|r\,{}^{(\mathbf{O})}p_4\|_{2,e}&\le\varepsilon_0 c,
\end{aligned}$$

which is assured by Proposition 8.2.1.

Also,

$$\|r^2\Lambda^3(\mathbf{O},W)\|_{2,e}\le c r_0^{-3/2}\Big(\|r^{7/2}\alpha\|_{\infty,e}\ {}^{(\mathrm{loc})}\|\tau_-\underline{\Theta}({}^{(\mathbf{O})}q)\|_{2,e}$$

$$+ \|r^3(\rho,\sigma)\|^{(\text{loc})}_{\infty,e}\|r(K,\Lambda)({}^{(\mathbf{O})}q)\|_{2,e}\Big)$$
$$\leq c\varepsilon_0 r_0^{-3/2}(\mathcal{Q}_1+\mathcal{Q}_2)^{1/2} \tag{8.3.1h}$$

provided that

$$^{(\text{loc})}\|\tau_-\underline{\Theta}({}^{(\mathbf{O})}q\|_{2,e}) \leq \varepsilon_0 c \tag{8.3.1i}$$
$$^{(\text{loc})}\|r(K,\Lambda)({}^{(\mathbf{O})}q)\|_{2,e} \leq \varepsilon_0 c, \tag{8.3.1j}$$

which holds in view of Proposition 8.2.1. Therefore, combining 8.3.1f–8.3.1h, we deduce

$$\|r^2\Lambda(\mathbf{O},W)\|_{2,e} \leq \varepsilon_0 c r_0^{-3/2}(\mathcal{Q}_1+\mathcal{Q}_2)^{1/2}. \tag{8.3.1k}$$

Proceeding in exactly the same manner, we find

$$\|r^2K(\mathbf{O},W)\|_{2,e} \leq \varepsilon_0 c r_0^{-3/2}(\mathcal{Q}_1+\mathcal{Q}_2)^{1/2}. \tag{8.3.2}$$

Therefore, back to 8.3.1c–8.3.1d, we conclude that the right-hand sides of both equations are bounded by $c\varepsilon_0 r_0^{-3/2}(\mathcal{Q}_1+\mathcal{Q}_2)$. Consequently, the first two terms on the right-hand side of 8.3.1b can be estimated by

$$\int_{V_t^e} |\phi\tau_+^4\rho(\hat{\mathcal{L}}_{\mathbf{O}}W)\Lambda(\mathbf{O},W)| \leq c\varepsilon_0\Big(\sup_{[0,t]}\mathcal{Q}_1+\sup_{[0,t]}\mathcal{Q}_2\Big) \tag{8.3.3}$$
$$\int_{V_t^e} \phi\tau_+^4\sigma(\hat{\mathcal{L}}_{\mathbf{O}}W)K(\mathbf{O},W) \leq c\varepsilon_0\Big(\sup_{[0,t]}\mathcal{Q}_1+\sup_{[0,t]}\mathcal{Q}_2\Big). \tag{8.3.4}$$

Proceeding in the same manner with the third integral in 8.3.1b, we run into a problem. Indeed, to estimate $\|r^2I(\mathbf{O},W)\|_{2,e}$, we write, according to Proposition 8.1.4,

$$\begin{aligned}
I^1(\mathbf{O},W) = {} & \mathrm{Qr}\left[{}^{(\mathbf{O})}\underline{\mathbf{m}}\,;\,\nabla\!\!\!/\beta\right] + \mathrm{Qr}\left[{}^{(\mathbf{O})}\underline{\mathbf{n}}\,;\,\beta_4\right] + \mathrm{Qr}\left[{}^{(\mathbf{O})}\mathbf{j}\,;\,\beta_3\right] \\
& + \mathrm{Qr}\left[{}^{(\mathbf{O})}\mathbf{i}\,;\,\nabla\!\!\!/(\rho,\sigma)\right] + \mathrm{Qr}\left[{}^{(\mathbf{O})}\underline{\mathbf{m}}\,;\,(\rho_4,\sigma_4)\right] \\
& + \boxed{\mathrm{Qr}\left[{}^{(\mathbf{O})}\mathbf{m}\,;\,(\rho_3,\sigma_3)\right]} \\
& + \mathrm{tr}\chi\left(\mathrm{Qr}\left[{}^{(\mathbf{O})}\underline{\mathbf{n}}\,;\,\beta\right] + \mathrm{Qr}\left[{}^{(\mathbf{O})}\underline{\mathbf{m}}\,;\,(\rho,\sigma)\right] + \mathrm{Qr}\left[{}^{(\mathbf{O})}\mathbf{i}\,;\,\underline{\beta}\right]\right) \\
& + \mathrm{tr}\underline{\chi}\Big(\mathrm{Qr}\left[{}^{(\mathbf{O})}\underline{\mathbf{m}}\,;\,\alpha\right] + \mathrm{Qr}\left[({}^{(\mathbf{O})}\mathbf{i},{}^{(\mathbf{O})}\mathbf{j})\,;\,\beta\right] \\
& + \boxed{\mathrm{Qr}\left[{}^{(\mathbf{O})}\mathbf{m}\,;\,(\rho,\sigma)\right]}\Big)
\end{aligned}$$

$$I^2(\mathbf{O},W) = \mathrm{Qr}\left[{}^{(\mathbf{O})}p_3\,;\,\beta\right] + \mathrm{Qr}\left[{}^{(\mathbf{O})}\not{p}\,;\,(\rho,\sigma)\right]$$

$$I^3(\mathbf{O}, W) = \boxed{\quad \mathrm{Qr}\left[\underline{\alpha}\,;\, \Xi({}^{(\mathbf{O})}q)\right] \quad} + \mathrm{Qr}\left[\underline{\beta}\,;\, (K, \Lambda, \Theta)({}^{(\mathbf{O})}q)\right]$$
$$+ \mathrm{Qr}\left[(\rho, \sigma)\,;\, (I, \underline{I})({}^{(\mathbf{O})}q)\right] + \mathrm{Qr}\left[\beta\,;\, (\underline{K}, \underline{\Lambda}, \underline{\Theta})({}^{(\mathbf{O})}q)\right]$$
$$+ \mathrm{Qr}\left[\alpha\,;\, \underline{\Xi}({}^{(\mathbf{O})}q)\right].$$

Now,

$$\begin{aligned}
\|r^2 I^1(\mathbf{O}, W)\|_{2,e} \leq\ & c r_0^{-2} \|r({}^{(\mathbf{O})}\mathbf{i}, {}^{(\mathbf{O})}\mathbf{j}, {}^{(\mathbf{O})}\underline{\mathbf{m}}, {}^{(\mathbf{O})}\underline{\mathbf{n}})\|_{\infty,e} \\
& \cdot \left(\|r^3 \nabla\!\!\!/ \beta\|_{2,e} + \|r^3 \beta_4\|_{2,e}\right. \\
& + \left.\|r^3 \beta_3\|_{2,e} + \|r^3 \nabla\!\!\!/ (\rho, \sigma)\|_{2,e} + \|r^3 (\rho_4, \sigma_4)\|_{2,e}\right) \\
& + c r_0^{-2} \|r({}^{(\mathbf{O})}\mathbf{i}, {}^{(\mathbf{O})}\mathbf{j})\|_{\infty,e} \|r\underline{\chi}\|_{\infty,e} \|r^2 \beta\|_{2,e} \\
& + c r_0^{-2} \|r{}^{(\mathbf{O})}\underline{\mathbf{m}}\|_{\infty,e} \|r\underline{\chi}\|_{\infty,e} \|r^2 \alpha\|_{2,e} \\
& + c r_0^{-2} \|r{}^{(\mathbf{O})}\underline{\mathbf{n}}\|_{\infty,e} \|r\chi\|_{\infty,e} \|r^2 \beta\|_{2,e} \\
& + c r_0^{-3/2} \|r{}^{(\mathbf{O})}\underline{\mathbf{m}}\|_{\infty,e} \|r\chi\|_{\infty,e} \|r^3 (\rho, \sigma)\|_{\infty,e} \\
& + c r (1+t)^{-1} \|r{}^{(\mathbf{O})}\mathbf{i}\|_{\infty,e} \|r\chi\|_{\infty,e} \|r\tau_- \underline{\beta}\|_{2,e} \\
\leq\ & c\varepsilon_0 r_0^{-3/2} (\mathcal{Q}_1 + \mathcal{Q}_2)^{1/2} + c\varepsilon_0 r_0^{-1} (\mathcal{Q}_1 + \mathcal{Q}_2)^{1/2}.
\end{aligned} \tag{8.3.5a}$$

Also,

$$\begin{aligned}
\|r^2 I^2(\mathbf{O}, W)\|_{2,e} \leq\ & c r_0^{-3/2} \|r^{7/2} \beta\|_{\infty,e} {}^{(\mathrm{loc})}\|\tau_-^{(\mathbf{O})} \underline{p}_3\|_{2,e} \\
& + c r_0^{-3/2} \|r^3 (\rho, \sigma)\|_{\infty,e} {}^{(\mathrm{loc})}\|r\, {}^{(\mathbf{O})}\not{p}\|_{2,e} \\
\leq\ & c\varepsilon_0 r_0^{-3/2} (\mathcal{Q}_1 + \mathcal{Q}_2)^{1/2}
\end{aligned} \tag{8.3.5b}$$

and

$$\begin{aligned}
\|r^2 I^3(\mathbf{O}, W)\|_{2,e} \leq\ & c r_0^{-3/2} \|r^{7/2} \alpha\|_{\infty,e} {}^{(\mathrm{loc})}\|\tau_- \underline{\Xi}({}^{(\mathbf{O})}q)\|_{2,e} \\
& + c r_0^{-3/2} \|r^{7/2} \beta\|_{\infty,e} {}^{(\mathrm{loc})}\|\,\|_{2,e} \\
& \cdot \tau_- (\underline{K}({}^{(\mathbf{O})}q), \underline{\Lambda}({}^{(\mathbf{O})}q), \underline{\Theta}({}^{(\mathbf{O})}q)) \\
& + c r_0^{-3/2} \|r^3 (\rho, \sigma)\|_{\infty,e} {}^{(\mathrm{loc})}\|r(I({}^{(\mathbf{O})}q), \underline{I}({}^{(\mathbf{O})}q))\|_{2,e} \\
& + c r_0^{-1} \|r^2 \tau_-^{3/2} \underline{\beta}\|_{\infty,e} {}^{(\mathrm{loc})}\|\,\|_{2,e} \\
& \cdot r(K({}^{(\mathbf{O})}q), \Lambda({}^{(\mathbf{O})}q), \Theta({}^{(\mathbf{O})}q)) \\
\leq\ & c\varepsilon_0 r_0^{-1} (\mathcal{Q}_1 + \mathcal{Q}_2)^{1/2} + c\varepsilon_0 r_0^{-3/2} (\mathcal{Q}_1 + \mathcal{Q}_2)^{1/2}.
\end{aligned} \tag{8.3.5c}$$

To avoid the presence of the term $O(r_0^{-1})$ on the right-hand side of 8.3.5a and 8.3.5c, which leads to divergence in 8.3.1b, we have to proceed differently.

First, observe that in the integral $\int_{V_t^e} \phi\tau_+^4|\beta(\hat{\mathcal{L}}_{\mathbf{O}}W)I^1(\mathbf{O},W)|$ the term that lead to divergence was given by the contribution to I^1 of $\mathrm{tr}\chi \mathrm{Qr}\left[{}^{(\mathbf{O})}\mathbf{i}\,;\,\underline{\beta}\right]$. Using our assumptions on $\mathrm{tr}\chi$ and ${}^{(\mathbf{O})}\mathbf{i}$,

$$\int_{V_t^e} \phi\tau_+^4\beta(\hat{\mathcal{L}}_{\mathbf{O}}W)\mathrm{tr}\chi\mathrm{Qr}\left[{}^{(\mathbf{O})}\mathbf{i}\,;\,\underline{\beta}\right] \leq c\varepsilon_0 \int_{V_t^e} \tau_+^2|\beta(\hat{\mathcal{L}}_{\mathbf{O}}W)||\underline{\beta}|.$$

Now according to the coarea formula of Lemma 8.1.2, we write

$$\begin{aligned}
\int_{V_t^e} \tau_+^2|\beta(\hat{\mathcal{L}}_{\mathbf{O}}W)||\underline{\beta}| &= \int_{u_0}^{\infty} du \int_{C_u} \tau_+^2|\beta(\hat{\mathcal{L}}_{\mathbf{O}}W)||\underline{\beta}| \\
&\leq \int_{u_0}^{\infty} \tau_-^{-2} du \left(\int_{C_u} \tau_+^4|\beta(\hat{\mathcal{L}}_{\mathbf{O}}W)|^2\right)^{1/2} \\
&\quad \cdot \left(\int_{C_u} \tau_-^4|\underline{\beta}(W)|^2\right)^{1/2}
\end{aligned}$$

where u_0 is the value of u that corresponds to the boundary of the interior region.

Since

$$\int_{C_u} \tau_-^4|\underline{\beta}(W)|^2 = \int_{t_0(u)}^{t} dt' \int_{S_{t',u}} \tau_-^4|\underline{\beta}(W)|^2$$

we can appeal to the property 7.5.12b of the rotation vectorfields (on page 179) to write

$$\int_{C_u} \tau_-^4|\underline{\beta}(W)|^2 \leq c\int_{C_u} \tau_-^4|\underline{\beta}(\hat{\mathcal{L}}_{\mathbf{O}}W)|^2.$$

On the other hand, by virtue of Proposition 8.1.2,

$$\begin{aligned}
\tau_+^4|\beta(\hat{\mathcal{L}}_{\mathbf{O}}W)|^2 &\leq Q(\hat{\mathcal{L}}_{\mathbf{O}}W)(\bar{K},\bar{K},T,e_4) \\
\tau_-^4|\underline{\beta}(\hat{\mathcal{L}}_{\mathbf{O}}W)|^2 &\leq Q(\hat{\mathcal{L}}_{\mathbf{O}}W)(\bar{K},\bar{K},T,e_4),
\end{aligned}$$

from which we infer that

$$\begin{aligned}
\int_{C_u} \tau_+^4|\beta(\hat{\mathcal{L}}_{\mathbf{O}}W)|^2 &\leq c \sup_{t\in[0,t_*]}\sup_{u\geq u_0(t)} \tilde{\mathcal{Q}}_1(W;u,t) \\
\int_{C_u} \tau_-^4|\underline{\beta}(\hat{\mathcal{L}}_{\mathbf{O}}W)|^2 &\leq c \sup_{t\in[0,t_*]}\sup_{u\geq u_0(t)} \tilde{\mathcal{Q}}_1(W;u,t)
\end{aligned}$$

with $\tilde{Q}_1(W;u,t)$ defined by the formula 8.1.4e.

In view of these and the definition 8.1.4g we conclude that

$$\int_{V_t^e} \tau_+^2 |\beta(\hat{\mathcal{L}}_{\mathbf{O}} W)||\underline{\beta}| \leq c\mathcal{Q}_1 *$$

and thus

$$\int_{V_t^e} |\phi \tau_+^4 \beta(\hat{\mathcal{L}}_{\mathbf{O}} W) \mathrm{tr}\chi \mathrm{Qr}\left[{}^{(\mathbf{O})}\mathbf{i}\,;\,\underline{\beta}\right]| \leq c\varepsilon_0 \mathcal{Q}_1 * . \tag{8.3.6a}$$

All the other terms of the integral $\int_{V_t^e} |\phi \tau_+^4 \beta(\hat{\mathcal{L}}_{\mathbf{O}} W) I^1(\mathbf{O}, W)|$ are treated exactly as before. Thus we find

$$\int_{V_t^e} |\phi \tau_+^4 \beta(\hat{\mathcal{L}}_{\mathbf{O}} W) I^1(\mathbf{O}, W)| \leq c\varepsilon_0 (\mathcal{Q}_1 * + \mathcal{Q}_2 *). \tag{8.3.6b}$$

Next we remark that the troublesome term of the integral corresponding to I^3, $\int_{V_t^e} |\phi \tau_+^4 \beta(\hat{\mathcal{L}}_{\mathbf{O}} W) I^3(\mathbf{O}, W)|$, is given by the contribution of the term $\mathrm{Qr}\left[\underline{\beta}\,;\,(K,\Lambda,\Theta)({}^{(\mathbf{O})}q)\right]$, which we estimate as follows:

$$\begin{aligned}\int_{V_t^e} \tau_+^4 \phi \beta(\hat{\mathcal{L}}_{\mathbf{O}} W) \cdot \mathrm{Qr}\,[\,;\,] &\leq \int_{u_0}^{\infty} du \int_{C_u} \tau_+^4 |\beta(\hat{\mathcal{L}}_{\mathbf{O}} W)||\mathrm{Qr}\,[\,;\,]| \\ &\leq \int_{u_0}^{\infty} du \left(\int_{C_u} \tau_+^4 |\beta(\hat{\mathcal{L}}_{\mathbf{O}} W)|^2 \right)^{1/2} \\ &\quad \cdot \left(\int_{C_u} \tau_+^4 |\mathrm{Qr}\,[\,;\,]|^2 \right)^{1/2} . \end{aligned} \tag{8.3.6c}$$

Now,

$$\begin{aligned}\int_{C_u} \tau_+^4 |\mathrm{Qr}\,[\,;\,]|^2 &= \int_{t_0(u)}^{t} dt' \int_{S_{u,t'}} \tau_+^4 |\mathrm{Qr}\,[\,;\,]|^2 \\ &\leq \int_{t_0(u)}^{t} dt' \left(\int_{S_{u,t'}} r^2 |\underline{\beta}|^4 \right)^{1/2} \\ &\quad \cdot \left(\int_{S_{u,t'}} r^6 |(K,\Lambda,\Theta)({}^{(\mathbf{O})}q)|^4 \right)^{1/2} \\ &\leq \tau_-^{-4} \sup_{[0,t]} |||r^{3/2}(K,\Lambda,\Theta)({}^{(\mathbf{O})}q)|||_{4,e}^2 |||r^{3/2}\tau_-^{3/2}\underline{\beta}|||_{4,e}^2 ,\end{aligned}$$

and taking into account the assumption 8.2.2e and the estimates 8.2.3d of Proposition 8.2.1, as well as the following immediate consequence of the global Sobolev inequalities and the comparison theorem,

$$|||r^{3/2}\tau_-^{3/2}\underline{\beta}|||_{4,e} \leq c\mathcal{Q}_1^{1/2},$$

we infer that

$$\left(\int_{C_u} \tau_+^4 |\mathrm{Qr}\,[\,;\,]\,|^2\right)^{1/2} \leq c\varepsilon_0\tau_-^{-2}\mathcal{Q}_1^{1/2}.$$

And finally, back to 8.3.6c,

$$\int_{V_t^e} \tau_+^4|\phi\beta(\hat{\mathcal{L}}_{\mathbf{O}}W)\cdot \mathrm{Qr}\,[\,;\,]\,| \leq \varepsilon_0\mathcal{Q}_1* . \tag{8.3.6d}$$

All the other terms of the integral $\int_{\Sigma_t^e} |\phi\tau_+^4\beta(\hat{\mathcal{L}}_{\mathbf{O}}W)I^3(\mathbf{O}, W)|$ are treated exactly as before. Consequently,

$$\int_{V_t^e} |\tau_+^4\phi\beta(\hat{\mathcal{L}}_{\mathbf{O}}W)\cdot I(\mathbf{O}, W)| \leq c\varepsilon_0(\mathcal{Q}_1* + \mathcal{Q}_2*). \tag{8.3.6e}$$

Going back to 8.3.1b, we conclude that

$$\int_\Sigma |\phi\tau_+^4 D(\mathbf{O}, W)_{344}| \leq c\varepsilon_0(\mathcal{Q}_1* + \mathcal{Q}_2 * . \tag{8.3.6f}$$

Proceeding in the same manner with all the other terms in 8.3.1a, we find

$$\mathrm{Integral}_A \leq c\varepsilon_0(\mathcal{Q}_1* + \mathcal{Q}_2*). \tag{8.3.6g}$$

B: Estimates for $\mathrm{Integral}_B = \int_{V_t} \phi\mathbf{Div}Q(\hat{\mathcal{L}}_T W)_{\beta\gamma\delta}\bar{K}^\beta\bar{K}^\gamma\bar{K}^\delta$

In view of formula 8.1.7b of Proposition 8.1.3 we write

$$\begin{aligned}\mathrm{Integral}_B = &\frac{1}{8}\int_{V_t} \phi\tau_+^6 D(T, W)_{444}\\ &+\frac{3}{8}\int_{V_t} \phi\tau_+^4\tau_-^2\, D(T, W)_{344}\\ &+\frac{3}{8}\int_{V_t} \phi\tau_+^2\tau_-^4\, D(T, W)_{334}\\ &+\frac{1}{8}\int_{V_t} \phi\tau_-^6\, D(T, W)_{333}.\end{aligned}$$

The most sensitive term to estimate now is $\int_{V_t} \phi\tau_+^6 D(T,W)_{444}$. As before, the estimates in the interior are completely straightforward and we concentrate only on the exterior part of the above integral. In view of the formula 8.1.7c of Proposition 8.1.3,

$$\int_{V_t^e} \phi\tau_+^6 D(T,W)_{444} = 4\int_{V_t^e} \phi\tau_+^6 \alpha(\hat{\mathcal{L}}_T W)\cdot\Theta(T,W) - 8\int_{V_t^e} \phi\tau_+^6 \beta(\hat{\mathcal{L}}_T W)\cdot\Xi(T,W). \quad (8.3.7a)$$

Proceeding as in 8.3.1c–8.3.1e, we write

$$\int_{\Sigma_t^e} |\phi\tau_+^6\alpha(\hat{\mathcal{L}}_T W)\cdot\Theta(T,W)| \le c\|r^3\alpha(\hat{\mathcal{L}}_T W)\|_{2,e}\|r^3\Theta(T,W)\|_{2,e} \le cQ_1^{1/2}\|r^3\Theta(T,W)\|_{2,e} \quad (8.3.7b)$$

$$\int_{\Sigma_t^e} |\phi\tau_+^6\beta(\hat{\mathcal{L}}_T W)\cdot\Xi(T,W)| \le c\|r^3\beta(\hat{\mathcal{L}}_T W)\|_{2,e}\|r^3\Xi(T,W)\|_{2,e} \le cQ_1^{1/2}\|r^3\Xi(T,W)\|_{2,e}. \quad (8.3.7c)$$

In view of the decomposition of J into $\frac{1}{2}(J^1+J^2+J^3)$ we write

$$\Theta(T,W) = \frac{1}{2}(\Theta^1(T,W)+\Theta^2(T,W)+\Theta^3(T,W))$$
$$\Xi(T,W) = \frac{1}{2}(\Xi^1(T,W)+\Xi^2(T,W)+\Xi^3(T,W)).$$

Now, in view of the formulas of Proposition 8.1.4 applied to the vectorfield T, we find

$$\begin{aligned}\Xi^1(T,W) = {} & \mathrm{Qr}\left[{}^{(T)}\mathbf{i}\,;\,\nabla\!\!\!/\alpha\right] + \mathrm{Qr}\left[{}^{(T)}\underline{\mathbf{m}}\,;\,\alpha_4\right] + \mathrm{Qr}\left[{}^{(T)}\mathbf{m}\,;\,\alpha_3\right] \\ & + \mathrm{Qr}\left[{}^{(T)}\mathbf{m}\,;\,\nabla\!\!\!/\beta\right] + \mathrm{Qr}\left[{}^{(T)}\mathbf{j}\,;\,\beta_4\right] + \mathrm{Qr}\left[{}^{(T)}\mathbf{n}\,;\,\beta_3\right] \\ & + \mathrm{tr}\chi\left(\mathrm{Qr}\left[{}^{(T)}\underline{\mathbf{m}}\,;\,\alpha\right] + \mathrm{Qr}\left[({}^{(T)}\mathbf{i},{}^{(T)}\mathbf{j})\,;\,\beta\right] + \mathrm{Qr}\left[T_m\,;\,(\rho,\sigma)\right]\right) \\ & + \mathrm{tr}\underline{\chi}\left(\mathrm{Qr}\left[{}^{(T)}\mathbf{m}\,;\,\alpha\right] + \mathrm{Qr}\left[{}^{(T)}\mathbf{n}\,;\,\beta\right]\right) \end{aligned} \quad (8.3.7d)$$

$$\Xi^2(T,W) = \mathrm{Qr}\left[{}^{(T)}p\!\!\!/\,;\,\alpha\right] + \mathrm{Qr}\left[{}^{(T)}p_4\,;\,\beta\right] \quad (8.3.7e)$$

$$\begin{aligned}\Xi^3(T,W) = {} & \mathrm{Qr}\left[\underline{\alpha}\,;\,(I,\underline{I})({}^{(T)}q)\right] \\ & + \mathrm{Qr}\left[\underline{\beta}\,;\,(\underline{K},\underline{\Lambda},\underline{\Theta})({}^{(T)}q)\right] \\ & + \mathrm{Qr}\left[(\rho,\sigma)\,;\,\underline{\Xi}({}^{(T)}q)\right]. \end{aligned} \quad (8.3.7f)$$

Now, taking into account exterior assumptions 8.2.2a for the null components of ${}^{(T)}\pi$ and also assumptions 7.5.8a on the Ricci coefficients, we derive

$$\begin{aligned}
\|r^3\Xi^1(T,W)\|_{2,e} &\le cr_0^{-1}\|r\tau_-{}^{(T)}\mathbf{i}\|_{\infty,e}\|r^3\not\nabla\alpha\|_{2,e}\\
&+cr_0^{-2}\|r^2\,{}^{(T)}\mathbf{j}\|_{\infty,e}\|r^3\alpha_4\|_{2,e}\\
&+cr_0^{-2}\|r^2({}^{(T)}\mathbf{m},\ {}^{(T)}\underline{\mathbf{m}},\ {}^{(T)}\mathbf{n},\ {}^{(T)}\underline{\mathbf{n}})\|_{\infty,e}\\
&\cdot\|r^3(\alpha_3,\not\nabla\beta,\beta_3,\beta_4)\|_{2,e}\\
&+cr_0^{-2}\|r^2({}^{(T)}\underline{\mathbf{m}},{}^{(T)}\mathbf{j})\|_{\infty,e}\|r\mathrm{tr}\chi\|_{\infty,e}\|r^2(\alpha,\beta)\|_{2,e}\\
&+cr_0^{-2}\|r^{2(T)}\mathbf{m}\|_{\infty,e}\|r\mathrm{tr}\chi\|_{\infty,e}\||r^3(\rho,\sigma)\||_{2,e}\\
&+cr_0^{-2}\|r^2({}^{(T)}\mathbf{m},\ {}^{(T)}\mathbf{n})\|_{\infty,e}\|r\mathrm{tr}\underline{\chi}\|_{\infty,e}\|r^2(\alpha,\beta)\|_{2,e}\\
&+cr_0^{-1}\|r\tau_-{}^{(T)}\mathbf{i}\|_{\infty,e}\|r\mathrm{tr}\chi\|_{\infty,e}\|r^2\beta\|_{2,e}\\
&\le c\varepsilon_0 r_0^{-2}\mathcal{Q}_1^{1/2}+c\varepsilon_0 r_0^{-1}\mathcal{Q}_1^{1/2}.
\end{aligned}\tag{8.3.7g}$$

Also, in view of estimates 8.2.3c of Proposition 8.1.1a,

$$\begin{aligned}
\|r^3\Xi^2(T,W)\|_{2,e} &\le \|r^{7/2}\alpha\|_{\infty,e}\|r^{-1/2}\,{}^{(T)}\not p\|_{2,e}\\
&+\|r^{7/2}\beta\|_{\infty,e}\|r^{-1/2}\,{}^{(T)}p_4\|_{2,e}\\
&\le c\varepsilon_0(1+t)^{-3/2}(\mathcal{Q}_1+\mathcal{Q}_2)
\end{aligned}\tag{8.3.7h}$$

and

$$\begin{aligned}
\|r^3\Xi^3(T,W)\|_{2,e} &\le \|r^{7/2}\alpha\|_{\infty,e}\|r^{-1/2}(I({}^{(T)}q),\underline{I}({}^{(T)}q))\|_{2,e}\\
&+\|r^{7/2}\beta\|_{\infty,e}\|r^{-1/2}(K({}^{(T)}q),\Lambda({}^{(T)}q),\Theta({}^{(T)}q))\|_{2,e}\\
&+\|r^3(\rho,\sigma)\|_{\infty,e}\|\Xi({}^{(T)}q)\|_{2,e}\\
&\le c\varepsilon_0(1+t)^{-3/2}(\mathcal{Q}_1+\mathcal{Q}_2).
\end{aligned}\tag{8.3.7i}$$

Similarly, we have

$$\begin{aligned}
\Theta^1(T,W) &= \mathrm{Qr}\left[{}^{(T)}\underline{\mathbf{m}}\,;\,\not\nabla\alpha\right]+\mathrm{Qr}\left[{}^{(T)}\underline{\mathbf{n}}\,;\,\alpha_4\right]+\mathrm{Qr}\left[{}^{(T)}\mathbf{j}\,;\,\alpha_3\right]\\
&+\mathrm{Qr}\left[{}^{(T)}\mathbf{i}\,;\,\not\nabla\beta\right]+\mathrm{Qr}\left[{}^{(T)}\underline{\mathbf{m}}\,;\,\beta_4\right]+\mathrm{Qr}\left[{}^{(T)}\mathbf{m}\,;\,\beta_3\right]\\
&+\mathrm{Qr}\left[{}^{(T)}\mathbf{m}\,;\,\not\nabla(\rho,\sigma)\right]+\mathrm{Qr}\left[{}^{(T)}\mathbf{j}\,;\,(\rho_4,\sigma_4)\right]\\
&+\mathrm{Qr}\left[{}^{(T)}\mathbf{n}\,;\,(\rho_3,\sigma_3)\right]\\
&+\mathrm{tr}\chi\left(\mathrm{Qr}\left[{}^{(T)}\underline{\mathbf{n}}\,;\,\alpha\right]+\mathrm{Qr}\left[{}^{(T)}\underline{\mathbf{m}}\,;\,\beta\right]\right.\\
&+\left.\mathrm{Qr}\left[({}^{(T)}\mathbf{i},{}^{(T)}\mathbf{j})\,;\,(\rho,\sigma)\right]+\mathrm{Qr}\left[{}^{(T)}\mathbf{m}\,;\,\underline{\beta}\right]\right)\\
&+\mathrm{tr}\underline{\chi}\left(\mathrm{Qr}\left[({}^{(T)}\mathbf{i},{}^{(T)}\mathbf{j})\,;\,\alpha\right]+\mathrm{Qr}\left[{}^{(T)}\mathbf{m}\,;\,\beta\right]\right.
\end{aligned}$$

$$+ \text{Qr}\left[{}^{(T)}\mathbf{n}\,;\,(\rho,\sigma)\right]\Big) \tag{8.3.8a}$$

$$\Theta^2(T,W) = \text{Qr}\left[{}^{(T)}p_3\,;\,\alpha\right] + \text{Qr}\left[{}^{(T)}\not{p}\,;\,\beta\right] + \text{Qr}\left[{}^{(T)}p_4\,;\,(\rho,\sigma)\right] \tag{8.3.8b}$$

$$\begin{aligned}\Theta^3(T,W) = {} & \text{Qr}\left[\alpha\,;\,\underline{K}({}^{(T)}q)\right] + \text{Qr}\left[\alpha\,;\,\underline{\Lambda}({}^{(T)}q)\right] \\ & + \boxed{\text{Qr}\left[\alpha\,;\,\underline{\Theta}({}^{(T)}q)\right]} \\ & + \text{Qr}\left[\beta\,;\,(I,\underline{I})({}^{(T)}q)\right] + \text{Qr}\left[(\rho,\sigma)\,;\,\Theta({}^{(T)}q)\right] \\ & + \boxed{\text{Qr}\left[\underline{\beta}\,;\,\Xi({}^{(T)}q)\right]}\,.\end{aligned} \tag{8.3.8c}$$

Hence, taking into account assumptions 8.2.2a,

$$\begin{aligned}\|r^3\Theta^1(T,W)\|_{2,e} \leq {} & r_0^{-1}\|r\tau_-{}^{(T)}\mathbf{i}\|_{\infty,e}\|r^3\not{\nabla}\beta\|_{2,e} \\ & + cr_0^{-2}\|r^2\,{}^{(T)}\mathbf{j}\|_{\infty,e}\|r^3(\alpha_3,\rho_4,\sigma_4)\|_{2,e} \\ & + cr_0^{-1}\|r^2\,{}^{(T)}\underline{\mathbf{n}}\|_{\infty,e}\|r^2\tau_-(\rho_3,\sigma_3)\|_{2,e} \\ & + cr_0^{-2}\|r^2({}^{(T)}\mathbf{m},\,{}^{(T)}\underline{\mathbf{m}},\,{}^{(T)}\mathbf{n},\,{}^{(T)}\underline{\mathbf{n}})\|_{\infty,e} \\ & \cdot\|r^3(\alpha_4,\not{\nabla}\alpha,\beta_3,\beta_4\not{\nabla}\rho,\not{\nabla}\sigma)\|_{2,e} \\ & + cr_0^{-2}\|r^2({}^{(T)}\underline{\mathbf{m}},{}^{(T)}\underline{\mathbf{n}})\|_{\infty,e}\|r\text{tr}\chi\|_{\infty,e}\|r^2(\alpha,\beta)\|_{2,e} \\ & + cr_0^{-2}\|r^2\,{}^{(T)}\mathbf{j}\|_{\infty,e}\|r\text{tr}\chi\|_{\infty,e}|||r^3(\rho,\sigma)|||_{2,e} \\ & + cr_0^{-2}\|r^2({}^{(T)}\mathbf{m},\,{}^{(T)}\mathbf{j})\|_{\infty,e}\|r\text{tr}\underline{\chi}\|_{\infty,e}\|r^2(\alpha,\beta)\|_{2,e} \\ & + cr_0^{-2}\|r^2\,{}^{(T)}\mathbf{n}\|_{\infty,e}\|r\text{tr}\underline{\chi}\|_{\infty,e}|||r^3(\rho,\sigma)|||_{2,e} \\ & + cr_0^{-1}\|r\tau_-\,{}^{(T)}\mathbf{i}\|_{\infty,e}\|r\text{tr}\chi\|_{\infty,e}|||r^3(\rho,\sigma)|||_{2,e} \\ & + cr_0^{-1}\|r\tau_-{}^{(T)}\mathbf{i}\|_{\infty,e}\|r\text{tr}\underline{\chi}\|_{\infty,e}\|r^2\alpha\|_{2,e} \\ & + cr_0^{-1}\|r^{2(T)}\mathbf{m}\|_{\infty,e}\|r\text{tr}\chi\|_{\infty,e}\|r\tau_-\underline{\beta}\|_{2,e} \\ \leq {} & c\varepsilon_0 r_0^{-2}Q_1^{1/2} + c\varepsilon_0 r_0^{-1}Q_1^{1/2}.\end{aligned} \tag{8.3.8d}$$

Also,

$$\begin{aligned}\|r^3\Theta^2(T,W)\|_{2,e} \leq {} & \|r^{7/2}\alpha\|_{\infty,e}\|r^{-1/2}\,{}^{(T)}p_3\|_{2,e} \\ & + \|r^{7/2}\beta\|_{\infty,e}\|r^{-1/2}\,{}^{(T)}\not{p}\|_{2,e} \\ & + \|r^3(\rho,\sigma)\|_{\infty,e}\|\,{}^{(T)}p_4\|_{2,e} \\ \leq {} & c\varepsilon_0(1+t)^{-3/2}(Q_1+Q_2)\end{aligned} \tag{8.3.8e}$$

and

$$\|r^3\Theta^3(T,W)\|_{2,e} \leq \|r^{7/2}\alpha\|_{\infty,e}\|r^{-1/2}\underline{K}({}^{(T)}q)\|_{2,e}$$

$$
\begin{aligned}
&+ \|r^{7/2}\alpha\|_{\infty,e}\|r^{-1/2}\underline{\Lambda}({}^{(T)}q)\|_{2,e} \\
&+ \|r^{7/2}\beta\|_{\infty,e}\|r^{-1/2}(I({}^{(T)}q), \underline{I}({}^{(T)}q))\|_{2,e} \\
&+ \|r^{3}(\rho,\sigma)\|_{\infty.e}\|\Theta({}^{(T)}q)\|_{2,e} \\
&\leq c\varepsilon_0(1+t)^{-3/2}(\mathcal{Q}_1+\mathcal{Q}_2).
\end{aligned} \tag{8.3.8f}
$$

Unfortunately, estimates 8.3.7g and 8.3.8d for $\Xi^1(T,W)$ and $\Theta^1(T,W)$ are not good enough due to the presence of the terms of order $O(r_0^{-1})$. We overcome this difficulty by proceeding in the same manner as for the estimates for Integral_A.

In doing this we have to focus only on those terms in the integrals on the right-hand side of 8.3.7a that lead to difficulties. In the case of the integral $\int_{\Sigma_t^e} \phi\tau_+^6\beta(\hat{\mathcal{L}}_T W)\cdot\Xi(T,W)$, these terms are related to the presence of ${}^{(T)}\mathbf{i}$ in the expression for Ξ^1. Let A_1, A_2 be the integrals

$$
A_1 = \int_{\Sigma_t^e} |\phi\tau_+^6\beta(\hat{\mathcal{L}}_T W)\cdot \mathrm{Qr}\left[{}^{(T)}\mathbf{i}\,;\,\nabla\!\!\!\!/\,\alpha\right]|
$$

and

$$
A_2 = \int_{\Sigma_t^e} |\phi\tau_+^6\mathrm{tr}\chi\beta(\hat{\mathcal{L}}_T W)\cdot \mathrm{Qr}\left[{}^{(T)}\mathbf{i}\,;\,\beta\right]|.
$$

To avoid the behavior of order $O(r_0^{-1})$ in the estimate 8.3.7g for which these terms were responsible, we proceed as follows:

$$
A_1 \leq c\sup_{[0,t]}\|r\tau_-{}^{(T)}\mathbf{i}\|_{\infty,e}\int_{r_0}^{\infty}\int_{V_t^e}\tau_+^5\tau_-^{-1}|\beta(\hat{\mathcal{L}}_T W)|\nabla\!\!\!\!/\,\alpha|. \tag{8.3.8g}
$$

Making use of the coarea formula of Lemma 8.1.2, we write

$$
\begin{aligned}
\int_{V_t^e}\tau_+^5\tau_-^{-1}|\beta(\hat{\mathcal{L}}_T W)|\nabla\!\!\!\!/\,\alpha| &= \int_{u_0}^{\infty}\tau_-^{-2}du\left(\int_{C_u}\tau_+^5\tau_-|\beta(\hat{\mathcal{L}}_T W)||\nabla\!\!\!\!/\,\alpha|\right) \\
&\leq \int_{u_0}^{\infty}\tau_-^{-2}du\left(\int_{C_u}\tau_+^4\tau_-^2|\beta(\hat{\mathcal{L}}_T W)|^2\right)^{1/2} \\
&\quad\cdot\left(\int_{C_u}\tau_+^6|\nabla\!\!\!\!/\,\alpha|^2\right)^{1/2}.
\end{aligned} \tag{8.3.8h}
$$

Now, by virtue of Proposition 8.1.2,

$$
\begin{aligned}
\tau_+^4\tau_-^2|\beta(\hat{\mathcal{L}}_T W)|^2 &\leq cQ(\hat{\mathcal{L}}_T W)(\bar{K},\bar{K},\bar{K},e_4) \\
\tau_+^6|\alpha(\hat{\mathcal{L}}_T W)|^2 &\leq cQ(\hat{\mathcal{L}}_T W)(\bar{K},\bar{K},\bar{K},e_4).
\end{aligned}
$$

Thus, recalling the definition 8.1.4e,

$$\int_{C_u} \tau_+^4 \tau_-^2 |\beta(\hat{\mathcal{L}}_T W)|^2 \leq \tilde{\mathcal{Q}}_1(W,u,t)$$

and, taking into account property 7.5.12a of the rotation vectorfields (see page 179), we deduce also that

$$\int_{C_u} r^6 |\nabla\!\!\!/ \alpha(W)|^2 \leq c\tilde{\mathcal{Q}}_1(W,u,t).$$

Therefore, going back to 8.3.8h and 8.3.8g, we conclude that

$$\begin{aligned} A_1 &= \int_{V_t^e} |\phi\tau_+^6 \beta(\hat{\mathcal{L}}_T W) \cdot \mathrm{Qr}\left[{}^{(T)}\mathbf{i}\,;\, \nabla\!\!\!/ \alpha\right]| \\ &\leq c\varepsilon_0 \sup_{t\in[0,t_*]} \sup_{u\geq u_0(t)} \tilde{\mathcal{Q}}_1(W;u,t) \int_{u_0}^{\infty} \tau_-^{-2} du \\ &\leq c\varepsilon_0 \mathcal{Q}_1 * . \end{aligned} \tag{8.3.8i}$$

Similarly,

$$A_2 \leq \|r\mathrm{tr}\chi\|_{\infty,e} \|r\tau_-^{(T)}\mathbf{i}\|_{\infty,e} \int_{V_t^e} \tau_+^4 \tau_-^{-1} |\beta(\hat{\mathcal{L}}_T W)||\beta| \tag{8.3.8j}$$

$$\begin{aligned} \int_{V_t^e} |\tau_+^4 \tau_-^{-1} |\beta(\hat{\mathcal{L}}_T W)||\beta| &\leq \int_{u_0}^{\infty} \tau_-^{-2} du \left(\int_{C_u} \tau_+^4 \tau_-^2 |\beta(\hat{\mathcal{L}}_T W)|^2 \right)^{1/2} \\ &\quad \cdot \left(\int_{C_u} \tau_+^4 |\beta|^2 \right)^{1/2} . \end{aligned}$$

Taking into account the property 7.5.12b of the rotation vectorfields, we can write

$$\int_{C_u} \tau_+^4 |\beta|^2 \leq c \int_{C_u} \tau_+^4 |\hat{\mathcal{L}}_{\mathbf{O}} \beta|^2,$$

and referring once again to Proposition 8.1.2, we deduce that

$$\int_{C_u} \tau_+^4 \tau_-^2 |\beta(\hat{\mathcal{L}}_T W)|^2 \leq c\tilde{\mathcal{Q}}_1(W,u,t)$$

$$\int_{C_u} \tau_+^4 |\beta|^2 \leq c\tilde{\mathcal{Q}}_1(W,u,t).$$

Therefore,

$$A_2 \leq \int_{V_t^e} |\phi\tau_+^6 \mathrm{tr}\chi \beta(\hat{\mathcal{L}}_T W) \cdot \mathrm{Qr}\left[{}^{(T)}\mathbf{i}\,;\, \beta\right]| \leq c\varepsilon_0 \mathcal{Q}_1 * . \tag{8.3.8k}$$

Finally, proceeding as before with all the other terms, we conclude that

$$\int_{V_t^e} |\phi\tau_+^6\beta(\hat{\mathcal{L}}_T W)(\text{remaining terms of } \Xi^1(T,W))| \le c\varepsilon_0 \mathcal{Q}_1*$$

and

$$\int_{V_t^e} |\phi\tau_+^6\beta(\hat{\mathcal{L}}_T W)\cdot \Xi^1(T,W)| \le c\varepsilon_0(\mathcal{Q}_1* + \mathcal{Q}_2*). \tag{8.3.8l}$$

Together with estimates 8.3.7h, 8.3.7i, and 8.3.7c, we infer that

$$\int_{V_t^e} |\phi\tau_+^6\beta(\hat{\mathcal{L}}_T W)\cdot \Xi(T,W)| \le c\varepsilon_0(\mathcal{Q}_1* + \mathcal{Q}_2*). \tag{8.3.8m}$$

Similarly, the terms in the integral $\int_{V_t^e} \phi\tau_+^6\alpha(\hat{\mathcal{L}}_T W)\cdot\Theta(T,W)$ that require special attention are

$$\begin{aligned}
B_1 &= \int_{V_t^e} |\phi\tau_+^6\alpha(\hat{\mathcal{L}}_T W)\cdot \mathrm{Qr}\left[{}^{(T)}\mathbf{i}\,;\,\nabla\!\!\!\!/\,\beta\right]| \\
B_2 &= \int_{V_t^e} |\phi\tau_+^6\alpha(\hat{\mathcal{L}}_T W)\cdot \mathrm{Qr}\left[{}^{(T)}\mathbf{n}\,;\,(\rho_3,\sigma_3)\right]| \\
B_3 &= \int_{V_t^e} |\phi\tau_+^6\mathrm{tr}\underline{\chi}\alpha(\hat{\mathcal{L}}_T W)\cdot \mathrm{Qr}\left[{}^{(T)}\mathbf{i}\,;\,\alpha\right]| \\
B_4 &= \int_{V_t^e} |\phi\tau_+^6\mathrm{tr}\chi\alpha(\hat{\mathcal{L}}_T W)\cdot \mathrm{Qr}\left[{}^{(T)}\mathbf{i}\,;\,(\rho,\sigma)\right]| \\
B_5 &= \int_{V_t^e} |\phi\tau_+^6\mathrm{tr}\chi\alpha(\hat{\mathcal{L}}_T W)\cdot \mathrm{Qr}\left[{}^{(T)}\mathbf{m}\,;\,\underline{\beta}\right]|.
\end{aligned}$$

For the first integral, we proceed precisely as in the derivation of 8.3.8i:

$$\begin{aligned}
B_1 &\le c\varepsilon_0 \int_{u_0}^{\infty} \tau_-^{-2} du \left(\int_{C_u} \tau_+^6 |\alpha(\hat{\mathcal{L}}_T W)|^2\right)^{1/2} \left(\int_{C_u} \tau_+^4\tau_-^2 |\nabla\!\!\!\!/\,\beta|^2\right)^{1/2} \\
&\le c\varepsilon_0 \mathcal{Q}_1*.
\end{aligned} \tag{8.3.8n}$$

For the third integral, we proceed as in 8.3.8k:

$$\begin{aligned}
B_3 &\le c\varepsilon_0 \int_{u_0}^{\infty} \tau_-^{-2} du \left(\int_{C_u} \tau_+^6 |\alpha(\hat{\mathcal{L}}_T W)|^2\right)^{1/2} \left(\int_{C_u} \tau_+^4 |\hat{\mathcal{L}}_O \alpha|^2\right)^{1/2} \\
&\le c\varepsilon_0 \mathcal{Q}_1*.
\end{aligned} \tag{8.3.8o}$$

For the second integral we proceed as follows:

$$B_2 \leq c \sup_{[0,t_*]} \|r^2 \, {}^{(T)}\mathbf{n}\|_{\infty,e}$$

$$\cdot \int_{u_0}^{\infty} \tau_-^{-2} du \left(\int_{C_u} \tau_+^6 |\alpha(\hat{\mathcal{L}}_T W)|^2 \right)^{1/2} \left(\int_{C_u} \tau_+^2 \tau_-^4 |(\rho_3, \sigma_3)|^2 \right)^{1/2}$$

$$\leq c\varepsilon_0 \mathcal{Q}_1 * \left(\int_{C_u} \tau_+^2 \tau_-^4 |(\rho_3, \sigma_3)|^2 \right)^{1/2}.$$

To estimate the integral on the right-hand side of 8.3.8p, *we refer back* to the equations 7.3.11e and 7.3.11g on page 161:

$$\rho_3 = -\text{d}\!\!/\text{iv} \, \underline{\beta} + E_3(\rho)$$

$$E_3(\rho) = -\frac{1}{2} \hat{\chi} \cdot \underline{\alpha} + \epsilon \cdot \underline{\beta} + 2(\underline{\xi} \cdot \beta - \zeta \cdot \underline{\beta})$$

$$\sigma_3 = -\text{c}\!\!/\text{url} \, \underline{\beta} + E_3(\sigma)$$

$$E_3(\sigma) = -\frac{1}{2} \hat{\chi} \cdot {}^*\underline{\alpha} + \epsilon \cdot {}^*\underline{\beta} - 2(\underline{\xi} \cdot {}^*\beta + \zeta \cdot {}^*\underline{\beta}).$$

Taking into account properties 7.5.12a and 7.5.12b of the rotation vectorfields together with Proposition 8.1.2, we have

$$\int_{C_u} r^2 \tau_-^4 |\nabla\!\!\!/ \underline{\beta}(W)|^2 \leq c\tilde{\mathcal{Q}}_1(W; u, t)$$

$$\int_{C_u} \tau_-^4 |\underline{\beta}(W)|^2 \leq c\tilde{\mathcal{Q}}_1(W; u, t)$$

$$\int_{C_u} \tau_+^4 |\beta(W)|^2 \leq c\tilde{\mathcal{Q}}_1(W; u, t).$$

Thus, taking into account assumptions 7.5.8a on page 176,

$$\int_{C_u} \tau_+^2 \tau_-^4 |(\rho_3, \sigma_3)|^2 \leq c\varepsilon_0 \left(\tilde{\mathcal{Q}}_1(W; u, t) + \int_{C_u} \tau_+^{-2} \tau_-^4 |\underline{\alpha}(W)|^2 \right).$$

Finally,

$$\int_{C_u} \tau_+^{-2} \tau_-^4 |\underline{\alpha}(W)|^2 \leq \int_{t_0(u)}^{t} dt' \int_{S_{u,t'}} \tau_+^{-2} \tau_-^4 |\underline{\alpha}(W)|^2$$

and, taking into account the degenerate form of the global Sobolev inequalities, as stated in Corollary 3.2.1.1 to Proposition 3.2.1, and the comparison lemma, we infer that

$$\int_{C_u} \tau_+^{-2} \tau_-^4 |\underline{\alpha}(W)|^2 \leq c \sup_{[0,t_*]} \mathcal{Q}_1(W, t).$$

Hence,

$$B_3 = \int_{V_t^\epsilon} |\phi\tau_+^6 \alpha(\hat{\mathcal{L}}_T W) \cdot \mathrm{Qr}\left[{}^{(T)}\mathbf{n}\,;\,(\rho_3, \sigma_3)\right]| \leq c\varepsilon_0 \mathcal{Q}_1 *. \quad (8.3.8p)$$

It remains to estimate the fourth and fifth integrals, B_4 and B_5:

$$B_4 \leq c\varepsilon_0 \int_{u_0}^{\infty} \tau_-^{-3/2} du \left(\int_{C_u} \tau_+^6 |\alpha(\hat{\mathcal{L}}_T W)|^2\right)^{1/2} \left(\int_{C_u} \tau_+^2 \tau_- |(\rho, \sigma)|^2\right)^{1/2}$$
$$\leq c\varepsilon_0 \mathcal{Q}_1 * \quad (8.3.8q)$$
$$B_5 \leq c\varepsilon_0 \int_{u_0}^{\infty} \tau_-^{-2} du \left(\int_{C_u} \tau_+^6 |\alpha(\hat{\mathcal{L}}_T W)|^2\right)^{1/2} \left(\int_{C_u} \tau_-^4 |\underline{\beta}|^2\right)^{1/2}$$
$$\leq c\varepsilon_0 \mathcal{Q}_1 *. \quad (8.3.8r)$$

Finally, treating the remaining terms precisely as before,

$$\int_{V_t^\epsilon} |\phi\tau_+^6 \alpha(\hat{\mathcal{L}}_T W)(\text{remaining terms of } \Theta^1(T, W))| \leq c \sup_{[0,t_*]} \mathcal{Q}_1(t, W),$$

we derive

$$\int_{V_t^\epsilon} |\phi\tau_+^6 \alpha(\hat{\mathcal{L}}_T W) \cdot \Theta^1(T, W)| \leq c\varepsilon_0(\mathcal{Q}_1 * + \mathcal{Q}_2 *). \quad (8.3.8s)$$

Together with estimates 8.3.8e, 8.3.8f, and 8.3.7b, we thus infer that

$$\int_{V_t^\epsilon} |\phi\tau_+^6 \alpha(\hat{\mathcal{L}}_T W) \cdot \Theta(T, W)| \leq c\varepsilon_0(\mathcal{Q}_1 * + \mathcal{Q}_2 *). \quad (8.3.9)$$

This, together with 8.3.8m, proves that

$$\int_{V_t^\epsilon} |\phi\tau_+^6 D(T, W)_{444}| \leq c\varepsilon_0(\mathcal{Q}_1 * + \mathcal{Q}_2 *). \quad (8.3.10)$$

Proceeding similarly with all the other terms, we obtain

$$\text{Integral}_B \leq c\varepsilon_0(\mathcal{Q}_1 * + \mathcal{Q}_2 *). \quad (8.3.11)$$

B: Estimates for $\text{Integral}_C = \int_{V_t} \phi Q(\hat{\mathcal{L}}_O W)_{\alpha\beta\gamma\delta}\, {}^{(\bar{K})}\pi^{\alpha\beta} \bar{K}^\gamma T^\delta$

We first decompose the integrand into null components, with the help of Lemma 7.3.2 on page 151, and observe that the most delicate terms to estimate are of the form

$$\tau_-^2 Q(\hat{\mathcal{L}}_O W)_{\alpha\beta 34}\, {}^{(\bar{K})}\hat{\pi}^{\alpha\beta} = \tau_+^2 \left(\frac{1}{4} Q_{4443}\, {}^{(\bar{K})}\hat{\pi}_{33} + \frac{1}{2} Q_{4433}\, {}^{(\bar{K})}\hat{\pi}_{34}\right.$$

$$
\begin{aligned}
&+ \frac{1}{4} Q_{3334}\,{}^{(\bar{K})}\hat{\pi}_{44} - Q_{A434}\,{}^{(\bar{K})}\hat{\pi}_{3A} \\
&- Q_{A334}\,{}^{(\bar{K})}\hat{\pi}_{4A} + Q_{AB34}\,{}^{(\bar{K})}\hat{\pi}_{AB}\Big) \\
&= \tau_+^2 \{ |\beta(\hat{\mathcal{L}}_{\mathbf{O}} W)|^2\,{}^{(\bar{K})}\underline{\mathbf{n}} + 2(|\rho(\hat{\mathcal{L}}_{\mathbf{O}} W)|^2 \\
&+ |\sigma(\hat{\mathcal{L}}_{\mathbf{O}} W)|^2)\,{}^{(\bar{K})}\mathbf{j} + |\underline{\beta}(\hat{\mathcal{L}}_{\mathbf{O}} W)|^2\,{}^{(\bar{K})}\mathbf{n} \\
&- 4\rho(\hat{\mathcal{L}}_{\mathbf{O}} W)\beta(\hat{\mathcal{L}}_{\mathbf{O}} W) \cdot {}^{(\bar{K})}\underline{\mathbf{m}} \\
&+ 4\sigma^\star\beta(\hat{\mathcal{L}}_{\mathbf{O}} W) \cdot {}^{(\bar{K})}\underline{\mathbf{m}} \\
&+ 4\rho(\hat{\mathcal{L}}_{\mathbf{O}} W)\underline{\beta}(\hat{\mathcal{L}}_{\mathbf{O}} W) \cdot {}^{(\bar{K})}\mathbf{m} \\
&+ 4\sigma(\hat{\mathcal{L}}_{\mathbf{O}} W)^\star\underline{\beta}(\hat{\mathcal{L}}_{\mathbf{O}} W) \cdot {}^{(\bar{K})}\mathbf{m} \\
&+ 2(\beta(\hat{\mathcal{L}}_{\mathbf{O}} W)\widehat{\otimes}\underline{\beta}(\hat{\mathcal{L}}_{\mathbf{O}} W)) \cdot {}^{(\bar{K})}\mathbf{i} \\
&+ 2(|\rho(\hat{\mathcal{L}}_{\mathbf{O}} W)|^2 + |\sigma(\hat{\mathcal{L}}_{\mathbf{O}} W)|^2)\mathrm{tr}\,{}^{(\bar{K})}\mathbf{i} \}.
\end{aligned}
$$

Estimating, with the help of assumptions 8.2.2c, we find

$$
\begin{aligned}
C :&= \int_{\Sigma_t^e} \tau_+^2 |\phi Q(\hat{\mathcal{L}}_{\mathbf{O}} W)_{\alpha\beta 34}\,{}^{(\bar{K})}\hat{\pi}^{\alpha\beta}| \\
&\leq c r_0^{-2} \|{}^{(\bar{K})}\underline{\mathbf{n}}\|_{\infty,e} \|\tau_+^2 \beta(\hat{\mathcal{L}}_{\mathbf{O}} W)\|_{2,e}^2 \\
&\quad + c r_0^{-2} \|{}^{(\bar{K})}\mathbf{j}\|_{\infty,e} \|\tau_+^2 (\rho(\hat{\mathcal{L}}_{\mathbf{O}} W), \sigma(\hat{\mathcal{L}}_{\mathbf{O}} W))\|_{2,e}^2 \\
&\quad + c r_0^{-2} \|r^2 \tau_-^{-2}\,{}^{(\bar{K})}\mathbf{n}\|_{\infty,e} \|\tau_+ \tau_-\, \underline{\beta}(\hat{\mathcal{L}}_{\mathbf{O}} W)\|_{2,e}^2 \\
&\quad + c r_0^{-2} \|{}^{(\bar{K})}\underline{\mathbf{m}}\|_{\infty,e} \|\tau_+^2 (\rho(\hat{\mathcal{L}}_{\mathbf{O}} W), \sigma(\hat{\mathcal{L}}_{\mathbf{O}} W))\|_{2,e} \\
&\quad \cdot \|\tau_+^2\, \beta(\hat{\mathcal{L}}_{\mathbf{O}} W)\|_{2,e}^2 \\
&\quad + c r_0^{-2} \|r^2 \tau_-^{-2}\,{}^{(\bar{K})}\mathbf{m}\|_{\infty,e} \|\tau_+^2 (\rho(\hat{\mathcal{L}}_{\mathbf{O}} W), \sigma(\hat{\mathcal{L}}_{\mathbf{O}} W))\|_{2,e} \\
&\quad \cdot \|\tau_+ \tau_-\, \underline{\beta}(\hat{\mathcal{L}}_{\mathbf{O}} W)\|_{2,e}^2 \\
&\quad + c r_0^{-2} \|{}^{(\bar{K})}\mathbf{i}\|_{\infty,e} \|\tau_+^2 (\rho(\hat{\mathcal{L}}_{\mathbf{O}} W), \sigma(\hat{\mathcal{L}}_{\mathbf{O}} W))\|_{2,e}^2 \\
&\quad + c r_0^{-1} \|{}^{(\bar{K})}\mathbf{i}\|_{\infty,e} \|\tau_+^2\, \beta(\hat{\mathcal{L}}_{\mathbf{O}} W)\|_{2,e} \\
&\quad c \|\tau_+ \tau_-\, \underline{\beta}(\hat{\mathcal{L}}_{\mathbf{O}} W)\|_{2,e} \\
&\leq c(\varepsilon_0 r_0^{-2} + \varepsilon_0 r_0^{-1}) \mathcal{Q}_1.
\end{aligned}
\tag{8.3.12a}
$$

Note that the only term that leads to divergence in the integral on V_t^e is

$$
C_1 = \int_{V_t^e} |\phi \tau_+^2 (\beta(\hat{\mathcal{L}}_{\mathbf{O}} W) \widehat{\otimes} \underline{\beta}(\hat{\mathcal{L}}_{\mathbf{O}} W)) \cdot {}^{(\bar{K})}\mathbf{i}|,
$$

which we treat as follows:

$$C_1 \leq c \sup_{[0,t]} \|{}^{(\bar{K})}\mathbf{i}\|_{\infty,e}$$
$$\cdot \int_{u_0}^{\infty} \tau_-^{-2} du \left(\int_{C_u} \tau_+^4 |\beta(\hat{\mathcal{L}}_O W|^2 \right)^{1/2} \left(\int_{C_u} \tau_-^4 |\underline{\beta}(\hat{\mathcal{L}}_O W|^2 \right)^{1/2}$$
$$\leq c\varepsilon_0 \mathcal{Q}_1 * . \tag{8.3.12b}$$

Thus,

$$\int_{V_t^e} \tau_+^2 |\phi Q(\hat{\mathcal{L}}_O W)_{\alpha\beta 34} {}^{(\bar{K})}\hat{\pi}^{\alpha\beta}| \leq c\varepsilon_0(\mathcal{Q}_1 * + \mathcal{Q}_2 *). \tag{8.3.12c}$$

All the other components of Integral$_C$ can be estimated in a straightforward manner, and we find

$$\text{Integral}_C \leq c\varepsilon_0(\mathcal{Q}_1 * + \mathcal{Q}_2 *) \tag{8.3.12d}$$

D: Estimates for Integral$_D$ = $\int_\Sigma \phi Q(\hat{\mathcal{L}}_T W)^{(\bar{K})}_{\alpha\beta\gamma\delta}\pi^{\alpha\beta}\bar{K}^\gamma \bar{K}^\delta$

Proceeding as in the estimates of Integral$_C$, we concentrate our attention on the terms that are most delicate to estimate, in the exterior region. Taking into account Lemma 7.3.2 on page 151, we find that these have the form

$$\begin{aligned}
\tau_+^4 \phi Q(\hat{\mathcal{L}}_T W)_{\alpha\beta 44} {}^{(\bar{K})}\hat{\pi}^{\alpha\beta} &= \tau_+^4 \Big(\frac{1}{4} Q_{4444} {}^{(\bar{K})}\hat{\pi}_{33} + \frac{1}{2} Q_{4443} {}^{(\bar{K})}\hat{\pi}_{34} \\
&\quad + \frac{1}{4} Q_{3344} {}^{(\bar{K})}\hat{\pi}_{44} - Q_{A444} {}^{(\bar{K})}\hat{\pi}_{3A} \\
&\quad - Q_{A344} {}^{(\bar{K})}\hat{\pi}_{4A} + Q_{AB44} {}^{(\bar{K})}\hat{\pi}_{AB} \Big) \\
&= \tau_+^4 \big(|\alpha(\hat{\mathcal{L}}_T W)|^2 {}^{(\bar{K})}\underline{\mathbf{n}} + 2|\beta(\hat{\mathcal{L}}_T W)|^2 {}^{(\bar{K})}\mathbf{j} \\
&\quad + (\rho(\hat{\mathcal{L}}_T W)^2 + \sigma(\hat{\mathcal{L}}_T W)^2)^2 {}^{(\bar{K})}\mathbf{n} \\
&\quad - 4\alpha(\hat{\mathcal{L}}_T W)_{AB} \beta(\hat{\mathcal{L}}_T W)_B^{(\bar{K})} \underline{\mathbf{m}}_A \\
&\quad - 4\rho(\hat{\mathcal{L}}_T W)\beta(\hat{\mathcal{L}}_T W) \cdot {}^{(\bar{K})}\mathbf{m} \\
&\quad + 4\sigma(\hat{\mathcal{L}}_T W)^* \beta(\hat{\mathcal{L}}_T W) \cdot {}^{(\bar{K})}\mathbf{m} \\
&\quad + 2|\beta(\hat{\mathcal{L}}_T W)|^2 \mathrm{tr}^{(\bar{K})}\mathbf{i} \\
&\quad + 2\rho(\hat{\mathcal{L}}_T W)\alpha(\hat{\mathcal{L}}_T W) \cdot {}^{(\bar{K})}\mathbf{i} \\
&\quad - 2\sigma(\hat{\mathcal{L}}_T W)^* \alpha(\hat{\mathcal{L}}_T W) \cdot {}^{(\bar{K})}\mathbf{i} \big).
\end{aligned}$$

Estimating, with the help of assumptions 8.2.2c, we find

$$\int_{\Sigma_t^e} |\phi \tau_+^4 Q(\hat{\mathcal{L}}_T W)_{\alpha\beta 44} {}^{(\bar{K})}\hat{\pi}^{\alpha\beta}| \leq c r_0^{-2} \|{}^{(\bar{K})}\underline{\mathbf{n}}\|_{\infty,e} \|\tau_+^3 \alpha(\hat{\mathcal{L}}_T W)\|_{2,e}^2$$

$$
\begin{aligned}
&+ cr_0^{-2} \| {}^{(\bar{K})}\mathbf{j} \|_{\infty,e} \| \tau_+^3 \beta(\hat{\mathcal{L}}_T W) \|_{2,e}^2 \\
&+ cr_0^{-2} \| \tau_+^2 \tau_-^{-2} \, {}^{(\bar{K})}\mathbf{n} \|_{\infty,e} \\
&\cdot \| \tau_+^2 \tau_- (\rho(\hat{\mathcal{L}}_T W), \sigma(\hat{\mathcal{L}}_T W)) \|_{2,e}^2 \\
&+ cr_0^{-2} \| {}^{(\bar{K})}\underline{\mathbf{m}} \|_{\infty,e} \| \tau_+^3 \alpha(\hat{\mathcal{L}}_T W) \|_{2,e} \\
&\cdot \| \tau_+^3 \beta(\hat{\mathcal{L}}_T W) \|_{2,e} \\
&+ cr_0^{-2} \| \tau_+^2 \tau_-^{-2} \, {}^{(\bar{K})}\mathbf{m} \|_{\infty,e} \| \tau_+^3 \beta(\hat{\mathcal{L}}_T W) \|_{2,e} \\
&\cdot \| \tau_+^2 \tau_- (\rho(\hat{\mathcal{L}}_T W), \sigma(\hat{\mathcal{L}}_T W)) \|_{2,e} \\
&+ cr_0^{-2} \| {}^{(\bar{K})}\mathbf{i} \|_{\infty,e} \| \tau_+^3 \beta(\hat{\mathcal{L}}_T W) \|_{2,e}^2 \\
&+ cr_0^{-1} \| {}^{(\bar{K})}\mathbf{i} \|_{\infty,e} \| \tau_+^3 \alpha(\hat{\mathcal{L}}_T W) \|_{2,e} \\
&\cdot \| \tau_+^2 \tau_- \left(\rho(\hat{\mathcal{L}}_T W), \sigma(\hat{\mathcal{L}}_T W) \|_{2,e} \right) \\
\leq\; & c(\varepsilon_0 r_0^{-2} + \varepsilon_0 r_0^{-1}) \mathcal{Q}_1 . \qquad (8.3.13a)
\end{aligned}
$$

On the other hand, concentrating only on the terms that lead to divergences,

$$
\begin{aligned}
\int_{V_t^e} |\phi \tau_+^4 \rho(\hat{\mathcal{L}}_T W) \alpha(\hat{\mathcal{L}}_T W) \cdot {}^{(\bar{K})}\mathbf{i}| &\leq c \sup_{[0,t]} \| {}^{(\bar{K})}\mathbf{i} \|_{\infty,e} \int_{u_0}^{\infty} \tau_-^{-2} du \\
&\cdot \left(\int_{C_u} \tau_+^6 |\alpha(\hat{\mathcal{L}}_T W)|^2 \right)^{1/2} \\
&\cdot \left(\int_{C_u} \tau_+^2 \tau_-^4 |\rho(\hat{\mathcal{L}}_T W)|^2 \right)^{1/2} \\
&\leq c\varepsilon_0 \mathcal{Q}_1 * . \qquad (8.3.13b)
\end{aligned}
$$

Thus,

$$
\int_{V_t^e} \phi \tau_+^4 Q(\hat{\mathcal{L}}_T W)_{\alpha\beta 44} \, {}^{(\bar{K})}\hat{\pi}^{\alpha\beta} \leq c\varepsilon_0 (\mathcal{Q}_1 * + \mathcal{Q}_2 *). \qquad (8.3.13c)
$$

All the other components of Integral_D can be estimated in a straightforward manner, and we find

$$
\text{Integral}_D \leq c\varepsilon_0 (\mathcal{Q}_1 * + \mathcal{Q}_2 *). \qquad (8.3.13d)
$$

E: Estimates for $\text{Integral}_E = \int_\Sigma \phi Q(\hat{\mathcal{L}}_O W)_{\alpha\beta\gamma\delta} \, {}^{(T)}\pi^{\alpha\beta} \bar{K}^\gamma \bar{K}^\delta$

As before, we concentrate our attention of the terms of Integral_E that have the highest weight in the wave zone. These have the form

$$
E =: \int_{V_t^e} \phi Q(\hat{\mathcal{L}}_O W)_{\alpha\beta 44} \, {}^{(T)}\hat{\pi}^{\alpha\beta}
$$

$$= \int_{\Sigma_t^e} \tau_+^4 \Big\{ \frac{1}{2} |\alpha(\hat{\mathcal{L}}_O W)|^2\, {}^{(T)}\underline{\mathbf{n}} + 2|\beta(\hat{\mathcal{L}}_O W)|^2\, {}^{(T)}\mathbf{j}$$
$$+ (\rho(\hat{\mathcal{L}}_O W)^2 + \sigma(\hat{\mathcal{L}}_O W)^2)^{(T)}\mathbf{n}$$
$$- 4\beta(\hat{\mathcal{L}}_O W) \cdot \alpha(\hat{\mathcal{L}}_O W) \cdot {}^{(T)}\underline{\mathbf{m}} - 4\rho(\hat{\mathcal{L}}_O W)\beta(\hat{\mathcal{L}}_O W) \cdot {}^{(T)}\mathbf{m}$$
$$+ 4\sigma(\hat{\mathcal{L}}_O W)^\star \beta(\hat{\mathcal{L}}_O W) \cdot {}^{(T)}\mathbf{m} - 2|\beta(\hat{\mathcal{L}}_O W)|^2 \mathrm{tr}\, {}^{(T)}\mathbf{i}$$
$$+ 2\rho(\hat{\mathcal{L}}_O W)\alpha(\hat{\mathcal{L}}_O W)^{(T)}\mathbf{i} - 2\sigma(\hat{\mathcal{L}}_O W)^\star \alpha(\hat{\mathcal{L}}_O W) \cdot {}^{(T)}\mathbf{i} \Big\}.$$

Now, using the assumptions 8.2.2a, we find

$$E \leq c r_0^{-2} \|\tau_+^2\, {}^{(T)}\underline{\mathbf{n}}\|_{\infty,e} \|\tau_+^2 \alpha(\hat{\mathcal{L}}_O W)\|_{2,e}^2$$
$$\leq c r_0^{-2} \|\tau_+^2\, {}^{(T)}\mathbf{j}\|_{\infty,e} \|\tau_+^2 \beta(\hat{\mathcal{L}}_O W)\|_{2,e}^2$$
$$+ c r_0^{-2} \|\tau_+^2\, {}^{(T)}\mathbf{n}\|_{\infty,e} \|\tau_+^2 (\rho(\hat{\mathcal{L}}_O W), \sigma(\hat{\mathcal{L}}_O W))\|_{2,e}^2 \qquad (8.3.14a)$$
$$+ c r_0^{-2} \|\tau_+^2\, {}^{(T)}\underline{\mathbf{m}}\|_{\infty,e} \|\tau_+^2 \alpha(\hat{\mathcal{L}}_O W)\|_{2,e} \|\tau_+^2 \beta(\hat{\mathcal{L}}_O W)\|_{2,e}$$
$$+ c r_0^{-2} \|\tau_+^2\, {}^{(T)}\mathbf{m}\|_{\infty,e} \|\tau_+^2 \beta(\hat{\mathcal{L}}_O W)\|_{2,e}$$
$$\cdot \|\tau_+^2 (\rho(\hat{\mathcal{L}}_O W), \sigma(\hat{\mathcal{L}}_O W))\|_{2,e}$$
$$+ c r_0^{-1} \|\tau_+ \tau_-\, {}^{(T)}\mathbf{i}\|_{\infty,e} \|\tau_+^2 \alpha(\hat{\mathcal{L}}_O W)\|_{2,e}$$
$$\cdot \|\tau_+^2 (\rho(\hat{\mathcal{L}}_O W), \sigma(\hat{\mathcal{L}}_O W))\|_{2,e}$$
$$+ c r_0^{-1} \|\tau_+ \tau_-\, {}^{(T)}\mathbf{i}\|_{\infty,e} \|\tau_+^2 \beta(\hat{\mathcal{L}}_O W)\|_{2,e}^2$$
$$\leq c(\varepsilon_0 r_0^{-2} + \varepsilon_0 r_0^{-1}) \mathcal{Q}_1. \qquad (8.3.14b)$$

The terms of order $O(r_0^{-1})$ in the estimates that lead to divergence when integrated on V_t^e can in fact be treated as follows:

$$E_1 =: \int_{V_t^e} |\phi \tau_+^4 \rho(\hat{\mathcal{L}}_O W) \alpha(\hat{\mathcal{L}}_O W)^{(T)}\mathbf{i}| \qquad (8.3.14c)$$
$$\leq \|\tau_+ \tau_-\, {}^{(T)}\mathbf{i}\|_{\infty,e} \int_{u_0}^{\infty} \tau_-^{-2} du$$
$$\cdot \left(\int_{C_u} \tau_+^4 |\alpha(\hat{\mathcal{L}}_O W)|^2 \right)^{1/2} \left(\int_{C_u} \tau_+^2 \tau_-^2 |\rho(\hat{\mathcal{L}}_O W)|^2 \right)^{1/2}$$
$$\leq c\varepsilon_0 \mathcal{Q}_1 *$$

and

$$E_2 =: \int_{V_t^e} |\phi \tau_+^4 \sigma(\hat{\mathcal{L}}_O W) \alpha(\hat{\mathcal{L}}_O W)^{(T)}\mathbf{i}| \qquad (8.3.14d)$$
$$\leq \|\tau_+ \tau_-\, {}^{(T)}\mathbf{i}\|_{\infty,e} \int_{u_0}^{\infty} \tau_-^{-2} du$$

$$+\cdot\left(\int_{C_u}\tau_+^4|\alpha(\hat{\mathcal{L}}_{\mathrm{O}}W)|^2\right)^{1/2}\left(\int_{C_u}\tau_+^2\tau_-^2|\sigma(\hat{\mathcal{L}}_{\mathrm{O}}W)|^2\right)^{1/2}$$
$$\leq c\varepsilon_0\mathcal{Q}_1*$$

and similarly for $E_3 = \int_{V_t^e}|\phi\tau_+^4|\beta(\hat{\mathcal{L}}_{\mathrm{O}}W)|^{2(T)}\mathbf{i}|$.

Hence,

$$\int_{V_t^e}|\phi\tau_-^4 Q(\hat{\mathcal{L}}_{\mathrm{O}}W)^{(T)}_{\alpha\beta 44}\hat{\pi}^{\alpha\beta}| \leq c\varepsilon_0(\mathcal{Q}_1* + \mathcal{Q}_2*). \tag{8.3.14e}$$

All the other components of $\mathrm{Integral}_E$ can be estimated in a straightforward manner, and we find

$$\mathrm{Integral}_E \leq c\varepsilon_0(\mathcal{Q}_1* + \mathcal{Q}_2*). \tag{8.3.14f}$$

Finally, we combine 8.3.6g, 8.3.11, 8.3.12d, 8.3.13d, and 8.3.14f and conclude that

$$\mathcal{E}_1(t) \leq c\varepsilon_0(\mathcal{Q}_1 * + \mathcal{Q}_2*), \tag{8.3.15}$$

which ends the proof of the first part of the theorem.

(ii.) **Estimates for $\mathcal{E}_2$:**

There are now nine integrals appearing in the definition of $\mathcal{E}_2$ (see equation 8.1.4d). As before, the estimates of these integrals in the interior region V_t^i offer no difficulty due to the uniformity of asymptotic behavior there. It thus suffices to estimate the eight integrals of 8.1.4d in the exterior region V_t^e. Among these eight integrals we can first convince ourselves that those which contain as second derivative $\hat{\mathcal{L}}_{\mathrm{O}}$, namely,

$$\int_{V_t^e}|\phi\mathbf{Div}Q(\hat{\mathcal{L}}_{\mathrm{O}}^2W)_{\beta\gamma\delta}\bar{K}^\beta\bar{K}^\gamma T^\delta|$$

$$\int_{V_t^e}|\phi\mathbf{Div}Q(\hat{\mathcal{L}}_{\mathrm{O}}\hat{\mathcal{L}}_T W)_{\beta\gamma\delta}\bar{K}^\beta\bar{K}^\gamma\bar{K}^\delta|$$

$$\int_{V_t^e}|\phi Q(\hat{\mathcal{L}}_{\mathrm{O}}^2W)^{(\bar{K})}_{\alpha\beta\gamma\delta}\pi^{\alpha\beta}\bar{K}^\gamma T^\delta|$$

$$\int_{V_t^e}|\phi Q(\hat{\mathcal{L}}_{\mathrm{O}}^2W)^{(T)}_{\alpha\beta\gamma\delta}\pi^{\alpha\beta}\bar{K}_\gamma\bar{K}_\delta$$

$$\int_{V_t^e}|\phi Q(\hat{\mathcal{L}}_{\mathrm{O}}\hat{\mathcal{L}}_T W)_{\alpha\beta\gamma\delta}{}^{(\bar{K})}\pi^{\alpha\beta}\bar{K}^\gamma\bar{K}^\delta|$$

are to be treated precisely as in the first part of the theorem. Indeed, this fact is due to the uniform behavior of all tangential derivatives of all null components of all covariant quantities appearing in these integrals. More precisely, each tangential derivative of any null component improves the asymptotic behavior by order $O(r^{-1})$ in the exterior region. We can thus repeat the arguments of the first part step by step to show that these four integrals are bounded.

Among the four remaining integrals we can easily dismiss the two that contain $\hat{\mathcal{L}}_T^2 W$, namely,

$$\int_{V_t^e} |\phi \mathbf{Div} Q(\hat{\mathcal{L}}_T^2 W)_{\beta\gamma\delta} \bar{K}^\beta \bar{K}^\gamma \bar{K}^\delta|$$

$$\int_{V_t^e} |\phi Q(\hat{\mathcal{L}}_T^2 W)_{\alpha\beta\gamma\delta} {}^{(\bar{K})}\pi^{\alpha\beta} \bar{K}^\gamma \bar{K}^\delta|.$$

Indeed, the weights of these integrals are the same as for the integrals containing only the first derivatives $\hat{\mathcal{L}}_W$ that were estimated in the first part of the theorem. Thus they are even easier to treat than the four integrals previously discussed.

The two remaining integrals are

$$\int_{V_t^e} |\phi \mathbf{Div} Q(\hat{\mathcal{L}}_S \hat{\mathcal{L}}_T W)_{\beta\gamma\delta} \bar{K}^\beta \bar{K}^\gamma \bar{K}^\delta|$$

$$\int_{V_t^e} |\phi Q(\hat{\mathcal{L}}_S \hat{\mathcal{L}}_T W)_{\alpha\beta\gamma\delta} {}^{(\bar{K})}\pi^{\alpha\beta} \bar{K}^\gamma \bar{K}^\delta|.$$

These integrals require the EA 2 assumptions 8.2.2g and 8.2.2h as well as 8.2.2l. In what follows we sketch the proof of the boundedness of these integrals.

F: Estimates for $\text{Integral}_F = \int_{V_t} \phi \mathbf{Div} Q(\hat{\mathcal{L}}_S \hat{\mathcal{L}}_T W)_{\beta\gamma\delta} \bar{K}^\beta \bar{K}^\gamma \bar{K}^\delta$

In view of formula 8.1.7b of Proposition 8.1.3 we write

$$\begin{aligned}
\text{Integral}_F = & \frac{1}{8} \int_{V_t} \phi \tau_+^6 D(T, S; W)_{444} \\
& + \frac{3}{8} \int_{V_t} \phi \tau_+^4 \tau_-^2 D(T, S; W)_{344} \\
& + \frac{3}{8} \int_{V_t} \phi \tau_+^2 \tau_-^2 D(T, S; W)_{334} \\
& + \frac{1}{8} \int_{V_t} \phi \tau_-^6 D(T, S; W)_{333}
\end{aligned}$$

where $D(T, S; W) = \mathbf{Div} Q(\hat{\mathcal{L}}_S \hat{\mathcal{L}}_T W)$.

The most sensitive term to estimate now is $\int_{V_t^e} \phi \tau_+^6 D(T, S; W)_{444}$. In view of the formula 8.1.7c of Proposition 8.1.3,

$$\int_{V_t^e} \phi \tau_+^6 D(T, S; W)_{444} = 4 \int_{V_t^e} \phi \tau_+^6 \alpha(\hat{\mathcal{L}}_S \hat{\mathcal{L}}_T W) \cdot \Theta(T, S; W)$$

$$- 8 \int_{V_t^e} \phi \tau_+^6 \beta(\hat{\mathcal{L}}_S \hat{\mathcal{L}}_T W) \cdot \Xi(T, S; W)$$

(8.3.16a)

where $\Theta(T, S; W), \Xi(T, S; W)$ are the corresponding null terms of the Weyl current $J(T, S; W)$ as defined by 8.1.2d.

Proceeding as in 8.3.7b–8.3.7c, we write

$$\begin{aligned}\int_{\Sigma_t^e} |\phi \tau_+^6 \alpha(\hat{\mathcal{L}}_S \hat{\mathcal{L}}_T W) \cdot \Theta(T, S; W)| &\leq c \|r^3 \alpha(\hat{\mathcal{L}}_S \hat{\mathcal{L}}_T W)\|_{2,e} \\ &\quad \cdot \|r^3 \Theta(T, S; W)\|_{2,e} \\ &\leq c Q_2^{1/2} \|r^3 \Theta(T, S; W)\|_{2,e}\end{aligned}$$

(8.3.16b)

$$\begin{aligned}\int_{\Sigma_t^e} |\phi \tau_+^6 \beta(\hat{\mathcal{L}}_S \hat{\mathcal{L}}_T W) \cdot \Xi(T, S; W)| &\leq c \|r^3 \beta(\hat{\mathcal{L}}_S \hat{\mathcal{L}}_T W)\|_{2,e} \\ &\quad \cdot \|r^3 \Xi(T, S; W)\|_{2,e} \\ &\leq c Q_2^{1/2} \|r^3 \Xi(T, S; W)\|_{2,e}.\end{aligned}$$

(8.3.16c)

In view of the decomposition 8.1.2d of $J(T, S; W)$ into J^0, J^1, J^2, and J^3 we write

$$\Theta(T, W) = \Theta^0(T, S; W) + \frac{1}{2}(\Theta^1(T, S; W) + \Theta^2(T, S; W) + \Theta^3(T, S; W))$$

$$\Xi(T, W) = \Xi^0(T, S; W) + \frac{1}{2}(\Xi^1(T, S; W) + \Xi^2(T, S; W) + \Xi^3(T, S; W)).$$

Now, for $i = 1, 2, 3$, we have

$$\begin{aligned}\Theta^i(T, S; W) &= \Theta^i(S, \hat{\mathcal{L}}_T W) \\ \Xi^i(T, S; W) &= \Xi^i(S, \hat{\mathcal{L}}_T W).\end{aligned}$$

In view of the formulas of Proposition 8.1.4 applied to the vectorfield T, we find

$$\begin{aligned}\Xi^1(S, \hat{\mathcal{L}}_T W) = &\operatorname{Qr}\left[{}^{(S)}\mathbf{i}\,;\, \nabla\!\!\!/ \alpha(\hat{\mathcal{L}}_T W)\right] + \operatorname{Qr}\left[{}^{(S)}\underline{\mathbf{m}}\,;\, \alpha_4(\hat{\mathcal{L}}_T W)\right] \\ &+ \operatorname{Qr}\left[{}^{(S)}\mathbf{m}\,;\, \alpha_3(\hat{\mathcal{L}}_T W)\right] + \operatorname{Qr}\left[{}^{(S)}\mathbf{m}\,;\, \nabla\!\!\!/ \beta(\hat{\mathcal{L}}_T W)\right] \\ &+ \operatorname{Qr}\left[{}^{(S)}\mathbf{j}\,;\, \beta_4(\hat{\mathcal{L}}_T W)\right] + \operatorname{Qr}\left[{}^{(S)}\mathbf{n}\,;\, \beta_3(\hat{\mathcal{L}}_T W)\right] \\ &+ \operatorname{tr}\chi \left(\operatorname{Qr}\left[{}^{(S)}\underline{\mathbf{m}}\,;\, \alpha(\hat{\mathcal{L}}_T W)\right] + \operatorname{Qr}\left[({}^{(S)}\mathbf{i}, {}^{(S)}\mathbf{j})\,;\, \beta(\hat{\mathcal{L}}_T W)\right]\right.\end{aligned}$$

$$+ \text{Qr}\left[{}^{(S)}\mathbf{m}\,;\,(\rho(\hat{\mathcal{L}}_T W), \sigma(\hat{\mathcal{L}}_T W))\right]\Big)$$
$$+ \text{tr}\underline{\chi}\left(\text{Qr}\left[{}^{(S)}\mathbf{m}\,;\,\alpha(\hat{\mathcal{L}}_T W)\right] + \text{Qr}\left[{}^{(S)}\mathbf{n}\,;\,\beta(\hat{\mathcal{L}}_T W)\right]\right)$$

$$\Xi^2(S, \hat{\mathcal{L}}_T W) = \text{Qr}\left[{}^{(S)}\not{p}\,;\,\alpha(\hat{\mathcal{L}}_T W)\right] + \text{Qr}\left[{}^{(S)}p_4\,;\,\beta(\hat{\mathcal{L}}_T W)\right]$$

$$\underline{\Xi}^3(S, \hat{\mathcal{L}}_T W) = \text{Qr}\left[\underline{\alpha}(\hat{\mathcal{L}}_T W)\,;\,(I, \underline{I})({}^{(S)}q)\right]$$
$$+ \text{Qr}\left[\underline{\beta}(\hat{\mathcal{L}}_T W)\,;\,(\underline{K}, \underline{\Lambda}, \underline{\Theta})({}^{(S)}q)\right]$$
$$+ \text{Qr}\left[(\rho(\hat{\mathcal{L}}_T W), \sigma(\hat{\mathcal{L}}_T W))\,;\,\underline{\Xi}({}^{(S)}q)\right].$$

Now, taking into account exterior assumptions 8.2.2g for the null components of ${}^{(S)}\pi$ and also estimates 8.2.5c of Proposition 8.2.1 for ${}^{(S)}q, {}^{(S)}p$, together with assumptions 7.5.8a on the Ricci coefficients and proceeding as in 8.3.7g–8.3.7i, we derive

$$\|r^3 \Xi^i(S, \hat{\mathcal{L}}_T W)\|_{2,e} \leq c\varepsilon_0 (1+t)^{-3/2}(\mathcal{Q}_1 + \mathcal{Q}_2) \tag{8.3.16d}$$

for all $i = 1, 2, 3$. Similarly, we have

$$\Theta^1(S, \hat{\mathcal{L}}_T W) = \text{Qr}\left[{}^{(S)}\underline{\mathbf{m}}\,;\,\nabla\!\!\!/\,\alpha(\hat{\mathcal{L}}_T W)\right] + \text{Qr}\left[{}^{(S)}\underline{\mathbf{n}}\,;\,\alpha_4(\hat{\mathcal{L}}_T W)\right]$$
$$+ \text{Qr}\left[{}^{(S)}\mathbf{j}\,;\,\alpha_3(\hat{\mathcal{L}}_T W)\right]$$
$$+ \text{Qr}\left[{}^{(S)}\mathbf{i}\,;\,\nabla\!\!\!/\,\beta(\hat{\mathcal{L}}_T W)\right] + \text{Qr}\left[{}^{(S)}\underline{\mathbf{m}}\,;\,\beta_4(\hat{\mathcal{L}}_T W)\right]$$
$$+ \text{Qr}\left[{}^{(S)}\mathbf{m}\,;\,\beta_3(\hat{\mathcal{L}}_T W)\right]$$
$$+ \text{Qr}\left[{}^{(S)}\mathbf{m}\,;\,\nabla\!\!\!/\,(\rho(\hat{\mathcal{L}}_T W), \sigma(\hat{\mathcal{L}}_T W))\right]$$
$$+ \text{Qr}\left[{}^{(S)}\mathbf{j}\,;\,(\rho_4(\hat{\mathcal{L}}_T W), \sigma_4(\hat{\mathcal{L}}_T W))\right]$$
$$+ \text{Qr}\left[{}^{(S)}\mathbf{n}\,;\,(\rho_3(\hat{\mathcal{L}}_T W), \sigma_3(\hat{\mathcal{L}}_T W))\right]$$
$$+ \text{tr}\chi\Big(\text{Qr}\left[{}^{(S)}\underline{\mathbf{n}}\,;\,\alpha(\hat{\mathcal{L}}_T W)\right] + \text{Qr}\left[{}^{(S)}\underline{\mathbf{m}}\,;\,\beta(\hat{\mathcal{L}}_T W)\right]$$
$$+ \text{Qr}\left[({}^{(S)}\mathbf{i}, {}^{(S)}\mathbf{j})\,;\,(\rho(\hat{\mathcal{L}}_T W), \sigma(\hat{\mathcal{L}}_T W))\right]$$
$$+ \text{Qr}\left[{}^{(S)}\mathbf{m}\,;\,\beta(\hat{\mathcal{L}}_T W)\right]\Big) + \text{tr}\underline{\chi}\Big(\text{Qr}\left[({}^{(S)}\mathbf{i}, {}^{(S)}\mathbf{j})\,;\,\alpha(\hat{\mathcal{L}}_T W)\right]$$
$$+ \text{Qr}\left[{}^{(S)}\mathbf{m}\,;\,\beta(\hat{\mathcal{L}}_T W)\right] + \text{Qr}\left[{}^{(T)}\mathbf{n}\,;\,(\rho(\hat{\mathcal{L}}_T W), \sigma(\hat{\mathcal{L}}_T W))\right]\Big)$$

$$\Theta^2(S, \hat{\mathcal{L}}_T W) = \text{Qr}\left[{}^{(S)}p_3\,;\,\alpha(\hat{\mathcal{L}}_T W)\right] + \text{Qr}\left[{}^{(S)}\not{p}\,;\,\beta(\hat{\mathcal{L}}_T W)\right]$$
$$+ \text{Qr}\left[{}^{(S)}p_4\,;\,(\rho(\hat{\mathcal{L}}_T W), \sigma(\hat{\mathcal{L}}_T W))\right]$$
$$\Theta^3(S, \hat{\mathcal{L}}_T W) = \text{Qr}\left[\alpha\,;\,\underline{K}({}^{(S)}q)\right] + \text{Qr}\left[\alpha(\hat{\mathcal{L}}_T W)\,;\,\underline{\Lambda}({}^{(S)}q)\right]$$

$$+\boxed{\mathrm{Qr}\left[\alpha(\hat{\mathcal{L}}_T W)\,;\,\underline{\Theta}({}^{(S)}q)\right]}$$
$$+\,\mathrm{Qr}\left[\beta(\hat{\mathcal{L}}_T W)\,;\,(I,\underline{I})({}^{(S)}q)\right]$$
$$+\,\mathrm{Qr}\left[(\rho(\hat{\mathcal{L}}_T W),\sigma(\hat{\mathcal{L}}_T W))\,;\,\Theta({}^{(S)}q)\right]$$
$$+\boxed{\mathrm{Qr}\left[\underline{\beta}(\hat{\mathcal{L}}_T W)\,;\,\Xi({}^{(S)}q)\right]}\,.$$

Hence, taking into account exterior assumptions 8.2.2g for the null components of ${}^{(S)}\pi$ and also the estimates 8.2.5c of Proposition 8.2.1 for ${}^{(S)}q$, ${}^{(S)}p$, together with assumptions 7.5.8a on the Ricci coefficients and proceeding as in 8.3.8d–8.3.8f, we derive

$$\|r^3\Theta^i(S,\hat{\mathcal{L}}_T W)\|_{2,e}\leq c\varepsilon_0(1+t)^{-3/2}(\mathcal{Q}_1+\mathcal{Q}_2) \tag{8.3.16e}$$

for all $i=1,2,3$.

It remains to estimate $\Xi^0(T,S;W)$, $\Theta^0(T,S;W)$. In view of the commutation properties of S with the null frame we have

$$\Xi^0(T,S;W)=\frac{1}{2}(\not{\mathcal{L}}_S\Xi^1(T,W)+\not{\mathcal{L}}_S\Xi^2(T,W)+\not{\mathcal{L}}_S\Xi^3(T,W))+\mathrm{l.o.t.}$$
$$\Theta^0(T,S;W)=\frac{1}{2}(\hat{\not{\mathcal{L}}}_S\Theta^1(T,W)+\hat{\not{\mathcal{L}}}_S\Theta^2(T,W)+\hat{\not{\mathcal{L}}}_S\Theta^3(T,W))+\mathrm{l.o.t.}$$

One can easily see, from formulas 8.3.7e–8.3.7f and 8.3.8b–8.3.8c, that

$$\|r^3\not{\mathcal{L}}_S\Xi^2(T,W)\|_{2,e},\|r^3\not{\mathcal{L}}_S\Xi^3(T,W)\|_{2,e}\leq c(1+t)^{-3/2}\varepsilon_0(\mathcal{Q}_1+\mathcal{Q}_2) \tag{8.3.16f}$$
$$\|r^3\hat{\not{\mathcal{L}}}_S\Theta^2(T,W)\|_{2,e},\|r^3\hat{\not{\mathcal{L}}}_S\Theta^3(T,W)\|_{2,e}\leq c(1+t)^{-3/2}\varepsilon_0(\mathcal{Q}_1+\mathcal{Q}_2). \tag{8.3.16g}$$

Moreover, the only terms in $\hat{\mathcal{L}}_S\Xi^1(T,W)$ that lead to a behavior of order $O((1+t)^{-1})$ when estimated in the same norm are those due to the presence of ${}^{(T)}\mathbf{i}$ in the expression 8.3.7d for $\Xi^1(T,W)$. The corresponding terms in the integral $\int_{V_t^e}\phi\tau_+^6\beta(\hat{\mathcal{L}}_S\hat{\mathcal{L}}_T W)\cdot\Xi(T,S;W)$ are then to be estimated precisely in the same way as the integrals

$$\int_{V_t^e}|\phi\tau_+^6\beta(\hat{\mathcal{L}}_T W)\cdot\mathrm{Qr}\left[{}^{(T)}\mathbf{i}\,;\,\nabla\!\!\!/\alpha\right]|$$

and

$$\int_{\Sigma_t^e}|\phi\tau_+^6\mathrm{tr}\chi\beta(\hat{\mathcal{L}}_T W)\cdot\mathrm{Qr}\left[{}^{(T)}\mathbf{i}\,;\,\beta\right]|$$

on page 244. Thus, finally,

$$\int_{V_t^e} |\phi\tau_+^6\beta(\hat{\mathcal{L}}_S\hat{\mathcal{L}}_T W)\cdot \Xi(T,S;W)| \le c\varepsilon_0(\mathcal{Q}_1* + \mathcal{Q}_2*). \qquad (8.3.16\text{h})$$

Similarly, the only terms in $\hat{\mathcal{L}}_S\Theta^1(T,W)$ that lead to the $O(t^{-1})$ behavior, when estimated in the norm $\|r^3\cdot\|_{2,e}$ are due to the presence of, on one hand, ${}^{(T)}\mathbf{i}$ and, on the other hand, ${}^{(T)}\mathbf{n}$ in the expression 8.3.8a for $\Theta^1(T,W)$. The corresponding terms in the integral $\int_{V_t^e}\phi\tau_+^6\alpha(\hat{\mathcal{L}}_S\hat{\mathcal{L}}_T W)\cdot\Theta(T,S;W)$ are then to be estimated precisely in the same way as the integrals

$$\int_{V_t^e} |\phi\tau_+^6\alpha(\hat{\mathcal{L}}_T W)\cdot \mathrm{Qr}\left[{}^{(T)}\mathbf{i}\,;\,\nabla\!\!\!/\beta\right]|$$

$$\int_{V_t^e} |\phi\tau_+^6\alpha(\hat{\mathcal{L}}_T W)\cdot \mathrm{Qr}\left[{}^{(T)}\mathbf{n}\,;\,(\rho_3,\sigma_3)\right]|$$

$$\int_{V_t^e} |\phi\tau_+^6\mathrm{tr}\underline{\chi}\alpha(\hat{\mathcal{L}}_T W)\cdot \mathrm{Qr}\left[{}^{(T)}\mathbf{i}\,;\,\alpha\right]|$$

$$\int_{V_t^e} |\phi\tau_+^6\mathrm{tr}\chi\alpha(\hat{\mathcal{L}}_T W)\cdot \mathrm{Qr}\left[{}^{(T)}\mathbf{i}\,;\,(\rho,\sigma)\right]|$$

on page 246. We derive

$$\int_{V_t^e} |\phi\tau_+^6\alpha(\hat{\mathcal{L}}_S\hat{\mathcal{L}}_T W)\cdot \Theta(T,S;W)| \le c\varepsilon_0(\mathcal{Q}_1* + \mathcal{Q}_2*). \qquad (8.3.16\text{i})$$

Thus,

$$\mathrm{Integral}_F \le c\varepsilon_0(\mathcal{Q}_1* + \mathcal{Q}_2*). \qquad (8.3.16\text{j})$$

Finally, the last integral, $\mathrm{Integral}_G = \int_{V_t} |\phi Q(\hat{\mathcal{L}}_S\hat{\mathcal{L}}_T W)_{\alpha\beta\gamma\delta}{}^{(\bar{K})}\pi^{\alpha\beta}\bar{K}^\gamma\bar{K}^\delta|$, is simpler and can be treated as $\mathrm{Integral}_D$.

We conclude that

$$\mathcal{E}_2(t) \le c\varepsilon_0(\mathcal{Q}_1* + \mathcal{Q}_2*), \qquad (8.3.17)$$

which ends the proof of the theorem.

Part III

Construction of Global Space-Times. Proof of the Main Theorem

CHAPTER 9

Construction of the Optical Function

In this chapter we shall assume that a space-time slab $\bigcup_{t\in[0,t_*]}\Sigma_t$ has been constructed, each Σ_t being a level set of the unique time function defined in the introduction, Σ_0 corresponding to the initial hypersurface, and Σ_{t_*} to the final hypersurface. Our main objective here is to construct an optical function u and use it in order to define the vectorfields K, S and the rotation vector fields Ω. The construction of this global optical function is obtained by matching an exterior optical function to an interior one. The exterior optical function is by far the more important one for us, and its construction is achieved in the following three steps.

9.1. Construction of the Exterior Optical Function

Step 1: First, to allow ourselves some room, we assume that a unit space-time slab $\bigcup_{t\in[-1,0]}\Sigma_t$ in the past of Σ_0 has also been constructed. We select an origin O_0 on the initial hypersurface Σ_0, and we consider the point O_{-1} on Σ_{-1}, which lies on the integral curve of T through O_0. Let C_0 be the future null geodesic cone with vertex at O_{-1}. Then C_0 shall be the level set $u=0$ of the desired optical function u.

Step 2: The cone C_0 intersects each maximal slice Σ_t on a 2-surface $S_{t,0}$. In particular, it intersects the final hypersurface Σ_{t_*} on the 2-surface $S_{t_*,0}$. In the second step we construct an appropriate radial extension of this 2-surface. More precisely, we construct function u_*, defined on Σ_{t_*}, such that $u_*=0$ on $S_{t_*,0}$ and each level set of u_* is a 2-surface diffeomorphic to S^2. For technical reasons we postpone the discussion of this step to Chapter 14. The main geometric ideas of the construction are exposed, however, in the first two pages of that chapter.

Step 3: In this step we extend the radial function u_*, defined in the previous step, to a bona fide optical function, namely a solution of the eikonal equation, defined in the whole space-time slab under consideration, with the exception of the interior of the timelike cylinder generated by the integral curves of T passing through each point of a geodesic ball of Σ_0 with center at O_0. We achieve this as follows:

Let $S_{t_*,\lambda}$ be an arbitrary level set of u_*, N its unit normal relative to Σ_{t_*}, and $e_4' = T + N$ its outgoing null normal. From each point p on Σ_{t_*} we construct, toward the past, the null geodesic in the direction of e_4. The set of these null geodesics generates a null hypersurface C_λ. We define u to be the function whose level hypersurfaces $u = \lambda$ coincide with C_λ. As we shall see, this construction defines the function u in the region exterior to the timelike cylinder. Clearly u, as uniquely defined by this construction, is an optical function, that is, it satisfies the following:

Eikonal Equation

$$\mathbf{g}^{\mu\nu}\frac{\partial u}{\partial x^\mu}\frac{\partial u}{\partial x^\nu} = 0.$$

We shall call u the *canonical exterior optical function of the space-time slab.*

We shall denote by $S_{t,u}$ the 2-surface that is the intersection of the null hypersurface C_u with the maximal hypersurface Σ_t:

$$S_{t,u} = C_u \bigcap \Sigma_t.$$

Let

$$l^\mu = \mathbf{g}^{\mu\nu}\frac{\partial u}{\partial x^\nu}.$$

The vector field l is the vectorfield tangent to the null geodesics that generate each C_u. We will write[1]

$$\frac{d}{ds} = \mathbf{D}_l$$

where the affine parameter function s is a solution of the equation

$$\mathbf{D}_l s = 1. \tag{9.1.1}$$

Clearly,

$$\frac{dl}{ds} = 0.$$

Let N be the vectorfield that is the outward unit normal of the surfaces $S_{t,u}$ relative to the hypersurfaces Σ_t. Then

$$e_4' = T + N$$

[1] Recall that $\mathbf{D}$ denotes the covariant derivative in space-time while ∇ stands for the induced covariant derivative on the slices Σ_t. Moreover, we use ∇ to denote the intrinsic covariant differentiation on each $S_{t,u}$

and

$$e_3' = T - N$$

are, respectively, the outgoing and incoming null normals of $S_{t,u}$.

The function a defined by the formula

$$a^{-1} = -\mathbf{D}_T u = |\nabla u| \tag{9.1.2}$$

is the lapse function of the foliation given by the null hypersurfaces C_u. We have

$$l = e_4 = a^{-1} e_4'.$$

We also introduce

$$\underline{l} = e_3 = a e_3'.$$

Then

$$\mathbf{g}(l, \underline{l}) = \mathbf{g}(e_3', e_4') = -2.$$

The tensor of projection to the surfaces $S_{t,u}$ is given by the formula

$$\Pi^\mu_\nu = \delta^\mu_\nu + \frac{1}{2}(e_3'^\mu e_{4\nu}' + e_4'^\mu e_{3\nu}') = \delta^\mu_\nu + \frac{1}{2}(l^\mu \underline{l}_\nu + \underline{l}^\mu l_\nu).$$

Let $(e_A : A = 1, 2)$ be a local orthonormal frame for $S_{t_*,u}$. We extend this frame to C_u according to the equations

$$\frac{de_A}{ds} = \underline{\zeta}_A l \tag{9.1.3a}$$

$$\mathbf{g}(T, e_A) = 0. \tag{9.1.3b}$$

Applying $\frac{d}{ds}$ to 9.1.3b and using 9.1.3a together with

$$\mathbf{g}(l, T) = -a^{-1}, \tag{9.1.3c}$$

we derive

$$\underline{\zeta}_A = a\mathbf{g}\left(e_A, \frac{dT}{ds}\right). \tag{9.1.3d}$$

Since l is a null geodesic vectorfield, the equation 9.1.3a implies that the frame e_A remains orthogonal to l. Thus, in view of 9.1.3b, it also remains orthogonal to N and $\underline{l}$. We therefore infer that

$$\begin{aligned}\frac{d}{ds}(\mathbf{g}(e_A, e_B)) &= \mathbf{g}\left(\frac{de_A}{ds}, e_B\right) + \mathbf{g}\left(e_A, \frac{de_B}{ds}\right)\\ &= \underline{\zeta}_A \mathbf{g}(l, e_B) + \underline{\zeta}_B \mathbf{g}(l, e_A) = 0.\end{aligned}$$

Consequently, since $\mathbf{g}(e_A, e_B) = \delta_{AB}$ on $S_{t_*,u}$, this holds everywhere on C_u. We conclude that $(e_A : A = 1, 2)$ is an orthonormal frame on $S_{t,u}$ for each

$t \in [0, t_*]$. Differentiating the equations $\mathbf{g}(\underline{l}, l) = -2$, $\mathbf{g}(\underline{l}, \underline{l}) = 0$, $\mathbf{g}(\underline{l}, e_A) = 0$ along l, we deduce that

$$\frac{d\underline{l}}{ds} = 2\underline{\zeta}_A e_A \tag{9.1.3e}$$

where $\underline{\zeta}$ is defined by 9.1.3c. We thus have two fundamental null tetrads: the null tetrad $(e_1, e_2, \underline{l} = e_3, l = e_4)$ and the normalized null tetrad (e_1, e_2, e_3', e_4'). Relative to these null tetrads, we can express the space-time metric $\mathbf{g}$ as

$$\begin{aligned} \mathbf{g}^{\mu\nu} &= -\frac{1}{2}(l^\mu \underline{l}^\nu + l^\nu \underline{l}^\mu) + \sum e_A^\mu e_B^\nu \\ &= -\frac{1}{2}(e_3'^\mu e_4'^\nu + e_3'^\nu e_4'^\mu) + \sum e_A^\mu e_B^\nu \end{aligned} \tag{9.1.4a}$$

and the tensor of projection to $S_{t,u}$ as

$$\Pi_\nu^\mu = \sum e_A^\mu e_B^\nu. \tag{9.1.4b}$$

Since

$$\partial_\mu t = -\phi^{-1} \mathbf{g}_{\mu\nu} T^\nu$$

where ϕ is the lapse function of the maximal foliation, we derive from 9.1.3c

$$\frac{dt}{ds} = \phi^{-1} a^{-1}. \tag{9.1.4c}$$

Also, applying $\frac{d}{ds}$ to equation 9.1.3c, we obtain

$$\frac{da}{ds} = -\nu \tag{9.1.4d}$$

where

$$\nu = -a^2 \mathbf{g}\left(l, \frac{dT}{ds}\right). \tag{9.1.4e}$$

In view of the fact that we can express

$$\mathbf{D}_\mu T_\nu = -T_\mu \phi^{-1} \nabla_\nu \phi - k_{\mu\nu} \tag{9.1.4f}$$

where $k_{\mu\nu}$ is the second fundamental form of the maximal foliation, we deduce

$$\begin{aligned} \mathbf{g}\left(\frac{dT}{ds}, e_A\right) &= a^{-1}(\mathbf{g}(e_A, \mathbf{D}_T T) + g(e_A, \mathbf{D}_N T)) \\ &= a^{-1}(\phi^{-1} \nabla_A \phi - \epsilon_A) \\ \mathbf{g}\left(\frac{dT}{ds}, l\right) &= a^{-2}(\mathbf{g}(N, \mathbf{D}_T T) + \mathbf{g}(N, \mathbf{D}_N T)) \\ &= a^{-2}(\phi^{-1} \mathbf{D}_N \phi - \delta) \end{aligned}$$

where

$$\epsilon_A = k_{\mu\nu} e^{\mu}_A N^{\nu} \tag{9.1.4g}$$
$$\delta = k_{\mu\nu} N^{\mu} N^{\nu}. \tag{9.1.4h}$$

Hence we deduce, from equations 9.1.3d and 9.1.4e,

$$\underline{\zeta}_A = \phi^{-1} \nabla\!\!\!/_A \phi - \epsilon_A \tag{9.1.4i}$$
$$\nu = -\phi^{-1} \nabla_N \phi + \delta. \tag{9.1.4j}$$

We define the normalized null second fundamental form of $S_{t,u}$ to be the 2-covariant tensorfield χ', tangential to $S_{t,u}$, given by the formula

$$\chi'_{\alpha\beta} = \Pi^{\mu}_{\alpha} \Pi^{\nu}_{\beta} \nabla_{\mu} e'_{4\nu}. \tag{9.1.5a}$$

Its components, relative to the orthonormal frame (e_A), are given by

$$\chi'_{AB} = \theta_{AB} - \eta_{AB} \tag{9.1.5b}$$

where θ_{AB} are the components of the second fundamental form of $S_{t,u}$ relative to Σ_t, while

$$\eta_{AB} = k_{\mu\nu} e^{\mu}_A e^{\nu}_B. \tag{9.1.5c}$$

Equation 9.1.5b shows that χ'_{AB} is symmetric.

We also define the second fundamental form of $S_{t,u}$ relative to the null geodesic normal l:

$$\chi_{\alpha\beta} = \Pi^{\mu}_{\alpha} \Pi^{\nu}_{\beta} \nabla_{\mu} l_{\nu} \tag{9.1.5d}$$

Clearly,

$$\chi_{\alpha\beta} = a^{-1} \chi'_{\alpha\beta}. \tag{9.1.5e}$$

The equation 9.1.5d can be written in the form

$$\chi_{AB} = e^{\mu}_A e^{\nu}_B \mathbf{D}_{\mu} l_{\nu}. \tag{9.1.5f}$$

Applying $\frac{d}{ds}$ to this equation and using 9.1.3a and the fact that l is a null geodesic, we obtain

$$\begin{aligned}
\frac{d\chi_{AB}}{ds} &= e^{\mu}_A e^{\nu}_B l^{\alpha} \mathbf{D}_{\alpha} (\mathbf{D}_{\mu} l_{\nu}) \\
&= e^{\mu}_A e^{\nu}_B l^{\alpha} (\mathbf{D}_{\mu} \mathbf{D}_{\alpha} l_{\nu} + R_{\nu\beta\alpha\mu} l^{\beta}) \\
&= -e^{\mu}_A e^{\nu}_B (g^{\alpha\beta} \mathbf{D}_{\mu} l_{\alpha} \mathbf{D}_{\beta} l_{\nu} + R_{\nu\beta\mu\alpha} l^{\beta} l^{\alpha}).
\end{aligned}$$

Hence, using the expression 9.1.4a, we obtain the following ordinary differential equation for χ_{AB} along the null geodesics generating C_u:

$$\frac{d\chi_{AB}}{ds} = -\sum \chi_{AC}\chi_{CB} - \alpha_{AB} \tag{9.1.6a}$$

where, we recall from 7.3.3c,

$$\alpha_{AB} = R_{A4B4}. \tag{9.1.6b}$$

Let $\hat{\chi}_{AB}$ be the traceless part of χ_{AB}. Recall that for any two symmetric, traceless 2-dimensional matrices A and B we have the algebraic identity

$$AB + BA = Itr(AB). \tag{9.1.6c}$$

In view of this fact we have

$$\sum \chi_{AC}\chi_{CB} = \delta_{AB}\left(\frac{1}{4}(tr\chi)^2 + \frac{1}{2}|\hat{\chi}|^2\right) + tr\chi\hat{\chi}_{AB}.$$

It follows that equation 9.1.6a decomposes into[2]

$$\frac{dtr\chi}{ds} = -\frac{1}{2}(tr\chi)^2 - |\hat{\chi}|^2 \tag{9.1.6d}$$

$$\frac{d\hat{\chi}_{AB}}{ds} = -tr\chi\hat{\chi}_{AB} - \alpha_{AB}. \tag{9.1.6e}$$

Next we define

$$\zeta_A = \frac{1}{2}e^{\mu}_{A}\underline{l}^{\nu}\mathbf{D}_{\mu}l_{\nu}. \tag{9.1.7a}$$

Since $l = a^{-1}(T + N)$, and $\underline{l} = a^2 l - 2aN$, we have

$$\begin{aligned}\zeta_A &= -ae^{\mu}_{A}N^{\nu}\mathbf{D}_{\mu}(a^{-1}(T_{\nu} + N_{\nu}))\\ &= a^{-1}e_A(a) - e^{\mu}_{A}N^{\nu}\mathbf{D}_{\mu}T_{\nu}.\end{aligned}$$

Hence, by virtue of 9.1.4f,

$$\zeta_A = a^{-1}e_A(a) + \epsilon_A. \tag{9.1.7b}$$

Applying $\frac{d}{ds}$ to 9.1.7a and using also 9.1.3a and 9.1.3e, we obtain

$$\begin{aligned}\frac{d\zeta_A}{ds} &= \frac{1}{2}e^{\mu}_{A}\underline{l}^{\nu}l^{\alpha}\mathbf{D}_{\alpha}(\mathbf{D}_{\mu}l_{\nu}) + e^{\mu}_{A}e^{\nu}_{B}\mathbf{D}_{\mu}l_{\nu}\zeta_B\\ &= \frac{1}{2}e^{\mu}_{A}\underline{l}^{\nu}l^{\alpha}(\mathbf{D}_{\mu}\mathbf{D}_{\alpha}l_{\nu} + R_{\nu\beta\alpha\mu}l^{\beta}) + \chi_{AB}\zeta_B\\ &= \frac{1}{2}e^{\mu}_{A}\underline{l}^{\nu}(g^{\alpha\beta}\mathbf{D}_{\mu}l_{\alpha}\mathbf{D}_{\beta}l_{\nu} + R_{\mu\alpha\nu\beta}l^{\alpha}l^{\beta}) + \chi_{AB}\zeta_B.\end{aligned}$$

[2] Compare these with the null structure equations derived in Proposition 7.4.1

Hence, in view of expression 9.1.4a[3],

$$\frac{d\zeta_A}{ds} = -\chi_{AB}\zeta_B + \chi_{AB}\underline{\zeta}_B - \beta_A \tag{9.1.8a}$$

where (see the definition in 7.3.3c)

$$\beta_A = \frac{1}{2}R_{A434}. \tag{9.1.8b}$$

Differentiating the equation $l^\mu\partial_\mu s = 1$ tangentially to $S_{t,u}$, we find, by 9.1.4a, 9.1.5f, and 9.1.7a,

$$e^\nu_A\frac{d}{ds}(\partial_\nu s) = -e^\nu_A\partial_\mu s\mathbf{D}_\nu l^\mu = -\chi_{AB}e_B(s) + \zeta_A.$$

Hence, in view of 9.1.3a,

$$\frac{d}{ds}(e_A(s)) = -\chi_{AB}e_B(s) + \zeta_A + \underline{\zeta}_A. \tag{9.1.9}$$

The Gauss curvature of the surface $S_{t,u}$ is given by 3.1.2d in terms of the embedding of $S_{t,u}$ in Σ_t. We recall from the general introduction (see formulas 1.0.14a and 1.0.14b) that the electric part of the space-time curvature is given in terms of the Ricci curvature and the second fundamental form of the maximal hypersurfaces Σ_t by

$$E_{ij} = R_{ij} - k_{im}k^m_j,$$

and the equation that expresses the vanishing of the trace of E is the constraint equation

$$R = |\, k\,|^2.$$

We have (see definition in Part II)

$$E_{NN} = \frac{1}{4}\mathbf{R}_{3434} = \rho. \tag{9.1.10}$$

Using these equations, we express

$$R - 2R_{NN} = |\eta|^2 - \delta^2 - 2\rho.$$

Substituting this expression in equation 3.1.2d and using 9.1.5b, we obtain

$$2K = \frac{1}{2}(tr\chi')^2 - |\,\hat{\chi}'\,|^2 - 2\hat{\eta}\cdot\hat{\chi}' - \delta tr\chi' - 2\rho \tag{9.1.11}$$

[3] Compare this with formula 7.4.2t in Proposition 7.4.1.

where $\hat{\eta}$ is the traceless part of η and we have used the fact that

$$tr\eta + \delta = trk = 0.$$

Let $A(t, u)$ be the area of the surface $S(t, u)$, and let $r(t, u)$ be defined by the formula

$$A = 4\pi r^2.$$

The rate of change of A along a given cone C_u is given by

$$\frac{\partial A}{\partial t} = \int_S \phi tr\chi' \tag{9.1.12a}$$

or, equivalently,

$$\frac{\partial r}{\partial t} = \frac{r}{2}\overline{\phi tr\chi'}. \tag{9.1.12b}$$

We denote

$$r_0(t) = r(t, 0).$$

Definition 9.1.1 The exterior region E of the space-time slab $\bigcup_{t\in[0,t_*]} \Sigma_t$ is the region defined by the inequality

$$r \geq \frac{r_0}{2}.$$

The interior region I is defined to be the union of the complement of E relative to the slab and M_0, the part of the interior of the cone C_0 that lies in the past of Σ_0.

We also define the extended exterior region E' as the region where

$$r \geq \frac{r_0}{4}.$$

In the following we shall confine attention to the exterior region E. We note that there is a $u_1 < 0$ such that

$$r(0, u_1) = \frac{r_0(0)}{2}.$$

We also define $t_0(u)$ to be the value of t at which the cone C_u intersects the boundary of E, that is, $t_0(u)$ is defined for $u < u_1$ to be the solution of the equation

$$r(t_0, u) = \frac{r_0(t_0)}{2},$$

while for $u \geq u_1$,

$$t_0 = 0.$$

We now introduce the following primitive hypothesis:

A₀:

$$1 + \frac{t}{2} \leq r_0 \leq 2 + 2t.$$

We also assume that the function $u_\star$, defined on the last slice $\Sigma_{t_\star}$ satisfies the following hypotheses:

A₁:

$$r \mid a_\star - 1 \mid \leq A$$

A₂:

$$\mid \frac{2}{tr\chi_\star} - r \mid \leq A, \qquad r^2 \mid \hat{\chi}_\star \mid \leq A.$$

$$r^2 a_\star^{-1} \mid \nabla\!\!\!/ a_\star \mid \leq A$$

The subscript $\star$ refers to quantities defined on the last slice $\Sigma_{t_\star}$. We assume further that the lapse function of the maximal foliation satisfies the following hypothesis:

B₁:

$$\inf \phi \geq \frac{1}{2}.$$

We note that by the maximum principle we have:

$$\sup \phi \leq 1.$$

We assume, in addition, that the components of the second fundamental form and the gradient of the lapse function of the maximal foliation satisfy in the exterior region E the following hypothesis:

B₂:

$$\sup_E r^2 \mid \delta \mid \leq B, \qquad \sup_E r^2 \mid \epsilon \mid \leq B, \qquad \sup_E r \mid \hat{\eta} \mid \leq B$$

$$\sup_E r^2 \phi^{-1} \mid \nabla\!\!\!/ \phi \mid \leq B, \qquad \sup_E r^2 \phi^{-1} \mid \nabla_N \phi \mid \leq B.$$

Finally, we assume that the components of the space-time curvature satisfy in E the following hypothesis:

B₃:

$$\sup_E r^{7/2} \mid \alpha \mid \leq B, \qquad r^{7/2} \mid \beta \mid \leq B, \qquad \sup_E r^3 \mid \rho \mid \leq B.$$

We note that these hypotheses imply that in the exterior region E the function ν (see 9.1.4j) satisfies

$$\mid \nu \mid \leq cBr^{-2}.$$

It then follows from equations 9.1.4d, 9.1.4c, and hypothesis $\mathbf{A_1}$ that if $A + B$ is sufficiently small,

$$\inf_E a \geq \frac{1}{2}, \qquad \sup_E a \leq 2. \tag{9.1.13}$$

We define a unique affine parameter function s by setting

$$s_\star = r$$

where $s_\star$ is the restriction of s to $\Sigma_{t_\star}$. For each cone C_u we define $t_1(u)$ to be the minimal value of $t' \geq t_0(u)$ such that

$$\frac{1}{2} < \frac{s}{r} < 2$$

on each $S_{t,u}$ for all $t \in (t', t_\star]$.

Proposition 9.1.1 *Under the hypotheses* $\mathbf{A_0}, \mathbf{A_1}, \mathbf{A_2}, \mathbf{B_1}, \mathbf{B_2}$, *and* $\mathbf{B_3}$ *there exist positive numerical constants* c_1 *and* c_2 *such that* $A + B < c_1$ *implies*

$$\mid \frac{2}{tr\chi} - s \mid < c_2(A + B),$$

$$s^2 \mid \hat{\chi} \mid < c_2(A + B),$$

for all $t \in [t_1, t_\star]$.

Proof 9.1.1 *Let*

$$h = \frac{2}{tr\chi}.$$

Then by equations 9.1.6d and 9.1.6e we have

$$\frac{dh}{ds} = 1 + \frac{1}{2}h^2 \mid \hat{\chi} \mid^2, \tag{9.1.14a}$$

$$\mid \frac{d \mid \hat{\chi} \mid}{ds} + 2h^{-1} \mid \hat{\chi} \mid\mid \leq \mid \alpha \mid, \tag{9.1.14b}$$

and by 9.1.13 and hypothesis $\mathbf{B_3}$ *we have*

$$\mid \alpha \mid \leq Bs^{-7/2}, \tag{9.1.14c}$$

and, according to hypothesis $\mathbf{A_2}$,

$$| h_\star - s_\star | \leq A, \qquad s_\star^2 | \hat{\chi}_\star | \leq A. \tag{9.1.14d}$$

We consider a given generator of the cone C_u. *Let* $k > 1$ *and let* Γ *be the property*

$$| h - s | < k(A + B), \qquad s^2 | \hat{\chi} | < k(A + B).$$

The property Γ *is continuous and true at* $s = s_\star$. *We shall show that* Γ *is inductive as long as* $t \geq t_1(u)$. *Let* Γ *hold in* $(s_1, s_\star]$. *Then, since by 9.1.14a*

$$h(s_1) - s_1 - h_\star + s_\star = -\frac{1}{2} \int_{s_1}^{s_\star} h^2 | \hat{\chi} |^2,$$

we obtain, in view of 9.1.14d,

$$-A - ck^2(A + B)^2 \leq h(s_1) - s_1 \leq A, \tag{9.1.15a}$$

provided that

$$k(A + B) \leq \frac{1}{8} \tag{9.1.15b}$$

(which guarantees that $s - k(A+B) \geq s/2$, *as* $s \geq r/2 \geq r_0/4 \geq 1/4$*). Also, by 9.1.14b and 9.1.14c,*

$$\frac{d | \hat{\chi} |}{ds} + \frac{2 | \hat{\chi} |}{s - k(A + B)} \geq -Bs^{-7/2},$$

which implies

$$(s_1 - k(A + B))^2 | \hat{\chi}(s_1) | - (s_\star - k(A + B))^2 | \hat{\chi}(s_\star) | \leq B \int_{s_1}^{s_\star} s^{-3/2} ds,$$

which, in view of 9.1.14d, in turn implies

$$s_1^2 | \hat{\chi}(s_1) | \leq c(A + B). \tag{9.1.15c}$$

We conclude from 9.1.15a and 9.1.15c that we can choose the constant k *so that if* $A + B$ *is sufficiently small* Γ *is true at* $s = s_1$. *Hence* Γ *is inductive and therefore true as long as* $t \geq t_1(u)$.

Proposition 9.1.2 *Let the hypotheses* $\mathbf{A_0}, \mathbf{A_1}, \mathbf{A_2}, \mathbf{B_1}, \mathbf{B_2}$, *and* $\mathbf{B_3}$ *hold. Then, if* $A + B$ *is sufficiently small, we have*

$$r | a - 1 | \leq c(A + B)$$

and

$$s^2 | \zeta | \leq c(A + B)$$

for all $t \in [t_1, t_\star]$.

Proof 9.1.2 *By 9.1.12b, 9.1.13, and Proposition 9.1.1, along a given cone C_u for $t \geq t_1(u)$ we have*

$$\frac{dr}{dt} = \frac{r}{2}\overline{\phi a tr\chi} \geq c,$$

where c is a positive numerical constant. Using r as a parameter along the generators of C_u, we obtain

$$\int_t^{t_*} \varphi \mid \nu \mid dt = \int_r^{r_*} \varphi \mid \nu \mid \left(\frac{dr}{dt}\right)^{-1} dr \leq cB\left(\frac{1}{r} - \frac{1}{r_*}\right),$$

which, in view of hypothesis $\mathbf{A_1}$ and the equation

$$\frac{da}{dt} = -\nu\phi a$$

(see 9.1.4c and 9.1.4d), implies the first part of the proposition provided that $A + B$ is sufficiently small.

From equation 9.1.8a we obtain

$$-\frac{d \mid \zeta \mid}{ds} - (h^{-1} + \mid \hat{\chi} \mid) \mid \zeta \mid \leq (h^{-1} + \mid \hat{\chi} \mid) \mid \underline{\zeta} \mid + \mid \beta \mid, \tag{9.1.16a}$$

and by hypotheses $\mathbf{B_2}$ and $\mathbf{B_3}$,

$$\mid \underline{\zeta} \mid \leq cBs^{-2}, \qquad \mid \beta \mid \leq Bs^{-7/2}. \tag{9.1.16b}$$

while by hypotheses $\mathbf{A_2}$ and $\mathbf{B_2}$,

$$s_*^2 \mid \zeta \mid \leq A + B. \tag{9.1.16c}$$

Substituting 9.1.16b as well as the results of Proposition 9.1.1 into inequality 9.1.16a we obtain

$$\begin{aligned}-\frac{d \mid \zeta \mid}{ds} - \left((s - c(A + B))^{-1} + c(A + B)s^{-2}\right) \mid \zeta \mid \leq \big((s - c(A + B))^{-1} \\ + c(A + B)s^{-2}\big) cBs^{-2} \\ + Bs^{-7/2}.\end{aligned}$$

which, upon integrating in the interval $[s, s_*]$ and using 9.1.16c, yields

$$s \mid \zeta \mid \leq c\left(\frac{(A + B)}{s_*} + B\left(\frac{1}{s} - \frac{1}{s_*}\right)\right).$$

which in turn implies the second part of the proposition.

Next we wish to prove that the oscillation on $S_{t,u}$ of the function $h = \frac{2}{\mathrm{tr}\chi}$ is bounded for all t and u. In view of Proposition 9.1.1 it suffices to show that the same is true for the oscillation of s on $S_{t,u}$. This may seem surprising in view of the equations 9.1.4c and 9.1.12b, which imply

$$\frac{dr}{ds} = \frac{r}{2}\phi^{-1}a^{-1}\overline{\phi a \mathrm{tr}\chi}.$$

The right-hand side is of order $1 + O(r^{-1})$, so at first glance it looks as if $s - r$ diverges logarithmically. Nevertheless, we have the following:

Proposition 9.1.3 *Let the hypotheses* $\mathbf{A_0}, \mathbf{A_1}, \mathbf{A_2}, \mathbf{B_1}, \mathbf{B_2}$, *and* $\mathbf{B_3}$ *hold. Then, if* $A + B$ *is sufficiently small, we have*

$$| \nabla\!\!\!/ s | \leq c(A + B)s^{-1}$$

and

$$| s - r | \leq c(A + B)$$

for all $t \in [t_1, t_*]$.

Proof 9.1.3 *We define*

$$\psi_A = s e_A(s) - s^2\zeta_A. \tag{9.1.17a}$$

By equations 9.1.9 and 9.1.8a we have

$$\frac{d\psi_A}{ds} + (\chi_{AB} - s^{-1}\delta_{AB})\psi_B = -s^2(\chi_{AB} - s^{-1}\delta_{AB})\underline{\zeta}_B + s^2\beta_A. \tag{9.1.17b}$$

It follows that

$$-\frac{d\,|\,\psi\,|}{ds} - (|\,h^{-1} - s^{-1}\,| + |\,\hat{\chi}\,|)\,|\,\psi\,| \\ \leq s^2(|\,h^{-1} - s^{-1}\,| + |\,\hat{\chi}\,|)\,|\,\underline{\zeta}\,| + s^2\,|\,\beta\,|.$$

Substituting the result of Proposition 9.1.1 and 9.1.16b, we then obtain the inequality

$$-\frac{d\,|\,\psi\,|}{ds} - c\frac{(A + B)}{s^2}\,|\,\psi\,| \leq \frac{cB}{s^{3/2}},$$

which, upon integrating in the interval $[s, s_*]$ *and using 9.1.16c and the fact that*

$$\nabla\!\!\!/ s_* = 0,$$

yields

$$|\,\psi\,| \leq c(A + B). \tag{9.1.17c}$$

By virtue of Proposition 9.1.2 this implies the first part of the proposition.

To prove the second part, we consider equation 9.1.11 for the Gauss curvature of the surfaces $S_{t,u}$. In view of hypotheses $\mathbf{B_2}$, $\mathbf{B_3}$, *and the results of Propositions 9.1.1 and 9.1.2, we conclude that, if $A+B$ is small, there are positive constants k_m and k_M such that*

$$k_m \leq s^2 K \leq k_M. \qquad (9.1.18a)$$

By the Bonnet-Myers theorem we then have

$$diam S \leq \pi k_m^{-\frac{1}{2}} s_M \qquad (9.1.18b)$$

where we denote by s_m and s_M, respectively, the minimum and maximum values of s on $S_{t,u}$. Integrating the inequality

$$|\nabla s| \leq c(A+B)s^{-1}$$

along a minimal geodesic of $S_{t,u}$, joining two points where these values are attained, and using 9.1.18b, we obtain

$$s_M - s_m \leq c(A+B)\frac{s_M}{s_m},$$

which if $A+B$ is sufficiently small implies

$$s_M - s_m \leq c(A+B). \qquad (9.1.18c)$$

Going back to equation9.1.11 and using the results of Propositions 9.1.1, 9.1.2, and hypotheses $\mathbf{B_2}$ *and* $\mathbf{B_3}$, *we also obtain*

$$|K - \frac{1}{s^2}| \leq \frac{c(A+B)}{s^3}.$$

Integrating this inequality on $S_{t,u}$ and using the fact that, by the Gauss-Bonnet theorem,

$$\overline{K} = \frac{1}{r^2},$$

we obtain that for $t \geq t_1$

$$\frac{1}{s_M^2} - \frac{c(A+B)}{r^3} \leq \frac{1}{r^2} \leq \frac{1}{s_m^2} + \frac{c(A+B)}{r^3},$$

which in view of 9.1.18c yields the second part of the proposition.

Finally, the second part of Proposition 9.1.3 implies the following:

Corollary 9.1.3.1 *If $A+B$ is sufficiently small, then*

$$t_1 = t_0.$$

We now define the function

$$\omega = \frac{1}{4}\underline{l}^\mu \underline{l}^\nu \mathbf{D}_\mu l_\nu. \tag{9.1.19a}$$

We have

$$\begin{aligned}\omega &= \frac{1}{2}\mathbf{D}'_3 a + \frac{1}{4} a e_3'^\mu e_3'^\nu \mathbf{D}_\mu e'_{4\nu} \\ &= \frac{1}{2} a \frac{da}{ds} - \nabla_N a + \frac{1}{2} a(-\mathbf{g}(\mathbf{D}_T T, N) + \mathbf{g}(\mathbf{D}_N T, N)).\end{aligned}$$

Thus, in view of equations 9.1.4d and 9.1.4f,

$$\omega = -\nabla_N a - a\delta. \tag{9.1.19b}$$

Applying $\frac{d}{ds}$ to 9.1.19a and using 9.1.3e, we obtain

$$\begin{aligned}\frac{d\omega}{ds} &= \frac{1}{4}\underline{l}^\mu \underline{l}^\nu l^\alpha \mathbf{D}_\alpha(\mathbf{D}_\mu l_\nu) + \frac{1}{2}\frac{d\underline{l}^\mu}{ds}\underline{l}^\nu \mathbf{D}_\mu l_\nu \\ &= \frac{1}{4}\underline{l}^\mu \underline{l}^\nu l^\alpha (\mathbf{D}_\mu \mathbf{D}_\alpha l_\nu + R_{\nu\beta\alpha\mu} l^\beta) + \underline{\zeta}_A e^\mu_A \underline{l}^\nu \mathbf{D}_\mu l_\nu \\ &= -\frac{1}{4}\underline{l}^\mu \underline{l}^\nu (\mathbf{g}^{\alpha\beta}\mathbf{D}_\mu l_\alpha \mathbf{D}_\beta l_\nu + R_{\mu\alpha\nu\beta} l^\alpha l^\beta) + 2\underline{\zeta}_A \zeta_A,\end{aligned}$$

hence, using expression 9.1.4a,

$$\frac{d\omega}{ds} = 2\underline{\zeta}\cdot\zeta - |\zeta|^2 - \rho. \tag{9.1.19c}$$

Remark 1 All the results of this subsection remain valid in the extended exterior region[4] E′, provided that, in addition to these assumptions, the assumptions $\mathbf{C}_1 - \mathbf{C}_3$ of the next subsection are satisfied.

9.2. Interior Construction of the Optical Function

Let P_0 be a point in the geodesic ball in Σ_0 of radius ε_1 with center at O_0. We denote by P the integral curve of T through P_0. The interior optical function u is defined by the following two conditions:

Condition 9.2.1 *On the timelike curve P, u is minus arc length on P as measured from P_{-1}.*

Condition 9.2.2 *The level sets C_u of u are the future null geodesic cones with vertices on P.*

[4] See the definition on page 268.

The interior optical function as defined by these conditions is of course also a solution of the Eikonal equation; however, the boundary conditions it satisfies are different from those of the exterior optical function. In particular, Condition 1 says that on P

$$T^\mu \partial_\mu u = -1. \tag{9.2.1a}$$

Defining as before the null geodesic vectorfield

$$l^\mu = \mathbf{g}^{\mu\nu} \partial_\nu u$$

we note that, since the integral curves of l are the generators of each cone C_u, at a point P_t of P l takes not a single value but rather a whole set of values, namely, the set of null vectors at P_t whose T component is equal to unity:

$$\mathbf{g}(l, T)|_P = -1. \tag{9.2.1b}$$

The natural boundary condition for the affine parameter function s is now

$$s|_P = 0.$$

The surfaces $S_{t,u}$, the vectorfields N and $\underline{l}$, and the lapse function a are defined as for the exterior optical function. By 9.2.1b we have

$$a|_P = 1.$$

The acceleration of the timelike curve P is the Σ-tangent vectorfield along P given by

$$\mathbf{b} = \mathbf{D}_T T = \phi^{-1} \nabla \phi. \tag{9.2.2}$$

By 9.2.1b on P we have

$$l = T + N.$$

At each point P_t of P the set of values of N is the unit sphere in the tangent space at P_t to Σ_t.

At each P_t and for each value of N we choose an orthonormal basis (e_A) for the plane in $T_{P_t}\Sigma_t$ orthogonal to that value. We then propagate (e_A) along the null geodesic generators of each cone C_u according to equation 9.1.3a. Then (e_A) remains orthonormal and orthogonal to both l and T, therefore (e_A) is an orthonormal basis for each $S_{t,u}$.

The null second fundamental form χ_{AB} is defined as for the exterior optical function and satisfies the equations 9.1.6a, 9.1.6d, and 9.1.6e. However, the boundary conditions are now the requirements that $tr\chi - (2/s)$ and $\hat{\chi}$ remain bounded as $s \to 0$. As we shall see, these imply that in fact

$$tr\chi = \frac{2}{s} + O(s^3), \qquad \hat{\chi}_{AB} = O(s).$$

The 1-form ζ is also defined as before and satisfies equation 9.1.8a. The boundary condition is now the requirement that ζ remain bounded as $s \to 0$.

Finally, the function ω is defined as for the exterior optical function and satisfies equation 9.1.19c. To obtain the boundary condition for ω, we consider the fact that

$$\omega = a^2 \mathbf{g}(\mathbf{D}_T l, T) = -a^2[\mathbf{D}_T(a^{-1}) + \mathbf{g}(\mathbf{D}_T T, l)] = \mathbf{D}_T a - \mathbf{b}_N a,$$

which implies that on P

$$\omega = -\mathbf{b}_N. \tag{9.2.3}$$

We define the space-time region

$$V = \bigcup_{t \in [0,t_*]} V_t$$

where V_t is the closed ball of radius

$$d_1(t) = (9/20)(1+t)$$

with center at P_t in Σ_t.

We now make the following assumption:

M: The region V is contained in the interior region I.

Remark 2 This assumption will be verified later in Corollary 15.0.5.1 in Chapter 15.

We introduce the following hypotheses on the lapse function and the second fundamental form of the maximal foliation:

$\mathbf{C}_1$:

$$\sup_I (2+t)|\phi - 1| \le B$$

$\mathbf{C}_2$:

$$\sup_I (2+t)^2 |k| \le B, \qquad \sup_V (2+t)^2 \phi^{-1} |\nabla \phi| \le B.$$

We also assume that the electric and magnetic parts of the space-time curvature satisfy in I :

$\mathbf{C}_3$:

$$\sup_I (2+t)^{7/2} |E| \le B, \qquad \sup_I (2+t)^{7/2} |H| \le B.$$

At each $t \in [0, t_*]$, let $u_M(t)$ be the greatest value of u such that the part of the cone C_u that lies in the past of Σ_t is contained in $V \bigcup M_0$. We then define the interior region I' to be the union of the set[5]

$$\bigcup_{t \in [0, t_*]} \mathrm{Int} S_{t, u_M(t)}$$

with the part of the cone $C_{u_M(0)}$ that lies in the past of Σ_0.

Let $u_m(t)$ be the value of u at P_t. By Condition 1 of the definition of u we have

$$u_m(t) = -\int_{-1}^{t} o|_P dt. \tag{9.2.4a}$$

Hence, for $-1 \leq t_1 \leq t_2 \leq t_*$

$$|u_m(t_2) - u_m(t_1) + t_2 - t_1| \leq cB \log\left(\frac{2 + t_2}{2 + t_1}\right). \tag{9.2.4b}$$

Hypothesis $\mathbf{C_2}$ implies that in V

$$|\nu| \leq cB(2 + t)^{-2}$$

and by equations 9.1.4c and 9.1.4d

$$\frac{da}{dt} = -\nu o a. \tag{9.2.5a}$$

As each point p in $I'_t = I' \bigcap \Sigma_t$ lies on a cone C_u such that the part of C_u in the past of Σ_t is contained in I, integrating equation 9.2.5a along the generator of C_u corresponding to p from the vertex, we obtain

$$\exp[-cB(2 + t_m)^{-1}] \leq a \leq \exp[cB(2 + t_m)^{-1}] \tag{9.2.5b}$$

where $t_m(u)$ is the value of t at the vertex of C_u (note that $t_m(u) = t$ is equivalent to $u_m(t) = u$. Now the fact that $S_{t, u_M(t)}$ is included in the closed ball of radius $d_1(t)$ with center at P_t in Σ_t implies that

$$\inf_{I \bigcap \Sigma_t} a \cdot (u_M(t) - u_m(t)) \leq d_1(t). \tag{9.2.6}$$

This follows from the the consideration that for any $p \in S_{t, u_1}$ the distance $d(p)$ from p to P_t is bounded from below by

$$d(p) \geq \int_{u_m}^{u_M} |\nabla u|^{-1} du \geq \int_{u_m}^{u_M} (\inf_{S_{t,u}} a) du.$$

[5] Where $\mathrm{Int} S_{t,u}$ is the interior of $S_{t,u}$ in Σ_t

From 9.2.4b, 9.2.5b, and 9.2.6 we obtain

$$\exp[-cB(2+t_0)^{-1}]\cdot\left(t-t_0-cB\log[(2+t)/(2+t_0)]\right)\leq d_1(t)$$

where

$$t_0(t)=t_m(u_M(t)).$$

This implies the following:

Lemma 9.2.1 *If B is sufficiently small, then, for all $t\in[0,t_*]$,*

$$2+t_0\geq\frac{1}{2}(2+t). \tag{9.2.7}$$

Note that $t_m(u)\geq t_0(t)$ for all $u\leq u_M(t)$.

Proposition 9.2.1 *Under the hypotheses* $\mathbf{C_1}$,$\mathbf{C_2}$, *and* $\mathbf{C_3}$ *with B sufficiently small, the following estimates hold in I':*

$$\begin{aligned}|\frac{2}{tr\chi}-s|&\leq cB^2s^5(2+t)^{-7},\\ |\hat{\chi}|&\leq cBs(2+t)^{-7/2}.\end{aligned}$$

Proof 9.2.1 *Let*

$$h=\frac{2}{tr\chi}$$

and consider a given generator of a given cone C_u. Let Δ be the property

$$|h-s|<s^2/(1+s)^2.$$

The property Δ is continuous. By the boundary conditions at the vertex of the cone

$$h=s+O(s^2),\qquad\hat{\chi}=O(1)$$

near $s=0$. By 9.1.14b we then have

$$s^2|\hat{\chi}(s)|\leq\int_0^s\exp\left[2\int_{s'}^{s}|(h(s''))^{-1}-s''^{-1}|ds''\right]\cdot|\alpha(s')|ds'. \tag{9.2.8a}$$

Hence,

$$\hat{\chi}=O(s).$$

Since by 9.1.14a

$$h(s)=s+\frac{1}{2}\int_0^s(h(s'))^2|\hat{\chi}(s')|^2ds', \tag{9.2.8b}$$

this implies

$$h = s + O(s^5),$$

therefore there is a $\varepsilon_1 > 0$ *such that for* $0 < s < \varepsilon_1$ Δ *is true. We shall show that* Δ *is inductive as long as* $t \leq t_M(u)$ *where* $t_M(u) = x$ *is equivalent to* $u_M(x) = u$, *that is, as long as the generator remains in* I'. *Let* Δ *hold in* $(0, s_1)$. *Then in* $(0, s_1)$ *we have*

$$|h^{-1} - s^{-1}| \leq c(1+s)^{-2},$$

which by 9.2.8a and hypothesis $\mathbf{C}_3$ *implies*

$$|\hat{\chi}| \leq cBs^3(2+t_m)^{-7/2}. \tag{9.2.8c}$$

Substituting this estimate as well as

$$|h| \leq cs: \quad s \in (0, s_1)$$

in 9.2.8b, we then obtain:

$$|h(s_1) - s_1| \leq cB^2 s_1^5 (2+t_m)^{-7}. \tag{9.2.8d}$$

In view of the lower bound for t_m *given by 9.2.7 and the fact that (by 9.1.4c)* $s \leq ct$ *in* I', *we conclude that if* B *is sufficiently small,* Δ *holds at* $s = s_1$. *Hence,* Δ *is inductive and therefore true as long as* $t \leq t_M(u)$. *The estimates 9.2.8c and 9.2.8d then hold throughout* I'.

Proposition 9.2.2 *Under the hypotheses* $\mathbf{C}_1$, $\mathbf{C}_2$, *and* $\mathbf{C}_3$ *with* B *sufficiently small,*

$$|a - 1| \leq cBs(2+t)^{-2}$$

and

$$|\zeta| \leq cB(2+t)^{-2}$$

hold in I'. *Furthermore,*

$$|\zeta_A - \frac{1}{s}\int_0^s \underline{\zeta}_A(s')ds'| \leq cBs(2+t)^{-7/2}.$$

Proof 9.2.2 *The estimate for* $a - 1$ *follows directly from equation 9.1.4d. To obtain the estimate for* ζ, *we consider equation 9.1.8a along a generator of a given cone deducing*

$$\frac{d|\zeta|}{ds} + (h^{-1} - |\hat{\chi}|)|\zeta| \leq (h^{-1} + |\hat{\chi}|)|\underline{\zeta}| + |\beta|. \tag{9.2.9a}$$

Integrating and using the boundary condition $\zeta = O(1)$ at the vertex, we obtain

$$s|\zeta(s)| \leq \int_0^s e(s') \exp\left[2\int_{s'}^s \left(|(h(s''))^{-1} - s''^{-1}| + |\hat{\chi}(s'')|\right) ds''\right] ds'$$
$$e(s') = s' \left[(h(s')^{-1} + |\hat{\chi}(s')|)|\underline{\zeta}(s')| + |\beta(s')|\right]. \tag{9.2.9b}$$

Using the results of Proposition 9.2.1 and the fact that by hypothesis $\mathbf{C_2}$

$$|\underline{\zeta}| \leq cB(2+t)^{-2}$$

in I', we conclude that for $t \leq t_M(u)$

$$|\zeta| \leq \frac{cB}{(2+t_m)^2}\left[1 + \frac{s}{(2+t_m)^{3/2}}\right],$$

which in view of the lower bound 9.2.7 for t_m yields the first two conclusions.

To obtain the last part of the proposition, we rewrite the equation 9.1.8a in the form

$$\frac{d}{ds}(s\zeta_A) - \underline{\zeta}_A = \left(\hat{\chi}_{AB} + (h^{-1} - s^{-1})\delta_{AB}\right)\left(s\underline{\zeta}_B - s\zeta_B\right)$$
$$- s\beta_A. \tag{9.2.10}$$

Integrating and using the previous conclusions yields the remainder of the proposition.

Proposition 9.2.3 *Under the hypotheses $\mathbf{C_1}$, $\mathbf{C_2}$, and $\mathbf{C_3}$ with B sufficiently small,*

$$|\nabla s| \leq cBs(2+t)^{-2}$$

and

$$|s - r| \leq cBr^2(2+t)^{-2}$$

hold in I'.

Proof 9.2.3 *Going back to equation 9.1.17b for ψ, we now deduce*

$$\frac{d|\psi|}{ds} \leq (|h^{-1} - s^{-1}| + |\hat{\chi}|)(|\psi| + s^2|\underline{\zeta}|) + s^2|\beta|. \tag{9.2.11a}$$

Integrating from the vertex and using the fact that ψ vanishes there, we obtain in I'

$$|\psi| \leq cBs^3(2+t)^{-7/2}. \tag{9.2.11b}$$

Hence, in view of Proposition 9.1.3e,

$$|\nabla s| \leq cBs(2+t)^{-2}. \tag{9.2.11c}$$

This implies that

$$s_M - s_m \leq cBs_M^2(2+t)^{-2}. \tag{9.2.11d}$$

Considering equation 9.1.11 for the Gauss curvature K of $S_{t,u}$ and using hypotheses $\mathbf{C_2}$, $\mathbf{C_2}$, *as well as the results of Propositions 9.2.1 and 9.2.2, we obtain*

$$|K - \frac{1}{s^2}| \leq \frac{cB}{s(2+t)^2}.$$

Integrating this inequality on $S_{t,u}$ and using the Gauss-Bonnet theorem, we obtain that for $t \leq t_M(u)$

$$\frac{1}{s_M^2} - \frac{cB}{s_m(2+t)^2} \leq \frac{1}{r^2} \leq \frac{1}{s_m^2} + \frac{cB}{s_m(2+t)^2},$$

which in view of 9.2.11d yields the second part of the proposition.

9.3. The Initial Cone C_0

In this section we establish an estimate for the normalized null second fundamental form χ' of the initial cone C_0, $\chi'_{AB} = \langle \mathbf{D}_A, e'_4 \rangle$, $e'_4 = T+N$. This differs from the estimates of the first section by the fact that the boundary condition for χ' on the last slice is replaced by the interior type boundary condition at the point O_{-1}.

Proposition 9.3.1 *Under the assumptions* $\mathbf{B_1}, \mathbf{B_2}, \mathbf{B_3}$ *and* $\mathbf{C_1}, \mathbf{C_2}, \mathbf{C_3}$, *the following hold:*

1. *In the interval $t \in [-1, 0]$ we have*

$$\sup_{S_{t,0}} |\mathrm{tr}\chi' - \frac{2}{r_0}| \leq cr_0 B$$

$$\sup_{S_{t,0}} |\hat{\chi}'| \leq cr_0 B.$$

2. *In the interval $t \in [0, t_*]$ we have*

$$\sup_{S_{t,0}} r_0^2 |\mathrm{tr}\chi' - \frac{2}{r_0}| \leq c\left(\sup_{S_{0,0}} |\mathrm{tr}\chi' - \frac{2}{r_0}| + \sup_{S_{0,0}} |\hat{\chi}'|\right) + cB$$

$$\sup_{S_{t,0}} r_0^2 |\hat{\chi}'| \leq c\left(\sup_{S_{0,0}} |\mathrm{tr}\chi' - \frac{2}{r_0}| + \sup_{S_{0,0}} |\hat{\chi}'|\right) + cB.$$

Proof 9.3.1 *The first part is identical to the interior estimates discussed in the previous section. To prove the second part, we set $a = 1$ on S_{0,t_*} and we propagate it down on C_0 with the help of the equation $\frac{da}{ds} = -\nu$ (see 9.1.4d).*

With this definition of a *we then set* $l = a^{-1}(T + N)$, $\underline{l} = a(T - N)$, *and* $\chi_{AB} = \langle \mathbf{D}_a l, e_B \rangle$. *We also set* $\zeta = \frac{1}{2}\langle \nabla\!\!\!/_A l, \underline{l} \rangle$ *and note that with this definition* $\zeta = \epsilon$ *on the last slice. We then proceed as in the first section with the boundary condition for* χ, *given on* $S_{0,0}$ *by*

$$\chi_{AB} = a^{-1}\chi'_{AB} \tag{9.3.1}$$

where $\chi'|_{S_{0,0}}$, *given by the first part of the proposition, while the boundary condition for* ζ *is given on* $S_{t_*,0}$ *by*

$$\zeta = \epsilon. \tag{9.3.2}$$

CHAPTER 10.

Third Version of the Main Theorem

10.1. Basic Notations, Norms

In this section we assume a space-time slab $\bigcup_{[0,t_*]} \Sigma_t$, foliated by a maximal time foliation t and an optical function u, and we introduce our basic norms involving the curvature tensor $\mathbf{R}$, second fundamental form k, and lapse function ϕ, as well as the optical quantities χ, ζ, ω.

Given an S-tangent tensorfield V we first define

$$|V|_{p,S}(t,u) = \left(\int_{S_{t,u}} |V|^p d\mu_\gamma \right)^{1/p} \quad \text{if } 1 \leq p < \infty \qquad (10.1.1a)$$
$$= \sup_{S_{t,u}} |V| \quad \text{if } p = \infty.$$

We also introduce the following norms defined in the interior and exterior regions Σ_t^i, Σ_t^e of each slice:

$$\|V\|_{p,i} = \left(\int_{\Sigma_t^i} |V|^p \right)^{1/p} \quad \text{if } 1 \leq p < \infty$$
$$= \sup_{\Sigma_t^i} |V| \quad \text{if } p = \infty$$

$$\|V\|_{p,e}(t) = \left(\int_{\Sigma_t^e} |V|^p \right)^{1/p} \quad \text{if } 1 \leq p < \infty \qquad (10.1.1b)$$
$$= \sup_{\Sigma_t^e} |V| \quad \text{if } p = \infty$$

$$|||V|||_{p,e}(t) = \sup_{u \geq u_0(t)} |V|_{p,S}(t,u) \quad \text{if } 1 \leq p < \infty$$
$$= \sup_{\Sigma_t^e} |V| \quad \text{if } p = \infty,$$

where we recall that $\Sigma_t^i = I$ consists of points for which $r \leq \frac{r_0(t)}{2}$ while $\Sigma_t^e = E$ consists of those for which $r \geq \frac{r_0(t)}{2}$, with $r_0(t)$ the value of r corresponding to the area of the surface of intersection between C_0 and Σ_t. Also recall that $u_1(t)$ is the value of u corresponding to $\frac{r_0(t)}{2}$.

We also introduce the local L^2 norm

$$^{(loc)}\|V\|_{2,e}(t) = \sup_{\lambda \geq u_0(t)} \left(\int_{D(\lambda,t)} |V|^2 \right)^{1/2} \tag{10.1.1c}$$

where $D(\lambda, t)$ is the annulus $\lambda \leq u \leq \lambda + 1$ on Σ_t.

Finally, we recall the definition of

$$\tau_- = \sqrt{1+u^2}$$
$$w_p = \min\{\tau_-^p r_0^{1/2}, r^p\}.$$

10.1.1. *Norms Involving the Curvature Tensor* **R**

Our main norm involving the curvatures was defined in the first chapter of Part II by formulas 7.6.1a–7.6.1f, which we recall here.

First we define the interior and exterior norms ${}^{\imath}\mathcal{R}_{[q]}$ and ${}^{e}\mathcal{R}_{[q]}$ and then set as our basic curvature norms

$$\mathcal{R}_{[q]} = \max({}^{e}\mathcal{R}_{[q]}, {}^{\imath}\mathcal{R}_{[q]}) \tag{10.1.2a}$$

for $q = 0, 1, 2$.

To define the interior norms, we set

$${}^{\imath}\mathcal{R}_q = r_0^{2+q}(\|\mathbf{D}^q E\|_{2,\imath} + \|\mathbf{D}^q H\|_{2,\imath}) \tag{10.1.2b}$$

where E, H are the electric and magnetic parts of the space-time curvature tensor **R**.

We then set

$${}^{\imath}\mathcal{R}_{[0]} = {}^{\imath}\mathcal{R}_0 \tag{10.1.2c}$$
$${}^{\imath}\mathcal{R}_{[1]} = {}^{\imath}\mathcal{R}_{[0]} + {}^{\imath}\mathcal{R}_1 \tag{10.1.2d}$$
$${}^{\imath}\mathcal{R}_{[2]} = {}^{\imath}\mathcal{R}_{[1]} + {}^{\imath}\mathcal{R}_2. \tag{10.1.2e}$$

We define the exterior quantities ${}^e\mathcal{R}_q$ as follows:[1]

$$
\begin{aligned}
{}^e\mathcal{R}_q(\underline{\alpha}) &= \|\tau_-^2 r^q \nabla\!\!\!/^q \underline{\alpha}\|_{2,e} & {}^e\mathcal{R}_q(\alpha) &= \|r^{q+2} \nabla\!\!\!/^q \alpha\|_{2,e} \\
{}^e\mathcal{R}_q(\underline{\alpha}_3) &= \|\tau_-^3 r^q \nabla\!\!\!/^q \underline{\alpha}_3\|_{2,e} & {}^e\mathcal{R}_q(\alpha_3) &= \|r^{q+3} \nabla\!\!\!/^q \alpha_3\|_{2,e} \\
{}^e\mathcal{R}_q(\underline{\alpha}_4) &= \|\tau_- r^{2+q} \nabla\!\!\!/^q \underline{\alpha}_4\|_{2,e} & {}^e\mathcal{R}_q(\alpha_4) &= \|r^{3+q} \nabla\!\!\!/^q \alpha_4\|_{2,e} \\
{}^e\mathcal{R}_q(\underline{\alpha}_{33}) &= \|\tau_-^4 r^q \nabla\!\!\!/^q \underline{\alpha}_{33}\|_{2,e} & {}^e\mathcal{R}_q(\alpha_{33}) &= \|r^{4+q} \nabla\!\!\!/^q \alpha_{33}\|_{2,e} \\
{}^e\mathcal{R}_q(\underline{\alpha}_{34}) &= \|\tau_-^2 r^{2+q} \nabla\!\!\!/^q \underline{\alpha}_{34}\|_{2,e} & {}^e\mathcal{R}_q(\alpha_{34}) &= \|r^{4+q} \nabla\!\!\!/^q \alpha_{34}\|_{2,e} \\
{}^e\mathcal{R}_q(\underline{\alpha}_{43}) &= \|\tau_-^2 r^{2+q} \nabla\!\!\!/^q \underline{\alpha}_{43}\|_{2,e} & {}^e\mathcal{R}_q(\alpha_{43}) &= \|r^{4+q} \nabla\!\!\!/^q \alpha_{43}\|_{2,e} \\
{}^e\mathcal{R}_q(\underline{\alpha}_{44}) &= \|\tau_- r^{3+q} \nabla\!\!\!/^q \underline{\alpha}_{44}\|_{2,e} & {}^e\mathcal{R}_q(\alpha_{44}) &= \|r^{4+q} \nabla\!\!\!/^q \alpha_{44}\|_{2,e} \\
{}^e\mathcal{R}_q(\underline{\beta}) &= \|\tau_- r^{q+1} \nabla\!\!\!/^q \underline{\beta}\|_{2,e} & {}^e\mathcal{R}_q(\beta) &= \|r^{q+2} \nabla\!\!\!/^q \beta\|_{2,e} \\
{}^e\mathcal{R}_q(\underline{\beta}_3) &= \|\tau_-^2 r^{q+1} \nabla\!\!\!/^q \underline{\beta}_3\|_{2,e} & {}^e\mathcal{R}_q(\beta_3) &= \|r^{q+3} \nabla\!\!\!/^q \beta_3\|_{2,e} \\
{}^e\mathcal{R}_q(\underline{\beta}_4) &= \|r^{3+q} \nabla\!\!\!/^q \underline{\beta}_4\|_{2,e} & {}^e\mathcal{R}_q(\beta_4) &= \|r^{3+q} \nabla\!\!\!/^q \beta_4\|_{2,e} \\
{}^e\mathcal{R}_q(\underline{\beta}_{33}) &= \|\tau_-^3 r^{q+1} \nabla\!\!\!/^q \underline{\beta}_{33}\|_{2,e} & {}^e\mathcal{R}_q(\beta_{33}) &= \|\tau_- r^{q+3} \nabla\!\!\!/^q \beta_{33}\|_{2,e} \\
{}^e\mathcal{R}_q(\underline{\beta}_{34}) &= \|\tau_- r^{3+q} \nabla\!\!\!/^q \underline{\beta}_{34}\|_{2,e} & {}^e\mathcal{R}_q(\beta_{34}) &= \|r^{4+q} \nabla\!\!\!/^q \beta_{34}\|_{2,e} \\
{}^e\mathcal{R}_q(\underline{\beta}_{43}) &= \|\tau_- r^{3+q} \nabla\!\!\!/^q \underline{\beta}_{43}\|_{2,e} & {}^e\mathcal{R}_q(\beta_{43}) &= \|r^{4+q} \nabla\!\!\!/^q \beta_{43}\|_{2,e} \\
{}^e\mathcal{R}_q(\underline{\beta}_{44}) &= \|r^{4+q} \nabla\!\!\!/^q \underline{\beta}_{44}\|_{2,e} & {}^e\mathcal{R}_q(\beta_{44}) &= \|r^{4+q} \nabla\!\!\!/^q \beta_{44}\|_{2,e} \\
{}^e\mathcal{R}_q(\rho - \bar{\rho}) &= \|r^{q+2} \nabla\!\!\!/^q (\rho - \bar{\rho})\|_{2,e} & {}^e\mathcal{R}_q(\sigma - \bar{\sigma}) &= \|r^{q+2} \nabla\!\!\!/^q (\sigma - \bar{\sigma})\|_{2,e} \\
{}^e\mathcal{R}_q(\rho_3) &= \|\tau_- r^{q+2} \nabla\!\!\!/^q \rho_3\|_{2,e} & {}^e\mathcal{R}_q(\sigma_3) &= \|\tau_- r^{q+2} \nabla\!\!\!/^q \sigma_3\|_{2,e} \\
{}^e\mathcal{R}_q(\rho_4) &= \|r^{3+q} \nabla\!\!\!/^q \rho_4\|_{2,e} & {}^e\mathcal{R}_q(\sigma_4) &= \|r^{3+q} \nabla\!\!\!/^q \sigma_4\|_{2,e} \\
{}^e\mathcal{R}_q(\rho_{33}) &= \|\tau_-^2 r^{q+2} \nabla\!\!\!/^q \rho_{33}\|_{2,e} & {}^e\mathcal{R}_q(\sigma_{33}) &= \|\tau_-^2 r^{q+2} \nabla\!\!\!/^q \sigma_{33}\|_{2,e} \\
{}^e\mathcal{R}_q(\rho_{34}) &= \|\tau_-^{\frac{1}{2}} r^{(\frac{7}{2}+q)} \nabla\!\!\!/^q \rho_{34}\|_{2,e} & {}^e\mathcal{R}_q(\sigma_{34}) &= \|\tau_-^{\frac{1}{2}} r^{(\frac{7}{2}+q)} \nabla\!\!\!/^q \sigma_{34}\|_{2,e} \\
{}^e\mathcal{R}_q(\rho_{43}) &= \|\tau_-^{\frac{1}{2}} r^{(\frac{7}{2}+q)} \nabla\!\!\!/^q \rho_{43}\|_{2,e} & {}^e\mathcal{R}_q(\sigma_{43}) &= \|\tau_-^{\frac{1}{2}} r^{(\frac{7}{2}+q)} \nabla\!\!\!/^q \sigma_{43}\|_{2,e} \\
{}^e\mathcal{R}_q(\rho_{44}) &= \|r^{4+q} \nabla\!\!\!/^q \rho_{44}\|_{2,e} & {}^e\mathcal{R}_q(\sigma_{44}) &= \|r^{4+q} \nabla\!\!\!/^q \sigma_{44}\|_{2,e}
\end{aligned}
\tag{10.1.2f}
$$

[1] Note that $\bar{\rho}, \bar{\sigma}$ are the mean values of ρ, σ on $S_{t,u}$.

where $\underline{\alpha}, \underline{\beta}, \rho, \sigma, \beta, \alpha$ is the null decomposition[2] of the curvature tensor $\mathbf{R}$ relative to the standard pair $e_3 = T - N$, $e_4 = T + N$. We now set

$$
\begin{aligned}
{}^e\mathcal{R}_0[\underline{\alpha}] &= {}^e\mathcal{R}_0(\underline{\alpha}) \\
{}^e\mathcal{R}_1[\underline{\alpha}] &= \left({}^e\mathcal{R}_1(\underline{\alpha})^2 + {}^e\mathcal{R}_0(\underline{\alpha}_3)^2 + {}^e\mathcal{R}_0(\underline{\alpha}_4)^2\right)^{1/2} \\
{}^e\mathcal{R}_2[\underline{\alpha}] &= \left({}^e\mathcal{R}_2(\underline{\alpha})^2 + {}^e\mathcal{R}_1(\underline{\alpha}_3)^2 + {}^e\mathcal{R}_1(\underline{\alpha}_4)^2\right. \qquad (10.1.2g)\\
&\quad \left. + {}^e\mathcal{R}_0(\underline{\alpha}_{33})^2 + {}^e\mathcal{R}_0(\underline{\alpha}_{34})^2 + {}^e\mathcal{R}_0(\underline{\alpha}_{43})^2 + {}^e\mathcal{R}_0(\underline{\alpha}_{44})^2\right)^{1/2},
\end{aligned}
$$

and similarly for all the other null components of $\mathbf{R}$.

The expressions ${}^e\mathcal{R}_q$ are thus defined, for all $q = 0, 1, 2$, by

$$
{}^e\mathcal{R}_q = \left({}^e\mathcal{R}_q[\underline{\alpha}]^2 + {}^e\mathcal{R}_q[\underline{\beta}]^2 + \ldots + {}^e\mathcal{R}_q[\alpha]^2\right)^{1/2}. \qquad (10.1.2h)
$$

We now define our basic exterior curvature norms by

$$
\begin{aligned}
{}^e\mathcal{R}_{[0]} &= {}^e\mathcal{R}_0 + \sup_{r \geq r_0/2} r^3 |\bar{\rho}, \bar{\sigma}| && (10.1.3a)\\
{}^e\mathcal{R}_{[1]} &= {}^e\mathcal{R}_{[0]} + {}^e\mathcal{R}_1 && (10.1.3b)\\
{}^e\mathcal{R}_{[2]} &= {}^e\mathcal{R}_{[1]} + {}^e\mathcal{R}_2. && (10.1.3c)
\end{aligned}
$$

For convenience, given v, one of the components of $\mathbf{R}$ in the exterior region, we denote by ${}^e\mathcal{R}_{[q]}(v)$ the part of $\mathcal{R}_{[q]}$ that contains only the angular derivatives of v.

[2] We recall that this decomposition as well as the definition of $\underline{\alpha}_3, \underline{\alpha}_4, \ldots$ was defined in the first chapter of Part II.

We will also make use of the following exterior norms:

$$
\begin{array}{llll}
\mathcal{R}_0^{p,S}(\underline{\alpha}) = |||r^{(1-\frac{2}{p})}\tau_-^{\frac{5}{2}}\underline{\alpha}|||_{p,e} & & \mathcal{R}_0^{p,S}(\alpha) = |||r^{(\frac{7}{2}-\frac{2}{p})}\alpha|||_{p,e} \\
\mathcal{R}_0^{p,S}(\underline{\beta}) = |||r^{(2-\frac{2}{p})}\tau_-^{\frac{3}{2}}\underline{\beta}|||_{p,e} & & \mathcal{R}_0^{p,S}(\beta) = |||r^{(\frac{7}{2}-\frac{2}{p})}\beta|||_{p,e} \\
\mathcal{R}_0^{p,S}(\rho) = |||r^{(3-\frac{2}{p})}\tau_-^{\frac{1}{2}}(\rho-\bar{\rho})|||_{p,e} & & \mathcal{R}_0^{p,S}(\sigma) = |||r^{(3-\frac{2}{p})}\tau_-^{\frac{1}{2}}(\sigma-\bar{\sigma})|||_{p,e} \\
\mathcal{R}_1^{p,S}(\underline{\alpha}) = |||r^{(2-\frac{2}{p})}\tau_-^{\frac{5}{2}}\not\nabla\underline{\alpha}|||_{p,e} & & \mathcal{R}_1^{p,S}(\alpha) = |||r^{(\frac{9}{2}-\frac{2}{p})}\not\nabla\alpha|||_{p,e} \\
\mathcal{R}_1^{p,S}(\underline{\alpha}_3) = |||r^{(1-\frac{2}{p})}\tau_-^{\frac{7}{2}}\underline{\alpha}_3|||_{p,e} & & \mathcal{R}_1^{p,S}(\alpha_3) = |||r^{(\frac{9}{2}-\frac{2}{p})}\alpha_3|||_{p,e} \\
\mathcal{R}_1^{p,S}(\underline{\alpha}_4) = |||r^{(3-\frac{2}{p})}\tau_-^{\frac{3}{2}}\underline{\alpha}_4|||_{p,e} & & \mathcal{R}_1^{p,S}(\alpha_4) = |||r^{(\frac{9}{2}-\frac{2}{p})}\alpha_4|||_{p,e} \\
\mathcal{R}_1^{p,S}(\underline{\beta}) = |||r^{(3-\frac{2}{p})}\tau_-^{\frac{3}{2}}\not\nabla\underline{\beta}|||_{p,e} & & \mathcal{R}_1^{p,S}(\beta) = |||r^{(\frac{9}{2}-\frac{2}{p})}\not\nabla\beta|||_{p,e} \\
\mathcal{R}_1^{p,S}(\underline{\beta}_3) = |||r^{(2-\frac{2}{p})}\tau_-^{\frac{5}{2}}\underline{\beta}_3|||_{p,e} & & \mathcal{R}_1^{p,S}(\beta_3) = |||r^{(4-\frac{2}{p})}\tau_-^{\frac{1}{2}}\beta_3|||_{p,e} \\
\mathcal{R}_1^{p,S}(\underline{\beta}_4) = |||r^{(4-\frac{2}{p})}\tau_-^{\frac{1}{2}}\underline{\beta}_4|||_{p,e} & & \mathcal{R}_1^{p,S}(\beta_4) = |||r^{(\frac{9}{2}-\frac{2}{p})}\beta_4|||_{p,e} \\
\mathcal{R}_1^{p,S}(\rho) = |||r^{(4-\frac{2}{p})}\tau_-^{\frac{1}{2}}\not\nabla\rho|||_{p,e} & & \mathcal{R}_1^{p,S}(\sigma) = |||r^{(4-\frac{2}{p})}\tau_-^{\frac{1}{2}}\not\nabla\sigma|||_{p,e} \\
\mathcal{R}_1^{p,S}(\rho_3) = |||r^{(3-\frac{2}{p})}\tau_-^{\frac{3}{2}}\rho_3|||_{p,e} & & \mathcal{R}_1^{p,S}(\sigma_3) = |||r^{(3-\frac{2}{p})}\tau_-^{\frac{3}{2}}\sigma_3|||_{p,e} \\
\mathcal{R}_1^{p,S}(\rho_4) = |||r^{(\frac{9}{2}-\frac{2}{p})}\rho_4|||_{p,e} & & \mathcal{R}_1^{p,S}(\sigma_4) = |||r^{(\frac{9}{2}-\frac{2}{p})}\sigma_4|||_{p,e}.
\end{array}
\tag{10.1.4a}
$$

We now set

$$
\begin{aligned}
\mathcal{R}_0^{p,S}[\underline{\alpha}] &= \mathcal{R}_0^{p,S}(\underline{\alpha}) \\
\mathcal{R}_1^{p,S}[\underline{\alpha}] &= \left(\mathcal{R}_1^{p,S}(\underline{\alpha})^2 + \mathcal{R}_0^{p,S}(\underline{\alpha}_3)^2 + \mathcal{R}_0^{p,S}(\underline{\alpha}_4)^2\right)^{1/2}
\end{aligned}
\tag{10.1.4b}
$$

and similarly for all the other null components of $\mathbf{R}$.

Finally we set, for all $q = 0, 1, 2$,

$$
\mathcal{R}_q^{p,S} = \left(\mathcal{R}_q^{p,S}[\underline{\alpha}]^2 + \mathcal{R}_q^{p,S}[\underline{\beta}]^2 + \ldots + \mathcal{R}_q^{p,S}[\alpha]^2\right)^{1/2}
\tag{10.1.4c}
$$

and

$$
\mathcal{R}_{[1]}^{p,S} = \left(\mathcal{R}_0^{p,S} + \mathcal{R}_1^{p,S}\right).
\tag{10.1.4d}
$$

Sometimes, for convenience, given v, one of the components of $\mathbf{R}$ in the exterior region, we will denote by $\mathcal{R}_{[1]}^{p,S}(v)$ the part of $\mathcal{R}_{[1]}^{p,S}$ that involves only the angular derivatives of v.

We will also use the L^∞ norms of the curvature defined as follows:

$$\mathcal{R}_0^\infty = \max\{{}^i\mathcal{R}_0^\infty, \mathcal{R}_0^{\infty,S}\} \tag{10.1.4e}$$

where

$${}^i\mathcal{R}_0^\infty = r_0^{7/2}(\|E\|_{\infty,i} + \|H\|_{\infty,i}). \tag{10.1.4f}$$

10.1.2. *Norms Involving the Second Fundamental Form of the Foliation*

We first define the expressions ${}^i\mathcal{K}_q^p$, ${}^e\mathcal{K}_q^p$ as interior and exterior weighted L^p norms of the q-covariant derivatives of the second fundamental form k. The full expression $\mathcal{K}_q^p$ is then defined by

$$\mathcal{K}_q^p = \max({}^i\mathcal{K}_q^p, {}^e\mathcal{K}_q^p). \tag{10.1.5a}$$

The interior expression ${}^i\mathcal{K}_q^p$, for $1 \leq p \leq \infty$, is simply

$${}^i\mathcal{K}_q^p = r_0^{2+q-\frac{3}{p}}\|\mathbf{D}^q k\|_{p,i}. \tag{10.1.5b}$$

To define ${}^e\mathcal{K}_q^p$, we first recall the decomposition of k relative to the radial foliation of u on Σ_t:

$$\begin{aligned} k_{NN} &= \delta \\ k_{AN} &= \epsilon_A \\ k_{AB} &= \eta_{AB}. \end{aligned} \tag{10.1.5c}$$

Moreover, we have decomposed η into its traceless part $\hat{\eta}$ and trace $\mathrm{tr}(\eta) = -\delta$. We also define

$$\begin{aligned} \delta_3 &= \not{D}_3\delta & \delta_4 &= \not{D}_4\delta \\ \epsilon_3 &= \not{D}_3\epsilon & \epsilon_4 &= \not{D}_4\epsilon \\ \hat{\eta}_3 &= \not{D}_3\hat{\eta} + \tfrac{1}{2}tr(\underline{\chi})\hat{\eta} & \hat{\eta}_4 &= \not{D}_4\hat{\eta} + \tfrac{1}{2}tr(\chi)\hat{\eta} \end{aligned} \tag{10.1.5d}$$

and set

$$\begin{aligned} {}^e\mathcal{K}_q^p(\delta) &= r_0^{-1/2}\|r^{(\frac{5}{2}-\frac{3}{p}+q)}\nabla\!\!\!/^q\delta\|_{p,e} \\ {}^e\mathcal{K}_q^p(\epsilon) &= r_0^{-1/2}\|r^{(\frac{5}{2}-\frac{3}{p}+q)}\nabla\!\!\!/^q\epsilon\|_{p,e} \\ {}^e\mathcal{K}_q^p(\hat{\eta}) &= r_0^{-1/2}\|r^{(1-\frac{2}{p}+q)}w_{(\frac{3}{2}-\frac{1}{p})}\nabla\!\!\!/^q\hat{\eta}\|_{p,e} \end{aligned} \tag{10.1.6a}$$

$$
\begin{aligned}
{}^e\mathcal{K}^p_{q+1}(\delta_4) &= r_0^{1/2}\|r^{(\frac{5}{2}-\frac{3}{p}+q)}\nabla\!\!\!/^q\delta_4\|_{p,e}\\
{}^e\mathcal{K}^p_{q+1}(\delta_3) &= r_0^{1/2}\|r^{(1-\frac{2}{p}+q)}w_{(\frac{3}{2}-\frac{1}{p})}\nabla\!\!\!/^q\delta_3\|_{p,e}\\
{}^e\mathcal{K}^p_{q+1}(\epsilon_4) &= r_0^{-1/2}\|r^{(\frac{7}{2}-\frac{3}{p}+q)}\nabla\!\!\!/^q\epsilon_4\|_{p,e}\\
{}^e\mathcal{K}^p_{q+1}(\epsilon_3) &= r_0^{-1/2}\|r^{(2-\frac{2}{p}+q)}w_{(\frac{3}{2}-\frac{1}{p})}\nabla\!\!\!/^q\epsilon_3\|_{p,e}\\
{}^e\mathcal{K}^p_{q+1}(\hat{\eta}_4) &= r_0^{-1/2}\|r^{(\frac{7}{2}-\frac{3}{p}+q)}\nabla\!\!\!/^q\hat{\eta}_4\|_{p,e}\\
{}^e\mathcal{K}^p_{q+1}(\hat{\eta}_3) &= r_0^{-1/2}\|r^{(1-\frac{2}{p}+q)}w_{(\frac{5}{2}-\frac{1}{p})}\nabla\!\!\!/^q\hat{\eta}_3\|_{p,e} \qquad (10.1.6b)\\
{}^e\mathcal{K}^p_{q+2}(\mathbf{D}_S\delta_4) &= r_0^{1/2}\|r^{(\frac{5}{2}-\frac{3}{p}+q)}\nabla\!\!\!/^q\mathbf{D}_S\delta_4\|_{p,e}\\
{}^e\mathcal{K}^p_{q+2}(\mathbf{D}_S\delta_3) &= r_0^{1/2}\|r^{(1-\frac{2}{p}+q)}w_{(\frac{3}{2}-\frac{1}{p})}\nabla\!\!\!/^q\mathbf{D}_S\delta_3\|_{p,e}\\
{}^e\mathcal{K}^p_{q+2}(\mathbf{D}\!\!\!\!/\,_S\epsilon_4) &= r_0^{-1/2}\|r^{(\frac{7}{2}-\frac{3}{p}+q)}\nabla\!\!\!/^q\mathbf{D}\!\!\!\!/\,_S\epsilon_4\|_{p,e}\\
{}^e\mathcal{K}^p_{q+2}(\mathbf{D}\!\!\!\!/\,_S\epsilon_3) &= r_0^{-1/2}\|r^{(2-\frac{2}{p}+q)}w_{(\frac{3}{2}-\frac{1}{p})}\nabla\!\!\!/^q\mathbf{D}\!\!\!\!/\,_S\epsilon_3\|_{p,e}\\
{}^e\mathcal{K}^p_{q+2}(\mathbf{D}\!\!\!\!/\,_S\hat{\eta}_4) &= r_0^{-1/2}\|r^{(\frac{7}{2}-\frac{3}{p}+q)}\nabla\!\!\!/^q\mathbf{D}\!\!\!\!/\,_S\hat{\eta}_4\|_{p,e}\\
{}^e\mathcal{K}^p_{q+2}(\mathbf{D}\!\!\!\!/\,_S\hat{\eta}_3) &= r_0^{-1/2}\|r^{(1-\frac{2}{p}+q)}w_{(\frac{5}{2}-\frac{1}{p})}\nabla\!\!\!/^q\mathbf{D}\!\!\!\!/\,_S\hat{\eta}_3\|_{p,e} \qquad (10.1.6c)
\end{aligned}
$$

where S is the scaling vectorfield $S = \frac{1}{2}(ve_4 - ue_3)$, with $v = 2r - u$.

We now set

$$
\begin{aligned}
{}^e\mathcal{K}^p_0[\delta] &= {}^e\mathcal{K}^p_0(\delta)\\
{}^e\mathcal{K}^p_1[\delta] &= {}^e\mathcal{K}^p_1(\delta) + {}^e\mathcal{K}^p_1(\delta_3) + {}^e\mathcal{K}^p_1(\delta_4)\\
{}^e\mathcal{K}^p_2[\delta] &= {}^e\mathcal{K}^p_2(\delta) + {}^e\mathcal{K}^p_2(\delta_3) + {}^e\mathcal{K}^p_2(\delta_4)\\
{}^e\mathcal{K}^p_3[\delta] &= {}^e\mathcal{K}^p_3(\delta) + {}^e\mathcal{K}^p_3(\delta_3) + {}^e\mathcal{K}^p_3(\delta_4) \qquad (10.1.6d)
\end{aligned}
$$

and similarly for $\epsilon, \hat{\eta}$. The exterior expressions ${}^e\mathcal{K}^p_q$ are thus defined, for all $q = 0, 1, 2, 3$, by

$$
{}^e\mathcal{K}^p_q = {}^e\mathcal{K}^p_q[\delta] + {}^e\mathcal{K}^p_q[\epsilon] + {}^e\mathcal{K}^p_q[\hat{\eta}]. \qquad (10.1.6e)
$$

For $p = 2$, which is the case of interest for us, we write simply ${}^e\mathcal{K}_q = {}^e\mathcal{K}^2_q$. We also set

$$
\begin{aligned}
{}^e\mathcal{K}^*_2 = {}&{}^e\mathcal{K}^2_2(\mathbf{D}_S\delta_4) + {}^e\mathcal{K}^2_2(\mathbf{D}_S\delta_3)\\
&+ {}^e\mathcal{K}^2_2(\mathbf{D}\!\!\!\!/\,_S\epsilon_4) + {}^e\mathcal{K}^2_2(\mathbf{D}\!\!\!\!/\,_S\epsilon_3)\\
&+ {}^e\mathcal{K}^2_2(\mathbf{D}\!\!\!\!/\,_S\hat{\eta}_4) + {}^e\mathcal{K}^2_2(\mathbf{D}\!\!\!\!/\,_S\hat{\eta}_3). \qquad (10.1.6f)
\end{aligned}
$$

We also define the following expressions, in the exterior region:

$$
\mathcal{K}^{p,S}_q(\delta) = r_0^{-1/2}|||r^{(\frac{5}{2}-\frac{2}{p}+q)}\nabla\!\!\!/^q\delta|||_{p,e}
$$

$$\mathcal{K}_q^{p,S}(\epsilon) = r_0^{-1/2}|||r^{(\frac{5}{2}-\frac{2}{p}+q)}\nabla^q\epsilon|||_{p,e}$$
$$\mathcal{K}_q^{p,S}(\hat{\eta}) = r_0^{-1/2}|||r^{(1-\frac{2}{p}+q)}w_{\frac{3}{2}}\nabla^q\hat{\eta}|||_{p,e} \tag{10.1.7a}$$

$$\mathcal{K}_{q+1}^{p,S}(\delta_4) = |||r^{(3-\frac{2}{p}+q)}\nabla^q\delta_4|||_{p,e}$$
$$\mathcal{K}_{q+1}^{p,S}(\delta_3) = |||r^{(\frac{3}{2}-\frac{2}{p}+q)}w_{\frac{3}{2}}\nabla^q\delta_3|||_{p,e}$$
$$\mathcal{K}_{q+1}^{p,S}(\epsilon_4) = r_0^{-1/2}|||r^{(\frac{7}{2}-\frac{2}{p}+q)}\nabla^q\epsilon_4|||_{p,e}$$
$$\mathcal{K}_{q+1}^{p,S}(\epsilon_3) = r_0^{-1/2}|||r^{(2-\frac{2}{p}+q)}w_{\frac{3}{2}}\nabla^q\epsilon_3|||_{p,e}$$
$$\mathcal{K}_{q+1}^{p,S}(\hat{\eta}_4) = r_0^{-1/2}|||r^{(\frac{7}{2}-\frac{2}{p}+q)}\nabla^q\hat{\eta}_4|||_{p,e}$$
$$\mathcal{K}_{q+1}^{p,S}(\hat{\eta}_3) = r_0^{-1/2}|||r^{(1-\frac{2}{p}+q)}w_{\frac{5}{2}}\nabla^q\hat{\eta}_3|||_{p,e} \tag{10.1.7b}$$

and set

$$\mathcal{K}_0^{p,S}[\delta] = \mathcal{K}_0^{p,S}(\delta)$$
$$\mathcal{K}_1^{p,S}[\delta] = \mathcal{K}_1^{p,S}(\delta) + \mathcal{K}_1^{p,S}(\delta_3) + \mathcal{K}_1^{p,S}(\delta_4) \tag{10.1.7c}$$

and similarly for $\epsilon, \hat{\eta}$. We then define, for $q = 0, 1$,

$$\mathcal{K}_q^{p,S} = \mathcal{K}_q^{p,S}[\delta] + \mathcal{K}_q^{p,S}[\epsilon] + \mathcal{K}_q^{p,S}[\hat{\eta}]. \tag{10.1.7d}$$

We are finally ready to define the basic space-time norms of the second fundamental form

$${}^{\imath}\mathcal{K}_{[0]} = {}^{\imath}\mathcal{K}_0 \tag{10.1.8a}$$
$${}^{\imath}\mathcal{K}_{[1]} = {}^{\imath}\mathcal{K}_{[0]} + {}^{\imath}\mathcal{K}_1 \tag{10.1.8b}$$
$${}^{\imath}\mathcal{K}_{[2]} = {}^{\imath}\mathcal{K}_{[1]} + {}^{\imath}\mathcal{K}_1 \tag{10.1.8c}$$
$${}^{\imath}\mathcal{K}_{[3]} = {}^{\imath}\mathcal{K}_{[2]} + {}^{\imath}\mathcal{K}_3 \tag{10.1.8d}$$
$${}^{e}\mathcal{K}_{[0]} = {}^{e}\mathcal{K}_0 \tag{10.1.8e}$$
$${}^{e}\mathcal{K}_{[1]} = {}^{e}\mathcal{K}_{[0]} + {}^{e}\mathcal{K}_1 + \mathcal{K}_0^{3,S} \tag{10.1.8f}$$
$${}^{e}\mathcal{K}_{[2]} = {}^{e}\mathcal{K}_{[1]} + {}^{e}\mathcal{K}_2 + \mathcal{K}_1^{3,S} \tag{10.1.8g}$$
$${}^{e}\mathcal{K}_{[3]} = {}^{e}\mathcal{K}_{[2]} + {}^{e}\mathcal{K}_3 + \mathcal{K}_2^{3,S} \tag{10.1.8h}$$
$${}^{e}\mathcal{K}_{[3]}^{*} = {}^{e}\mathcal{K}_{[3]} + {}^{e}\mathcal{K}_2{}^{*} \tag{10.1.8i}$$
$$\mathcal{K}_{[q]} = {}^{\imath}\mathcal{K}_{[q]} + {}^{e}\mathcal{K}_{[q]} \tag{10.1.8j}$$

For convenience, given v, one of the components of k in the exterior region, we denote by ${}^{e}\mathcal{K}_{[q]}(v)$ the part of ${}^{e}\mathcal{K}_{[q]}$ that contains only angular derivatives of v.

We will also make use of the L^∞ norms:

$$\mathcal{K}_0^\infty = \max\{{}^\imath\mathcal{K}_0^\infty, \mathcal{K}_0^{\infty,S}\} \tag{10.1.8k}$$

$$\mathcal{K}_1^\infty = \max\{{}^\imath\mathcal{K}_1^\infty, \mathcal{K}_1^{\infty,S}\}. \tag{10.1.8l}$$

10.1.3. *Norms Involving the Lapse Function ϕ*

As for the curvature and second fundamental form, we first define the expressions ${}^\imath\mathcal{L}_q^p$, ${}^e\mathcal{L}_q^p$ as interior and exterior weighted L^p norms of the q+1-covariant derivatives of the logarithm φ of the lapse function ϕ. The full expression $\mathcal{L}_q^p$ is then defined by

$$\mathcal{L}_q^p = \max({}^\imath\mathcal{L}_q^p, {}^e\mathcal{L}_q^p). \tag{10.1.9a}$$

The interior expression ${}^\imath\mathcal{L}_q^p$ is simply

$${}^\imath\mathcal{L}_q^p = r_0^{2+q-\frac{3}{p}} \|\mathbf{D}^{q+1}(\varphi)\|_{p,\imath}. \tag{10.1.9b}$$

To define ${}^e\mathcal{L}_q^p$, we first introduce the following decomposition of $\nabla\varphi$:

$$\not{\phi}_A = \nabla\!\!\!/_A\varphi \tag{10.1.9c}$$

$$\varphi_N = \nabla_N\varphi. \tag{10.1.9d}$$

We now set

$$\begin{aligned}
{}^e\mathcal{L}_q^p(\not{\phi}) &= r_0^{-1/2}\|r^{(\frac{5}{2}-\frac{3}{p}+q)}\nabla\!\!\!/^q\not{\phi}\|_{p,e}\\
{}^e\mathcal{L}_q^p(\varphi_N) &= r_0^{1/2}\|r^{(\frac{3}{2}-\frac{3}{p}+q)}\nabla\!\!\!/^q\varphi_N\|_{p,e}\\
{}^e\mathcal{L}_{q+1}^p(\mathbf{D}\!\!\!\!/_4\not{\phi}) &= r_0^{-1/2}\|r^{(\frac{7}{2}-\frac{3}{p}+q)}\nabla\!\!\!/^q\mathbf{D}\!\!\!\!/_4\not{\phi}\|_{p,e}\\
{}^e\mathcal{L}_{q+1}^p(\mathbf{D}\!\!\!\!/_3\not{\phi}) &= r_0^{-1/2}\|r^{(\frac{7}{2}-\frac{3}{p}+q)}\nabla\!\!\!/^q\mathbf{D}\!\!\!\!/_3\not{\phi}\|_{p,e}\\
{}^e\mathcal{L}_{q+1}^p(\mathbf{D}_4\varphi_N) &= r_0^{3/2}\|r^{(\frac{3}{2}-\frac{3}{p}+q)}\nabla\!\!\!/^q\mathbf{D}_4\varphi_N\|_{p,e}\\
{}^e\mathcal{L}_{q+1}^p(\mathbf{D}_3\varphi_N) &= r_0^{3/2}\|r^{(\frac{3}{2}-\frac{3}{p}+q)}w_{(\frac{3}{2}-\frac{1}{p})}\nabla\!\!\!/^q\mathbf{D}_3\varphi_N\|_{p,e}
\end{aligned} \tag{10.1.9e}$$

$$\begin{aligned}
{}^e\mathcal{L}_{q+1}^p(\mathbf{D}\!\!\!\!/_S\not{\phi}) &= r_0^{-1/2}\|r^{(\frac{5}{2}-\frac{3}{p}+q)}\nabla\!\!\!/^q\not{\phi}\|_{p,e}\\
{}^e\mathcal{L}_{q+1}^p(\mathbf{D}_S\varphi_N) &= r_0^{1/2}\|r^{(\frac{3}{2}-\frac{3}{p}+q)}\nabla\!\!\!/^q\varphi_N\|_{p,e}\\
{}^e\mathcal{L}_{q+2}^p(\mathbf{D}\!\!\!\!/_S\mathbf{D}\!\!\!\!/_4\not{\phi}) &= r_0^{-1/2}\|r^{(\frac{7}{2}-\frac{3}{p}+q)}\nabla\!\!\!/^q\mathbf{D}\!\!\!\!/_4\not{\phi}\|_{p,e}\\
{}^e\mathcal{L}_{q+2}^p(\mathbf{D}\!\!\!\!/_S\mathbf{D}\!\!\!\!/_3\not{\phi}) &= r_0^{-1/2}\|r^{(\frac{7}{2}-\frac{3}{p}+q)}\nabla\!\!\!/^q\mathbf{D}\!\!\!\!/_3\not{\phi}\|_{p,e}\\
{}^e\mathcal{L}_{q+2}^p(\mathbf{D}_S\mathbf{D}_4\varphi_N) &= r_0^{3/2}\|r^{(\frac{3}{2}-\frac{3}{p}+q)}\nabla\!\!\!/^q\mathbf{D}_4\varphi_N\|_{p,e}\\
{}^e\mathcal{L}_{q+2}^p(\mathbf{D}_S\mathbf{D}_3\varphi_N) &= r_0^{3/2}\|r^{(\frac{3}{2}-\frac{3}{p}+q)}w_{(\frac{3}{2}-\frac{1}{p})}\nabla\!\!\!/^q\mathbf{D}_3\varphi_N\|_{p,e}.
\end{aligned} \tag{10.1.9f}$$

We now set

$$
\begin{aligned}
{}^e\mathcal{L}_0^p[\not\phi] &= {}^e\mathcal{L}_0^p(\not\phi) \\
{}^e\mathcal{L}_1^p[\not\phi] &= {}^e\mathcal{L}_1^p(\not\phi) + {}^e\mathcal{L}_1^p(\not\mathbf{D}_3\not\phi) + {}^e\mathcal{L}_1^p(\not\mathbf{D}_4\not\phi) \\
{}^e\mathcal{L}_2^p[\not\phi] &= {}^e\mathcal{L}_2^p(\not\phi) + {}^e\mathcal{L}_2^p(\not\mathbf{D}_3\not\phi) + {}^e\mathcal{L}_2^p(\not\mathbf{D}_4\not\phi) \\
{}^e\mathcal{L}_3^p[\not\phi] &= {}^e\mathcal{L}_3^p(\not\phi) + {}^e\mathcal{L}_3^p(\not\mathbf{D}_3\not\phi) + {}^e\mathcal{L}_3^p(\not\mathbf{D}_4\not\phi)
\end{aligned}
\tag{10.1.9g}
$$

and similarly for φ_N.

The exterior expressions ${}^e\mathcal{L}_q^p$ are thus defined, for all $q = 0, 1, 2, 3$, by

$$
{}^e\mathcal{L}_q^p = {}^e\mathcal{L}_q^p[\not\phi] + {}^e\mathcal{L}_q^p[\varphi_N]. \tag{10.1.9h}
$$

For $p = 2$, which is the case of interest for us, we write simply ${}^e\mathcal{L}_q = {}^e\mathcal{L}_q^2$.

We also set

$$
\begin{aligned}
{}^e\mathcal{L}_2^* = {}^e\mathcal{L}_2^2(\not\mathbf{D}_S\not\mathbf{D}_4\not\phi) + {}^e\mathcal{L}_2^2(\not\mathbf{D}_S\not\mathbf{D}_3\not\phi) \\
+ {}^e\mathcal{L}_2^2(\mathbf{D}_S\mathbf{D}_4\varphi_N) + {}^e\mathcal{L}_2^2(\mathbf{D}_S\mathbf{D}_3\varphi_N).
\end{aligned}
\tag{10.1.9i}
$$

We also define the following norms:

$$
\mathcal{L}_q^{p,S}(\not\phi) = r_0^{-1/2}|||r^{(\frac{5}{2}-\frac{2}{p}+q)}\not\nabla^q\not\phi|||_{p,e} \tag{10.1.10a}
$$

$$
\mathcal{L}_q^{p,S}(\varphi_N) = |||r^{(2-\frac{2}{p}+q)}\not\nabla^q\varphi_N|||_{p,e} \tag{10.1.10b}
$$

$$
\mathcal{L}_q^{p,S}(\not\mathbf{D}_4\not\phi) = r_0^{-1/2}|||r^{(\frac{7}{2}-\frac{2}{p}+q)}\not\nabla^q\not\mathbf{D}_4\not\phi|||_{p,e} \tag{10.1.10c}
$$

$$
\mathcal{L}_q^{p,S}(\not\mathbf{D}_3\not\phi) = r_0^{-1/2}|||r^{(\frac{7}{2}-\frac{2}{p}+q)}\not\nabla^q\not\mathbf{D}_3\not\phi|||_{p,e} \tag{10.1.10d}
$$

$$
\mathcal{L}_q^{p,S}(\mathbf{D}_4\varphi_N) = r_0|||r^{(2-\frac{2}{p}+q)}\not\nabla^q\mathbf{D}_4\varphi_N|||_{p,e} \tag{10.1.10e}
$$

$$
\mathcal{L}_q^{p,S}(\mathbf{D}_3\varphi_N) = r_0|||r^{(\frac{1}{2}-\frac{2}{p}+q)}w_{\frac{3}{2}}\not\nabla^q\mathbf{D}_3\varphi_N|||_{p,e}, \tag{10.1.10f}
$$

and set

$$
\begin{aligned}
\mathcal{L}_0^{p,S}[\not\phi] &= \mathcal{L}_0^{p,S}(\not\phi) \\
\mathcal{L}_1^{p,S}[\not\phi] &= \mathcal{L}_1^{p,S}(\not\phi) + \mathcal{L}_1^{p,S}(\not\mathbf{D}_3\not\phi) + \mathcal{L}_1^{p,S}(\not\mathbf{D}_4\not\phi),
\end{aligned}
\tag{10.1.10g}
$$

and similarly for φ_N. We then define, for $q = 0, 1$,

$$
\mathcal{L}_q^{p,S} = \mathcal{L}_q^{p,S}[\not\phi] + \mathcal{L}_q^{p,S}[\varphi_N]. \tag{10.1.10h}
$$

We are finally ready to define the basic space-time norms of the lapse function:

$$
{}^\imath\mathcal{L}_{[0]} = {}^\imath\mathcal{L}_{[0]}
$$

$$
\begin{aligned}
{}^{\imath}\mathcal{L}_{[1]} &= {}^{\imath}\mathcal{L}_{[0]} + {}^{\imath}\mathcal{L}_1 \\
{}^{\imath}\mathcal{L}_{[2]} &= {}^{\imath}\mathcal{L}_{[1]} + {}^{\imath}\mathcal{L}_2 \\
{}^{\imath}\mathcal{L}_{[3]} &= {}^{\imath}\mathcal{L}_{[2]} + {}^{\imath}\mathcal{L}_3
\end{aligned}
\tag{10.1.11a}
$$

$$
\begin{aligned}
{}^{e}\mathcal{L}_{[0]} &= {}^{e}\mathcal{L}_{[0]} \\
{}^{e}\mathcal{L}_{[1]} &= {}^{e}\mathcal{L}_{[0]} + {}^{e}\mathcal{L}_1 + \mathcal{L}_0^{3,S} \\
{}^{e}\mathcal{L}_{[2]} &= {}^{e}\mathcal{L}_{[1]} + {}^{e}\mathcal{L}_2 + \mathcal{L}_1^{3,S} \\
{}^{e}\mathcal{L}_{[3]} &= {}^{e}\mathcal{L}_{[2]} + {}^{e}\mathcal{L}_3 \\
{}^{e}\mathcal{L}^*_{[3]} &= {}^{e}\mathcal{L}_{[3]} + {}^{e}\mathcal{L}^*_2
\end{aligned}
\tag{10.1.11b}
$$

and

$$
\mathcal{L}_{[q]} = {}^{\imath}\mathcal{L}_{[q]} + {}^{e}\mathcal{L}_{[q]}. \tag{10.1.11c}
$$

Also, we will make use of the L^∞ norms

$$
\mathcal{L}_0^\infty = \max\{{}^{\imath}\mathcal{L}_0^\infty, \mathcal{L}_0^{\infty,S}\} \tag{10.1.11d}
$$

$$
\mathcal{K}_1^\infty = \max\{{}^{\imath}\mathcal{L}_1^\infty, \mathcal{L}_1^{\infty,S}\}. \tag{10.1.11e}
$$

10.1.4. *Norms Involving the Hessian of the Optical Function*

Let U be the Hessian of the optical function u, $\mathbf{D}^2u$. As before, we define the interior and exterior expressions ${}^{\imath}\mathcal{O}_q, {}^{e}\mathcal{O}_q$ and set

$$
\mathcal{O}_q = \max\{{}^{e}\mathcal{O}_q, {}^{\imath}\mathcal{O}_q\}. \tag{10.1.12}
$$

In the interior region the components of $\mathbf{D}^2u$ can be described, relative to the standard null frame, in terms of $\chi' = a\chi, \zeta' = \zeta, \omega' = a^{-1}\omega$, and $\underline{\Delta} = -\frac{1}{2}g^{\mu\nu}\frac{\partial u}{\partial x^\mu}\frac{\partial u}{\partial x^\nu}$. Remark that $\underline{\Delta} = 0$ everywhere except the matching region.
We now define

$$
\begin{aligned}
{}^{\imath}\mathcal{O}_q^p(\mathrm{tr}\chi' - \overline{\mathrm{tr}\chi'}) &= r_0^{\frac{3}{2}+q-\frac{3}{p}} \sum_{\imath+\jmath+k=q} \|\nabla\!\!\!/^{\imath}\mathbf{D}_3^{\jmath}\mathbf{D}_4^k(\mathrm{tr}\chi' - \overline{\mathrm{tr}\chi'})\|_{p,\imath} \\
{}^{\imath}\mathcal{O}_q^p(\hat{\chi}') &= r_0^{\frac{3}{2}+q-\frac{3}{p}} \sum_{\imath+\jmath+k=q} \|\nabla\!\!\!/^{\imath}\mathbf{D}\!\!\!\!/_3^{\jmath}\mathbf{D}\!\!\!\!/_4^k\hat{\chi}'\|_{p,\imath} \\
{}^{\imath}\mathcal{O}_q^p(\zeta') &= r_0^{\frac{3}{2}+q-\frac{3}{p}} \sum_{\imath+\jmath+k=q} \|\nabla\!\!\!/^{\imath}\mathbf{D}_3^{\jmath}\mathbf{D}_4^k\zeta'\|_{p,\imath} \\
{}^{\imath}\mathcal{O}_q^p(\omega') &= r_0^{\frac{3}{2}+q-\frac{3}{p}} \sum_{\imath+\jmath+k=q} \|\nabla\!\!\!/^{\imath}\mathbf{D}_3^{\jmath}\mathbf{D}_4^k\omega\|_{p,\imath} \\
{}^{\imath}\mathcal{O}_{q+1}^p(\underline{\Delta}) &= r_0^{\frac{3}{2}+q-\frac{3}{p}} \sum_{\imath+\jmath+k=q+1} \|\nabla\!\!\!/^{\imath}\mathbf{D}_3^{\jmath}\mathbf{D}_4^k\underline{\Delta}\|_{p,\imath}.
\end{aligned}
\tag{10.1.13a}
$$

$$
{}^{\imath}\mathcal{O}^p_q = {}^{\imath}\mathcal{O}^p_q(\mathrm{tr}\chi' - \overline{\mathrm{tr}\chi'}) + {}^{\imath}\mathcal{O}^p_q(\hat{\chi}') + {}^{\imath}\mathcal{O}^p_q(\zeta')
+ {}^{\imath}\mathcal{O}^p_{q+1}(\underline{\Delta}). \tag{10.1.13b}
$$

As usual, we write ${}^{\imath}\mathcal{O}_q = {}^{\imath}\mathcal{O}^2_q$.

To define the exterior norms of the optical function, we use, instead of the standard null pair, the l-pair and the corresponding null frame. We recall that in the exterior region the Hessian of the optical function was decomposed, relative to the l-null frame, into the components $\mathrm{tr}\chi$, $\hat{\chi}$, ζ, and ω. We also introduce the following:

$$
\begin{aligned}
\mathrm{tr}\chi_3 &= \mathbf{D}_3\mathrm{tr}\chi + \tfrac{1}{2}\mathrm{tr}\underline{\chi}\mathrm{tr}\chi & \mathrm{tr}\chi_4 &= \mathbf{D}_4\mathrm{tr}\chi + \tfrac{1}{2}\mathrm{tr}\chi\mathrm{tr}\chi \\
\hat{\chi}_3 &= \not{\mathbf{D}}_3\hat{\chi} + tr\underline{\chi}\hat{\chi} & \hat{\chi}_4 &= \not{\mathbf{D}}_4\hat{\chi} + tr\chi\hat{\chi} \\
\zeta_3 &= \not{\mathbf{D}}_3\zeta + tr\underline{\chi}\zeta & \hat{\chi}_4 &= \not{\mathbf{D}}_4\hat{\chi} + \tfrac{1}{2}tr\chi\zeta \\
\omega_3 &= \mathbf{D}_3\omega & \omega_4 &= \mathbf{D}_4\omega.
\end{aligned} \tag{10.1.14a}
$$

We now define

$$
\begin{aligned}
\mathcal{O}^{p,S}_q(tr\chi - \overline{tr\chi}) &= |||r^{(2+q-\frac{2}{p})}\nabla^q(\mathrm{tr}\chi - \overline{tr\chi})|||_{p,e} \\
\mathcal{O}^{p,S}_q(\hat{\chi}) &= |||r^{(2+q-\frac{2}{p})}\nabla^q(\hat{\chi})|||_{p,e} \\
\mathcal{O}^{p,S}_q(\chi) &= \max\{\mathcal{O}^{p,S}_q(tr\chi - \overline{tr\chi}),\ \mathcal{O}^{p,S}_q(\hat{\chi})\} \\
\mathcal{O}^{p,S}_q(\zeta) &= |||r^{(2+q-\frac{2}{p})}\nabla^q\zeta|||_{p,e} \\
\mathcal{O}^{p,S}_q(\omega) &= |||r^{(\frac{1}{2}+q-\frac{2}{p})}w_{3/2}\nabla^q\omega|||_{p,e}
\end{aligned} \tag{10.1.14b}
$$

$$
\begin{aligned}
\mathcal{O}^{p,S}_{q+1}(\mathrm{tr}\chi_3) &= |||r^{(\frac{3}{2}+q-\frac{2}{p})}w_{3/2}\nabla^q(\mathrm{tr}\chi_3)|||_{p,e} \\
\mathcal{O}^{p,S}_{q+1}(\mathrm{tr}\chi_4) &= |||r^{(3+q-\frac{2}{p})}\nabla^q(\mathrm{tr}\chi_4)|||_{p,e} \\
\mathcal{O}^{p,S}_{q+1}(\hat{\chi}_3) &= |||r^{(3+q-\frac{2}{p})}\nabla^q\hat{\chi}_3|||_{p,e} \\
\mathcal{O}^{p,S}_{q+1}(\hat{\chi}_4) &= |||r^{(3+q-\frac{2}{p})}\nabla^q\hat{\chi}_4|||_{p,e} \\
\mathcal{O}^{p,S}_{q+1}(\zeta_3) &= |||r^{(\frac{3}{2}+q-\frac{2}{p})}w_{3/2}\nabla^q\zeta_3|||_{p,e} \\
\mathcal{O}^{p,S}_{q+1}(\zeta_4) &= |||r^{(3+q-\frac{2}{p})}\nabla^q\zeta_4|||_{p,e} \\
\mathcal{O}^{p,S}_{q+1}(\omega_3) &= |||r^{(\frac{1}{2}+q-\frac{2}{p})}w_{5/2}\nabla^q\mathbf{D}_3\omega|||_{p,e} \\
\mathcal{O}^{p,S}_{q+1}(\omega_4) &= |||r^{(3+q-\frac{2}{p})}\nabla^q\omega_4|||_{p,e}.
\end{aligned} \tag{10.1.14c}
$$

We now set

$$
\begin{aligned}
\mathcal{O}_0^{p,S}[\mathrm{tr}\chi] &= \mathcal{O}_0^{p,S}(\mathrm{tr}\chi)\\
\mathcal{O}_1^{p,S}[\mathrm{tr}\chi] &= \mathcal{O}_1^{p,S}(\mathrm{tr}\chi) + \mathcal{O}_1^{p,S}(\mathrm{tr}\chi_3) + \mathcal{O}_1^{p,S}(\mathrm{tr}\chi_4)\\
\mathcal{O}_2^{p,S}[\mathrm{tr}\chi] &= \mathcal{O}_2^{p,S}(\mathrm{tr}\chi) + \mathcal{O}_2^{p,S}(\mathrm{tr}\chi_3) + \mathcal{O}_2^{p,S}(\mathrm{tr}\chi_4)
\end{aligned}
\tag{10.1.14d}
$$

and similarly for $\hat{\chi}$, ζ, and ω.

The exterior expressions $\mathcal{O}_q^{p,S}$, $q = 0, 1, 2$, are now defined by

$$
\mathcal{O}_q^{p,S} = \mathcal{O}_q^{p,S}[\mathrm{tr}\chi] + \mathcal{O}_q^{p,S}[\hat{\chi}] + \mathcal{O}_q^{p,S}[\zeta] + \mathcal{O}_q^{p,S}[\omega]. \tag{10.1.14e}
$$

We also introduce the following expressions for $q = 1, 2, 3$:

$$
\begin{aligned}
{}^e\mathcal{O}_q(\chi) &= {}^{(\mathrm{loc})}\|r^{1+q}\nabla\!\!\!/^q\chi\|_{2,e}\\
{}^e\mathcal{O}_q(\zeta) &= {}^{(\mathrm{loc})}\|r^{1+q}\nabla\!\!\!/^q\zeta\|_{2,e},
\end{aligned}
\tag{10.1.15a}
$$

and, for $q = 1, 2$,

$$
{}^e\mathcal{O}_q(\omega) = {}^{(\mathrm{loc})}\|r^{-1/2+q}w_{3/2}\nabla\!\!\!/^q\omega\|_{2,e}, \tag{10.1.15b}
$$

while for $q = 3$,

$$
{}^e\mathcal{O}_3(\omega) = r_0^{-1/2}\|r^3 w_1\nabla\!\!\!/^3\omega\|_{2,e}^2. \tag{10.1.15c}
$$

Also, for $q = 0, 1, 2$,

$$
\begin{aligned}
{}^e\mathcal{O}_{q+1}(\mathrm{tr}\chi_3) &= {}^{(\mathrm{loc})}\|r^{(\frac{1}{2}+q)}w_{3/2}\nabla\!\!\!/^q(\mathrm{tr}\chi_3)\|_{2,e}\\
{}^e\mathcal{O}_{q+1}(\mathrm{tr}\chi_4) &= {}^{(\mathrm{loc})}\|r^{2+q}\nabla\!\!\!/^q(\mathrm{tr}\chi_4)\|_{2,e}\\
{}^e\mathcal{O}_{q+1}(\hat{\chi}_3) &= {}^{(\mathrm{loc})}\|r^{2+q}\nabla\!\!\!/^q(\mathrm{tr}\chi_3)\|_{2,e}\\
{}^e\mathcal{O}_{q+1}(\mathrm{tr}\chi_4) &= {}^{(\mathrm{loc})}\|r^{2+q}\nabla\!\!\!/^q(\mathrm{tr}\chi_4)\|_{2,e},
\end{aligned}
\tag{10.1.15d}
$$

and for $q = 0, 1$,

$$
\begin{aligned}
{}^e\mathcal{O}_{q+1}(\zeta_3) &= {}^{(\mathrm{loc})}\|r^{(\frac{1}{2}+q)}w_{3/2}\nabla\!\!\!/^q\zeta_3\|_{2,e}\\
{}^e\mathcal{O}_{q+1}(\zeta_4) &= {}^{(\mathrm{loc})}\|r^{2+q}\nabla\!\!\!/^q\zeta_4\|_{2,e}\\
{}^e\mathcal{O}_{q+1}(\omega_3) &= {}^{(\mathrm{loc})}\|r^{(-\frac{1}{2}+q)}w_{5/2}\nabla\!\!\!/^q\mathbf{D}_3\omega\|_{2,e}\\
{}^e\mathcal{O}_{q+1}(\omega_4) &= {}^{(\mathrm{loc})}\|r^{2+q}\nabla\!\!\!/^q\omega_4\|_{2,e},
\end{aligned}
\tag{10.1.15e}
$$

while for $q = 2$,

$$
\begin{aligned}
{}^e\mathcal{O}_3(\zeta_3) &= r_0^{-1/2}\|r^3 w_1 \nabla\!\!\!/^2 \zeta_3\|_{2,e}^2 & {}^e\mathcal{O}_3(\zeta_4) &= {}^{(\mathrm{loc})}\|r^{2+q}\nabla\!\!\!/^2\zeta_4\|_{2,e} \\
{}^e\mathcal{O}_3(\omega_3) &= r_0^{-1/2}\|r^2 w_2 \nabla\!\!\!/^2 \omega_3\|_{2,e}^2 & {}^e\mathcal{O}_3(\omega_4) &= {}^{(\mathrm{loc})}\|r^4\nabla\!\!\!/^2\omega_4\|_{2,e}.
\end{aligned}
\tag{10.1.15f}
$$

We now set

$$
\begin{aligned}
{}^e\mathcal{O}_0[\mathrm{tr}\chi] &= {}^e\mathcal{O}_0(\mathrm{tr}\chi) \\
{}^e\mathcal{O}_1[\mathrm{tr}\chi] &= {}^e\mathcal{O}_1(\mathrm{tr}\chi) + {}^e\mathcal{O}_1(\mathrm{tr}\chi_3) + {}^e\mathcal{O}_1(\mathrm{tr}\chi_4) \\
{}^e\mathcal{O}_2[\mathrm{tr}\chi] &= {}^e\mathcal{O}_2(\mathrm{tr}\chi) + {}^e\mathcal{O}_2(\mathrm{tr}\chi_3) + {}^e\mathcal{O}_2(\mathrm{tr}\chi_4) \\
{}^e\mathcal{O}_3[\mathrm{tr}\chi] &= {}^e\mathcal{O}_3(\mathrm{tr}\chi) + {}^e\mathcal{O}_3(\mathrm{tr}\chi_3) + {}^e\mathcal{O}_3(\mathrm{tr}\chi_4)
\end{aligned}
\tag{10.1.15g}
$$

and similarly for $\hat{\chi}$, ζ, and ω.

The exterior expressions ${}^e\mathcal{O}_q$, $q = 0, 1, 2$ are defined by

$$
{}^e\mathcal{O}_q = {}^e\mathcal{O}_q[\mathrm{tr}\chi] + {}^e\mathcal{O}_q[\hat{\chi}] + {}^e\mathcal{O}_q[\zeta] + {}^e\mathcal{O}_q[\omega]. \tag{10.1.15h}
$$

We are finally ready to define our basic norms involving $\mathbf{D}^2 u$:

$$
\begin{aligned}
{}^i\mathcal{O}_{[0]}^{\infty} &= {}^i\mathcal{O}_0^{\infty} + r_0^{3/2}\sup_{\Sigma^i}|\overline{\mathrm{tr}\chi} - \frac{2}{r}| + r_0^{1/2}\sup_{I}|a-1| \\
{}^i\mathcal{O}_{[1]}^{\infty} &= {}^i\mathcal{O}_{[0]}^{\infty} + {}^i\mathcal{O}_1^{\infty} \\
{}^i\mathcal{O}_{[2]} &= {}^i\mathcal{O}_2 + {}^i\mathcal{O}_{[0]}^{\infty} \\
{}^i\mathcal{O}_{[3]} &= {}^i\mathcal{O}_{[2]} + {}^i\mathcal{O}_3 + {}^i\mathcal{O}_{[1]}^{\infty}
\end{aligned}
\tag{10.1.16a}
$$

$$
\begin{aligned}
{}^e\mathcal{O}_{[0]}^{\infty} &= {}^e\mathcal{O}_0^{\infty} + \sup_{r\geq\frac{r_0}{2}} r^2|\overline{\mathrm{tr}\chi} - \frac{2}{r}| + \sup_{r\geq\frac{r_0}{2}} r|a-1| \\
{}^e\mathcal{O}_{[1]}^{\infty} &= {}^e\mathcal{O}_{[0]}^{\infty} + {}^e\mathcal{O}_1^{\infty} \\
{}^e\mathcal{O}_{[2]} &= {}^e\mathcal{O}_2 + {}^e\mathcal{O}_{[0]}^{\infty} + \mathcal{O}_1^{4,S} \\
{}^e\mathcal{O}_{[3]} &= {}^e\mathcal{O}_{[2]} + {}^e\mathcal{O}_3 + {}^e\mathcal{O}_{[1]}^{\infty} + \mathcal{O}_2^{4,S}
\end{aligned}
\tag{10.1.16b}
$$

$$
\begin{aligned}
\mathcal{O}_{[q]}^{\infty} &= {}^i\mathcal{O}_{[q]}^{\infty} + {}^e\mathcal{O}_{[q]}^{\infty} \\
\mathcal{O}_{[2]} &= {}^i\mathcal{O}_{[2]} + {}^e\mathcal{O}_{[2]}
\end{aligned}
\tag{10.1.16c}
$$

where

$$
{}^e\mathcal{O}_q^{\infty} = \mathcal{O}_q^{\infty,S}.
$$

For convenience, given v, one of the components of U in the exterior region, we will denote by $\mathcal{O}^{\infty}_{[q]}(v), \mathcal{O}_{[q]}(v)$ the parts of $\mathcal{O}^{\infty}_{[q]}, \mathcal{O}_{[q]}$ that contain only angular derivatives of v.

10.2. Statement and Proof of the Main Theorem

We are now ready to state the final and complete version[3] of our main theorem:

Theorem 10.2.1 (Third Version of the Main Theorem:) Any strongly asymptotically flat, maximal initial data set that satisfies the global smallness assumption, stated in the introduction, leads to a unique, globally hyperbolic, smooth, and geodesically complete solution of the Einstein-Vacuum equations, which is foliated by a normal maximal time function t, defined for all[4] $t \geq -1$. Moreover, there exists a global, smooth, exterior optical function u, namely a solution of the Eikonal equation defined everywhere in the exterior region $r \geq r_0/2$, with $r_0(t)$ representing the radius of the 2-surfaces of intersection between the hypersurfaces Σ_t and a fixed null cone C_0 [5] with vertex at a point on Σ_{-1}, with respect to which

$$
\begin{gathered}
{}^e\mathcal{R}_0^{\infty}, {}^e\mathcal{O}_0^{\infty}, {}^e\mathcal{K}_0^{\infty}, {}^e\mathcal{L}_0^{\infty} \leq \varepsilon_0 \\
{}^e\mathcal{R}_{[2]}, {}^e\mathcal{O}_{[3]}, {}^e\mathcal{K}_{[3]}, {}^e\mathcal{L}_{[3]} \leq \varepsilon_0 .
\end{gathered}
$$

Moreover, in the complement of the exterior region,

$$
\begin{gathered}
{}^i\mathcal{R}_0^{\infty}, {}^i\mathcal{K}_0^{\infty}, {}^i\mathcal{L}_0^{\infty} \leq \varepsilon_0 \\
{}^i\mathcal{R}_{[2]}, {}^i\mathcal{K}_{[3]}, {}^i\mathcal{L}_{[3]} \leq \varepsilon_0 .
\end{gathered}
$$

In what follows we give a detailed account of the main steps in the proof.

Proof 10.2.1 (the Main Theorem) *In view of Theorem 5.0.1 we may assume that the initial data set Σ, g, k verifies the smallness assumptions, relative to the global radial foliation induced by the distance function d_0 from a point O in Σ:*

$$
\begin{aligned}
\left(\int_{\Sigma} \sum_{l=0}^{3} \sigma_0^{2l+2} |\nabla^l k|^2 \right)^{1/2} &\leq \varepsilon \\
\mathcal{D}_0 &\leq \varepsilon \qquad (10.2.1)
\end{aligned}
$$

[3] The first two versions appear in the introduction.
[4] The level hypersurface $t = 0$ corresponds to the initial data set.
[5] Namely, the level hypersurface $u = 0$.

where $\sigma_0 = (1 + d_0^2)^{1/2}$ *and*

$$\begin{aligned}
\mathcal{D}_0^2 = & \|\sigma_0^2(Q - \bar{Q})\|_2^2 + \|\sigma_0^3 Q_N\|_2^2 + \|\sigma_0^3 \nabla\!\!\!\!/\, Q\|_2^2 \\
& + \|\sigma_0^4 \nabla_N Q_N\|_2^2 + \|\sigma_0^4 \nabla\!\!\!\!/\, Q_N\|_2^2 + \|\sigma_0^4 \nabla\!\!\!\!/^2 Q\|_2^2 \\
& + \|\sigma_0^2 P\|_2^2 + \|\sigma_0^3 \nabla\!\!\!\!/\, P\|_2^2 + \|\sigma_0^3 \nabla\!\!\!\!/_N P\|_2^2 \\
& + \|\sigma_0^4 \nabla\!\!\!\!/^2 P\|_2^2 + \|\sigma_0^4 \nabla\!\!\!\!/\, \nabla\!\!\!\!/_N P\|_2^2 + \|\sigma_0^4 \nabla\!\!\!\!/_N^2 P\|_2^2 \\
& + \|\sigma_0^2 \hat{S}\|_2^2 + \|\sigma_0^3 \nabla\!\!\!\!/\, \hat{S}\|_2^2 + \|\sigma_0^3 \nabla\!\!\!\!/_N \hat{S}\|_2^2 \\
& + \|\sigma_0^4 \nabla\!\!\!\!/^2 \hat{S}\|_2^2 + \|\sigma_0^4 \nabla\!\!\!\!/\, \nabla\!\!\!\!/_N \hat{S}\|_2^2 + \|\sigma_0^4 \nabla\!\!\!\!/_N^2 \hat{S}\|_2^2 + (\sup_{\Sigma} \sigma_0^3 \bar{Q})^2
\end{aligned}$$

with

$$Q = \widehat{Ric}_{NN}, \qquad P_A = \widehat{Ric}_{AN},$$
$$S_{AB} = \widehat{Ric}_{AB}, \qquad S_{AB} = \hat{S}_{AB} - \frac{1}{2} Q \gamma_{AB}$$

being the decomposition of $\widehat{Ric}$ *relative to the foliation and* r *being the function defined by the condition that* $4\pi r^2$ *is the area of the level surfaces of the foliation.*

We next state the following form of the well-known local existence theorem:

Theorem 10.2.2 *Let* (Σ, g_0, k_0) *be an initial data set verifying the following conditions:*

1. (Σ, g_0) *is a complete Riemannian manifold diffeomorphic to* $\Re^3$.
2. *The isoperimetric constant* $I(\Sigma, g_0)$ *is finite, where* I *is defined to be*

$$\sup_S \frac{V(S)}{A(S)^{3/2}}$$

with S *an arbitrary surface in* Σ, $A(S)$ *its area, and* $V(S)$ *the enclosed volume.*

3. *The Ricci curvature* $Ric(g_0)$ *verifies, relative to the distance function* d_0 *from a given point* O,

$$Ric(g_0) \in H_{2,1}(\Sigma, g_0).$$

4. k *is a 2-covariant symmetric trace-free tensorfield on* Σ *verifying*

$$k \in H_{3,1}(\Sigma, g_0)$$

where for a given tensorfield h, $\|h\|_{H_{s,\delta}}(\Sigma, g_0)$ *denotes the norm*

$$\|h\|_{H_{s,\delta}(\Sigma, g_0)} = \left(\sum_{i=0}^{s} \int \sigma_0^{2\delta+2i} |\nabla_0^i h|^2 d\mu_{g_0} \right)^{1/2}$$

and $\sigma_0 = \sqrt{1 + d_0^2}$.

5. (g_0, k_0) verify the constraint equations on Σ.

Then there exists a unique, local-in-time smooth development, foliated by a normal, maximal time foliation t with range in some interval $[0, t_*]$ and with $t = 0$ corresponding to the initial slice Σ. Moreover,

$$g(t) - g_0 \in C^1([0, t_*]\,;\, H_{3,1}(\Sigma, g_0))$$
$$k(t) \in C^0([0, t_*]\,;\, H_{3,1}(\Sigma, g_0)).$$

Furthermore,

$$Ric(t) \in C^0([0, t_*]\,;\, H_{2,1}(\Sigma, g_0)).$$

On the other hand, if Condition 2 is strengthened to

$$\sigma_0^3 |Ric(g_0)| \text{ is bounded}$$
$$Ric(g_0) \in H_{2,1}(\Sigma, g_0)$$
$$B(g_0) \in H_{1,3}(\Sigma, g_0)$$

where B is the Bach tensor, then this solution also satisfies

$$\sigma_0^3 |Ric(g(t))| \text{ is bounded}$$
$$B(g(t)) \in H_{1,3}(\Sigma, g_0).$$

An outline of the proof of the local existence theorem is provided at the end of this section. We now return to the proof of the main theorem.

We define $\mathcal{S}$ to be the set of all $t \geq 0$ such that there exists a space-time slab $\bigcup_{t' \in [0,t]} \Sigma_{t'}$ endowed with a canonical exterior optical function[6] with respect to which the following bootstrap assumptions hold:

BA₀: For all $t' \in [0, t]$,

$$\frac{1}{2}(1 + t') \leq r_0(t') \leq \frac{3}{2}(1 + t').$$

BA₁: For all $t' \in [0, t]$,

$${}^e\mathcal{O}_0^\infty, \mathcal{R}_0^\infty, \mathcal{K}_0^\infty, \mathcal{L}_0^\infty \leq \varepsilon_0.$$

BA₂: For all $t' \in [0, t]$,

$${}^e\mathcal{O}_{[3]}, \mathcal{R}_{[2]}, \mathcal{K}_{[3]}^*, \mathcal{L}_{[3]}^* \leq \varepsilon_0.$$

[6] Namely, an optical function defined starting from the "last slice" Σ_t and initial cone C_0 according to the unique prescription outlined in the previous chapter.

Auxiliary Assumption:

On the last slice Σ_t,

$$\sup_{\Sigma_t} r_0^{-1/2} r^{5/2} |\nabla\!\!\!/ \log a| \leq \varepsilon_0.$$

Step 1: We show that the set $\mathcal{S}$ is not empty, as it contains at least $t = 0$. Indeed, in view of the local existence theorem we can construct the past slab $\bigcup_{t' \in [-1,0]} \Sigma_{t'}$ and the initial cone C_0 with vertex at a point on Σ_{-1}. We then construct an exterior optical function u on $t = 0$ by solving the inverse lapse problem, starting on the 2-surface of intersection between C_0 and Σ_0, according to the results of Chapter 14. Using the comparison results of Chapter 14, in the case when the background radial function u' is the distance function d from the 2-surface $S_{0,0}$ and for small ε in 10.2.1, we then check that the norms ${}^e\mathcal{O}_0^\infty, \mathcal{R}_0^\infty, \mathcal{K}_0^\infty, \mathcal{L}_0^\infty$ and $\mathcal{R}_{[2]}, \mathcal{K}_{[3]}^*, \mathcal{L}_{[3]}^*, {}^e\mathcal{O}_{[3]}$ can be made arbitrarily small and thus check assumptions $\mathbf{BA_1}$, and $\mathbf{BA_2}$. The auxiliary assumption is a consequence of the last conclusion of Proposition 14.0.1.

Step 2: Let $t_* = \sup \mathcal{S}$. If $t_* = \infty$, global existence is proved. Assume that $t_* < \infty$. Then, clearly, $t_* \in \mathcal{S}$. We first extend the exterior optical function u, with which the space-time slab $\bigcup_{t' \in [0,t_*]} \Sigma_{t'}$ comes equipped, to the interior region by using the results of Chapters 9, 13, and 15. Since

$$\begin{aligned} {}^i\mathcal{O}_0^\infty &\leq c(\mathcal{R}_0^\infty + \mathcal{K}_0^\infty + \mathcal{L}_0^\infty) \\ {}^i\mathcal{O}_{[3]} &\leq c(\mathcal{R}_{[2]} + \mathcal{K}_{[3]} + \mathcal{L}_{[3]}), \end{aligned}$$

we infer that

$$ {}^i\mathcal{O}_0^\infty, {}^i\mathcal{O}_{[3]} \leq c\varepsilon_0. \tag{10.2.2}$$

We remark that the extension of u to the interior region is not unique; in fact, it depends on the choice of a Lebesgue point. The construction of the interior optical function is first defined in Chapter 9, and is further discussed in Chapters 13 and 15.

Step 3: This is the main step of the proof. We show, using the bootstrap assumptions $\mathbf{BA_0}, \mathbf{BA_1}, \mathbf{BA_2}$, and the result of step 2, that the size of the norms ${}^e\mathcal{O}_0^\infty, \mathcal{R}_0^\infty, \mathcal{K}_0^\infty, \mathcal{L}_0^\infty$ and ${}^e\mathcal{O}_{[3]}, \mathcal{R}_{[2]}, \mathcal{K}_{[3]}^*, \mathcal{L}_{[3]}^*$ cannot exceed a constant multiple of the size of the data at $t = 0$. We therefore can choose ε and ε_0 sufficiently small such that

$$\begin{aligned} {}^e\mathcal{O}_0^\infty, \mathcal{R}_0^\infty, \mathcal{K}_0^\infty, \mathcal{L}_0^\infty &\leq \frac{1}{2}\varepsilon_0 \\ {}^e\mathcal{O}_{[3]}, \mathcal{R}_{[2]}, \mathcal{K}_{[3]}^*, \mathcal{L}_{[3]}^* &\leq \frac{1}{2}\varepsilon_0. \end{aligned} \tag{10.2.3}$$

We achieve this as follows:

Step 3_a: We use the bootstrap assumptions $\mathbf{BA_0}$, $\mathbf{BA_1}$, $\mathbf{BA_2}$, as well as the properties of the interior optical function 10.2.2 to check all the assumptions of comparison theorem 7.6.1 as well as the main theorem of Part II, Theorem 8.2.1. Assumptions 0, 1, 2 of Theorem 7.6.1 are a direct consequence of bootstrap assumptions $\mathbf{BA_0}$, $\mathbf{BA_1}$, $\mathbf{BA_2}$, and 10.2.2. Assumptions **IA0–IA2** and **EA0–EA2** for vectorfields $T, S, \bar{K}$ are easy to check in view of the formulas 7.5.4a, and 7.5.4b for ${}^{(T)}\pi$; formulas 7.5.6c for ${}^{(T)}\hat{\pi}$; and formulas 7.5.7 for ${}^{(\bar{K})}\hat{\pi}$. On the other hand, to check Properties 0, 1, 2 of Proposition 7.5.3 as well as assumptions **IA0–IA2** and **EA0–EA2** for the rotation vectorfields $\mathbf{O}$, we need the results of Chapter 16 and Theorems 16.1.1, 16.1.2, and 16.2.1.

According to Theorem 8.2.1, we have

$$\mathcal{Q}_1* + \mathcal{Q}_2* \leq c(\mathcal{Q}_1(0) + \mathcal{Q}_2(0)).$$

Thus by the Comparison Theorem 7.6.1, we conclude that, for all $t \in [0, t_*]$,

$$\mathcal{R}_{[2]}(t) \leq c\mathcal{R}_{[2]}(0).$$

Moreover, in view of the definition of $\mathcal{Q}_1*$, $\mathcal{Q}_2*$, we derive the additional information, on the null hypersurface C_0, for the null components α, β of the curvature tensor

$$\mathcal{A} = \left[\int_{C_0}\left(r^4|\alpha|^2 + |r^6|\nabla\!\!\!/\alpha|^2 + r^8|\nabla\!\!\!/^2\alpha|^2\right)\right]^{1/2} \leq c\mathcal{R}_{[2]}(0) \quad (10.2.4a)$$

$$\mathcal{B} = \left[\int_{C_0}\left(r^4|\beta|^2 + |r^6|\nabla\!\!\!/\beta|^2 + r^8|\nabla\!\!\!/^2\beta|^2\right)\right]^{1/2} \leq c\mathcal{R}_{[2]}(0). \quad (10.2.4b)$$

Next we show, using once again the comparison results of Lemma 14.0.1 as well as Proposition 14.0.1[7] and Proposition 14.0.2, applied to the distance function d and the restriction of the optical function u to the initial slice, to conclude that $\mathcal{R}_{[2]}(0)$ can be bounded by a constant multiple of ε. Hence,

$$\mathcal{R}_{[2]} + \mathcal{A} + \mathcal{B} \leq c\varepsilon.$$

We then use bootstrap assumption $\mathbf{BA_1}$, which enables us to apply the Sobolev inequalities of Chapter 3 in order to conclude that

$$\mathcal{R}_0^\infty \leq c\varepsilon.$$

Choosing ε sufficiently small in 10.2.1, we then infer that

$$\mathcal{R}_{[2]} + \mathcal{A} + \mathcal{B} \leq \frac{1}{2}\varepsilon_0 \quad (10.2.5)$$

$$\mathcal{R}_0^\infty \leq \frac{1}{2}\varepsilon_0. \quad (10.2.6)$$

[7] See equations 14 0.16–14.0.20c.

Step 3_b: We show that bootstrap assumptions $\mathbf{BA_0}$, $\mathbf{BA_1}$, $\mathbf{BA_2}$, and the result of step 2, 10.2.2, imply

$$\mathcal{K}^*_{[3]}, \mathcal{L}^*_{[3]} \leq c\mathcal{R}_{[2]}.$$

Then, by the Sobolev inequalities, we conclude that

$$\mathcal{K}^\infty_0, \mathcal{L}^\infty_0 \leq c\mathcal{R}_{[2]},$$

and therefore, choosing ε sufficiently small,

$$\mathcal{K}^\infty_0, \mathcal{L}^\infty_0 \leq \frac{1}{2}\varepsilon_0 \tag{10.2.7}$$

$$\mathcal{K}^*_{[3]}, \mathcal{L}^*_{[3]} \leq \frac{1}{2}\varepsilon_0. \tag{10.2.8}$$

This is achieved in Chapters 11 and 12.

Step 3_c: We show that bootstrap assumptions $\mathbf{BA_0}$, $\mathbf{BA_1}$, and $\mathbf{BA_2}$ imply that

$$^e\mathcal{O}^\infty_0 \leq c\varepsilon$$

$$^e\mathcal{O}_{[3]} \leq c\varepsilon,$$

and therefore, if ε is sufficiently small,

$$^e\mathcal{O}^\infty_0 \leq \frac{1}{2}\varepsilon_0 \tag{10.2.9}$$

$$^e\mathcal{O}_{[3]} \leq \frac{1}{2}\varepsilon_0. \tag{10.2.10}$$

This is achieved in Chapters 9, 13, and 14.

As a consequence of the last conclusion of Proposition 14.0.1, we also show that

$$\sup_{\Sigma_{t_*}} r_0^{-1/2} r^{5/2} |\nabla \log a| \leq \frac{1}{2}\varepsilon_0. \tag{10.2.11}$$

Step 4: We use the result of the previous step and the local existence theorem, with data at t_*, to extend our space-time up from t_* to $t_* + \delta$. We also extend the exterior optical function u, with which the slab $\bigcup_{t\in[0,t_*]} \Sigma_t$ comes equipped, by continuing the null geodesic generators of the hypersurfaces C_u into the future up to $t_* + \delta$. We pick δ sufficiently small so that the size of the norms $^e\mathcal{O}^\infty_0, \mathcal{R}^\infty_0, \mathcal{K}^\infty_0, \mathcal{L}^\infty_0$ as well as of the norms $^e\mathcal{O}_{[3]}, \mathcal{R}_{[2]}, \mathcal{K}^*_{[3]}, \mathcal{L}^*_{[3]}$ remains strictly smaller than ε_0. Also, the quantity $\sup_{\Sigma_{t_*+\delta}} r_0^{-1/2} r^{5/2} |\nabla \log a|$, appearing in the auxiliary assumption, is strictly smaller than ε_0.

We now start with $\Sigma_{t_*+\delta}$, as last slice, and its cut with C_0, as initial 2-surface, to solve the propagation equation and construct a new exterior optical function on $\Sigma_{t_*+\delta}$, using the results of Chapter 14.[8] This is then extended to the past according to the construction described in Chapters 9 and 13. In view of the continuity properties of the propagation equations we infer that the new norms ${}^e\mathcal{O}_0^\infty, \mathcal{R}_0^\infty, \mathcal{K}_0^\infty, \mathcal{L}_0^\infty$ and ${}^e\mathcal{O}_{[3]}, \mathcal{R}_{[2]}, \mathcal{K}^*_{[3]}, \mathcal{L}^*_{[3]}$ can be made arbitrarily close to the previous ones and therefore check that bootstrap assumptions $\mathbf{BA_1}, \mathbf{BA_2}$, as well as the auxiliary assumption, still hold. The assumption $\mathbf{BA_0}$ is also checked from Lemma 15.0.1. Therefore $t_* + \delta \in \mathcal{S}$, which contradicts our assumption that $t_* < \infty$.

Step 5: To complete the proof of the main theorem, we show that the exterior optical function ${}^{(t)}u$ defined on the slab $\bigcup_{t'\in[0,t]} \Sigma_{t'}$, starting from the last slice Σ_t, converges as $t \to \infty$ to a global exterior optical function u. Let ${}^{(t)}C_\lambda$ be the $\lambda-$ level set of ${}^{(t)}u$. Now, the 0- level set of ${}^{(t)}u$ is the part of the fixed cone C_0 contained in the slab $\bigcup_{t'\in[0,t]} \Sigma_{t'}$. For $t_2 > t_1 > 0$, consider the functions ${}^{(t_2)}u$ and ${}^{(t_1)}u$ on Σ_{t_1}. Since their 0-level sets coincide, we can again apply the comparison results of Chapter 14 to conclude that the difference of the functions in any region of the form $\left(\bigcup_{\lambda\in[\lambda_1,\lambda_2]} {}^{(t_1)}C_\lambda\right) \bigcap \Sigma_{t_1}, \lambda_1, \lambda_2$ fixed but arbitrary, tends to zero as $t_1 \to \infty$. As a consequence, in any space-time region $\bigcup_{\lambda\in[\lambda_1,\lambda_2]} {}^{(t_1)}C_\lambda$ the difference of ${}^{(t_2)}u$ from ${}^{(t_1)}u$ tends to zero as $t_1 \to \infty$. The convergence of ${}^{(t)}u$ to u then follows. For completeness we give in what follows a short description of the proof of the local existence theorem.

Proof 10.2.2 *The Local Existence Theorem*

The proof of the theorem follows along the main lines of the proof of the well-known existence result of Choque-Bruhat [Br1], with certain simple modifications. For this reason we describe the main steps in the proof.

We recall from the introduction that relative to the maximal foliation and its associated normal flow the Einstein-Vacuum equations take the following form:

Constraint Equations of a Maximal Foliation

$$trk = 0 \qquad (10.2.12a)$$

$$\nabla^j k_{ji} = 0 \qquad (10.2.12b)$$

$$R = |k|^2. \qquad (10.2.12c)$$

[8] Applied now to the case when u' is the natural extension of the old optical function and u is the new one.

Evolution Equations of a Maximal Foliation

$$\partial_t g_{ij} = -2\phi k_{ij} \tag{10.2.13a}$$
$$\partial_t k_{ij} = -\nabla_i \nabla_j \phi + \phi(R_{ij} - 2k_{ia}k^a{}_j). \tag{10.2.13b}$$

Lapse Equation of a Maximal Foliation

$$\triangle \phi = |k|^2 \phi. \tag{10.2.14}$$

As it stands, the evolution equations are not in standard hyperbolic form. This is due to the fact that the Ricci curvature R_{ij}, expressed relative to the metric g_{ij}, is not an elliptic differential operator. To overcome this difficulty, we differentiate the evolution equation for k_{ij} with respect to t, and, making use of the constraint equations, we derive instead a second-order hyperbolic system.

Let us introduce the following notation:

$$\begin{aligned} A &= \mathrm{tr} k \\ B &= R - |k|^2 \\ C_i &= (div k)_i - \frac{1}{2}\nabla_i \mathrm{tr} k \\ D_{ij} &= \phi^{-1}\partial_t k_{ij} + \phi^{-1}\nabla_i\nabla_j\phi - R_{ij} + 2k_{im}k^m_j. \end{aligned} \tag{10.2.15}$$

To compute the time derivative of D_{ij}, we first calculate the time derivative of R_{ij}

$$\partial_t R_{ij} = \nabla_a \dot{\Gamma}^a_{ij} - \nabla_i \dot{\Gamma}^a_{aj}$$

where $\dot{\Gamma}^a_{ij}$, the time derivative of the connection, is the following tensorfield:

$$\begin{aligned} \dot{\Gamma}^a_{ij} &= \frac{1}{2} g^{ab}(\nabla_i \dot{g}_{jb} + \nabla_j \dot{g}_{ib} - \nabla_b \dot{g}_{ij}) \\ &= -\nabla_i(\phi k^a{}_j) - \nabla_j(\phi k^a{}_i) + \nabla^a(\phi k_{ij}). \end{aligned}$$

Hence,

$$\begin{aligned} \partial_t R_{ij} = &-\nabla_a\nabla_i(\phi k^a{}_j) - \nabla_a\nabla_j(\phi k^a{}_i) + \nabla_i\nabla_j(\phi \mathrm{tr} k) \\ &+ \Delta(\phi k_{ij}), \end{aligned} \tag{10.2.16a}$$

or, in terms of the notation 10.2.15,

$$\phi^{-1}\partial_t R_{ij} = \triangle k_{ij} - \nabla_i C_j - \nabla_j C_i + H_{ij} \tag{10.2.16b}$$

$$\begin{aligned}\phi H_{ij} &= \phi I_{ij} + \nabla^m\phi(2\nabla_m k_{ij} - \nabla_i k_{mj} - \nabla_j k_{mi}) \\ &\quad - \nabla_i\phi\nabla^m k_{mj} - \nabla_j\phi\nabla^m k_{mi} + \nabla_i\phi\nabla_j \mathrm{tr}k + \nabla_j\phi\nabla_i \mathrm{tr}k \\ &\quad - k_i{}^m\nabla_j\nabla_m\phi - k_j{}^m\nabla_i\nabla_m\phi + \mathrm{tr}k\nabla_i\nabla_j\phi + k_{ij}\triangle\phi \\ I_{ij} &= -3(k_i{}^m R_{jm} + k_j{}^m R_{im}) + 2g_{ij}k^{mn}R_{mn} \\ &\quad + 2\mathrm{tr}kR_{ij} + (k_{ij} - g_{ij}\mathrm{tr}k)R.\end{aligned}$$

We then calculate

$$\begin{aligned}\phi^{-1}\partial_t D_{ij} &= (\phi^{-1}\partial_t)^2 k_{ij} - \triangle k_{ij} + \nabla_i C_j + \nabla_j C_i \\ &\quad + N_{ij}\end{aligned} \tag{10.2.16c}$$

$$N_{ij} = L_{ij} - H_{ij}$$

$$\begin{aligned}\phi^2 L_{ij} &= \nabla_i\nabla_j\dot\phi - \phi^{-1}\dot\phi\nabla_i\nabla_j\phi - \dot\Gamma^m_{ij}\nabla_m\phi \\ &\quad + 2\phi(k_i{}^m\partial_t k_{jm} + k_j{}^m\partial_t k_{im}) + 4\phi^2 k^{mn}k_{im}k_{jn}.\end{aligned} \tag{10.2.16d}$$

Therefore, the Einstein equations imply the following second-order hyperbolic system for k_{ij}, on which we will base our proof of local existence:

$$-(\phi^{-1}\partial_t)^2 k_{ij} + \triangle k_{ij} = N_{ij}. \tag{10.2.17a}$$

This equation is to be considered in conjunction with the equation

$$-\phi^{-1}\partial_t g_{ij} = 2k_{ij} \tag{10.2.17b}$$

as well as the lapse equation

$$\triangle\phi = |k|^2\phi. \tag{10.2.17c}$$

In other words the lapse ϕ is defined to be the solution of the elliptic equation 10.2.17c on each Σ_t, with the boundary condition $\phi \to 1$ at ∞. From the lapse equation we derive the following equation for $\dot\phi$:

$$\begin{aligned}\triangle\dot\phi - |k|^2\dot\phi &= g^{ij}\dot\Gamma^m_{ij}\nabla_m\phi - 2\phi k^{ij}\nabla_i\nabla_j\phi \\ &\quad + 2k^{ij}\partial_t k_{ij} + 4\phi k^{mn}k_{im}k^i{}_n.\end{aligned} \tag{10.2.17d}$$

Lemma 10.2.1 *Any solution of the coupled elliptic-hyperbolic system 10.2.17a–10.2.17c whose initial data,* $g_{ij}, k_{ij}, \dot{k}_{ij}$*, verify the original constraint equations* $A = 0, B = 0, C_i = 0$*, as well as* $D_{ij} = 0$*,*[9] *is a solution of the original system 10.2.12a–10.2.14.*

To prove this, one derives a system of evolution equations for $A, B, C, and D$, which holds for any given solution of the system 10.2.17a–10.2.17c:

$$\phi^{-1}\partial_t A = F := B + \mathrm{tr}D \tag{10.2.18a}$$

$$\phi^{-1}\partial_t F = \triangle A - 4\nabla\phi \cdot C + 2\nabla\phi \cdot \nabla A + 2\phi|k|^2 A \tag{10.2.18b}$$

$$\begin{aligned}\phi^{-1}\partial_t C_i = {} & \nabla^j D_{ij} - \frac{1}{2}\nabla_i \mathrm{tr}D \\ & + \phi^{-1}\nabla^j \phi D_{ij} - \frac{1}{2}\phi^{-1}\nabla_i \phi F \\ & - k_i{}^j \nabla_j A - \phi^{-1} k_i{}^j \nabla_j \phi A \end{aligned} \tag{10.2.18c}$$

$$\phi^{-1}\partial_t D_{ij} = \nabla_i C_j + \nabla_j C_i. \tag{10.2.18d}$$

We remark that A, F, and C satisfy wave equations of the type

$$-(\phi^{-1}\partial_t)^2 A + \triangle A = M$$
$$-(\phi^{-1}\partial_t)^2 F + \triangle F = M'$$
$$-(\phi^{-1}\partial_t)^2 C_i + \triangle C_i = M_i''$$

where M, M', M'' are linear expressions in the quantities

$$A, \nabla A, \nabla^2 A, F, \nabla F, C, \nabla C, D, \nabla D.$$

We then prove the uniqueness of the system 10.2.18a–10.2.18d by deriving a homogeneous differential inequality for the quantity

$$\begin{aligned}\mathcal{E} = \int_{\Sigma_t} & (|A|^2 + |\nabla A|^2 + |F|^2 + |\nabla F|^2 + |\phi^{-1}\partial_t F|^2 \\ & + |C|^2 + |\nabla C|^2 + |\phi^{-1}\partial_t C|^2 + |D|^2).\end{aligned}$$

It remains to prove local existence for the system 10.2.17a–10.2.17b–10.2.17c for initial data g_0, k_0 as in the assumptions of the theorem and $\dot{k}_0$, verifying the assumption

$$\dot{k}_0 \in H_{2,1}(\Sigma, g_0).$$

[9] This equation expresses $\dot{k}_{ij}$ in terms of g_{ij} and k_{ij}.

Before proceeding we remark that for any Riemannian manifold (Σ, g_0) verifying the assumptions of the theorem we have the following global Sobolev inequalities, for any given tensorfield f:

$$\|\sigma_0 f\|_{L^6(\Sigma,g_0)} \leq c\|f\|_{H_{1,0}(\Sigma,g_0)} \tag{10.2.19}$$

$$\sup_\Sigma |\sigma_0^{3/2} f| \leq c\|f\|^{1/2}_{H_{1,0}(\Sigma,g_0)}\|f\|^{1/2}_{H_{2,0}(\Sigma,g_0)}. \tag{10.2.20}$$

It is well known that the isoperimetric inequality is equivalent to the Sobolev inequality

$$\|f\|_{L^{3/2}(\Sigma,g_0)} \leq I^{2/3}\|\nabla_0 f\|_{L^1(\Sigma,g_0)}$$

for functions f with compact support on Σ. This inequality immediately implies

$$\|f\|_{L^6(\Sigma,g_0)} \leq c\|\nabla_0 f\|_{L^2(\Sigma,g_0)},$$

and, by the Moser iteration scheme (see [Gil-Tr]),

$$\sup_\Sigma |f| \leq c\|f\|^{1/2}_{L^6(\Sigma,g_0)}\|\nabla_0 f\|^{1/2}_{L^6(\Sigma,g_0)}.$$

The desired inequalities, 10.2.19–10.2.20, are then direct consequences.

We now define the following iteration with respect to the triplets $(g_n, k_n, \dot{k}_n)$ defined on the product manifold $[0, t_*] \times \Sigma$. First we associate to such a triplet the lapse function ϕ_n to be the solution of

$$\triangle_n \phi_n = |k_n|_n^2 \phi_n \tag{10.2.21a}$$

tending to 1 at ∞ on Σ at each t. Next we define g_{n+1} by

$$\partial_t g_{n+1} = -2\phi_n k_n \tag{10.2.21b}$$

and $g_{n+1}|_{t=0}$ given by the prescribed initial data for g. We then define k_{n+1} to be the solution of the wave equation

$$-(\phi_n^{-1}\partial_t)^2 k_{n+1} + \triangle_n k_{n+1} = N_n \tag{10.2.21c}$$

such that $k_{n+1}|_{t=0}$, $\dot{k}_{n+1}|_{t=0}$ agree with the given initial data for k and $\dot{k}$. The inhomogeneous term N_n corresponds to the term N on the right-hand side of 10.2.17a with g, k, ϕ replaced by g_n, k_n, ϕ_n.

We then show that we can find a suitable large positive constant E such that the sequence of triplets $(g_n, k_n, \dot{k}_n)$ verify the following conditions:

1. The smallest eigenvalue of g_n relative to g_0 is not less than E^{-1}, and the largest eigenvalue is not greater than E.

2. The following estimates hold:

$$\|g_n - g_0\|_{H_{3,1}(\Sigma,g_0)} \leq E \tag{10.2.22a}$$
$$\|k_n\|_{H_{3,1}(\Sigma,g_0)} \leq E \tag{10.2.22b}$$
$$\|\dot{k}_n\|_{H_{2,1}(\Sigma,g_0)} \leq E \tag{10.2.22c}$$

provided that the time interval $[0, t_*]$ is sufficiently small.

These conditions imply that the associated sequence ϕ_n verifies

$$\begin{aligned}\|\phi_n - 1\|_{H_{4,-1}(\Sigma,g_0)} &\leq E' \\ \|\dot{\phi}_n\|_{H_{4,-1}(\Sigma,g_0)} &\leq E' \\ \inf \phi_n &\geq E'^{-1}\end{aligned} \tag{10.2.22d}$$

where E' is a positive constant depending on E. In fact, we show this by first using standard weighted L^2 theory for the Poisson equation to deduce

$$\begin{aligned}\|\phi_n - 1\|_{H_{4,-1}(\Sigma,g_n)} &\leq E'' \\ \|\dot{\phi}_n\|_{H_{4,-1}(\Sigma,g_n)} &\leq E'' \\ \inf \phi_n &\geq E'^{-1}.\end{aligned} \tag{10.2.23}$$

The estimates 10.2.22d then follow in view of the equivalence of the metrics $g_n(t)$ and g_0 implied by conditions 1 and 2.

Next, assuming that conditions 1 and 2 are satisfied for the triplet $(g_n, k_n, \dot{k}_n)$, we show that the same holds true for the triplet $(g_{n+1}, k_{n+1}, \dot{k}_{n+1})$. First, the estimate for g_{n+1} follow easily from those for k_n, ϕ_n. Then, using standard wave equation type energy estimates corresponding to the vectorfield $\sigma_0^2\partial_t$ applied to 10.2.21c, we show that $\sigma_0\partial_t k_{n+1}, \sigma_0\nabla k_{n+1} \in L^2(\Sigma, g_n)$. This result is weak as far as the decay at ∞ of ∇k_{n+1} is concerned. This is due to the presence on the right-hand side of 10.2.21c of the term $-\nabla_i\nabla_j\dot{\phi}_{n+1}$, which belongs only to $H_{2,1}$. We overcome this difficulty by considering the induced wave equations for $\operatorname{div} k_{n+1}, \operatorname{curl} k_{n+1}$ for which the corresponding term on the right-hand side behaves manifestly better at infinity. Thus, by standard energy estimates, we establish that $\nabla\operatorname{div} k_{n+1}, \nabla\operatorname{curl} k_{n+1} \in H_{1,3}$ and $\partial_t\operatorname{div} k_{n+1}, \partial_t\operatorname{curl} k_{n+1} \in H_{1,3}$. On the other hand, by integrating $\partial_t k_{n+1}$, we have $k_{n+1} \in H_{0,1}$. Finally, in view of the standard elliptic estimates

$$\|k_{n+1}\|_{H_{s,\delta}} \leq c(\|k_{n+1}\|_{H_{0,\delta}} + \|\operatorname{div} k_{n+1}\|_{H_{s-1,\delta+1}} + \|\operatorname{curl} k_{n+1}\|_{H_{s-1,\delta+1}}),$$

we deduce the correct estimates for k_{n+1} and $\partial_t k_{n+1}$.

The estimates for g_{n+1} are obtained from k_{n+1} by integration in time. This establishes the fact that all the triplets $(g_n, k_n, \dot{k}_n)$ satisfy the conditions 1 and 2. By a standard contraction argument we then show that the corresponding sequence converges to a solution of the system 10.2.17a–10.2.17b–10.2.17c. In view of Lemma 10.2.1 we thus obtain a solution of the original system. By the equation 10.2.13b we then obtain

$$\mathrm{Ric}(g) \in H_{2,1}(\Sigma, g_0).$$

Also remark that, in view of the fact that $g - g_0 \in H_{3,1}$, we also have $\mathrm{Ric}(g) - \mathrm{Ric}(g_0) \in H_{1,3}$. Therefore, using the inequality 10.2.20, we infer that

$$\sup_{\Sigma} \sigma_0^3 |\mathrm{Ric}(g) - \mathrm{Ric}(g_0)|$$

is bounded.

Finally, to prove the last statement of the theorem, we consider the curl of the equation 10.2.17a, which induces a wave equation for curl k, to show that $\partial_t k \in H_{1,3}$. Then the curl of the equation 10.2.13b implies that $B(g)$ belongs to $H_{1,3}$.

CHAPTER 11

Second Fundamental Form

11.1. Preliminaries

The aim of this section is to derive the estimates for the second fundamental form of the time foliation k needed in Step 3_b of the proof of the main theorem. We recall (see 1.0.14b, 1.0.14c) that k satisfies the elliptic system, on each slice Σ_t,

$$\begin{aligned} \mathrm{tr} k &= 0 \\ \mathrm{curl} k &= H \\ \mathrm{div} k &= 0. \end{aligned} \tag{11.1.1a}$$

Also,

$$R_{ij} = k_{ia} k^a_j + E_{ij} \tag{11.1.1b}$$

where E, H are the electric and magnetic parts of the space-time curvature relative to the time foliation. The Bianchi identities of the space-time imply that, in particular, E and H verify, on Σ_t, the following divergence equations:[1]

$$\mathrm{div} E = k \wedge H \tag{11.1.1c}$$

$$\mathrm{div} H = -k \wedge E. \tag{11.1.1d}$$

We will show that equations 11.1.1a and 11.1.1b completely determine the metric g and the second fundamental form of each Σ_t, in terms of E and H. We specify the metric g relative to the radial foliation induced, on Σ_t, by the extended optical function u, as defined in Step 2 of the proof of the main theorem. More precisely, we describe it relative to

1. The lapse function a.
2. The second fundamental form θ.
3. The Gauss curvature K.

[1] Given two symmetric traceless tensors A, B on Σ, we write $A \wedge B_i = \in_i{}^{mn} A_m^k B_{kn}$.

Recall that (see 9.1.5b, 9.1.7a, 9.1.19b)

$$\theta = \chi' + \eta$$
$$a^{-1}\nabla a = \zeta' - \epsilon \qquad (11.1.1e)$$
$$a^{-1}\nabla_N a = -(\omega' + \delta') \qquad (11.1.1f)$$

where χ', ζ', ω' are the components of the Hessian of the optical function u decomposed relative to the standard frame, while η, ϵ, δ are the components of the decomposition of k relative to the radial foliation. For convenience, throughout this section, we drop the primes in the definition of χ', ζ', ω'; we write them simply χ, ζ, ω.

According to Proposition 4.4.3 of Part I, the equations 11.1.1a can be decomposed, relative to the radial foliation, according to the formulas

$$\not{d}\text{iv}\,\epsilon = -\nabla_N\delta - \frac{3}{2}\text{tr}\theta\delta + \hat{\eta}\cdot\hat{\theta} - 2(a^{-1}\not\nabla a)\cdot\epsilon \qquad (11.1.2a)$$
$$\text{c}\not{u}\text{rl}\,\epsilon = \sigma + \hat{\theta}\wedge\hat{\eta} \qquad (11.1.2b)$$
$$\not\nabla_N\epsilon + \text{tr}\theta\epsilon = \frac{1}{2}(\beta - \underline{\beta}) + \not\nabla\delta - 2\hat{\theta}\cdot\epsilon + \frac{3}{2}(a^{-1}\not\nabla a)\cdot\delta - \hat{\eta}\cdot(a^{-1}\not\nabla a) \qquad (11.1.2c)$$

$$\not{d}\text{iv}\,\hat{\eta} = \frac{1}{2}(-\underline{\beta} + \beta) - \frac{1}{2}\not\nabla\delta + \hat{\theta}\cdot\epsilon - \frac{1}{2}\text{tr}\theta\epsilon \qquad (11.1.2d)$$
$$\not\nabla_N\hat{\eta} + \frac{1}{2}\text{tr}\theta\hat{\eta} = \frac{1}{4}(-\underline{\alpha} + \alpha) + \frac{1}{2}\not\nabla\widehat{\otimes}\epsilon + \frac{3}{2}\delta\hat{\theta} + (a^{-1}\not\nabla a)\widehat{\otimes}\epsilon. \qquad (11.1.2e)$$

We will assume that the main bootstrap assumptions made in Chapter 9 hold here, and we estimate the first three derivatives of the second fundamental form k with respect to the space-time curvature tensor $\mathbf{R}$. To be more precise, we write down a simplified set of hypotheses that ensure the proof of the desired results. Before doing this, we introduce, using the notation of Chapter 10, the following norms for the second fundamental form k and space-time curvature $\mathbf{R}$:[2]

Norms for k

In the interior region I, $r \leq \frac{r_0}{2}$, we define, for all $1 \leq p \leq \infty$,

$$\overline{{}^{\imath}\mathcal{K}_q^p} = r_0^{2+q-\frac{3}{p}}\|\nabla^q k\|_{p,\imath}. \qquad (11.1.3a)$$

[2] Unlike the corresponding norms of Chapter 10, these norms involve only differentiation on Σ_t.

In the exterior region E, $r \geq \frac{r_0}{2}$, we define the following:

$$\overline{{}^e\mathcal{K}_q^p}(\delta) = r_0^{-1/2}\Big\{ \|r^{(\frac{5}{2}-\frac{3}{p}+q)} \not\nabla^q \delta\|_{p,e}^p \tag{11.1.3b}$$

$$+ \sum_{i+j=q\,,\,j\geq 1} \|r^{(2-\frac{2}{p}+i)} w_{(\frac{1}{2}-\frac{1}{p}+j)} \not\nabla^i \nabla_N^j \delta\|_{p,e}^p \Big\}^{1/p}$$

$$\overline{{}^e\mathcal{K}_q^p}(\epsilon) = r_0^{-1/2}\Big\{ \|r^{(\frac{5}{2}-\frac{3}{p}+q)} \not\nabla^q \epsilon\|_{p,e}^p \tag{11.1.3c}$$

$$+ \sum_{i+j=q\,,\,j\geq 1} \|r^{(2-\frac{2}{p}+i)} w_{(\frac{1}{2}-\frac{1}{p}+j)} \not\nabla^i \not\nabla_N^j \epsilon\|_{p,e}^p \Big\}^{1/p}$$

$$\overline{{}^e\mathcal{K}_q^p}(\hat{\eta}) = r_0^{-1/2}\Big\{ \sum_{i+j=q} \|r^{(1-\frac{2}{p}+i)} w_{(\frac{3}{2}-\frac{1}{p}+j)} \not\nabla^i \not\nabla_N^j \hat{\eta}\|_{p,e}^p \Big\}^{1/p}. \tag{11.1.3d}$$

We now set

$$\overline{{}^e\mathcal{K}_q^p} = \overline{{}^e\mathcal{K}_q^p}(\delta) + \overline{{}^e\mathcal{K}_q^p}(\epsilon) + \overline{{}^e\mathcal{K}_q^p}(\hat{\eta}) \tag{11.1.3e}$$

and define the expressions

$$\overline{\mathcal{K}_q^p} = \max(\overline{{}^i\mathcal{K}_q^p}, \overline{{}^e\mathcal{K}_q^p}). \tag{11.1.3f}$$

For the particular case $p = 2$ we write simply

$$\overline{\mathcal{K}_q} = \overline{\mathcal{K}_q^2}.$$

We also define the following:

$$\overline{\mathcal{K}_q^{p,S}}(\delta) = r_0^{-1/2}\Big\{ |||r^{(\frac{5}{2}-\frac{2}{p}+q)} \not\nabla^q \delta|||_{p,e}^p \tag{11.1.4a}$$

$$+ \sum_{i+j=q\,,\,j\geq 1} |||r^{(2-\frac{2}{p}+i)} w_{1/2+j} \not\nabla^i \nabla_N^j \delta|||_{p,e}^p \Big\}^{1/p}$$

$$\overline{\mathcal{K}_q^{p,S}}(\epsilon) = r_0^{-1/2}\Big\{ |||r^{(\frac{5}{2}-\frac{2}{p}+q)} \not\nabla^q \delta|||_{p,e}^p \tag{11.1.4b}$$

$$+ \sum_{i+j=q\,,\,j\geq 1} |||r^{(2-\frac{2}{p}+i)} w_{1/2+j} \not\nabla^i \not\nabla_N^j \epsilon|||_{p,e}^p \Big\}^{1/p}$$

$$\overline{\mathcal{K}_q^{p,S}}(\hat{\eta}) = r_0^{-1/2}\Big\{ \sum_{i+j=q} |||r^{(1-\frac{2}{p}+i)} w_{3/2+j} \not\nabla^i \not\nabla_N^j \hat{\eta}|||_{p,e}^p \Big\}^{1/p} \tag{11.1.4c}$$

and set

$$\overline{\mathcal{K}_q^{p,S}} = \overline{\mathcal{K}_q^{p,S}}(\delta) + \overline{\mathcal{K}_q^{p,S}}(\epsilon) + \overline{\mathcal{K}_q^{p,S}}(\hat{\eta}). \tag{11.1.4d}$$

Finally, we introduce the basic norms of the second fundamental form:

$$\overline{\mathcal{K}_{[0]}} = \overline{\mathcal{K}_0} \tag{11.1.5a}$$

$$\overline{\mathcal{K}_{[1]}} = \overline{\mathcal{K}_{[0]}} + \overline{\mathcal{K}_1} + \overline{\mathcal{K}_0^{3,S}} \tag{11.1.5b}$$

$$\overline{\mathcal{K}_{[2]}} = \overline{\mathcal{K}_{[1]}} + \overline{\mathcal{K}_2} + \overline{\mathcal{K}_1^{3,S}} \tag{11.1.5c}$$

$$\overline{\mathcal{K}_{[3]}} = \overline{\mathcal{K}_{[2]}} + \overline{\mathcal{K}_3} + \overline{\mathcal{K}_2^{3,S}}. \tag{11.1.5d}$$

Norms for **R**

In the interior region I we define

$$\overline{{}^{i}\mathcal{R}_q} = r_0^{2+q}\|\mathbf{D}^q\mathbf{R}\|_{2,i}. \tag{11.1.6a}$$

In the exterior region E we define

$$\overline{{}^{e}\mathcal{R}_0}^2 = \int_{\Sigma_t^e} \tau_-^4|\underline{\alpha}|^2 + \int_{\Sigma_t^e} \tau_-^2 r^2|\underline{\beta}|^2 + \int_{\Sigma_t^e} r^4|\rho - \bar{\rho}|^2 \tag{11.1.6b}$$
$$+ \int_{\Sigma_t^e} r^4|\sigma - \bar{\sigma}|^2 + \int_{\Sigma_t^e} r^4|\beta|^2 + \int_{\Sigma} r^4|\alpha|^2$$

$$\overline{{}^{e}\mathcal{R}_1}^2 = \int_{\Sigma_t^e} r^2\tau_-^4|\nabla\underline{\alpha}|^2 + \int_{\Sigma_t^e} r^4\tau_-^2|\nabla\underline{\beta}|^2 + \int_{\Sigma_t^e} r^6|\nabla\rho|^2 \tag{11.1.6c}$$
$$+ \int_{\Sigma_t^e} r^6|\nabla\sigma|^2 + \int_{\Sigma_t^e} r^6|\nabla\beta|^2 + \int_{\Sigma_t^e} r^6|\nabla\alpha|^2$$
$$+ \int_{\Sigma_t^e} \tau_-^6|\nabla_N\underline{\alpha}|^2 + \int_{\Sigma_t^e} \tau_-^4 r^2|\nabla_N\underline{\beta}|^2$$
$$+ \int_{\Sigma} \tau_-^2 r^4|\rho_N|^2 + \int_{\Sigma_t^e} \tau_-^2 r^4|\sigma_N|^2$$
$$+ \int_{\Sigma_t^e} r^6|\nabla_N\beta|^2 + \int_{\Sigma_t^e} r^6|\nabla_N\alpha|^2$$

$$\overline{{}^{e}\mathcal{R}_2}^2 = \int_{\Sigma_t^e} r^4\tau_-^4|\nabla^2\underline{\alpha}|^2 + \int_{\Sigma_t^e} r^6\tau_-^2|\nabla^2\underline{\beta}|^2 + \int_{\Sigma_t^e} r^8|\nabla^2\rho|^2 \tag{11.1.6d}$$
$$+ \int_{\Sigma_t^e} r^8|\nabla^2\sigma|^2 + \int_{\Sigma_t^e} r^8|\nabla^2\beta|^2 + \int_{\Sigma_t^e} r^8|\nabla^2\alpha|^2$$

$$+\int_{\Sigma_t^e} r^2\tau_-^6|\nabla\!\!\!/\nabla\!\!\!/_N\underline{\alpha}|^2 + \int_{\Sigma_t^e} \tau_-^4 r^4|\nabla\!\!\!/\nabla\!\!\!/_N\underline{\beta}|^2 + \int_{\Sigma_t^e} \tau_-^2 r^6|\nabla\!\!\!/\rho_N|^2$$
$$+\int_{\Sigma_t^e} \tau_-^2 r^6|\nabla\!\!\!/\sigma_N|^2 + \int_{\Sigma_t^e} r^8|\nabla\!\!\!/\nabla\!\!\!/_N\beta|^2 + \int_{\Sigma_t^e} r^8|\nabla\!\!\!/\nabla\!\!\!/_N\alpha|^2$$
$$+\int_{\Sigma_t^e} \tau_-^8|\nabla\!\!\!/_N^2\underline{\alpha}|^2 + \int_{\Sigma_t^e} \tau_-^6 r^2|\nabla\!\!\!/_N^2\underline{\beta}|^2 + \int_{\Sigma_t^e} \tau_-^4 r^4|\nabla_N\rho_N|^2$$
$$+\int_{\Sigma_t^e} \tau_-^4 r^4|\nabla_N\sigma_N|^2 + \int_{\Sigma_t^e} \tau_-^2 r^6|\nabla\!\!\!/_N^2\beta|^2 + \int_{\Sigma_t^e} r^8|\nabla\!\!\!/_N^2\alpha|^2$$

where

$$\rho_N = \nabla_N\rho + \frac{3}{2}\text{tr}\theta\rho, \qquad \sigma_N = \nabla_N\sigma + \frac{3}{2}\text{tr}\theta\sigma.$$

We now introduce, for $q = 0, 1, 2$,

$$\overline{\mathcal{R}_q}(W) = \max(\overline{{}^e\mathcal{R}_q}, \overline{{}^i\mathcal{R}_q}) \tag{11.1.6e}$$

and our main norms

$$\overline{\mathcal{R}_{[1]}} = \overline{\mathcal{R}_0} + \overline{\mathcal{R}_1} \tag{11.1.6f}$$
$$\overline{\mathcal{R}_{[2]}} = \overline{\mathcal{R}_{[1]}} + \overline{\mathcal{R}_2}. \tag{11.1.6g}$$

We now introduce the following assumptions:

Assumption 0: The fundamental constants[3] ${a_m}^{-1}, a_M, h_m^{-1}, h_M, k_m^{-1}, k_M$ of the foliation induced by the optical function u on the t-slices Σ_t are bounded:

$${a_m}^{-1}, a_M, h_m^{-1}, h_M, k_m^{-1}, k_M \leq c. \tag{11.1.7a}$$

Assumption 1: The constants[4] $\varsigma, a_1, \tilde{a}_1, h_1, a_2$ are bounded according to

$$\varsigma, a_1, \tilde{a}_1, h_1, a_2 \leq c. \tag{11.1.7b}$$

Assumption 2: The space-time curvature norm $\overline{\mathcal{R}_{[1]}}$ is small:

$$\overline{\mathcal{R}_{[1]}} \leq \varepsilon_0. \tag{11.1.7c}$$

Next we give a necessary set of simplified assumptions that have to be satisfied by the Hessian of the optical function:

Assumption 3: The components χ, ζ, ω of the Hessian of the extended optical function u verify the following:

[3] See definition 3.1.3 on page 56

[4] See 3.1.3, 4.1.2c, 4.1.2d, and 4 1.2b.

Interior Assumptions for the Extended Optical Function

$$\begin{aligned} r_0^{3/2} \sup_I |\mathrm{tr}\chi - \overline{\mathrm{tr}\chi}| &\leq \varepsilon_0 \\ r_0^{3/2} \sup_I |\hat{\chi}| &\leq \varepsilon_0 \\ r_0^{3/2} \sup_I |\zeta| &\leq \varepsilon_0 \\ r_0^{3/2} \sup_I |\omega| &\leq \varepsilon_0 \end{aligned} \tag{11.1.8a}$$

Also, for the first derivatives,

$$r_0^{1/2} \|r^{1/2} \nabla\!\!\!/ (\chi, \zeta, \omega)\|_{2,\imath} \leq \varepsilon_0. \tag{11.1.8b}$$

Exterior Assumptions for the Optical Function

$$\begin{aligned} \sup_E r^2 |\mathrm{tr}\chi - \overline{\mathrm{tr}\chi}| &\leq \varepsilon_0 \\ \sup_E r^2 |\hat{\chi}| &\leq \varepsilon_0 \\ \sup_E r^2 |\zeta| &\leq \varepsilon_0 \\ \sup_E r^{1/2} w_{3/2} |\omega| &\leq \varepsilon_0 \end{aligned} \tag{11.1.8c}$$

also

$$r_0^{1/2} \|r^{-1} w_1 \omega\|_{2,e} \leq \varepsilon_0 \tag{11.1.8d}$$

and, for the first derivatives,

$$\begin{aligned} r_0^{1/2} \|r \nabla\!\!\!/ (\chi, \zeta)\|_{2,e} &\leq \varepsilon_0 \\ r_0^{1/2} \|w_1 \nabla\!\!\!/ \omega\|_{2,e} &\leq \varepsilon_0. \end{aligned} \tag{11.1.8e}$$

Finally, we write down the following bootstrap assumptions for the second fundamental form k:

Bootstrap Assumption 1: $\overline{\mathcal{K}_0^\infty} \leq \varepsilon_0, \qquad \overline{\mathcal{K}_{[1]}} \leq \varepsilon_0$

Bootstrap Assumption 2: $\|r^{3/2} \tau_-^{1/2} \nabla\!\!\!/ \hat{\eta}\|_{4,e} \leq \varepsilon_0.$

Next we display a list of estimates for the fundamental parameters of the radial foliation $a, \mathrm{tr}\theta, \hat{\theta}$, as well as for the following scalars (see 3.1.1i and 4.1.3d):

$$\lambda = \frac{dr}{du} = \frac{r}{2}\overline{a\mathrm{tr}\theta} \tag{11.1.9}$$

$$\kappa = ra^{-1}(a\mathrm{tr}\theta - \overline{a tr\theta}). \tag{11.1.10}$$

Lemma 11.1.1 *Assumptions 0, 1, 2, and 3 together with bootstrap assumptions 1 and 2 imply the estimates*

$$r_0^{3/2} \sup_I |\mathrm{tr}\theta - \overline{\mathrm{tr}\theta}| \leq c\varepsilon_0$$
$$r_0^{3/2} \sup_I r^{-1}|\kappa| \leq c\varepsilon_0$$
$$r_0^{3/2} \sup_I |\hat{\theta}| \leq c\varepsilon_0$$
$$r_0^{3/2} \sup_I |a^{-1}\nabla\!\!\!/ a| \leq c\varepsilon_0$$
$$r_0^{3/2} \sup_I |a^{-1}\nabla_N a| \leq c\varepsilon_0$$

$$r_0^{1/2}\|r^{1/2}\nabla\!\!\!/\mathrm{tr}\theta\|_{2,\imath} \leq c\varepsilon_0$$
$$r_0^{1/2}\|r^{-1/2}\nabla\!\!\!/\kappa\|_{2,\imath} \leq c\varepsilon_0$$
$$r_0^{1/2}\|r^{1/2}\nabla\!\!\!/\hat{\theta}\|_{2,\imath} \leq c\varepsilon_0$$
$$r_0^{1/2}\|r^{1/2}a^{-1}\nabla\!\!\!/^2 a\|_{2,\imath} \leq c\varepsilon_0$$
$$r_0^{1/2}\|r^{1/2}a^{-1}\nabla_N a\|_{2,\imath} \leq c\varepsilon_0;$$

and in the exterior

$$\sup_E r^2|\mathrm{tr}\theta - \overline{\mathrm{tr}\theta}| \leq c\varepsilon_0$$
$$\sup_E r|\kappa| \leq c\varepsilon_0$$
$$\sup_E r^{1/2}w_{3/2}|\hat{\theta}| \leq c\varepsilon_0$$
$$\sup_E r^2|a^{-1}\nabla\!\!\!/ a| \leq c\varepsilon_0$$
$$\sup_E r^{1/2}w_{3/2}|a^{-1}\nabla_N a| \leq c\varepsilon_0$$

$$r_0^{1/2}\|r\nabla\!\!\!/\mathrm{tr}\theta\|_{2,e} \leq c\varepsilon_0$$

$$
\begin{aligned}
r_0^{1/2}\|\nabla\!\!\!/ \kappa\|_{2,e} &\le c\varepsilon_0\\
r_0^{1/2}\|w_1\nabla\!\!\!/ \hat{\theta}\|_{2,e} &\le c\varepsilon_0\\
r_0^{1/2}\|r^{-1}w_1 a^{-1}\nabla_N a\|_{2,e} &\le c\varepsilon_0\\
r_0^{1/2}\|r a^{-1}\nabla\!\!\!/^2 a\|_{2,e} &\le c\varepsilon_0\\
r_0^{1/2}\|w_1 a^{-1}\nabla_N a\|_{2,e} &\le c\varepsilon_0.
\end{aligned}
$$

The proof of the lemma follows easily from our assumptions, the formulas 11.1.1e, and the definitions 11.1.10. In the interior we also make use of the following formulas, which are immediate consequences of 11.1.1a:

$$
\begin{aligned}
\nabla\!\!\!/_i\delta &= \Pi_i{}^a N^b N^c \nabla_a k_{bc} + 2\theta_i{}^a\epsilon_a\\
\nabla\!\!\!/_i\epsilon_j &= \Pi_i{}^a\Pi_j{}^b N^c\nabla_a k_{bc} + \theta_i{}^a\hat{\eta}_{aj} - \frac{3}{2}\theta_{ij}\delta\\
\nabla\!\!\!/_l\hat{\eta}_{ij} &= \Pi_i{}^a\Pi_j{}^b\Pi_l{}^c\nabla_c k_{ab} - \theta_{il}\epsilon_j - \theta_{jl}\epsilon_i + \gamma_{ij}\theta_l{}^m\epsilon_m
\end{aligned}
$$

where Π is the projection operator of the radial foliation (see 3.1.1b) and $\gamma = \Pi_i^m\Pi_j^n g_{mn}$ is the projected metric.

In the next lemma we record a list of estimates that involve normal derivatives of the quantities $r\mathrm{tr}\chi$, $\hat{\chi}$, ζ as well as $r\mathrm{tr}\theta$, $\hat{\theta}$, κ and that follow easily from the bootstrap assumptions.

Lemma 11.1.2 *Assumptions 0, 1, 2, and 3 and bootstrap assumptions 1 and 2 imply the estimates*

$$
\begin{aligned}
r_0^{1/2}\|r^{-1/2}\nabla_N(r\mathrm{tr}\chi)\|_{2,i} &\le c(\varepsilon_0 + \overline{\mathcal{R}_{[1]}})\\
r_0^{1/2}\|r^{1/2}\nabla\!\!\!/_N\hat{\chi}\|_{2,i} &\le c(\varepsilon_0 + \overline{\mathcal{R}_{[1]}})\\
r_0^{1/2}\|r^{1/2}\nabla\!\!\!/_N\zeta\|_{2,i} &\le c(\varepsilon_0 + \overline{\mathcal{R}_{[1]}})
\end{aligned}
$$

and, in the exterior,

$$
\begin{aligned}
r_0^{1/2}\|\nabla_N(r\mathrm{tr}\chi)\|_{2,e} &\le c(\varepsilon_0 + \overline{\mathcal{R}_{[1]}})\\
r_0^{1/2}\|w_1\nabla\!\!\!/_N\hat{\chi}\|_{2,e} &\le c(\varepsilon_0 + \overline{\mathcal{R}_{[1]}})\\
r_0^{1/2}\|w_1\nabla\!\!\!/_N\zeta\|_{2,e} &\le c(\varepsilon_0 + \overline{\mathcal{R}_{[1]}}).
\end{aligned}
$$

Similarly,

$$
r_0^{1/2}\|r^{-1/2}\nabla_N(r\mathrm{tr}\theta)\|_{2,i} \le c(\varepsilon_0 + \overline{\mathcal{R}_{[1]}})
$$

$$r_0^{1/2}\|r^{-1/2}\nabla\!\!\!/_N\kappa\|_{2,i} \le c(\varepsilon_0 + \overline{\mathcal{R}_{[1]}})$$
$$r_0^{1/2}\|r^{1/2}\nabla\!\!\!/_N\hat{\theta}\|_{2,i} \le c(\varepsilon_0 + \overline{\mathcal{R}_{[1]}}),$$

while in the exterior,

$$r_0^{1/2}\|r^{-1}w_1\nabla_N(r\mathrm{tr}\theta)\|_{2,e} \le c(\varepsilon_0 + \overline{\mathcal{R}_{[1]}})$$
$$r_0^{1/2}\|r^{-1}w_1\nabla_N\kappa\|_{2,e} \le c(\varepsilon_0 + \overline{\mathcal{R}_{[1]}})$$
$$r_0^{1/2}\|r^{-1}w_2\nabla\!\!\!/_N\hat{\theta}\|_{2,e} \le c(\varepsilon_0 + \overline{\mathcal{R}_{[1]}}).$$

The first part of the lemma is an immediate consequence of our assumptions as well as the following formulas:

$$\nabla_N\mathrm{tr}\chi + \frac{1}{2}(\mathrm{tr}\chi)^2 = -\frac{1}{2}\mathrm{tr}\chi\delta - \hat{\chi}\cdot\hat{\eta} - |\hat{\chi}|^2 - \mathrm{d}\!\!/\mathrm{iv}\,\zeta - |\zeta|^2 - \rho$$
$$\nabla_N(r\mathrm{tr}\chi) = r(\nabla_N\mathrm{tr}\chi + \frac{1}{2}(\mathrm{tr}\chi)^2) - 2\mathrm{tr}\chi(r\mathrm{tr}\chi - 2a^{-1}\lambda)$$
$$\nabla\!\!\!/_N\hat{\chi}_{ij} = -\frac{1}{2}\mathrm{tr}\chi\hat{\eta}_{ij} - (\mathrm{tr}\chi + \frac{1}{2}\delta)\hat{\chi}_{ij} - \frac{1}{2}\alpha_{ij}$$
$$-\frac{1}{2}(\nabla\!\!\!/_i\zeta_j + \nabla\!\!\!/_j\zeta_i - \gamma_{ij}\mathrm{d}\!\!/\mathrm{iv}\,\zeta) - \zeta_i\zeta_j + \frac{1}{2}\gamma_{ij}|\zeta|^2$$
$$\nabla\!\!\!/_N\zeta_i = -\nabla\!\!\!/_i\omega - \mathrm{tr}\chi\zeta_i - \hat{\chi}_{ij}\zeta_j - 2\hat{\eta}\cdot\zeta - \omega\zeta_i$$
$$+\,\omega\epsilon_i - \hat{\chi}_{ij}\epsilon_j + \delta\epsilon_i + \frac{1}{2}\underline{\beta}_i - \frac{1}{2}\beta_i.$$

Although assumption 2 for the space-time curvature norm does not contain information on $\bar{\rho}$ and $\bar{\sigma}$, we show now that this can be obtained as follows:

One can in fact easily estimate $\bar{\sigma}$ from the curl equation in 11.1.1a. Indeed, the normal-normal component of that equation (see also 11.1.2a–11.1.2e) is

$$\begin{aligned} \mathrm{c}\!\!/\mathrm{url}\,\epsilon &= \sigma + \hat{\theta}\wedge\hat{\eta} \\ &= \sigma + \hat{\chi}\wedge\hat{\eta}. \end{aligned} \tag{11.1.11}$$

Integrating 11.1.11 on B_r, the interior of the sphere S_r, we infer that

$$\bar{\sigma} = -\overline{\hat{\chi}\wedge\hat{\eta}} \tag{11.1.12a}$$

and consequently, in view of our assumptions,

$$\sup_E(r^3 w_{3/2}|\bar{\sigma}|) \le c\varepsilon_0 r_0^{1/2}\overline{\mathcal{K}_{[1]}}. \tag{11.1.12b}$$

On the other hand, the situation with $\bar{\rho}$ is completely different. To estimate it, we have to appeal, instead of to 11.1.1a, to the Bianchi identity 11.1.1c. Contracting the left-hand side of that equation with the position vectorfield Z (see 4.1.3a) and integrating on B_r, the interior of the spheres S_r, we find

$$\int_{S_r} E(Z,N) = \frac{1}{2}\int_{B_r} \pi \cdot E + \int_{B_r} Z \cdot (k \wedge E)$$

where π is the deformation tensor of Z. Taking into account formula 4.1.3b for the traceless part of π and the fact that, in view of the definition of κ, $\int_{S_u} \kappa = 0$ and therefore by the coarea formula, $\int_{B_r} a^{-1}\kappa\rho = \int_{B_r} a^{-1}\kappa(\rho - \bar{\rho})$, we deduce the following equation for $\bar{\rho}$:

$$\begin{aligned} 4\pi r^3 \bar{\rho} = &-\int_{B_r} a^{-1}\kappa(\rho - \bar{\rho}) \\ &+ \frac{1}{2}\int_{B_r} r\Big((a^{-1}\nabla\!\!\!/ a + \epsilon)\cdot \underline{\beta} + (a^{-1}\nabla\!\!\!/ a - \epsilon) \\ &\cdot \beta + \frac{1}{2}\hat{\chi}\cdot \underline{\alpha} - \frac{1}{2}\underline{\hat{\chi}}\cdot \alpha\Big). \end{aligned} \tag{11.1.12c}$$

Estimating the left-hand side of this equation, we find

$$\sup_{E\bigcap\{u\le 0\}} r^3 w_{3/2}|\bar{\rho}| \le c\varepsilon_0 r_0^{1/2}\overline{\mathcal{R}_{[1]}} \tag{11.1.12d}$$

$$\sup_{E\bigcap\{u>0\}} r^3|\bar{\rho}| \le c\varepsilon_0\,\overline{\mathcal{R}_{[1]}}. \tag{11.1.12e}$$

We also record here the following inequalities concerning $\nabla_N\lambda$:

$$\sup_I r_0^{3/2}|\nabla_N\lambda| \le c(\varepsilon_0 + \overline{\mathcal{R}_{[1]}})$$

$$\sup_E r^{1/2} w_{3/2}|\nabla_N\lambda| \le c(\varepsilon_0 + \overline{\mathcal{R}_{[1]}}). \tag{11.1.13a}$$

These follow easily from the calculation

$$\nabla_N\lambda = \frac{1}{2} r a^{-1}\left(-\frac{1}{2}r^{-2}\overline{a^2\kappa^2} - \overline{a^2|\hat{\theta}|^2} + \overline{|\nabla\!\!\!/ a|^2} + \overline{\mathrm{tr}\theta a\nabla_N a} - \overline{a^2 R_{NN}}\right). \tag{11.1.13b}$$

11.2. The Exterior Estimates

Our strategy here is to estimate successively the components δ, ϵ, $\hat{\eta}$ of the second fundamental form. The first and most complicated step is to estimate δ.[5] To

[5] Remark that, in view of the equations 11.1.2a–11.1.2e, once δ, $\nabla_N\delta$ are estimated, ϵ, $\hat{\eta}$ follow easily by solving Hodge systems on the 2-surfaces of the radial foliation.

begin, we define, with the help of the position vectorfield Z introduced in 4.1.3a,

$$(\mathbf{i}_Z k)_i = k_{ij} Z^j = r(\epsilon_i + \delta N_i). \tag{11.2.1a}$$

Using the equations 11.1.1a, we deduce

$$\operatorname{curl}(\mathbf{i}_Z k) = F + G \tag{11.2.1b}$$

$$\operatorname{div}(\mathbf{i}_Z k) = \frac{1}{2}\hat{\pi}^{ij} k_{ij} \tag{11.2.1c}$$

where

$$F_i = H_{ij} Z^j \tag{11.2.1d}$$

$$G_i = \frac{1}{2} \in_i{}^{mn} \aleph_m^{\ j} k_{nj} \tag{11.2.1e}$$

and $\aleph$ is the expression

$$\aleph_m^{\ j} = \hat{\pi}_m^{\ j} - ra^{-1}(N_m \nabla^j a - N^j \nabla_m a) \tag{11.2.1f}$$

defined in terms of the traceless part of the deformation tensor of Z (see 4.2.4).

Remark 1 *In view of formula 4.2.4, taking into account the assumptions and results of the previous paragraph, we find that the term $\hat{\pi}^{ij} k_{ij}$, which appears on the right-hand side of 11.2.1c, has a very large contribution in the wave zone. Indeed,*

$$\frac{1}{2}\hat{\pi}^{ij} k_{ij} = r\hat{\theta}^{ij}\hat{\eta}_{ij} + l.o.t$$
$$= O(r^{-1}); \quad along\ u = const.$$

This a purely nonlinear effect. More precisely, this is a nonlinear effect by comparison to the behavior of k in flat space. Indeed, if we assume that k is a solution of the equations 11.1.1a in flat space, with H the magnetic part of a solution to the Bianchi equations (see definition on page 3) enjoying the usual decay properties, we find the following asymptotic results for the components $\epsilon, \delta, \hat{\eta}$:

$$\delta = O(r^{-5/2})$$
$$\epsilon = O(r^{-2}\tau_-^{-1/2})$$
$$\hat{\eta} = O(r^{-1}\tau_-^{-3/2}).$$

As a matter of fact, with Z the position vectorfield in $\Re^3$, we find

$$curl(\mathbf{i}_Z k) = i_Z H$$
$$div(\mathbf{i}_Z k) = 0.$$

In components, the second equation reads

$$\not{d}iv\,\epsilon + \nabla_N \delta + \frac{3}{2}\mathrm{tr}\theta \cdot \delta = 0.$$

Using this equation, we find that in the linear theory $\not{d}iv\,\epsilon, \nabla_N\delta = O(r^{-3})$ *in the wave zone. In contrast, in our case we have the equation*

$$\not{d}iv\,\epsilon + \nabla_N \delta + \frac{3}{2}\mathrm{tr}\theta \cdot \delta = -(2a^{-1}\epsilon \cdot \not{\nabla} a - \hat{\eta} \cdot \theta)$$

where we see that the term $\hat{\eta} \cdot \theta = O(r^{-2})$ *dominates the contribution, predicted by the linear theory, of each term on the left-hand side.*

To circumvent this problem, we decompose $\mathbf{i}_Z k$ according to

$$\mathbf{i}_Z k = \widetilde{\mathbf{i}_Z k} + \nabla\psi \tag{11.2.2a}$$

where

$$\triangle\psi = r|\hat{\eta}|^2. \tag{11.2.2b}$$

In view of 11.2.1b and 11.2.1c we deduce

$$\mathrm{curl}(\widetilde{\mathbf{i}_Z k}) = F + G \tag{11.2.2c}$$

$$\mathrm{div}(\widetilde{\mathbf{i}_Z k}) = I \tag{11.2.2d}$$

where

$$\begin{aligned} I &= \frac{1}{2}\hat{\pi}^{ij}k_{ij} - r|\hat{\eta}|^2 \\ &= r\hat{\chi} \cdot \hat{\eta} - \kappa\delta - 2r^{-1}a^{-1}\not{\nabla} a \cdot \epsilon. \end{aligned} \tag{11.2.2e}$$

We obtain a Poisson equation for $\widetilde{\mathbf{i}_Z k}$ by applying to 11.2.2c and 11.2.2d the following formula, which holds for an arbitrary 1-form v:

$$\mathrm{curl}(\mathrm{curl}\,v)_i = R_i{}^j v_j - \triangle v_i + \nabla_i(\mathrm{div}\,v). \tag{11.2.2f}$$

Thus

$$\triangle(\widetilde{\mathbf{i}_Z k})_i = -\mathrm{curl}(F+G)_i + R_i{}^j(\widetilde{\mathbf{i}_Z k})_j + \nabla_i I. \tag{11.2.2g}$$

Contracting once again with the position vectorfield Z, we find, for the scalar function

$$p = Z \cdot \widetilde{\mathbf{i}_Z k} = r(r\delta - \nabla_N\psi), \tag{11.2.2h}$$

the following:

$$\begin{aligned}\triangle p &= Z \cdot \triangle(\widetilde{\mathbf{i}_Z k}) + 2\nabla^j Z^i \nabla_j(\widetilde{\mathbf{i}_Z k_i}) + \triangle Z \cdot \widetilde{\mathbf{i}_Z k} \\ &= Z \cdot \triangle(\widetilde{\mathbf{i}_Z k}) + \operatorname{curl} Z \cdot \operatorname{curl}(\widetilde{\mathbf{i}_Z k}) \\ &\quad + \hat{\pi}^{ij}(\nabla_j(\widetilde{\mathbf{i}_Z k})_i + \nabla_i \mathbf{i}_Z k_j - \frac{2}{3} g_{ij} \operatorname{div}(\widetilde{\mathbf{i}_Z k})) \\ &\quad + \frac{1}{3}\operatorname{tr}\pi \operatorname{div}(\widetilde{\mathbf{i}_Z k}) + \widetilde{\mathbf{i}_Z k} \cdot \triangle Z. \end{aligned} \tag{11.2.2i}$$

Therefore, in view of 11.2.2g and taking also into account the formula

$$\triangle Z^i + R_i{}^j Z^j = \nabla_j \hat{\pi}^{ij} - \frac{1}{6}\nabla^i(\operatorname{tr}\pi),$$

we derive

$$\begin{aligned}\triangle p = &-Z \cdot \operatorname{curl}(F + G) + \nabla_Z I + \aleph^{ij}\nabla_i(\mathbf{i}_Z k_j) \\ &+ \frac{1}{3}\operatorname{tr}\pi I - \hat{\pi}^{ij}\nabla_i\nabla_j\psi \\ &+ (\nabla_j \hat{\pi}^{ij} - \frac{1}{6}\nabla^i(\operatorname{tr}\pi))((\mathbf{i}_Z k)_i - \nabla_i\psi)\end{aligned} \tag{11.2.3}$$

where F, G are given by formulas 11.2.1d and 11.2.1e, and I by 11.2.2e.

We now observe that for any 1-form A on Σ we have

$$Z \cdot \operatorname{curl} A = r \, c\!\!\!/url \, A\!\!\!/$$

where $A\!\!\!/ = \Pi_i{}^j A_j$. Applying this fact to F, we find

$$Z \cdot \operatorname{curl} F = \frac{1}{2} r^2 (d\!\!\!/iv \, \underline{\beta} - d\!\!\!/iv \, \beta).$$

Remark 2 *The term $r^2 d\!\!\!/iv \, \underline{\beta}$ still gives a large contribution in the wave zone. Yet we could attempt at this point to apply to 11.2.3 the degenerate elliptic estimates derived in Propositions 4.2.1ii and 4.2.2ii. Indeed, in view of our bootstrap assumptions for the space-time curvature we can check that $\|\tau_- \triangle p\|_2$ is finite. Unfortunately, however, the degenerate elliptic estimates, which we have, require one more degree of differentiability in the angular directions than the usual nondegenerate estimates. This will immediately take us out of the range of differentiability that we allow for the space-time curvature. The troublesome term $r^2 d\!\!\!/iv \, \underline{\beta}$ can, however, be eliminated with the help of the Bianchi identities. Indeed, we recall*

$$div E_i - \in_i{}^{ab} \, k_a{}^c H_c{}^b = 0,$$

which, after contracting with N_i, yields

$$\frac{1}{2}(\not{d}\mathrm{iv}\,\underline{\beta} + \not{d}\mathrm{iv}\,\beta) = \nabla_N \rho - \frac{1}{4}\hat{\chi}\cdot\underline{\alpha} + \frac{1}{4}\underline{\hat{\chi}}\cdot\alpha + \frac{3}{2}\mathrm{tr}\theta\rho - (a^{-1}\not{\nabla} a + \frac{1}{2}\epsilon)\cdot\underline{\beta} - (a^{-1}\not{\nabla} a - \frac{1}{2}\epsilon)\cdot\beta. \tag{11.2.4}$$

We can therefore express the term $r^2\not{d}\mathrm{iv}\,\underline{\beta}$, which occurs on the right-hand side of 11.2.3, in terms of $\nabla_Z\rho$. We can then try to define a function q to be the solution of $\triangle q = -r\rho$ that tends to zero at infinity and then derive a Poisson equation for the difference $p - \nabla_Z q$. The hope is that this procedure will allow us to apply the nondegenerate elliptic estimates for this residual equation as well as the equation satisfied by q. This doesn't quite work, for two reasons. First, in view of the decay rate of ρ toward spacelike infinity, there is no solution q of $\triangle q = -r\rho$ tending to zero at infinity on Σ_t. The other difficulty is the presence of the slowly decaying term $\hat{\chi}\cdot\underline{\alpha}$ on the right-hand side of 11.2.4. The first difficulty can be overcome by replacing ρ by $\rho - \bar{\rho}$ in the definition of q. We also surmount the second difficulty by replacing ρ by the following scalar function, which we call the approximate mass aspect function:[6]

$$\mu = -\rho - \hat{\chi}\cdot\hat{\eta}. \tag{11.2.5}$$

We first calculate the normal derivative of μ,

$$\nabla_N(r^3\mu) + \frac{3}{2}\kappa r^2\mu = -r^3\{\frac{1}{2}(\not{d}\mathrm{iv}\,\underline{\beta} + \not{d}\mathrm{iv}\,\beta) + \frac{1}{2}a^{-1}\not{\nabla} a\cdot(\underline{\beta}+\beta) - r^{-1}a^{-1}\lambda|\hat{\eta}|^2 + J\}, \tag{11.2.6a}$$

where

$$J + a^{-1}\hat{\eta}\cdot\not{\nabla}^2 a = -\frac{1}{2}r^{-1}\kappa|\hat{\eta}|^2 + \frac{1}{2}\hat{\chi}\cdot\alpha + \frac{1}{2}\delta(3|\hat{\chi}|^2 - |\hat{\eta}|^2) + \frac{1}{2}(a^{-1}\not{\nabla} a + \epsilon)\cdot\underline{\beta} + \frac{1}{2}(a^{-1}\not{\nabla} a - \epsilon)\cdot\beta - \epsilon\cdot\hat{\eta}\cdot\epsilon + (\hat{\chi} - \hat{\eta})\cdot\not{\nabla}\epsilon + 2a^{-1}\not{\nabla} a\cdot(\hat{\chi} - \hat{\eta})\cdot\epsilon. \tag{11.2.6b}$$

This follows with the help of the equations

$$\not{\nabla}_N\hat{\eta}_{ij} = -\frac{1}{2}\mathrm{tr}\theta\hat{\eta}_{ij} - \frac{1}{4}\underline{\alpha}_{ij} + \frac{1}{4}\alpha_{ij} + \frac{1}{2}(\not{\nabla}_i\epsilon_j + \not{\nabla}_j\epsilon_i - \gamma_{ij}\not{d}\mathrm{iv}\,\epsilon) \tag{11.2.6c}$$

[6] The correct definition will be given in Chapter 17.

$$+ a^{-1}(\epsilon_i \nabla\!\!\!/_j a + \epsilon_j \nabla\!\!\!/_i a - \gamma_{ij}\epsilon \cdot \nabla\!\!\!/ a) + \frac{3}{2}\hat{\theta}_{ij}\delta$$

$$\begin{aligned}\nabla\!\!\!/_N \hat{\chi}_{ij} = &-\mathrm{tr}\theta(\hat{\chi}_{ij} + \frac{1}{2}\hat{\eta}_{ij}) - \delta(\frac{3}{2}\hat{\chi}_{ij} + \frac{1}{2}\hat{\eta}_{ij}) \\ &- a^{-1}(\nabla\!\!\!/_i \nabla\!\!\!/_j a - \frac{1}{2}\gamma_{ij}\Delta\!\!\!/ a) - \frac{1}{2}\alpha_{ij} \\ &- a^{-1}(\epsilon_i \nabla\!\!\!/_j a + \epsilon_j \nabla\!\!\!/_i a - \gamma_{ij}\epsilon \cdot \nabla\!\!\!/ a) \\ &- \frac{1}{2}(\nabla\!\!\!/_i \epsilon_j - \nabla\!\!\!/_j \epsilon_i - \gamma_{ij}\,\mathrm{d}\!\!\!/\mathrm{iv}\,\epsilon) - \epsilon_i\epsilon_j + \frac{1}{2}\gamma_{ij}\epsilon^2.\end{aligned} \tag{11.2.6d}$$

The first equation is 11.1.2e, while the second is an immediate consequence of the null structure equations for $\mathbf{D}\!\!\!\!/_4\hat{\chi}, \mathbf{D}\!\!\!\!/_3\hat{\chi}$ derived in Part II.

Now, using formula 3.1.11, we infer from 11.2.6a that the average of μ over the spheres $S_{t,u}$ verifies

$$\frac{dr^3(\bar{\mu})}{du} = \lambda r^2\overline{|\hat{\eta}|^2} - r^2\overline{a\kappa(\mu - \overline{\mu})} - r^3\overline{aJ}. \tag{11.2.6e}$$

Subtracting this last equation from 11.2.6a, we find

$$-\frac{1}{2}r^2(\mathrm{d}\!\!\!/\mathrm{iv}\,\underline{\beta} + \mathrm{d}\!\!\!/\mathrm{iv}\,\beta) = r^{-1}\nabla_N[r^3(\mu - \bar{\mu})] - ra^{-1}\lambda(|\hat{\eta}|^2 - \overline{|\hat{\eta}|^2}) + \mathrm{Error}, \tag{11.2.6f}$$

with the error terms given by

$$\begin{aligned}\mathrm{Error} = &\frac{1}{2}r^2a^{-1}\nabla a \cdot (\underline{\beta} + \beta) + r^2a^{-1}(aJ - \overline{aJ}) \\ &+ \frac{1}{2}a^{-1}[3a\kappa\mu - \overline{a\kappa(\mu - \overline{\mu})}].\end{aligned} \tag{11.2.6g}$$

Finally, substituting this expression into 11.2.2e, we find

$$\begin{aligned}\Delta p = &\nabla_Z[(\mu - \bar{\mu}) + I] - ra^{-1}\lambda(|\hat{\eta}|^2 - \overline{|\hat{\eta}|^2}) \\ &+ r^2\,\mathrm{d}\!\!\!/\mathrm{iv}\,\beta + \frac{1}{3}\mathrm{tr}\pi[(\mu - \bar{\mu}) + I] + \mathrm{l.o.t}\end{aligned} \tag{11.2.6h}$$

where the lower order terms, l.o.t, will appear explicitly in the formula 11.2.7d.

We have thus succeeded in incorporating the troublesome terms into the derivative with respect to Z of a function with proper decay properties.

There is one more hurdle, however: the term $ra^{-1}\lambda(|\hat{\eta}|^2 - \overline{|\hat{\eta}|^2})$ decays in the wave zone no faster then the terms we have eliminated, although with one more degree of differentiability. In fact this term is of the same type as the term $r|\hat{\eta}|^2$, which had previously appeared on the right-hand side of 11.2.1c and which we have isolated in equation 11.2.2b. We overcome this problem in the same fashion, subtracting from p an additional scalar function ψ'. More precisely, we introduce the functions q and ψ' to be the solutions, vanishing at infinity, to the equations

$$\Delta q = r(\mu - \bar{\mu}) + I \tag{11.2.7a}$$
$$\Delta \psi' = -ra^{-1}\lambda(|\hat{\eta}|^2 - \overline{|\hat{\eta}|^2}), \tag{11.2.7b}$$

and we define

$$q' = p - \nabla_Z q - \psi'. \tag{11.2.7c}$$

In view of formulas 11.2.3, 11.2.6f, and 11.2.6g, and using also the commutation formulas 4.2.4 and 4.1.3b, we check that q' verifies the equation

$$\Delta q' = r^2 \mathrm{d\!\!/iv}\, \beta + S \tag{11.2.7d}$$

with S the precise error term

$$\begin{aligned} S = &-\hat{\pi} \cdot \nabla^2(q + \psi) - (\nabla_j \hat{\pi}^{ij} - \frac{1}{6}\nabla^i(\mathrm{tr}\pi))(\nabla_i q - \mathbf{i}_Z k_i + \nabla_i \psi) \\ &- r\mathrm{c\!\!/url}\, \not{G} + r^2 a^{-1}(aJ - \overline{aJ}) + \aleph^{ij}\nabla_i(\mathbf{i}_Z k)_j \\ &+ \frac{1}{2} r a^{-1}\left[3a\kappa\mu - \frac{4}{3}a\kappa(\mu - \bar{\mu}) + \overline{a\kappa(\mu - \bar{\mu})}\right] + \frac{1}{2}r^2 a^{-1} \nabla\!\!\!/ a \cdot (\underline{\beta} + \beta). \end{aligned} \tag{11.2.7e}$$

We collect all the previous formulas into the following:

Lemma 11.2.1 *The component $\delta = k(N, N)$ of the second fundamental form can be decomposed according to the formula*

$$\delta = r^{-2}(q' + \nabla_Z q + \psi' + \nabla_Z \psi)$$

where the scalars ψ, q, q', ψ' are the solutions, vanishing at infinity, of the Poisson equations

$$\Delta \psi = r|\hat{\eta}|^2 \tag{11.2.8a}$$
$$\Delta q = r(\mu - \bar{\mu}) + I \tag{11.2.8b}$$
$$\Delta \psi' = -ra^{-1}\lambda(|\hat{\eta}|^2 - \overline{|\hat{\eta}|^2}) \tag{11.2.8c}$$
$$\Delta q' = r^2 \mathrm{d\!\!/iv}\, \beta + S \tag{11.2.8d}$$

with the error terms I and S given, respectively, by formulas 11.2.2e and 11.2.7e.

We now start estimating ψ. In view of our bootstrap assumptions, we easily check the global estimates

$$\begin{aligned}\|\tau_-\Delta\psi\|_2 &\le c\varepsilon_0\overline{\mathcal{K}_{[2]}}\\ \|\tau_- r\nabla\!\!\!/\Delta\psi\|_2 &\le c\varepsilon_0\overline{\mathcal{K}_{[2]}}\\ \|\tau_- r^2\nabla\!\!\!/^2\Delta\psi\|_2 &\le c\varepsilon_0\overline{\mathcal{K}_{[2]}},\end{aligned} \tag{11.2.9a}$$

noting that the last inequality is derived with the help of bootstrap assumption 2.

Moreover, in the exterior,

$$\begin{aligned}\|\tau_-^2\nabla_N\Delta\psi\|_{2,e} &\le c\varepsilon_0\overline{\mathcal{K}_{[2]}}\\ |||r^{1/3}\tau_-^{3/2}\Delta\psi|||_{3,e} &\le c\varepsilon_0\overline{\mathcal{K}_{[2]}}.\end{aligned} \tag{11.2.9b}$$

The same estimates are also valid for $\Delta\psi'$. We are thus in a position to apply the degenerate elliptic estimates of Propositions 4.2.1ii, 4.2.2ii, 4.2.3ii, and 4.2.4 to equations 11.2.8a and 11.2.8b. We derive the following:

Proposition 11.2.1 *Under the assumptions made at the beginning of this chapter, there exists a constant c such that the scalar ψ verifies the estimates*

$$\begin{aligned}r_0^{-1/2}\|\nabla\psi\|_2 &\le c\varepsilon_0\overline{\mathcal{K}_{[2]}}\\ r_0^{-1/2}\left\{\|r\nabla\!\!\!/^2\psi\|_2^2+\|r\nabla\!\!\!/\nabla_N\psi\|_2^2+\|w_1\nabla_N^2\psi\|_2\right\}^{1/2} &\le c\varepsilon_0\overline{\mathcal{K}_{[2]}}\\ r_0^{-1/2}\left\{\|r^2\nabla\!\!\!/^3\psi\|_2^2+\|r^2\nabla\!\!\!/^2\nabla_N\psi\|_2^2+\|w_1\nabla\!\!\!/\nabla_N^2\psi\|_2\right\}^{1/2} &\le c\varepsilon_0\overline{\mathcal{K}_{[2]}}\\ r_0^{-1/2}\|w_2\nabla_N^3\psi\|_{2,e} &\le c\varepsilon_0\overline{\mathcal{K}_{[2]}}.\end{aligned}$$

Also,

$$\begin{aligned}r_0^{-1/2}|||r^{5/6}\nabla\psi|||_3 &\le c\varepsilon_0\overline{\mathcal{K}_{[2]}}\\ r_0^{-1/2}\left\{|||r^{11/6}\nabla\!\!\!/^2\psi|||_3^3+|||r^{11/6}\nabla\!\!\!/\nabla_N\psi|||_3^3\right\}^{1/3} &\le c\varepsilon_0\overline{\mathcal{K}_{[2]}}\\ r_0^{-1/2}|||r^{1/3}w_{3/2}\nabla_N^2\psi|||_{3,e} &\le c\varepsilon_0\overline{\mathcal{K}_{[2]}}.\end{aligned}$$

As a consequence, we also have

$$r_0^{-1/2}\sup_E(r^{3/2}|\nabla\psi|)\le c\varepsilon_0\overline{\mathcal{K}_{[2]}}\,.$$

The scalar ψ' verifies precisely the same estimates.

We next estimate the scalars q and q'. We first check the global estimates

$$\begin{aligned}\|r\Delta q\|_2 &\le c(\overline{\mathcal{R}_{[1]}}+\varepsilon_0\overline{\mathcal{K}_{[2]}})\\ \|r^2\nabla\!\!\!/\Delta q\|_2 &\le c(\overline{\mathcal{R}_{[1]}}+\varepsilon_0\overline{\mathcal{K}_{[2]}}).\end{aligned} \tag{11.2.10a}$$

Also, in the exterior region,

$$\|r\tau_- \nabla\!\!\!/_N \Delta q\|_{2,e} \leq c(\overline{\mathcal{R}_{[1]}} + \varepsilon_0 \overline{\mathcal{K}_{[2]}})$$
$$|||r^{4/3}\tau_-^{1/2} \Delta q|||_{3,e} \leq c(\overline{\mathcal{R}_{[1]}} + \varepsilon_0 \overline{\mathcal{K}_{[2]}}). \qquad (11.2.10b)$$

Therefore we can apply the nondegenerate elliptic estimates of Propositions 4.2.1i, 4.2.2i, and 4.2.3i to deduce the following:

Proposition 11.2.2 *Under the assumptions made at the beginning of this section there exists a constant c such that*

$$\left\{ \|\nabla q\|_2^2 + \|r\nabla^2 q\|_2^2 + \sum_{i+j=3,\, i\geq 1} \|r^2 \nabla\!\!\!/^i \nabla_N^j q\|_2^2 \right\}^{1/2} \leq c(\overline{\mathcal{R}_{[1]}} + \varepsilon_0 \overline{\mathcal{K}_{[2]}})$$
$$\|r\tau_- \nabla_N^3 q\|_{2,e} \leq c(\overline{\mathcal{R}_{[1]}} + \varepsilon_0 \overline{\mathcal{K}_{[2]}}).$$

Also,

$$\left\{ |||r^{5/6} \nabla q|||_3^3 + |||r^{11/6} \nabla\!\!\!/^2 q|||_3^3 + |||r^{11/6} \nabla\!\!\!/ \nabla_N q|||_3^3 \right\}^{1/3} \leq c(\overline{\mathcal{R}_{[1]}} + \varepsilon_0 \overline{\mathcal{K}_{[2]}})$$
$$|||r^{4/3}\tau_-^{1/2} \nabla_N^2 q|||_{3,e} \leq c(\overline{\mathcal{R}_{[1]}} + \varepsilon_0 \overline{\mathcal{K}_{[2]}}).$$

As a consequence, we also have

$$\sup_E (r^{3/2} \nabla q) \leq c(\overline{\mathcal{R}_{[1]}} + \varepsilon_0 \overline{\mathcal{K}_{[2]}}).$$

Next we make use of the results of the previous two propositions, as well as the bootstrap assumptions, to estimate

$$\|r\Delta q'\|_2 \leq c(\overline{\mathcal{R}_{[1]}} + \varepsilon_0 \overline{\mathcal{K}_{[2]}}). \qquad (11.2.11)$$

This is obtained in a straightforward manner with the exception of the terms $-r\,\mathrm{c}\!\!/\mathrm{url}\, \mathcal{G}\!\!\!/ + r^2 a^{-1}(aJ - \overline{aJ})$ in the expression 11.2.7e for S. These terms are estimated as follows. By 11.2.1e we have

$$\begin{aligned}
-\mathrm{c}\!\!/\mathrm{url}\, \mathcal{G}\!\!\!/ - ra^{-1}\hat{\eta} \cdot \nabla\!\!\!/^2 a = {} & r\epsilon \cdot \mathrm{d}\!\!/\mathrm{iv}\,\hat{\theta} + r\hat{\theta} \cdot \nabla\!\!\!/\epsilon + \frac{1}{2}\epsilon \cdot \nabla\!\!\!/\kappa + \frac{1}{2}\kappa\, \mathrm{d}\!\!/\mathrm{iv}\,\epsilon \\
& + ra^{-1}\nabla\!\!\!/ a \cdot \mathrm{d}\!\!/\mathrm{iv}\,\hat{\eta} - ra^{-2}\nabla\!\!\!/ a \cdot \hat{\eta} \cdot \nabla\!\!\!/ a \\
& - \frac{1}{2} ra^{-1} \Delta\!\!\!/ a \delta \\
& - \frac{1}{2} ra^{-1} \nabla\!\!\!/ a \cdot \nabla\!\!\!/\delta + \frac{1}{2} ra^{-2} |\nabla\!\!\!/ a|^2 \delta
\end{aligned}$$

and we can estimate

$$\|r^2(\mathrm{c}\nabla\!\!\!/\mathrm{rl}\, \not{G} + ra^{-1}\hat{\eta} \cdot \nabla\!\!\!/^2 a)\|_2 \leq c\varepsilon_0 \overline{\mathcal{K}_{[2]}}. \tag{11.2.12a}$$

Also from 11.2.6b we find

$$\|r^3(J + a^{-1}\hat{\eta} \cdot \nabla\!\!\!/^2 a)\|_2 \leq c(\overline{\mathcal{R}_{[1]}} + \varepsilon_0 \overline{\mathcal{K}_{[2]}}) \tag{11.2.12b}$$

and we have

$$\|r^3 a^{-1}\overline{aJ}\|_2 \leq c(\|r^3(J + a^{-1}\hat{\eta} \cdot \nabla\!\!\!/^2 a)\|_2 + \|r^3 a^{-1}\overline{\hat{\eta} \cdot \nabla\!\!\!/^2 a}\|_2).$$

On the other hand,

$$\overline{\hat{\eta} \cdot \nabla\!\!\!/^2 a} = -\frac{1}{4\pi r^2} \int_{S_{t,u}} \mathrm{d}\!\!\!/\mathrm{iv}\, \hat{\eta} \cdot \nabla\!\!\!/ a.$$

It follows that in the exterior region

$$\|r^3 a^{-1}\overline{\hat{\eta} \cdot \nabla\!\!\!/^2 a}\|_{2,e} \leq c\varepsilon_0 \|r\nabla\!\!\!/\hat{\eta}\|_{2,e} \leq c\varepsilon_0 \overline{\mathcal{K}_{[2]}},$$

while in the interior region

$$\|r^3 a^{-1}\overline{\hat{\eta} \cdot \nabla\!\!\!/^2 a}\|_{2,i} \leq c\varepsilon_0 \overline{\mathcal{K}_{[2]}}.$$

Hence,

$$\|r^3 a^{-1}\overline{aJ}\|_2 \leq c(\overline{\mathcal{R}_{[1]}} + \varepsilon_0 \overline{\mathcal{K}_{[2]}}). \tag{11.2.12c}$$

From 11.2.12a, 11.2.12b, and 11.2.12c, in view of the cancellation of the term $a^{-1}\hat{\eta} \cdot \nabla\!\!\!/^2 a$, we deduce

$$\|r^2(-\mathrm{c}\nabla\!\!\!/\mathrm{rl}\, \not{G} + ra^{-1}(aJ - \overline{aJ}))\|_2 \leq c(\overline{\mathcal{R}_{[1]}} + \varepsilon_0 \overline{\mathcal{K}_{[2]}}).$$

The estimate 11.2.11 follows. Applying the nondegenerate elliptic theory then yields the following:

Proposition 11.2.3 *Under the general assumptions made at the beginning of this section the following estimates are valid for q':*

$$\left\{\|r^{-1}q'\|_2^2 + \|\nabla q'\|_2^2 + \|r\nabla^2 q'\|_2^2\right\}^{1/2} \leq c(\overline{\mathcal{R}_{[1]}} + \varepsilon_0 \overline{\mathcal{K}_{[2]}})$$
$$\left\{|||r^{1/6}q'|||_3^3 + |||r^{7/6}\nabla q'|||_3^3\right\}^{1/3} \leq c(\overline{\mathcal{R}_{[1]}} + \varepsilon_0 \overline{\mathcal{K}_{[2]}}).$$

Consequently,

$$\sup_E (r^{1/2}|q'|) \leq c(\overline{\mathcal{R}_{[1]}} + \varepsilon_0 \overline{\mathcal{K}_{[2]}}).$$

Combining the results of the last three propositions, we derive

Corollary 11.2.0.1 *Under the same assumptions, there exists a positive constant c such that*

$$\overline{{}^{e}\mathcal{K}_{[2]}}(\delta) \le c(\overline{\mathcal{R}_{[1]}} + \varepsilon_0 \overline{\mathcal{K}_{[2]}}).$$

Next, using the estimates we have already derived for δ, we estimate the component ϵ. According to equations 11.1.2a–11.1.2e we have

$$\not{d}\mathrm{iv}\,\epsilon = -\nabla_N \delta + \hat{\theta} \cdot \hat{\eta} - \frac{3}{2}\mathrm{tr}\theta\delta - 2a^{-1}\not{\nabla} a \cdot \epsilon \tag{11.2.13a}$$

$$\begin{aligned} \mathrm{c}\not{u}\mathrm{rl}\,\epsilon &= \sigma + \hat{\theta} \wedge \hat{\eta} \\ &= \sigma + \hat{\chi} \wedge \hat{\eta}. \end{aligned} \tag{11.2.13b}$$

Also,

$$\begin{aligned} \nabla_N \epsilon + \mathrm{tr}\theta \cdot \epsilon &= \not{\nabla}\delta - 2\hat{\theta} \cdot \epsilon - a^{-1}\not{\nabla} a \cdot \hat{\eta} \\ &\quad + \frac{3}{2}a^{-1}\not{\nabla} a \cdot \delta + \frac{1}{2}\beta - \frac{1}{2}\underline{\beta}. \end{aligned} \tag{11.2.13c}$$

Now, as discussed in Remark 1, the first two terms on the right-hand side of 11.2.13a are slowly decaying in the wave zone. However, in view of the decomposition $\delta = r^{-2}p + r^{-1}\nabla_N\psi$, $p = q' + \nabla_Z q + \psi'$ given by Lemma 11.2.1,[7] and since, according to 4.2.5, $\nabla_N^2 = \triangle - \mathrm{tr}\theta\nabla_N - \not{\triangle} - a^{-1}\not{\nabla} a \cdot \not{\nabla}$, we can write[8]

$$\begin{aligned} \nabla_N \delta - \hat{\theta} \cdot \hat{\eta} &= -2r^{-3}(\nabla_N r)p + r^{-2}\nabla_N p - r^{-2}(\nabla_N r)\nabla_N \psi \\ &\quad + r^{-1}\nabla_N^2 \psi \\ &= -\hat{\chi} \cdot \hat{\eta} - r^{-1}\not{\triangle}\psi - r^{-2}(r\mathrm{tr}\theta + a^{-1}\lambda)\nabla_N \psi \\ &\quad - r^{-1}a^{-1}\not{\nabla} a \cdot \not{\nabla}\psi + r^{-2}\nabla_N p - 2r^{-3}a^{-1}\lambda p. \end{aligned} \tag{11.2.13d}$$

Then, with the help of the estimates for ψ, ψ', q, q' in Propositions 11.2.1 and 11.1.1b and our assumptions, we find the following:

$$\begin{aligned} \|r^2(\nabla_N \delta - \theta \cdot \hat{\eta})\|_{2,e} &\le c r_0^{1/2}(\overline{\mathcal{R}_{[1]}} + \varepsilon_0 \overline{\mathcal{K}_{[2]}}) \\ \|r^3 \not{\nabla}(\nabla_N \delta - \theta \cdot \hat{\eta})\|_{2,e} &\le c r_0^{1/2}(\overline{\mathcal{R}_{[1]}} + \varepsilon_0 \overline{\mathcal{K}_{[2]}}) \\ |||r^{17/6}(\nabla_N \delta - \theta \cdot \hat{\eta})|||_{3,e} &\le c r_0^{1/2}(\overline{\mathcal{R}_{[1]}} + \varepsilon_0 \overline{\mathcal{K}_{[2]}}). \end{aligned}$$

[7] See also 11.2.2h.

[8] Recall also that, $\nabla_N r = \frac{r}{2a}\overline{a\mathrm{tr}\theta}$ (see 3.1.1j)

Consequently, also making use of the exterior estimate 11.1.12b for $\bar{\sigma}$,

$$r_0^{-1/2}\|r^2 \mathrm{d}\!\!/\mathrm{iv}\,\epsilon\|_{2,e} \le c(\overline{\mathcal{R}_{[1]}} + \varepsilon_0 \overline{\mathcal{K}_{[2]}})$$
$$r_0^{-1/2}\|r^2 \mathrm{c}\!\!/\mathrm{url}\,\epsilon\|_{2,e} \le c(\overline{\mathcal{R}_{[1]}} + \varepsilon_0 \overline{\mathcal{K}_{[2]}})$$
$$r_0^{-1/2}\|r^3 \nabla\!\!\!/\, \mathrm{d}\!\!/\mathrm{iv}\,\epsilon\|_{2,e} \le c(\overline{\mathcal{R}_{[1]}} + \varepsilon_0 \overline{\mathcal{K}_{[2]}})$$
$$r_0^{-1/2}\|r^3 \nabla\!\!\!/\, \mathrm{c}\!\!/\mathrm{url}\,\epsilon\|_{2,e} \le c(\overline{\mathcal{R}_{[1]}} + \varepsilon_0 \overline{\mathcal{K}_{[2]}})$$

and also

$$r_0^{-1/2}|||r^{17/6} \mathrm{d}\!\!/\mathrm{iv}\,\epsilon|||_{3,e} \le c(\overline{\mathcal{R}_{[1]}} + \varepsilon_0 \overline{\mathcal{K}_{[2]}})$$
$$r_0^{-1/2}|||r^{17/6} \mathrm{c}\!\!/\mathrm{url}\,\epsilon|||_{3,e} \le c(\overline{\mathcal{R}_{[1]}} + \varepsilon_0 \overline{\mathcal{K}_{[2]}}).$$

Finally, we appeal to the 2-dimensional elliptic estimates developed in Chapter 2 (see Proposition 2.2.2 and Proposition 2.3.1 applied to a Hodge system of type $\mathbf{H}_1$) and derive, without any further subtleties,

$$\overline{{}^e\mathcal{K}_{[2]}}(\epsilon) \le c(\overline{\mathcal{R}_{[1]}} + \varepsilon_0 \overline{\mathcal{K}_{[2]}}). \tag{11.2.13e}$$

It thus only remains to estimate $\hat{\eta}$. This can be done using the system

$$\mathrm{d}\!\!/\mathrm{iv}\,\hat{\eta} = -\frac{1}{2}\underline{\beta} + \frac{1}{2}\beta - \frac{1}{2}\nabla\!\!\!/\,\delta - \frac{1}{2}\mathrm{tr}\theta\epsilon + \hat{\theta}\cdot\epsilon. \tag{11.2.14a}$$

Also,

$$\nabla\!\!\!/_N \hat{\eta}_{ij} + \frac{1}{2}\mathrm{tr}\theta\cdot\hat{\eta}_{ij} = -\frac{1}{4}\underline{\alpha} + \frac{1}{4}\alpha + \frac{1}{2}(\nabla\!\!\!/_i \epsilon_j + \nabla\!\!\!/_j \epsilon_i - \gamma_{ij}\mathrm{d}\!\!/\mathrm{iv}\,\epsilon)$$
$$+ a^{-1}(\epsilon_i \nabla\!\!\!/_j a + \epsilon_j \nabla\!\!\!/_i a - \gamma_{ij}\epsilon\nabla\!\!\!/ a) + \frac{3}{2}\hat{\theta}_{ij}\delta. \tag{11.2.14b}$$

We thus find

$$r_0^{-1/2}\|r w_1 \mathrm{d}\!\!/\mathrm{iv}\,\hat{\eta}\|_{2,e} \le c(\overline{\mathcal{R}_{[1]}} + \varepsilon_0 \overline{\mathcal{K}_{[2]}})$$
$$r_0^{-1/2}\|r^2 w_1 \nabla\!\!\!/\, \mathrm{d}\!\!/\mathrm{iv}\,\hat{\eta}\|_{2,e} \le c(\overline{\mathcal{R}_{[1]}} + \varepsilon_0 \overline{\mathcal{K}_{[2]}})$$
$$r_0^{-1/2}|||r^{4/3} w_{3/2} \mathrm{d}\!\!/\mathrm{iv}\,\hat{\eta}|||_{3,e} \le c(\overline{\mathcal{R}_{[1]}} + \varepsilon_0 \overline{\mathcal{K}_{[2]}})$$

and therefore, with the help of the 2-dimensional elliptic theory developed in Chapter 2 (see Proposition 2.2.2 and Proposition 2.3.1 applied to a Hodge system of type $\mathbf{H}_2$) we infer, in a straightforward manner,

$$\overline{{}^e\mathcal{K}_{[2]}}(\hat{\eta}) \le c(\overline{\mathcal{R}_{[1]}} + \varepsilon_0 \overline{\mathcal{K}_{[2]}}). \tag{11.2.14c}$$

Combining Corollary 11.2.0.1 together with 11.2.13e and 11.2.14c, we complete the estimates in the exterior for the first derivatives of the second fundamental form. We summarize our result in the following:

Proposition 11.2.4 *Under our general assumptions, there exists a constant c such that*

$$\overline{{}^e\mathcal{K}_{[2]}} \leq c(\overline{\mathcal{R}_{[1]}} + \varepsilon_0 \overline{\mathcal{K}_{[2]}}).$$

11.3. The Interior Estimates

In this section we will provide the interior estimates for the first two derivatives of k. We first recall, from page 65, the definition of the truncation function

$$f(x) = h(r(x)/r_0) \tag{11.3.1a}$$

where x is an arbitrary point on Σ, $r(x)$ is the radius of the sphere u = const passing through it, and h is a C^∞ decreasing function defined on the real line such that

$$h(s) = \begin{cases} 1 & : \ s \leq 1/2 \\ 0 & : \ s \geq 3/4. \end{cases} \tag{11.3.1b}$$

Recalling Lemma 3.2.2 and taking into account the estimate 11.1.13a for $\nabla_N \lambda$, we have the following:

Lemma 11.3.1 *The truncation function f is supported in the set Int($S_{3r_0/4}$), while its gradient, ∇f, is supported in $E \cap$ Int($S_{3r_0/4}$). Moreover, there exists a numerical, positive constant c such that*

$$\tau_- \geq r_0/c$$

on the support of f and

$$\sup_\Sigma |\nabla f| \leq c r_0^{-1}$$
$$\sup_\Sigma |\nabla^2 f| \leq c r_0^{-2}.$$

Using the truncation function f, we define

$$\overset{\circ}{k} = f k. \tag{11.3.2a}$$

Thus $\overset{\circ}{k}$ has compact support included in Int($S_{3r_0/4}$) and satisfies the equations

$$\text{curl } \overset{\circ}{k} = \overset{\circ}{H}$$
$$\text{div } \overset{\circ}{k} = \overset{\circ}{F} \tag{11.3.2b}$$
$$\text{tr } \overset{\circ}{k} = 0 \tag{11.3.2c}$$

where

$$\overset{\circ}{H}_{ij} = fH_{ij} + \frac{1}{2}(\in_i{}^{ab}\nabla_a fk_{bj} + \in_j{}^{ab}\nabla_a fk_{bi}) \tag{11.3.2d}$$

$$\overset{\circ}{F}_j = \nabla^i fk_{ij}. \tag{11.3.2e}$$

We now apply to the Hodge system 11.3.2b the integral identity of Proposition 11.2.2g,

$$\int_\Sigma \left(|\nabla \overset{\circ}{k}|^2 + 3R_{ij}\overset{\circ}{k}{}^i{}_m \overset{\circ}{k}{}^j{}_m - \frac{1}{2}R|\overset{\circ}{k}|^2 \right) = \int_\Sigma (|\overset{\circ}{H}|^2 + \frac{3}{2}|\overset{\circ}{F}|^2). \tag{11.3.2f}$$

Now, in view of 11.1.1b, we have $R_{ij} = E_{ij} + k^2_{ij}$, where $k^2_{ij} = k^m_i k_{mj}$ and $R = |k|^2 = \mathrm{tr}k^2$. Therefore,

$$\begin{aligned} 3R_{ij}\overset{\circ}{k}{}^i{}_m \overset{\circ}{k}{}^j{}_m - \frac{1}{2}R|\overset{\circ}{k}|^2 &= f^2 \left(3E_{ij}k^m_i k_{mj} + 3\mathrm{tr}[(k^2)^2] - \frac{1}{2}(\mathrm{tr}k^2)^2 \right) \\ &\geq f^2 \left(3E_{ij}k^m_i k_{mj} + \frac{1}{2}(\mathrm{tr}k^2)^2 \right) \\ &\geq f^2 \left(-3|E||k|^2 + \frac{1}{2}|k|^4 \right) \geq -\frac{9}{2}f^2|E|^2. \end{aligned}$$

And from

$$|\overset{\circ}{H}| \leq c(f^2|H|^2 + |\nabla f|^2|k|^2) \tag{11.3.2g}$$

$$|\overset{\circ}{F}| \leq c|\nabla f|^2|k|^2, \tag{11.3.2h}$$

we derive the following:

$$\int_\Sigma |\nabla \overset{\circ}{k}|^2 \leq c\left(\int_\Sigma f^2(|E|^2 + |H|^2) + \int_\Sigma |\nabla f|^2|k|^2 \right). \tag{11.3.2i}$$

Now, in view of the definition of the norm ${}^i\mathcal{R}_{[1]}$, the integral of $|E|^2+|H|^2$ on the interior region $r \leq r_0/2$ is $\leq cr_0^{-3}\,{}^i\mathcal{R}_{[1]}$ while, in view of the definition of ${}^e\mathcal{R}_{[1]}$ complemented by the exterior estimates 11.1.12b and 11.1.12d for $\bar{\sigma}, \bar{\rho}$, the same integral on the region $r_0/2 \leq r \leq 3r_0/4$ is $\leq cr_0^{-3}({}^i\mathcal{R}_{[1]} + \varepsilon_0\overline{\mathcal{K}_{[1]}})$. Therefore,

$$\int_{\mathrm{supp(f)}} (|E|^2 + |H|^2) \leq cr_0^{-3}(\overline{\mathcal{R}_{[1]}} + \varepsilon_0\overline{\mathcal{K}_{[1]}}).$$

On the other hand, by virtue of Lemma 11.3.1 and the exterior estimates of the previous section (see 11.2.14c) we have

$$\int_\Sigma |\nabla f|^2 |k|^2 \leq c r_0^{-3}\, \overline{{}^e\mathcal{K}_{[1]}}$$
$$\leq c r_0^{-3}(\overline{\mathcal{R}_{[1]}} + \varepsilon_0 \overline{\mathcal{K}_{[2]}}),$$

and, combining them, we infer that

$$r_0^{3/2} \|\nabla \overset{\circ}{k}\|_2 \leq (\overline{\mathcal{R}_{[1]}} + \varepsilon_0 \overline{\mathcal{K}_{[2]}}). \tag{11.3.2j}$$

To estimate the second derivatives of $\overset{\circ}{k}$, we appeal to Proposition 4.2.7d. Thus we infer that

$$\int_\Sigma |\nabla^2 \overset{\circ}{k}|^2 \leq$$
$$c\left(\int_\Sigma (|\nabla \overset{\circ}{H}|^2 + |\nabla \overset{\circ}{F}|^2) + \int_\Sigma |\mathrm{Ric}||\nabla \overset{\circ}{k}|^2 + \int_\Sigma |\mathrm{Ric}|^2 |\overset{\circ}{k}|^2\right).$$

Now

$$\int_\Sigma |\mathrm{Ric}||\nabla \overset{\circ}{k}|^2 \leq \left(\int_{\mathrm{Int}(S_{3r_0/4})} |\mathrm{Ric}|^3\right)^{1/3} \left(\int_\Sigma |\nabla \overset{\circ}{k}|^2\right)^{1/2} \left(\int_\Sigma |\nabla \overset{\circ}{k}|^6\right)^{1/6}$$
$$\int_\Sigma |\mathrm{Ric}|^2 |\overset{\circ}{k}|^2 \leq \left(\int_{\mathrm{Int}(S_{3r_0/4})} |\mathrm{Ric}|^3\right)^{2/3} \left(\int_\Sigma |\overset{\circ}{k}|^6\right)^{1/3},$$

and, using the classical Sobolev inequalities (see Corollary 3.2.1.2), we obtain

$$\|\nabla^2 \overset{\circ}{k}\|_2 \leq c\left(\|\nabla \overset{\circ}{H}\|_2 + \|\nabla \overset{\circ}{F}\|_2 + \left(\int_{\mathrm{Int}(S_{3r_0/4})} |\mathrm{Ric}|^3\right)^{1/3} \|\nabla \overset{\circ}{k}\|_2\right). \tag{11.3.3a}$$

On the other hand,

$$|\nabla \overset{\circ}{H}| \leq c\left(f|\nabla f| + |\nabla f||H| + |\nabla f||\nabla \overset{\circ}{k}| + |\nabla^2 f||k|\right)$$
$$|\nabla \overset{\circ}{F}| \leq c\left(|\nabla f||\nabla \overset{\circ}{k}| + |\nabla^2 f||k|\right),$$

therefore

$$\|\nabla \overset{\circ}{H}\|_2 + \|\nabla \overset{\circ}{F}\|_2 \leq c r_0^{-5/2}(\overline{\mathcal{R}_{[1]}} + \varepsilon_0 \overline{\mathcal{K}_{[2]}}). \tag{11.3.3b}$$

Indeed, this is immediate from the definition of the interior and exterior norms ${}^{i}\mathcal{R}_{[1]}, {}^{e}\mathcal{R}_{[1]}$ and from the exterior estimates for $\bar{\sigma}, \nabla_N\bar{\sigma}$, which follow from formula 11.1.12a, that is, 11.1.12b as well as

$$\|r^{3/2}w_2\nabla_N\bar{\sigma}\|_{2,e} \leq c\varepsilon_0(\overline{\mathcal{R}_{[1]}} + \overline{\mathcal{K}_{[2]}}) ,$$

which implies

$$\left(\int_{\mathrm{Int}(S_{3r_0/4})}|\nabla_N\bar{\sigma}|^2\right)^{1/2} \leq cr_0^{-7/2}\varepsilon_0(\overline{\mathcal{R}_{[1]}} + \overline{\mathcal{K}_{[2]}}).$$

It remains to estimate $(\int_{\mathrm{Int}(S_{3r_0/4})}|\mathrm{Ric}|^3)^{1/3}$. Making use once again of formula 11.1.1b, the assumption $\overline{\mathcal{R}_{[1]}} \leq \varepsilon_0$, we derive

$$\begin{aligned}\left(\int_{\mathrm{Int}(S_{3r_0/4})}|\mathrm{Ric}|^3\right)^{1/3} &\leq \left(\int_{\mathrm{Int}(S_{3r_0/4})}|E|^3\right)^{1/3} + \left(\int_{\mathrm{Int}(S_{3r_0/4})}|k|^6\right)^{1/3}\\ &\leq cr_0^{-2}(\overline{\mathcal{R}_{[1]}} + \varepsilon_0\overline{\mathcal{K}_{[2]}}).\end{aligned} \tag{11.3.3c}$$

Substituting these estimates in 11.3.3a, we obtain

$$r_0^{5/2}\|\nabla^2 \overset{\circ}{k}\|_2 \leq c(\overline{\mathcal{R}_{[1]}} + \varepsilon_0\overline{\mathcal{K}_{[2]}}),$$

which in conjunction with 11.3.2j yields the following:

Proposition 11.3.1 *Under the assumptions made at the beginning of this chapter, there exists a numerical constant c such that*

$$\overline{{}^{i}\mathcal{K}_{[2]}} \leq c(\overline{\mathcal{R}_{[1]}} + \varepsilon_0\overline{\mathcal{K}_{[2]}}).$$

Combining this with the Proposition 11.2.4, we conclude that

$$\overline{\mathcal{K}_{[2]}} \leq c(\overline{\mathcal{R}_{[1]}} + \varepsilon_0\overline{\mathcal{K}_{[2]}}),$$

from which we read the first part of the following:

Theorem 11.3.1 *Under the assumptions made at the beginning of this chapter, if ε_0 is chosen sufficiently small, we have, with a numerical constant c,*

$$\overline{\mathcal{K}_{[2]}} \leq c\overline{\mathcal{R}_{[1]}}.$$

Moreover,

$$\|r^{3/2}\tau_-^{3/4}\nabla\hat{\eta}\|_{4,e} \leq c\overline{\mathcal{K}_{[2]}}.$$

The second statement of the theorem allows us to recover bootstrap assumption 2. To prove it, we set $f = r\nabla\hat{\eta}$ and check, in view of the definition of $\overline{\mathcal{K}_{[2]}}$,

$$r_0^{-1/2}\left(\int w_1^2|f|^2 + r^2 w_1^2|\nabla f|^2 + w_2^2|\nabla_N f|^2\right)^{1/2} \leq c\overline{\mathcal{K}_{[2]}}.$$

Now, $w_1 = \min\{r_0^{1/2}\tau_-, r\} \geq c r_0^{1/2}\tau_-^{1/2}$, $w_2 = \min\{r_0^{1/2}\tau_-^2, r^2\} \geq c r_0^{1/2}\tau_-^{3/2}$. Hence,

$$\left(\int \tau_-|f|^2 + r^2\tau_-|\nabla f|^2 + \tau_-^3|\nabla_N f|^2\right)^{1/2} \leq c\overline{\mathcal{K}_{[2]}}$$

and, according to the degenerate form of the global Sobolev inequality 3.2.1 applied to $h = \tau_-^{1/2} f$, we infer that

$$\|r^{1/2}\tau_-^{1/4}h\|_{4,e} \leq c\|h\|_{2,e}^{1/4}\|r^{2/3}\tau_-^{1/3}h\|_{6,e}^{3/4} \leq c\overline{\mathcal{K}_{[2]}}$$

and therefore

$$\|r^{3/2}\tau_-^{3/4}\nabla\hat{\eta}\|_{4,e} \leq c\overline{\mathcal{K}_{[2]}}.$$

Finally, it remains to estimate the third derivatives of k. This is now quite straightforward, following the same steps as before, but easier since it does not require any bootstrap argument. Assumptions 0, 1, 2, and 3 have to be strengthened, however, to include the second derivatives of the optical quantities χ, ζ, ω.

Assumption 4: The components χ, ζ, ω of the hessian of the extended optical function u verify the following:

In the Interior

$$r_0^{1/2}\|r^{3/2}\nabla^2(\chi, \zeta, \omega)\|_{2,\imath} \leq \varepsilon_0. \tag{11.3.4}$$

In the Exterior

$$r_0^{1/2}\|r^{-1}w_2\nabla_N\omega\|_{2,e} \leq \varepsilon_0 \tag{11.3.5}$$

and

$$\begin{aligned} r_0^{1/2}\|r^2\nabla^2(\chi, \zeta)\|_{2,e} &\leq \varepsilon_0 \\ r_0^{1/2}\|r w_1\nabla^2\omega\|_{2,e} &\leq \varepsilon_0 \\ r_0^{1/2}\|w_2\nabla\nabla_N\omega\|_{2,e} &\leq \varepsilon_0. \end{aligned} \tag{11.3.6}$$

Theorem 11.3.2 *Under assumptions 0–4, if ε_0 is chosen sufficiently small, we have, with a numerical constant c,*

$$\overline{\mathcal{K}_{[3]}} \leq c\overline{\mathcal{R}_{[2]}}.$$

11.4. Estimates for the Time Derivatives

We recall (see the introduction of Part I) that relative to the product parametrization of the space-time[9] and arbitrary coordinates on Σ_0 we have

$$\partial_t k_{ij} = -\nabla_i \nabla_j \phi + \phi(R_{ij} - 2k_{ia}k^a{}_j). \tag{11.4.1a}$$

This equation can be rewritten in the form

$$\begin{aligned} \bar{\mathbf{D}}_T k_{ij} &= -\phi^{-1}\nabla_i\nabla_j\phi + R_{ij} \\ &= -\phi^{-1}\nabla_i\nabla_j\phi + E_{ij} + k_{ia}k^a{}_j \end{aligned} \tag{11.4.1b}$$

where $\bar{\mathbf{D}}_T$ is the projection of $\mathbf{D}_T$ to Σ_t.

In view of the radial decomposition of k the equation 11.4.1b takes the form[10]

$$\mathbf{D}_T\delta = -\phi^{-1}\nabla_N^2\phi + \rho + \delta^2 - \zeta\cdot\underline{\zeta} + \zeta\cdot\epsilon - \underline{\zeta}\cdot\epsilon \tag{11.4.2a}$$

$$\begin{aligned} \not{\mathbf{D}}_T\epsilon = &-\phi^{-1}\not\nabla\nabla_N\phi + \frac{1}{2}(\underline{\beta}+\beta) + (a^{-1}\not\nabla a)\phi^{-1}\nabla_N\phi \\ &- \frac{3}{2}(\zeta - \phi^{-1}\not\nabla\phi)\delta + (\zeta - \phi^{-1}\not\nabla\phi + \epsilon)\cdot\hat{\eta} + \frac{1}{2}\delta\epsilon \end{aligned} \tag{11.4.2b}$$

$$\begin{aligned} \not{\mathbf{D}}_T\hat{\eta} = &-\phi^{-1}\hat{\not\nabla}^2\phi + \frac{1}{4}(\underline{\alpha}+\alpha) \\ &- \delta\hat{\eta} + \epsilon\widehat{\otimes}\epsilon - (\zeta - \phi^{-1}\not\nabla\phi)\widehat{\otimes}\epsilon. \end{aligned} \tag{11.4.2c}$$

We next use 11.4.2a–11.4.2c to derive equations for $\mathbf{D}_4\delta, \not{\mathbf{D}}_4\epsilon, \not{\mathbf{D}}_4\hat{\eta}$.

To derive the equation for $\delta_4 = \mathbf{D}_4\delta$, we recall that $\nu = -\phi^{-1}\nabla_N\phi + \delta$ and write

[9] Namely, the parametrization obtained by following the integral curves of $\mathbf{D}T$ starting on the initial slice.

[10] The calculations can be easily checked by considering an orthonormal frame e_1, e_2, N, with N the unit normal to the radial foliation, and using the projection formulas $\bar{\mathbf{D}}_T e_A = \not{\mathbf{D}}_T e_A - (\zeta_A - \phi^{-1}\not\nabla_A\phi)N$ and $\bar{\mathbf{D}}_T N = (\zeta_A - \phi^{-1}\not\nabla_A\phi)e_A$.

$$\begin{aligned}\delta_4 &= \mathbf{D}_T\delta + \mathbf{D}_N\delta \\ &= \nabla_N(-\phi^{-1}\nabla_N\phi + \delta) - \phi^{-2}|\nabla_N\phi|^2 + \rho + \delta^2 - \zeta\cdot\underline{\zeta} \\ &\quad + \zeta\cdot\epsilon - \underline{\zeta}\cdot\epsilon.\end{aligned}$$

Hence,

$$\delta_4 = \nabla_N\nu - \phi^{-2}|\nabla_N\phi|^2 + \rho + \delta^2 - \zeta\cdot\underline{\zeta} + \zeta\cdot\epsilon - \underline{\zeta}\cdot\epsilon. \tag{11.4.3a}$$

On the other hand, in view of the polar decomposition formula 4.2.5 and Lemma 11.2.1,

$$\begin{aligned}\nabla_N\nu - \phi^{-2}|\nabla_N\phi|^2 &= -\phi^{-1}\nabla_N^2\phi + \nabla_N\delta \\ &= -\phi^{-1}(\Delta\phi - \not\Delta\phi - \mathrm{tr}\theta\nabla_N\phi - a^{-1}\not\nabla a\cdot\not\nabla\phi) \\ &\quad + \nabla_N(r^{-2}q' + r^{-1}\nabla_N q + r^{-2}\psi' + r^{-1}\nabla_N\psi).\end{aligned}$$

Henceforth, in view of the lapse equation $\Delta\phi = \phi|k|^2$,

$$\nabla_N\nu - \phi^{-2}|\nabla_N\phi|^2 = -|k|^2 + r^{-1}\nabla_N^2\psi + \phi^{-1}\not\Delta\phi + E \tag{11.4.3b}$$

$$\begin{aligned}E &= \phi^{-1}(\mathrm{tr}\theta\nabla_N\phi + a^{-1}\not\nabla a\cdot\not\nabla\phi) \\ &\quad + \nabla_N(r^{-2}q' + r^{-1}\nabla_N q + r^{-2}\psi') - r^{-2}a^{-1}\lambda\nabla_N\psi.\end{aligned} \tag{11.4.3c}$$

Finally, using equation 11.2.2b, $\Delta\psi = r|\hat{\eta}|^2$, and once again the polar decomposition formula 4.2.5, we derive

$$\nabla_N\nu - \phi^{-2}|\nabla_N\phi|^2 = -|k|^2 + |\hat{\eta}|^2 + \phi^{-1}\not\Delta\phi + E' \tag{11.4.3d}$$

$$E' = E - r^{-1}(\mathrm{tr}\theta\nabla_N\psi + a^{-1}\not\nabla a\cdot\not\nabla\psi).$$

Therefore, in view of 11.4.3a,

$$\begin{aligned}\delta_4 &= \phi^{-1}\not\Delta\phi + \rho + E'' \\ E'' &= E' - \zeta\cdot\underline{\zeta} + \zeta\cdot\epsilon - \underline{\zeta}\cdot\epsilon - 2|\epsilon|^2 - \frac{1}{2}|\delta|^2.\end{aligned} \tag{11.4.4}$$

To derive equations for $\not\mathbf{D}_4\epsilon, \not\mathbf{D}_4\hat{\eta}$, we recall equations 11.2.13c and 11.2.14b:

$$\not\nabla_N\epsilon + \mathrm{tr}\theta\cdot\epsilon = \not\nabla\delta - 2\hat{\theta}\cdot\epsilon - a^{-1}\not\nabla a\cdot\hat{\eta}$$

$$+\frac{3}{2}a^{-1}\nabla\!\!\!/ a\cdot\delta+\frac{1}{2}\beta-\frac{1}{2}\underline{\beta} \tag{11.4.5}$$

$$\nabla\!\!\!/_N\hat{\eta}+\frac{1}{2}\mathrm{tr}\theta\cdot\hat{\eta}=-\frac{1}{4}\underline{\alpha}+\frac{1}{4}\alpha+\frac{1}{2}(\nabla\!\!\!/\widehat{\otimes}\epsilon)$$
$$+a^{-1}(\nabla\!\!\!/ a\widehat{\otimes}\epsilon)+\frac{3}{2}\hat{\theta}\delta$$

Henceforth, together with 11.4.2b–11.4.2c, and recalling the notation $\hat{\eta}_4=\mathbf{D}_4\hat{\eta}+\frac{1}{2}\mathrm{tr}\chi\hat{\eta}$, $\epsilon_4=\mathbf{D}\!\!\!/_4\epsilon$, we infer that

$$\epsilon_4=-\phi^{-1}\nabla\!\!\!/\nabla_N\phi+\nabla\!\!\!/\delta+\beta \tag{11.4.6a}$$
$$+(a^{-1}\nabla\!\!\!/ a)\phi^{-1}\nabla_N\phi+\frac{3}{2}(a^{-1}\nabla\!\!\!/ a-\zeta+\phi^{-1}\nabla\!\!\!/\phi)\delta \tag{11.4.6b}$$
$$+(\zeta-\phi^{-1}\nabla\!\!\!/\phi-a^{-1}\nabla\!\!\!/ a)\cdot\hat{\eta}+(\hat{\eta}-2\hat{\theta})\cdot\epsilon+(\frac{1}{2}\delta-tr\theta)\epsilon$$

$$\hat{\eta}_4=-\phi^{-1}\widehat{\nabla\!\!\!/^2}\phi+\frac{1}{2}(\nabla\!\!\!/\widehat{\otimes}\epsilon)+\frac{1}{2}\alpha \tag{11.4.6c}$$
$$-\frac{1}{2}\delta\hat{\eta}+a^{-1}(\nabla\!\!\!/ a\widehat{\otimes}\epsilon)+\frac{3}{2}\hat{\theta}\delta+\epsilon\widehat{\otimes}\epsilon-(\zeta-\phi^{-1}\nabla\!\!\!/\phi)\widehat{\otimes}\epsilon.$$

We are now ready to state our main results.

Proposition 11.4.1 *With the notation of Chapter 10, if bootstrap assumptions* $\mathbf{BA}_1$ *and* $\mathbf{BA}_2$ *are verified, there exists a constant c such that for all* $0\le q\le 2$,

$$^{e}\mathcal{K}_{q+1}(\delta_4)\le c(^{e}\mathcal{L}_{q+1}(\phi\!\!\!/)+{}^{e}\mathcal{R}_{[q]})$$
$$^{e}\mathcal{K}_{q+1}(\epsilon_4)\le c(^{e}\mathcal{L}_{q+1}(\varphi_N)+{}^{e}\mathcal{R}_{[q]})$$
$$^{e}\mathcal{K}_{q+1}(\hat{\eta}_4)\le c(^{e}\mathcal{L}_{q+1}(\phi\!\!\!/)+{}^{e}\mathcal{R}_{[q]}),$$

and for all $0\le q\le 1$,

$$\mathcal{K}^{4,S}_{q+1}(\delta_4)\le c(\mathcal{L}^{4,S}_{q+1}(\phi\!\!\!/)+{}^{e}\mathcal{R}_{[q+1]})$$
$$\mathcal{K}^{4,S}_{q+1}(\epsilon_4)\le c(\mathcal{L}^{4,S}_{q+1}(\varphi_N)+{}^{e}\mathcal{R}_{[q+1]})$$
$$\mathcal{K}^{4,S}_{q+1}(\hat{\eta}_4)\le c(\mathcal{L}^{4,S}_{q+1}(\phi\!\!\!/)+{}^{e}\mathcal{R}_{[q+1]}).$$

Also, for all $0\le q\le 2$,

$$^{e}\mathcal{K}_{q+1}(\delta_3)\le c(^{e}\mathcal{L}_{q+1}(\phi\!\!\!/)+{}^{e}\mathcal{R}_{[q]})$$
$$^{e}\mathcal{K}_{q+1}(\epsilon_3)\le c(^{e}\mathcal{L}_{q+1}(\varphi_N)+{}^{e}\mathcal{R}_{[q]})$$
$$^{e}\mathcal{K}_{q+1}(\hat{\eta}_3)\le c(^{e}\mathcal{L}_{q+1}(\phi\!\!\!/)+{}^{e}\mathcal{R}_{[q]}),$$

and for all $0 \leq q \leq 1$,

$$\mathcal{K}^{4,S}_{q+1}(\delta_3) \leq c(\mathcal{L}^{4,S}_{q+1}(\not{\phi}) + {}^e\mathcal{R}_{[q+1]})$$
$$\mathcal{K}^{4,S}_{q+1}(\epsilon_3) \leq c(\mathcal{L}^{4,S}_{q+1}(\varphi_N) + {}^e\mathcal{R}_{[q+1]})$$
$$\mathcal{K}^{4,S}_{q+1}(\hat{\eta}_3) \leq c(\mathcal{L}^{4,S}_{q+1}(\not{\phi}) + {}^e\mathcal{R}_{[q+1]}).$$

The proof of the first part of the proposition follows easily from equations 11.4.4, 11.4.6a, and 11.4.6c, together with Theorems 11.3.1 and 11.3.2. The second part is a straightforward consequence of 11.4.2a–11.4.2c and Theorems 11.3.1 and 11.3.2.

The following proposition is also an easy consequence of 11.4.4, 11.4.6a, 11.4.6c, and 11.4.2a–11.4.2c, together with Theorems 11.3.1 and 11.3.2.

Proposition 11.4.2 *Under the same assumptions as in Proposition 11.4.1 we have, for* $0 \leq q \leq 1$,

$${}^e\mathcal{K}_{q+2}(\mathbf{D}_S\delta_4) \leq c({}^e\mathcal{L}_{q+2}(\not{\mathbf{D}}_S\not{\phi}) + {}^e\mathcal{R}_{[q]})$$
$${}^e\mathcal{K}_{q+2}(\not{\mathbf{D}}_S\epsilon_4) \leq c({}^e\mathcal{L}_{q+2}(\mathbf{D}_S\varphi_N) + {}^e\mathcal{R}_{[q]})$$
$${}^e\mathcal{K}_{q+2}(\not{\mathbf{D}}_S\hat{\eta}_4) \leq ({}^e\mathcal{L}_{q+2}(\not{\mathbf{D}}_S\not{\phi}) + {}^e\mathcal{R}_{[q]}).$$

Also,

$${}^e\mathcal{K}_{q+2}(\mathbf{D}_S\delta_3) \leq c({}^e\mathcal{L}_{q+2}(\not{\mathbf{D}}_S\not{\phi}) + {}^e\mathcal{R}_{[q]})$$
$${}^e\mathcal{K}_{q+2}(\not{\mathbf{D}}_S\epsilon_3) \leq c({}^e\mathcal{L}_{q+2}(\mathbf{D}_S\varphi_N) + {}^e\mathcal{R}_{[q]})$$
$${}^e\mathcal{K}_{q+2}(\not{\mathbf{D}}_S\hat{\eta}_3) \leq ({}^e\mathcal{L}_{q+2}(\not{\mathbf{D}}_S\not{\phi}) + {}^e\mathcal{R}_{[q]}).$$

CHAPTER 12

The Lapse Function

In this chapter we make use of the estimates for the Poisson equation of Chapter 4 to estimate the lapse function ϕ. We recall (see 1.0.13) that ϕ verifies the lapse equation

$$\triangle\phi = |k|^2\phi \tag{12.0.1a}$$

and $\phi \rightarrow 1$ at infinity on each Σ_t. In view of the maximum principle and Harnack inequality we infer that $0 < \phi \leq 1$. Throughout this chapter we assume that the bootstrap assumptions $\mathbf{BA_1}$ and $\mathbf{BA_2}$ hold and that the estimates for the second fundamental form k have already been derived. In other words, we have, with the notation of the previous chapter,

$$\overline{\mathcal{K}_{[2]}}(t) \leq c\mathcal{R}_{[1]}(t) \leq c\varepsilon_0 \tag{12.0.1b}$$

$$\overline{\mathcal{K}_{[3]}}(t) \leq c\mathcal{R}_{[2]}(t) \leq c\varepsilon_0. \tag{12.0.1c}$$

The estimates of the lapse function depend not only on the division between the interior region $I = \Sigma^i$, where $r \leq \frac{r_0}{2}$, and the exterior region $E = \Sigma^e$, where $r \geq \frac{r_0}{2}$, but also on the "far exterior region" FE, which we define as the set of points where $r \geq 2r_0$. We will denote by WZ the wave zone region, complementary to both interior and far exterior regions, given by the set of points for which $\frac{r_0}{2} \leq r \leq 2r_0$.

We now let $f = |k|^2\phi$ in 12.0.1a and check

$$\left(\int_{\Sigma_t} \tau_-^2|f|^2\right)^{1/2} \leq \frac{c\varepsilon_0^2}{r_0}. \tag{12.0.2a}$$

Therefore, we can appeal to the degenerate estimates for the Poisson equation developed in Chapter 4. More precisely, using Proposition 4.2.1ii, we derive

$$\|\nabla\phi\|_2 \leq cr_0^{-1/2}\varepsilon_0^2. \tag{12.0.2b}$$

In view of 12.0.1b we thus infer that

$$\left(\int_{\Sigma_t} r^2\tau_-^2|\nabla\!\!\!/ f|^2\right)^{1/2} \leq \frac{c\varepsilon_0^2}{r_0}, \tag{12.0.2c}$$

and therefore, using the degenerate estimates of Proposition 4.2.2ii, we deduce

$$\|r\nabla\!\!\!/^2\phi\|_2 + \|r\nabla\!\!\!/\nabla_N\phi\|_2 + \|w_1\nabla_N^2\phi\|_2 \le cr_0^{-1/2}\varepsilon_0^2. \tag{12.0.2d}$$

Repeating the same argument once more, with the help of the degenerate estimates of Proposition 4.2.3ii, we infer that

$$\|r^2\nabla\!\!\!/^3\phi\|_2 + \|r^2\nabla\!\!\!/^2\nabla_N\phi\|_2 + \|w_1\nabla\!\!\!/\nabla_N^2\phi\|_2 \le cr_0^{-1/2}\varepsilon_0^2. \tag{12.0.2e}$$

Also taking one normal derivative of the equation, we conclude the following:

Proposition 12.0.1 *Under the assumptions made at the beginning of this chapter, there exists a constant c depending on $\mathcal{R}_{[1]}, \mathcal{R}_{[2]}$ such that the lapse function ϕ verifies the following estimates:*

$$r_0^{1/2}\|\nabla\phi\|_{2,e} \le c\varepsilon_0^2$$
$$r_0^{1/2}\left\{\|r\nabla\!\!\!/^2\phi\|_{2,e} + \|r\nabla\!\!\!/\nabla_N\phi\|_{2,e} + \|w_1\nabla_N^2\phi\|_{2,e}\right\} \le c\varepsilon_0^2$$
$$r_0^{1/2}\left\{\|r^2\nabla\!\!\!/^3\phi\|_{2,e} + \|r^2\nabla\!\!\!/^2\nabla_N\phi\|_{2,e} + \|rw_1\nabla\!\!\!/\nabla_N^2\phi\|_{2,e}\right\} \le c\varepsilon_0^2$$
$$r_0^{1/2}\|w_2\nabla_N^3\phi\|_{2,e} \le c\varepsilon_0^2.$$

Also,

$$r_0^{1/2}|||r^{5/6}\nabla\phi|||_{3,e} \le c\varepsilon_0^2$$
$$r_0^{1/2}\left\{|||r^{11/6}\nabla\!\!\!/^2\phi|||_{3,e}^3 + |||r^{11/6}\nabla\!\!\!/\nabla_N\phi|||_{3,e}^3\right\}^{1/3} \le c\varepsilon_0^2.$$

As a consequence of these and of Proposition 3.2.3 we also have

$$r_0^{1/2}\sup_E r^{1/2}|1-\phi| \le c\varepsilon_0^2$$
$$r_0^{1/2}\sup_E r^{3/2}|\nabla\phi| \le c\varepsilon_0^2.$$

On the other hand, in the interior region I there exists a numerical constant c, which depends only on $\mathcal{R}_{[1]}, \mathcal{R}_{[2]}$ such that, for all $1 \le q \le 3$,

$$r_0^{\frac{5}{2}+q}\|\nabla^q(\phi-1)\|_{2,\iota} \le c\varepsilon_0^2.$$

In particular, we infer that

$$r_0\sup_I|1-\phi| \le c\varepsilon_0^2$$
$$r_0^2\sup_I|\nabla\phi| \le c\varepsilon_0^2.$$

The interior estimates for ϕ are obtained by considering the cut-off function f introduced on page 65 (see 11.3.1a) and deriving an equation for $f(\phi - 1)$. Using standard L^2-estimates,[1] and taking into account 12.0.1b and the exterior estimates of Proposition 12.0.1, we arrive at the desired results.

We next estimate ϕ in the far exterior region F of Σ_t. To achieve this, we consider the truncation function

$$e(x) = \begin{cases} 1 & : r \geq 2r_0 \\ 0 & : r \leq \frac{3r_0}{2} \end{cases}. \tag{12.0.3a}$$

Clearly, ∇e is supported in the region $\frac{3r_0}{2} \leq r \leq 2r_0$ and

$$\sup_\Sigma |\nabla e| \leq c r_0^{-1}$$
$$\sup_\Sigma |\nabla^2 e| \leq c r_0^{-2}.$$

Let $\phi^e = (\phi - 1)e$. It satisfies the equation

$$\triangle \phi^e = f^e \tag{12.0.3b}$$
$$f^e = \underbrace{|k|^2 \phi e}_{f_1} + \underbrace{2\nabla\phi \cdot \nabla e + (\phi - 1)\triangle e}_{f_0}.$$

Remark that

$$P[\phi^e] = P[\phi] = -\frac{1}{4\pi}\int_\Sigma \phi |k|^2 \leq c\varepsilon_0^2 \tag{12.0.3c}$$

where $P[\phi]$ is the charge of ϕ defined according to 4.1.4. Now, taking into account the assumptions 12.0.1b and Proposition 12.0.1,

$$\int_\Sigma r^4 |f_1|^2 \leq c\varepsilon_0^4 r_0^{-1}$$
$$\int_\Sigma r^4 |f_0|^2 \leq c \int_{\frac{3r_0}{2} \leq r \leq 2r_0} (r_0^2 |\nabla\phi|^2 + |\phi - 1|^2) \leq c r_0 \varepsilon_0^4,$$

whence

$$\int_\Sigma r^4 |f^e|^2 \leq c r_0 \varepsilon_0^4. \tag{12.0.3d}$$

We are thus in a position to apply the results of Proposition 4.3.1 to the equation 12.0.3b and deduce the following:

[1] See the remark following Proposition 4.4.2.

Proposition 12.0.2 *Under the assumptions made at the beginning of this chapter there exists a constant c depending on $\mathcal{R}_{[1]}, \mathcal{R}_{[2]}$ such that the lapse function ϕ verifies the following estimates:*

$$\|\phi - 1 - \frac{1}{r}P[\phi]\|_{2,fe} + \|\nabla(r\phi)\|_{2,fe} + \|r^2\nabla^2(r\phi)\|_{2,fe} \leq cr_0^{1/2}\varepsilon_0^2$$

$$\|r^3\nabla^3(r\phi)\|_{2,fe} \leq cr_0^{1/2}\varepsilon_0^2$$

where $\| \, \|_{2,fe}$ denotes the L^2-norm in the exterior region FE. Also,

$$P[\phi] \leq c\varepsilon_0^2. \tag{12.0.4}$$

As a consequence, using the nondegenerate form of the Sobolev inequalities, we infer that

$$\sup_{FE} |r(\phi - 1) - P[\phi]| \leq cr_0^{1/2}\frac{1}{r^{1/2}}\varepsilon_0^2$$

$$\sup_{FE} |\nabla[r(\phi) - 1)]| \leq cr_0^{1/2}\frac{1}{r^{3/2}}\varepsilon_0^2.$$

We next consider time derivatives of ϕ. To obtain good estimates in the wave zone WZ, we need to introduce the space-time scaling operator

$$S = \underline{t}T + Z \tag{12.0.5a}$$

where $\underline{t} = (r - u)$ and Z is the position vectorfield (see page 80) acting on Σ_t,

$$Z = rN.$$

Computing the commutator between ∂_t and $\triangle_g$ with the help of 1.0.11a–1.0.12b we find

$$[\partial_t, \triangle_g]\phi = 2k^{ij}(\phi\nabla_i\nabla_j\phi + \nabla_i\phi\nabla_j\phi). \tag{12.0.5b}$$

Also,

$$\partial_t|k|^2 = 2k^{ij}(-\nabla_i\nabla_j\phi + \phi R_{ij}).$$

Hence

$$\triangle_g\frac{\partial\phi}{\partial t} - |k|^2\frac{\partial\phi}{\partial t} = -2k^{ij}\left(\phi\nabla_i\nabla_j\phi + \nabla_i\phi\nabla_j\phi - \phi^2R_{ij}\right). \tag{12.0.5c}$$

Therefore also

$$\triangle(\underline{t}\frac{\partial\phi}{\partial t}) - |k|^2\underline{t}\frac{\partial\phi}{\partial t} = -2\underline{t}k^{ij}\left(\phi\nabla_i\nabla_j\phi + \nabla_i\phi\nabla_j\phi - R_{ij}\right) \tag{12.0.5d}$$
$$+ \frac{2}{\underline{t}}\nabla\underline{t}\cdot\nabla(\underline{t}\frac{\partial\phi}{\partial t}) + \underline{t}\frac{\partial\phi}{\partial t}(\underline{t}^{-1}\triangle\underline{t} - 2\underline{t}^{-2}|\nabla\underline{t}|^2).$$

On the other hand, recalling commutation formula 4.2.4, we infer that

$$\triangle \nabla_Z \phi - |k|^2 \nabla_Z \phi = 2\phi k^{ij} \nabla_Z k_{ij} + \hat{\pi}^{ij} \nabla_i \nabla_j \phi + \frac{1}{3} \mathrm{tr}\pi |k|^2 \phi + \nabla_j \hat{\pi}^{ij} \nabla_i \phi - \frac{1}{6} \nabla^i \mathrm{tr}\pi \nabla_i \phi. \tag{12.0.5e}$$

Combining 12.0.5d with 12.0.5e, we deduce the formula

$$\begin{aligned}
\triangle(S\phi) - |k|^2 S\phi = {} & 2\phi k^{ij}(\phi \underline{t} R_{ij} + r \nabla_N k_{ij}) \\
& + 2(\underline{t}^{-1} \nabla \underline{t} - \phi^{-1} \nabla \phi) \cdot \nabla(S\phi - Z\phi) \\
& + (2\underline{t}^{-1} \phi^{-1} \nabla \underline{t} \cdot \nabla \phi - |k|^2 \\
& + \underline{t}^{-1} \triangle \underline{t} - 2\underline{t}^{-2} |\nabla \underline{t}|^2) \cdot (S\phi - Z\phi) \\
& + (\hat{\pi}^{ij} - 2\phi \underline{t} k^{ij}) \nabla_i \nabla_j \phi - 2\underline{t} k^{ij} \phi^{-1} \nabla_i \phi \nabla_j \phi \\
& + \frac{1}{3} \mathrm{tr}\pi |k|^2 \phi + \nabla_j \hat{\pi}^{ij} \nabla_i \phi - \frac{1}{6} \nabla^i \mathrm{tr}\pi \nabla_i \phi.
\end{aligned} \tag{12.0.5f}$$

We now recall, from 11.2.19c, the formula

$$\nabla\!\!\!/_N \hat{\eta}_{ij} = -\frac{1}{2} \mathrm{tr}\theta \hat{\eta}_{ij} - \frac{1}{4} \underline{\alpha}_{ij} + \dots \tag{12.0.5g}$$

Hence,

$$k^{ij} \nabla_N k_{ij} = \hat{\eta} \cdot (-\frac{1}{2} \mathrm{tr}\theta \hat{\eta} - \frac{1}{4} \underline{\alpha}_{ij}) + \text{l.o.t.} \tag{12.0.5h}$$

Also,

$$R_{ij} = E_{ij} + k_{ia} k^a{}_j,$$

and taking into account 7.3.3e, we infer that

$$k^{ij} R_{ij} = \frac{1}{4} \hat{\eta} \cdot \underline{\alpha} + \text{l.o.t.} \tag{12.0.5i}$$

The lower-order terms denoted by l.o.t., in both 12.0.5h and 12.0.5i, are of order $\max(r^{-4} \tau_-^{-3/2}, r^{-5})$ or better in the wave zone. In view of these formulas and the fact that $\underline{t}$ and r differ only by a term of order τ_- in the wave zone we infer that

$$\left(\int_\Sigma \tau_-^2 |\phi k \cdot (\phi \underline{t} Ric + r \nabla_N k)|^2 \right)^{1/2} \leq \frac{c\varepsilon_0}{r_0}.$$

We also note that

$$\begin{aligned}
\nabla \underline{t} &= a^{-1}(\lambda - 1)N \\
\triangle \underline{t} &= a^{-1} \nabla_N \lambda - (\lambda - 1) a^{-1} (a^{-1} \nabla_N a - \mathrm{tr}\theta).
\end{aligned}$$

Thus, taking into account 11.1.13a, as well as the result of the previous proposition, the right-hand side of equation 12.0.5f satisfies the estimates 12.0.2a and therefore, proceeding exactly as in the derivation of the degenerate estimate of Proposition 4.2.1ii,[2] we first obtain

$$r_0^{1/2}\|\nabla S(\phi)\|_2 \leq c\varepsilon_0^2 + c\varepsilon_0\|\nabla S(\phi)\|_2,$$

and consequently, for small ε_0,

$$r_0^{1/2}\|\nabla S(\phi)\|_2 \leq c\varepsilon_0^2. \tag{12.0.5j}$$

Then, writing 12.0.5f in the form

$$\triangle(S\phi) = f,$$

we easily check that

$$\left(\int_{\Sigma_t} \tau_-^2 |f|^2\right)^{1/2} + \left(\int_{\Sigma_t} r^2\tau_-^2 |\nabla\!\!\!/ f|^2\right)^{1/2} \leq \frac{c\varepsilon_0^2}{r_0} + c\varepsilon_0 r_0^{-1/2}(\|r\nabla\!\!\!/^2(S\phi)\|_2 + \|r\nabla\!\!\!/\nabla_N(S\phi))\|.$$

Therefore, in view of the degenerate estimates of Proposition 4.2.2ii, for small ε_0

$$r_0^{1/2}\left(\|r\nabla\!\!\!/^2(S\phi)\|_2 + \|r\nabla\!\!\!/\nabla_N(S\phi)\| + \|w_1\nabla_N^2(S\phi)\|\right) \leq c\varepsilon_0^2 \tag{12.0.5k}$$

with c a constant depending on $\mathcal{R}_{[1]}, \mathcal{R}_{[2]}$.

We thus obtain the following:

Proposition 12.0.3 *Under the assumptions made at the beginning of this chapter there exists a constant c depending on $\mathcal{R}_{[1]}, \mathcal{R}_{[2]}$ such that the lapse function ϕ verifies the following estimates:*

$$r_0^{1/2}\|\nabla(S\phi)\|_2 \leq c\varepsilon_0^2$$

$$r_0^{1/2}\left\{\|r\nabla\!\!\!/^2(S\phi)\|_2 + \|r\nabla\!\!\!/\nabla_N(S\phi)\| + \|w_1\nabla_N^2(S\phi)\|\right\} \leq c\varepsilon_0^2.$$

Also,

$$r_0^{1/2}|||r^{5/6}\nabla(S\phi)|||_{3,e} \leq c\varepsilon_0^2.$$

As a consequence of these and of Proposition 3.2.3 we also have

$$r_0^{1/2}\sup_E r^{1/2}|(S\phi)| \leq c\varepsilon_0^2.$$

[2] We also take into account that the contribution of the term $|k|^2 S(\phi)$ on the left-hand side of 12 0.5f has favorable sign and can thus be discarded.

On the other hand, in the interior region I there exists a numerical constant c, which depends only on $\mathcal{R}_{[1]}, \mathcal{R}_{[2]}$ such that, for all $1 \leq q \leq 2$,

$$r_0^{\frac{5}{2}+q}\|\nabla^q(S\phi)\|_{2,\imath} \leq c\varepsilon_0^2.$$

In particular, we infer that

$$r_0 \sup_I |(S\phi)| \leq c\varepsilon_0^2.$$

To obtain estimates in the far exterior region FE, we start with the equation 12.0.5c, which we now localize with the help of the truncation function 12.0.3a. Thus, introducing $\psi^e = e\frac{\partial\phi}{\partial t}$, we find

$$\triangle_g\psi^e - |k|^2\psi^e = h^e \tag{12.0.6a}$$

where

$$\begin{aligned} h^e = &-2\underbrace{ek^{\imath\jmath}\left(\phi\nabla_\imath\nabla_\jmath\phi + \nabla_\imath\phi\nabla_\jmath\phi - \phi^2R_{\imath\jmath}\right)}_{h_1^e} \\ &+\underbrace{2\left(\nabla\frac{\partial\phi}{\partial t}\right)\cdot\nabla e + \frac{\partial\phi}{\partial t}\Delta e}_{h_2^e}. \end{aligned} \tag{12.0.6b}$$

We first remark that, by the results of the previous two propositions,

$$\left(\int_\Sigma r^2|h^e|^2\right)^{1/2} \leq c\varepsilon_0^2 r_0^{-3/2}.$$

Multiplying the equation 12.0.6a by ψ^e and proceeding precisely as in the proof of the nondegenerate estimate of Proposition 4.2.1, we infer that

$$\|\nabla\psi^e\|_2 \leq \|rh^e\|_2 \leq c\varepsilon_0^2 r_0^{-3/2}. \tag{12.0.6c}$$

We now remark that

$$\begin{aligned} P[\psi^e] = P\left[\frac{\partial\phi}{\partial t}\right] &= -\frac{1}{4\pi}\int_{\Sigma_t}|k|^2\frac{\partial\phi}{\partial t} \\ &+\frac{1}{4\pi}\int_{\Sigma_t}2k^{\imath\jmath}(\phi\nabla_\imath\nabla_\jmath\phi + \nabla_\imath\phi\nabla_\jmath\phi - \phi^2R_{\imath\jmath}). \end{aligned} \tag{12.0.6d}$$

Now, in view of 12.0.5g and 12.0.5i, the bad term $k^{\imath\jmath}R_{\imath\jmath}$ can be written in the form

$$k^{ij}R_{ij} = -\frac{1}{2}(\nabla_N|\hat{\eta}|^2 + \mathrm{tr}\theta|\hat{\eta}|^2) + \mathrm{l.o.t.} \tag{12.0.6e}$$
$$= -\frac{1}{2}\mathrm{div}(N|\hat{\eta}|^2) + \mathrm{l.o.t.} \tag{12.0.6f}$$

where the error terms l.o.t. are of order max $(r^{-4}\tau_-^{-3/2}, r^{-5})$ or better in the wave zone.

Thus, by integration,

$$\int_{\Sigma_t} 2k^{ij}(\phi\nabla_i\nabla_j\phi + \nabla_i\phi\nabla_j\phi - \phi^2 R_{ij}) \leq c\varepsilon_0^2 r_0^{-2}. \tag{12.0.6g}$$

On the other hand,

$$\int_{\Sigma_t} |k|^2\frac{\partial\phi}{\partial t} = \int_{FE} |k|^2\frac{\partial\phi}{\partial t} + \int_{WZ} |k|^2\frac{\partial\phi}{\partial t} + \int_I |k|^2\frac{\partial\phi}{\partial t}.$$

According to the sup-norm estimates of Propositions 12.0.1 and 12.0.2,

$$\sup_{WZ}|\frac{\partial\phi}{\partial t}| \leq \varepsilon_0^2 r_0^{-2}$$
$$\sup_I|\frac{\partial\phi}{\partial t}| \leq \varepsilon_0^2 r_0^{-2}.$$

Therefore,

$$|\int_E |k|^2\frac{\partial\phi}{\partial t}| \leq c\varepsilon_0^2 r_0^{-2}.$$

Also, using 12.0.6c, we easily deduce that

$$|\int_{FE} |k|^2\frac{\partial\phi}{\partial t}| = |\int_{FE} |k|^2\psi^e| \leq c\varepsilon_0^2 r_0^{-2}.$$

Consequently,

$$|\int_{\Sigma_t} |k|^2\frac{\partial\phi}{\partial t}| \leq c\varepsilon_0^2 r_0^{-2} \tag{12.0.6h}$$

and thus, combining 12.0.6h with 12.0.6g, we conclude that

$$|P[\psi^e]| = |P[\frac{\partial\phi}{\partial t}]| \leq c\varepsilon_0^2 r_0^{-2}. \tag{12.0.6i}$$

Now, going back to 12.0.6a, we estimate, with the help of the results of Propositions 12.0.1 and 12.0.2,

$$\left(\int_\Sigma r^4|h_1^e|^2\right)^{1/2} \leq c\varepsilon_0^2 r_0^{-2}$$
$$\left(\int_\Sigma r^4|h_2^e|^2\right)^{1/2} \leq c\left(\int_{\frac{3r_0}{2}\leq r\leq 2r_0} r_0^2|\nabla\frac{\partial\phi}{\partial t}|^2+|\frac{\partial\phi}{\partial t}|^2\right)^{1/2} \leq c\varepsilon_0^2 r_0^{-1}.$$

We are thus in a position to apply the results of Proposition 4.2.4 to the equation 12.0.3b and deduce the following:

Proposition 12.0.4

$$\|\frac{\partial\phi}{\partial t}-\frac{1}{r}P[\frac{\partial\phi}{\partial t}]\|_{2,fe}+\|\nabla(r\frac{\partial\phi}{\partial t})\|_{2,fe}+\|r^2\nabla^2(r\frac{\partial\phi}{\partial t})\|_{2,fe} \leq cr_0^{-1/2}\varepsilon_0^2 \tag{12.0.6j}$$

and

$$P[\frac{\partial\phi}{\partial t}] \leq c\varepsilon_0^2 r_0^{-2}. \tag{12.0.6k}$$

As a consequence, using the nondegenerate form of the Sobolev inequalities, we infer that

$$\sup_{FE}|r\frac{\partial\phi}{\partial t}-P[\frac{\partial\phi}{\partial t}]| \leq cr_0^{-1/2}\frac{1}{r^{1/2}}\varepsilon_0^2. \tag{12.0.6l}$$

We are now ready to state our main results.

Theorem 12.0.1 *Assume that the space-time slab $\bigcup_{t'\in[0,t]}\Sigma_{t'}$ satisfies the basic bootstrap assumptions* $\mathbf{BA_0}$, $\mathbf{BA_1}$, *and* $\mathbf{BA_2}$. *There exists a constant c such that*

$$\mathcal{L}_{[3]} \leq c\varepsilon_0. \tag{12.0.7}$$

Using Theorem 12.0.1 together with Proposition 11.4.1, we obtain

Theorem 12.0.2 *Assume that the space-time slab $\bigcup_{t'\in[0,t]}\Sigma_{t'}$ satisfies the basic bootstrap assumptions* $\mathbf{BA_0}$, $\mathbf{BA_1}$, *and* $\mathbf{BA_2}$. *There exists a constant c such that*

$$\mathcal{K}_{[3]} \leq c\varepsilon_0 \tag{12.0.8}$$
$$\mathcal{K}^*_{[3]} \leq c\varepsilon_0. \tag{12.0.9}$$

Finally, using the result of Theorem 12.0.2 and following the same steps as in the derivation of Propositions 12.0.2 and 12.0.4 we also prove

Theorem 12.0.3 *Assume that the space-time slab $\bigcup_{t' \in [0,t]} \Sigma_{t'}$ satisfies the basic bootstrap assumptions* $\mathbf{BA_0}$, $\mathbf{BA_1}$, *and* $\mathbf{BA_2}$. *There exists a constant c such that*

$$\mathcal{L}^*_{[3]} \leq c\epsilon_0^2. \tag{12.0.10}$$

CHAPTER 13

Derivatives of the Optical Function

13.1. Higher Derivatives of the Exterior Optical Function

The aim of this section is to derive estimates in the exterior for the higher derivatives of the optical function u.

Throughout this section we use as null pair the l-null pair of (t, u), that is,

$$l = -\mathbf{g}^{\mu\nu}\mathbf{D}_\nu u = a^{-1}(T+N) \qquad \underline{l} = a(T-N)$$

where a is the lapse function of the foliation on Σ_t given by the level hypersurfaces of u,

$$a^{-1} = |\nabla u| = \mathbf{D}_T u.$$

Relative to the l-pair, the Ricci coefficients defined by formulas 7.3.1b in Chapter 7, take the form

$$\begin{aligned} H &= \chi & \underline{H} &= \underline{\chi} \\ Z &= \zeta & \underline{Z} &= \underline{\zeta} \\ Y &= 0 & \underline{Y} &= \underline{\xi} \\ \Omega &= 0 & \underline{\Omega} &= -\omega \\ V &= \zeta. \end{aligned} \tag{13.1.1a}$$

Thus, for a corresponding null frame $e_4 = l$, $e_3 = \underline{l}$, $(e_A)_{A=1,2}$, we have

$$\begin{aligned} \mathbf{D}_A e_3 &= \underline{\chi}_{AB} e_B + \zeta_A e_3 & \mathbf{D}_A e_4 &= \chi_{AB} e_B - \zeta_A e_4 \\ \mathbf{D}_3 e_3 &= 2\underline{\xi}_A e_A + 2\omega e_3 & \mathbf{D}_3 e_4 &= 2\zeta_A e_A - 2\omega e_4 \\ \mathbf{D}_4 e_3 &= 2\underline{\zeta}_A e_A & \mathbf{D}_4 e_4 &= 0. \end{aligned} \tag{13.1.1b}$$

Also,

$$\begin{aligned}
\mathbf{D}_B e_A &= \nabla\!\!\!/_B e_A + \frac{1}{2}\chi_{AB} e_3 + \frac{1}{2}\underline{\chi}_{AB} e_4 \\
\mathbf{D}_3 e_A &= \mathbf{D}\!\!\!/_3 e_A + \zeta_A e_3 + \underline{\xi}_A e_4 \qquad (13.1.1c) \\
\mathbf{D}_4 e_A &= \mathbf{D}\!\!\!/_4 a_A + \underline{\zeta}_A e_4.
\end{aligned}$$

The Ricci coefficients $\chi, \underline{\chi}, \zeta, \underline{\zeta}, \underline{\xi}, \omega$ of the l-frame are connected to the Ricci coefficients $\chi', \underline{\chi}', \zeta', \underline{\zeta}', \underline{\xi}', \nu, \underline{\nu}$ of the standard null frame $e_4' = T + N, e_3' = T - N$ through the formulas

$$\begin{aligned}
\chi &= a^{-1}\chi' & \underline{\chi} &= a\underline{\chi}' \\
\zeta &= \zeta' & \underline{\zeta} &= \underline{\zeta}' \qquad (13.1.1d) \\
\underline{\xi} &= a^2\underline{\xi}' & \omega &= \tfrac{1}{2}a(a^{-1}\mathbf{D}_3 a - \underline{\nu}).
\end{aligned}$$

Moreover, we recall, from Proposition 7.5.1, that the Ricci coefficients of the standard frame are given by the formulas

$$\begin{aligned}
\chi'_{AB} &= \theta_{AB} - k_{AB} \\
\underline{\chi}'_{AB} &= -\theta_{AB} - k_{AB} \\
\underline{\xi}'_A &= \phi^{-1}\nabla\!\!\!/_A\phi - a^{-1}\nabla\!\!\!/_A a \\
\underline{\zeta}'_A &= \phi^{-1}\nabla\!\!\!/_A\phi - \epsilon_A \\
\zeta'_A &= a^{-1}\nabla\!\!\!/_A a + \epsilon_A \\
\nu &= -\phi^{-1}\nabla\!\!\!/_N\phi + \delta \\
\underline{\nu} &= \phi^{-1}\nabla\!\!\!/_N\phi + \delta. \qquad (13.1.1e)
\end{aligned}$$

We also recall from Chapter 9 that the Hessian of u decomposes into the null second fundamental form χ_{AB}, the 1-form ζ_A, and the function ω, which satisfy along the generators of each level set C_u of u the equations

$$\frac{d\chi_{AB}}{ds} = -\chi_{AC}\chi_{CB} + \alpha_{AB} \qquad (13.1.2a)$$

$$\frac{d\zeta_A}{ds} = -\chi_{AB}\zeta_B + \chi_{AB}\underline{\zeta}_B - \beta_A \qquad (13.1.2b)$$

and

$$\frac{d\omega}{ds} = 2\underline{\zeta}\cdot\zeta - |\zeta|^2 - \rho. \qquad (13.1.2c)$$

Equation 13.1.2a is equivalent to

$$\frac{d\mathrm{tr}\chi}{ds} + \frac{1}{2}(\mathrm{tr}\chi)^2 = -|\hat{\chi}|^2 \tag{13.1.2d}$$

$$\frac{d\hat{\chi}_{AB}}{ds} + \mathrm{tr}\chi\hat{\chi}_{AB} = \alpha_{AB}. \tag{13.1.2e}$$

Moreover, the general null-structure equations, discussed in Chapter 7, applied to the l-null pair give us, in addition, the following:

$$\frac{d}{ds}\mathrm{tr}\underline{\chi} + \frac{1}{2}\mathrm{tr}\chi\mathrm{tr}\underline{\chi} = 2\mathrm{d}\!\!/\mathrm{iv}\,\underline{\zeta} - \hat{\chi}\cdot\underline{\hat{\chi}} + 2|\underline{\zeta}|^2 + 2\rho \tag{13.1.2f}$$

$$\frac{d}{ds}\underline{\hat{\chi}} + \frac{1}{2}\mathrm{tr}\chi\underline{\hat{\chi}} = \nabla\!\!\!/\widehat{\otimes}\underline{\zeta} - \frac{1}{2}\mathrm{tr}\underline{\chi}\hat{\chi} + \underline{\zeta}\widehat{\otimes}\underline{\zeta} \tag{13.1.2g}$$

and

$$\mathbf{D}_3\mathrm{tr}\chi + \frac{1}{2}\mathrm{tr}\underline{\chi}\mathrm{tr}\chi + 2\omega\mathrm{tr}\chi = 2\mathrm{d}\!\!/\mathrm{iv}\,\zeta - \hat{\chi}\cdot\underline{\hat{\chi}} + 2|\zeta|^2 + 2\rho \tag{13.1.2h}$$

$$\mathrm{c}\!\!/\mathrm{url}\,\zeta = \sigma - \frac{1}{2}\hat{\chi}\wedge\underline{\hat{\chi}} \tag{13.1.2i}$$

$$\mathbf{D}\!\!\!/_3\hat{\chi} + \frac{1}{2}\mathrm{tr}\underline{\chi}\hat{\chi} + 2\omega\hat{\chi} = (\nabla\!\!\!/\widehat{\otimes}\zeta) - \frac{1}{2}\mathrm{tr}\chi\underline{\hat{\chi}} + \zeta\widehat{\otimes}\zeta \tag{13.1.2j}$$

$$\mathbf{D}\!\!\!/_3\zeta + \mathrm{tr}\underline{\chi}\zeta = 2\nabla\!\!\!/\omega - 2\underline{\hat{\chi}}\cdot\zeta + \frac{1}{2}\mathrm{tr}\chi\underline{\xi} + \hat{\chi}\cdot\underline{\xi} - \underline{\beta}. \tag{13.1.2k}$$

Also, we have the null-Codazzi equation

$$(\mathrm{d}\!\!/\mathrm{iv}\,\hat{\chi})_A + \hat{\chi}_{AB}\zeta_B = \frac{1}{2}(\nabla\!\!\!/_A\mathrm{tr}\chi + \zeta_A\mathrm{tr}\chi) - \beta_A, \tag{13.1.2l}$$

the conjugate null-Codazzi equation

$$(\mathrm{d}\!\!/\mathrm{iv}\,\underline{\hat{\chi}})_A - \underline{\hat{\chi}}_{AB}\zeta_B = \frac{1}{2}(\nabla\!\!\!/_A\mathrm{tr}\underline{\chi} - \zeta_A\mathrm{tr}\underline{\chi}) + \underline{\beta}_A, \tag{13.1.2m}$$

and the Gauss equations for the Gauss curvature K of $S_{t,u}$

$$K = -\frac{1}{4}\mathrm{tr}\chi\mathrm{tr}\underline{\chi} + \frac{1}{2}\hat{\chi}\cdot\underline{\hat{\chi}} - \rho. \tag{13.1.2n}$$

We will derive estimates for the derivatives of χ_{AB}, ζ_A, and ω of up to the third order. Given V an arbitrary S-tangent tensorfield, we will use the following norms:

$$|V|_{p,S}(t,u) = \left(\int_{S_{t,u}} |V|^p\right)^{1/p}$$

and

$$^{(loc)}||V||(t,\kappa) = \left(\int_{D(t,\kappa)} |V|^2\right)^{1/2}$$

where $D(t,\kappa)$ is the annular region $\kappa \leq u \leq \kappa + 1$ on Σ_t. We also define $^{(\mathrm{loc})}||V||_{2,e}(t)$ to be the supremum over $\kappa \leq u_1$ of the norms $^{(\mathrm{loc})}||V||(t,\kappa)$.[1]

The following lemma will be used repeatedly in the sequel:

Lemma 13.1.1 *Let $U_{A_1 \ldots A_k}$ and $F_{A_1 \ldots A_k}$ be k-covariant tensorfields tangential to the surfaces $S_{t,u}$ that verify*

$$\frac{dU_{A_1\ldots A_k}}{ds} + \lambda_0 tr\chi U_{A_1\ldots A_k} = F_{A_1\ldots A_k},$$

where λ_0 is a nonnegative real number. We let: $\lambda_1 = 2(\lambda_0 - \frac{1}{p})$. Then

1. *on each cone C_u*

$$|r^{\lambda_1} U|_{p,S}(t,u) \leq c\left(|r^{\lambda_1} U|_{p,S}(t_*,u) + \int_t^{t_*} |r^{\lambda_1} F|_{p,S}(t',u)dt'\right)$$

   where c is a constant independent of u.

2. *for each κ:*

$$^{(loc)}||r^{\lambda_1} U||(t,\kappa) \leq c\left(^{(loc)}||r^{\lambda_1} U||(t_*,\kappa) + \int_t^{t_*} {}^{(loc)}||r^{\lambda_1} F||(t',\kappa)dt'\right)$$

   where c is a constant independent of κ.

Proof 13.1.1 *Recalling from Chapter 9 that along each generator of C_u*

$$\frac{dt}{ds} = a^{-1}\phi^{-1}, \tag{13.1.3}$$

we deduce, for any function f, the formula

$$\frac{d}{dt}\left(\int_{S_{t,u}} f d\mu_\gamma\right) = \int_{S_{t,u}} \left(\frac{df}{ds} + tr\chi f\right) a\phi d\mu_\gamma. \tag{13.1.4}$$

[1] We recall that $u_1 = u_1(t)$ is the value of u corresponding to $r_0/2$.

In particular, setting $f = 1$, we obtain

$$\frac{dr}{dt} = \frac{r}{2}\overline{a\phi tr\chi}. \tag{13.1.5a}$$

Hence for any function f and any real number λ, we have

$$\frac{d}{dt}\left(\int_{S_{t,u}} r^{\lambda} f d\mu_{\gamma}\right) = \int_{S_{t,u}} r^{\lambda}\left(\frac{df}{ds} + (1 + \frac{\lambda}{2})tr\chi f\right) a\phi - \frac{\lambda}{2}\int_{S_{t,u}} r^{\lambda} f(a\phi tr\chi - \overline{a\phi tr\chi}). \tag{13.1.5b}$$

Now the equation satisfied by the tensorfield U implies

$$\frac{d}{ds}|U|^p + \lambda_0 p tr\chi|U|^p \leq |F||U|^{p-1}.$$

Setting

$$f = r^{\lambda_1 p}|U|^p$$

where

$$1 + \frac{\lambda_1 p}{2} = \lambda_0 p$$

in formula 13.1.5b and using this inequality, we obtain

$$\frac{d}{dt}\left(\int_{S_{t,u}} r^{\lambda_1 p}|U|^p\right) \leq p\int_{S_{t,u}} r^{\lambda_1 p}|F||U|^{p-1} - \frac{\lambda_1 p}{2}\int_{S_{t,u}} r^{\lambda_1 p}|U|^p(a\phi tr\chi - \overline{a\phi tr\chi}). \tag{13.1.5c}$$

Now, by the Holder inequality,

$$\int_{S_{t,u}} r^{\lambda_1 p}|F||U|^{p-1} \leq \left(\int_{S_{t,u}} r^{\lambda_1 p}|F|^p\right)^{\frac{1}{p}}\left(\int_{S_{t,u}} r^{\lambda_1 p}|U|^p\right)^{\frac{p-1}{p}},$$

whereas by Propositions 9.1.1, 9.1.2, and 9.1.3,

$$|a\phi tr\chi - \overline{a\phi tr\chi}| \leq cr^{-2},$$

provided that the quantities A and B are sufficiently small. Substituting in 13.1.5c, we obtain the inequality

$$\frac{d}{dt}|r^{\lambda_1}U|_{p,S} \leq c\left(|r^{\lambda_1}F|_{p,S} + r^{-2}|r^{\lambda_1}U|_{p,S}\right),$$

which upon integration in the interval $[t, t_*]$ $t \geq t_0$ *yields conclusion (i). In view of the fact that by the coarea formula*

$$^{(loc)}||V||(t,\kappa) = \left(\int_\kappa^{\kappa+1}\left(\int_{S(t,u)} a|V|^2\right) du\right)^{1/2}$$

we have

$$a_M^{-1/2(loc)}||V||(t,\kappa) \leq \left(\int_\kappa^{\kappa+1} |V|^2_{2,S}(t,u)\right)^{1/2} \leq a_M^{-1/2(loc)}||V||(t,\kappa).$$

Taking the L^2 *norm of (i) in the interval* $\kappa \leq u \leq \kappa+1$, *we obtain conclusion (ii).*

The following commutation lemma will also be frequently used in this section.

Lemma 13.1.2 *Let* $U_{\underline{A}} = U_{A_1\ldots A_k}$ *be a k-covariant tensorfield in space-time that is tangent to the surfaces* $S_{t,u}$ *at each point. Then*

$$\begin{aligned}
(i) \quad \not{D}_4\nabla\!\!\!/_B U_{\underline{A}} - \nabla\!\!\!/_B\not{D}_4 U_{\underline{A}} &= -\chi_{BC}\nabla\!\!\!/_C U_{\underline{A}} + (\zeta_B + \underline{\zeta}_B)\not{D}_4 U_{\underline{A}} \\
&\quad + \sum_i (\chi_{A_iB}\underline{\zeta}_C - \chi_{BC}\underline{\zeta}_{A_i} \\
&\qquad + \in_{A_iC}{}^*\beta_B) U_{A_1\ldots\check{C}\ldots A_k} \\
(ii) \quad \not{D}_4\not{D}_3 U_{\underline{A}} - \not{D}_3\not{D}_4 U_{\underline{A}} &= 2(\underline{\zeta}_B - \zeta_B)\nabla\!\!\!/_B U_{\underline{A}} + 2\omega\not{D}_4 U_{\underline{A}} \\
&\quad + 2\sum_i (\zeta_{A_i}\underline{\zeta}_B - \underline{\zeta}_{A_i}\zeta_B - \sigma \in_{A_iB}) \\
&\quad \times U_{A_1\ldots\check{B}\ldots A_k} \\
(iii) \quad \not{D}_3\nabla\!\!\!/_B U_{\underline{A}} - \nabla\!\!\!/_B\not{D}_3 U_{\underline{A}} &= -\underline{\chi}_{BC}\nabla\!\!\!/_C U_{\underline{A}} + \underline{\xi}_B\not{D}_4 U_{\underline{A}} \\
&\quad + \sum_i \Big(\chi_{A_iB}\underline{\xi}_C - \chi_{BC}\underline{\xi}_{A_i} \\
&\qquad + \underline{\chi}_{A_iB}\zeta_C - \underline{\chi}_{BC}\zeta_{A_i} \\
&\qquad - \in_{A_iC}{}^*\underline{\beta}_B\Big) U_{A_1\ldots\check{C}\ldots A_k}.
\end{aligned}$$

Proof 13.1.2 *It suffices to consider the case where* $U_{\underline{A}}$ *is a 1-form. Relation (i) results from the fact that*

$$\begin{aligned}
\mathbf{D}^2_{4B}U_A &= \not{D}_4\nabla\!\!\!/_B U_A + \chi_{BC}\underline{\zeta}_A U_C - \underline{\zeta}_B\not{D}_4 U_A, \\
\mathbf{D}^2_{B4}U_A &= \nabla\!\!\!/_B\not{D}_4 U_A + \chi_{AB}\underline{\zeta}_C U_C - \chi_{BC}\nabla\!\!\!/_C U_A + \zeta_B\not{D}_4 U_A,
\end{aligned}$$

and

$$\mathbf{D}^2_{4B}U_A - \mathbf{D}^2_{B4}U_A = \mathbf{R}_{AC4B}U_C = \in_{AC}{}^\star\beta_B U_C.$$

Relation (ii) results from the fact that

$$\mathbf{D}^2_{43}U_A = \not{\mathbf{D}}_4\not{\mathbf{D}}_3 U_A - 2\underline{\zeta}_B \nabla\!\!\!/_B U_A + 2\underline{\zeta}_A \zeta_B U_B,$$
$$\mathbf{D}^2_{34}U_A = \not{\mathbf{D}}_3\not{\mathbf{D}}_4 U_A - 2\zeta_B \nabla\!\!\!/_B U_A + 2\zeta_A \underline{\zeta}_B U_B + 2\omega\not{\mathbf{D}}_4 U_A,$$

and

$$\mathbf{D}^2_{43}U_A - \mathbf{D}^2_{34}U_A = \mathbf{R}_{AB43}U_B = -2 \in_{AB} \sigma U_B.$$

Finally, relation (iii) results from the fact that

$$\mathbf{D}^2_{3B}U_A = \not{\mathbf{D}}_3\nabla\!\!\!/_B U_A + (\zeta_A\underline{\chi}_{BC} + \underline{\xi}_A \chi_{BC})U_C - \zeta_B\not{\mathbf{D}}_3 U_A - \underline{\xi}_B\not{\mathbf{D}}_4 U_A,$$
$$\mathbf{D}^2_{B3}U_A = \nabla\!\!\!/_B\not{\mathbf{D}}_3 U_A + (\chi_{AB}\underline{\xi}_C + \underline{\chi}_{AB}\zeta_C)U_C - \underline{\chi}_{BC}\nabla\!\!\!/_C U_A - \zeta_B\not{\mathbf{D}}_3 U_A,$$

and

$$\mathbf{D}^2_{3B}U_A - \mathbf{D}^2_{B3}U_A = \mathbf{R}_{AC3B}U_C = - \in_{AC} {}^\star\underline{\beta}_B U_C.$$

Corollary 13.1.2.1 *If U is a scalar satisfying*

$$\frac{dU}{ds} = F,$$

then

$$(i) \quad \frac{d}{ds}\nabla\!\!\!/_A U + \chi_{AB}\nabla\!\!\!/_B U = \nabla\!\!\!/_A F + (\zeta_A + \underline{\zeta}_A)F$$
$$(ii) \quad \frac{d}{ds}\mathbf{D}_3 U = \mathbf{D}_3 F + 2\omega F + 2(\underline{\zeta}_A - \zeta_A)\nabla\!\!\!/_A U.$$

Corollary 13.1.2.2 *If U_A is a 1-form in space-time that is tangential to the surfaces $S_{t,u}$ at each point and satisfies*

$$\frac{dU_A}{ds} = F_A,$$

then

$$\frac{d}{ds}\not{d}iv\, U + \chi_{AB}\nabla\!\!\!/_A U_B = \not{d}iv\, F + (\zeta + \underline{\zeta})\cdot F + (\frac{1}{2}tr\chi\underline{\zeta}_A - \hat{\chi}_{AB}\underline{\zeta}_B + \beta_A)U_A.$$

Corollary 13.1.2.3 *If U is a scalar satisfying*

$$\frac{dU}{ds} = F,$$

then

$$\begin{aligned}\frac{d}{ds}\Delta\!\!\!/ U + 2\chi_{AB}\nabla\!\!\!/_A\nabla\!\!\!/_B U &= \Delta\!\!\!/ F + 2(\zeta_A + \underline{\zeta}_A)\nabla\!\!\!/_A F \\ &+ [\frac{1}{2}tr\chi(\zeta_A + \underline{\zeta}_A) + \hat{\chi}_{AB}(\zeta_B - \underline{\zeta}_B) \\ &+ \nabla\!\!\!/_A tr\chi + tr\chi\zeta_A]\nabla\!\!\!/_A U \\ &+ (d\!\!\!/ iv\zeta + d\!\!\!/ iv\underline{\zeta} + |\zeta + \underline{\zeta}|^2)F.\end{aligned}$$

Corollary 13.1.2.4 *If U_A is a 1-form tangential to $S_{t,u}$ satisfying the Hodge system*

$$d\!\!\!/ iv U = F$$

$$c\!\!\!/ url U = G,$$

then $D\!\!\!\!/_3 U$ satisfies the Hodge system

$$\begin{aligned}d\!\!\!/ iv D\!\!\!\!/_3 U &= D\!\!\!\!/_3 F + \frac{1}{2}tr\underline{\chi}F + \hat{\chi}_{AB}\nabla\!\!\!/_B U_A - \underline{\xi}_A D\!\!\!\!/_4 U_A \\ &- (\frac{1}{2}tr\chi\underline{\xi}_A + \frac{1}{2}tr\underline{\chi}\zeta_A - \hat{\chi}_{AB}\underline{\xi}_B - \hat{\underline{\chi}}_{AB}\zeta_B - \underline{\beta}_A)U_A\end{aligned}$$

$$\begin{aligned}c\!\!\!/ url D\!\!\!\!/_3 U &= D\!\!\!\!/_3 G + \frac{1}{2}tr\underline{\chi}G + {}^*\hat{\underline{\chi}}_{AB}\nabla\!\!\!/_B U_A + {}^*\underline{\xi}_A D\!\!\!\!/_4 U_A \\ &+ (\frac{1}{2}tr\chi{}^*\underline{\xi}_A + \frac{1}{2}tr\underline{\chi}{}^*\zeta_A + \hat{\chi}_{AB}{}^*\underline{\xi}_B + \hat{\underline{\chi}}_{AB}{}^*\zeta_B - {}^*\underline{\beta}_A)U_A.\end{aligned}$$

Corollary 13.1.2.5 *If U is a scalar satisfying*

$$\Delta\!\!\!/ U = F,$$

then $\mathbf{D}_3 U$ satisfies

$$\begin{aligned}\Delta\!\!\!/ \mathbf{D}_3 U &= \mathbf{D}_3 F + tr\underline{\chi}F + \hat{\underline{\chi}}_{AB}\nabla\!\!\!/_B\nabla\!\!\!/_A U - \underline{\xi}_A D\!\!\!\!/_4\nabla\!\!\!/_A U \\ &- (\frac{1}{2}tr\chi\underline{\xi}_A + \frac{1}{2}tr\underline{\chi}\zeta_A - \hat{\chi}_{AB}\underline{\xi}_B - \hat{\underline{\chi}}_{AB}\zeta_B - \underline{\beta}_A)\nabla\!\!\!/_A U.\end{aligned}$$

At first glance the equations 13.1.2a–13.1.2c seem to imply that the components χ, ζ, ω of the optical function have the same differentiability properties as the components α, β, ρ of the curvature tensor. In the case of χ we circumvent this difficulty by considering the trace equation 13.1.2d together with the

Codazzi equations 13.1.2l. Thus differentiating 13.1.2d tangentially to $S_{t,u}$ and using Corollary 13.1.2.1, we obtain

$$\frac{d}{ds}(\nabla\!\!\!/_A tr\chi) + \frac{3}{2}tr\chi\nabla\!\!\!/_A tr\chi = -\hat{\chi}_{AB}\nabla\!\!\!/_B tr\chi - 2\hat{\chi}_{BC}\nabla\!\!\!/_A\hat{\chi}_{BC}$$
$$- (\zeta_A + \underline{\zeta}_A)(|\hat{\chi}|^2 + \frac{1}{2}(tr\chi)^2). \quad (13.1.6a)$$

The term of $\nabla\!\!\!/\hat{\chi}$, on the right-hand side of 13.1.6a, can be eliminated after integration on $S_{t,u}$ with the help of the Codazzi equation, which, according to the results of Chapter 2, can be viewed as an elliptic Hodge system relative to $\hat{\chi}$. Thus our strategy is to apply Lemma 13.1.1 to the equation 13.1.6a and then eliminate $\nabla\!\!\!/\hat{\chi}$ using the L^p theory for the Hodge systems of type $\mathbf{H_2}$ developed in Chapter 2.

There is one more difficulty, however. The term $(\zeta_A + \underline{\zeta}_A)(tr\chi)^2$ on the right-hand side of 13.1.6a decays only like r^{-4} along each generator of C_u. This leads to a logarithmic divergence when we apply Lemma 13.1.1. To avoid this difficulty, we introduce the 1-form $\not{\kappa}$,

$$\not{\kappa}_A = \nabla\!\!\!/_A tr\chi + tr\chi\zeta_A. \quad (13.1.6b)$$

Taking into account equation 13.1.2b, we deduce the following propagation equation for $\not{\kappa}$

$$\frac{d}{ds}\not{\kappa}_A + \frac{3}{2}tr\chi\not{\kappa}_A = -\hat{\chi}_{AB}\not{\kappa}_B - 2\hat{\chi}_{BC}\nabla\!\!\!/_A\hat{\chi}_{BC}$$
$$- tr\chi\beta_A + tr\chi\hat{\chi}_{AB}\underline{\zeta}_B - 2|\hat{\chi}|^2\zeta_A. \quad (13.1.6c)$$

Remark now that the slowest decaying term on the right-hand side of 13.1.6c is $-tr\chi\beta_A$, which decays like $r^{-9/2}$. This is fine at this level; however, it will present difficulties later on when we estimate the top derivatives of χ.

We are now ready to prove the following:

Proposition 13.1.1 *Assume that, for a given* $2 \leq p < \infty$,

$$|r^{2-2/p}\epsilon|_{p,S}, \ |r^{2-2/p}\nabla\!\!\!/\phi|_{p,S} \leq c$$

everywhere in the exterior region. Then the 1-form $\not{\kappa} = \nabla\!\!\!/\mathrm{tr}\chi + tr\chi\zeta$ *and the tensor* $\nabla\!\!\!/\hat{\chi}$ *verify the estimates*

$$|r^{3-2/p}(\not{\kappa}, \nabla\!\!\!/\hat{\chi})|_{p,S}(t,u) \leq c\left(|r^{3-2/p}\not{\kappa}|_{p,S}(t_*,u) + \mathcal{R}_0^{p,S}(\beta) + {}^e\mathcal{O}_0^\infty(\chi,\zeta)\right).$$

Proof 13.1.3 *We rewrite propagation equation 13.1.6c in the form*

$$\frac{d}{ds}\not{k}_A + \frac{3}{2}tr\chi\not{k}_A = F_A \tag{13.1.7a}$$

where

$$F_A = -\hat{\chi}_{AB}\not{k}_B - 2\hat{\chi}_{BC}\nabla\!\!\!/_A\hat{\chi}_{BC} + F'_A \tag{13.1.7b}$$

and

$$F'_A = -tr\chi\beta_A + tr\chi\hat{\chi}_{AB}\underline{\zeta}_B - 2|\hat{\chi}|^2\zeta_A. \tag{13.1.7c}$$

In terms of $\not{k}$ *the null Codazzi equation 13.1.21 reads*

$$\nabla\!\!\!/_B\hat{\chi}_{AB} + \zeta_B\hat{\chi}_{AB} = \frac{1}{2}\not{k}_A - \beta_A. \tag{13.1.8}$$

We now apply Lemma 13.1.1, with $\lambda_0 = 3/2, \lambda_1 = 3 - 2/p$, *to equation 13.1.7a to obtain*

$$|r^{3-2/p}\not{k}|_{p,S}(t,u) \le c\left(|r^{3-2/p}\not{k}|_{p,S}(t_*,u) + \int_t^{t_*} |r^{3-2/p}F|_{p,S}(t',u)dt'\right). \tag{13.1.9a}$$

From 13.1.7b we have

$$|r^{3-2/p}F|_{p,S} \le c\left(r^{-2}(|r^{3-2/p}\not{k}|_{p,S} + |r^{3-2/p}\nabla\!\!\!/\hat{\chi}|_{p,S}) + |r^{3-2/p}F'|_{p,S}\right), \tag{13.1.9b}$$

and from 13.1.7c we have

$$|r^{3-2/p}F'|_{p,S} \le c(r^{-3/2}|r^{7/2-2/p}\beta|_{p,S} + {}^e\mathcal{O}_0^\infty(\chi,\zeta)r^{-2}). \tag{13.1.9c}$$

We now apply the elliptic L^p *estimates of Proposition 2.3.1 (on page 45) to the null Codazzi equation 13.1.8 to estimate*

$$|r^{3-2/p}\nabla\!\!\!/\hat{\chi}|_{p,S} \le c(|r^{3-2/p}\not{k}|_{p,S} + r^{-1/2}|r^{7/2-2/p}\beta|_{p,S} + {}^e\mathcal{O}_0^\infty(\chi,\zeta)r^{-1}). \tag{13.1.9d}$$

Substituting estimates 13.1.9b, 13.1.9c, and 13.1.9d in 13.1.9a, we obtain the following integral inequality for $|r^{3-2/p}\not{k}|_{p,S}$*:*

$$\begin{aligned}|r^{3-2/p}\not{k}|_{p,S}(t,u) \le c\{&|r^{3-2/p}\not{k}|_{p,S}(t_*,u)\\ &+\int_t^{t_*} r(t',u)^{-2}|r^{3-2/p}\not{k}|_{p,S}(t',u)dt'\\ &+r(t,u)^{-1/2}\mathcal{R}_0^{p,S}(\beta)\\ &+r(t,u)^{-1}{}^e\mathcal{O}_0^\infty(\chi,\zeta)\}.\end{aligned}$$

Applying the Gronwall lemma to this inequality yields the conclusion of the proposition regarding $\not{k}$*. The estimate 13.1.9d then yields the conclusion regarding* $\nabla\!\!\!/\hat{\chi}$*.*

We now attack our second problem, that of how to estimate the first tangential derivatives of ζ. We have to avoid differentiating equation 13.1.2b, as this leads to derivatives of the curvature β.

Consider instead equations 13.1.2f and 13.1.2h, which we rewrite as follows:

$$\frac{d}{ds}\mathrm{tr}\underline{\chi} + \frac{1}{2}\mathrm{tr}\chi\mathrm{tr}\underline{\chi} = -2\underline{\mu} + 2|\underline{\zeta}|^2 \quad (13.1.10a)$$

$$\mathbf{D}_3\mathrm{tr}\chi + \frac{1}{2}\mathrm{tr}\underline{\chi}\mathrm{tr}\chi + 2\omega\mathrm{tr}\chi = -2\mu + 2|\zeta|^2 \quad (13.1.10b)$$

where

$$\mu = -\mathrm{d\!\!/iv}\,\zeta + \frac{1}{2}\hat{\chi}\cdot\underline{\hat{\chi}} - \rho \quad (13.1.10c)$$

$$\underline{\mu} = -\mathrm{d\!\!/iv}\,\underline{\zeta} + \frac{1}{2}\hat{\chi}\cdot\underline{\hat{\chi}} - \rho. \quad (13.1.10d)$$

Our strategy is to derive a propagation equation for μ and use it in conjunction with the following divergence-curl system for ζ:

$$\mathrm{d\!\!/iv}\,\zeta = -\mu + \frac{1}{2}\hat{\chi}\cdot\underline{\hat{\chi}} - \rho \quad (13.1.10e)$$

$$\mathrm{c\!\!/url}\,\zeta = \sigma - \frac{1}{2}\hat{\chi}\wedge\underline{\hat{\chi}}. \quad (13.1.10f)$$

First, applying Corollary 13.1.2.1 to equation 13.1.2d, we obtain the following propagation equation for $\mathbf{D}_3 tr\chi$:

$$\frac{d}{ds}\mathbf{D}_3 tr\chi + tr\chi\mathbf{D}_3 tr\chi = -2\hat{\chi}_{AB}\mathbf{D\!\!\!/}_3\hat{\chi}_{AB} + 2(\underline{\zeta}_A - \zeta_A)\nabla\!\!\!\!/_A tr\chi - 2\omega\left(|\hat{\chi}|^2 + \frac{1}{2}(tr\chi)^2\right)$$

The term $-\omega(tr\chi)^2$ decays only like r^{-3} along each generator of C_u. This, when integrated along C_u according to Lemma 13.1.1, will lead to divergence. We eliminate it by considering the expression $\mathbf{D}'_3\mathrm{tr}\chi = \mathbf{D}_3 tr\chi + 2\omega\mathrm{tr}\chi$. In view of 13.1.2c and 13.1.2d,

$$\frac{d}{ds}(\omega tr\chi) + \omega(tr\chi)^2 = tr\chi(-|\zeta|^2 + 2\underline{\zeta}\cdot\zeta - \rho) - \omega\left(|\hat{\chi}|^2 - \frac{1}{2}(tr\chi)^2\right).$$

Thus,

$$\frac{d}{ds}\mathbf{D}'_3 tr\chi + tr\chi\mathbf{D}'_3 tr\chi = -2\hat{\chi}\cdot\mathbf{D}_3\hat{\chi} - 4\omega|\hat{\chi}|^2 + 2(\underline{\zeta} - \zeta)\cdot\nabla\!\!\!\!/\mathrm{tr}\chi + 2\mathrm{tr}\chi(-|\zeta|^2 + 2\zeta\cdot\underline{\zeta} - \rho). \quad (13.1.10g)$$

On the other hand, according to equations 13.1.10a and 13.1.2d,

$$\frac{d}{ds}(\mathrm{tr}\chi\mathrm{tr}\underline{\chi}) + \mathrm{tr}\chi(\mathrm{tr}\chi\mathrm{tr}\underline{\chi}) = \mathrm{tr}\chi(-2\underline{\mu} + 2|\underline{\zeta}|^2) - \mathrm{tr}\underline{\chi}|\hat{\chi}|^2. \tag{13.1.10h}$$

Combining 13.1.10g together with 13.1.10h, we derive

$$\begin{aligned}\frac{d}{ds}\mu' + tr\chi\mu' = {} & \hat{\chi}\cdot\mathbf{D}_3\hat{\chi} + 2\omega|\hat{\chi}|^2 + \frac{1}{4}\mathrm{tr}\underline{\chi}|\hat{\chi}|^2 \\ & + (\zeta - \underline{\zeta})\cdot\nabla\!\!\!/\,\mathrm{tr}\chi \\ & - \frac{1}{2}\mathrm{tr}\chi(-\underline{\mu} - 2|\zeta|^2 + |\underline{\zeta}|^2 + 4\zeta\cdot\underline{\zeta} - 2\rho)\end{aligned}$$

where $\mu' = -\frac{1}{2}(\mathbf{D}_3'\mathrm{tr}\chi + \frac{1}{2}\mathrm{tr}\chi\mathrm{tr}\underline{\chi})$. In view of 13.1.2j,

$$\hat{\chi}\cdot\mathbf{D}\!\!\!\!/_3\hat{\chi} + 2\omega|\hat{\chi}|^2 + \frac{1}{2}\mathrm{tr}\underline{\chi}|\hat{\chi}|^2 = -\frac{1}{2}tr\chi\hat{\chi}\cdot\underline{\hat{\chi}} + \hat{\chi}\cdot(\nabla\!\!\!/\,\widehat{\otimes}\zeta) + 2\zeta\cdot\hat{\chi}\cdot\zeta\,.$$

Hence

$$\begin{aligned}\frac{d}{ds}\mu' + tr\chi\mu' = {} & \hat{\chi}\cdot(\nabla\!\!\!/\,\widehat{\otimes}\zeta) + (\zeta - \underline{\zeta})\cdot\nabla\!\!\!/\,\mathrm{tr}\chi - \frac{1}{4}\mathrm{tr}\underline{\chi}|\hat{\chi}|^2 \\ & - \frac{1}{2}\mathrm{tr}\chi(\mathrm{d}\!\!\!/\mathrm{iv}\,\underline{\zeta} + \frac{1}{2}\hat{\chi}\cdot\underline{\hat{\chi}} - \rho + |\underline{\zeta}|^2 - 2|\zeta|^2 + 4\zeta\cdot\underline{\zeta}) \\ & + 2\zeta\cdot\hat{\chi}\cdot\zeta.\end{aligned} \tag{13.1.10i}$$

Finally, using 13.1.2b, we have

$$\frac{d}{ds}|\zeta|^2 + \mathrm{tr}\chi|\zeta|^2 = \mathrm{tr}\chi(\zeta\cdot\underline{\zeta}) - 2\zeta\cdot\hat{\chi}\cdot\zeta + 2\zeta\cdot\hat{\chi}\cdot\underline{\zeta} - 2\zeta\cdot\beta,$$

and therefore, since by 13.1.10b $\mu = \mu' + |\zeta|^2$,

$$\begin{aligned}\frac{d}{ds}\mu + tr\chi\mu = {} & \hat{\chi}\cdot(\nabla\!\!\!/\,\widehat{\otimes}\zeta) + (\zeta - \underline{\zeta})\cdot\nabla\!\!\!/\,\mathrm{tr}\chi - \frac{1}{4}\mathrm{tr}\underline{\chi}|\hat{\chi}|^2 \\ & + \frac{1}{2}\mathrm{tr}\chi\Big(-\mathrm{d}\!\!\!/\mathrm{iv}\,\underline{\zeta} - \frac{1}{2}\hat{\chi}\cdot\underline{\hat{\chi}} + \rho - |\underline{\zeta}|^2 + 2|\zeta|^2 - 2\zeta\cdot\underline{\zeta}\Big) \\ & + 2\zeta\cdot\hat{\chi}\cdot\underline{\zeta} - 2\zeta\cdot\beta\end{aligned}$$

or, taking into account the definition of $\not{k}$,

$$\begin{aligned}\frac{d}{ds}\mu + tr\chi\mu = {} & \hat{\chi}\cdot(\nabla\!\!\!/\,\widehat{\otimes}\zeta) + (\zeta - \underline{\zeta})\cdot\not{k} - \frac{1}{4}\mathrm{tr}\underline{\chi}|\hat{\chi}|^2 \\ & + \frac{1}{2}\mathrm{tr}\chi\Big(-\mathrm{d}\!\!\!/\mathrm{iv}\,\underline{\zeta} - \frac{1}{2}\hat{\chi}\cdot\underline{\hat{\chi}} + \rho - |\underline{\zeta}|^2\Big) + 2\zeta\cdot\hat{\chi}\cdot\underline{\zeta} - 2\zeta\cdot\beta.\end{aligned} \tag{13.1.11}$$

This is the desired propagation equation for μ. The term $\hat{\chi} \cdot (\nabla\!\!\!/ \widehat{\otimes} \zeta)$ on the right-hand side of 13.1.11 can be eliminated, after integration on $S_{t,u}$, with the help of the system 13.1.10e–13.1.10f, which can be viewed as an elliptic Hodge system in ζ.

The quantity μ plays a central role for the Einstein equations.[2] We call it the *mass aspect function* of the corresponding surface $S_{t,u}$. Despite the fact that μ is such an important quantity, intimately connected with the Hawking and Bondi mass, for technical reasons it is not quite appropriate for our estimates here. Indeed, the presence of $d\!\!\!/\mathrm{iv}\,\underline{\zeta}$ on the right-hand side of the propagation equation 13.1.11 requires more differentiability assumptions on the components of the second fundamental form k. We want, in fact, a propagation equation that does not require derivatives of k. With this in mind we introduce the following *modified mass aspect function*:

$$\tilde{\mu} = -\frac{1}{2}\mathbf{D}_3 \mathrm{tr}\chi + \frac{1}{4}a^2(\mathrm{tr}\chi)^2 - \omega \mathrm{tr}\chi. \tag{13.1.12a}$$

Noting that $\tilde{\mu} = \mu' + \frac{1}{2}a\delta\mathrm{tr}\chi$, we then derive the propagation equation

$$\begin{aligned} \frac{d}{ds}\tilde{\mu} + tr\chi\tilde{\mu} &= 2\hat{\chi}\cdot\nabla\!\!\!/\zeta + (\zeta - \underline{\zeta})\cdot k\!\!\!/ \\ &\quad -\frac{1}{2}tr\chi(\hat{\chi}\cdot\underline{\hat{\chi}} - 2\rho + 2\underline{\zeta}\cdot\zeta) + 2\zeta\cdot\hat{\chi}\cdot\zeta \\ &\quad -a\delta|\hat{\chi}|^2 - \frac{1}{2}a\nu(\mathrm{tr}\chi)^2 \end{aligned} \tag{13.1.12b}$$

in which no derivatives of k appear. We couple 13.1.12b with the following Hodge system for ζ:

$$\begin{aligned} d\!\!\!/\mathrm{iv}\,\zeta &= -\tilde{\mu} - \rho - |\zeta|^2 + \frac{1}{2}\hat{\chi}\cdot\underline{\hat{\chi}} + \frac{1}{2}a\delta\mathrm{tr}\chi \\ c\!\!\!/\mathrm{url}\,\zeta &= \sigma - \frac{1}{2}\hat{\chi}\wedge\underline{\hat{\chi}}. \end{aligned} \tag{13.1.12c}$$

Similarly to our treatment of χ, our strategy in deriving estimates for $\nabla\!\!\!/\zeta$ is to apply Lemma 13.1.1 to equation 13.1.12b and then eliminate $\nabla\!\!\!/\zeta$ using the L^p theory for the Hodge system 13.1.12c.[3] We are thus ready to prove the following:

[2] So does the dual quantity $\underline{\mu}$, which we use in Chapter 17 to derive the Bondi radiation law.

[3] This system is of type $\mathbf{H}_1$, developed in Chapter 2

Proposition 13.1.2 *Consider the modified mass aspect function* $\tilde{\mu} = -\frac{1}{2}\mathbf{D}_3\mathrm{tr}\chi + \frac{a^2}{4}(\mathrm{tr}\chi)^2 - \omega\mathrm{tr}\chi$ *and assume that, in addition to the assumptions of Proposition 13.1.1,*

$$|r^{2-2/p}\nabla_N\phi|_{p,S},\ |r^{2-2/p}\delta|_{p,S},\ |r^{1-2/p}\hat{\eta}|_{p,S} \leq c.$$

Then, for all $2 \leq p < \infty$,

$$\begin{aligned}|r^{3-2/p}(\tilde{\mu}, \nabla\!\!\!/\zeta)|_{p,S}(t,u) \leq\ & c|r^{3-2/p}\underline{\hat{\chi}}|_{p,S}(t_*,u)\\ &+ c|r^{3-2/p}\tilde{\mu}|_{p,S}(t_*,u)\\ &+ c\left(\mathcal{R}_0^{p,S}(\beta,\rho,\sigma) + \mathcal{K}_0^{p,S} + \mathcal{L}_0^{p,S} + {}^{e}\mathcal{O}_0^{\infty}(\chi,\zeta)\right).\end{aligned}$$

Proof 13.1.4 *We rewrite propagation equation 13.1.11 in the form*

$$\frac{d}{ds}\tilde{\mu} + tr\chi\tilde{\mu} = 2\hat{\chi}\cdot(\nabla\!\!\!/\hat{\otimes}\zeta) + G \tag{13.1.13a}$$

where

$$\begin{aligned}G = &(\zeta - \underline{\zeta})\cdot\underline{\hat{\chi}} - \frac{1}{2}tr\chi(\hat{\chi}\cdot\underline{\hat{\chi}} - 2\rho + 2\underline{\zeta}\cdot\zeta)\\ &+ 2\zeta\cdot\hat{\chi}\cdot\zeta - a\delta|\hat{\chi}|^2 - \frac{1}{2}a\nu(\mathrm{tr}\chi)^2.\end{aligned} \tag{13.1.13b}$$

We consider 13.1.13a coupled together with the Hodge system 13.1.12c on $S_{t,u}$ *and first apply Lemma 13.1.2a, with* $\lambda_0 = 1, \lambda_1 = 2 - 2/p$.

$$\begin{aligned}|r^{2-2/p}\tilde{\mu}|_{p,S}(t,u) \leq c\Big\{&(|r^{2-2/p}\tilde{\mu}|_{p,S}(t_*,u)\\ &+ \int_t^{t_*}(r(t',u))^{-2}|r^{2-2/p}\nabla\!\!\!/\zeta|_{p,S}(t',u)dt'\\ &+ \int_t^{t_*}|r^{2-2/p}G|_{p,S}(t',u)dt'\Big\}.\end{aligned} \tag{13.1.14a}$$

Next we apply the L^p estimates of Proposition 2.3.1 to the Hodge system for ζ to estimate

$$\begin{aligned}|r^{2-2/p}\nabla\!\!\!/\zeta|_{p,S} \leq\ & c|r^{2-2/p}\tilde{\mu}|_{p,S}\\ &+ r^{-1}\left(|r^{3-2/p}(\rho,\sigma)|_{p,S} + |r^{2-2/p}\delta|_{p,S} + {}^{e}\mathcal{O}_0^{\infty}(\chi,\zeta)\right).\end{aligned} \tag{13.1.14b}$$

We also estimate

$$\begin{aligned}|r^{2-2/p}G|_{p,S} \leq\ & c(r^{-3}|r^{3-2/p}\underline{\hat{\chi}}|_{p,S} + r^{-2}|r^{3-2/p}\rho|_{p,S} + \mathcal{K}_0^{p,S}\\ &+ \mathcal{L}_0^{p,S} + {}^{e}\mathcal{O}_0^{\infty}(\chi,\zeta)).\end{aligned} \tag{13.1.14c}$$

Substituting in 13.1.16c the estimate for $\not{k}$ given by Proposition 13.1.2a, and then using the resulting estimate together with 13.1.14b into 13.1.14a, we obtain the following integral inequality for $|r^{2-2/p}\tilde{\mu}|_{p,S}$:

$$
\begin{aligned}
|r^{2-2/p}\tilde{\mu}|_{p,S}(t,u) \leq\ & c|r^{2-2/p}\tilde{\mu}|_{p,S}(t_*,u) \\
& + c\int_t^{t_*} r(t',u)^{-2}|r^{2-2/p}\tilde{\mu}|_{p,S}(t',u)dt' \\
& + c\int_t^{t_*} r(t',u)^{-2}|r^{3-2/p}\not{k}|_{p,S}(t_*,u) \\
& + c\int_t^{t_*} r(t',u)^{-2}\big(\mathcal{R}_0^{p,S}(\beta,\rho,\sigma) + \mathcal{K}_0^{p,S} \\
& + \mathcal{L}_0^{p,S} + {}^{e}\mathcal{O}_0^{\infty}(\chi,\zeta)dt'\big).
\end{aligned}
$$

Applying the Gronwall lemma to this inequality yields the conclusion of the proposition regarding $\tilde{\mu}$. The estimate 13.1.14b then yields the conclusion regarding $\not{\nabla}\zeta$.

We next estimate the second angular derivatives of χ.

Proposition 13.1.3 *Let*

$$2 \leq p \leq 4.$$

If in addition to the assumptions of Proposition 13.1.2 we have

$$|r^{3-2/p}\not{\nabla}(\epsilon, \not{\nabla}\phi)|_{p,S} \leq c$$

and

$$\mathcal{R}_0^{2p,S}(\beta), \mathcal{K}_0^{2p,S}, \mathcal{L}_0^{2p,S} \leq c,$$

then

$$
\begin{aligned}
|r^{4-2/p}(\not{\nabla}\not{k}, \not{\nabla}^2\hat{\chi})|_{p,S}(t,u) \leq c\big(&|r^{4-2/p}\not{\nabla}\not{k}|_{p,S}(t_*,u) \\
&+ \mathcal{R}_{[1]}^{p,S}(\beta) + \mathcal{O}_{[1]}^{2p,S}(\chi,\zeta)\big).
\end{aligned}
$$

Proof 13.1.5 *We apply Commutation Lemma 13.1.2 to equation 13.1.7a to obtain*

$$
\begin{aligned}
\frac{d}{ds}\not{\nabla}_B\not{k}_A + 2tr\chi\not{\nabla}_B\not{k}_A = & -\hat{\chi}_{AC}\not{\nabla}_B\not{k}_C - \hat{\chi}_{BC}\not{\nabla}_C\not{k}_A \\
& - 2\hat{\chi}_{CD}\not{\nabla}_B\not{\nabla}_A\hat{\chi}_{CD} + \not{F}_{AB} \quad (13.1.15a)
\end{aligned}
$$

where

$$\begin{aligned}\not{F}_{AB} = &- \nabla_B \hat{\chi}_{AC} \not{\kappa}_C - 2\nabla_B \hat{\chi}_{CD} \nabla_A \hat{\chi}_{CD} + \nabla_B F'_A + (\zeta_B + \underline{\zeta}_B) F'_A \\ &+ (\chi_{AB} \underline{\zeta}_C - \chi_{BC} \underline{\zeta}_A + \in_{AC} {}^*\beta_B) \not{\kappa}_C \\ &- \frac{3}{2} \not{\kappa}_A (\nabla_B tr\chi + (\underline{\zeta}_B + \zeta_B) tr\chi). \end{aligned} \quad (13.1.15b)$$

We then apply Lemma 13.1.1, with $\lambda_0 = 2$, $\lambda_1 = 4 - 2/p$, *to equation 13.1.15a to obtain*

$$\begin{aligned}|r^{4-2/p} \nabla \not{\kappa}|_{p,S}(t,u) \leq c\Big(&|r^{4-2/p} \nabla \not{\kappa}|_{p,S}(t_*,u) + \\ &\int_t^{t_*} (r(t',u))^{-2} |r^{4-2/p} \nabla^2 \hat{\chi}|_{p,S}(t',u) dt' \\ &+ \int_t^{t_*} |r^{4-2/p} \not{F}|_{p,S}(t',u) dt' \Big). \end{aligned} \quad (13.1.15c)$$

We now appeal once again to the L^p *estimates of Proposition 2.3.1 applied to the null Codazzi equation 13.1.8 to estimate*

$$|r^{4-2/p} \nabla^2 \hat{\chi}|_{p,S} \leq c \left(|r^{4-2/p} \nabla \not{\kappa}|_{p,S} + r^{-1/2} \mathcal{R}_{[1]}^{p,S}(\beta) + r^{-1} \mathcal{O}_{[1]}^{2p,S}(\chi, \zeta) \right). \quad (13.1.16)$$

Using our assumptions, we find

$$|r^{4-2/p} \not{F}|_{p,S} \leq c \left(|r^{4-2/p} \nabla F'|_{p,S} + r^{-1} |r^{3-2/p} F'|_{p,S} + r^{-2} \mathcal{O}_{[1]}^{2p,S}(\chi, \zeta) \right) \quad (13.1.17a)$$

and

$$|r^{4-2/p} \nabla F'|_{p,S} \leq c \left(r^{-3/2} \mathcal{R}_{[1]}^{p,S}(\beta) + r^{-2} \mathcal{O}_{[1]}^{2p,S}(\chi, \zeta) \right). \quad (13.1.17b)$$

Substituting estimates 13.1.16, 13.1.17a, 13.1.17b, and 13.1.9c in 13.1.15c, we obtain the following integral inequality for $|r^{4-2/p} \nabla \not{\kappa}|_{p,S}$:

$$\begin{aligned}|r^{4-2/p} \nabla \not{\kappa}|_{p,S}(t,u) \leq c\Big\{&|r^{4-2/p} \nabla \not{\kappa}|_{p,S}(t_*,u) \\ &+ \int_t^{t_*} (r(t',u))^{-2} |r^{4-2/p} \nabla \not{\kappa}|_{p,S}(t',u) dt' \\ &+ \int_t^{t_*} (r(t',u))^{-3/2} \mathcal{R}_{[1]}^{p,S}(\beta)(t') dt' \\ &+ (r(t,u))^{-1} \mathcal{O}_{[1]}^{2p,S}(\chi, \zeta) \Big\}. \end{aligned}$$

Applying the Gronwall lemma to this inequality yields the conclusion of the proposition regarding $\nabla \not{\kappa}$. *The estimate 13.1.16 then yields the conclusion regarding* $\nabla^2 \hat{\chi}$.

Proposition 13.1.4 *Let*

$$2 \leq p \leq 4.$$

If in addition to the assumptions of Proposition 13.1.2 we have

$$|r^{3-2/p}\nabla\!\!\!/ \nabla_N \phi|_{p,S},\ |r^{3-2/p}\nabla\!\!\!/ \delta|_{p,S},\ |r^{2-2/p}\nabla\!\!\!/ \hat{\eta}|_{p,S} \leq c$$

and

$$\mathcal{R}_0^{2p,S}(\rho,\sigma), \mathcal{K}_0^{2p,S}, \mathcal{L}_0^{2p,S} \leq c$$

then

$$\begin{aligned}|r^{4-2/p}(\nabla\!\!\!/ \tilde{\mu}, \nabla\!\!\!/^2 \zeta)|_{p,S}(t,u) \leq c \Big(&|r^{4-2/p}\nabla\!\!\!/ \hat{\chi}\!\!\!/|_{p,S}(t_*,u) + |r^{4-2/p}\nabla\!\!\!/ \tilde{\mu}|_{p,S}(t_*,u) \\ &+ \mathcal{R}_{[1]}^{p,S}(\beta,\rho,\sigma) + \mathcal{O}_{[1]}^{2p,S}(\chi,\zeta) \\ &+ \mathcal{K}_{[1]}^{p,S} + \mathcal{L}_{[1]}^{p,S}\Big).\end{aligned}$$

Proof 13.1.6 *We apply Corollary 13.1.2.1 of Commutation Lemma 13.1.2 to equation 13.1.13a to obtain*

$$\begin{aligned}\frac{d}{ds}\nabla\!\!\!/_A \tilde{\mu} + \frac{3}{2} tr\chi \nabla\!\!\!/_A \tilde{\mu} = &-\hat{\chi}_{AB}\nabla\!\!\!/_B \tilde{\mu} + 2\hat{\chi}_{BC}\nabla\!\!\!/_A \nabla\!\!\!/_B \zeta_C \\ &+ 2\nabla\!\!\!/_A \hat{\chi}_{BC}\nabla\!\!\!/_B \zeta_C + \mathcal{G}\!\!\!/_A \qquad (13.1.18a)\end{aligned}$$

where

$$\begin{aligned}\mathcal{G}\!\!\!/_A = &\ 2\nabla\!\!\!/_A \hat{\chi}_{BC}\nabla\!\!\!/_B \zeta_C + \nabla\!\!\!/_A G + (\underline{\zeta}_A + \zeta_A)(G + 2\hat{\chi}_{BC}\nabla\!\!\!/_B \zeta_C) \\ &- \tilde{\mu}(\nabla\!\!\!/_A tr\chi + (\underline{\zeta}_A + \zeta_A)tr\chi). \qquad (13.1.18b)\end{aligned}$$

We then apply Lemma 13.1.1, with $\lambda_0 = 3/2$, $\lambda_1 = 3-2/p$, to equation 13.1.18a:

$$\begin{aligned}|r^{3-2/p}\nabla\!\!\!/ \tilde{\mu}|_{p,S}(t,u) \leq c \Big(&|r^{3-2/p}\nabla\!\!\!/ \tilde{\mu}|_{p,S}(t_*,u) \\ &+ \int_t^{t_*} (r(t',u))^{-2} |r^{3-2/p}\nabla\!\!\!/^2 \zeta|_{p,S}(t',u)dt' \\ &+ \int_t^{t_*} |r^{3-2/p}\mathcal{G}\!\!\!/|_{p,S}(t',u)dt'\Big). \qquad (13.1.19)\end{aligned}$$

Then we apply Proposition 2.3.1 to the Hodge system 13.1.12c for ζ to estimate

$$\begin{aligned}|r^{3-2/p}\nabla\!\!\!/^2 \zeta|_{p,S} \leq c\Big\{ &|r^{3-2/p}\nabla\!\!\!/ \tilde{\mu}|_{p,S} + r^{-1}\mathcal{R}_{[1]}^{p,S}(\rho,\sigma) \\ &+ r^{-1}(\mathcal{K}_{[1]}^{p,S}(\delta) + \mathcal{O}_{[1]}^{2p,S}(\chi,\zeta))\Big\}. \qquad (13.1.20)\end{aligned}$$

We also estimate

$$|r^{3-2/p}\not{G}|_{p,S} \leq c\left(|r^{3-2/p}\not{\nabla} G|_{p,S} + r^{-1}|r^{2-2/p}G|_{p,S} + r^{-3}\mathcal{O}^{2p,S}_{[1]}(\chi,\zeta)\right) \tag{13.1.21a}$$

and

$$\begin{aligned} |r^{3-2/p}\not{\nabla} G|_{p,S} \leq c\Big(&r^{-3}|r^{4-2/p}\not{\nabla}\not{k}|_{p,S} \\ &+ r^{-2}(\mathcal{R}^{p,S}_{[1]}(\rho) \\ &+ \mathcal{K}^{p,S}_{[1]} + \mathcal{L}^{p,S}_{[1]} + \mathcal{O}^{2p,S}_{[1]}(\chi,\zeta))\Big). \end{aligned} \tag{13.1.21b}$$

Substituting in 13.1.21b, the estimate for $\not{\nabla}\not{k}$ given by Proposition 13.1.2c, and substituting in turn the resulting estimate as well as estimates 13.1.20, 13.1.21a, and 13.1.16c into 13.1.19, we obtain the following integral inequality for $|r^{3-2/p}\not{\nabla}\tilde{\mu}|_{p,S}$:

$$\begin{aligned} |r^{3-2/p}\not{\nabla}\tilde{\mu}|_{p,S}(t,u) \leq c\Big\{&|r^{3-2/p}\not{\nabla}\tilde{\mu}|_{p,S}({}_*t,u) \\ &+ \int_t^{t_*}(r(t',u))^{-2}|r^{3-2/p}\not{\nabla}\tilde{\mu}|_{p,S}(t',u)dt' \\ &+ (r(t,u))^{-1}(|r^{4-2/p}\not{\nabla}\not{k}|_{p,S}(t_*,u) + \mathcal{R}^{p,S}_{[1]}(\beta,\rho,\sigma) \\ &+ \mathcal{K}^{p,S}_{[1]} + \mathcal{L}^{p,S}_{[1]} + \mathcal{O}^{2p,S}_{[1]}(\chi,\zeta))\Big\}. \end{aligned} \tag{13.1.21c}$$

Applying the Gronwall lemma to this inequality yields the conclusion of the proposition regarding $\not{\nabla}\tilde{\mu}$. The estimate 13.1.20 then yields the conclusion regarding $\not{\nabla}^2\zeta$.

We are now concentrating our attention on the highest tangential derivatives for χ, ζ. The estimates for the third derivatives of ζ proceed precisely in the same manner as those for the second derivatives. The estimates for the the third derivatives of χ present, a new difficulty, however. This is because the propagation equation for the second derivatives of $\not{k}$ requires the second derivatives of β. These are to be estimated in the local L^2 norms ${}^{(loc)}\|\ \|_{2,e}$ introduced at the beginning of this section. Yet the term ${}^{(loc)}\|r^4 \mathrm{tr}\chi\not{\nabla}^2\beta\|_{2,e}$, which appears on the right-hand side of the propagation equation, decays only like r^{-1} and thus leads to a logarithmic divergence. To avoid this we introduce the function

$$\not{k} = \not{d}\mathrm{iv}\not{k} + tr\chi\rho \tag{13.1.22}$$

and derive a propagation equation for it as follows:

First we go back to equation 13.1.7a and apply Corollary 13.1.2.2[4] to obtain

$$\frac{d}{ds}(\not{d}\mathrm{iv}\not{k}) + 2tr\chi(\not{d}\mathrm{iv}\not{k}) = -2\hat{\chi}_{AB}\not{\nabla}_B\not{k}_A - 2\hat{\chi}_{AB}\not{\Delta}\hat{\chi}_{AB} - tr\chi\not{d}\mathrm{iv}\,\beta + E \tag{13.1.23a}$$

where

$$\begin{aligned} E = &-2|\hat{\chi}|^2\not{d}\mathrm{iv}\,\zeta + tr\chi\hat{\chi}_{AB}\not{\nabla}_B\underline{\zeta}_A - 2|\not{k}|^2 \\ &- 2\not{\nabla}_A\hat{\chi}_{BC}(\not{\nabla}_A\hat{\chi}_{BC} + \hat{\chi}_{BC}(3\zeta_A + \underline{\zeta}_A)) \\ &+ \beta\cdot\not{k} - 2tr\chi\beta\cdot\underline{\zeta} - \frac{1}{2}tr\chi\underline{\zeta}\cdot\not{k} - \hat{\chi}_{AB}\not{k}_A\underline{\zeta}_B \\ &+ tr\chi\hat{\chi}_{AB}(\underline{\zeta}_A - \zeta_A)\underline{\zeta}_B - 2|\hat{\chi}|^2(|\zeta|^2 + \zeta\cdot\underline{\zeta}). \end{aligned} \tag{13.1.23b}$$

Taking into account the identity

$$\hat{\chi}_{AB}\not{\Delta}\hat{\chi}_{AB} = 2K|\hat{\chi}|^2 + 2\hat{\chi}_{AB}\not{\nabla}_B(-\zeta_C\hat{\chi}_{AC} + \frac{1}{2}\not{k}_A + \beta_A), \tag{13.1.23c}$$

which follows by differentiating the Codazzi equation 13.1.21, the equation 13.1.23a reduces to

$$\frac{d}{ds}(\not{d}\mathrm{iv}\not{k}) + 2tr\chi(\not{d}\mathrm{iv}\not{k}) = -4\hat{\chi}_{AB}\not{\nabla}_B\not{k}_A - tr\chi\not{d}\mathrm{iv}\,\beta - 4\hat{\chi}_{AB}\not{\nabla}_B\beta_A + E' \tag{13.1.23d}$$

where

$$\begin{aligned} E' = &\, tr\chi\hat{\chi}_{AB}\not{\nabla}_B\underline{\zeta}_A - 2|\not{k}|^2 + 4\hat{\chi}_{AB}\zeta_C\not{\nabla}_B\hat{\chi}_{AC} \\ &- 2\not{\nabla}_A\hat{\chi}_{BC}(\not{\nabla}_A\hat{\chi}_{BC} + \hat{\chi}_{BC}(3\zeta_A + \underline{\zeta}_A)) - \beta\cdot\not{k} - \frac{1}{2}tr\chi\underline{\zeta}\cdot\not{k} \\ &- \hat{\chi}_{AB}\not{k}_A\underline{\zeta}_B + tr\chi\hat{\chi}_{AB}(\underline{\zeta}_A - \zeta_A)\underline{\zeta}_B - 4K|\hat{\chi}|^2. \end{aligned} \tag{13.1.23e}$$

Our aim is to eliminate the troublesome term $tr\chi\not{d}\mathrm{iv}\,\beta$, which appears on the right-hand side of 13.1.23d. We achieve this with the help of the following consequence of the Bianchi identities, proved in Part II (Proposition 7.3.2, formula 7.3.11d):

$$\frac{d\rho}{ds} = \not{d}\mathrm{iv}\,\beta - \frac{3}{2}tr\chi\rho - \frac{1}{2}\underline{\hat{\chi}}\cdot\alpha + (\zeta + 2\underline{\zeta})\cdot\beta.$$

Using this, we arrive at

$$\frac{d}{ds}\not{k} + 2tr\chi\not{k} = H \tag{13.1.24a}$$

[4] We also apply the Codazzi equations 13.1.21.

where

$$H = -4\hat{\chi}_{AB}\nabla_B \not{k}_A - 4\hat{\chi}_{AB}\nabla_B\beta_A + E' - \frac{1}{2}tr\chi\hat{\underline{\chi}}\cdot\alpha + tr\chi(\zeta + 2\underline{\zeta})\cdot\beta - |\hat{\chi}|^2\rho. \qquad (13.1.24b)$$

Remark that $\mathrm{tr}\chi \mathrm{d}\!\!/\mathrm{iv}\,\beta$ has disappeared; we have in its place $\hat{\chi}\cdot\nabla\beta$, which has better decay. The slowest decaying term now on the right-hand side of 13.1.24a is $tr\chi\hat{\underline{\chi}}\cdot\alpha$. After differentiating equation 13.1.24a tangentially to $S_{t,u}$ and applying Corollary 13.1.2.1, we obtain an equation in which the slowest decaying term, ${}^{(\mathrm{loc})}\|r^4\mathrm{tr}\chi\hat{\underline{\chi}}\cdot\nabla\alpha\|_{2,e}$, can be estimated in terms of the $|\ |_{2,S}$ norm and decays like $r^{-3/2}$. More precisely, we have

$$\frac{d}{ds}\nabla_A \not{k} + \frac{5}{2}tr\chi\nabla_A\not{k} = \not{H} \qquad (13.1.25a)$$

where

$$\not{H} = -\hat{\chi}_{AB}\nabla_B\not{k} + \nabla_A H + (\zeta_A + \underline{\zeta}_A)H - 2(\not{\chi} + tr\chi\underline{\zeta}_A)\not{k}. \qquad (13.1.25b)$$

We also note that the 1-form $\not{\chi}$ satisfies the following Hodge system:

$$\mathrm{d}\!\!/\mathrm{iv}\,\not{\chi} = \not{k} - \rho tr\chi \qquad (13.1.26a)$$

$$\mathrm{c}\!\!/\mathrm{url}\,\not{\chi} = tr\chi\,\mathrm{c}\!\!/\mathrm{url}\,\zeta + \zeta\cdot{}^*\not{\chi}. \qquad (13.1.26b)$$

Proposition 13.1.5 *Suppose that the assumptions of Proposition 13.1.3 are satisfied for p = 4. Let, in addition,*

$${}^{(loc)}\|r^3\nabla^2(\epsilon,\nabla\phi)\|(t,\kappa) \le c,$$

and consider the function

$$\not{k} = \mathrm{d}\!\!/\mathrm{iv}\,\not{\chi} + tr\chi\rho.$$

Then

$${}^{(loc)}\|r^4(\nabla\not{k},\nabla^2\not{\chi},\nabla^3\chi)\|(t,\kappa) \le c\left({}^{(loc)}\|r^4\nabla\not{k}\|(t_*,\kappa) + {}^e\mathcal{R}_2(\alpha,\beta) + {}^e\mathcal{O}_{[2]}(\chi,\zeta) + {}^e\mathcal{R}_{[1]}\right).$$

Proof 13.1.7 *We apply part (ii) of Lemma 13.1.1, with $\lambda_0 = 5/2$, $\lambda_1 = 4$, and $p = 2$, to equation 13.1.25a. Thus*

$${}^{(loc)}\|r^4\nabla\not{k}\|(t,\kappa) \le c\left({}^{(loc)}\|r^4\nabla\not{k}\|(t_*,\kappa) + \int_t^{t_*}{}^{(loc)}\|r^4\not{H}\|(t',\kappa)dt'\right). \qquad (13.1.27a)$$

Using the Sobolev inequality on $S_{t,u}$ and the previous results of this section, we have

$$|r^3 \not{k}|_{\infty,S}(t,u) \leq c\mathcal{O}^{4,S}_{[2]}(\chi) \leq c. \tag{13.1.27b}$$

Therefore we can estimate

$$^{(loc)}||r^4 \not{H}||(t,\kappa) \leq c^{(loc)}||r^4 \nabla H||(t,\kappa) + c(r(t,\kappa))^{-1(loc)}||r^3 H||(t,\kappa) + c(r(t,\kappa))^{-2}\left(^{(loc)}||r^4 \nabla \not{k}||(t,\kappa) +^{(loc)} ||r^3 \not{k}||(t,\kappa)\right) \tag{13.1.27c}$$

The worst decaying term, $tr\chi \hat{\underline{\chi}} \cdot \nabla\alpha$, can be estimated as follows:

$$^{(loc)}||tr\chi\hat{\underline{\chi}} \cdot \nabla\alpha||(t,\kappa) \leq c(r(t,\kappa))^{-3/2}\mathcal{R}^{2,S}_1(\alpha) \leq c(r(t,\kappa))^{-3/2\,e}\mathcal{R}_{[2]}(\alpha)(t).$$

Consequently,

$$^{(loc)}||r^4 \nabla H||(t,\kappa) \leq c\left(^{(loc)}||r^4 \nabla E'||(t,\kappa) + (r(t,\kappa))^{-3/2\,e}\mathcal{R}_{[2]}(\alpha,\beta)(t) + r(t,\kappa)^{-2}(^{(loc)}||r^4 \nabla^2 \not{k}||(t,\kappa) + {}^e\mathcal{O}_{[2]}(\chi)(t))\right). \tag{13.1.27d}$$

We apply the L^2-estimates of Proposition 2.2.2 to the Hodge system 13.1.26a, 13.1.26b for $\not{k}$ to estimate, taking into account 13.1.27b,

$$^{(loc)}||r^4 \nabla^2 \not{k}||(t,\kappa) \leq c\left(^{(loc)}||r^4 \nabla \not{k}||(t,\kappa) + {}^e\mathcal{O}_{[2]}(\chi,\zeta) + {}^e\mathcal{R}_{[1]}\right). \tag{13.1.27e}$$

Also, using the fact that, in view of Gauss equation 13.1.2n and Proposition 13.1.2c,

$$^{(loc)}||r^3 \nabla K|| \leq c,$$

we can estimate

$$^{(loc)}||r^4 \nabla E'||(t,\kappa) \leq cr(t,\kappa)^{-2}({}^e\mathcal{O}_{[2]}(\chi,\zeta) + {}^e\mathcal{R}_{[1]}). \tag{13.1.27f}$$

Substituting 13.1.27e and 13.1.27f in 13.1.27d and the resulting estimate in turn in 13.1.27c, we obtain

$$^{(loc)}||r^4 \not{H}||(t,\kappa) \leq c\Big(r(t,\kappa)^{-2}\left(^{(loc)}||r^4 \nabla \not{k}||(t,\kappa) + {}^e\mathcal{O}_{[2]}(\chi,\zeta) + {}^e\mathcal{R}_{[1]}\right) + \left(r(t,\kappa)\right)^{-3/2\,e}\mathcal{R}_{[2]}(\alpha,\beta)(t)\Big). \tag{13.1.27g}$$

Substituting this estimate into 13.1.27a yields the following integral inequality for $^{(loc)}||r^4 \nabla \not{k}||$:

$$^{(loc)}||r^4 \nabla \not{k}||(t,\kappa) \leq c\Big(^{(loc)}||r^4 \nabla \not{k}||(t_*,\kappa)$$

$$+ \int_t^{t_*} (r(t', \kappa))^{-2 (loc)} \| r^4 \nabla \hat{\chi} \| (t', \kappa) dt'$$
$$+ \int_t^{t_*} r(t', \kappa)^{-3/2} {}^e\mathcal{R}_{[2]}(\alpha, \beta)(t') dt' \Big)$$
$$+ c r(t, \kappa)^{-1} ({}^e\mathcal{O}_{[2]}(\chi, \zeta) + {}^e\mathcal{R}_{[1]}).$$

Applying the Gronwall lemma to this inequality yields the conclusion regarding $\nabla \hat{\chi}$. The estimate for $\nabla^2 \hat{\chi}$ then follows from 13.1.27e, while the estimate for $\nabla^3 tr\chi$ follows from that for $\nabla^2 \hat{\chi}$ by applying the elliptic estimates of Chapter 2, to the identity

$$\Delta tr\chi = \text{div}\, \hat{\chi} - tr\chi\, \text{div}\, \zeta - \zeta \cdot tr\chi.$$

Finally, the conclusion of the proposition regarding $\nabla^3 \hat{\chi}$ follows from the estimate for $\nabla^3 tr\chi$ by applying once again the elliptic estimates of Chapter 2 to the null Codazzi equation.

By contrast, the estimates for the top derivatives of $\tilde{\mu}$ do not involve any additional complications. We proceed as in the proof of Proposition 13.1.4, differentiating once more the propagation equation for $\tilde{\mu}$, and then following the same steps. We thus prove the following:

Proposition 13.1.6 *If in addition to the assumptions of Proposition13.1.4*

$${}^{(loc)}\| r^3 \nabla^2 \nabla_N \phi \|(t, \kappa),\ {}^{(loc)}\| r^3 \nabla^2 \delta \|(t, \kappa),\ {}^{(loc)}\| r^2 \nabla^2 \hat{\eta} \|(t, \kappa) \le c,$$

then

$${}^{(loc)}\| r^4 (\nabla^2 \tilde{\mu}, \nabla^3 \zeta) \|(t, \kappa) \le c \big({}^{(loc)}\| r^4 \nabla \hat{\chi} \|(t_*, \kappa) + {}^{(loc)}\| r^4 \nabla^2 \tilde{\mu} \|(t_*, \kappa)$$
$$+ {}^e\mathcal{R}_{[2]} + {}^e\mathcal{K}_{[2]} + {}^e\mathcal{L}_{[2]} + {}^e\mathcal{O}_{[2]}(\chi, \zeta) \big).$$

This completes our estimates for χ, ζ. We next turn our attention to ω. Unlike χ, ζ, the gain of differentiation becomes manifest only when we take its second angular derivatives. We first prove, by a straightforward differentiation of the equation 13.1.2c, the following:

Proposition 13.1.7 *Assume that, in addition of the assumptions to 13.1.2, we have, for $2 \le p \le 4$,*

$$|r^{2-2/p}(\epsilon, \nabla \phi)|_{p,S},\ |r^{3-2/p} \nabla (\epsilon, \nabla \phi)|_{p,S} \le c.$$

Then

$$|r^{3/2-2/p} w_{3/2} \nabla \omega|_{p,S}(t, u) \le c \Big(|r^{3/2-2/p} w_{3/2} \nabla \omega|_{p,S}(t_*, u)$$
$$+ \mathcal{R}^{p,S}_{[1]}(\rho) + \mathcal{O}^{p,S}_1(\zeta) + {}^e\mathcal{O}^\infty_0(\zeta) \Big).$$

Proof 13.1.8 *Differentiating equation 13.1.2c tangentially to $S_{t,u}$ and using Corollary 13.1.2.1, we obtain*

$$\frac{d}{ds}\nabla\!\!\!/_A\omega + \frac{1}{2}tr\chi\nabla\!\!\!/_A\omega = -\hat{\chi}_{AB}\nabla\!\!\!/_B\omega - \nabla\!\!\!/_A\rho + E_A \qquad (13.1.28a)$$

where

$$E_A = 2\zeta_B\nabla\!\!\!/_A\underline{\zeta}_B + 2(\underline{\zeta}_B - \zeta_B)\nabla\!\!\!/_A\zeta_B + (\zeta_A + \underline{\zeta}_A)(-|\zeta|^2 + 2\underline{\zeta}\cdot\zeta - \rho). \qquad (13.1.28b)$$

We then apply Lemma 13.1.1, with $\lambda_0 = 1/2$, $\lambda_1 = 1 - 2/p$, to obtain

$$\begin{aligned}|r^{1-2/p}\nabla\!\!\!/\omega|_{p,S}(t,u) \leq c\Big(&|r^{1-2/p}\nabla\!\!\!/\omega|_{p,S}(t_*,u)\\ &+(\tau_-(u))^{-1/2}\int_t^{t_*}(r(t',u))^{-3}\\ &\cdot|r^{4-2/p}\tau_-^{1/2}\nabla\!\!\!/\rho|_{p,S}(t',u)dt'\\ &+\int_t^{t_*}|r^{1-2/p}E|_{p,S}(t',u)dt'\Big).\end{aligned} \qquad (13.1.29)$$

As

$$|r^{4-2/p}\tau_-^{1/2}\nabla\!\!\!/\rho|_{p,S}(t',u) \leq \mathcal{R}_1^{p,S}(\rho)$$

and we can estimate

$$|r^{1-2/p}E|_{p,S} \leq r^{-4}(\mathcal{R}_0^{p,S}(\rho) + \mathcal{O}_1^{p,S}(\zeta) + {}^e\mathcal{O}_0^\infty(\zeta)),$$

we conclude

$$\begin{aligned}|r^{1-2/p}\nabla\!\!\!/\omega|_{p,S}(t,u) \leq c(&|r^{1-2/p}\nabla\!\!\!/\omega|_{p,S}(t_*,u) + r^{-2}\tau_-^{-1/2}\mathcal{R}_1^{p,S}(\rho)\\ &+ r^{-3}(\mathcal{R}_0^{p,S}(\rho) + \mathcal{O}_1^{p,S}(\zeta) + {}^e\mathcal{O}_0^\infty(\zeta))),\end{aligned}$$

from which the result follows.[5]

We are now ready to estimate the second angular derivatives of ω. Here we introduce the crucial quantity

$$\psi\!\!\!/ = \Delta\!\!\!/\omega - d\!\!\!/iv\,\underline{\beta}, \qquad (13.1.30)$$

which allows us to gain back a derivative. This is seen as follows:

[5] In view of the last equation it looks as if we could derive a stronger result. The weaker result is necessary because of the last slice.

We first apply the operator $\not{\mathrm{d}}\mathrm{iv}$ to equation 13.1.28a and, using Corollary 13.1.2.2 and the null Codazzi equation, we obtain

$$\begin{aligned}\frac{d\not\Delta\omega}{ds} + tr\chi\not\Delta\omega = {} & -2\hat{\chi}_{AB}\not\nabla_A\not\nabla_B\omega - \not\Delta\rho + \not{\mathrm{d}}\mathrm{iv}\, E + (\zeta_A + \underline{\zeta}_A) \\ & \cdot(-\not\nabla_A\rho + E_A) + (-\not{k}_A + 2\beta_A - 2\underline{\zeta}_B\hat{\chi}_{AB})\not\nabla_A\omega.\end{aligned} \tag{13.1.31a}$$

We next appeal to the consequence of the Bianchi equations,

$$\frac{d\underline{\beta}_A}{ds} + tr\chi\underline{\beta}_A = -\not\nabla_A\rho + \in_{AB}\not\nabla_B\sigma + 2\hat{\underline{\chi}}_{AB}\beta_B - 3\rho\underline{\zeta}_A + 3\sigma{}^*\underline{\zeta}_A, \tag{13.1.31b}$$

which was proved in Part II (Proposition 7.3.2, formula 7.3.11h). Applying to it the operator $\not{\mathrm{d}}\mathrm{iv}$, with the help of Corollary 13.1.2.2, as well as the conjugate null Codazzi equation

$$\not\nabla_B\underline{\chi}_{AB} - \not\nabla_A tr\underline{\chi} - \zeta_B\underline{\chi}_{AB} + \zeta_A tr\underline{\chi} = \underline{\beta}_A, \tag{13.1.31c}$$

we infer

$$\begin{aligned}\frac{d}{ds}(\not{\mathrm{d}}\mathrm{iv}\,\underline{\beta}) + \frac{3}{2}tr\chi\not{\mathrm{d}}\mathrm{iv}\,\underline{\beta} = {} & -\not\Delta\rho - \hat{\chi}_{AB}\not\nabla_A\underline{\beta}_B + 2\hat{\underline{\chi}}_{AB}\not\nabla_A\beta_B \\ & - (\zeta + 4\underline{\zeta})\cdot\not\nabla\rho - ({}^*\zeta - 2{}^*\underline{\zeta})\cdot\not\nabla\sigma \\ & - 3(\not{\mathrm{d}}\mathrm{iv}\,\underline{\zeta} + \underline{\zeta}\cdot\zeta + |\underline{\zeta}|^2)\rho + 3(\mathrm{c}\not{\mathrm{u}}\mathrm{rl}\,\underline{\zeta} + {}^*\underline{\zeta}\cdot\zeta)\sigma \\ & + \left(-\not{k}_A - \hat{\chi}_{AB}\underline{\zeta}_B + \frac{1}{2}tr\chi\underline{\zeta}_A\right)\underline{\beta}_A \\ & + \left(\not{\underline{k}}_A + (4\zeta_B + 2\underline{\zeta}_B)\hat{\underline{\chi}}_{AB}\right)\beta_A + 3\beta\cdot\underline{\beta}\end{aligned} \tag{13.1.31d}$$

where

$$\not{\underline{k}}_A = \not\nabla_A tr\underline{\chi} - \zeta_A \mathrm{tr}\underline{\chi}.$$

We now make the observation that by subtracting 13.1.31d from 13.1.31a, the harmful term $\not\Delta\rho$, involving the second derivative of the curvature, is eliminated. We thus conclude that the function $\not\psi$ satisfies the propagation equation

$$\frac{d\not\psi}{ds} + tr\chi\not\psi = -2\hat{\chi}_{AB}\not\nabla_A\not\nabla_B\omega + \frac{1}{2}tr\chi\not{\mathrm{d}}\mathrm{iv}\,\underline{\beta} + \not{E} \tag{13.1.32a}$$

where

$$\begin{aligned}\not{E} = {} & \not{\mathrm{d}}\mathrm{iv}\, E + (\zeta + \underline{\zeta})\cdot E + \hat{\chi}\cdot\not\nabla\underline{\beta} - 2\hat{\underline{\chi}}\cdot\not\nabla\beta + 3\underline{\zeta}\cdot\not\nabla\rho + ({}^*\zeta - 2{}^*\underline{\zeta})\cdot\not\nabla\sigma \\ & + \left(\not{k} + \underline{\zeta}\cdot\hat{\chi} - \frac{1}{2}tr\chi\underline{\zeta}\right)\cdot\underline{\beta} - \left(\not{\underline{k}} + (4\zeta + 2\underline{\zeta})\cdot\hat{\underline{\chi}}\right)\cdot\beta - 3\beta\cdot\underline{\beta} \\ & + 3(\not{\mathrm{d}}\mathrm{iv}\,\underline{\zeta} + \underline{\zeta}\cdot\zeta + |\underline{\zeta}|^2)\rho - 3(\mathrm{c}\not{\mathrm{u}}\mathrm{rl}\,\underline{\zeta} + {}^*\underline{\zeta}\cdot\zeta)\sigma \\ & + (-\not{k} + 2\beta - 2\underline{\zeta}\cdot\hat{\chi})\cdot\not\nabla\omega.\end{aligned} \tag{13.1.32b}$$

This allows us to prove the following:

Proposition 13.1.8 *If in addition to the assumptions of Proposition 13.1.7 we have*

$$|r^{3-1/p}\nabla\!\!\!/^2(\epsilon, \nabla\!\!\!/\phi)|_{p,S} \leq c$$

for $2 \leq p \leq 4$ *and*

$${}^e\mathcal{R}_0^\infty \leq c,$$

then

$$\begin{aligned}|r^{5/2-2/p}w_{3/2}(\psi\!\!\!/, \nabla\!\!\!/^2\omega)|_{p,S}(t,u) \leq c\Big(&|r^{5/2-2/p}w_{3/2}\psi\!\!\!/|_{p,S}(t_*,u)\\ &+ \mathcal{R}_1^{p,S} + {}^e\mathcal{R}_0^\infty\\ &+ \mathcal{O}_2^{p,S}(\zeta) + {}^e\mathcal{O}_{[1]}^\infty(\chi,\zeta)\Big)\end{aligned}$$

where $\psi\!\!\!/ = \Delta\!\!\!/\omega - d\!\!\!/iv\,\underline{\beta}$.

Proof 13.1.9 *We first observe that, according to the previous proposition,* $\mathcal{O}_1^{p,S}(\omega) \leq c$. *Thus, using our assumptions as well as the results of the previous propositions for* χ, ζ, *we can estimate*

$$|r^{5-2/p}E\!\!\!/|_{p,S} \leq c(\mathcal{R}_1^{p,S} + {}^e\mathcal{R}_0^\infty + \mathcal{O}_2^{p,S}(\zeta) + {}^e\mathcal{O}_{[1]}^\infty(\chi,\zeta)). \tag{13.1.33a}$$

Now we apply Lemma 3.1.1, with $\lambda_0 = 1, \lambda_1 = 2-2/p$, *to equation 13.1.32a,*

$$\begin{aligned}|r^{2-2/p}\psi\!\!\!/|_{p,S}(t,u) \leq\ & c|r^{2-2/p}\psi\!\!\!/|_{p,S}(t_*,u)\\ &+ c\int_t^{t_*}(r(t',u))^{-2}|r^{2-2/p}\nabla\!\!\!/^2\omega|_{p,S}(t',u)dt'\\ &+ c(\tau_-(u))^{-3/2}\\ &\cdot\int_t^{t_*}(r(t',u))^{-2}\\ &\cdot|r^{3-2/p}c\tau_-^{3/2}\nabla\!\!\!/\underline{\beta}|_{p,S}(t',u)dt'\\ &+ c\int_t^{t_*}(r(t',u))^{-3}|r^{5-2/p}E\!\!\!/|_{p,S}(t',u)dt'.\end{aligned} \tag{13.1.33b}$$

We now apply Corollary 2.3.1.1 of Proposition 2.3.1 to the equation

$$\Delta\!\!\!/\omega = \psi\!\!\!/ + d\!\!\!/iv\,\underline{\beta}$$

to estimate

$$|r^{2-2/p}\nabla\!\!\!/^2\omega|_{p,S} \leq c(|r^{2-2/p}\psi\!\!\!/|_{p,S}+r^{-1}\tau_-^{-3/2}|r^{3-2/p}\tau_-^{3/2}\nabla\!\!\!/\underline{\beta}|_{p,S}). \quad (13.1.33c)$$

Substituting estimates 13.1.33a and 13.1.33c in 13.1.33b yields the following integral inequality for $|r^{2-2/p}\psi\!\!\!/|_{p,S}$*:*

$$\begin{aligned}|r^{2-2/p}\psi\!\!\!/|_{p,S}(t,u) \leq c\Big(&|r^{2-2/p}\psi\!\!\!/|_{p,S}(t_*,u)\\ &+\int_t^{t_*}(r(t',u))^{-2}|r^{2-2/p}\psi\!\!\!/|_{p,S}(t',u)dt'\\ &+\mathcal{R}_1^{p,S}r^{-1}\tau_-^{-3/2}\\ &+r^{-2}(\mathcal{R}_1^{p,S}+{}^e\mathcal{R}_0^\infty+\mathcal{O}_2^{p,S}(\zeta)+{}^e\mathcal{O}_{[1]}^\infty(\chi,\zeta))\Big).\end{aligned}$$

The Gronwall lemma then yields the conclusion regarding $\psi\!\!\!/$*. In view of 13.1.33c the conclusion regarding* $\nabla\!\!\!/^2\omega$ *then follows.*

The third tangential derivative of ω can now be estimated without any additional complications. This requires taking the tangential derivatives of the propagation equation for $\psi\!\!\!/$ and then proceeding as in the derivation of the third derivatives of χ or ζ. We prove the following:

Proposition 13.1.9 *Suppose that the assumptions of Proposition 13.1.7 are satisfied for* $p=4$*. Assume in addition that*

$$r_0^{-1/2}\|r^4\nabla\!\!\!/^3(\epsilon,\nabla\!\!\!/\phi)\|_{2,e} \leq c.$$

Then

$$\begin{aligned}r_0^{-1/2}(t)\|r^3w_1\nabla\!\!\!/^3\omega\|_{2,e}(t) \leq c\big(&r_0^{-1/2}(t_*)\|r^3w_1\nabla\!\!\!/^3\omega\|_{2,e}(t_*)\\ &+{}^e\mathcal{R}_{[2]}+{}^e\mathcal{O}_{[1]}^\infty+{}^e\mathcal{O}_{[1]}\big).\end{aligned}$$

We next estimate $\mathbf{D}_3\omega$:

Proposition 13.1.10 *If in addition to the assumptions of Proposition 13.1.7 we have, for* $2\leq p\leq 4$*,*

$$|r^{2-2/p}\mathbf{D}\!\!\!\!/_3(\epsilon,\nabla\!\!\!/\phi)|_{p,S} \leq c$$

as well as

$$\mathcal{K}_0^{p,S} \leq c,$$

then

$$\begin{aligned}|r^{1/2-2/p}w_{5/2}\mathbf{D}_3\omega|_{p,S}(t,u) \leq c\Big(&|r^{1/2-2/p}w_{5/2}\mathbf{D}_3\omega|_{p,S}(t_*,u)\\ &+|r^{3/2-2/p}w_{3/2}\nabla\!\!\!/\omega|_{p,S}(t_*,u)\\ &+\mathcal{R}_1^{p,S}+\mathcal{R}_0^{p,S}+{}^e\mathcal{O}_0^\infty\Big).\end{aligned}$$

Proof 13.1.10 *Applying* $\mathbf{D}_3$ *to equation 13.1.2c and using Corollary 13.1.2.1, we obtain*

$$\frac{d\mathbf{D}_3\omega}{ds} = -\mathbf{D}_3\rho + 2\zeta\cdot\not{\mathbf{D}}_3\underline{\zeta} + 2(\underline{\zeta}-\zeta)\cdot\not{\mathbf{D}}_3\zeta + 2(\underline{\zeta}-\zeta)\cdot\not{\nabla}\omega - 2\omega(\rho+|\zeta|^2-2\zeta\cdot\underline{\zeta}).$$

Now we have

$$\not{\mathbf{D}}_3\zeta = 2\not{\nabla}\omega - \underline{\beta} + \chi\cdot\underline{\xi} - 2\underline{\chi}\cdot\zeta. \tag{13.1.34}$$

Hence,

$$\frac{d\mathbf{D}_3\omega}{ds} = -\mathbf{D}_3\rho + 6(\underline{\zeta}-\zeta)\cdot\not{\nabla}\omega + I \tag{13.1.35a}$$

where

$$I = 2\zeta\cdot\not{\mathbf{D}}_3\underline{\zeta} + 2(\underline{\zeta}-\zeta)\cdot(-\underline{\beta}+\chi\cdot\underline{\xi}-2\underline{\chi}\cdot\zeta) - 2\omega(\rho+|\zeta|^2-2\zeta\cdot\underline{\zeta}). \tag{13.1.35b}$$

We then apply Lemma 13.1.1, with $\lambda_0 = 0$, $\lambda_1 = -2/p$, *to 13.1.35a and obtain*

$$\begin{aligned}|r^{-2/p}\mathbf{D}_3\omega|_{p,S}(t,u) \leq c\Big(&|r^{-2/p}\mathbf{D}_3\omega|_{p,S}(t_*,u)\\ &+\int_t^{t_*}|r^{-2/p}\mathbf{D}_3\rho|_{p,S}(t',u)dt'\\ &+\int_t^{t_*}(r(t',u))^{-4}|r^{2-2/p}\not{\nabla}\omega|_{p,S}(t',u)dt'\\ &+\int_t^{t_*}|r^{-2/p}I|_{p,S}(t',u)dt'\Big).\end{aligned} \tag{13.1.36}$$

We need an estimate of $\mathbf{D}_3\rho$ *in* $L^p(S_{t,u})$. *We have*

$$|r^{-2/p}\mathbf{D}_3(\rho-\overline{\rho})|_{p,S} \leq r^{-3}\tau_-^{-3/2}\mathcal{R}_1^{p,S}(\rho). \tag{13.1.37a}$$

To estimate $\mathbf{D}_3\overline{\rho}$, *we consider the Bianchi identity*

$$\mathbf{D}_3\rho = -\not{d}iv\,\underline{\beta} - \frac{3}{2}tr\underline{\chi}\rho - \frac{1}{2}\hat{\chi}\cdot\underline{\alpha} + 2\underline{\zeta}\cdot\beta - \zeta\cdot\underline{\beta}, \tag{13.1.37b}$$

which was proved in Part II (formula 7.3.11e). Remark also that, for an arbitrary function f,

$$\mathbf{D}_3\left(\int_{S_{t,u}} f d\bar{\mu}_\gamma\right) = a\phi^{-1}\int_{S_{t,u}}(\mathbf{D}_3 f + tr\underline{\chi}f)\phi a^{-1}d\bar{\mu}_\gamma. \tag{13.1.37c}$$

Setting $f = 1$ in equation 13.1.37c yields

$$\mathbf{D}_3 A = a\phi^{-1}\int_{S_{t,u}} \phi a^{-1}\mathrm{tr}\underline{\chi}\, d\bar{\mu}_\gamma \tag{13.1.37d}$$

where $A(t,u)$ is the area of $S_{t,u}$. Since, by definition, $A = 4\pi r^2$, we may write 13.1.37d in the form

$$\mathbf{D}_3 r = a\phi^{-1}\frac{r}{2}\overline{\phi a^{-1}\mathrm{tr}\underline{\chi}}. \tag{13.1.37e}$$

Setting $f = r\rho$ in 13.1.37c and using 13.1.37e, we obtain

$$\begin{aligned} a^{-1}\phi\mathbf{D}_3(4\pi r^3\overline{\rho}) = \frac{r}{2}\int_{S_{t,u}} (\phi a^{-1}\mathrm{tr}\underline{\chi} - \overline{\phi a^{-1}\mathrm{tr}\underline{\chi}})(\rho - \overline{\rho}) \\ + r\int_{S_{t,u}} \phi a^{-1}(-\frac{1}{2}\hat{\chi}\cdot\underline{\alpha} + 2\underline{\xi}\cdot\beta - \zeta\cdot\underline{\beta}) \end{aligned} \tag{13.1.37f}$$

from which it follows that

$$|\mathbf{D}_3(r^3\overline{\rho})|_{\infty,S} \le c\tau_-^{-5/2}\mathcal{R}_0^{2,S}. \tag{13.1.37g}$$

Combining 13.1.37a and 13.1.37g, we obtain

$$|r^{-2/p}\mathbf{D}_3\rho|_{p,S} \le c(r^{-3}\tau_-^{-3/2}\mathcal{R}_1^{p,S} + r^{-4}\mathcal{R}_0^{p,S}).$$

Substituting this estimate together with

$$|r^{-2/p}I|_{p,S} \le c(\mathcal{R}_0^{p,S} + {}^e\mathcal{O}_0^{\infty}),$$

as well as the result of Proposition 13.1.6, in 13.1.36 we obtain

$$\begin{aligned} |r^{-2/p}\mathbf{D}_3\omega|_{p,S}(t,u) \le c|r^{-2/p}\mathbf{D}_3\omega|_{p,S}(t_*,u) + cr^{-2}\tau_-^{-3/2}\mathcal{R}_1^{p,S} \\ + cr^{-3}\Big(|r^{3/2-2/p}w_{3/2}\nabla\!\!\!/\,\omega|_{p,S}(t_*,u) \\ + \mathcal{R}_1^{p,S} + \mathcal{R}_0^{p,S} + {}^e\mathcal{O}_0^{\infty}\Big), \end{aligned}$$

from which the result follows.

Next we estimate the angular derivatives of $\mathbf{D}_3\omega$. To do this, we appeal once again to the quantity $\psi\!\!\!/ = \Delta\!\!\!/\,\omega - \mathrm{d}\!\!\!/\mathrm{iv}\,\underline{\beta}$. Differentiating the propagation equation 13.1.32a relative to e_3, we derive, with the help of Corollary 13.1.2.1,

$$\frac{d\mathbf{D}_3\psi\!\!\!/}{ds} + tr\chi\mathbf{D}_3\psi\!\!\!/ = -2\hat{\chi}_{AB}\nabla\!\!\!/_A\nabla\!\!\!/_B\mathbf{D}_3\omega + \frac{1}{2}tr\chi\,\mathrm{d}\!\!\!/\mathrm{iv}\,\mathbf{D}\!\!\!\!/_3\underline{\beta} + F\!\!\!/_3 \tag{13.1.38a}$$

where

$$\not{E}_3 = \mathbf{D}_3\not{E} - \mathbf{D}_3\mathrm{tr}\chi\not{\psi} - 2\not{\mathbf{D}}_3\hat{\chi}\cdot\not{\nabla}^2\omega \qquad (13.1.38b)$$
$$- 2\hat{\chi}\cdot[\not{\mathbf{D}}_3, \not{\nabla}^2]\omega + \mathrm{tr}\chi[\not{\mathbf{D}}_3, \not{\mathrm{div}}]\underline{\beta} + \frac{1}{2}\mathbf{D}_3\mathrm{tr}\chi\not{\mathrm{div}}\,\underline{\beta}.$$

According to formula 13.1.32b, using Commutation Lemma 13.1.2 and its corollaries and dropping the lower-order terms, we write

$$\mathbf{D}_3\not{E} = \not{\mathrm{div}}\,\mathbf{D}_3E + \hat{\chi}\cdot\not{\nabla}\not{\mathbf{D}}_3\underline{\beta} - 2\underline{\hat{\chi}}\cdot\not{\nabla}\mathbf{D}_3\beta$$
$$+ 3\underline{\zeta}\not{\nabla}\mathbf{D}_3\rho + (\zeta^* - 2\underline{\zeta}^*)\not{\nabla}\mathbf{D}_3\sigma + \text{l.o.t.}$$

On the other hand, we recall from the null Bianchi equations proved in Proposition 7.3.2,

$$\not{\mathbf{D}}_3\underline{\beta} + 2\mathrm{tr}\underline{\chi}\underline{\beta} = -\not{\mathrm{div}}\,\underline{\alpha} + \omega\underline{\beta} + \zeta\cdot\underline{\alpha} + 3(-\underline{\xi}\rho + \underline{\xi}^*\sigma)$$
$$\not{\mathbf{D}}_3\rho + \frac{3}{2}\mathrm{tr}\underline{\chi}\rho = -\not{\mathrm{div}}\,\underline{\beta} - \frac{1}{2}\hat{\chi}\cdot\underline{\alpha} + \zeta\cdot\underline{\beta} + (2\underline{\xi}\cdot\beta - \zeta\cdot\underline{\beta})$$
$$\not{\mathbf{D}}_3\sigma + \frac{3}{2}\mathrm{tr}\underline{\chi}\sigma = -\not{\mathrm{curl}}\,\underline{\beta} - \frac{1}{2}\underline{\hat{\chi}}\cdot^*\underline{\alpha} + \zeta\cdot^*\underline{\beta} - (2\underline{\xi}\cdot^*\beta + \zeta\cdot^*\underline{\beta})$$
$$\frac{d\underline{\beta}_A}{ds} + tr\chi\underline{\beta}_A = -\not{\nabla}_A\rho + \in_{AB}\not{\nabla}_B\sigma + 2\underline{\hat{\chi}}_{AB}\beta_B - 3\rho\underline{\zeta}_A + 3\sigma^*\underline{\zeta}_A.$$

Hence, dropping again the lower-order terms and recalling the definition of E in 13.1.28b together with formula 13.1.2k for $\not{\mathbf{D}}_3\zeta$, we write

$$\mathbf{D}_3\not{E} = \mathrm{Qr}\left[\hat{\chi}\,;\,\not{\nabla}\underline{\alpha}\right] + \mathrm{Qr}\left[(\zeta,\underline{\zeta})\,;\,\not{\nabla}^2\underline{\beta}\right] + \mathrm{Qr}\left[\underline{\hat{\chi}}\,;\,\not{\nabla}^2(\rho,\sigma)\right]$$
$$+ \mathrm{Qr}\left[(\zeta,\underline{\zeta})\,;\,\not{\nabla}^3\omega\right] + \mathrm{Qr}\left[(\zeta,\underline{\zeta})\,;\,\not{\nabla}^2\not{\mathbf{D}}_3\underline{\zeta}\right] + \text{l.o.t.}$$

where we denote by Qr [;] any quadratic form with coefficients that depend only on the induced metric and area form of $S_{t,u}$. Calculating also the commutators $[\not{\mathbf{D}}_3, \not{\nabla}^2]\omega$, $[\not{\mathbf{D}}_3, \not{\mathrm{div}}]\underline{\beta}$, we write

$$\not{E}_3 = \mathrm{Qr}\left[\hat{\chi}\,;\,\not{\nabla}\underline{\alpha}\right] + \mathrm{Qr}\left[(\zeta,\underline{\zeta})\,;\,\not{\nabla}^2\underline{\beta}\right] + \mathrm{Qr}\left[\underline{\hat{\chi}}\,;\,\not{\nabla}^2(\rho,\sigma)\right]$$
$$+ \mathrm{Qr}\left[(\zeta,\underline{\zeta})\,;\,\not{\nabla}^3\omega\right] + \mathrm{Qr}\left[(\zeta,\underline{\zeta})\,;\,\not{\nabla}^2\not{\mathbf{D}}_3\underline{\zeta}\right] + \text{l.o.t.} \qquad (13.1.38c)$$

Hence, back to 13.1.38a, we can write

$$\frac{d\mathbf{D}_3\not{\psi}}{ds} + tr\chi\mathbf{D}_3\not{\psi} = -2\hat{\chi}_{AB}\not{\nabla}_A\not{\nabla}_B\mathbf{D}_3\omega - \frac{1}{2}tr\chi\not{\mathrm{div}}\,\not{\mathrm{div}}\,\underline{\alpha} \qquad (13.1.38d)$$
$$+ \mathrm{Qr}\left[\hat{\chi}\,;\,\not{\nabla}\underline{\alpha}\right] + \mathrm{Qr}\left[(\zeta,\underline{\zeta})\,;\,\not{\nabla}^2\underline{\beta}\right] + \mathrm{Qr}\left[\underline{\hat{\chi}}\,;\,\not{\nabla}^2(\rho,\sigma)\right]$$
$$+ \mathrm{Qr}\left[(\zeta,\underline{\zeta})\,;\,\not{\nabla}^3\omega\right] + \mathrm{Qr}\left[(\zeta,\underline{\zeta})\,;\,\not{\nabla}^2\not{\mathbf{D}}_3\underline{\zeta}\right] + \text{l.o.t.}$$

Remark 1 *The terms denoted by l.o.t. are strictly lower-order terms as far as differentiability is concerned. In other words, they contain only terms that have at most the level of differentiability of the first derivatives of the curvature tensor. Moreover, using the null Bianchi equations, we can arrange that all first derivatives of the null components of the curvature be tangential. We can also use formulas 13.1.2a–13.1.2k to express the* $\not{D}_3, \not{D}_4$ *derivatives of* χ, ζ *in terms of tangential derivatives of* ζ, ω. *Finally, we observe that the rate of decay of the l.o.t. terms is at least equal to that of the terms included on the right-hand side of 13.1.38c.*

On the other hand, differentiating the equation $\not{\Delta}\omega = \not{\psi} + \not{d}iv\, \underline{\beta}$ *relative to* $\mathbf{D}_3$ *and using Corollary 13.1.2.5 of the Commutation Lemma 13.1.2 together with*

$$[\not{D}_3, \not{d}iv\,]\underline{\beta} = -\frac{1}{2}\mathrm{tr}\underline{\chi}\not{d}iv\,\underline{\beta} - \hat{\underline{\chi}} \cdot \not{\nabla}\underline{\beta} + l.o.t. \tag{13.1.38e}$$

and $\not{D}_3\underline{\beta} = -\not{d}iv\,\underline{\alpha} + l.o.t.$ *we derive*

$$\not{\Delta}\mathbf{D}_3\omega = \mathbf{D}_3\not{\psi} - \not{d}iv\,\not{d}iv\,\underline{\alpha} + l.o.t.' \tag{13.1.38f}$$

Remark 2 *Once again, the terms that we denote by l.o.t.′ contain only terms that have at most the level of differentiability of the first derivatives of the curvature tensor. Moreover, all first derivatives of the null components of the curvature are tangential. Also, the rate of decay of the l.o.t.′ terms is at least equal to that of the terms included on the right-hand side of 13.1.38f.*

We now proceed as in the proof of Propositions 13.1.7 and 13.1.8 and derive the following:

Proposition 13.1.11 *We make the assumption that, with the notation of Chapter 9,*

$${}^{e}\mathcal{K}_{[3]}, {}^{e}\mathcal{L}_{[3]} \leq c.$$

Then

$$\begin{aligned} r_0^{-1/2}(t)\|rw_2\not{\nabla}\mathbf{D}_3\omega\|_{2,e}(t) \leq c\Big(&r_0^{-1/2}(t_*)(\|rw_2\not{\nabla}\mathbf{D}_3\omega\|_{2,e}(t_*) \\ &+ \|r^2w_2\not{\nabla}^2\mathbf{D}_3\omega\|_{2,e}(t_*) \\ &+ \|r^2w_1\not{\nabla}^3\omega\|_{2,e}(t_*)) \\ &+ {}^{e}\mathcal{R}_{[2]} + {}^{e}\mathcal{O}^{\infty}_{[1]} + {}^{e}\mathcal{O}_{[1]}\Big) \end{aligned}$$

$$\begin{aligned} r_0^{-1/2}(t)\|r^2w_2\not{\nabla}^2\mathbf{D}_3\omega\|_{2,e}(t) \leq c\Big(&r_0^{-1/2}(t_*)(\|rw_2\not{\nabla}\mathbf{D}_3\omega\|_{2,e}(t_*) \\ &+ \|r^2w_2\not{\nabla}^2\mathbf{D}_3\omega\|_{2,e}(t_*) \end{aligned}$$

$$+ \|r^3 w_1 \nabla\!\!\!/^3 \omega\|_{2,e}(t_*))$$
$$+ {}^e\mathcal{R}_{[2]} + {}^e\mathcal{O}^{\infty}_{[1]} + {}^e\mathcal{O}_{[1]}\big).$$

We are now ready to state the theorem that summarizes the main results of this section. But before doing this, we introduce the following quantity defined on the last slice Σ_{t_*}:

$$\begin{aligned}
{}^e\mathcal{O}_{[3]}{}^* &= |r(r\mathrm{tr}\chi - 2)|_{\infty,S}(t_*, u) + |r^2\hat{\chi}|_{\infty,S}(t_*, u) \\
&+ |r^{5/2}\hat{\chi}\!\!\!/|_{4,S}(t_*, u) + |r^{7/2}\nabla\!\!\!/\hat{\chi}\!\!\!/|_{4,S}(t_*, u) \\
&+ |r^2\zeta|_{\infty,S}(t_*, u) + |r^{5/2}\tilde{\mu}|_{4,S}(t_*, u) + |r^{7/2}\nabla\!\!\!/\tilde{\mu}|_{4,S}(t_*, u) \\
&+ |r^{1/2}w_{3/2}\omega|_{\infty,S}(t_*, u) + |rw_{3/2}\nabla\!\!\!/\omega|_{4,S}(t_*, u) \\
&+ |r^2 w_{3/2}\nabla\!\!\!/^2\omega|_{4,S}(t_*, u) + |w_{5/2}\nabla_N\omega|_{4,S}(t_*, u) \\
&+{}^{(loc)}\|r^4\nabla\!\!\!/^2\hat{\chi}\!\!\!/\|_{2,e}(t_*) + {}^{(loc)}\|r^4\nabla\!\!\!/^2\tilde{\mu}\|_{2,e}(t_*) \\
&+ r_0^{-1/2}\|rw_2\nabla\!\!\!/\nabla_N\omega\|_{2,e}(t_*) + r_0^{-1/2}\|r^3 w_1\nabla\!\!\!/^3\omega\|_{2,e}(t_*) \\
&+ r_0^{-1/2}\|r^2 w_2\nabla\!\!\!/^2\nabla_N\omega\|_{2,e}(t_*).
\end{aligned} \tag{13.1.38g}$$

Theorem 13.1.1 *Assume that the space-time slab $\bigcup_{t\in[0,t_*]}\Sigma_t$ satisfies the bootstrap assumptions* $\mathbf{BA}_0$, $\mathbf{BA}_1$, *and* $\mathbf{BA}_2$. *Moreover, we assume that*[6]

$${}^e\mathcal{O}^{\infty}_0 \le c.$$

Then there exists another constant c such that

$$\begin{aligned}
{}^e\mathcal{O}_{[3]} \le c\big({}^e\mathcal{O}_{[3]}{}^* &+ {}^e\mathcal{R}_{[2]} \\
&+ {}^e\mathcal{K}_{[2]} + {}^e\mathcal{L}_{[2]} + {}^e\mathcal{O}^{\infty}_0\big).
\end{aligned}$$

To prove the theorem, we need only remark that the normal derivatives ∇_N can be expressed in terms of the derivatives in the e_3 and e_4 directions. Whereas the e_3 derivatives of our optical quantities have been already estimated throughout this chapter, the e_4 derivatives can be trivially estimated directly from the corresponding propagation equations.

Remark 3 The results of Theorem 13.1.1 remain valid in the extended exterior region E'.[7]

[6] These estimates were in fact proved in Chapter 9.

[7] See the definition on page 268.

In the next proposition we establish an estimate for the tangential derivatives of the trace of the normalized null second fundamental form χ' of the initial cone C_0. We recall that the initial cone was constructed starting with the point O_{-1} on the slice Σ_{-1}. We may assume that estimates for the tangential derivatives $\mathrm{tr}\chi'$, $\chi'_{AB} = \langle \mathbf{D}_A, e'_4\rangle$, $e'_4 = (T+N)$, on $S_{0,0}$ have already been established.[8] *We recall from Chapter 9 that on the initial cone χ' was renormalized with the help of the lapse function as defined by taking $a = 1$ on S_{0,t_*} and propagating it down on C_0 with the help of the equation $\frac{da}{ds} = -\nu$ (see 9.1.4d). With this definition of a we have set $l = a^{-1}(T+N)$, $\underline{l} = a(T-N)$, and $\chi_{AB} = \langle \mathbf{D}_a l, e_B\rangle$. We have also set $\zeta = \frac{1}{2}\langle \nabla\!\!\!/_A l, \underline{l}\rangle$. We recall that with this definition $\zeta = \epsilon$ on the last slice.*

We shall now estimate the following quantity on the last slice Σ_{t_*}:

$$\mathcal{S}(0)(t) = |r_0^{5/2}\nabla\!\!\!/ tr\chi'|_{4,S_{t,0}} + |r_0^{7/2}\Delta\!\!\!/ tr\chi'|_{4,S_{t,0}} + |r_0^4\nabla\!\!\!/\Delta\!\!\!/ tr\chi'|_{2,S_{t,0}} \quad (13.1.39)$$

Using the quantities $\mathcal{A}, \mathcal{B}$ introduced in 10.2.4a, we prove

Proposition 13.1.12 *Suppose that the assumptions of Proposition 13.1.4 are satisfied for $p = 4$ and that, in addition,*

$$|r^3\nabla\!\!\!/^2(\epsilon, \nabla\!\!\!/\phi)|_{2,S}(t,0) \leq c.$$

Then

$$\mathcal{S}(0)(t_*) \leq c\left(\mathcal{S}(0)(0) + \mathcal{A} + \mathcal{B} + {}^e\mathcal{R}_{[2]} + {}^e\mathcal{K}_{[3]} + {}^e\mathcal{L}_{[3]}\right).$$

Proof 13.1.11 *The proof follows precisely the same steps as described in Propositions 13.1.1–13.1.5. The only subtlety comes at the top level, when we estimate $\nabla\!\!\!/\Delta\!\!\!/\chi'$, which is done through $\nabla\!\!\!/\not{k}$.*

We apply part (ii) of Lemma 13.1.1, with $\lambda_0 = 5/2, \lambda_1 = 4$ and $p = 2$, to equation 13.1.25a. Instead of 13.1.27a, we have on the cone C_0

$$|r^4\nabla\!\!\!/\not{k}|_{2,S}(t,0) \leq c\left(|r^4\nabla\!\!\!/\not{k}|_{2,S}(0,0) + \int_0^t |r^4\not{H}|_{2,S}(t',0)dt'\right). \quad (13.1.40a)$$

We then simply remark that the terms involving $\nabla\!\!\!/^2\beta$ in the integral $\int_0^t |r^4\not{H}|_{2,S}(t',0)dt'$ can be estimated in terms of quantities $\mathcal{A} + \mathcal{B}$, while all other terms can be treated exactly as before.

Finally, in the end of this section we prove a proposition concerning the top derivatives of ω relative to $\mathbf{D}_3$. Though these are not present in the definition

[8] The estimates from O_{-1} to $S_{0,0}$ will be discussed in the next paragraph as part of the interior estimates for the optical function.

of the norm ${}^e\mathcal{O}_{[3]}$, *they need to be estimated in the matching region, where the distinction between tangential and transversal derivatives becomes blurred. In order to do this we introduce the following notation on the last slice* Σ_{t^*}:

$$
{}^e\mathcal{N}^* = r_0^{-1/2}\big(\|rw_3\nabla_N^2\omega\|_{2,e}(t_*) \\
+ \|r^2 w_3 \nabla\!\!\!/ \nabla_N^2\omega\|_{2,e}(t_*) + \|rw_4\nabla_N^3\omega\|_{2,e}(t_*)\big). \quad (13.1.40b)
$$

We will prove the following:

Proposition 13.1.13 *Assume that the assumptions and results of Theorem 13.1.1 hold true. Then there exists another constant c such that*

$$
r_0^{-1/2}(t)\|rw_3\mathbf{D}_3^2\omega\|_{2,e}(t) \le c\left({}^e\mathcal{N}^* + {}^e\mathcal{R}_{[2]} + {}^e\mathcal{K}_{[2]} + {}^e\mathcal{L}_{[2]} + {}^e\mathcal{O}_0^\infty\right)
$$
$$
r_0^{-1/2}(t)\|r^2 w_3\nabla\!\!\!/\mathbf{D}_3^2\omega\|_{2,e}(t) \le c\left({}^e\mathcal{N}^* + {}^e\mathcal{R}_{[2]} + {}^e\mathcal{K}_{[2]} + {}^e\mathcal{L}_{[2]} + {}^e\mathcal{O}_0^\infty\right)
$$
$$
r_0^{-1/2}(t)\|rw_4\mathbf{D}_3^3\omega\|_{2,e}(t) \le c\left({}^e\mathcal{N}^* + {}^e\mathcal{R}_{[2]} + {}^e\mathcal{K}_{[2]} + {}^e\mathcal{L}_{[2]} + {}^e\mathcal{O}_0^\infty\right).
$$

Proof 13.1.12 *The issue here is the differentiability*[9] *of* $\mathbf{D}_3^2\omega$, $\nabla\!\!\!/\mathbf{D}_3^2\omega$, *and* $\mathbf{D}_3^3\omega$, *which is not at all obvious. To see this, we consider again our basic quantity* $\psi\!\!\!/ = \Delta\!\!\!/\omega + d\!\!\!/iv\,\underline{\beta}$. *Differentiating once again the equation 13.1.38d in the direction of* $\mathbf{D}_3$, *we derive equations of the form*

$$
\frac{d\mathbf{D}_3^2\psi\!\!\!/}{ds} + tr\chi\mathbf{D}_3^2\psi\!\!\!/ = -2\hat{\chi}_{AB}\nabla\!\!\!/_A\nabla\!\!\!/_B\mathbf{D}_3^2\omega + F_{33} \quad (13.1.41a)
$$
$$
\frac{d\mathbf{D}_3^3\psi\!\!\!/}{ds} + tr\chi\mathbf{D}_3^3\psi\!\!\!/ = -2\hat{\chi}_{AB}\nabla\!\!\!/_A\nabla\!\!\!/_B\mathbf{D}_3^3\omega + F_{333}. \quad (13.1.41b)
$$

Remark that the error terms F_{33}, F_{333} *contain respectively three and four derivatives of the curvature tensor, which looks unacceptable in view of our construction. Yet we recall from our remark on page 380 that all derivatives of the null components of the curvature present on the right-hand side of 13.1.38d are tangential. Thus* F_{33} *contains at worst second tangential derivatives of arbitrary first derivatives of the curvature, while* F_{333} *contains at worst second tangential derivatives of arbitrary second derivatives of the curvature. This fact is crucial for us; the two tangential derivatives can be recovered with the help of the elliptic equation on* $S_{t,u}$:

$$
\Delta\!\!\!/\mathbf{D}_3^2\omega = \mathbf{D}_3^2\psi\!\!\!/ - d\!\!\!/iv\,d\!\!\!/iv\,\mathbf{D}_3\underline{\alpha} + G_{33} \quad (13.1.41c)
$$
$$
\Delta\!\!\!/\mathbf{D}_3^3\omega = \mathbf{D}_3^3\psi\!\!\!/ - d\!\!\!/iv\,d\!\!\!/iv\,\mathbf{D}_3^2\underline{\alpha} + G_{333}. \quad (13.1.41d)
$$

[9] Once the differentiability is established, the asymptotic behavior, which is completely consistent with all our other results, follows easily from the formulas.

Once again we remark that on the right-hand side of the equation 13.1.41c the worst terms contain at most two tangential derivatives of arbitrary first derivatives of the curvature, while on the right-hand side of the equation 13.1.41d the worst terms contain at most two tangential derivatives of arbitrary second derivatives of the curvature. Thus $\mathbf{D}_3^2\omega$ *should have the order of differentiability of the first derivatives of the curvature while* $\nabla\mathbf{D}_3^2\omega$ *and* $\mathbf{D}_3^3\omega$ *should have the order of differentiability of the second derivatives of the curvature. This argument can be made precise as follows:*

First we introduce $\phi_{t_*,t,u}$ *to be the diffeomorphisms of* $S_{t_*,u}$ *onto* $S_{t,u}$ *generated by* l. *Pulling back the equations 13.1.38c–13.1.38f on* $S_{t_*,u}$, *we obtain equations of the type*

$$\frac{d(r^2\mathbf{D}_3^2\phi)}{dt} = L(\nabla^2\mathbf{D}_3^2\omega) + r^2F_{33} \qquad (13.1.41e)$$

$$\frac{d(r^2\mathbf{D}_3^3\phi)}{ds} = L(\nabla^2\mathbf{D}_3^3\omega) + r^2F_{333} \qquad (13.1.41f)$$

where L *is a linear operator with continuous bounded coefficients. On* $S = S_{t_*,u}$ *we consider the Sobolev spaces of currents* $W^{-s,p}$. *We recall that, for smooth* ϕ *and positive exponents* s,

$$|\phi|_{W^{s,p}(S)} = \sum_{i=0}^{s} r^{i-s}|\nabla^i\phi|_{L^p(S)}.$$

A current U *belongs to* $W^{-s,p}(S)$ *if there exists a constant* c *such that for every smooth* ϕ

$$|\langle U,\phi\rangle| \le c|\phi|_{W^{s,q}(S)}$$

where q *is the exponent conjugate to* p. *The norm* $U_{W^{-s,p}(S)}$ *is defined as the smallest* c *with this property. We remark that if* $U \in W^{-s,p}(S)$ *for some positive integer* s, *then the current* ∇U, *defined by* $\langle \nabla U,\phi\rangle = \langle U,\nabla\phi\rangle$, *is in* $W^{-s-1,p}(S)$ *and*

$$r|\nabla U|_{W^{-s-1,p}(S)} \le |U|_{W^{-s,p}(S)}.$$

In what follows, $H_s(S_u) = W^{s,2}(S_{t_*,u})$, and we denote by $|\ |_{H_s(S_u)}$ the corresponding norms. We remark that, for all $u_1(t_*) \le u_1 \le u_2$ and appropriate weights[10] on the left-hand side, we have

$$\int_{u_1}^{u_2} ar^2|F_{33}|^2_{H_{-1}(S'_u)}du' \le c\left({}^e\mathcal{R}_{[2]} + {}^e\mathcal{K}_{[2]} + {}^e\mathcal{L}_{[2]} + {}^e\mathcal{O}_0^\infty\right)$$

[10] The issue being one of differentiability, we neglect the precise weights.

$$\int_{u_1}^{u_2} ar^2|\nabla F_{33}|^2_{H_{-2}(S'_u)}du' \le c\left({}^e\mathcal{R}_{[2]} + {}^e\mathcal{K}_{[2]} + {}^e\mathcal{L}_{[2]} + {}^e\mathcal{O}_0^\infty\right)$$
$$\int_{u_1}^{u_2} ar^2|F_{333}|^2_{H_{-2}(S'_u)}du' \le c\left({}^e\mathcal{R}_{[2]} + {}^e\mathcal{K}_{[2]} + {}^e\mathcal{L}_{[2]} + {}^e\mathcal{O}_0^\infty\right). \tag{13.1.41g}$$

We now make use of the following:

Lemma 13.1.3 *Consider the following equation on $S_{t_*,u}$:*

$$\frac{dU}{dt} = H.$$

Then

$$|U(t)|_X \le |U(t_*)|_X + \int_t^{t_*} |H|_X$$

where X is one of the distribution spaces $H_s(S_u)$ with negative s.

Applying the Lemma to 13.1.41e, we derive

$$r^2|\mathbf{D}_3^2\not{\kappa}(t)|_{H_{-1}(S_u)} \le cr^2|\mathbf{D}_3^2\not{\kappa}(t_*)|_{H_{-1}(S_u)} + c\int_t^{t_*} |\mathbf{D}_3^2\omega|_{H_1(S_u)}$$
$$+ c\int_t^{t_*} r^2|F_{33}|_{H_{-1}(S_u)}. \tag{13.1.41h}$$

On the other hand, applying Corollary 2.3.1.2 of Proposition 2.3.1 (see Chapter 2) to the Poisson equation on $S_{t_*,u}$ obtained by pulling back the equation 13.1.41c, we derive

$$|(\mathbf{D}_3^2\omega - \overline{\mathbf{D}_3^2\omega})|_{H_1(S_u)} \le |\mathbf{D}_3\not{\kappa}(t)|_{H_{-1}(S_u)} + |\nabla\mathbf{D}_3^2\underline{\alpha}|_{H_0(S_u)} + |G_{33}|_{H_{-1}(S_u)}. \tag{13.1.41i}$$

Remark that, neglecting the appropriate weights,

$$\int_{u_1}^{u_2} a|G_{33}|^2_{H_{-1}(S'_u)}du' \le c\left({}^e\mathcal{R}_{[2]} + {}^e\mathcal{K}_{[2]} + {}^e\mathcal{L}_{[2]} + {}^e\mathcal{O}_0^\infty\right). \tag{13.1.41j}$$

Therefore, inserting 13.1.41h,

$$r^2|\mathbf{D}_3^2\not{\kappa}(t)|_{H_{-1}(S_u)} \le cr^2|\mathbf{D}_3^2\not{\kappa}(t_*)|_{H_{-1}(S_u)} + c\int_t^{t_*} |\mathbf{D}_3^2\not{\kappa}(t)|_{H_{-1}(S_u)}$$
$$+ c\int_t^{t_*} r^2|F_{33}|_{H_{-1}(S_u)} + c\int_t^{t_*} |\nabla^2\mathbf{D}_3\underline{\alpha}|_{H_{-1}(S_u)}$$
$$+ c\int_t^{t_*} |G_{33}|_{H_{-1}(S_u)}$$

and, applying the Gronwall inequality, we obtain

$$r^2|\mathbf{D}_3^2\not{\psi}(t)|_{H_{-1}(S_u)} \leq cr^2|\mathbf{D}_3^2\not{\psi}(t_*)|_{H_{-1}(S_u)} + c\int_t^{t_*} r^2|F_{33}|_{H_{-1}(S_u)}$$
$$+ c\int_t^{t_*} |\nabla\!\!\!/^2\mathbf{D}_3\underline{\alpha}|_{H_{-1}(S_u)} + c\int_t^{t_*} |G_{33}|_{H_{-1}(S_u)}. \tag{13.1.41k}$$

Finally, integrating in u inequalities 13.1.41c and 13.1.41i and taking into account 13.1.41g and 13.1.41j, we infer that $(\mathbf{D}_3^2\omega - \overline{\mathbf{D}_3^2\omega}) \in L^2(\Sigma_t^e)$. To show that $\overline{\mathbf{D}_3^2\omega}$ is also L^2 integrable, we go back to equation 13.1.38a, which we differentiate once more with respect to $\mathbf{D}_3$. Keeping track of the appropriate weights, we derive the first conclusion of the proposition. The second conclusion follows in the same manner by differentiating the equation 13.1.41a in the tangential direction. To derive the third conclusion, we first estimate $\|(\mathbf{D}_3^3\omega - \overline{\mathbf{D}_3^3\omega})\|_{2,e}$ from equations 13.1.41b and 13.1.41d, proceeding precisely as above. It then remains to estimate $\overline{\mathbf{D}_3^3\omega}$. This follows from differentiating the equation 13.1.38a twice in the direction of $\mathbf{D}_3$. The corresponding propagation equation $\overline{\mathbf{D}_3^3\omega}$ involves the integral on $S_{t,u}$ of $\mathbf{D}_3^3\rho$. Using the null Bianchi equation for $\mathbf{D}_3\rho$, we can express this in terms of $\text{d}\!\!\!/\text{iv}\,\mathbf{D}\!\!\!\!/_3^2\underline{\beta}$ whose integral on $S_{t,u}$ is zero. All the other terms can be estimated in a straightforward manner. The third conclusion follows in the same manner from 13.1.41b and 13.1.41d.

13.2. Derivatives of the Interior Optical Function

In this section we estimate the higher derivatives of the interior optical function. We recall from Chapter 9 that the interior optical function, which we also denote by u, is another solution of the Eikonal equation; its construction differs from that of the exterior optical function only through its boundary conditions, which are now imposed on the central line P. Thus our procedure to estimate the higher derivatives of the the interior optical function follows precisely the same lines outlined in the previous section; we need only to take special care to describe the behavior on the central line. We recall from Chapter 9 that the surface $S_{t,u_M(t)}$ is the boundary of the interior region I_t' in Σ_t and that $u_m(t)$ is the value of u at the center P_t. We set

$$J(t) = [u_m(t), u_M(t)].$$

Regarding χ we prove the following:

Proposition 13.2.1 *Assuming that* ${}^{\iota}\mathcal{O}_0^\infty \le c$, *we have the following estimates:*

$$\sup_{u\in J(t)} |\not{\chi}, \nabla\!\!\!\!/\,\chi|_{4,S}(t,u) \le c(2+t)^{-3}({}^{\iota}\mathcal{R}_{[1]} + {}^{\iota}\mathcal{K}_{[1]} + {}^{\iota}\mathcal{L}_{[1]} + {}^{\iota}\mathcal{O}_0^\infty)$$

$$\|r^{1/2}(\not{\chi}, \nabla\!\!\!\!/\,\chi)\|_{I_t'} \le c(2+t)^{-3/2}({}^{\iota}\mathcal{R}_{[0]} + {}^{\iota}\mathcal{K}_{[0]} + {}^{\iota}\mathcal{L}_{[0]} + {}^{\iota}\mathcal{O}_0^\infty).$$

Also,

$$\sup_{u\in J(t)} |r(\nabla\!\!\!\!/\,\not{\chi}, \nabla\!\!\!\!/^2\chi)|_{4,S}(t,u) \le c(2+t)^{-3}({}^{\iota}\mathcal{R}_{[2]} + {}^{\iota}\mathcal{K}_{[2]} + {}^{\iota}\mathcal{L}_{[2]} + {}^{\iota}\mathcal{O}_0^\infty)$$

$$\|r^{3/2}(\nabla\!\!\!\!/\,\not{\chi}, \nabla\!\!\!\!/^2\chi)\|_{I_t'} \le c(2+t)^{-3/2}({}^{\iota}\mathcal{R}_{[1]} + {}^{\iota}\mathcal{K}_{[1]} + {}^{\iota}\mathcal{L}_{[1]} + {}^{\iota}\mathcal{O}_0^\infty).$$

Finally,

$$\|r^{5/2}(\nabla\!\!\!\!/\,\not{\chi}, \nabla\!\!\!\!/^2\not{\chi}, \nabla\!\!\!\!/^3\chi)\|_{I_t'} \le c(2+t)^{-3/2}({}^{\iota}\mathcal{R}_{[2]} + {}^{\iota}\mathcal{K}_{[2]} + {}^{\iota}\mathcal{L}_{[2]} + {}^{\iota}\mathcal{O}_0^\infty).$$

Proof 13.2.1 *We recall from Chapter 9 that* $t_m(u)$ *is the value of* t *at the vertex of a given* C_u *on the central line* P. *Thus* $r(u, t_m(u)) = 0$ *and* $\not{\chi}$ *verifies the boundary condition*

$$\lim_{t\to t_m(u)} |r^{3-2/p}\not{\chi}|_{p,S}(t,u) = 0. \tag{13.2.1}$$

Now $\not{\chi}$ *satisfies the propagation equation 13.1.7a. Therefore it also verifies the estimate 13.1.9a where, in view of 13.2.1, the boundary term is zero. More precisely,*

$$|r^{3-2/p}\not{\chi}|_{p,S}(t,u) \le c\int_{t_m}^t |r^{3-2/p}F|_{p,S}(t',u)dt'. \tag{13.2.2a}$$

Now, from the definition of F *in 13.1.7b,*

$$\begin{aligned} |r^{3-2/p}F|_{p,S} \le\ & c(2+t)^{-2}(|r^{3-2/p}\not{\chi}|_{p,S} + c|r^{3-2/p}\nabla\!\!\!\!/\,\hat{\chi}|_{p,S}) \\ & + c|r^{2-2/p}\beta|_{p,S} + c(2+t)^{-2}|r^{2-2/p}\underline{\zeta}|_{p,S} \\ & + c(2+t)^{-6}r^{3}\,{}^{\iota}\mathcal{O}_0^\infty. \end{aligned} \tag{13.2.2b}$$

On the other hand, from the elliptic estimates on the Codazzi equation 13.1.8,

$$|r^{3-2/p}\nabla\!\!\!\!/\,\hat{\chi}|_{p,S} \le c\left(|r^{3-2/p}\not{\chi}|_{p,S} + |r^{3-2/p}\beta|_{p,S} + {}^{\iota}\mathcal{O}_0^\infty r^3(2+t)^{-4}\right). \tag{13.2.2c}$$

Using 13.2.2b–13.2.2c, we infer from 13.2.2a

$$\begin{aligned} |r^{3-2/p}\not{\chi}|_{p,S}(t,u) \le\ & c\int_{t_m}^t |r^{2-2/p}\beta|_{p,S}(t',u) \\ & + c(2+t)^{-2}\int_{t_m}^t |r^{2-2/p}\underline{\zeta}|_{p,S} + c(2+t)^{-5}r^{3}\,{}^{\iota}\mathcal{O}_0^\infty. \end{aligned} \tag{13.2.2d}$$

Now take $p = 4$ in 13.2.2d. According to the Sobolev inequalities,

$$|\beta|_{4,S}(t,u) \leq c(\|\nabla E\|_{I'_t} + \|\nabla H\|_{I'_t}) + cr_0^{-1}(\|E\|_{I'_t} + \|H\|_{I'_t}) \\ \leq c(2+t)^{-3}{}^{i}\mathcal{R}_{[1]}.$$

Similarly,

$$|\underline{\zeta}|_{4,S}(t,u) \leq c(2+t)^{-3/2}({}^{i}\mathcal{K}_{[1]} + {}^{i}\mathcal{L}_{[1]}).$$

Making use of these in 13.2.2d and taking also into account that

$$t - t_m(u) \leq cr,$$

we infer that

$$|r^{5/2}\underline{\hat{\chi}}|_{4,S}(t,u) \leq cr^{5/2}(2+t)^{-3}{}^{i}\mathcal{R}_{[1]} \\ + cr^{5/2}(2+t)^{-7/2}({}^{i}\mathcal{K}_{[1]} + {}^{i}\mathcal{L}_{[1]}) + c(2+t)^{-5}r^{3}{}^{i}\mathcal{O}_0^{\infty}.$$

Finally, dividing by $r^{5/2}$, we obtain the desired statement of the proposition.
Next we set $p = 2$ in 13.2.2d and multiply by $r^{-3/2}$. We derive

$$|r^{1/2}\underline{\hat{\chi}}|_{2,S}(t,u) \leq cr^{-3/2}\int_{t_m}^{t}|r\beta|_{2,S}(t',u)dt' \\ + cr^{-3/2}(2+t)^{-2}\int_{t_m}^{t}|r\underline{\zeta}|_{2,S}dt' \\ + c(2+t)^{-5}r^{3/2}{}^{i}\mathcal{O}_0^{\infty}. \tag{13.2.2e}$$

Now, using $t - t_m \leq cr$, we obtain the inequality

$$r^{-3}\left(\int_{t_m}^{t}|rf|_{2,S}(t',u)dt'\right)^2 \leq c\int_{t_m}^{t}|f|^2_{2,S}(t',u)dt'.$$

Thus, squaring 13.2.2e and integrating the result with respect to u over $J(t)$, we infer, recalling from Chapter 9 that $t_m(u_M(t)) = t_0(t)$,

$$\int_{u_m(t)}^{u_M(t)}|r^{1/2}\underline{\hat{\chi}}|^2_{2,S}(t,u)du \leq c\int_{t_0(t)}^{t}\left(\int_{I'_{t'}}|\beta|^2\right)dt' \\ + c(2+t)^{-4}\int_{t_0(t)}^{t}\left(\int_{I'_{t'}}|\underline{\zeta}|^2\right)dt' \\ + c(2+t)^{-6}({}^{i}\mathcal{O}_0^{\infty})^2.$$

According to the definition of the interior norms ${}^i\mathcal{R}_{[0]}$, ${}^i\mathcal{K}_{[0]}$, *and* ${}^i\mathcal{L}_{[0]}$, *we have*

$$\int_{t_0(t)}^{t}\left(\int_{I'_{t'}}|\beta|^2\right)dt' \le c(2+t)^{-3}({}^i\mathcal{R}_{[0]})^2$$
$$\int_{t_0(t)}^{t}\left(\int_{I'_{t'}}|\underline{\zeta}|^2\right)dt' \le c\Big(({}^i\mathcal{K}_{[0]})^2+({}^i\mathcal{L}_{[0]})^2\Big).$$

Hence,

$$\|r^{1/2}\underline{\chi}\|^2_{I'_t} \le c(2+t)^{-3}({}^i\mathcal{R}_{[0]})^2 + c(2+t)^{-4}\Big(({}^i\mathcal{K}_{[0]})^2+({}^i\mathcal{L}_{[0]})^2\Big) + c(2+t)^{-6}({}^i\mathcal{O}_0^\infty)^2.$$

The remaining estimates follow in the same manner.

Proposition 13.2.2 *Assuming that* ${}^i\mathcal{O}_0^\infty, {}^i\mathcal{K}_0^\infty, {}^i\mathcal{L}_0^\infty \le c$, *the following estimates hold:*

$$\sup_{u\in J(t)}|r^{1/2}(\tilde{\mu},\nabla\!\!\!/\zeta)|_{4,S}(t,u) \le c(2+t)^{-2}\big({}^i\mathcal{R}_{[1]}+{}^i\mathcal{K}_{[1]}+{}^i\mathcal{L}_{[1]} + {}^i\mathcal{K}_0^\infty + {}^i\mathcal{L}_0^\infty + {}^i\mathcal{O}_0^\infty\big)$$

$$\|r^{1/2}(\tilde{\mu},\nabla\!\!\!/\zeta)\|_{I'_t} \le c(2+t)^{-1}({}^i\mathcal{R}_{[0]}+{}^i\mathcal{K}_{[1]}+{}^i\mathcal{L}_{[1]}+{}^i\mathcal{O}_0^\infty).$$

Also,

$$\sup_{u\in J(t)}|r^{3/2}(\nabla\!\!\!/\tilde{\mu},\nabla\!\!\!/^2\zeta)|_{4,S}(t,u) \le c(2+t)^{-2}\big({}^i\mathcal{R}_{[2]}+{}^i\mathcal{K}_{[2]}+{}^i\mathcal{L}_{[2]} + {}^i\mathcal{K}_{[1]}^\infty + {}^i\mathcal{L}_{[1]}^\infty + {}^i\mathcal{O}_{[0]}^\infty\big)$$

$$\|r^{3/2}(\nabla\!\!\!/\tilde{\mu},\nabla\!\!\!/^2\zeta)\|_{I'_t} \le c(2+t)^{-1}({}^i\mathcal{R}_{[1]}+{}^i\mathcal{K}_{[2]}+{}^i\mathcal{L}_{[2]}+{}^i\mathcal{O}_0^\infty).$$

Finally,

$$\|r^{5/2}(\nabla\!\!\!/^2\tilde{\mu},\nabla\!\!\!/^3\zeta)\|_{I'_t} \le c(2+t)^{-1}({}^i\mathcal{R}_{[2]}+{}^i\mathcal{K}_{[3]}+{}^i\mathcal{L}_{[3]}+{}^i\mathcal{O}_0^\infty).$$

Proof 13.2.2 *Recall that*

$$\tilde{\mu} = -\frac{1}{2}\Big(\mathbf{D}_3\mathrm{tr}\chi - \frac{a^2}{2}(\mathrm{tr}\chi)^2 + 2\omega\mathrm{tr}\chi\Big) \tag{13.2.3}$$
$$= -\mathrm{d}\!\!\!/\mathrm{iv}\,\zeta - \rho + \frac{1}{2}\hat{\chi}\cdot\underline{\hat{\chi}} - |\zeta|^2 + \frac{1}{2}a\mathrm{tr}\chi\delta. \tag{13.2.4}$$

$\tilde{\mu}$ satisfies the following boundary condition

$$\lim_{t\to t_m(u)} |r^{2-2/p}\tilde{\mu}|_{p,S}(t,u) = 0. \tag{13.2.5}$$

On the other hand, it satisfies the propagation equation 13.1.13a and therefore also the following analogue of the inequality 13.1.14a:

$$|r^{2-2/p}\tilde{\mu}|_{p,S}(t,u) \le c\int_{t_m(u)}^{t} (2+t)^{-2}|r^{2-2/p}\nabla\!\!\!/\,\zeta|_{p,S}(t',u)dt' + c\int_{t_m(u)}^{t} |r^{2-2/p}G|_{p,S}(t',u)dt'.$$

From the Hodge system 13.1.12c we deduce

$$|r^{2-2/p}\nabla\!\!\!/\,\zeta|_{p,S} \le c|r^{2-2/p}\tilde{\mu}|_{p,S} + c|r^{2-2/p}(\rho,\sigma)|_{p,S} \tag{13.2.6}$$

$$+ c|r^{1-2/p}\delta|_{p,S} + c\,{}^{i}\mathcal{O}_0^{\infty} r^2(2+t)^{-4}. \tag{13.2.7}$$

Now, in 13.2.6,

$$|r^{2-2/p}G|_{p,S}(t',u) \le c(2+t)^{-2}|r^{2-2/p}\not{\chi}|_{p,S} + c|r^{1-2/p}\rho|_{p,S} + c|r^{-2/p}\nu|_{p,S} + c(2+t)^{-4}r\,{}^{i}\mathcal{O}_0^{\infty}. \tag{13.2.8}$$

Substituting, we find

$$|r^{2-2/p}\tilde{\mu}|_{p,S}(t,u) \le c\int_{t_m(u)}^{t} |r^{1-2/p}(\rho,\sigma)|_{p,S}dt' + c(2+t)^{-2}\int_{t_m(u)}^{t} |r^{2-2/p}\not{\chi}|_{p,S}dt' + c\int_{t_m(u)}^{t} |r^{-2/p}\nu|_{p,S}dt' + cr^2(2+t)^{-4}\,{}^{i}\mathcal{O}_0^{\infty}. \tag{13.2.9}$$

Setting $p = 4$ and using the fact that

$$|(\rho,\sigma)|_{4,S} \le c(2+t)^{-3}\,{}^{i}\mathcal{R}_{[1]}$$

and

$$|r^{-1/2}\nu|_{4,S} \le c(2+t)^{-2}({}^{i}\mathcal{K}_0^{\infty} + {}^{i}\mathcal{L}_0^{\infty})$$

as well as the results of the previous proposition, we infer that

$$|r^{1/2}\tilde{\mu}|_{4,S}(t,u) \le c(2+t)^{-2}({}^{i}\mathcal{R}_{[1]} + {}^{i}\mathcal{K}_{[1]} + {}^{i}\mathcal{L}_{[1]} + {}^{i}\mathcal{K}_0^{\infty} + {}^{i}\mathcal{L}_0^{\infty} + {}^{i}\mathcal{O}_0^{\infty}).$$

On the other hand, setting $p = 2$ in 13.2.9, multiplying the resulting inequality by $r^{-1/2}$, and squaring and integrating with respect to u over $J(t)$, we obtain

$$\int_{u_m(t)}^{u_M(t)} |r^{1/2}\tilde{\mu}|^2_{2,S}(t,u)du \leq c\int_{t_0(t)}^{t}\left(\int_{I'_{t'}} |(\rho,\sigma)|^2\right)dt' + c(2+t)^{-4}\int_{t_0(t)}^{t}\left(\int_{I'_{t'}} |r\not{\chi}|^2\right)dt' + c\int_{t_0(t)}^{t}\left(\int_{I'_{t'}} |r^{-1}\nu|^2\right)dt' + c(2+t)^{-4}({}^i\mathcal{O}_0^\infty)^2.$$

Now, by the generalized Poincaré inequality,

$$\|r^{-1}\nu\|_{I'_t} \leq c(2+t)^{-2}({}^i\mathcal{K}_{[1]} + {}^i\mathcal{L}_{[1]}).$$

Hence, also using the results of the previous proposition,

$$\|r^{1/2}\tilde{\mu}\|_{I'_t} \leq c(2+t)^{-1}({}^i\mathcal{R}_{[0]} + {}^i\mathcal{K}_{[1]} + {}^i\mathcal{L}_{[1]} + {}^i\mathcal{O}_0^\infty).$$

The first two estimates of the proposition follow. The remaining estimates follow in a similar manner.

Before proceeding to estimate the derivatives of the function ω, we need to establish estimates for the angular and time derivatives of $r^{-1}(\zeta - \underline{\zeta})$ near the central line. The 1-form $\zeta - \underline{\zeta}$ appears in the expression for the commutator

$$[\mathbf{D}_4, \mathbf{D}_3]f = 2(\zeta - \underline{\zeta})\cdot \nabla\!\!\!\!/ f - 2\omega \mathbf{D}_4 f$$

applied to functions f. By Proposition 9.2.2, $\zeta - \underline{\zeta}$ vanishes on the central line. The regularity of the angular and time derivatives of $\mathbf{D}_T\omega$ on the central line requires that the corresponding derivatives of $\zeta - \underline{\zeta}$ vanish there as well.

Proposition 13.2.3 *Under the same assumptions as in the previous proposition we have*

$$\sup_{u\in J(t)} |r^{-1/2}\nabla\!\!\!\!/(\zeta - \underline{\zeta})|_{4,S}(t,u) \leq c(2+t)^{-3}({}^i\mathcal{R}_{[2]} + {}^i\mathcal{K}_{[3]} + {}^i\mathcal{L}_{[3]} + {}^i\mathcal{O}_0^\infty),$$

and for each $q > 0$

$$\|r^{q-1/2}\not{\Delta}(\zeta - \underline{\zeta})\|_{I'_t} \leq c(q)(2+t)^{q-3}({}^i\mathcal{R}_{[2]} + {}^i\mathcal{K}_{[3]} + {}^i\mathcal{L}_{[3]} + {}^i\mathcal{O}_0^\infty).$$

Proof 13.2.3 *Differentiating the propagation equation for ζ, equation 13.1.2b, tangentially to $S_{t,u}$ and applying commutation lemma 13.1.2, we obtain*

$$\frac{d}{ds}\nabla\!\!\!\!/_A\zeta_B + \mathrm{tr}\chi\nabla\!\!\!\!/_A\zeta_B = -\hat{\chi}_{BC}\nabla\!\!\!\!/_A\zeta_C - \hat{\chi}_{AC}\nabla\!\!\!\!/_C\zeta_B + \frac{1}{2}\mathrm{tr}\chi\nabla\!\!\!\!/_A\underline{\zeta}_B - \nabla\!\!\!\!/_A\beta_A + F_{AB} \qquad (13.2.10a)$$

where

$$
\begin{aligned}
F_{AB} = {} & \hat{\chi}_{BC}\nabla\!\!\!/_A\underline{\zeta}_C - \frac{1}{2}\chi\!\!\!/_A(\zeta_B - \underline{\zeta}_B) - \nabla\!\!\!/_A\hat{\chi}_{BC}(\zeta_C - \underline{\zeta}_C) \\
& + \frac{1}{2}\mathrm{tr}\chi\zeta_A(\zeta_B - \underline{\zeta}_B) - (\zeta_A + \underline{\zeta}_A)(\chi_{BC}(\zeta_C - \underline{\zeta}_C) + \beta_B) \\
& + (\chi_{AB}\underline{\zeta}_C - \chi_{AC}\underline{\zeta}_B + \in_{BC}{}^*\beta_A)\zeta_C.
\end{aligned}
\tag{13.2.10b}
$$

We now introduce

$$
\psi\!\!\!/_{AB} = s^2\nabla\!\!\!/_A\zeta_B - \int_0^s s'\nabla\!\!\!/_A\underline{\zeta}_B ds'. \tag{13.2.10c}
$$

Then $\psi\!\!\!/$ satisfies the propagation equation

$$
\begin{aligned}
\frac{d}{ds}\psi\!\!\!/_{AB} = s^2\Big[& \Big(s^{-1} - \frac{1}{2}\mathrm{tr}\chi\Big)(2\nabla\!\!\!/_A\zeta_B - \nabla\!\!\!/_A\underline{\zeta}_B) - \hat{\chi}_{AC}\nabla\!\!\!/_C\zeta_B \\
& - \hat{\chi}_{BC}\nabla\!\!\!/_A\zeta_C + F_{AB} - \nabla\!\!\!/_A\beta_B\Big].
\end{aligned}
\tag{13.2.10d}
$$

Applying Lemma 13.1.1, in the case $p = 4, \lambda_0 = 1/2$, to this equation, noting that

$$
\lim_{t\to t_m(u)} |r^{-1/2}\psi\!\!\!/|_{4,S}(t,u) \to 0,
$$

we obtain

$$
\begin{aligned}
|r^{-1/2}\psi\!\!\!/|_{4,S}(t,u) \le {} & c(2+t)^{-7/2}\int_{t_m(u)}^t |r^{5/2}(\nabla\!\!\!/\zeta, \nabla\!\!\!/\underline{\zeta})|_{4,S}dt' \\
& + c\int_{t_m(u)}^t |r^{3/2}F|_{4,S}dt' + c\int_{t_m(u)}^t |r^{3/2}\nabla\!\!\!/\beta|_{4,S}dt'.
\end{aligned}
\tag{13.2.10e}
$$

We have

$$
\begin{aligned}
|r^{1/2}\nabla\!\!\!/\beta|_{4,S} & \le c\Big[r^{1/2}(|\nabla E|_{4,S} + |\nabla H|_{4,S}) + |E|_{\infty,S} + |H|_{\infty,S}\Big] \\
& \le c(2+t)^{-7/2}{}^{\imath}\mathcal{R}_{[2]}.
\end{aligned}
$$

Using the results of the previous propositions, which imply in particular that

$$
|r^{1/2}F|_{4,S} \le c(2+t)^{-4}({}^{\imath}\mathcal{R}_{[1]} + {}^{\imath}\mathcal{K}_{[2]} + {}^{\imath}\mathcal{L}_{[2]} + {}^{\imath}\mathcal{O}_0^{\infty}),
$$

we then deduce

$$
|r^{-1/2}\psi\!\!\!/_{4,S} \le cr^2(2+t)^{-7/2}({}^{\imath}\mathcal{R}_{[2]} + {}^{\imath}\mathcal{K}_{[2]} + {}^{\imath}\mathcal{L}_{[2]} + {}^{\imath}\mathcal{O}_0^{\infty}). \tag{13.2.10f}
$$

We decompose

$$\nabla\!\!\!/_A \underline{\zeta}_B = \underline{\zeta}'_{AB} + \underline{\zeta}_{AB} \tag{13.2.11a}$$

where

$$\underline{\zeta}'_{AB} = \nabla_A \mathbf{b}_B - \nabla_A k_{BN}, \tag{13.2.11b}$$

$$\begin{aligned}\underline{\zeta}_{AB} = &-\frac{1}{2}\mathbf{b}_N \mathrm{tr}\theta\delta_{AB} - \mathbf{b}_N\hat{\theta}_{AB}\\ &-\frac{1}{2}(\hat{\eta}_{AB} - \frac{3}{2}\delta\delta_{AB})\mathrm{tr}\theta - \hat{\eta}_{CB}\hat{\theta}_{AC} + \frac{3}{2}\delta\hat{\theta}_{AB}.\end{aligned} \tag{13.2.11c}$$

We can estimate

$$\begin{aligned}\left| r^{-1/2}\frac{d}{ds}(s\underline{\zeta})\right|_{4,S} \leq c(2+t)^{-3}({}^{\imath}\mathcal{K}^{\infty}_{[1]} + {}^{\imath}\mathcal{L}^{\infty}_{[1]}\\ + {}^{\imath}\mathcal{R}^{\infty}_{0} + {}^{\imath}\mathcal{O}^{\infty}_{0}).\end{aligned} \tag{13.2.11d}$$

It follows that

$$\begin{aligned}\left| r^{-1/2}\left(s\underline{\zeta}_{AB} - \frac{1}{s}\int_0^s s'\underline{\zeta}_{AB}ds'\right)\right|_{4,S} \leq cr(2+t)^{-3}({}^{\imath}\mathcal{K}^{\infty}_{[1]} + {}^{\imath}\mathcal{L}^{\infty}_{[1]}\\ + {}^{\imath}\mathcal{R}^{\infty}_{0} + {}^{\imath}\mathcal{O}^{\infty}_{0}).\end{aligned} \tag{13.2.11e}$$

On the other hand,

$$|r^{-1/2}\underline{\zeta}'|_{4,S} \leq c(2+t)^{-3}({}^{\imath}\mathcal{K}^{\infty}_{[1]} + {}^{\imath}\mathcal{L}^{\infty}_{[1]}). \tag{13.2.11f}$$

In view of 13.2.10f, 13.2.11e, and 13.2.11f, the first part of the proposition follows. To obtain the second part, we first derive from equation 13.2.10a, using Lemma 13.1.2b, the following propagation equation for $\Delta\!\!\!/\zeta$*:*

$$\begin{aligned}\frac{d}{ds}\Delta\!\!\!/\zeta_A + \frac{3}{2}\mathrm{tr}\chi\Delta\!\!\!/\zeta_A = &-2\hat{\chi}_{BC}\nabla\!\!\!/_C\nabla\!\!\!/_B\zeta_A - \hat{\chi}_{AC}\Delta\!\!\!/\zeta_C\\ &+\frac{1}{2}\mathrm{tr}\chi\Delta\!\!\!/\underline{\zeta}_A + F\!\!\!/_A - \Delta\!\!\!/\beta_A\end{aligned} \tag{13.2.12a}$$

where

$$\begin{aligned}F\!\!\!/_A = &\nabla\!\!\!/_B F_{BA} - \frac{1}{2}\nabla\!\!\!/_B\mathrm{tr}\chi(2\nabla\!\!\!/_B\zeta_A - \nabla\!\!\!/_B\underline{\zeta}_A)\\ &-\nabla\!\!\!/_B\hat{\chi}_{AC}\nabla\!\!\!/_B\zeta_C - \nabla\!\!\!/_B\hat{\chi}_{BC}\nabla\!\!\!/_C\zeta_A\\ &+(\zeta_B + \underline{\zeta}_B)(-\mathrm{tr}\chi\nabla\!\!\!/_B\zeta_A - \hat{\chi}_{AC}\nabla\!\!\!/_B\zeta_C - \hat{\chi}_{BC}\nabla\!\!\!/_C\zeta_A\\ &+\frac{1}{2}\mathrm{tr}\chi\nabla\!\!\!/_B\underline{\zeta}_A + F_{BA} - \nabla\!\!\!/_B\beta_A)\\ &+(tr\chi\underline{\zeta}_C - \chi_{BC}\underline{\zeta}_B + \epsilon_{BC}\,{}^*\beta_B)\nabla\!\!\!/_C\zeta_A\\ &+(\chi_{AB}\underline{\zeta}_C - \chi_{BC}\underline{\zeta}_A + \epsilon_{AC}\,{}^*\beta_B)\nabla\!\!\!/_B\zeta_C.\end{aligned} \tag{13.2.12b}$$

We now define

$$\psi_A = s^3 \not\Delta\zeta_A - \int_0^s s'^2 \not\Delta\underline{\zeta}_A ds'. \tag{13.2.12c}$$

Then ψ satisfies the propagation equation

$$\frac{d}{ds}\psi = s^3 \Big[(s^{-1} - \frac{1}{2}\mathrm{tr}\chi)(3\not\Delta\zeta_A - \not\Delta\underline{\zeta}_A) \\ - 2\hat{\chi}_{BC}\not\nabla_C\not\nabla_B\zeta_A - \hat{\chi}_{AC}\not\Delta\zeta_C + \not F_A - \not\Delta\beta_A\Big]. \tag{13.2.12d}$$

We then apply Lemma 13.1.1 to obtain

$$|r^{-1}\psi|_{2,S}(t,u) \leq c(2+t)^{-7/2}\int_{t_m(u)}^t |r^3(\not\nabla^2\zeta, \not\nabla^2\underline{\zeta})|_{2,S}dt' \tag{13.2.12e}$$
$$+ c\int_{t_m(u)}^t |r^2 \not F|_{2,S}dt' + c\int_{t_m(u)}^t |r^2\not\Delta\beta|_{2,S}dt'.$$

We have

$$|r\not\Delta\beta|_{2,S} \leq c(|r\nabla^2 E|_{2,S} + |r\nabla^2 H|_{2,S}) + c(|\nabla E|_{2,S} + |\nabla H|_{2,S}) \\ + c(|E|_{\infty,S} + |H|_{\infty,S}) \\ \leq cr(|\nabla^2 E|_{2,S} + |\nabla^2 H|_{2,S}) + c(2+t)^{-7/2}\,{}^{\imath}\mathcal{R}_{[2]}.$$

Using the results of the previous propositions, which imply in particular that

$$|r\not F|_{2,S} \leq c(2+t)^{-4}({}^{\imath}\mathcal{R}_{[2]} + {}^{\imath}\mathcal{K}_{[3]} + {}^{\imath}\mathcal{L}_{[3]} + {}^{\imath}\mathcal{O}_0^\infty),$$

we then obtain

$$|r^{-1}\psi|_{2,S}(t,u) \leq cr^2(2+t)^{-7/2}({}^{\imath}\mathcal{R}_{[2]} + {}^{\imath}\mathcal{K}_{[3]} + {}^{\imath}\mathcal{L}_{[3]} + {}^{\imath}\mathcal{O}_0^\infty) \\ + c\int_{t_m(u)}^t (|r^2\nabla^2 E|_{2,S} + |r^2\nabla^2 H|_{2,S})dt'.$$

Multiplying this inequality by $r^{q-5/2}$, $q > 0$, and squaring and integrating the result with respect to u over $J(t)$ yields

$$\|r^{q-7/2}\psi\|_{I'_t} \leq c(2+t)^{q-7/2}({}^{\imath}\mathcal{R}_{[2]} + {}^{\imath}\mathcal{K}_{[3]} + {}^{\imath}\mathcal{L}_{[3]} + {}^{\imath}\mathcal{O}_0^\infty). \tag{13.2.12f}$$

We now decompose

$$\not\Delta\underline{\zeta}_A = \not\underline{\zeta}_A + \not\underline{\zeta}_A \tag{13.2.13a}$$

where

$$
\begin{aligned}
\underline{\not{L}}'_A &= \nabla^2_{BB}\mathbf{b}_A - \nabla^2_{BB}k_{AN} \\
&\quad - \mathrm{tr}\theta(\nabla_N\mathbf{b}_A - \nabla_N k_{AN}) - 2\theta_{AB}(\nabla_B\mathbf{b}_N - \nabla_B k_{NN}) \\
&\quad - 2\theta_{BC}\nabla_B k_{AC} - \not{d}iv\,\theta_A(\mathbf{b}_N - \delta) - \not{d}iv\,\theta_B\eta_{AB}, \qquad (13.2.13b) \\
\underline{\not{L}}_A &= -\theta_{AB}\theta_{BC}(\mathbf{b}_C - \epsilon_C) + |\theta|^2\epsilon_A. \qquad (13.2.13c)
\end{aligned}
$$

By equations 13.1.2d and 13.1.2e we can estimate

$$
\left| r^{-1}\frac{d}{ds}(s^2\underline{\not{L}})\right|_{2,S} \leq c(2+t)^{-3}({}^{\iota}\mathcal{K}^{\infty}_{[1]} + {}^{\iota}\mathcal{L}^{\infty}_{[1]} + {}^{\iota}\mathcal{R}^{\infty}_{0} + {}^{\iota}\mathcal{O}^{\infty}_{0}). \qquad (13.2.13d)
$$

This implies

$$
\begin{aligned}
\left| r^{-1}\left(s^2\underline{\not{L}}_A - \frac{1}{s}\int_0^s s'^2\underline{\not{L}}_A ds'\right)\right|_{2,S} &\leq cr(2+t)^{-3}({}^{\iota}\mathcal{K}^{\infty}_{[1]} + {}^{\iota}\mathcal{L}^{\infty}_{[1]} \\
&\quad + {}^{\iota}\mathcal{R}^{\infty}_{0} + {}^{\iota}\mathcal{O}^{\infty}_{0}).
\end{aligned}
\qquad (13.2.13e)
$$

On the other hand,

$$
|\underline{\not{L}}'|_{2,S} \leq c(2+t)^{-3}({}^{\iota}\mathcal{K}_{[3]} + {}^{\iota}\mathcal{L}_{[3]}). \qquad (13.2.13f)
$$

The estimates 13.2.12f, 13.2.13e, and 13.2.13f taken together yield the remainder of the proposition.

Proposition 13.2.4 *Under the same assumptions as in the previous proposition we have*

$$
\sup_{u\in J(t)} |r^{-1}\not{D}_T(\zeta - \underline{\zeta})|_{4,S}(t,u) \leq c(2+t)^{-7/2}({}^{\iota}\mathcal{R}_{[2]} + {}^{\iota}\mathcal{K}_{[3]} + {}^{\iota}\mathcal{L}_{[3]} + {}^{\iota}\mathcal{O}^{\infty}_{0}).
$$

Proof 13.2.4 *By commutation lemma 13.1.2 we have, for any 1-form U tangential to $S_{t,u}$,*

$$
\begin{aligned}
[\not{D}_4, \not{D}_T]U_A &= a^{-1}(\underline{\zeta}_B - \zeta_B)\not{\nabla}_B U_A \\
&\quad + (a^{-1}\omega - \nu)\not{D}_4 U_A + a^{-1}\nu\not{D}_T U_A. \qquad (13.2.14)
\end{aligned}
$$

We apply $\not{D}_T$ to the propagation equation for ζ and then apply the above formula to obtain the following propagation equation for $\not{D}_T\zeta$:

$$
\begin{aligned}
\frac{d}{ds}\not{D}_T\zeta_A + \frac{1}{2}\mathrm{tr}\chi\not{D}_T\zeta_A - a^{-1}\nu\not{D}_T\zeta_A &= \frac{1}{2}\mathrm{tr}\chi\not{D}_T\underline{\zeta}_A - \hat{\chi}_{AB}\not{D}_T\zeta_B \\
&\quad - \not{D}_T\beta_A + L_A \qquad (13.2.15a)
\end{aligned}
$$

where

$$
\begin{aligned}
L_A = & -\frac{1}{2}(\zeta_A - \underline{\zeta}_A)\mathbf{D}_T \mathrm{tr}\chi - (\zeta_B - \underline{\zeta}_B)\not{\mathbf{D}}_T \hat{\chi}_{AB} \\
& - a^{-1}(\zeta_B - \underline{\zeta}_B)\not{\nabla}_B \zeta_A + \hat{\chi}_{AB}\not{\mathbf{D}}_T \underline{\zeta}_B \\
& - (a^{-1}\omega - \nu)(\chi_{AB}(\zeta_B - \underline{\zeta}_B) + \beta_A).
\end{aligned} \tag{13.2.15b}
$$

We apply Lemma 13.1.1, in the case $p = 4, \lambda_0 = 1/2$, to the above equation, noting that

$$
\lim_{t \to t_m(u)} |r^{1/2}\not{\mathbf{D}}_T \zeta|_{4,S}(t,u) \to 0,
$$

to obtain

$$
\begin{aligned}
|r^{1/2}\not{\mathbf{D}}_T \zeta|_{4,S}(t,u) \leq & \, c\int_{t_m(u)}^{t} |r^{-1/2}\not{\mathbf{D}}_T \underline{\zeta}|_{4,S} dt' \\
& + c\int_{t_m(u)}^{t} |r^{1/2} L|_{4,S} dt' + c\int_{t_m(u)}^{t} |r^{1/2}\not{\mathbf{D}}_T \beta|_{4,S} dt'.
\end{aligned} \tag{13.2.16}
$$

We have

$$
\begin{aligned}
(\not{\mathbf{D}}_T \beta)_\mu = & \frac{1}{2}\Pi_\mu^\nu \mathbf{D}_T(\mathbf{R}_{\gamma\delta\kappa\lambda}\Pi_\nu^\gamma l^\delta \underline{l}^\kappa l^\lambda) \\
= & \frac{1}{2}\Pi_\mu^\gamma l^\delta \underline{l}^\kappa l^\lambda \mathbf{D}_T(\mathbf{R}_{\gamma\delta\kappa\lambda}) + \frac{1}{2} l^\delta \underline{l}^\kappa l^\lambda \Pi_\mu^\nu \mathbf{D}_T \Pi_\nu^\gamma \\
& + \Pi_\mu^\gamma(\underline{l}^\kappa l^\lambda \mathbf{D}_T l^\delta + l^\delta l^\lambda \mathbf{D}_T \underline{l}^\kappa + l^\delta \underline{l}^\kappa \mathbf{D}_T l^\lambda).
\end{aligned}
$$

Now, formulas 13.1.1b give

$$
\mathbf{D}_T l_\mu = a^{-1}(\zeta_\mu - \omega l_\mu) \tag{13.2.17a}
$$

$$
\mathbf{D}_T \underline{l}_\mu = a\underline{\zeta}_\mu + a^{-1}(\underline{\xi}_\mu + \omega \underline{l}_\mu) \tag{13.2.17b}
$$

$$
\begin{aligned}
\mathbf{D}_T \Pi_\nu^\mu = & \frac{1}{2}a^{-1}(\zeta^\mu \underline{l}_\nu + \underline{l}^\mu \zeta_\nu + \underline{\xi}^\mu l_\nu + l^\mu \underline{\xi}_\nu) \\
& + \frac{1}{2}a(\underline{\zeta}^\mu l_\nu + l^\mu \underline{\zeta}_\nu).
\end{aligned} \tag{13.2.17c}
$$

In view of this we can estimate

$$
|\not{\mathbf{D}}_T \beta|_{4,S} \leq c(2+t)^{-4\imath}\mathcal{R}_{[2]}. \tag{13.2.18a}
$$

Similarly,

$$
|\not{\mathbf{D}}_T \underline{\zeta}|_{4,S} \leq c(2+t)^{-5/2}({}^{\imath}\mathcal{K}_{[2]} + {}^{\imath}\mathcal{L}_{[2]}). \tag{13.2.18b}
$$

Also, from equations 13.1.2d and 13.1.2h, using the results of the previous proposition, we can estimate

$$|r^{1/2}(\mathbf{D}_T \mathrm{tr}\chi, \not{\mathbf{D}}_T\hat{\chi})|_{4,S} \leq c(2+t)^{-2}({}^{\imath}\mathcal{R}_{[1]} + {}^{\imath}\mathcal{K}_{[1]} + {}^{\imath}\mathcal{L}_{[1]} + {}^{\imath}\mathcal{K}_0^{\infty} + {}^{\imath}\mathcal{L}_0^{\infty} + {}^{\imath}\mathcal{O}_0^{\infty}). \tag{13.2.18c}$$

Taking into account the fact that by Proposition 9.2.2,

$$\sup_{u\in J(t)} |r^{-1}(\zeta - \underline{\zeta})|_{\infty,S} \leq c(2+t)^{-3}, \tag{13.2.18d}$$

we find

$$|r^{-1/2}L_A|_{4,S} \leq c(2+t)^{-5}({}^{\imath}\mathcal{R}_{[2]}+{}^{\imath}\mathcal{K}_{[1]}+{}^{\imath}\mathcal{L}_{[1]}+{}^{\imath}\mathcal{K}_0^{\infty}+{}^{\imath}\mathcal{L}_0^{\infty}+{}^{\imath}\mathcal{O}_0^{\infty}). \tag{13.2.18e}$$

Substituting the estimates 13.2.18a, 13.2.18b, and 13.2.18e in 13.2.16, we obtain

$$\sup_{u\in J(t)} |\not{\mathbf{D}}_T\zeta|_{4,S} \leq c(2+t)^{-5/2}({}^{\imath}\mathcal{R}_{[2]} + {}^{\imath}\mathcal{K}_{[2]} + {}^{\imath}\mathcal{L}_{[2]} + {}^{\imath}\mathcal{O}_0^{\infty}). \tag{13.2.19}$$

We now introduce

$$(\psi_T)_A = s\not{\mathbf{D}}_T\zeta_A - \int_0^s \not{\mathbf{D}}_T\zeta_A ds', \tag{13.2.20}$$

which satisfies the propagation equation

$$\frac{d}{ds}(\psi_T)_A = s\Big[\Big(s^{-1} - \frac{1}{2}\mathrm{tr}\chi\Big)\big(\not{\mathbf{D}}_T\zeta_A - \not{\mathbf{D}}_T\underline{\zeta}_A\big) + a^{-1}\nu\not{\mathbf{D}}_T\zeta_A - \hat{\chi}_{AB}\not{\mathbf{D}}_T\underline{\zeta} - \not{\mathbf{D}}_T\beta_A + L_A\Big].$$

Applying Lemma 13.1.2a, in the case p = 4, $\lambda_0 = 0$, to this equation and using the estimates 13.2.18a, 13.2.18b, and 13.2.18e as well as 13.2.19, we conclude that

$$|r^{-1/2}\psi_T|_{4,S} \leq cr^{3/2}(2+t)^{-4}({}^{\imath}\mathcal{R}_{[2]} + {}^{\imath}\mathcal{K}_{[2]} + {}^{\imath}\mathcal{L}_{[2]} + {}^{\imath}\mathcal{O}_0^{\infty}). \tag{13.2.21}$$

In view of the fact that

$$|r^{-1/2}(\not{\mathbf{D}}_T\underline{\zeta}_A - \frac{1}{s}\int_0^s \not{\mathbf{D}}_T\zeta_A)|_{4,S}(t,u) \leq cr^{1/2} \sup_{t'\in[t_m(u),t]} |\not{\mathbf{D}}_4\not{\mathbf{D}}_T\underline{\zeta}|_{4,S}(t',u),$$

while

$$\sup_{u\in J(t)} |\not{\mathbf{D}}_4\not{\mathbf{D}}_T\underline{\zeta}|_{4,S}(t,u) \leq c(2+t)^{-7/2}({}^{\imath}\mathcal{K}_{[3]} + {}^{\imath}\mathcal{L}_{[3]}),$$

the proposition follows.

We now proceed to estimate the derivatives of the function ω.

Proposition 13.2.5 *Under the same assumptions as in Proposition 13.2.2,*

$$\sup_{u\in J(t)} |r^{1/2}\nabla\!\!\!/\omega|_{4,S}(t,u) \leq c(2+t)^{-2}({}^{\iota}\mathcal{R}_{[2]} + {}^{\iota}\mathcal{K}_{[2]} + {}^{\iota}\mathcal{L}_{[2]} + {}^{\iota}\mathcal{O}_0^{\infty}).$$

Also,

$$\sup_{u\in J(t)} |r^{3/2}(\not{\psi}, \nabla\!\!\!/^2\omega)|_{4,S} \leq c(2+t)^{-2}({}^{\iota}\mathcal{R}_{[2]} + {}^{\iota}\mathcal{K}_{[3]} + {}^{\iota}\mathcal{L}_{[3]} + {}^{\iota}\mathcal{O}_0^{\infty})$$

$$\|r(\nabla\!\!\!/\not{\psi}, \nabla\!\!\!/^2\omega)\|_{I'_t} \leq c(2+t)^{-3/2}({}^{\iota}\mathcal{R}_{[1]} + {}^{\iota}\mathcal{K}_{[2]} + {}^{\iota}\mathcal{L}_{[2]} + {}^{\iota}\mathcal{O}_0^{\infty}).$$

Finally,

$$\|r^2(\nabla\!\!\!/^2\not{\psi}, \nabla\!\!\!/^3\omega)\|_{I'_t} \leq c(2+t)^{-3/2}({}^{\iota}\mathcal{R}_{[2]} + {}^{\iota}\mathcal{K}_{[3]} + {}^{\iota}\mathcal{L}_{[3]} + {}^{\iota}\mathcal{O}_0^{\infty}).$$

Proof 13.2.5 *We first recall, from Chapter 9, the limit of ω as we approach the central line P, at P_t, along the null generator initiating at P_t in the direction of $T + N$, where N belongs to the unit 2-sphere in the tangent space at P_t to Σ_t, is given by minus the component of $\mathbf{b} = \phi^{-1}\nabla\phi$ in the direction of N. We write simply*

$$\omega|_P = -\mathbf{b}_N.$$

Thus, we consider $\omega|_P$ as a function defined on the unit 2-sphere in the tangent space at $P_t \in P$ to Σ_t. If we denote by ϑ the angle between $\mathbf{b}$ and N in the tangent space of Σ_t at P_t,

$$\omega|_P = -|\mathbf{b}|\cos\vartheta.$$

Now remark that the induced metric γ_{AB} on the surfaces $S_{t,u}$, when rescaled by r^{-2}, converges as $t \longrightarrow t_m(u)$ to the metric $\overset{\circ}{\gamma}_{AB}$ of the unit sphere. More precisely, we denote by $f_{t,u}$ the mapping $f_{t,u} : S^2 \longrightarrow S_{t,u}$, which takes a vector N in the unit sphere S^2 in the tangent space to $\Sigma_{t_m(u)}$ at $P_{t_m(u)}$, that is, the vertex of C_u, to the point of intersection of the corresponding null generator of C_u with Σ_t. We have $\omega|P = \lim_{t\to t_m(u)} f^*_{t,u}(\omega|_{S_{t,u}})$. We then consider the metric on S^2 given by

$$\tilde{\gamma} = f^*_{t,u}(r^{-2}\gamma). \tag{13.2.22}$$

This metric converges to $\overset{\circ}{\gamma}$ as $t \longrightarrow t_m(u)$.

Also $\tilde{\nabla\!\!\!/}$ converges to $\overset{\circ}{\nabla\!\!\!/}$. Thus

$$|r\nabla\!\!\!/\omega|_\gamma \to |\overset{\circ}{\nabla\!\!\!/}(\omega|P)|_{\overset{\circ}{\gamma}}$$

and we have $\overset{\circ}{\nabla\!\!\!/}(\omega|P) = -\Pi\mathbf{b}$, *where* Π *is the projection to the plane orthogonal to* N *in the tangent space at* P_t *to* Σ_t.

It follows that

$$\lim_{t\to t_m(u)} |r^{1-2/p}\nabla\!\!\!/\omega|_{p,S}(t,u) = c_p|\mathbf{b}| \tag{13.2.23}$$

where $c_p = (\int_{S^2}(1-\cos^2\vartheta)^{p/2})^{1/p}$.

We now recall that $\nabla\!\!\!/\omega$ *verifies, in the interior region, the equation 13.1.28a. Thus, proceeding as in the derivation of 13.1.29, we infer that*

$$|r^{1-2/p}\nabla\!\!\!/\omega|_{p,S}(t,u) \le c|\mathbf{b}(t_m(u))| + c\int_{t_m(u)}^{t}|r^{1-2/p}\nabla\!\!\!/\rho|_{p,S}dt' + c\int_{t_m(u)}^{t}|r^{1-2/p}E|_{p,S}dt', \tag{13.2.24}$$

with E *the error term defined in 13.1.28b. Remarking that* $\rho = E_{NN}$, *we have, for* $p = 4$,

$$|r^{1/2}\nabla\!\!\!/\rho|_{4,S} \le c(|r^{1/2}|\nabla E|_{4,S} + |E|_{\infty,S}) \le c(2+t)^{-7/2}{}^{\imath}\mathcal{R}_{[2]}.$$

We next estimate the error term. According to the formula 13.1.28b we have

$$|r^{1-2/p}E|_{p,S} \le c(2+t)^{-2}(|r^{1-2/p}\nabla\!\!\!/\underline{\zeta}|_{p,S} + |r^{1-2/p}\nabla\!\!\!/\zeta|_{p,S}) + cr(2+t)^{-11/2}({}^{\imath}\mathcal{R}_0^\infty + {}^{\imath}\mathcal{O}_0^\infty).$$

For $p = 4$

$$|r^{1/2}\nabla\!\!\!/\underline{\zeta}|_{4,S} \le c(2+t)^{-2}({}^{\imath}\mathcal{K}_{[2]} + {}^{\imath}\mathcal{L}_{[2]})$$
$$|r^{1/2}\nabla\!\!\!/\zeta|_{4,S} \le c(2+t)^{-2}({}^{\imath}\mathcal{R}_{[1]} + {}^{\imath}\mathcal{K}_0^\infty + {}^{\imath}\mathcal{L}_0^\infty + {}^{\imath}\mathcal{O}_0^\infty).$$

Also,

$$|\mathbf{b}(t_m(u))| \le c(2+t)^{-2}{}^{\imath}\mathcal{L}_0^\infty.$$

Hence, back to 13.2.24, we deduce

$$|r^{1/2}\nabla\!\!\!/\omega|_{p,S}(t,u) \le c(2+t)^{-2}\left({}^{\imath}\mathcal{R}_{[2]} + {}^{\imath}\mathcal{K}_{[2]} + {}^{\imath}\mathcal{L}_{[2]} + {}^{\imath}\mathcal{O}_0^\infty\right). \tag{13.2.25}$$

We next estimate the quantity $\psi\!\!\!/$ *introduced in 13.2.28a. First remark that the following limit exists in the* L^p *norm on* S^2:

$$\lim_{t\to t_m(u)} f_{t,u}^*(r^2\Delta\!\!\!/\omega|_{S_{t,u}}) = \overset{\circ}{\Delta\!\!\!/}(\omega|_P) = -2\omega|_P$$

while $\lim_{r\to 0} r^2 \mathrm{d\!\!/iv}\,\underline{\beta} = 0$. *Hence*

$$\lim_{t\to t_m(u)} f^*_{t,u}(r^2 \psi\!\!\!/|_{S_{t,u}}) = 2\mathbf{b}_N. \tag{13.2.26}$$

Now considering the propagation equation 13.1.32a and proceeding as in the derivation of the inequality 13.1.33b, we derive

$$\begin{aligned} |r^{2-2/p}\psi\!\!\!/|_{p,S}(t,u) \leq c|\mathbf{b}(t_m(u))| & \\ + c\int_{t_m(u)}^{t} (2+t')^{-2}|r^{2-2/p}\nabla\!\!\!\!/^2\psi\!\!\!/|_{p,S}dt' & \\ + c\int_{t_m(u)}^{t} |r^{1-2/p}\nabla\!\!\!\!/\underline{\beta}|_{p,S}dt' & \\ + c\int_{t_m(u)}^{t} |r^{2-2/p}E\!\!\!/|_{p,S}dt'. & \end{aligned} \tag{13.2.27}$$

We now apply Corollary 2.3.1.1 of Proposition 2.3.1 to the equation

$$\Delta\!\!\!\!/\omega = \psi\!\!\!/ + \mathrm{d\!\!/iv}\,\underline{\beta}$$

to estimate

$$|r^{2-2/p}\nabla\!\!\!\!/^2\omega|_{p,S} \leq c(|r^{2-2/p}\nabla\!\!\!\!/^2\omega|_{p,S} + |r^{2-2/p}\nabla\!\!\!\!/\underline{\beta}|_{p,S}).$$

For $p = 4$ *we have, as before for* ρ,

$$\begin{aligned} |r^{1/2}\nabla\!\!\!\!/\underline{\beta}|_{4,S} &\leq c(|r^{1/2}|\nabla E|_{4,S} + |E|_{\infty,S}) \\ &\leq c(2+t)^{-7/2}\,{}^{\imath}\mathcal{R}_{[2]}. \end{aligned}$$

From 13.1.32b,

$$|r^{3/2}E\!\!\!/|_{4,S} \leq c(2+t)^{-7/2}({}^{\imath}\mathcal{R}_{[2]} + {}^{\imath}\mathcal{K}_{[3]} + {}^{\imath}\mathcal{L}_{[3]} + {}^{\imath}\mathcal{O}_0^\infty).$$

Substituting, we derive

$$|r^{3/2}(\psi\!\!\!/, \nabla\!\!\!\!/^2\omega)|_{4,S} \leq c(2+t)^{-7/2}({}^{\imath}\mathcal{R}_{[2]} + {}^{\imath}\mathcal{K}_{[3]} + {}^{\imath}\mathcal{L}_{[3]} + {}^{\imath}\mathcal{O}_0^\infty).$$

The $L^2(I'_t)$ *estimate for* $\psi\!\!\!/$, $\nabla\!\!\!\!/^2\omega$ *follows using the fact that*

$$\|\nabla\!\!\!\!/\underline{\beta}\|_{I'_t} \leq c(2+t)^{-3}\,{}^{\imath}\mathcal{R}_{[1]}.$$

The remaining estimates follow.

Proposition 13.2.6 *Assuming that*

$$ {}^{i}\mathcal{O}_0^\infty, {}^{i}\mathcal{R}_{[2]}, {}^{i}\mathcal{K}_{[3]}, {}^{i}\mathcal{L}_{[3]} \le c, $$

the following estimates hold:

$$ \sup_{u\in J(t)} |r^{-1/2}\mathbf{D}_T\omega|_{4,S}(t,u) \le c(2+t)^{-3}({}^{i}\mathcal{R}_{[2]} + {}^{i}\mathcal{K}_{[2]} + {}^{i}\mathcal{L}_{[3]} + {}^{i}\mathcal{O}_0^\infty). $$

Also,

$$ \sup_{u\in J(t)} |r^{1/2}\nabla\!\!\!/\,\mathbf{D}_T\omega|_{4,S} \le c(2+t)^{-3}({}^{i}\mathcal{R}_{[2]} + {}^{i}\mathcal{K}_{[3]} + {}^{i}\mathcal{L}_{[3]} + {}^{i}\mathcal{O}_0^\infty). $$

and

$$ \|r\nabla\!\!\!/^2\mathbf{D}_T\omega)\|_{I'_t} \le c(2+t)^{-5/2}({}^{i}\mathcal{R}_{[2]} + {}^{i}\mathcal{K}_{[3]} + {}^{i}\mathcal{L}_{[3]} + {}^{i}\mathcal{O}_0^\infty). $$

Proof 13.2.6 *From equations 13.1.2c and 13.1.35a we deduce the following propagation equation for* $\mathbf{D}_T\omega$*:*

$$ \frac{d}{ds}\mathbf{D}_T\omega - a^{-1}\nu\mathbf{D}_T\omega = -\mathbf{D}_T\rho + 3a^{-1}(\underline{\zeta} - \zeta)\cdot\nabla\!\!\!/\,\omega + I' \tag{13.2.28a} $$

where

$$ \begin{aligned} I' = {} & 2\zeta\cdot \mathbf{D}\!\!\!\!/\,_T\underline{\zeta} \qquad (13.2.28b) \\ & +(\underline{\zeta} - \zeta)\cdot(a(-\chi\cdot\zeta + \chi\cdot\underline{\zeta} - \beta) + a^{-1}(-\underline{\beta} + \chi\cdot\underline{\xi} - 2\underline{\chi}\cdot\zeta)) \\ & +(a^{-1}\omega - \nu)(-\rho + 2\underline{\zeta}\cdot\zeta - |\zeta|^2) \end{aligned} $$

To obtain the boundary condition for $\mathbf{D}_T\omega$*, we recall that we can express*

$$ \omega = \mathbf{D}_T a - \mathbf{b}_N a $$

and, from equations 13.1.1b,

$$ \mathbf{D}_T N = (\zeta_A - \mathbf{b}_A)e_A + \mathbf{b}_N T, $$

hence

$$ \mathbf{D}_T\omega = \mathbf{D}_T^2 a - a\mathbf{g}(\mathbf{D}_T\mathbf{b}, N) - a(\mathbf{b}\cdot\zeta + |\mathbf{b}_N|^2 - |\mathbf{b}|^2). \tag{13.2.29} $$

Taking into account the fact that by Proposition 9.2.2 $\zeta_A \to \underline{\zeta}_A = \mathbf{b}_A - \epsilon_A$ *as we approach the vertex of* C_u *on* P*, we obtain*

$$ \mathbf{D}_T\omega|_P = -(\mathbf{D}_T\mathbf{b})_N + \epsilon\cdot\mathbf{b}. \tag{13.2.30} $$

We now proceed as in the derivation of 13.1.36 to deduce from 13.2.28a

$$|r^{-2/p}\mathbf{D}_T\omega|_{p,S} \leq c|\mathbf{D}_T\mathbf{b}(t_m(u))| + c|\epsilon(t_m(u))||\mathbf{b}(t_m(u))| + c\int_{t_m(u)}^{t} |r^{-2/p}\mathbf{D}_T\rho|_{p,S}dt' + c\int_{t_m(u)}^{t} |r^{-2/p}(\underline{\zeta} - \zeta)\cdot \nabla\!\!\!/\,\omega|_{p,S}dt' + c\int_{t_m(u)}^{t} |r^{-2/p}I'|_{p,S}dt'. \tag{13.2.31}$$

Taking $p = 4$, we have

$$|\mathbf{D}_T\rho|_{4,S} \leq c(2+t)^{-4}\,{}^{1}\mathcal{R}_{[2]}.$$

In view of the fact that by Proposition 9.2.2

$$\sup_{u\in J(t)} |r^{-1}(\zeta - \underline{\zeta})|_{\infty,S} \leq c(2+t)^{-3},$$

using the results of the previous proposition, we can estimate

$$|r^{-1/2}(\underline{\zeta} - \zeta)\cdot \nabla\!\!\!/\,\omega|_{4,S} \leq c(2+t)^{-5}({}^{1}\mathcal{R}_{[2]} + {}^{1}\mathcal{K}_{[2]} + {}^{1}\mathcal{L}_{[2]} + {}^{1}\mathcal{O}_0^{\infty}).$$

Also,

$$|I'|_{4,S} \leq c(2+t)^{-9/2}({}^{1}\mathcal{R}_{[1]} + {}^{1}\mathcal{K}_{[2]} + {}^{1}\mathcal{L}_{[2]} + {}^{1}\mathcal{O}_0^{\infty})$$

and

$$|\mathbf{D}_T\mathbf{b}(t_m(u))| \leq c(2+t)^{-3}\,{}^{1}\mathcal{L}_{[1]}^{\infty} \leq c(2+t)^{-3}\,{}^{1}\mathcal{L}_{[3]}.$$

Substituting this in 13.2.31, we deduce the first estimate of the proposition.
To obtain the remaining estimates, we consider the function

$$\psi\!\!\!/_T = \Delta\!\!\!/\,\mathbf{D}_T\omega - d\!\!\!/iv\,\mathbf{D}\!\!\!\!/_T\underline{\beta}. \tag{13.2.32a}$$

We have

$$\frac{d}{ds}\psi\!\!\!/_T = \Delta\!\!\!/\left(\frac{d\mathbf{D}_T\omega}{ds}\right) - d\!\!\!/iv\left(\mathbf{D}\!\!\!\!/_T\frac{d\underline{\beta}}{ds}\right) - d\!\!\!/iv\,C_1 + C_2 - C_3 \tag{13.2.32b}$$

where

$$C_1 = [\mathbf{D}\!\!\!\!/_4, \mathbf{D}\!\!\!\!/_T]\underline{\beta}$$
$$C_2 = [\mathbf{D}_4, \Delta\!\!\!/]\mathbf{D}_T\omega$$
$$C_3 = [\mathbf{D}\!\!\!\!/_4, d\!\!\!/iv\,]\mathbf{D}\!\!\!\!/_T\underline{\beta}.$$

By Commutation Lemma 13.1.2 and Corollaries 13.1.2.2 and 13.1.2.3 we have

$$(C_1)_A = a^{-1}(\underline{\zeta}_B - \zeta_B)\nabla\!\!\!/_B\underline{\beta}_A + +a^{-1}\nu D\!\!\!\!/_T\underline{\beta}_A + (a^{-1}\omega - \nu)\frac{d\underline{\beta}_A}{ds}$$
$$+ 2(\zeta_A\underline{\zeta}_B - \underline{\zeta}_A\zeta_B - \sigma \in_{AB})\underline{\beta}_B$$
$$C_2 = -\mathrm{tr}\chi\triangle\!\!\!\!/\mathbf{D}_T\omega - 2\hat{\chi}\cdot\nabla\!\!\!/^2\mathbf{D}_T\omega + C_2',$$
$$C_2' = 2(\zeta + \underline{\zeta})\cdot\nabla\!\!\!/\left(\frac{d\mathbf{D}_T\omega}{ds}\right)$$
$$+ [\frac{1}{2}\mathrm{tr}\chi(\zeta + \underline{\zeta}) + \hat{\chi}\cdot(\zeta - \underline{\zeta}) + k\!\!\!/]\cdot\nabla\!\!\!/\mathbf{D}_T\omega$$
$$+ (d\!\!\!/iv\zeta + d\!\!\!/iv\underline{\zeta} + |\zeta + \underline{\zeta}|^2)\frac{d\mathbf{D}_T\omega}{ds}$$
$$C_3 = -\frac{1}{2}\mathrm{tr}\chi d\!\!\!/iv D\!\!\!\!/_T\underline{\beta} - \hat{\chi}\cdot\nabla\!\!\!/D\!\!\!\!/_T\underline{\beta} + C_3',$$
$$C_3' = (\zeta + \underline{\zeta})\cdot\frac{dD\!\!\!\!/_T\underline{\beta}}{ds} + \left(\frac{1}{2}\mathrm{tr}\chi\underline{\zeta} - \hat{\chi}\cdot\underline{\zeta} + \beta\right)\cdot D\!\!\!\!/_T\underline{\beta}. \tag{13.2.32c}$$

According to equation 13.2.28a

$$\frac{d\mathbf{D}_T\omega}{ds} = -\mathbf{D}_T\rho + I_0$$

where

$$I_0 = a^{-1}\nu\mathbf{D}_T\omega + 3a^{-1}(\underline{\zeta} - \zeta)\cdot\nabla\!\!\!/\omega + I'. \tag{13.2.32d}$$

Differentiating tangentially to $S_{t,u}$, we obtain

$$\nabla\!\!\!/\left(\frac{d\mathbf{D}_T\omega}{ds}\right) = -\nabla\!\!\!/\mathbf{D}_T\rho + a^{-1}\nu\nabla\!\!\!/\mathbf{D}_T\omega + I_1, \tag{13.2.32e}$$
$$I_1 = \nabla\!\!\!/(a^{-1}\nu)\mathbf{D}_T\omega + 3a^{-1}(\underline{\zeta} - \zeta)\cdot\nabla\!\!\!/^2\omega$$
$$+ 3\nabla\!\!\!/(a^{-1}(\underline{\zeta} - \zeta))\cdot\nabla\!\!\!/\omega + \nabla\!\!\!/I'$$

$$\triangle\!\!\!\!/\left(\frac{d\mathbf{D}_T\omega}{ds}\right) = -\triangle\!\!\!\!/\mathbf{D}_T\rho + a^{-1}\nu\triangle\!\!\!\!/\mathbf{D}_T\omega + 2\nabla\!\!\!/(a^{-1}\nu)\cdot\nabla\!\!\!/\mathbf{D}_T\omega$$
$$+ 3a^{-1}(\underline{\zeta} - \zeta)\cdot\nabla\!\!\!/\triangle\!\!\!\!/\omega + I_2,$$
$$I_2 = \triangle\!\!\!\!/(a^{-1}\nu)\mathbf{D}_T\omega + 3Ka^{-1}(\underline{\zeta} - \zeta)\cdot\nabla\!\!\!/\omega + 6\nabla\!\!\!/(a^{-1}(\underline{\zeta} - \zeta))\cdot\nabla\!\!\!/^2\omega$$
$$+ 3\triangle\!\!\!\!/(a^{-1}(\underline{\zeta} - \zeta))\nabla\!\!\!/\omega + \triangle\!\!\!\!/I'. \tag{13.2.32f}$$

Using 13.2.32e, we can express

$$C_2' = \left[2a^{-1}\nu(\underline{\zeta} - \zeta) + \frac{1}{2}\mathrm{tr}\chi(\zeta + \underline{\zeta}) + \hat{\chi}\cdot(\zeta - \underline{\zeta}) + k\!\!\!/\right]$$

$$
\begin{aligned}
&\cdot \nabla\!\!\!/ \mathbf{D}_T\omega + C_2'',\\
C_2'' &= 2(\zeta + \underline{\zeta})\cdot(I_1 - \nabla\!\!\!/ \mathbf{D}_T\rho)\\
&\quad + (d\!\!/iv\,\zeta + d\!\!/iv\,\underline{\zeta} + |\zeta + \underline{\zeta}|^2)(I_0 - \mathbf{D}_T\rho). \qquad (13.2.32g)
\end{aligned}
$$

On the other hand, applying $D\!\!\!/_T$ to the Bianchi equation 13.1.31b, we obtain

$$
\begin{aligned}
D\!\!\!/_T\left(\frac{d\beta}{ds}\right)_A &= -\nabla\!\!\!/_A\mathbf{D}_T\rho + \in_{AB}\nabla\!\!\!/_B\mathbf{D}_T\sigma\\
&\quad - \mathrm{tr}\chi D\!\!\!/_T\underline{\beta}_A - \mathbf{D}_T\mathrm{tr}\chi\underline{\beta}_A + (M_T)_A \qquad (13.2.32h)
\end{aligned}
$$

where

$$
\begin{aligned}
(M_T)_A &= -[\mathbf{D}_T, \nabla\!\!\!/_A]\rho + \in_{AB}[\mathbf{D}_T, \nabla\!\!\!/_B]\sigma + D\!\!\!/_T M_A,\\
M_A &= 2\hat{\underline{\chi}}_{AB}\beta_B - 3\rho\underline{\zeta}_A + 3\sigma{}^*\underline{\zeta}_A \qquad (13.2.32i)
\end{aligned}
$$

and, by Lemma 13.1.2, for any function f

$$
[D\!\!\!/_T, \nabla\!\!\!/_A]f = -\frac{1}{2}\delta\nabla\!\!\!/_A f + \hat{\eta}_{AB}\nabla\!\!\!/_B f + a\mathbf{b}_A\mathbf{D}_4 f.
$$

Hence,

$$
\begin{aligned}
d\!\!/iv\,D\!\!\!/_T\left(\frac{d\beta}{ds}\right) &= -\Delta\!\!\!/\mathbf{D}_T\rho - \mathrm{tr}\chi\, d\!\!/iv\,(D\!\!\!/_T\underline{\beta}) - \mathbf{D}_T\mathrm{tr}\chi\, d\!\!/iv\,\underline{\beta}\\
&\quad - \nabla\!\!\!/\mathbf{D}_T\mathrm{tr}\chi\cdot\underline{\beta} + d\!\!/iv\,M_T \qquad (13.2.32j)
\end{aligned}
$$

Substituting 13.2.32f and 13.2.32j in 13.2.32b, we obtain the following propagation equation for $\psi\!\!\!/_T$:

$$
\begin{aligned}
\frac{d}{ds}\psi\!\!\!/_T + \mathrm{tr}\chi\psi\!\!\!/_T &= -2\hat{\chi}\cdot\nabla\!\!\!/^2\mathbf{D}_T\omega + a^{-1}\nu\Delta\!\!\!/\mathbf{D}_T\omega + J\cdot\nabla\!\!\!/\mathbf{D}_T\omega\\
&\quad + 3a^{-1}(\underline{\zeta} - \zeta)\cdot\nabla\!\!\!/\Delta\!\!\!/\omega + \frac{1}{2}\mathrm{tr}\chi\, d\!\!/iv\,(D\!\!\!/_T\underline{\beta}) + L
\end{aligned}
\qquad (13.2.32k)
$$

where

$$
\begin{aligned}
J &= 2\nabla\!\!\!/(a^{-1}\nu) + 2a^{-1}\nu(\underline{\zeta} - \zeta)\\
&\quad + \frac{1}{2}\mathrm{tr}\chi(\zeta + \underline{\zeta}) + \hat{\chi}\cdot(\zeta - \underline{\zeta}) + \chi\!\!\!/ \qquad (13.2.32l)
\end{aligned}
$$

and

$$
\begin{aligned}
L &= \hat{\chi}\cdot\nabla\!\!\!/(D\!\!\!/_T\underline{\beta}) - \mathbf{D}_T\mathrm{tr}\chi\, d\!\!/iv\,\underline{\beta} - \nabla\!\!\!/(\mathbf{D}_T\mathrm{tr}\chi)\cdot\underline{\beta}\\
&\quad + d\!\!/iv\,M_T - d\!\!/iv\,C_1 + I_2 + C_2'' - C_3'. \qquad (13.2.32m)
\end{aligned}
$$

The boundary condition for $\not\psi_T$ is

$$\not\psi_T|_P = 2(\mathbf{D}_T\mathbf{b})_N - 4\epsilon\cdot\mathbf{b} + 4\delta\mathbf{b}_N. \tag{13.2.33}$$

To obtain the last estimate of the proposition, we apply Lemma 13.1.2a in the case $p = 2$, $\lambda_0 = 1$, in view of the fact that by Proposition 13.2.1

$$|J|_{\infty,S} \leq cr^{-1}(2+t)^{-2},$$

to deduce

$$\begin{aligned}|r\not\psi_T|_{2,S} \leq\ & c|\mathbf{D}_T\mathbf{b}(t_m(u))| + c|(\epsilon,\delta)(t_m(u))||\mathbf{b}(t_m(u))| \\ & + (2+t)^{-2}c\int_{t_m(u)}^{t}(|r\not\nabla^2\mathbf{D}_T\omega|_{2,S} + |\not\nabla\mathbf{D}_T\omega|_{2,S})dt' \\ & + c(2+t)^{-3}\int_{t_m(u)}^{t}|r^2\not\nabla\not\Delta\omega|_{2,S}dt' \\ & + c\int_{t_m(u)}^{t}|\not{d}iv(\not{\mathbf{D}}_T\underline{\beta})|_{2,S}dt' + c\int_{t_m(u)}^{t}|rL|_{2,S}dt'.\end{aligned} \tag{13.2.34}$$

Corollary 2.3.1.1 of Proposition 2.3.1 applied to equation 13.2.32a yields

$$|r\not\nabla^2\mathbf{D}_T\omega|_{2,S} + |\not\nabla\mathbf{D}_T\omega|_{2,S} \leq c|r\not\psi_T|_{2,S} + c|r\not{d}iv(\not{\mathbf{D}}_T\underline{\beta})|_{2,S}.$$

Hence, by substituting and using the Gronwall lemma we obtain

$$\begin{aligned}|r\not\psi_T|_{2,S} \leq\ & c(2+t)^{-3}\,{}^{\imath}\mathcal{L}^{\infty}_{[1]} + c(2+t)^{-3}\int_{t_m(u)}^{t}|r^2\not\nabla\not\Delta\omega|_{2,S}dt' \\ & + c\int_{t_m(u)}^{t}|\not{d}iv(\not{\mathbf{D}}_T\underline{\beta})|_{2,S}dt' \\ & + c\int_{t_m(u)}^{t}|rL|_{2,S}dt'.\end{aligned} \tag{13.2.35}$$

Squaring and integrating the result over $J(t)$, we infer

$$\begin{aligned}\|r(\not\psi_T, \not\nabla^2\mathbf{D}_T\omega)\|^2_{I'_t} \leq\ & c(2+t)^{-5}({}^{\imath}\mathcal{L}^{\infty}_{[1]})^2 \\ & + c(2+t)^{-3}\int_{t_0(t)}^{t}\|r^2\not\nabla\not\Delta\omega\|^2_{I'_{t'}}dt' \\ & + c(2+t)\int_{t_0(t)}^{t}\|\not{d}iv(\not{\mathbf{D}}_T\underline{\beta})\|^2_{I'_{t'}}dt' \\ & + c(2+t)\int_{t_0(t)}^{t}\|rL\|^2_{I'_{t'}}dt'.\end{aligned} \tag{13.2.36}$$

We can estimate

$$\|\not{d}iv(\not{\mathcal{D}}_T\underline{\beta})\|_{I_t} \leq c(2+t)^{-4}{}^{\imath}\mathcal{R}_{[2]}.$$

Also, using the results of Proposition 13.2.3 and the fact that by Proposition 13.2.2 and equations 13.1.2d and 13.1.2h

$$|r\mathbf{D}_T \mathrm{tr}\chi|_{\infty,S} \leq c(2+t)^{-2},$$
$$|r^{3/2}\not{\nabla}\mathbf{D}_T \mathrm{tr}\chi|_{4,S} \leq c(2+t)^{-2},$$

we obtain

$$\|rL\|_{I'_t} \leq c(2+t)^{-9/2}({}^{\imath}\mathcal{R}_{[2]} + {}^{\imath}\mathcal{K}_{[3]} + {}^{\imath}\mathcal{L}_{[3]} + {}^{\imath}\mathcal{O}_0^{\infty}).$$

Substituting this result as well as the result of Proposition 13.2.5 in 13.2.36 yields the last conclusion of the proposition.

To obtain the second conclusion, we define

$$U = f^*_{t,u}(r^2\not{\phi}_T|_{S_{t,u}}), \qquad (13.2.37a)$$

and we write equation 13.2.32k as an equation on S^2:

$$\frac{dU}{dt} = F.$$

We then have

$$\|U(t,u)\|_{W^4_{-1}(S^2)} \leq \|U(t_m(u),u)\|_{W^4_{-1}(S^2)} + \int_{t_m(u)}^{t} \|F(t',u)\|_{W^4_{-1}(S^2)}dt'. \qquad (13.2.37b)$$

The principal term in F is

$$f^*_{t,u}\left(\frac{1}{2}\mathrm{tr}\chi\not{d}iv(\not{\mathcal{D}}_T\underline{\beta})|_{S_{t,u}}\right) = r^{-1}H\not{d}iv_{\overset{\circ}{\gamma}}V$$

where

$$H = f^*_{t,u}(\frac{1}{2}r\mathrm{tr}\chi|_{S_{t,u}}),$$
$$V = f^*_{t,u}(\not{\mathcal{D}}_T\underline{\beta}|_{S_{t,u}}).$$

Now for any $\varphi \in C^{\infty}(S^2)$ we have

$$|\langle H\not{d}iv_{\overset{\circ}{\gamma}}V, \varphi\rangle| \leq c\|V\|_{L^4(S^2)}\|\varphi\|_{W_1^{4/3}(S^2)}.$$

Hence

$$\|r^{-1}H\not{d}iv_{\overset{\circ}{\gamma}}V\|_{W^4_{-1}(S^2)} \leq cr^{-1}\|V\|_{L^4(S^2)}.$$

On the other hand,

$$\|V\|_{L^4(S^2)} \leq cr^{1/2}|\not{D}_T\underline{\beta}|_{4,S} \leq cr^{1/2}(2+t)^{-4}\,{}^{\imath}\mathcal{R}_{[2]}.$$

Therefore the contribution of the principal term in F to the integral in 13.2.37b is bounded by

$$cr^{1/2}(2+t)^{-4}\,{}^{\imath}\mathcal{R}_{[2]}.$$

Thus the remaining conclusion of the proposition follows.

We must finally estimate $\mathbf{D}_T^2\omega$. Differentiating 13.2.28a along T and applying the commutation formula

$$[\frac{d}{ds}, \mathbf{D}_T]f = a^{-1}\nu\mathbf{D}_T f + (a^{-1}\omega - \nu)\mathbf{D}_4 f + a^{-1}(\underline{\zeta} - \zeta)\cdot \not{\nabla} f$$

for functions f, we obtain the following propagation equation for $\mathbf{D}_T^2\omega$:

$$\frac{d}{ds}\mathbf{D}_T^2\omega - 2a^{-1}\nu\mathbf{D}_T^2\omega = -\mathbf{D}_T^2\rho + 4a^{-1}(\underline{\zeta} - \zeta)\cdot \not{\nabla}\mathbf{D}_T\omega + I_T \tag{13.2.38a}$$

where

$$\begin{aligned} I_T = {} & \mathbf{D}_T I' + (a^{-1}\omega - \nu)(I' - \mathbf{D}_T\rho) \\ & + 3[\mathbf{D}_T(a^{-1}(\underline{\zeta} - \zeta)) + (a^{-1}\omega - \nu)a^{-1}(\underline{\zeta} - \zeta)]\cdot \not{\nabla}\omega \\ & + [\mathbf{D}_T(a^{-1}\nu) + (a^{-1}\omega - \nu)a^{-1}\nu]\mathbf{D}_T\omega \\ & + 3a^{-1}(\underline{\zeta} - \zeta)\cdot [\not{D}_T, \not{\nabla}]\omega. \end{aligned} \tag{13.2.38b}$$

The boundary condition for $\mathbf{D}_T^2\omega$ is obtained by differentiating 13.2.29 along T. Taking into account the fact that by Proposition 13.2.4 $\not{D}_T\zeta \to \not{D}_T\underline{\zeta}$, as we approach the line P we find

$$\begin{aligned} \mathbf{D}_T^2\omega|_P = {} & -(\mathbf{D}_T^2\mathbf{b})_N + (2\epsilon - \mathbf{b}_N N - \mathbf{b})\cdot \mathbf{D}_T\mathbf{b} \\ & + (|\mathbf{b}|^2 + 2\mathbf{b}\cdot\epsilon - |\underline{\zeta}|^2 + \mathbf{b}\cdot\underline{\zeta})\mathbf{b}_N + \mathbf{b}\cdot \not{D}_T\underline{\zeta}. \end{aligned} \tag{13.2.38c}$$

We then apply Lemma 13.1.2a in the case $p = 2$, $\lambda_0 = 0$, to equation 13.2.38a. Taking into account the fact that by Propositions 9.2.2 and 13.2.2c

$$\begin{aligned} |r^{-1}(\underline{\zeta} - \zeta)\cdot \not{\nabla}\mathbf{D}_T\omega|_{2,S} & \leq c(2+t)^{-3}|r^{1/2}\not{\nabla}\mathbf{D}_T\omega|_{4,S} \\ & \leq c(2+t)^{-6}({}^{\imath}\mathcal{R}_{[2]} + {}^{\imath}\mathcal{K}_{[3]} + {}^{\imath}\mathcal{L}_{[3]} + {}^{\imath}\mathcal{O}_0^{\infty}), \end{aligned}$$

while using Proposition 13.2.4, we can estimate

$$\begin{aligned} |r^{-1}I_T|_{2,S} & \leq c|r^{-1/2}I_T|_{4,S} \\ & \leq cr^{-1/2}(2+t)^{-11/2}({}^{\imath}\mathcal{R}_{[2]} + {}^{\imath}\mathcal{K}_{[3]} + {}^{\imath}\mathcal{L}_{[3]} + {}^{\imath}\mathcal{O}_0^{\infty}). \end{aligned}$$

We deduce

$$|r^{-1}\mathbf{D}_T^2\omega|_{2,S}(t,u) \leq c|\mathbf{D}_T^2\mathbf{b}(t_m(u),u)| + c\int_{t_m(u)}^t |r^{-1}\mathbf{D}_T^2\rho|_{2,S}dt' \leq c(2+t)^{-5}({}^{\imath}\mathcal{R}_{[2]} + {}^{\imath}\mathcal{K}_{[3]} + {}^{\imath}\mathcal{L}_{[3]} + {}^{\imath}\mathcal{O}_0^{\infty}). \tag{13.2.38d}$$

Squaring and integrating with respect to u over $J(t)$, we obtain, for any given $q > 0$,

$$\|r^{-1}\mathbf{D}_T^2\omega\|_{I_t'}^2 \leq c\int_{t_0}^t |\mathbf{D}_T^2\mathbf{b}(P_t')|^2 dt' + c(2+t)^q \int_{t_0}^t \left(\int_{I_{t'}'} r^{-1-q}|\mathbf{D}_T^2 E|^2\right)dt' \leq c(2+t)^{-9}({}^{\imath}\mathcal{R}_{[2]} + {}^{\imath}\mathcal{K}_{[3]} + {}^{\imath}\mathcal{L}_{[3]} + {}^{\imath}\mathcal{O}_0^{\infty})^2. \tag{13.2.38e}$$

The two integrals on the right are defined for almost all choices of the central line P. However, only for suitable choices of P are they bounded in terms of the quantities ${}^{\imath}\mathcal{R}_{[2]}, {}^{\imath}\mathcal{K}_{[3]}, {}^{\imath}\mathcal{L}_{[3]}$. To be more precise, let us consider the space-time slab $\bigcup_{t\in[-1,t_*]}\Sigma_t$ as a product $[-1,t_*]\times\Sigma_0$, the product structure being given by the normal flow of the maximal foliation. We remark that maximality implies that volumes are preserved in time: $d\mu_{g_t} = d\mu_{g_0}$. We define on Σ_0 the functions

$$A(x) = \int_{-2/3}^{t_*} (2+t)^{2(3-q)}|\mathbf{D}_T^2\mathbf{b}(t,x)|^2 dt \tag{13.2.39a}$$

$$B(x) = \int_{-2/3}^{t_*}\int_{I_t'} \frac{(2+t)^{7-q}}{d^{1+q}(x,y)}|\mathbf{D}_T^2 E(t,y)|^2 d\mu_{g_0}(y)dt \quad (q>0) \tag{13.2.39b}$$

where $d(x,y)$ is the distance of the points x, y in Σ_0. Let B_{ε_1} be the ball in Σ_0 with center at O_0 and of radius ε_1. We then have

$$\int_{B_{\varepsilon_1}} A(x)d\mu_{g_0}(x) \leq (Vol(B_{\varepsilon_1}))^{2/3}\int_{-2/3}^{t_*}(2+t)^{2(3-q)}\|\mathbf{D}_T^2\mathbf{b}(t)\|_{L^6(B_{\varepsilon_1})}^2 dt \leq c\varepsilon_1^2\int_{-2/3}^{t_*}(2+t)^{2(3-q)}\left(\|\nabla\mathbf{D}_T^2\mathbf{b}\|_{I_t}^2 + (2+t)^{-2}\|\mathbf{D}_T^2\mathbf{b}\|_{I_t}^2\right)dt \leq c\varepsilon_1^2({}^{\imath}\mathcal{L}_{[3]})^2 \tag{13.2.39c}$$

where

$$I_t = I \bigcap \Sigma_t.$$

Also, in view of the fact that

$$\int_{B_{\epsilon_1}} \frac{d\mu_{g_0}(x)}{d^{1+q}(x,y)} \leq c\varepsilon_1^{2-q},$$

$$\int_{B_{\epsilon_1}} B(x)d\mu_{g_0}(x) \leq c\varepsilon_1^{2-q} \int_{-2/3}^{t_*} (2+t)^{7-q} \|\mathbf{D}_T^2 E(t)\|_{I_t}^2 dt$$
$$\leq c\varepsilon_1^{2-q} ({}^{\imath}\mathcal{R}_{[2]})^2. \tag{13.2.39d}$$

We now select the point $P_0 \in B_{\varepsilon_1}$ such that

Condition 1

$$A(P_0) \leq \frac{3}{Vol(B_{\varepsilon_1})} \int_{B_{\epsilon_1}} A(x)d\mu_{g_0}(x)$$

Condition 2

$$B(P_0) \leq \frac{3}{Vol(B_{\varepsilon_1})} \int_{B_{\epsilon_1}} B(x)d\mu_{g_0}(x).$$

We note that the measure of the set of such points is at least $(1/3)Vol(B_{\varepsilon_1})$. With P_0 selected in this way, by 13.2.39c we have

$$\left(\int_{t_0}^{t} |\mathbf{D}_T^2 \mathbf{b}(P_t')|^2 dt'\right)^{1/2} \leq c\varepsilon_1^{-1/2} (2+t)^{q-3\imath} \mathcal{L}_{[3]}.$$

Also since for any $P_t' \in S_{t,u}$

$$d(P_0, P_0') \leq c d(P_t, P_t') \leq c r(t,u),$$

by 13.2.39d we have

$$\left(\int_{t_0}^{t} \left(\int_{I_{t'}'} r^{-1-q} |\mathbf{D}_T^2 E|^2\right) dt'\right)^{1/2} \leq c\varepsilon_1^{-1/2-q/2} (2+t)^{q/2-7/2} {}^{\imath}\mathcal{R}_{[2]}.$$

We conclude with the following:

Proposition 13.2.7 *Let*

$$^{\imath}\mathcal{R}_{[2]}, {}^{\imath}\mathcal{K}_{[3]}, {}^{\imath}\mathcal{L}_{[3]} \leq c.$$

Then there exists a point $P_0 \in B_{\varepsilon_1}$ *such that the interior optical function based on the corresponding line* P *satisfies, for some* $q > 0$:

$$\|r^{-1}\mathbf{D}_T^2\omega\|_{I_t'}^2 \leq c\varepsilon_1^{-(1+q)/2}(2+t)^{q-3}({}^{\imath}\mathcal{R}_{[2]} + {}^{\imath}\mathcal{K}_{[3]} + {}^{\imath}\mathcal{L}_{[3]}).$$

CHAPTER 14

The Last Slice

In this chapter we construct the function u on the last slice Σ_{t_*}. Given the surface $S_{t_*,0}$, the intersection of the standard cone C_0 with Σ_{t_*}, we define u_* to be the solution of the following inverse lapse problem:

$$|\nabla u_*|^{-1} = a, \qquad u_*|_{S_{t_*,0}} = 0 \tag{14.0.1a}$$

where a satisfies on each level surface S_{t_*,u_*} of u_* the equation

$$\not{\Delta} \log a = f - \overline{f} - \not{d}\mathrm{iv}\,\epsilon, \qquad \overline{\log a} = 0 \tag{14.0.1b}$$

with

$$f = K - \frac{1}{4}(tr\chi)^2. \tag{14.0.1c}$$

In this chapter the optical quantities and the curvature components are expressed relative to the normalized null normals

$$e_3' = T - N = a^{-1}e_3, \qquad e_4' = T + N = ae_4,$$

N being the outward unit normal to S_{t_*,u_*} relative to Σ_{t_*}. However, for convenience of notation we have dropped the primes. We shall show in the following that if the first and second fundamental forms of Σ_{t_*} satisfy certain estimates, then this inverse lapse problem has a solution that is global in the exterior of the surface $S_{t_*,0}$ and extends in the interior a distance that is a given fraction of r_0.

Equation 14.0.1b can be expressed in the form

$$\not{d}\mathrm{iv}\,\zeta = f - \overline{f}, \tag{14.0.2a}$$

which is to be considered in conjunction with equation 13.1.2i, that is,

$$\not{c}\mathrm{url}\,\zeta = \sigma - \hat{\eta} \wedge \hat{\chi}. \tag{14.0.2b}$$

Combining equations 13.1.2d and 13.1.2h, we obtain the following equation for $\nabla_N tr\chi$ on each Σ_t, in particular on Σ_{t_*}:

$$\nabla_N tr\chi = -\frac{1}{2}(tr\chi)^2 - \frac{1}{2}\delta tr\chi - |\hat{\chi}|^2 - \hat{\eta} \cdot \hat{\chi} - |\zeta|^2 - \not{d}\mathrm{iv}\,\zeta - \rho.$$

Recalling equation 13.1.2n for the Gauss curvature,

$$K = \frac{1}{4}(tr\chi)^2 - \frac{1}{2}|\hat{\chi}|^2 - \hat{\eta}\cdot\hat{\chi} - \frac{1}{2}\delta tr\chi - \rho, \tag{14.0.3}$$

and using 14.0.2a, we deduce the following propagation equation for $tr\chi$ on Σ_{t_*}:

$$\nabla_N tr\chi + \frac{1}{2}(tr\chi)^2 = -F_0 - \overline{f} \tag{14.0.4a}$$

where

$$F_0 = \frac{1}{2}|\hat{\chi}|^2 + |\zeta|^2. \tag{14.0.4b}$$

We then apply the commutation formula of Corollary 3.2.3.1, for the commutator $[\nabla\!\!\!/_N, \nabla\!\!\!/]$ applied to functions to obtain the propagation equation for $\nabla\!\!\!/ tr\chi$,

$$\nabla\!\!\!/_N(\nabla\!\!\!/ tr\chi) + \frac{3}{2}tr\theta\nabla\!\!\!/ tr\chi = -F_1, \tag{14.0.5a}$$

where

$$\begin{aligned} F_1 &= (\hat{\chi} + \hat{\eta})\cdot\nabla\!\!\!/ tr\chi + \delta\nabla\!\!\!/ tr\chi + \nabla\!\!\!/ F_0 \\ &\quad + \left(\frac{1}{2}(tr\chi)^2 + F_0 + \overline{f}\right)(\zeta - \epsilon). \end{aligned} \tag{14.0.5b}$$

This equation is to be considered in conjunction with the (normalized) null Codazzi equation

$$d\!\!\!/iv\,\hat{\chi} - \frac{1}{2}\nabla\!\!\!/ tr\chi + \epsilon\cdot\hat{\chi} - \frac{1}{2}\epsilon tr\chi = -\beta. \tag{14.0.6}$$

In order to prove the global existence of solutions to the inverse lapse problem 14.0.1a, 14.0.1b, we will need to assume estimates for the space-time curvature on the last slice. We cannot assume estimates for the components of the curvature with respect to a null frame adapted to the surfaces S_{t_*,u_*} since these surfaces, with the exception of $S_{t_*,0}$, have not been constructed yet. We can, however, assume estimates for the curvature components relative to a null frame defined by the level surfaces of the distance function on Σ_{t_*} from the surface $S_{t_*,0}$. In translating from one null frame to the other we shall make use of the following lemma:

Lemma 14.0.1 *Let Σ be a 3-dimensional Riemannian manifold and let S_0 be a surface in Σ diffeomorphic to S^2. Let u' and u be smooth functions without critical points defined in tubular neighborhoods U' and U, $U' \supset U$, of S_0 and vanishing on S_0. We can take*

$$U' = \bigcup_{u'\in I'} S'_{u'}, \qquad U = \bigcup_{u\in I} S_u$$

where I', I are open real intervals including 0, and $S'_{u'}$ and S_u are the level surfaces of u' and u respectively. With

$$r' = (Area(S'_{u'})/4\pi)^{1/2}, \qquad r = (Area(S_u)/4\pi)^{1/2}$$

let

$$diam(S'_{u'}) \leq 2\pi r', \qquad diam(S_u) \leq 2\pi r,$$

$$1/2 \leq a' \leq 2, \qquad 1/2 \leq a \leq 2,$$

where $a' = |\nabla u'|^{-1}$, $a = |\nabla u|^{-1}$, and let

$$\inf_{S'_{u'}}(r'/2)tr\theta' \geq 1/2, \qquad \sup_{S'_{u'}}(r'/2)tr\theta' \leq 2,$$

$$\inf_{S_u}(r/2)tr\theta \geq 1/2, \qquad \sup_{S_u}(r/2)tr\theta \leq 2,$$

hold for all $u' \in I'$ and all $u \in I$ where θ' and θ are the second fundamental forms of $S'_{u'}$ and S_u respectively. Then there is a numerical constant c such that in U:

$$c^{-1}u \leq u' \leq cu, \qquad c^{-1}r' \leq r \leq cr'.$$

Suppose also that

$$\sup_{S'_{u'}} |tr\theta' - \overline{tr\theta'}| \leq cr'^{-2}$$

and

$$\sup_{S'_{u'}} |\hat{\theta}'| \leq c(r'^{-1}\tau_-'^{-3/2} + r'^{-2})$$

hold for all $u' \in I'$, where $\tau_-' = (1 + u'^2)^{1/2}$. We denote

$$A' = r_0^{-1/2} \sup_{u' \in I'} \sup_{S'_{u'}} r'^{5/2} |\nabla\!\!\!/' \log a'|,$$

$$A = r_0^{-1/2} \sup_{u \in I} \sup_{S_u} r^{5/2} |\nabla\!\!\!/ \log a|$$

where $\nabla\!\!\!/'$ and $\nabla\!\!\!/$ are the derivatives tangential to $S'_{u'}$ and S_u, respectively. Consider the unit normal vectorfields N' and N of the foliations $\{S'_{u'}\}$ and $\{S_u\}$ respectively, and consider the angle φ between N and N'. Then if A and A' are sufficiently small, $\varphi < \pi/2$ in U and there is a numerical constant c such that

$$\sin\varphi \leq c(A + A')|u'|r'^{-2}.$$

Finally, denoting by

$$B' = r_0^{-1/2} \sup_{u' \in I'} \sup_{S'_{u'}} r'^{3/2} |\nabla_{N'} \log a'|,$$

for all $u \in I$ we have

$$\begin{aligned} osc_{S_u} u' &\leq c(A + A' + B')|u|r^{-1}, \\ osc_{S_u} r' &\leq c(A + A' + B')|u|r^{-1}, \\ \sup_{S_u} |r - r'| &\leq c(A + A' + B')|u|r^{-1}, \end{aligned}$$

provided that $A, A', B' \leq 1$.

Proof 14.0.1 *Since $\varphi = 0$ on S_0, by continuity $\varphi < \pi/2$ in a neighborhood of 0. Let $\hat{I}$ be the maximal subinterval of I where $\varphi < \pi/2$. We shall show later on that $\hat{I} = I$. We first compare the functions u and u' in*

$$\hat{U} = \bigcup_{u \in \hat{I}} S_u.$$

Let $p \in \hat{U}$ be a point on S_u and let Γ be the integral curve of N joining p to S_0. Denoting by s arc length on Γ, we have $ds/du = a$, hence

$$u'(p) = \int_\Gamma \nabla_N u' ds \leq \int_\Gamma |\nabla u'| ds = \int_0^u (a/a')|_\Gamma.$$

Let Γ' be the integral curve of N' joining p to S_0. Then $\Gamma' \subset \hat{U}$. Using the integral curves of N to construct a diffeomorphism of U onto $I \times S_0$, we can express the metric in U in the form

$$g_{ij} dx^i dx^j = a^2 du^2 + \gamma_{AB} d\omega^A d\omega^B$$

where (ω^1, ω^2) are local coordinates on S_0. Then Γ' is given by an equation of the form $\omega^A = \omega^A(u)$. Let s' denote arc length on Γ'. Then

$$\begin{aligned} u'(p) &= \int_{\Gamma'} \nabla_{N'} u' ds' = \int_{\Gamma'} (1/a')|_{\Gamma'} ds' \\ &= \int_0^u \frac{1}{a'} \left(a^2 + \gamma_{AB} \frac{d\omega^A}{du} \frac{d\omega^B}{du} \right)^{1/2}_{|_{\Gamma'}} du \geq \int_0^u (a/a')|_{\Gamma'} du. \end{aligned}$$

Therefore

$$\int_0^u \inf_{S_u} (a/a') du \leq u' \leq \int_0^u \sup_{S_u} (a/a') du \tag{14.0.7a}$$

and

$$osc_{S_u} u' \leq \int_0^u osc_{S_u}(a/a')du. \tag{14.0.7b}$$

The assumptions of the lemma then imply that

$$c^{-1}u \leq u' \leq cu. \tag{14.0.8a}$$

The assumptions also imply

$$\frac{1}{4} \leq \frac{dr'}{du'} = \frac{1}{2}\overline{a'r'tr\theta'} \leq 4, \qquad \frac{1}{4} \leq \frac{dr}{du} = \frac{1}{2}\overline{artr\theta} \leq 4.$$

Then by 14.0.8a and the fact that

$$r'|_{u'=0} = r|_{u=0} = r_0$$

we obtain

$$c^{-1}r' \leq r \leq cr'. \tag{14.0.8b}$$

To derive a more precise comparison of the areas of S_u and $S'_{u'}$, we use the integral curves of N' to construct a diffeomorphism of U' onto $I' \times S_0$ and express the metric in U' in the form

$$g_{ij}dx^i dx^j = a'^2 du'^2 + \gamma'_{AB} d\omega^A d\omega^B.$$

Then a surface S_u is given by an equation of the form

$$u' = f(u, \omega),$$

and the metric induced on S_u is given by

$$\gamma_{AB} = \gamma'_{AB} + a'^2 \frac{\partial f}{\partial \omega^A}\frac{\partial f}{\partial \omega^B}.$$

Since

$$det\gamma = det\gamma' \left(1 + a'^2 \gamma'^{AB} \frac{\partial f}{\partial \omega^A}\frac{\partial f}{\partial \omega^B}\right)$$

we have

$$Area(S_u) \geq \int_{S_0} (det\gamma'(f(u,\omega),\omega))^{1/2} d^2\omega.$$

On the other hand,

$$\frac{\partial det\gamma'}{\partial u'} = a' tr\theta' det\gamma' > 0.$$

Therefore, setting

$$u'_m(u) = \inf_{S_u} u', \qquad u'_M(u) = \sup_{S_u} u',$$

we have

$$det\gamma'(f(u,\omega),\omega) \geq det\gamma'(u'_m(u),\omega).$$

Hence

$$Area(S_u) \geq Area(S'_{u'_m(u)}).$$

Similarly, interchanging the roles of the functions u' *and* u *and using the fact that also* $tr\theta > 0$, *we obtain*

$$Area(S'_{u'}) \geq Area(S_{u_m(u')})$$

where

$$u_m(u') = \inf_{S'_{u'}} u, \qquad u_M(u') = \sup_{S'_{u'}} u.$$

Now $u_m(c') = c$ *implies that* $u'_M(c) = c'$. *This is seen as follows: Let* $p \in S'_{c'}$ *achieve* $\inf_{S'_{c'}} u = c$. *Then for every* $q \in S_c$ *we have* $u'(q) \leq c'$, *for if there is a* $q \in S_c$ *such that* $u'(q) > c'$, *then* q *belongs to the exterior of* $S'_{c'}$. *In view of the fact that*

$$\nabla_{N'} u = a^{-1} \cos\varphi > 0,$$

following the integral curve of N' *through* q *inward, we obtain a point* $q' \in S'_{c'}$ *where* $u(q') < c$, *contradicting the definition of* c. *We conclude that*

$$Area(S'_{u'_M(u)}) \geq Area(S_u) \geq Area(S'_{u'_m(u)}),$$

or, equivalently,

$$r'(u'_M(u)) \geq r(u) \geq r'(u'_m(u)). \tag{14.0.9a}$$

It follows that

$$\sup_{S_u} |r - r'| \leq osc_{S_u} r'. \tag{14.0.9b}$$

We have

$$N' = \cos\varphi N + Y, \qquad N = \cos\varphi N' + Y' \tag{14.0.10}$$

where Y *and* Y' *are vectorfields tangential to* S_u *and* $S'_{u'}$ *respectively, and*[1]

$$|Y| = |Y'| = \sin\varphi.$$

In view of the expressions (see 3.1.1c)

$$\nabla_i N_j = \theta_{ij} - N_i \nabla\!\!\!\!/_j \log a, \qquad \nabla_i N'_j = \theta'_{ij} - N'_i \nabla\!\!\!\!/'_j \log a',$$

[1] Note that we are restricting attention to $\hat{U}$.

differentiating the first of 14.0.10 we obtain:

$$\theta'_{ij} - N'_i \nabla'_j \log a' = \cos\varphi(\theta_{ij} - N_i \nabla_j \log a) + \nabla_i \cos\varphi N_j + \nabla_i Y_j. \qquad (14.0.11)$$

Contracting this equation with $N^i N^j$ and using the facts that by 14.0.10

$$\theta'(N,N) = \theta'(Y',Y')$$
$$(N, \nabla' \log a') = \nabla_{Y'} \log a'$$
$$(\nabla_N Y, N) = -(Y, \nabla_N N) = \nabla_Y \log a$$

we obtain

$$\nabla_N \cos\varphi = \theta'(Y',Y') - \nabla_Y \log a - \cos\varphi \nabla_{Y'} \log a'. \qquad (14.0.12)$$

Letting $\hat{Y} = Y/|Y|$, $\hat{Y}' = Y'/|Y'|$, we can write this equation in the form

$$\nabla_N \varphi + (\frac{1}{2} tr\theta' + \hat{\theta}(\hat{Y},\hat{Y})) \sin\varphi = \cos\varphi \nabla_{\hat{Y}'} \log a' + \nabla_{\hat{Y}} \log a. \qquad (14.0.13)$$

Now

$$\nabla_N r' = \lambda' \cos\varphi$$

where

$$\lambda' = \nabla_{N'} r' = (r'/2a')\overline{a' tr\theta'}.$$

Hence, using r' as a parameter along the integral curves of N and setting

$$\gamma = r' \sin\varphi,$$

we obtain

$$|\lambda' \frac{d\gamma}{dr'} + [(1/2a')(a' tr\theta' - \overline{a' tr\theta'}) + \hat{\theta}'(\hat{Y}',\hat{Y}')]\gamma| \leq r'(|\nabla' \log a'| + |\nabla \log a|).$$

Integrating and taking into account 14.0.8b, the assumptions, and the fact that $\gamma(r_0) = 0$ yields

$$\gamma \leq \exp\left[\int_{r_0}^{r'} \left(\frac{1}{2a'}|a' tr\theta' - \overline{a' tr\theta'}| + |\hat{\theta}'|\right) \frac{dr'}{\lambda'}\right] \cdot \left(\int_{r_0}^{r'} r'(|\nabla \log a| + |\nabla' \log a'|)\frac{dr'}{\lambda'}\right)$$
$$\leq c(A + A')(1 - r_0/r'),$$

from which the estimate for $\sin\varphi$ *follows for the region* $\hat{U}$. *Now if* $\hat{I} \neq I$, *then* $\hat{I}$ *has a boundary point* $u_1 \in I$ *and at some* $p \in S_{u_1}$ *we have* $\varphi(p) = \pi/2$. *But this contradicts the estimate if* A *and* A' *are sufficiently small. Therefore* $\hat{I} = I$.

For any vector X *tangent to* S_u *at a point we have*

$$X = X' + (X, Y)N'$$

where X' *is the projection of* X *to the surface* $S'_{u'}$ *through that point. It follows that*

$$|\nabla\!\!\!/ \log a'| \leq |\nabla\!\!\!/' \log a'| + |Y||\nabla_{N'} \log a'|.$$

Thus the estimate on $\sin\varphi = |Y|$, *together with the assumptions of the lemma and 14.0.8b, implies that*

$$|\nabla\!\!\!/ \log a'| \leq c(A' + B')r_0^{1/2} r^{-5/2},$$

which in view of the assumption on the diameter of S_u *yields*

$$\log(\sup_{S_u} a' / \inf_{S_u} a') \leq c(A' + B')r_0^{1/2} r^{-3/2}. \tag{14.0.14a}$$

Since

$$\log(\sup_{S_u} a / \inf_{S_u} a) \leq cAr_0^{1/2} r^{-3/2}, \tag{14.0.14b}$$

the estimate for the oscillation of u' *on* S_u *follows. The estimate for the oscillation of* r' *on* S_u *follows from the fact that*

$$osc_{S_u} r' = \int_J \frac{dr'}{du'} du' \leq c\, osc_{S_u} u'$$

where J *is the interval*

$$J = [\inf_{S_u} u', \sup_{S_u} u'], \qquad |J| = osc_{S_u} u'.$$

Finally, the estimate for $r - r'$ *follows by virtue of 14.0.9b.*

In this chapter we shall also make use of the following lemma, which is analogous to Lemma 13.1.1:

Lemma 14.0.2 *Let* Σ, S_0, u, *and* U *be as in Lemma 14.0.1, and suppose that for all* $u \in I$:

$$diam(S_u) \leq 2\pi r, \qquad a \leq 2,$$

$$\inf_{S_u}(r/2)tr\theta \geq 1/2, \qquad \sup_{S_u}(r/2)tr\theta \leq 2,$$

and

$$\sup_{S_u} |tr\theta - \overline{tr\theta}| \leq cr^{-2}, \qquad \sup_{u \in I} \sup_{S_u} r^2 |\nabla\!\!\!/ \log a| \leq c.$$

Let V and F be k-covariant tensorfields on Σ tangent to the surfaces S_u, which verify in U:

$$\nabla\!\!\!/_N V + \lambda_0 tr\theta V = F,$$

where λ_0 is a nonnegative real. Then, given any $p \geq 2$, setting

$$\lambda_1 = 2(\lambda_0 - \frac{1}{p})$$

for all $u \in I$, we have

$$|r^{\lambda_1} V|_{p,S}(u) \leq c \left(|r^{\lambda_1} V|_{p,S}(0) + | \int_0^u |r^{\lambda_1} F|_{p,S}(u')du' | \right).$$

Proof 14.0.2 *The proof is along the lines of that of 3.1.1c, using the identity*

$$\begin{aligned}\frac{d}{du}\left(\int_{S_u} r^{\lambda_1 p}|V|^p\right) = \int_{S_u} & a r^{\lambda_1 p}(\nabla_N |V|^p + \lambda_0 p tr\theta |V|^p) \\ & - \frac{\lambda_1 p}{2}\int_{S_u} r^{\lambda_1 p}|V|^p(atr\theta - \overline{atr\theta}),\end{aligned}$$

which follows from the fact that for any function f defined in U

$$\frac{d}{du}\left(\int_{S_u} f\right) = \int_{S_u} a(\nabla_N f + tr\theta f).$$

We are now ready to prove the following:

Proposition 14.0.1 *Let $p > 2$. Suppose that the surface $S_0 = S_{t_*,0}$ satisfies:*

S_0:

$$r_0^2 \inf_{S_0} K > \frac{1}{2}, \qquad r_0^2 \sup_{S_0} K < 2,$$

$$\frac{r_0}{2} \inf_{S_0} tr\chi > \frac{1}{2}, \qquad \frac{r_0}{2} \sup_{S_0} tr\chi < 2,$$

and

S_1:

$$y(0) = |r_0^{3-2/p} \nabla\!\!\!/ tr\chi|_{p,S_0} \leq \varepsilon_0.$$

Let u' be a smooth function without critical points, defined in a tubular neighborhood U' of S_0, which includes the whole exterior of S_0 and extends in the interior up to a level surface $S'_{u'_0}$ of u' of area $A'_0 = 4\pi(r_0/4)^2$. Let u' vanish on S_0 and tend to infinity at infinity. We have

$$U' = \bigcup_{u' \in (u'_0, \infty)} S'_{u'},$$

where $S'_{u'}$ are the level surfaces of u'. With $r' = (Area(S'_{u'})/4\pi)^{1/2}$, we make the following set of assumptions:

U'_0:

For all $u' \in (u'_0, \infty)$,

$$diam(S'_{u'}) \leq 2\pi r', \qquad 1/2 \leq a' \leq 2.$$

U'_1:

For all $u' \in (u'_0, \infty)$,

$$\inf_{S'_{u'}} (r'/2) tr\theta' \geq 1/2, \qquad \sup_{S'_{u'}} (r'/2) tr\theta' \leq 2,$$

$$\sup_{S'_{u'}} |tr\chi' - \overline{tr\chi'}| \leq c r'^{-2}, \qquad \sup_{S'_{u'}} |\hat{\chi}'| \leq c r'^{-2},$$

$$\sup_{S'_{u'}} r'^{5/2} |\nabla\!\!\!/' \log a'| \leq c r_0^{1/2}, \qquad \sup_{S'_{u'}} r'^{3/2} |\nabla_{N'} \log a'| \leq c r_0^{1/2}$$

where $\tau'_- = (1 + u'^2)^{1/2}$.

We denote by $(\delta', \epsilon', \hat{\eta}')$ the decomposition of the second fundamental form of Σ and by $(\underline{\alpha}', \underline{\beta}', \rho', \sigma', \beta', \alpha')$ the null decomposition of the space-time curvature on Σ relative to the surfaces $S'_{u'}$. We denote by

$$\overline{\mathcal{K}_0'^{\infty}}$$

the maximum of the three quantities:

$$r_0^{-1/2} \sup_{U'} r'^{5/2} |\delta'|, \qquad r_0^{-1/2} \sup_{U'} r'^{5/2} |\epsilon'|,$$

$$r_0^{-1/2} \sup_{U'} r' [r_0^{-1/2} \tau_-'^{-3/2} + r'^{-3/2}]^{-1} |\hat{\eta}'|$$

and by

$$\overline{\mathcal{R}_0'^{\infty}}$$

the maximum of the 10 quantities

$$\sup_{U'} r' \tau_-'^{5/2} |\underline{\alpha}'|, \qquad \sup_{U'} r'^2 \tau_-'^{3/2} |\underline{\beta}'|,$$

$$\sup_{U'} r'^3 \tau_-'^{1/2} |\rho' - \bar{\rho}'|, \qquad \sup_{U'} r'^3 \tau_-'^{1/2} |\sigma' - \bar{\sigma}'|,$$

$$\sup_{U'} r'^3 |\bar{\rho}'|, \qquad \sup_{U'} r'^3 |\bar{\sigma}'|,$$

$$\sup_{(u_0', \infty)} r'^3 \tau_-' |\frac{d\bar{\rho}'}{du'}|, \qquad \sup_{(u_0', \infty)} r'^3 \tau_-' |\frac{d\bar{\sigma}'}{du'}|,$$

$$\sup_{U'} r'^{7/2} |\beta'|, \qquad \sup_{U'} r'^{7/2} |\alpha'|$$

where $\bar{\rho}', \bar{\sigma}'$ are the mean values on $S_{u'}'$ of ρ', σ', respectively.

Suppose that also, in addition to the assumptions $\mathbf{S}_0, \mathbf{S}_1$ *and* $\mathbf{U}_0, \mathbf{U}_1$, *we have:*

$\mathbf{H}_0$:

$$\overline{\mathcal{K}_0'^{\infty}} \leq \varepsilon_0, \qquad \overline{\mathcal{R}_0'^{\infty}} \leq \varepsilon_0.$$

Then, if ε_0 is sufficiently small, the inverse lapse problem 14.0.1a, 14.0.1b has a solution that is global in the exterior of S_0 and extends in the interior at least up to a level surface S_{u_0} of u of area $A_0 = 4\pi(r_0/4)^2$.

Also, letting $r(u) = (Area(S_u)/4\pi)^{1/2}$, for all $u \in (u_0, \infty)$ the level surfaces S_u of the function u satisfy

$$|Kr^2 - 1|, |(r/2)tr\chi - 1| \leq cr^{-1}(y(0) + \overline{\mathcal{K}_0'^{\infty}} + \overline{\mathcal{R}_0'^{\infty}}),$$

$$|r^{2-2/p}\hat{\chi}|_{p,S} + |r^{3-2/p}\nabla\hat{\chi}|_{p,S} + |r^{3-2/p}\nabla tr\chi|_{p,S} \leq c(y(0) + \overline{\mathcal{K}_0'^{\infty}} + \overline{\mathcal{R}_0'^{\infty}}),$$

where $r(u) = (Area(S_u)/4\pi)^{1/2}$ and c is a numerical constant. Furthermore

$$r_0^{-1/2}(|r^{5/2-2/p}\zeta|_{p,S} + |r^{7/2-2/p}\nabla\zeta|_{p,S}) \leq c(y(0) + \overline{\mathcal{K}_0'^{\infty}} + \overline{\mathcal{R}_0'^{\infty}}).$$

Proof 14.0.3 *Let us denote by $P_0(u)$ the following property of S_u:*

$$\inf_{S_u} r^2 K > 1/2, \qquad \sup_{S_u} r^2 K < 2$$

and

$$\inf_{S_u}(r/2)tr\theta > 1/2, \qquad \sup_{S_u}(r/2)tr\theta < 2,$$

$$\sup_{S_u}|tr\theta - \overline{tr\theta}| < r^{-2}, \qquad \sup_{S_u}|\hat{\theta}| < r^{-1}\tau_-^{-3/2} + r^{-2}$$

and

$$\inf_{S_u} a > 1/2, \qquad \sup_{S_u} a < 2, \qquad \sup_{S_u}|\nabla\!\!\!/ \log a| \leq r_0^{1/2} r^{-5/2}.$$

Let us also denote by $P_1(u)$ the property

$$y(u) = |r^{3-2/p}\nabla\!\!\!/ tr\chi|_{p,S}(u) < 1$$

and let P be the property P_0 and P_1. We first show that, under our hypotheses, P is true at $u = 0$. Indeed, since $S_0' = S_0$ we have

$$\sup_{S_0}(|\delta|, |\epsilon|) \leq \varepsilon_0 r_0^{-2}, \qquad \sup_{S_0}|\hat{\eta}| \leq \varepsilon_0 r_0^{-1},$$

$$\sup_{S_0}(|\rho - \bar{\rho}|, |\sigma - \bar{\sigma}|) \leq \varepsilon_0 r_0^{-3}, \qquad \sup_{S_0}|\beta| \leq \varepsilon_0 r_0^{-7/2}.$$

The only parts of $P(0)$ that are not obvious are those concerning $\hat{\theta}$,a and $\nabla\!\!\!/ \log a$. Applying the 2-dimensional elliptic estimates of Chapter 2, to the null Codazzi equation14.0.6 on S_0, we obtain, in view of our assumption on the Gauss curvature of S_0 and on $y(0)$,

$$|r^{2-2/p}\hat{\chi}|_{p,S}(0) + |r^{3-2/p}\nabla\!\!\!/ \hat{\chi}|_{p,S}(0) \leq c\varepsilon_0$$

and by the Sobolev inequality on S_0

$$\sup_{S_0}|\hat{\chi}| \leq c\varepsilon_0 r_0^{-2}.$$

The part concerning $\hat{\theta}$ follows. We then consider the system 14.0.2a, 14.0.2b on S_0. By equation 14.0.3 we have

$$f = -\rho - \frac{1}{2}|\hat{\chi}|^2 - \hat{\eta}\cdot\hat{\chi} - \frac{1}{2}\delta tr\chi. \tag{14.0.15}$$

Using this we find

$$\sup_{S_0}|f - \bar{f}| \leq c\varepsilon_0 r_0^{-3}.$$

Hence

$$|r^{3-2/p}\not{d}iv\,\zeta|_{p,S}(0) \leq c\varepsilon_0.$$

Similarly

$$|r^{3-2/p}c\not{u}rl\,\zeta|_{p,S}(0) \leq c\varepsilon_0.$$

Applying the 2-dimensional L^p elliptic estimates of Chapter 2, we then obtain

$$|r^{2-2/p}\zeta|_{p,S}(0) + |r^{3-2/p}\not{\nabla}\zeta|_{p,S}(0) \leq c\varepsilon_0,$$

hence also

$$\sup_{S_0}|\zeta| \leq c\varepsilon_0 r_0^{-2}.$$

In view of the fact that by our assumption on the Gauss curvature of S_0 we have

$$diam(S_0) < \pi 2^{1/2} r_0,$$

the remaining parts of property P at $u = 0$ follow. By the local existence theorem and continuity, P remains true in an open interval containing 0. Now, either the inverse lapse problem 14.0.1a, 14.0.1b has a global solution in the exterior of S_0 and P is true for all $u \geq 0$, or there is a $u_1 > 0$ such that P is true for all $u \in [0, u_1)$ but P is false at $u = u_1$. We shall show that the second alternative is impossible. Denoting

$$I = [0, u_1), \qquad U = \bigcup_{u \in I} S_u,$$

the fact that P holds in I allows us to apply Lemma 14.0.1a in I to conclude that

$$c^{-1}u \leq u' \leq cu, \qquad c^{-1}r' \leq r \leq cr', \tag{14.0.16}$$

$$osc_{S_u} u' \leq c|u|r^{-1} \tag{14.0.17}$$

and

$$\sin\varphi \leq cur^{-2}. \tag{14.0.18}$$

Using formulas 14.0.10, we can express the components of the second fundamental form of Σ relative to N in terms of the components relative to N'. For example,

$$\delta = \cos^2\varphi\delta' + 2\cos\varphi\epsilon'(Y') + \eta'(Y', Y').$$

We can also express the components of the space-time curvature on Σ relative to $l = T + N, \underline{l} = T - N$, in terms of the components relative to $l' = T + N'$, $\underline{l}' = T - N'$. For example,

$$\rho = \cos^2\varphi\rho' - \cos\varphi(\beta'(Y') + \underline{\beta}'(Y')) + \frac{1}{4}\alpha'(Y', Y') + \frac{1}{4}\underline{\alpha}'(Y', Y').$$

Using the estimate 14.0.18 together with 14.0.16, we conclude that in U

$$r_0^{-1/2} \sup_U r^{5/2}(|\delta|, |\epsilon|) \leq c\overline{\mathcal{K}_0'^{\infty}}, \tag{14.0.19a}$$

$$r_0^{-1/2} \sup_U r\left[r_0^{-1/2}\tau_-^{-3/2} + r^{-3/2}\right]^{-1} |\hat{\eta}| \leq c\overline{\mathcal{K}_0'^{\infty}} \tag{14.0.19b}$$

and

$$\sup_U r^{7/2}|\beta| \leq c\overline{\mathcal{R}_0'^{\infty}}, \tag{14.0.20a}$$

$$\sup_U r^3(|\rho|, |\sigma|) \leq c\overline{\mathcal{R}_0'^{\infty}}. \tag{14.0.20b}$$

To estimate $\rho - \bar{\rho}$ and $\sigma - \bar{\sigma}$, where $\bar{\rho}$ and $\bar{\sigma}$ are the mean values of ρ and σ on S_u, we remark that if f is an arbitrary function and $F(u')$ is the mean value of f on $S'_{u'}$,

$$F(u') = (Area(S'_{u'}))^{-1} \int_{S'_{u'}} f,$$

then the difference of $F \circ u'$ from its mean value on S_u is bounded by

$$|F \circ u' - (Area(S_u))^{-1} \int_{S_u} F \circ u'| \leq osc_J F \leq osc_{S_u} u' \cdot \sup_J |\frac{dF}{du'}|$$

where J is the interval

$$J = [\inf_{S_u} u', \sup_{S_u} u'].$$

Using 14.0.17, we obtain:

$$\sup_U r^3 \tau_-^{1/2}(|\rho - \bar{\rho}|, |\sigma - \bar{\sigma}|) \leq c\overline{\mathcal{R}_0'^{\infty}}. \tag{14.0.20c}$$

Let

$$x(u) = |r^{2-2/p}\hat{\chi}|_{p,S}(u) + |r^{3-2/p}\nabla\!\!\!\!/\,\hat{\chi}|_{p,S}(u). \tag{14.0.21}$$

We now apply 2-dimensional L^p elliptic estimates of Chapter 2 (see Proposition 2.3.1) to the null Codazzi equation 14.0.6 on S_u for each $u \in I$ to obtain, in view of 14.0.19a and 14.0.20a,

$$x(u) \leq c(y(u) + \overline{\mathcal{K}_0'^{\infty}} + \overline{\mathcal{R}_0'^{\infty}}) \tag{14.0.22a}$$

and, by the Sobolev inequality on S_u,

$$\sup_{S_u} r^2|\hat{\chi}| \leq cx(u). \tag{14.0.22b}$$

Using 14.0.19a, 14.0.19b, 14.0.20c, and 14.0.22b, we estimate (see 14.0.15)

$$\sup_{S_u} |f - \overline{f}| \leq c r_0^{1/2} r^{-7/2} (y(u) + \overline{\mathcal{K}_0'^{\infty}} + \overline{\mathcal{R}_0'^{\infty}}). \tag{14.0.23a}$$

Also, using 14.0.20b,

$$\sup_{S_u} |f| \leq c r^{-3} (y(u) + \overline{\mathcal{K}_0'^{\infty}} + \overline{\mathcal{R}_0'^{\infty}}). \tag{14.0.23b}$$

It follows that

$$|r^{7/2-2/p}(\not{d}iv\,\zeta, c\not{u}rl\,\zeta)|_{p,S}(u) \leq c r_0^{1/2} (y(u) + \overline{\mathcal{K}_0'^{\infty}} + \overline{\mathcal{R}_0'^{\infty}}).$$

Applying once again the 2-dimensional L^p elliptic estimates of Chapter 2 (see Proposition 2.3.1), we then obtain

$$\begin{aligned} z(u) &= r_0^{-1/2} \left(|r^{5/2-2/p}\zeta|_{p,S}(u) + |r^{7/2-2/p}\not{\nabla}\zeta|_{p,S}(u) \right) \\ &\leq c(y(u) + \overline{\mathcal{K}_0'^{\infty}} + \overline{\mathcal{R}_0'^{\infty}}) \end{aligned} \tag{14.0.24a}$$

and, by the Sobolev inequality on S_u,

$$r_0^{-1/2} \sup_{S_u} |r^{5/2}\zeta| \leq c z(u). \tag{14.0.24b}$$

This allows us to estimate, for all $u \in I$,

$$\begin{aligned} |r^{3-2/p} F_1|_{p,S}(u) \leq c \Big(& r^{-1/2}(r_0^{-1/2} \tau_-^{-3/2} + r^{-3/2}) y(u) \\ & + r_0^{1/2} r^{-3/2} (y(u) + \overline{\mathcal{K}_0'^{\infty}} + \overline{\mathcal{R}_0'^{\infty}}) \Big). \end{aligned} \tag{14.0.25}$$

In view of the fact that property P holds true in I we can then apply Lemma 14.0.2 with $\lambda_0 = 3/2$, $\lambda_1 = 3 - 2/p$ to equation 14.0.5a to obtain

$$\begin{aligned} y(u) \leq c(y(0) + & \int_0^u du' (r_0^{-1/2} r^{-1/2} \tau_-^{-3/2} + r_0^{1/2} r^{-3/2}) y(u') \\ & + \int_0^u du' r_0^{1/2} r^{-3/2} (\overline{\mathcal{K}_0'^{\infty}} + \overline{\mathcal{R}_0'^{\infty}})). \end{aligned} \tag{14.0.26}$$

By the Gronwall lemma and the fact that by virtue of P $1/4 < dr/du < 4$, we conclude that for all $u \in \overline{I}$

$$y(u) \leq c(y(0) + \overline{\mathcal{K}_0'^{\infty}} + \overline{\mathcal{R}_0'^{\infty}}), \tag{14.0.27a}$$

hence, by 14.0.22a and 14.0.24a,

$$x(u), z(u) \leq c(y(0) + \overline{\mathcal{K}_0'^{\infty}} + \overline{\mathcal{R}_0'^{\infty}}). \tag{14.0.27b}$$

These inequalities imply

$$\begin{aligned}
\sup_{S_u} r^2 |tr\chi - \overline{tr\chi}| &\leq c(y(0) + \overline{\mathcal{K}_0'^{\infty}} + \overline{\mathcal{R}_0'^{\infty}}) \\
\sup_{S_u} r^2 |\hat{\chi}| &\leq c(y(0) + \overline{\mathcal{K}_0'^{\infty}} + \overline{\mathcal{R}_0'^{\infty}}) \\
r_0^{-1/2} \sup_{S_u} r^{5/2} |\zeta| &\leq c(y(0) + \overline{\mathcal{K}_0'^{\infty}} + \overline{\mathcal{R}_0'^{\infty}}).
\end{aligned} \tag{14.0.27c}$$

In particular, at $u = u_1$,

$$x(u_1), y(u_1), z(u_1) \leq c\varepsilon_0, \tag{14.0.27d}$$

$$\sup_{S_{u_1}} r^2 |tr\chi - \overline{tr\chi}| \leq c\varepsilon_0, \qquad \sup_{S_{u_1}} r^2 |\hat{\chi}| \leq c\varepsilon_0, \qquad r_0^{-1/2} \sup_{S_{u_1}} r^{5/2} |\zeta| \leq c\varepsilon_0, \tag{14.0.27e}$$

and, in view of 14.0.19a and 14.0.19b,

$$\begin{aligned}
\sup_{S_{u_1}} |tr\theta - \overline{\mathrm{tr}\theta}| &\leq c\varepsilon_0 r^{-2} \\
\sup_{S_{u_1}} |\hat{\theta}| &\leq c\varepsilon_0 (r^{-1}\tau_-^{-3/2} + r^{-2}) \\
\sup_{S_{u_1}} |\nabla \log a| &\leq c\varepsilon_0 r_0^{1/2} r^{-5/2}.
\end{aligned} \tag{14.0.28}$$

As we are requiring $\overline{\log a} = 0$, the last implies

$$\sup_{S_{u_1}} |\log a| \leq c\varepsilon_0 r_0^{1/2} r^{-3/2}. \tag{14.0.29}$$

Now, by the Gauss-Bonnet formula (see 14.0.1c)

$$\overline{f} = \frac{1}{r^2} - \frac{1}{4}\overline{(tr\chi)^2},$$

hence

$$\frac{4}{r^2} - (\overline{tr\chi})^2 = 4\overline{f} + \frac{1}{4\pi r^2} \int_{S_u} (tr\chi - \overline{tr\chi})^2.$$

In view of the estimates 14.0.23b and 14.0.27c, we then obtain

$$\sup_{S_u} r^2 |\frac{2}{r} - tr\chi| \leq c(y(0) + \overline{\mathcal{K}_0'^{\infty}} + \overline{\mathcal{R}_0'^{\infty}}) \tag{14.0.30}$$

and, since

$$K = (1/4)(tr\chi)^2 + f,$$

$$\sup_{S_u} r^3 \left|\frac{1}{r^2} - K\right| \leq c(y(0) + \overline{\mathcal{K}_0'^{\infty}} + \overline{\mathcal{R}_0'^{\infty}}). \tag{14.0.31}$$

This implies in particular that

$$\sup_{S_{u_1}} \left|\frac{2}{r} - tr\theta\right| \leq c\varepsilon_0, \qquad \sup_{S_{u_1}} \left|\frac{1}{r^2} - K\right| \leq c\varepsilon_0. \tag{14.0.32}$$

From 14.0.27d, 14.0.28, 14.0.29, and 14.0.32, we conclude that if ε_0 is sufficiently small, then property P holds also at $u = u_1$. It follows that the inverse lapse problem 14.0.1a, 14.0.1b has a global solution in the exterior of S_0 and P is true for all $u \in [0, \infty)$. Therefore the estimates 14.0.32, 14.0.27a, and 14.0.27b also hold for all $u \in [0, \infty)$. The proof for $u \in (u_0, 0)$ is similar, only simpler.

We proceed to estimate the higher tangential derivatives of χ and ζ.

Proposition 14.0.2 *Let the hypotheses of Proposition 14.0.1 be satisfied for $p = 4$. Suppose that the foliation given by the function u' satisfies in addition*

U′₂:

For all $u' \in (u_0', \infty)$,

$$|r'^2 \nabla\!\!\!/' tr\chi'|_{2,S'}(u') \leq c,$$

$$|r'^2 \nabla\!\!\!/'^2 \log a'|_{2,S'}(u') \leq c, \qquad |r' \tau_-' \nabla\!\!\!/_{N'} \nabla\!\!\!/' \log a'|_{2,S'}(u') \leq c.$$

Suppose also that the components of the second fundamental form of Σ relative to N' satisfy

$\mathbf{H}_1$:

$$\overline{\mathcal{K}_{[1]}'^{\infty}} \leq c$$

where

$$\overline{\mathcal{K}_{[1]}'^{\infty}} = \overline{\mathcal{K}_0'^{\infty}} + \overline{\mathcal{K}_1'^{\infty}}$$

and $\overline{\mathcal{K}_1'^{\infty}}$ is the maximum of the six quantities

$$r_0^{-1/2} \sup_{U'} r'^{7/2} |\nabla\!\!\!/' \delta'|, \qquad r_0^{-1/2} \sup_{U'} w_{'3/2} r'^2 |\nabla_{N'} \delta'|,$$

$$r_0^{-1/2}\sup_{U'} r'^{7/2}|\nabla\!\!\!/' \epsilon'|, \qquad r_0^{-1/2}\sup_{U'} w_{'3/2} r'^2|\nabla\!\!\!/_{N'}\epsilon'|,$$

$$r_0^{-1/2}\sup_{U'} w_{'3/2} r'^2|\nabla\!\!\!/' \hat{\eta}'|, \qquad r_0^{-1/2}\sup_{U'} w_{'5/2} r'|\nabla\!\!\!/'_{N'}\hat{\eta}'|.$$

Here

$$w'_q = [r_0^{-1/2}\tau_-'^{-q} + r'^{-q}]^{-1}.$$

Let

$$r_0(\overline{\mathcal{K}'_0})^2 = \int_{U'}\{r'^2(|\delta'|^2 + |\epsilon'|^2) + w_1'^2|\hat{\eta}'|^2\},$$

$$r_0(\overline{\mathcal{K}'_1})^2 = \int_{U'}\{r'^4(|\nabla\!\!\!/'\delta'|^2 + |\nabla\!\!\!/'\epsilon'|^2) + r'^2 w_1'^2(|\nabla_{N'}\delta'|^2 + |\nabla\!\!\!/'_{N'}\epsilon'|^2)\}$$
$$+ \int_{U'}\{r'^2 w_1'^2|\nabla\!\!\!/'\hat{\eta}'|^2 + w_2'^2|\nabla\!\!\!/'_{N'}\hat{\eta}'|^2\},$$

$$r_0(\overline{\mathcal{K}'_2})^2 = \int_{U'} r'^6(|\nabla\!\!\!/'^2\delta'|^2 + |\nabla\!\!\!/'^2\epsilon'|^2)$$
$$+ \int_{U'}\{r'^4 w_1'^2(|\nabla\!\!\!/'\nabla_{N'}\delta'|^2 + |\nabla\!\!\!/'\nabla\!\!\!/'_{N'}\epsilon'|^2)$$
$$+ r'^2 w_2'^2(|\nabla^2_{N'}\delta'|^2 + |\nabla\!\!\!/'^2_{N'}\epsilon'|^2)\}$$
$$+ \int_{U'}\{r'^4 w_1'^2|\nabla\!\!\!/'^2\hat{\eta}'|^2 + r'^2 w_2'^2|\nabla\!\!\!/'\nabla\!\!\!/'_{N'}\hat{\eta}'|^2 + w_3'^2|\nabla\!\!\!/'^2_{N'}\hat{\eta}'|^2\},$$

and let

$$\overline{\mathcal{K}'_{[2]}} = \overline{\mathcal{K}'_0} + \overline{\mathcal{K}'_1} + \overline{\mathcal{K}'_2}.$$

Let also

$$(\overline{\mathcal{R}'_0})^2 = \int_{U'}\{\tau_-'^4|\underline{\alpha}'|^2 + \tau_-'^2 r'^2|\underline{\beta}'|^2$$
$$+ r'^4(|\rho' - \bar{\rho}'|^2 + |\sigma' - \bar{\sigma}'|^2 + |\beta'|^2 + |\alpha'|^2)\}$$

$$(\overline{\mathcal{R}'_1})^2 = \int_{U'}\{\tau_-'^4 r'^2|\nabla\!\!\!/'\underline{\alpha}'|^2 + \tau_-'^6|\nabla\!\!\!/'_{N'}\underline{\alpha}'|^2\}$$
$$+ \int_{U'}\{\tau_-'^2 r'^4|\nabla\!\!\!/'\underline{\beta}'|^2 + \tau_-'^4 r'^2|\nabla\!\!\!/'_{N'}\underline{\beta}'|^2\}$$
$$+ \int_{U'}\{r'^6(|\nabla\!\!\!/'\rho'|^2 + |\nabla\!\!\!/'\sigma'|^2) + r'^4\tau_-'^2(|\rho'_{N'}|^2 + |\sigma'_{N'}|^2)\}$$
$$+ \int_{U'} r'^6\{|\nabla\!\!\!/'\beta'|^2 + |\nabla\!\!\!/'_{N'}\beta'|^2 + |\nabla\!\!\!/'\alpha'|^2 + |\nabla\!\!\!/'_{N'}\alpha'|^2\}$$

$$(\overline{\mathcal{R}'_2})^2 = \int_{U'}\{\tau_-'^4 r'^4|\nabla\!\!\!/'^2\underline{\alpha}'|^2 + \tau_-'^6 r'^2|\nabla\!\!\!/'\nabla\!\!\!/'_{N'}\underline{\alpha}'|^2 + \tau_-'^8|\nabla\!\!\!/'^2_{N'}\underline{\alpha}'|^2\}$$

$$+\int_{U'}\{\tau_-'^2 r'^6|\nabla'^2\underline{\beta}'|^2+\tau_-'^4 r'^4|\nabla'\nabla_{N'}\underline{\beta}'|^2+\tau_-'^6 r'^2|\nabla_{N'}'^2\underline{\beta}'|^2\}$$

$$+\int_{U'}\{r'^8(|\nabla'^2\rho'|^2+|\nabla'^2\sigma'|^2)$$

$$+r'^6\tau_-'^2(|\nabla'\rho_{N'}'|^2+|\nabla'\sigma_{N'}'|^2)+r'^4\tau_-'^4(|\nabla_{N'}\rho_{N'}'|^2+|\nabla_{N'}\sigma_{N'}'|^2)\}$$

$$+\int_{U'}\{r'^8(|\nabla'^2\beta'|^2+|\nabla'\nabla_{N'}'\beta'|^2)+r'^6\tau_-'^2|\nabla_{N'}'^2\beta'|^2\}$$

$$+\int_{U'}r'^8\{|\nabla'^2\alpha'|^2+|\nabla'\nabla_{N'}'\alpha'|^2+|\nabla_{N'}'^2\alpha'|^2\},$$

where

$$\rho_{N'}'=\nabla_{N'}\rho'+(3/2)tr\theta'\rho',\qquad \sigma_{N'}'=\nabla_{N'}\sigma'+(3/2)tr\theta'\sigma'$$

and with

$$\overline{\mathcal{R}_{[2]}'}=\sup_{U'}r'^3|\bar{\rho}'|+\sup_{U'}r'^3|\bar{\sigma}'|+\overline{\mathcal{R}_0'}+\overline{\mathcal{R}_1'}+\overline{\mathcal{R}_2'}.$$

We assume

H$_2$:

$$\overline{\mathcal{R}_{[2]}'}\leq c.$$

If the surface S_0 satisfies in addition

$$\nabla^2 tr\chi\in L^4(S_0),$$

then, for all $u\in(u_0,\infty)$, the surfaces S_u satisfy

$$|r^{7/2}\nabla^2\hat{\chi}|_{4,S}(u)+|r^{7/2}\nabla^2 tr\chi|_{4,S}(u)\leq c\Big(|r_0^{7/2}\Delta tr\chi|_{4,S_0}+|r_0^{5/2}\nabla tr\chi|_{4,S_0}$$
$$+\overline{\mathcal{K}_{[1]}'^{\infty}}+\overline{\mathcal{R}_{[2]}'}\Big)$$

and

$$r_0^{-1/2}|r^4\nabla^2\zeta|_{4,S}(u)\leq c\Big(|r_0^{7/2}\Delta tr\chi|_{4,S_0}+|r_0^{5/2}\nabla tr\chi|_{4,S_0}+\overline{\mathcal{K}_{[1]}'^{\infty}}+\overline{\mathcal{R}_{[2]}'}\Big).$$

If also

$$\nabla^3 tr\chi\in L^2(S_0)$$

and

$\mathbf{S}_2$:

$$|r_0^{7/2} \not{\Delta} tr\chi|_{4,S_0} \leq c,$$

then for all $u \in (u_0, \infty)$

$$|r^4 \not{\nabla}^3 tr\chi|_{2,S}(u) \leq c\Big(|r_0^4 \not{\nabla} \not{\Delta} tr\chi|_{2,S_0} + |r_0^{7/2} \not{\Delta} tr\chi|_{4,S_0} + |r_0^{5/2} \not{\nabla} tr\chi|_{4,S_0} + \overline{\mathcal{K}'^{\infty}_{[1]}} + \overline{\mathcal{K}'_{[2]}} + \overline{\mathcal{R}'_{[2]}} \Big).$$

Proof 14.0.4 *We first derive equations for the derivatives of* $\cos\varphi$ *and* Y, *supplementing equation 14.0.12. Contracting 14.0.11 with* $N^i \Pi^j_m$, *where* $\Pi^j_m = \delta^j_m - N^j N_m$ *is the projection to the surfaces* S_u, *we obtain*

$$\not{\nabla}_N Y_m = \cos\varphi(\not{\nabla}_m \log a - \Pi^j_m \not{\nabla}' \log a') + Y'^i \Pi^j_m \theta'_{ij}. \qquad (14.0.33)$$

On the other hand, contracting 14.0.11 with $\Pi^i_m N^j$, *and using the fact that* $\Pi^i_m N^j \nabla_i Y_j = -\Pi^i_m Y^j \nabla_i N_j = -\theta_{mj} Y^j$, *we obtain*

$$\not{\nabla}_m \cos\varphi = \theta_{mj} Y^j + \Pi^i_m Y'^j \theta'_{ij} - Y_m \nabla_{Y'} \log a'. \qquad (14.0.34)$$

Finally, contracting with $\Pi^i_m \Pi^j_n$ *yields*

$$\not{\nabla}_m Y_n = -\cos\varphi\theta_{mn} + \Pi^i_m \Pi^j_n \theta'_{ij} - Y_m \Pi^j_n \not{\nabla}'_j \log a'. \qquad (14.0.35)$$

Using the estimate given by Lemma 14.0.1 for $\sin\varphi = |Y| = |Y'|$, *we find that in* U

$$|\nabla_N \cos\varphi| \leq c|u|r^{-4}, \qquad |\not{\nabla}_N Y| \leq cr^{-2}. \qquad (14.0.36)$$

To estimate $\not{\nabla}\cos\varphi$, *we write*

$$\theta_{mj} Y^j + \Pi^i_m Y'^j \theta'_{ij} = \chi_{mj} Y^j + \Pi^i_m Y'^j \chi'_{ij} + \eta_{mj} Y^j + \Pi^i_m Y'^j \eta'_{ij}.$$

Using the fact that

$$\Pi^i_m \Pi'^n_i = \Pi^n_m - \cos\varphi Y_m N^n - Y_m Y^n \qquad (14.0.37a)$$

and

$$\Pi^i_m Y'_i = -\cos\varphi Y_m, \qquad (N, Y') = \sin^2\varphi, \qquad Y' = -\cos\varphi Y + \sin^2\varphi N, \qquad (14.0.37b)$$

we obtain

$$\eta_{mj} Y^j + \Pi^i_m Y'^j \eta'_{ij} = (1 - \cos\varphi)\eta_{mj} Y^j + \sin^2\varphi\epsilon_m - \cos\varphi\sin^2\varphi\delta Y_m + (\cos^2\varphi - \sin^2\varphi) Y_m \epsilon_j Y^j + \cos\varphi Y_m \eta_{ij} Y^i Y^j.$$

We can also express

$$\chi_{mj}Y^j + \Pi^i_m Y'^j \chi'_{ij} = (1/2)(tr\chi - \cos\varphi tr\chi')Y_m + \hat{\chi}_{mj}Y^j + \Pi^i_m Y'^j \hat{\chi}'_{ij}.$$

Now the assumptions of the proposition imply

$$\sup_{S'_{u'}} r'^2 \left|\frac{2}{r'} - tr\chi'\right| \leq c. \tag{14.0.38}$$

This is seen by using the formula for the Gauss curvature of the surfaces $S'_{u'}$,

$$K' = (1/4)(tr\chi')^2 - (1/2)|\hat{\chi}'|^2 - \hat{\eta}' \cdot \hat{\chi}' - (1/2)\delta' tr\chi' - \rho',$$

and applying the Gauss-Bonnet theorem as in the last step of the proof of Proposition 14.0.1. In view of the estimate

$$\sup_{S_u} |r - r'| \leq c|u|r^{-1} \tag{14.0.39}$$

given by Lemma 14.0.1, we then know that

$$|tr\chi - tr\chi'| \leq cr^{-2} \tag{14.0.40}$$

holds in U. *This then implies*

$$|\nabla\!\!\!/ \cos\varphi| \leq c|u|r^{-4}. \tag{14.0.41}$$

To estimate $\nabla\!\!\!/ Y$, *we write*

$$-\cos\varphi\theta_{mn} + \Pi^i_m\Pi^j_n\theta'_{ij} = -\cos\varphi\chi_{mn} + \Pi^i_m\Pi^j_n\chi'_{ij} - \cos\varphi\eta_{mn} + \Pi^i_m\Pi^j_n\eta'_{ij}.$$

Using 14.0.37a and 14.0.37b we obtain

$$\begin{aligned}-\cos\varphi\eta_{mn} + \Pi^i_m\Pi^j_n\eta'_{ij} = (1 - \cos\varphi)\eta_{mn} - \cos\varphi(\epsilon_m Y_n + \epsilon_n Y_m)\\ + \cos^2\varphi Y_m Y_n \delta - (\eta_{mi}Y_n + \eta_{ni}Y_m)Y^i\\ + 2\cos\varphi Y_m Y_n \epsilon_i Y^i + Y_m Y_n \eta_{ij} Y^i Y^j.\end{aligned}$$

Also, using the fact that

$$\Pi^i_m\Pi^j_n\gamma'_{ij} = \gamma_{mn} - Y_m Y_n, \tag{14.0.42}$$

we express

$$\begin{aligned}-\cos\varphi\chi_{mn} + \Pi^i_m\Pi^j_n\chi'_{ij} = (1/2)(-\cos\varphi tr\chi + tr\chi')\gamma_{mn} - (1/2)tr\chi' Y_m Y_n\\ - \cos\varphi\hat{\chi}_{mn} + \Pi^i_m\Pi^j_n\hat{\chi}'_{ij}.\end{aligned}$$

In view of 14.0.40 we conclude that in U

$$|\nabla\!\!\!/ Y| \leq cr^{-2}. \tag{14.0.43}$$

Making use of the estimates 14.0.36, 14.0.41, and 14.0.43, as well as the estimate for $\sin\varphi$ *given by Lemma 14.0.1a, we can show that there is a constant c such that the components of the second fundamental form of* Σ *relative to N satisfy*

$$\overline{\mathcal{K}^{\infty}_{[1]}} \leq c\overline{\mathcal{K}'^{\infty}_{[1]}} \tag{14.0.44}$$

where

$$\overline{\mathcal{K}^{\infty}_{[1]}} = \overline{\mathcal{K}^{\infty}_{0}} + \overline{\mathcal{K}^{\infty}_{1}}$$

and $\overline{\mathcal{K}^{\infty}_{0}}$ *is the maximum of the three quantities:*

$$r_0^{-1/2} \sup_U r^{5/2}|\delta|, \qquad r_0^{-1/2} \sup_U r^{5/2}|\epsilon|,$$

$$r_0^{-1/2} \sup_U w_{3/2} r|\hat{\eta}|,$$

while $\overline{\mathcal{K}^{\infty}_{1}}$ *is the maximum of the six quantities:*

$$r_0^{-1/2} \sup_U r^{7/2}|\nabla\!\!\!/\delta|, \qquad r_0^{-1/2} \sup_U w_{3/2} r^2|\nabla_N \delta|,$$

$$r_0^{-1/2} \sup_U r^{7/2}|\nabla\!\!\!/\epsilon|, \qquad r_0^{-1/2} \sup_U w_{3/2} r^2|\nabla\!\!\!/_N \epsilon|,$$

$$r_0^{-1/2} \sup_U w_{3/2} r^2|\nabla\!\!\!/\hat{\eta}|, \qquad r_0^{-1/2} \sup_U w_{5/2} r|\nabla\!\!\!/_N \hat{\eta}|.$$

Here

$$w_q = [r_0^{-1/2}\tau_-^{-q} + r^{-q}]^{-1}.$$

Differentiating equations 14.0.12, 14.0.33, 14.0.34, and 14.0.35 tangentially to S_u *and making use of 14.0.44, the results of Proposition 14.0.1 and the assumptions of the present proposition, we then obtain, for all* $u \in (u_0, \infty)$,

$$\begin{aligned} {}^{(loc)}\|r^3\nabla\!\!\!/\nabla_N \cos\varphi\|(u) \leq c, \qquad & {}^{(loc)}\|r^3\nabla\!\!\!/^2 \cos\varphi\|(u) \leq c \\ {}^{(loc)}\|r^2\nabla\!\!\!/\nabla\!\!\!/_N Y\|(u) \leq c, \qquad & {}^{(loc)}\|r^2\nabla\!\!\!/^2 Y\|(u) \leq c \end{aligned} \tag{14.0.45}$$

where for any $\Sigma-$ *tangent tensorfield V*

$${}^{(loc)}\|V\|(\kappa) = \left(\int_{D(\kappa)} |V|^2\right)^{1/2}$$

with $D(\kappa)$ the annular region $\kappa < u < \kappa + 1$. The estimate for $\sin\varphi$ *together with the estimates 14.0.36, 14.0.41, 14.0.43, and 14.0.45 then yield*

$$\overline{\mathcal{K}_{[1]}} + \int_U r^6(|\nabla\!\!\!/^2\delta|^2 + |\nabla\!\!\!/^2\epsilon|^2) + \int_U \{r^4 w_1^2(|\nabla\!\!\!/\nabla_N\delta|^2 + |\nabla\!\!\!/\nabla\!\!\!/_N\epsilon|^2)\} + \int_U \{r^4 w_1^2|\nabla\!\!\!/^2\hat{\eta}|^2 + r^2 w_2^2|\nabla\!\!\!/\nabla\!\!\!/_N\hat{\eta}|^2\} \leq c\overline{\mathcal{K}'_{[2]}} \tag{14.0.46}$$

where

$$\overline{\mathcal{K}_{[1]}} = \overline{\mathcal{K}_0} + \overline{\mathcal{K}_1},$$

$$r_0(\overline{\mathcal{K}_0})^2 = \int_U \{r^2(|\delta|^2 + |\epsilon|^2) + w_1^2|\hat{\eta}|^2\},$$

$$r_0(\overline{\mathcal{K}_1})^2 = \int_U \{r^4(|\nabla\!\!\!/\delta|^2 + |\nabla\!\!\!/\epsilon|^2) + r^2 w_1^2(|\nabla_N\delta|^2 + |\nabla\!\!\!/_N\epsilon|^2)\} + \int_U \{r^2 w_1^2|\nabla\!\!\!/\hat{\eta}|^2 + w_2^2|\nabla\!\!\!/_N\hat{\eta}|^2\},$$

and also

$$\begin{aligned}\overline{\mathcal{R}_{[1]}} &+ \int_U \{\tau_-^4 r^4|\nabla\!\!\!/^2\underline{\alpha}|^2 + \tau_-^6 r^2|\nabla\!\!\!/\nabla\!\!\!/_N\underline{\alpha}|^2\} \\ &+ \int_U \{\tau_-^2 r^6|\nabla\!\!\!/^2\underline{\beta}|^2 + \tau_-^4 r^4|\nabla\!\!\!/\nabla\!\!\!/_N\underline{\beta}|^2\} \\ &+ \int_U \{r^8(|\nabla\!\!\!/^2\rho|^2 + |\nabla\!\!\!/^2\sigma|^2) + r^6\tau_-^2(|\nabla\!\!\!/\rho_N|^2 + |\nabla\!\!\!/\sigma_N|^2)\} \\ &+ \int_U r^8\{|\nabla\!\!\!/^2\beta|^2 + |\nabla\!\!\!/\nabla\!\!\!/_N\beta|^2 + |\nabla\!\!\!/^2\alpha|^2 + |\nabla\!\!\!/\nabla\!\!\!/_N\alpha|^2\} \leq c\overline{\mathcal{R}'_{[2]}}\end{aligned} \tag{14.0.47}$$

where

$$\overline{\mathcal{R}_{[1]}} = \overline{\mathcal{R}_0} + \overline{\mathcal{R}_1},$$

$$(\overline{\mathcal{R}_0})^2 = \int_U \{\tau_-^4|\underline{\alpha}|^2 + \tau_-^2 r^2|\underline{\beta}|^2 + r^4(|\rho - \bar{\rho}|^2 + |\sigma - \bar{\sigma}|^2 + |\beta|^2 + |\alpha|^2)\},$$

$$(\overline{\mathcal{R}_1})^2 = \int_U \{\tau_-^4 r^2|\nabla\!\!\!/\underline{\alpha}|^2 + \tau_-^6|\nabla\!\!\!/_N\underline{\alpha}|^2\} + \int_U \{\tau_-^2 r^4|\nabla\!\!\!/\underline{\beta}|^2 + \tau_-^4 r^2|\nabla\!\!\!/_N\underline{\beta}|^2\} + \int_U \{r^6(|\nabla\!\!\!/\rho|^2 + |\nabla\!\!\!/\sigma^2) + r^4\tau_-^2(|\rho_N|^2 + |\sigma_N|^2)\} + \int_U r^6\{|\nabla\!\!\!/\beta|^2 + |\nabla\!\!\!/_N\beta|^2 + |\nabla\!\!\!/\alpha|^2 + |\nabla\!\!\!/_N\alpha|^2\},$$

and

$$\rho_N = \nabla_N \rho + (3/2)tr\theta\rho, \qquad \sigma_N = \nabla_N \sigma + (3/2)tr\theta\sigma.$$

By the global Sobolev inequalities we then have, for all $u \in (u_0, \infty)$,

$$\begin{aligned}&|r^{3/2}\tau_-^{5/2}\nabla\!\!\!/\,\underline{\alpha}|_{4,S} + |r^{5/2}\tau_-^{3/2}\nabla\!\!\!/\,\underline{\beta}|_{4,S} + |r^{7/2}\tau_-^{1/2}\nabla\!\!\!/\,\rho|_{4,S}\\ &+|r^{7/2}\tau_-^{1/2}\nabla\!\!\!/\,\sigma|_{4,S} + |r^4\nabla\!\!\!/\,\beta|_{4,S} + |r^4\nabla\!\!\!/\,\alpha|_{4,S} \leq c\overline{\mathcal{R}'_{[2]}}. \end{aligned} \tag{14.0.48}$$

We now go back to the propagation equation for $tr\chi$ (equation 14.0.4a) and by applying the commutation formula of Corollary 3.2.3.1 to the commutator $[\nabla_N, \Delta\!\!\!/]$ we deduce

$$\nabla_N \Delta\!\!\!/ tr\chi + 2tr\theta \Delta\!\!\!/ tr\chi = -F_2 \tag{14.0.49a}$$

where

$$\begin{aligned}F_2 = {} & 2(\hat{\chi}+\hat{\eta})\cdot \hat{\nabla\!\!\!/}^2 tr\chi + \delta\Delta\!\!\!/ tr\chi + 2|\nabla\!\!\!/ tr\chi|^2\\ &+(2P - \nabla\!\!\!/\delta)\cdot\nabla\!\!\!/ tr\chi + 2tr\chi(\zeta-\epsilon)\cdot\nabla\!\!\!/ tr\chi + 2(\zeta-\epsilon)\cdot(\hat{\chi}+\hat{\eta})\cdot\nabla\!\!\!/ tr\chi\\ &+(f-\overline{f}+|\zeta-\epsilon|^2)(\frac{1}{2}(tr\chi)^2 + F_0 + \overline{f}) + 2(\zeta-\epsilon)\cdot\nabla\!\!\!/ F_0 + \Delta\!\!\!/ F_0\end{aligned} \tag{14.0.49b}$$

and

$$P_m = \Pi^i_m N^j R_{ij} = -(1/2)(\underline{\beta}_m + \beta_m) + \hat{\eta}_m{}^n\epsilon_n + (1/2)\delta\epsilon_m.$$

Since

$$\Delta\!\!\!/ F_0 = \hat{\chi}\cdot\Delta\!\!\!/\hat{\chi} + |\nabla\!\!\!/\hat{\chi}|^2 + 2\zeta\cdot\Delta\!\!\!/\zeta + 2|\nabla\!\!\!/\zeta|^2,$$

equation 14.0.49a is to be considered in conjunction with the equations

$$\Delta\!\!\!/\hat{\chi} - 2K\hat{\chi} = \hat{\nabla\!\!\!/}^2 tr\chi + 2\mathcal{D}\!\!\!/_2^*\left(\beta + \epsilon\cdot\hat{\chi} - \frac{1}{2}\epsilon tr\chi\right) \tag{14.0.50}$$

$$\Delta\!\!\!/\zeta - K\zeta = -\nabla\!\!\!/ f - {}^*\nabla\!\!\!/(\sigma - \hat{\eta}\wedge\hat{\chi}). \tag{14.0.51}$$

The first of these equations is obtained by applying the operator $\mathcal{D}\!\!\!/_2^*$ to the null Codazzi equation 14.0.6 and using the identity

$$\mathcal{D}\!\!\!/_2^*\mathcal{D}\!\!\!/_2 = -\frac{1}{2}\Delta\!\!\!/ + K,$$

while the second is obtained by applying the operator $\mathcal{D}\!\!\!/_1^*$ to the $\zeta-$ system 14.0.2a, 14.0.2b and using the identity

$$\mathcal{D}\!\!\!/_1^*\mathcal{D}\!\!\!/_1 = -\Delta\!\!\!/ + K.$$

Lemma 14.0.2 with $p = 4, \lambda_0 = 2, \lambda_1 = \frac{7}{2}$ *applied to 14.0.49a yields*

$$|r^{7/2} \not\Delta tr\chi|_{4,S}(u) \leq c\left(|r_0^{7/2} \not\Delta tr\chi|_{4,S_0} + \left|\int_0^u |r^{7/2} F_2|_{4,S}(u')du'\right|\right). \tag{14.0.52}$$

From 14.0.50 and 14.0.51 and Proposition 2.2.2 we obtain:

$$|r^{7/2} \not\nabla^2 tr\chi|_{4,S} \leq c|r^{7/2} \not\Delta tr\chi|_{4,S}, \tag{14.0.53a}$$

$$|r^{7/2} \not\Delta \hat{\chi}|_{4,S} \leq c(|r^2 \hat{\chi}|_{\infty,S} + |r^{7/2} \not\Delta tr\chi|_{4,S} + \overline{\mathcal{K}'^{\infty}_{[1]}} + \overline{\mathcal{R}'_{[2]}}), \tag{14.0.53b}$$

$$r_0^{-1/2}|r^4 \not\Delta \zeta|_{4,S} \leq c(r_0^{-1/2}|r^{5/2}\zeta|_{\infty,S} + r_0^{-1/2}|r^4 \not\nabla f|_{4,S} + \overline{\mathcal{K}'^{\infty}_{[1]}} + \overline{\mathcal{R}'_{[2]}}), \tag{14.0.53c}$$

and from 14.0.15

$$r_0^{-1/2}|r^4 \not\nabla f|_{4,S} \leq c\left(|r^{5/2} \not\nabla tr\chi|_{4,S} + |r^{5/2} \not\nabla \hat{\chi}|_{4,S} + \overline{\mathcal{K}'^{\infty}_{[1]}} + \overline{\mathcal{R}'_{[2]}}\right). \tag{14.0.53d}$$

Using the above estimates as well as the results of Proposition 14.0.1 and the fact that since by the Sobolev inequality on S_u

$$|r^3 \not\nabla tr\chi|_{\infty,S} \leq c\left(|r^{5/2} \not\nabla tr\chi|_{4,S} + |r^{7/2} \not\nabla^2 tr\chi|_{4,S}\right),$$

we have

$$|r^{7/2}|\not\nabla tr\chi|^2|_{4,S} \leq cr^{-2}|r^{5/2} \not\nabla tr\chi|_{4,S}\left(r^{5/2} \not\nabla tr\chi|_{4,S} + |r^{7/2} \not\Delta tr\chi|_{4,S}\right),$$

we obtain

$$\begin{aligned} |r^{7/2} F_2|_{4,S} \leq\ & c\left(r^{-1}\tau_-^{-3/2} + r^{-2}\right)|r^{7/2} \not\Delta tr\chi|_{4,S} \\ & + c\left(r^{-1}\tau_-^{-3/2} + r^{-2} + r_0^{1/2} r^{-3/2}\right)|r_0^{5/2} \not\nabla tr\chi|_{4,S_0} \\ & + c\left(r^{-1}\tau_-^{-3/2} + r^{-2} + r_0^{1/2} r^{-3/2}\right)\left(\overline{\mathcal{K}'^{\infty}_{[1]}} + \overline{\mathcal{R}'_{[2]}}\right). \end{aligned} \tag{14.0.54}$$

Substituting this estimate in 14.0.52 and applying the Gronwall lemma to the resulting integral inequality, we obtain the conclusion regarding $|r^{7/2} \not\Delta tr\chi|_{4,S}$. *The conclusions regarding*

$$|r^{7/2} \not\nabla^2 tr\chi|_{4,S}, \qquad |r^{7/2} \not\nabla^2 \hat{\chi}|_{4,S}, \qquad r_0^{-1/2}|r^4 \not\nabla^2 \zeta|_{4,S}$$

then follow from 14.0.53a, 14.0.53b, 14.0.53c, 14.0.53d, and the results of Chapter 2.

We now differentiate equation 14.0.49a tangentially to S_u to obtain

$$\nabla\!\!\!/_N \nabla\!\!\!/ \Delta\!\!\!/ tr\chi + \frac{5}{2}\nabla\!\!\!/ \Delta\!\!\!/ tr\chi = -F_3 \tag{14.0.55a}$$

where

$$\begin{aligned}F_3 = (\hat{\chi} + \hat{\eta}) \cdot \nabla\!\!\!/ \Delta\!\!\!/ tr\chi + 2(\nabla\!\!\!/ tr\chi - \nabla\!\!\!/ \delta)\Delta\!\!\!/ tr\chi \\ + (\zeta - \epsilon)(2tr\theta \Delta\!\!\!/ tr\chi + F_2) + \nabla\!\!\!/ F_2.\end{aligned} \tag{14.0.55b}$$

Applying to 14.0.55a Lemma 14.0.2 with $p = 2$, $\lambda_0 = \frac{5}{2}$, $\lambda_1 = 4$ yields

$$|r^4 \nabla\!\!\!/ \Delta\!\!\!/ tr\chi|_{2,S}(u) \leq c\left(|r_0^4 \nabla\!\!\!/ \Delta\!\!\!/ tr\chi|_{2,S_0} + \left|\int_0^u |r^4 F_3|_{2,S}(u')du'\right|\right). \tag{14.0.55c}$$

By 14.0.50, 14.0.51, Proposition 2.2.2, and the previous step,

$$|r^4 \nabla\!\!\!/^3 tr\chi|_{2,S} \leq c(|r^4 \nabla\!\!\!/ \Delta\!\!\!/ tr\chi|_{2,S} + |r^3 \Delta\!\!\!/ tr\chi|_{2,S}) \tag{14.0.56a}$$

$$\begin{aligned}|r^4 \nabla\!\!\!/ \Delta\!\!\!/ \hat{\chi}|_{2,S} \leq c\Big(&|r^4 \nabla\!\!\!/ \Delta\!\!\!/ tr\chi|_{2,S} + |r^3 \nabla\!\!\!/^2 \epsilon|_{2,S} + |r^4 \nabla\!\!\!/^2 \beta|_{2,S} \\ &+ |r_0^{7/2} \Delta\!\!\!/ tr\chi|_{4,S_0} + |r_0^{5/2} \nabla\!\!\!/ tr\chi|_{4,S_0} \\ &+ \overline{\mathcal{K}'^{\infty}_{[1]}} + \overline{\mathcal{R}'_{[2]}}\Big)\end{aligned} \tag{14.0.56b}$$

$$\begin{aligned}|r^4 \nabla\!\!\!/ \Delta\!\!\!/ \zeta|_{2,S} \leq c\Big(&|r^4 \nabla\!\!\!/^2 f|_{2,S} + |r^2 \nabla\!\!\!/^2 \hat{\eta}|_{2,S} + |r^4 \nabla\!\!\!/^2 \sigma|_{2,S} \\ &+ |r_0^{7/2} \Delta\!\!\!/ tr\chi|_{4,S_0} + |r_0^{5/2} \nabla\!\!\!/ tr\chi|_{4,S_0} \\ &+ \overline{\mathcal{K}'^{\infty}_{[1]}} + \overline{\mathcal{R}'_{[2]}}\Big);\end{aligned} \tag{14.0.56c}$$

and from 14.0.15,

$$\begin{aligned}|r^4 \nabla\!\!\!/^2 f|_{2,S} \leq c\Big(&|r^3 \nabla\!\!\!/^2 \delta|_{2,S} + |r^2 \nabla\!\!\!/^2 \hat{\eta}|_{2,S} + |r^4 \nabla\!\!\!/^2 \rho|_{2,S} \\ &+ |r_0^{7/2} \Delta\!\!\!/ tr\chi|_{4,S_0} + |r_0^{5/2} \nabla\!\!\!/ tr\chi|_{4,S_0} \\ &+ \overline{\mathcal{K}'^{\infty}_{[1]}} + \overline{\mathcal{R}'_{[2]}}\Big).\end{aligned} \tag{14.0.56d}$$

Using this, we can estimate:

$$\begin{aligned}|r^4 F_3|_{2,S} \leq c\big[&(r^{-1}\tau_-^{-3/2} + r^{-2})|r^4 \nabla\!\!\!/ \Delta\!\!\!/ tr\chi|_{2,S} + r^{-2}G^{1/2} + r^{-2}H^{1/2} \\ &+ (r^{-1}\tau_-^{-3/2} + r_0^{1/2} r^{-3/2})(|r_0^{7/2} \Delta\!\!\!/ tr\chi|_{4,S_0} + |r_0^{5/2} \nabla\!\!\!/ tr\chi|_{4,S_0} \\ &+ \overline{\mathcal{K}'^{\infty}_{[1]}} + \overline{\mathcal{R}'_{[2]}})\big]\end{aligned} \tag{14.0.57}$$

where

$$G = |r^3 \nabla\!\!\!/^2 \delta|_{2,S} + |r^3 \nabla\!\!\!/^2 \epsilon|_{2,S} + |r^2 \nabla\!\!\!/^2 \hat{\eta}|_{2,S}$$
$$H = |r^4 \nabla\!\!\!/^2 \beta|_{2,S} + |r^4 \nabla\!\!\!/^2 \rho|_{2,S} + |r^4 \nabla\!\!\!/^2 \sigma|_{2,S}.$$

Substituting this estimate in 14.0.55c and applying the Gronwall lemma to the resulting integral inequality, we obtain

$$|r^4 \nabla\!\!\!/ \triangle\!\!\!/ tr\chi|_{2,S} \leq c\Big\{ |r_0^4 \nabla\!\!\!/ \triangle\!\!\!/ tr\chi|_{2,S_0} + \int_{u_0}^{\infty} r^{-2} G^{1/2} du + \int_{u_0}^{\infty} r^{-2} H^{1/2} du$$
$$+ |r_0^{7/2} \triangle\!\!\!/ tr\chi|_{4,S_0} + |r_0^{5/2} \nabla\!\!\!/ tr\chi|_{4,S_0}$$
$$+ \overline{\mathcal{K}_{[1]}^{\prime\infty}} + \overline{\mathcal{R}_{[2]}^{\prime}} \Big\}. \tag{14.0.58}$$

In view of the fact that, by 14.0.46, 14.0.47,

$$\int_{u_0}^{\infty} r^{-2} G^{1/2} du \leq c\overline{\mathcal{K}_{[2]}^{\prime}}, \qquad \int_{u_0}^{\infty} r^{-2} H^{1/2} du \leq c\overline{\mathcal{R}_{[2]}^{\prime}},$$

the remaining conclusions of the proposition follow.

We shall now derive an equation for the function ω on the last slice Σ_{t_*}. As

$$\omega = -\nabla_N \log a - \delta,$$

we have to apply ∇_N to equation 14.0.1b. It is advantageous, however, to apply instead $a\nabla_N = \partial/\partial u$, since the commutator $[a\nabla_N, \triangle\!\!\!/]$ applied to functions ϕ has the simple expression

$$[a\nabla_N, \triangle\!\!\!/]\phi = -2a\hat{\theta} \cdot \nabla\!\!\!/^2 \phi - atr\theta \triangle\!\!\!/ \phi - 2 d\!\!\!/ iv(a\hat{\theta}) \cdot \nabla\!\!\!/ \phi. \tag{14.0.59}$$

Also if U is an S-tangent 1- form,

$$a\nabla_N(d\!\!\!/ iv U) - d\!\!\!/ iv(a\nabla\!\!\!/_N U) = -2a\hat{\theta} \cdot \nabla\!\!\!/ U - atr\theta d\!\!\!/ iv U - 2 d\!\!\!/ iv(a\hat{\theta}) \cdot U. \tag{14.0.60}$$

To compute $\nabla_N f$ we use the following general formula for the normal variation of the Gauss curvature of a surface:

$$a(\nabla_N K + tr\theta K) = d\!\!\!/ iv\, d\!\!\!/ iv(a\hat{\theta}) - \frac{1}{2} \triangle\!\!\!/ (atr\theta). \tag{14.0.61}$$

Together with the propagation equation, 14.0.4a this gives:

$$\nabla_N f + tr\theta f = a^{-1}(d\!\!\!/ iv\, d\!\!\!/ iv(a\hat{\theta}) - \frac{1}{2} \triangle\!\!\!/ (atr\theta)) + h$$

where

$$h = \frac{1}{2}tr\chi\Big(\frac{1}{2}\delta tr\chi + \frac{1}{2}|\hat{\chi}|^2 + |\zeta|^2 + \overline{f}\Big). \tag{14.0.62}$$

On the other hand, from the equation

$$a\nabla\!\!\!/_N\epsilon = -\mathrm{d}\!\!\!/\mathrm{iv}\,(a\hat{\eta}) + \frac{1}{2}\nabla\!\!\!/(a\delta) + \delta\nabla\!\!\!/ a - a\hat{\theta}\cdot\epsilon - \frac{3}{2}atr\theta\epsilon$$

(see 11.2.13c, 11.2.14a) and the commutation formula 14.0.60, we obtain

$$\nabla\!\!\!/_N\mathrm{d}\!\!\!/\mathrm{iv}\,\epsilon + tr\theta\mathrm{d}\!\!\!/\mathrm{iv}\,\epsilon = a^{-1}\Big(-\mathrm{d}\!\!\!/\mathrm{iv}\,\mathrm{d}\!\!\!/\mathrm{iv}\,(a\hat{\eta}) + \frac{1}{2}\Delta\!\!\!\!/(a\delta)\Big) + a^{-1}\mathrm{d}\!\!\!/\mathrm{iv}\,aA$$

where

$$A = \delta\nabla\!\!\!/\log a - 3\hat{\theta}\cdot\epsilon - \frac{3}{2}atr\theta\epsilon.$$

Hence,

$$\begin{aligned}\nabla_N(f - \mathrm{d}\!\!\!/\mathrm{iv}\,\epsilon) + tr\theta(f - \mathrm{d}\!\!\!/\mathrm{iv}\,\epsilon) &= -a^{-1}\Big(\mathrm{d}\!\!\!/\mathrm{iv}\,\mathrm{d}\!\!\!/\mathrm{iv}\,(a\underline{\hat{\chi}}) - \frac{1}{2}\Delta\!\!\!\!/(atr\underline{\chi})\Big)\\ &\quad - a^{-1}\Delta\!\!\!\!/(a\delta) - a^{-1}\mathrm{d}\!\!\!/\mathrm{iv}\,aA + h.\end{aligned}$$

We now use the conjugate null Codazzi equation

$$\mathrm{d}\!\!\!/\mathrm{iv}\,\underline{\hat{\chi}} - \frac{1}{2}\nabla\!\!\!/ tr\underline{\chi} - \epsilon\cdot\underline{\hat{\chi}} + \frac{1}{2}\epsilon tr\underline{\chi} = \underline{\beta}$$

to express

$$\mathrm{d}\!\!\!/\mathrm{iv}\,(a\underline{\hat{\chi}}) - \frac{1}{2}\nabla\!\!\!/(atr\underline{\chi}) = a\Big(\underline{\beta} + \underline{\hat{\chi}}\cdot\zeta - \frac{1}{2}tr\underline{\chi}\zeta\Big),$$

and, substituting, we obtain

$$\nabla_N(f - \mathrm{d}\!\!\!/\mathrm{iv}\,\epsilon) + tr\theta(f - \mathrm{d}\!\!\!/\mathrm{iv}\,\epsilon) = -a^{-1}\Delta\!\!\!\!/(a\delta) - a^{-1}\mathrm{d}\!\!\!/\mathrm{iv}\,aA' + h$$

where

$$A' = \underline{\beta} + \underline{\hat{\chi}}\cdot\zeta - \frac{1}{2}tr\underline{\chi}\zeta + A.$$

Also, in view of 14.0.4a and the fact that

$$\overline{f} = \frac{1}{r^2} - \frac{1}{4}\overline{(tr\chi)^2},$$

we have

$$\frac{d\overline{f}}{du} + \overline{atr\theta f} = -\overline{ah}.$$

Applying ∇_N to equation 14.0.1b, making use of the commutation formula 14.0.59, and taking into account the previous expressions, we obtain the following equation for the function ω on Σ_{t_*}:

$$\Delta\!\!\!/(a\omega) = (atr\theta - \overline{atr\theta})\overline{f} - (ah - \overline{ah}) + d\!\!\!/iv(aB) \tag{14.0.63a}$$

where

$$B = \underline{\beta} - 3\hat{\chi}\cdot\zeta - 4\hat{\eta}\cdot\zeta + \frac{1}{2}tr\chi\zeta$$
$$- \hat{\chi}\cdot\epsilon - \hat{\eta}\cdot\epsilon - \frac{3}{2}tr\chi\epsilon + \frac{1}{2}\delta\epsilon. \tag{14.0.63b}$$

We note that

$$\overline{a\omega} = \overline{(atr\theta - \overline{atr\theta})\log a} - \overline{a\delta}. \tag{14.0.64}$$

Proposition 14.0.3 *If the assumptions of Proposition 14.0.1 are satisfied, then, for all $u \in (u_0, \infty)$,*

$$|r^{1/2-2/p}w_{3/2}a\omega|_{p,S}(u) \le c\Big(|r_0^{3-2/p}\nabla\!\!\!/ tr\chi|_{p,S_0} + \overline{\mathcal{K}_0'^{\infty}} + \overline{\mathcal{R}_0'^{\infty}}\Big)$$
$$|r^{3/2-2/p}w_{3/2}\nabla\!\!\!/(a\omega)|_{p,S}(u) \le c\Big(|r_0^{3-2/p}\nabla\!\!\!/ tr\chi|_{p,S_0} + \overline{\mathcal{K}_0'^{\infty}} + \overline{\mathcal{R}_0'^{\infty}}\Big).$$

Moreover, if the assumptions of Proposition 14.0.2 are satisfied as well, then, for all $u \in (u_0, \infty)$,

$$|r^2 w_{3/2}\nabla\!\!\!/^2(a\omega)|_{4,S}(u) \le c\Big(|r_0^{5/2}\nabla\!\!\!/ tr\chi|_{4,S_0} + \overline{\mathcal{K}_{[1]}'^{\infty}} + \overline{\mathcal{R}_{[2]}'}\Big)$$

and

$$r_0^{-1/2}||r^3 w_1\nabla\!\!\!/^3(a\omega)||_{L^2(U)} \le c\Big(|r_0^{5/2}\nabla\!\!\!/ tr\chi|_{4,S_0} + |r_0^3\Delta\!\!\!/ tr\chi|_{2,S_0}$$
$$+ \overline{\mathcal{K}_{[1]}'^{\infty}} + \overline{\mathcal{K}_{[2]}'} + \overline{\mathcal{R}_{[2]}'}\Big).$$

Proof 14.0.5 *In view of equation 14.0.63b, the first assertion follows by applying Corollary 2.3.1.2 to $a\omega - \overline{a\omega}$ and using 14.0.44 and the results of Proposition 14.0.1. The remaining assertions follow by applying Corollary 2.3.1.1 and using in addition 14.0.46, 14.0.47 as well as the results of Proposition 14.0.2.*

Differentiating equations 14.0.12 and 14.0.33 normally to S_u and making use of the first part of Proposition 14.0.3 we obtain

$$^{(loc)}||r^3\nabla_N^2 \cos\varphi||(u) \le c, \qquad ^{(loc)}||r\tau_-\nabla\!\!\!/_N^2 Y||(u) \le c, \tag{14.0.65}$$

which, together with 14.0.45, 14.0.46, and 14.0.47, yield

$$\overline{\mathcal{K}_{[2]}} \le c\overline{\mathcal{K}_{[2]}'}, \qquad \overline{\mathcal{R}_{[2]}} \le c\overline{\mathcal{R}_{[2]}'}. \tag{14.0.66}$$

Finally differentiating equation 14.0.63b normally to S_u, using the commutation formula 14.0.59, and taking into account the formulas

$$\nabla\!\!\!/_N \hat{\chi}_{ij} + tr\theta\hat{\chi}_{ij} = -\frac{1}{2}(\nabla\!\!\!/_i\zeta_j + \nabla\!\!\!/_j\zeta_i - \gamma_{ij} d\!\!/iv\zeta) - \zeta_i\zeta_j + \frac{1}{2}\gamma_{ij}|\zeta|^2 \\ - \frac{3}{2}\delta\hat{\chi}_{ij} - \frac{1}{2}tr\chi\hat{\eta}_{ij} - \frac{1}{2}\alpha_{ij},$$

$$\nabla\!\!\!/_N \zeta_i + tr\theta\zeta_i = -\nabla\!\!\!/_i\omega - \zeta_i\omega + \epsilon_i\omega - \hat{\chi}_{ij}\zeta_j - 2\hat{\eta}_{ij}\zeta_j - \delta\zeta_i \\ - \hat{\chi}_{ij}\epsilon_j + \delta\epsilon_i + \frac{1}{2}\underline{\beta}_i - \frac{1}{2}\beta_i,$$

we deduce the following:

Proposition 14.0.4 *If the assumptions of Proposition 14.0.2 are satisfied, then for all $u \in (u_0, \infty)$*

$$|w_{5/2}\nabla_N(a\omega)|_{4,S}(u) + |w_{5/2}r\nabla\!\!\!/\nabla_N(a\omega)|_{4,S}(u) \le c|r_0^{5/2}\nabla\!\!\!/ tr\chi|_{4,S_0} \\ + c\overline{\mathcal{K}'^{\infty}_{[1]}} + \overline{\mathcal{R}'_{[2]}}$$

and

$$r_0^{-1/2}\|r^2 w_2 \nabla\!\!\!/^2\nabla_N(a\omega)\|_{L^2(U)} \le c\Big(|r_0^{5/2}\nabla\!\!\!/ tr\chi|_{4,S_0} + |r_0^3 \Delta\!\!\!/ tr\chi|_{2,S_0} \\ + \overline{\mathcal{K}'^{\infty}_{[1]}} + \overline{\mathcal{K}'_{[2]}} + \overline{\mathcal{R}'_{[2]}}\Big).$$

We now summarize the results of this chapter in the following:

Theorem 14.0.1 *Suppose that the surface $S_0 = S_{t_*,0}$ satisfies the assumptions* $\mathbf{S}_0, \mathbf{S}_1, \mathbf{S}_2$ *of Propositions 14.0.1, 14.0.2.*

Let u' be a smooth function that vanishes on S_0 and has no critical points, defined in the tubular neighborhood $U' = \bigcup_{u' \in (u'_0, \infty)} S'_{u'}$ of S_0 which includes the whole exterior of S_0 and extends in the interior up to a level surface $S'_{u'_0}$ of u' of area $A'_0 = 4\pi(r_0/4)^2$. Assume that u' verifies the hypotheses $\mathbf{U'}_0, \mathbf{U'}_1, \mathbf{U'}_2$ *of Propositions 14.0.1, 14.0.2. Assume also, with the notation introduced there,*[2]

$$\overline{\mathcal{K}'^{\infty}_0} \le \varepsilon_0, \qquad \overline{\mathcal{R}'^{\infty}_0} \le \varepsilon_0$$

and

$$\overline{\mathcal{K}'^{\infty}_{[1]}} \le c, \qquad \overline{\mathcal{R}'^{\infty}_{[2]}} \le c.$$

[2] These are precisely the assumptions $\mathbf{H}_0, \mathbf{H}_1, \mathbf{H}_2$ of Propositions 14.0.1 and 14.0.2.

Then

$$ {}^e\mathcal{O}_{[3]}{}^* \leq c\left(\mathcal{S}(0)^* + \overline{\mathcal{R}'_{[2]}} + \overline{\mathcal{K}'_{[2]}} + \overline{\mathcal{K}'^{\infty}_{[1]}}\right) $$

where ${}^e\mathcal{O}_{[3]}{}^*$ *is the norm introduced in Theorem 13.1.1 and*[3]

$$ \mathcal{S}(0)^* = |r_0^{5/2}\nabla\!\!\!/\, tr\chi|_{4,S_0} + |r_0^{7/2}\triangle\!\!\!/\, tr\chi|_{4,S_0} + |r_0^4\nabla\!\!\!/\,\triangle\!\!\!/\, tr\chi|_{2,S_0}. $$

We note that the quantity $\tilde{\mu}$, which enters the norm ${}^e\mathcal{O}_{[3]}{}^*$, is, in view of the first of equations 13.1.12c together with 14.0.2a, given by

$$ \tilde{\mu} = \delta\mathrm{tr}\chi - |\zeta|^2 + \overline{f}. $$

Thus the required estimates for this quantity are immediate consequences of Propositions 14.0.1 and 14.0.2.

Finally, in the end of this chapter we prove a proposition concerning the top derivatives of $a\omega$ relative to ∇_N. This is necessary in order to estimate the quantity ${}^e\mathcal{N}^*$ introduced in Proposition 13.1.12. We will prove the following:

Proposition 14.0.5 *If all the assumptions of Theorem 14.0.1 are satisfied, then, for all* $u \in (u_0, \infty)$,

$$ r_0^{-1/2}\|rw_3\nabla_N^2(a\omega)\|_{L^2(U)} \leq c\left(\mathcal{S}(0)^* + \overline{\mathcal{R}'_{[2]}} + \overline{\mathcal{K}'_{[2]}} + \overline{\mathcal{K}'^{\infty}_{[1]}}\right) $$
$$ r_0^{-1/2}\|r^2w_3\nabla\!\!\!/\,\nabla_N^2(a\omega)\|_{L^2(U)} \leq c\left(\mathcal{S}(0)^* + \overline{\mathcal{R}'_{[2]}} + \overline{\mathcal{K}'_{[2]}} + \overline{\mathcal{K}'^{\infty}_{[1]}}\right) $$
$$ r_0^{-1/2}\|rw_4\nabla_N^3(a\omega)\|_{L^2(U)} \leq c\left(\mathcal{S}(0)^* + \overline{\mathcal{R}'_{[2]}} + \overline{\mathcal{K}'_{[2]}} + \overline{\mathcal{K}'^{\infty}_{[1]}}\right). $$

Proof 14.0.6 *The proof is similar to, and much simpler than, Proposition 13.1.13. We have to differentiate the equation 14.0.63a up to three times in the normal direction. The only dangerous terms are generated at the top level, when we take three derivatives in the normal direction. This leads to an equation of the form*

$$ \triangle\!\!\!/\,\nabla_N^3(a\omega) = a\,\mathrm{d}\!\!\!/\mathrm{iv}\,(\nabla\!\!\!/_N^3\underline{\beta}) + l.o.t. $$

The terms that we denote by l.o.t. involve, at worst, terms that involve four derivatives of $\hat{\eta}, \epsilon, \delta$, *at least one of which is in the tangential direction. On the other hand, in view of the null Bianchi equations 7.3.11h, 7.3.11i (see Proposition 7.3.2) we have*

$$ \underline{\beta}_4 = \mathcal{D}\!\!\!\!/_1^*(\rho, \sigma) + l.o.t. $$
$$ \underline{\beta}_3 = -\mathrm{d}\!\!\!/\mathrm{iv}\,\underline{\alpha} + l.o.t. $$

[3] Note that $\mathcal{S}(0)^* = \mathcal{S}(0)(t_*)$ where the latter was introduced in Proposition 13.1.12.

Hence, we can express $\nabla\!\!\!/_N \underline{\beta}$ *in terms of the tangential derivatives of* $\underline{\alpha}$ *and* ρ, σ. *Therefore, at the top level* $\nabla_N^3(a\omega)$ *satisfies an equation of the type*

$$\Delta\!\!\!/ \nabla_N^3(a\omega) = F$$

where F *contains, at worst, second tangential derivatives of two derivatives of the curvature and one tangential derivative of three derivatives of the second fundamental form. Using the estimates of Corollary 2.3.1.2, we derive an estimate for the* L^2 *norm of* $\nabla_N^3(a\omega) - \overline{\nabla_N^3(a\omega)}$. *We then estimate* $\overline{\nabla_N^3(a\omega)}$ *by differentiating the equation 14.0.64 three times in the normal direction.*

CHAPTER 15

The Matching

In this chapter we shall match the exterior optical function u_E with the interior optical function u_I to obtain a globally defined function u. Let h_1 be a C^∞ function on $\Re$ such that

$$h_1 = \begin{cases} = 1; & t \leq \frac{7}{20} \\ = 0; & t \geq \frac{8}{20}. \end{cases} \tag{15.0.1}$$

Let

$$\vartheta = \frac{u_E + 1 + t}{1 + t}, \qquad f = h \circ \vartheta \tag{15.0.2a}$$

and let

$$u = (1 - f_1)u_E + f_1 u_I. \tag{15.0.2b}$$

As the support of $1 - f_1$ is included in the domain of definition of u_E and the support of f_1 is included in the domain of definition of u_I, the function u is defined globally. To estimate the Hessian of u, we have to compare u_E and u_I in the matching region

$$u_E \in [-(13/20)(1 + t), -(12/20)(1 + t)].$$

We do this by comparing each of u_E,u_I to the distance function from the point O_t. This is relatively easy for u_I but considerably more difficult for u_E.

We begin with the following preliminary lemma:

Lemma 15.0.1 *Suppose that along the cone C_0 and for all $t \in [-1, t_*]$ the lapse function ϕ of the maximal foliation and the restriction η to $S_{t,0}$ of the second fundamental form of the maximal hypersurfaces satisfy*

$$\sup_{S_{t,0}} |\phi - 1| \leq B(2 + t)^{-1},$$

$$\sup_{S_{t,0}} |\hat{\eta}| \leq B(2 + t)^{-1}, \qquad \sup_{S_{t,0}} |\delta| \leq B(2 + t)^{-2},$$

while the sectional curvature of space-time in the plane tangent to $S_{t,0}$, that is, $-\rho$, satisfies

$$\sup_{S_{t,0}} |\rho| \leq B(2+t)^{-3}.$$

Suppose also that for all $t \in [-1, t_*]$ the null second fundamental form of $S_{t,0}$ satisfies

$$osc_{S_{t,0}} tr\chi \leq A(2+t)^{-2}, \qquad \sup_{S_{t,0}} |\hat{\chi}| \leq A(2+t)^{-2}.$$

Then, setting

$$r_0(t) = \sqrt{Area(S_{t,0})/4\pi}$$

for all $t \in [-1, t_*]$, we have

$$|r_0(t) - 1 - t| \leq c(A+B)\log(2+t)$$

provided that A and B are sufficiently small.

Proof 15.0.1 *We recall the Gauss equation 13.1.2n of the embedding of a surface, lying in Σ_t, in space-time:*

$$K - \frac{1}{4}(tr\chi)^2 + \frac{1}{2}\delta tr\chi = -\frac{1}{2}|\hat{\chi}|^2 - \hat{\eta} \cdot \hat{\chi} - \rho. \tag{15.0.3}$$

Our assumptions for the surface $S_{t,0}$ then imply

$$K - \frac{1}{4}(tr\chi)^2_M \leq \frac{cB}{(2+t)^2}(tr\chi)_M + \frac{c(A+B)}{(2+t)^3}$$

$$\frac{1}{4}(tr\chi)^2_m - K \leq \frac{cB}{(2+t)^2}(tr\chi)_M + \frac{c(A+B)}{(2+t)^3}$$

where $(tr\chi)_m$ and $(tr\chi)_M$ are respectively the minimum and maximum values of $tr\chi$ on $S_{t,0}$. Since

$$osc_{S_{t,0}} tr\chi = (tr\chi)_M - (tr\chi)_m \leq A(2+t)^{-2}$$

and by the Gauss-Bonnet formula

$$\overline{K} = \frac{1}{r_0^2},$$

these imply

$$\frac{1}{4}\left[(tr\chi)_M - \frac{A}{(2+t)^2}\right]^2 - \frac{cB}{(2+t)^2}(tr\chi)_M - \frac{c(A+B)}{(2+t)^3} \leq \frac{1}{r_0^2},$$

$$\frac{1}{4}(tr\chi)^2_M + \frac{cB}{(2+t)^2}(tr\chi)_M + \frac{c(A+B)}{(2+t)^3} \geq \frac{1}{r_0^2},$$

which, provided that A and B are sufficiently small, yield

$$|tr\chi - \frac{2}{r_0}| \leq \frac{c(A+B)}{(2+t)^2}.$$

The result then follows by integrating the formula

$$\frac{dr_0}{dt} = \frac{r_0}{2}\overline{\phi tr\chi},$$

using the assumption for ϕ.

In the following, we denote by U the interior of the cone C_0 in the past of the final slice Σ_{t_*} and by

$$U_t = U \bigcap \Sigma_t.$$

We denote by O the integral curve of T through the vertex of C_0 and by O_t the point $O \bigcap \Sigma_t$ on this curve.

Lemma 15.0.2 *Let the exterior optical function satisfy*

$$\sup_{U_t \bigcap E} |\zeta| \leq A(2+t)^{-2}$$

for all $t \in [0, t_*]$. Suppose that the lapse function of the maximal foliation satisfies in U

$$\sup_U (2+t)|\phi - 1| \leq B, \qquad \sup_U (2+t)^2|\nabla\phi| \leq B$$

and

$$\phi \leq 1$$

everywhere.[1] *Suppose also that the components, relative to the exterior optical function, of the second fundamental form of the maximal foliation satisfy in $U_t \bigcap E$ for all $t \in [0, t_*]$*

$$|\delta|, |\epsilon| \leq B(2+t)^{-2}, \qquad |\hat{\eta}| \leq B[\tau_-^{-3/2}(2+t)^{-1} + (2+t)^{-2}],$$

while in $U \setminus E$ we have

$$\sup_{U\setminus E} (2+t)^2|k| \leq B.$$

[1] This assumption is verified in our case in view of the maximum principle applied to the lapse equation (see Chapter 12).

Then, provided that A and B are sufficiently small, for each $t \in [0, t_*]$ the surface $S_{t,0}$ is contained within the geodesic ball in Σ_t of radius

$$d_0(t) = 1 + t + cB \log(2 + t)$$

centered at O_t.

Proof 15.0.2 *Let f_τ be the 1-parameter group of diffeomorphisms generated by the vectorfield $\frac{\partial}{\partial t} = \phi T$. Then $f_\tau|_{\Sigma_t}$ is a diffeomorphism of Σ_t onto $\Sigma_{t+\tau}$. Consider a given generator Γ of C_0 and let, for each $t_1 \in [-1, t_*]$, ${}^{t_1}\Gamma'$ be the projection to Σ_{t_1}, given by the integral curves of $\frac{\partial}{\partial t}$ of the segment of Γ that lies in the past of Σ_{t_1}:*

$${}^{t_1}\Gamma' = \{f_{t_1 - t}(\Gamma_t) : t \in [-1, t_1]\}$$

where Γ_t is the point $\Gamma \bigcap \Sigma_t$. Let Γ be parametrized by t and let V_t be the projection to Σ_t of the tangent vector of Γ at Γ_t. We have

$$V = \phi N.$$

The arc length of ${}^{t_1}\Gamma'$ is given by

$$L(t_1) = \int_{-1}^{t_1} \sqrt{G(t_1, t)} dt \tag{15.0.4a}$$

where

$$G(t_1, t) = g_{t_1}(f'_{t_1 - t} V_t, f'_{t_1 - t} V_t) \tag{15.0.4b}$$

and g_{t_1} is the induced metric on Σ_{t_1}. Differentiating with respect to t_1 we obtain:

$$\frac{dL}{dt_1} = \phi(\Gamma_{t_1}) - \int_{-1}^{t_1} \phi(f_{t_1 - t}(\Gamma_t)) \frac{K(t_1, t)}{\sqrt{G(t_1, t)}} dt \tag{15.0.4c}$$

where

$$K(t_1, t) = k_{t_1}(f'_{t_1 - t} V_t, f'_{t_1 - t} V_t) \tag{15.0.4d}$$

where k_{t_1} is the second fundamental form of Σ_{t_1}. Since $\phi \leq 1$ we have

$$\frac{dL}{dt_1} \leq 1 + \int_{-1}^{t_1} \frac{|K(t_1, t)|}{\sqrt{G(t_1, t)}} dt. \tag{15.0.4e}$$

For some $t_0 \in [0, t_1]$, the point $f_{t_1 - t_0}(\Gamma_{t_0})$ on the arc ${}^{t_1}\Gamma'$ lies on the boundary of the exterior region in Σ_{t_1}. We have

$$|K(t_1, t)| \leq |k_{t_1}|_{g_{t_1}} |f'_{t_1 - t} V_t|^2_{g_{t_1}}$$

while

$$G(t_1,t) = |f'_{t_1-t}V_t|^2_{g_{t_1}}.$$

Hence, we can estimate the interior contribution to the integral in 15.0.4e by

$$\int_{-1}^{t_0} \frac{|K(t_1,t)|}{\sqrt{G(t_1,t)}} dt \le B(2+t_1)^{-2} \int_{-1}^{t_0} |f'_{t_1-t}V_t|_{g_{t_1}} dt. \tag{15.0.4f}$$

Since by our hypotheses in $U_{t_1} \bigcap E$ it holds that

$$|k_{t_1}|_{g_{t_1}} \le cB\left[(1+|u_E|)^{-3/2}(2+t_1)^{-1} + (2+t_1)^{-2}\right],$$

the exterior contribution can be estimated by

$$\begin{aligned}\int_{t_0}^{t_1} \frac{|K(t_1,t)|}{\sqrt{G(t_1,t)}} dt \le cB\big[(2+t_1)^{-1} &\int_{t_0}^{t_1} (1+|u_E(f_{t_1-t}(\Gamma_t))|)^{-3/2} |f'_{t_1-t}V_t|_{g_{t_1}} dt \\ &+ (2+t_1)^{-2} \int_{t_0}^{t_1} |f'_{t_1-t}V_t|_{g_{t_1}} dt\big].\end{aligned} \tag{15.0.4g}$$

To proceed further, we need estimates of the eigenvalues λ_1, λ_2, λ_3 of $f^\star_{t-t_1}g_t$ at an arbitrary point p of U_{t_1} relative to g_{t_1}. Let

$$\lambda = \min \lambda_1, \lambda_2, \lambda_3, \quad \Lambda = \max \lambda_1, \lambda_2, \lambda_3.$$

The fact that

$$\frac{\partial}{\partial t} \det_{g_{t_1}} f^\star_{t-t_1} g_t(p) = -2(\phi trk)(f_{t-t_1}(p)) \det_{g_{t_1}} f^\star_{t-t_1} g_t(p) = 0$$

implies

$$\lambda_1\lambda_2\lambda_3 = \det_{g_{t_1}} f^\star_{t-t_1} g_t(p) = 1. \tag{15.0.5a}$$

Hence

$$\lambda\Lambda^2 \ge 1. \tag{15.0.5b}$$

Therefore, it suffices to obtain an upper bound for Λ. Let X be any vector tangent to Σ_{t_1} at p of unit length

$$|X|_{g_{t_1}} = 1.$$

Then

$$\frac{\partial}{\partial t} f^\star_{t-t_1} g_t(X,X) = -2\phi(f_{t-t_1}(p)) f^\star_{t-t_1} k_t(X,X). \tag{15.0.5c}$$

Now,

$$|f^{\star}_{t-t_1}k_t(X,X)| \leq |f^{\star}_{t-t_1}k_t|_{g_{t_1}}(p)\cdot|X|^2_{g_{t_1}} = |f^{\star}_{t-t_1}k_t|_{g_{t_1}}(p)$$

and

$$|k_t|^2_{g_t}(f_{t-t_1}(p)) = |f^{\star}_{t-t_1}k_t|^2_{f^{\star}_{t-t_1}g_t} \geq \Lambda^{-2}|f^{\star}_{t-t_1}k_t|^2_{g_{t_1}}(p).$$

Therefore, substituting in 15.0.5c, integrating, and taking the supremum over all X in S_p, the unit sphere in the tangent space to Σ_{t_1} at p, we obtain the integral inequality

$$\Lambda(t) = \sup_{X\in S_p} f^{\star}_{t-t_1}g_t(X,X) \leq 1+2\int_t^{t_1}|k_{t'}|_{g_{t'}}(f_{t'-t_1}(p))\Lambda(t')dt'.$$

The Gronwall lemma then yields

$$\Lambda(t) \leq \exp\left(2\int_t^{t_1}|k_{t'}|_{g_{t'}}(f_{t'-t_1}(p))dt'\right) \leq c \qquad (15.0.5d)$$

provided that $B \leq c$.

In view of 15.0.5b, 15.0.5d, and the fact that

$$|V_t|_{g_t} = \phi(\Gamma_t) \leq 1,$$

we can estimate

$$|f'_{t_1-t}V_t|^2_{g_{t_1}} \leq \lambda^{-1}|f'_{t_1-t}V_t|^2_{f^{\star}_{t-t_1}g_t} \leq \Lambda^2|V_t|^2_{g_t} \leq c. \qquad (15.0.6a)$$

We have

$$\frac{d}{dt}u_E(f_{t_1-t}(\Gamma_t)) = \nabla u_E(f'_{t_1-t}V_t) = (a(f_{t_1-t}(\Gamma_t)))^{-1}\phi(\Gamma_t)\iota(f_{t_1-t}(\Gamma_t),t) \qquad (15.0.6b)$$

where for any $p\in U_{t_1}\bigcap E$

$$\iota(p,t) = g_{t_1}(N_p, f'_{t_1-t}N_{f_{t-t_1}(p)}). \qquad (15.0.6c)$$

As

$$\frac{\partial}{\partial t}f'_{t_1-t}N_{f_{t-t_1}(p)} = f'_{t_1-t}(\mathcal{L}_{\frac{\partial}{\partial t}}N)_{f_{t-t_1}(p)},$$

it holds that

$$\left|\frac{\partial\iota(p,t)}{\partial t}\right| \leq |f'_{t_1-t}(\mathcal{L}_{\frac{\partial}{\partial t}}N)_{f_{t-t_1}(p)}|_{g_{t_1}} \leq c|(\mathcal{L}_{\frac{\partial}{\partial t}}N)_{f_{t-t_1}(p)}|_{g_t}. \qquad (15.0.6d)$$

Now

$$\mathcal{L}_{\frac{\partial}{\partial t}}N = \phi(\zeta - \underline{\zeta} + \delta N).$$

The hypotheses of the lemma imply that in $U_t \bigcap E$

$$|\mathcal{L}_{\frac{\partial}{\partial t}} N|_{g_t} \leq cB(2+t)^{-2}. \tag{15.0.6e}$$

Hence, in view of the fact that for any $p \in U_{t_1} \bigcap E$ *the segment of the integral curve of* $\frac{\partial}{\partial t}$ *through* p *in the past of* Σ_{t_1} *in* U *is contained in* $U \bigcap E$*, substituting 15.0.6e in 15.0.6d and integrating, noting that* $\iota(p, t_1) = 1$*, we obtain*

$$\iota(f_{t_1 - t}(\Gamma_t), t) \geq 1 - cB\left[(2+t)^{-1} - (2+t_1)^{-1}\right]. \tag{15.0.6f}$$

Therefore if B *is sufficiently small, we have:*

$$\iota(f_{t_1 - t}(\Gamma_t), t) \geq \frac{1}{2},$$

which by 15.0.6b implies

$$\frac{d}{dt} u_E(f_{t_1 - t}(\Gamma_t)) \geq \frac{1}{c}.$$

It follows that

$$\int_{t_0}^{t_1} (1 + |u_E(f_{t_1 - t}(\Gamma_t))|)^{-3/2} dt \leq c.$$

Substituting this together with the estimate 15.0.6a in 15.0.4f, and 15.0.4g, we conclude that the integral in inequality 15.0.4e is bounded by $cB(2 + t_1)^{-1}$*, hence:*

$$\frac{dL}{dt_1} \leq 1 + cB(2+t_1)^{-1}.$$

Integrating from $t_1 = -1$ *then yields*

$$L(t_1) \leq 1 + t_1 + cB \log(2 + t_1),$$

which, in view of the fact that any point of the surface $S_{t_1, u}$ *is joined to the point* O_{t_1} *by an arc* ${}^{t_1}\Gamma'$ *lying in* Σ_{t_1}*, implies the lemma.*

Corollary 15.0.2.1 *Let the hypotheses of Lemma 15.0.2 hold and let the exterior optical function satisfy in addition*

$$\inf_{S_{t,u_E}} a \geq 1 - A(2+t)^{-1}$$

for all $t \in [0, t_*]$, $u_E \in [-(13/20)(1+t), 0]$. *Then the surface* S_{t,u_E} *is contained within the geodesic ball in* Σ_t *of radius*

$$d_{E,M}(t, u_E) = 1 + t + u_E + c(A + B) \log(2 + t)$$

centered at O_t.

Proof 15.0.3 *Consider an arbitrary point p on S_{t,u_E}. Since $p \in U$ and C_0 is a Cauchy surface for U, the integral curve of T drawn toward the past intersects C_0 at a point p_0. Let Γ be the generator of C_0 through p_0 and ${}^t\Gamma'$ the projection to Σ_t of the segment of Γ that lies in the past of Σ_t. Then the arc ${}^t\Gamma'$ passes through p and by the proof of Lemma 15.0.2 its length is at most*

$$1 + t + cB\log(2 + t).$$

On the other hand, the segment of this arc between p and the end point on $S_{t,0}$ has length at least

$$\int_{u_E}^{0} \inf_{S_{t,u}} a du \geq [1 - A(2 + t)^{-1}]|u_E|.$$

Therefore the length of the segment joining O_t and p is at most $d_{E,M}(t, u_E)$.

To proceed further, we have to show that the surface S_{t,u_E} contains a geodesic ball in Σ_t of radius

$$d_{E,m}(t, u_E) = 1 + t + u_E - c(A + B)\log(2 + t).$$

This shall be done with the help of the following two lemmas.

Lemma 15.0.3 *Let Σ be a 3-dimensional Riemannian manifold diffeomorphic to $\Re^3$ and let*

$$\{S_u : u \geq 0\}$$

be a foliation of the exterior of S_0 by surfaces S_u diffeomorphic to S^2 and of positive-definite second fundamental form θ_{AB}. Suppose that the Ricci curvature of Σ in I_0, the interior of S_0, satisfies

$$\sup_{I_0} |Ric| \leq B(diam(S_0))^{-3}.$$

Then, provided that B is sufficiently small, for any point $O \in I_0$ the exponential map $\exp_O$, with its range restricted to I_0, is a diffeomorphism. In particular, each geodesic from O is minimizing in I_0.

Proof 15.0.4 *We first show that I_0 is geodesically convex. Suppose that P_1 and P_2 are two points in I_0 for which there exists a minimal geodesic segment Γ_{12} joining them which is not contained in I_0. We can project the portion of Γ_{12} in the exterior of S_0 back to S_0 by following the normal flow of the foliation $\{S_u : u \geq 0\}$. The fact that θ_{AB} is positive-definite implies that this projection contracts all lengths. We thus obtain a curve Γ'_{12} joining P_1 and P_2 of smaller length.*

Thus, if O is an arbitrary point in I_0, any other point in I_0 can be joined to it by a minimal geodesic lying in I_0. We consider the star-shaped domain $\mathbf{D}$ *in* $T_O(\Sigma)$, *which consists of all ray segments from the origin whose image through the exponential map is both minimizing and contained in I_0. Then* $\exp_O$ *maps* $\mathbf{D}$ *onto I_0. Those boundary points of* $\mathbf{D}$ *that are not mapped to S_0 are cut points of* $\exp_O$. *The assumption on the curvature of Σ in I_0 implies by a standard argument (see Chapter 5) that at all points of the closure* $\overline{\mathbf{D}}$ *the derivative of* $\exp_O$ *is nonsingular, hence locally a diffeomorphism. We can thus extend* $\mathbf{D}$ *through any of the cut points to obtain a covering map from an open, star shaped domain $\Omega \supset \overline{\mathbf{D}}$ onto I_0. Since I_0 is diffeomorphic to $\Re^3$ this map must be a diffeomorphism. This implies in particular that the boundary* $\overline{\mathbf{D}}$ *consists only of those points that are mapped to S_0.*

Lemma 15.0.4 *Let Σ be a 3-dimensional Riemannian manifold diffeomorphic to $\Re^3$ and let S be a surface, diffeomorphic to S^2, embedded in Σ. Suppose that the second fundamental form of S is positive-definite and*

$$osc_S tr\theta \leq Ar^{-2}, \qquad \sup_S |\hat{\theta}| \leq Ar^{-2}$$

where

$$r = \sqrt{Area(S)/4\pi}.$$

Suppose also that the Ricci curvature of Σ in I, the interior of S, satisfies

$$\sup_I |Ric| \leq Br^{-3}.$$

Then there are numerical constants c_1 and c_2 such that if

$$A + B \leq c_1 r,$$

S contains a geodesic ball of radius at least

$$r - c_2(A + B).$$

Proof 15.0.5 *The Gauss equation for S reads*

$$K - \frac{1}{4}(tr\theta)^2 = -\frac{1}{2}|\hat{\theta}|^2 + \frac{1}{2}R - R_{NN}.$$

Our assumptions therefore imply

$$|K - \frac{1}{4}(tr\theta)^2| \leq \varepsilon_1 r^{-2} \tag{15.0.7a}$$

where

$$\varepsilon_1 = \frac{3}{2}\frac{B}{r} + \frac{1}{2}\frac{A^2}{r^2}. \tag{15.0.7b}$$

Since by the Gauss-Bonnet formula

$$\overline{K} = \frac{1}{r^2},$$

15.0.7a allows us to conclude that

$$\frac{(tr\theta)_m}{2} \leq \frac{\sqrt{1+\varepsilon_1}}{r}, \qquad \frac{(tr\theta)_M}{2} \geq \frac{\sqrt{1-\varepsilon_1}}{r} \tag{15.0.7c}$$

provided that $\varepsilon_1 \leq 1$. *Here* $(tr\theta)_m$ *and* $(tr\theta)_M$ *are respectively the minimum and maximum values of* $tr\theta$ *on* S. *Our assumption*

$$osc_S tr\theta = (tr\theta)_M - (tr\theta)_m \leq Ar^{-2}$$

then yields

$$\frac{2}{r}(\sqrt{1-\varepsilon_1} - \varepsilon_2) \leq (tr\theta)_m \leq (tr\theta)_M \leq \frac{2}{r}(\sqrt{1+\varepsilon_1} + \varepsilon_2). \tag{15.0.7d}$$

If $\varepsilon_1 + \varepsilon_2^2 < 1$, *setting*

$$h = \frac{2}{tr\theta}$$

we can express 15.0.7d in the form

$$\frac{r}{\sqrt{1+\varepsilon_1} + \varepsilon_2} \leq h \leq \frac{r}{\sqrt{1-\varepsilon_1} - \varepsilon_2}. \tag{15.0.7e}$$

It follows that for every $k > 1$ *there is a* $c_0 > 0$ *such that if* $A + B \leq c_0$ *then*

$$|h - r| \leq \frac{3}{4}k(A + B). \tag{15.0.7f}$$

We now construct the surfaces S_s, $s \leq 0$ *by going distance* $|s|$ *inward along each geodesic normal to* S:

$$S_s = \{\exp_p(sN_p) \ : p \in S\}.$$

We shall first show that for each $s \in [s_2, 0]$,

$$s_2 = -r + c_2(A + B),$$

the surface S_s *is immersed in* Σ *with positive-definite second fundamental form. Picking a local orthonormal frame* $e_A : A = 1, 2$ *for* S *and parallel propagating it along the geodesic normals, we obtain a local orthonormal frame for* S_s *at* *each s. In terms of this frame the second fundamental form θ of S_s satisfies the equations*

$$\frac{dtr\theta}{ds} = -\frac{1}{2}(tr\theta)^2 - |\hat{\theta}|^2 - R_{NN} \tag{15.0.8a}$$

$$\frac{d\hat{\theta}_{AB}}{ds} = -tr\theta\hat{\theta}_{AB} - \hat{S}_{AB} \tag{15.0.8b}$$

where S_{AB} are the components of the projection to S_s of the Ricci curvature of Σ. Setting again

$$h = \frac{2}{tr\theta}$$

equation 15.0.8a takes the form

$$\frac{dh}{ds} = 1 + \frac{1}{2}h^2(|\hat{\theta}|^2 + R_{NN}), \tag{15.0.8c}$$

while equation 15.0.8b implies

$$|\frac{d|\hat{\theta}|}{ds} + 2h^{-1}|\hat{\theta}|| \leq |\hat{S}|. \tag{15.0.8d}$$

Let now Γ be the property

$$\Gamma : \quad |h - r - s| < k(A + B).$$

Γ is a continuous property. By 15.0.7f, Γ is true at $s = 0$. We shall show that Γ is inductive as long as $s \geq s_2$. For let Γ hold in $(s_1, 0]$. Integrating 15.0.8d along a normal geodesic and using our assumption on the Ricci curvature of Σ, we obtain

$$\hat{\theta}(s)| \leq |\hat{\theta}(0)| \exp\left[2\int_s^0 (h(s'))^{-1}ds'\right] + r^{-3}B\int_s^0 \exp\left[2\int_s^{s'} (h(s''))^{-1}ds''\right] ds'. \tag{15.0.8e}$$

In $(s_1, 0]$ we have

$$h > l + s, \qquad h < L + s \tag{15.0.8f}$$

where

$$l = r - k(A + B), \qquad L = r + k(A + B).$$

The constant c_2 is to be chosen greater than k so that we have $|s_2| < l$. We then obtain

$$\int_s^{s'} (h(s''))^{-1} ds'' \le \log\left(\frac{l+s'}{l+s}\right),$$

$$\int_s^0 \exp\left[2\int_s^{s'} (h(s''))^{-1} ds''\right] ds' \le \frac{l^3}{(l+s)^2}.$$

Substituting these estimates and using our hypothesis on $\hat{\theta}$ of S, we conclude that for all $s \in [s_1, 0]$

$$|\hat{\theta}(s)| \le \frac{A+B}{(l+s)^2}. \tag{15.0.8g}$$

Integrating 15.0.8c along a normal geodesic gives

$$h(s) = h(0) + s - \frac{1}{2}\int_s^0 h^2(|\hat{\theta}|^2 + R_{NN}) ds'. \tag{15.0.8h}$$

Substituting 15.0.8f as well as 15.0.8g we obtain that in $[s_1, 0]$

$$|h(s) - h(0) - s| \le E(s) \tag{15.0.8i}$$

where

$$E(s) = \frac{1}{2}\int_s^0 (L+s')^2 \left[\frac{(A+B)^2}{(l+s)^4} + \frac{B}{r^3}\right]. \tag{15.0.8j}$$

Now setting

$$c_2 = 3k$$

we have $l + s \ge 2k(A+B)$, $L + s \le 2(l+s)$ and we can estimate

$$E \le \left(\frac{1}{k} + \frac{2}{3}\right)(A+B).$$

Hence taking k to satisfy

$$\frac{1}{k} + \frac{2}{3} < \frac{k}{4},$$

we have

$$E < \frac{k}{4}(A+B),$$

which by 15.0.7f and 15.0.8h implies that for all $s \in [s_1, 0]$, in particular at $s = s_1$,

$$|h(s) - r - s| \le |h - h(0) - s| + |h(0) - r| < k(A+B).$$

We conclude that Γ is inductive and thus true for all $s \in [s_2, 0]$. Therefore 15.0.8g holds in $[s_2, 0]$. It follows that θ, the second fundamental form of S_s, is positive-definite for all $s \in [s_2, 0]$.

We finally show that the surfaces S_s are actually embedded in Σ. Either this is true for all $s \in [s_2, 0]$ or there is a $s_0 \in [s_2, 0]$ such that S_s is embedded for all $s \in (s_0, 0]$ but S_{s_0} has self-intersections. Let q be a self-intersection point of S_{s_0}. There are distinct points $p_1, p_2 \in S$ such that

$$\exp_{p_1}(s_0 N_{p_1}) = \exp_{p_2}(s_0 N_{p_2}) = q.$$

Taking disjoint neighborhoods U_1 and U_2 of p_1 and p_2 respectively, we consider the corresponding sheets V_1 and V_2 of S_{s_0}:

$$V_1 = \{\exp_p(s_0 N_p) \quad : p \in U_1\}, \qquad V_2 = \{\exp_p(s_0 N_p) \quad : p \in U_2\}.$$

Then V_1 and V_2 have a contact point at q, that is, the tangent planes to V_1 and V_2 at q coincide and the outward normals N_1 and N_2 are opposite. For if this is not so, then for all s in a small enough neighborhood of s_0 the surface S_s has a self-intersection point (in the direction of $N_1 + N_2$ from q), thus contradicting the hypothesis that the surfaces S_s are embedded for $s > s_0$. On the other hand, the fact that the second fundamental form of S_{s_0} is positive-definite implies that S_{s_0} is locally a convex graph over the tangent plane at each of its points in the direction of the inward normal. Thus V_1 is a convex graph over the tangent plane at q with respect to $-N_1 = N_2$, and V_2 is a convex graph over the tangent plane at q with respect to $-N_2 = N_1$. But this implies that for some $\varepsilon > 0$ and all $s \in [s_0, s_0 + \varepsilon]$ the surfaces S_s have self-intersections, contradicting again the hypothesis that S_{s_0} is the first surface with self-intersections. We conclude that the surfaces S_s are embedded in Σ for all $s \in [s_2, 0]$. It follows that the S_s are the level sets of the distance function from S. Hence each point on S_{s_2} or its interior has distance from S at least s_2, which gives the conclusion of the lemma.

We now apply the lemmas to prove the following:

Lemma 15.0.5 *Let the hypotheses of Lemmas 15.0.1 and 15.0.2 hold and also let*

$$\sup_{U \setminus E} (2 + t)^3 (|E|, |H|) \leq B$$

where E and H are the electric and magnetic components of the space-time curvature. Suppose that the exterior optical function u_E satisfies in addition

$$\sup_{S_{t, u_E}} |a - 1| \leq A r^{-1},$$

$$osc_{S_{t,u_E}} tr\chi \leq Ar^{-2}, \qquad \sup_{S_{t,u_E}} |\hat{\chi}| \leq Ar^{-2}$$

for all $t \in [0, t_*]$, $u_E \in [-(13/20)(1+t), 0]$, *where*

$$r(t, u_E) = \sqrt{Area(S_{t,u_E})/4\pi}.$$

Then, provided that A and B are sufficiently small,
(i) For all $t \in [0, t_*]$ *and* $u_E \in [-(13/20)(1+t), 0]$

$$\sup_{S_{t,u_E}} |tr\chi - \frac{2}{r}| \leq \frac{c(A+B)}{r^2},$$

and

$$|r(t, u_E) - 1 - t - u_E| \leq c(A+B)\log(2+t).$$

(ii) For each $t \in [0, t_*]$ *and* $u_E \in [-(13/20)(1+t), -(12/20)(1+t)]$ *the surface* S_{t,u_E} *contains a geodesic ball in* Σ_t *of radius*

$$d_{E,m}(t, u_E) = 1 + t + u_E - c(A+B)\log(2+t)$$

centered at O_t.

Proof 15.0.6 *Since*

$$r(t, u_E) \leq r(t, 0) \leq c(1+t),$$

the assumptions imply, by the same argument as in the proof of Lemma 15.0.1,

$$|tr\chi - \frac{2}{r}| \leq \frac{c(A+B)}{r^2}.$$

The other assumptions then imply

$$|\frac{\partial r}{\partial u_E} - 1| = |\frac{r}{2}\overline{atr\theta} - 1| \leq \frac{c(A+B)}{r}.$$

Hence, integrating, we obtain that for $u_E \in [-(13/20)(1+t), 0]$

$$|r(t, u_E) - r_0(t) - u_E| \leq c(A+B),$$

which upon substituting the result of Lemma 15.0.1 yields conclusion (i).

In the region $u_E \in [-(13/20)(1+t), -(12/20)(1+t)]$ *in each* Σ_t *our assumptions then imply*

$$osc_{S_{t,u_E}} tr\theta \leq c(A+B)r^{-2}, \qquad \sup_{S_{t,u_E}} |\hat{\theta}| \leq c(A+B)r^{-2}$$

and

$$|Ric| \leq cBr^{-3}.$$

The hypotheses of Lemmas 15.0.3 and 15.0.4 are therefore verified for each surface S_{t,u_E} in this region. We conclude from Lemma 15.0.3 that the exponential map based at O_t is a diffeomorphism when its range is restricted to the interior of S_{t,u_E} and that each geodesic from O_t is minimizing in the interior of S_{t,u_E}. We conclude from Lemma 15.0.4 that S_{t,u_E} contains a geodesic ball $B_{d_m}(p)$ of radius

$$d_m(t, u_E) = r(t, u_E) - c(A + B) \geq 1 + t + u_E - c(A + B)\log(2 + t)$$

centered at some point p. Consider now the geodesic from O_t to p produced up to the boundary of S_{t,u_E}. Let the geodesic intersect the boundary of $B_{d_m}(p)$ at q_1 and the boundary of S_{t,u_E} at q_2. Then, since this geodesic is minimizing and by Corollary 15.0.2 the surface S_{t,u_E} is contained in the ball $B_{d_{E,M}}(O_t)$ of radius $d_{E,M}$ centered at O_t, we have

$$d_{E,M} \geq dist(q_2, O_t) = dist(p, O_t) + dist(p, q_1) + dist(q_2, q_1), dist(p, q_1) = d_m.$$

Hence

$$dist(p, O_t) \leq d_{E,M} - d_m.$$

Then, by the triangle inequality, $B_{d_m}(p)$ includes a ball of radius

$$2d_m - d_{E,M}.$$

Conclusion (ii) follows.

Corollary 15.0.5.1 *Under the hypotheses of Lemma 15.0.5, the interior region I_t in each Σ_t, $t \in [0, t_*]$ contains a ball of radius*

$$\frac{1}{2}(1 + t) - c(A + B)\log(2 + t)$$

centered at O_t. In particular, if A, B, and ε_1 are sufficiently small the region V, defined in Chapter 9, is contained in I.

We recall from Chapter 9 that the point P_0, which defines the line P on which the construction of the interior optical function is based, is contained in the ball in $\sum_0$ with center at O_0 and of radius ε_1.

Lemma 15.0.6 *Let the hypotheses of Lemma 15.0.5 hold and let φ be the angle between the normal at a given point to the surface S_{t,u_E} and the tangent to the geodesic from O_t to that point. Then for each $t \in [0, t_*]$ and $u_E \in [-(13/20)(1 + t), -(12/20)(1 + t)]$*

$$\varphi \leq c(A + B)\frac{\log(2 + t)}{(2 + t)}.$$

Proof 15.0.7 *We confine our attention to the ball of radius*

$$d_2(t) = d_{E,M}(t, -(12/20)(1+t))$$

centered at O_t in which all surfaces under consideration are contained. Let N be the unit outward normal to S_{t,u_E} and N' the unit tangent vectorfield of the geodesic congruence from O_t. Let us denote by s the distance function in Σ_t from O_t. We have

$$N' = N\cos\varphi + Y, \qquad N = N'\cos\varphi + Y',$$

where Y is tangent to S_{t,u_E}, Y' is tangent to $\partial B_s(O_t)$, and

$$|Y| = |Y'| = \sin\varphi.$$

At an extremum of the function s on S_{t,u_E} the angle φ vanishes, since at such a point $\nabla\!\!\!/ s = 0$, which implies $N' = \nabla s = N$. By formula 14.0.34 of the previous chapter,

$$\begin{aligned}\nabla\!\!\!/_m \cos\varphi &= \theta_{mj}Y^j + \Pi^i_m Y'^j\theta'_{ij}\\ &= \frac{1}{2}(tr\theta - \cos\varphi tr\theta')Y_m + \hat{\theta}_{mj}Y^j + \Pi^i_m Y'^j\hat{\theta}'_{ij}\end{aligned} \tag{15.0.9a}$$

where θ' is the second fundamental form of the geodesic spheres $\partial B_s(O_t)$. Since in $B_{d_2(t)}(O_t)$ we have

$$|Ric| \le cB(2+t)^{-3},$$

it is easy to show that

$$|tr\theta' - \frac{2}{s}| \le \frac{cB}{(2+t)^2}, \qquad |\hat{\theta}'| \le \frac{cB}{(2+t)^2}$$

for all $s \in (0, d_2(t)]$. Now let p be an arbitrary point on S_{t,u_E}. We join p to an extremal point e of s on S_{t,u_E} by a minimal geodesic Γ of S_{t,u_E}. Let us denote by v arc length on Γ. Then by 14.0.12 along Γ we have

$$\frac{d\varphi}{dv} \le |\nabla\!\!\!/\varphi| \le \frac{1}{2}tr\theta'(1-\cos\varphi) + \frac{1}{2}|tr\theta - tr\theta'| + |\hat{\theta}| + |\hat{\theta}'|. \tag{15.0.9b}$$

The fact that on S_{t,u_E}

$$d_{E,m}(t,u_E) \le s \le d_{E,M}(t,u_E),$$

together with the estimate for $r(t,u_E)$ given by Lemma 15.0.5, gives

$$\sup_{S_{t,u_E}} |r - s| \le c(A+B)\log(2+t).$$

This, in conjunction with the above estimates for θ' *and the estimates for* θ *given in the proof of Lemma 15.0.5, yield*

$$|tr\theta - tr\theta'| \leq c(A+B)r^{-2}\log(2+t), \qquad |\hat{\theta}|, |\hat{\theta}'| \leq c(A+B)r^{-2}.$$

Hence

$$\frac{d\varphi}{dv} \leq cr^{-1}(1-\cos\varphi) + c(A+B)r^{-2}\log(2+t). \tag{15.0.9c}$$

Integrating on Γ *and using the fact that* φ *vanishes at e, we obtain the conclusion of the lemma.*

Lemma 15.0.7 *Let*

$$\sup_U (2+t)|\phi - 1| \leq B, \qquad \sup_I (2+t)^2|k| \leq B$$

$$\sup_I (2+t)^3(|E|, |H|) \leq B,$$

I being the interior region defined in the first chapter of Part III, and let the interior optical function u_I *satisfy*

$$\sup_{S_{t,u_I}} |a-1| \leq A(2+t)^{-1}$$

$$osc_{S_{t,u_I}} tr\chi \leq A(2+t)^{-2}, \qquad \sup_{S_{t,u_I}} |\hat{\chi}| \leq A(2+t)^{-2}$$

for all $u_I \in [u_{I,m}(t), u_{I,M}(t)]$ *and* $t \in [0, t_*]$. *Then, provided that A and B are sufficiently small,*

(i) For all $u_I \in [u_{I,m}(t), u_{I,M}(t)]$ *and* $t \in [0, t_*]$

$$\sup_{S_{t,u_I}} |tr\chi - \frac{2}{r}| \leq c(A+B)(2+t)^{-2}$$

and

$$|r(t,u_I) - 1 - t - u_I| \leq c(A+B)\log(2+t).$$

(ii) For each $u_I \in [u_{I,m}(t), u_{I,M}(t)]$ *and* $t \in [0, t_*]$ *the surface* S_{t,u_I} *is contained in the geodesic ball in* Σ_t *of radius*

$$d_{I,M} = 1 + t + u_I + c(A+B)\log(2+t) + c\varepsilon_1$$

centered at O_t *and contains the geodesic ball of radius*

$$d_{I,m} = 1 + t + u_I - c(A+B)\log(2+t) - c\varepsilon_1$$

centered at O_t.
(iii) For each $u_I \in [u_{I,m}(t), u_{I,M}(t)]$ *and* $t \in [0, t_*]$, *the angle* φ *between the normal at the given point to the surface* S_{t,u_I} *and the tangent to the geodesic from* O_t *to that point, is bounded by*

$$\varphi \le c(A + B + \varepsilon_1)\frac{\log(2+t)}{(2+t)}.$$

We note that the point P_0 *is contained in the ball in* Σ_0 *of radius* ε_1 *centered at* O_0.

Proof 15.0.8 *The first statement in (i) follows as in the proof of Lemma 15.0.1. We then have*

$$|\frac{\partial r}{\partial u_I} - 1| = |\frac{r}{2}\overline{atr\theta} - 1| \le c(A+B)(2+t)^{-1}.$$

As $r(t, u_{I,m}(t)) = 0$, *integrating in* $[u_{I,m}(t), u_I]$, *we obtain*

$$|r(t,u_I) - u_I + u_{I,m}(t)| \le c(A+B).$$

In view of the fact that by 9.2.4b

$$|u_{I,m}(t) + 1 + t| \le cB\log(2+t), \qquad (15.0.10)$$

the remaining statement in (i) follows. To prove (ii), we observe that for any $p \in S_{t,u_I}$

$$\begin{aligned} dist(p, P_t) &\le \int_{u_{I,m}(t)}^{u_I} \sup_{S_{t,u}} a du \le [1 + A(2+t)^{-1}](u_I - u_{I,m}(t)) \\ &\le 1 + t + u_I + c(A+B)\log(2+t) \end{aligned}$$

and

$$\begin{aligned} dist(p, P_t) &\ge \int_{u_{I,m}(t)}^{u_I} (\inf_{S_{t,u}} a) du \ge [1 - A(2+t)^{-1}](u_I - u_{I,m}(t)) \\ &\ge 1 + t + u_I - c(A+B)\log(2+t), \end{aligned}$$

using 15.0.10. On the other hand, it is easy to show that

$$dist(P_t, O_t) \le c\,dist(P_0, O_0) \le c\varepsilon_1$$

The triangle inequality then yields (ii). Conclusion (iii) then follows by the argument of Lemma 15.0.6.

Let

$$l^\mu = \partial^\mu u,$$

where u is the global optical function defined by 15.0.2b. We have

$$l = a^{-1}N + b^{-1}T$$

where N is the unit normal to the surfaces $S_{t,u}$ in Σ_t. We denote by χ'_{AB} the second fundamental form of $S_{t,u}$ with respect to the normalized null normal $e_4 = T + N$, $\zeta'_A = \nabla\!\!\!/_A \log a + \epsilon_A$, $\omega' = -\nabla_N \log a - \delta$. We are now ready to prove the following:

Proposition 15.0.1 *Let the hypotheses of Lemmas 15.0.5 and 15.0.7 be satisfied. In addition let the exterior optical function u_E satisfy*

$$\sup_{S_{t,u_E}} |\zeta, \omega| \leq A r^{-2}$$

for all $t \in [0, t_*]$, $u_E \in [-(13/20)(1+t), -(12/20)(1+t)]$, and let the interior optical function u_I satisfy

$$\sup_{S_{t,u_I}} |\zeta, \omega| \leq A(2+t)^{-2}$$

for all $u_I \in [u_{I,m}(t), u_{I,M}(t)]$ and $t \in [0, t_*]$.

Then, in the matching region M_1 defined to be the set of points p where $u_E(p) \in [-(13/20)(1+t), -(12/20)(1+t)]$, we have

1. $|u_E - u_I| \leq c(A+B)\log(2+t) + c\varepsilon_1$.
2. $|r_E - r_I| \leq c(A+B)\log(2+t) + c\varepsilon_1$.
3. $|r(t,u) - 1 - t - u| \leq c(A+B+\varepsilon_1)\log(2+t)$.
4. *The angle $\varphi_{E,I}$ between the vectors N_E and N_I[2] is bounded by $c(A+B+\varepsilon_1)\frac{\log(2+t)}{2+t}$. Also,*

$$|\langle l_E, l_I\rangle| \leq c(A+B+\varepsilon_1)^2\left(\frac{\log(2+t)}{2+t}\right)^2$$

   and, with $l_\mu = \partial_\mu u$,

$$|\langle l, l\rangle| \leq c(A+B+\varepsilon_1)\frac{\log(2+t)}{2+t}.$$

5. $|a-1|, |b-1| \leq c(A+B+\varepsilon_1)\frac{\log(2+t)}{2+t}$.

[2] Namely, the unit normals to S_{t,u_E} and S_{t,u_I}, respectively, in Σ_t.

6.
$$\sup_{S_{t,u}} |tr\chi' - \frac{2}{r}| \le c(A+B+\varepsilon_1)\frac{\log(2+t)}{(2+t)^2},$$
$$\sup_{S_{t,u}} |\hat{\chi}'| \le c(A+B+\varepsilon_1)\frac{\log(2+t)}{(2+t)^2}.$$

7. $\sup_{S_{t,u}} |\zeta', \omega'| \le c(A+B+\varepsilon_1)\frac{\log(2+t)}{(2+t)^2}$.
8. $|\nabla \log b| \le c(A+B+\varepsilon_1)\frac{\log(2+t)}{(2+t)^2}$.
9. $|\mathbf{D}_l l| \le c(A+B+\varepsilon_1)\frac{\log(2+t)}{(2+t)^2}$.

Proof 15.0.9 *To prove the first conclusion, consider a level surface S_{t,u_E} of u_E. Let v_M and v_m be respectively the maximum and minimum values of u_I on S_{t,u_E}. Now by Corollary 15.0.2.1, S_{t,u_E} is contained within the geodesic ball in Σ_t of radius $d_{E,M}(u_E)$ centered at O_t, while by Lemma 15.0.7, S_{t,u_I} contains the geodesic ball of radius $d_{I,m}(u_I)$ centered at O_t. Therefore*

$$d_{E,M}(u_E) \ge d_{I,m}(v_M),$$

which is equivalent to

$$v_M - u_E \le c(A+B)\log(2+t) + c\varepsilon_1.$$

On the other hand, by Lemma 15.0.5, S_{t,u_E} contains the geodesic ball in Σ_t of radius $d_{E,m}(u_E)$ centered at O_t, while by Lemma 15.0.7, S_{t,u_I} is contained within the geodesic ball of radius $d_{I,M}(u_I)$ centered at O_t. Therefore

$$d_{I,M}(v_m) \ge d_{E,m}(u_E),$$

which is equivalent to

$$u_E - v_m \le c(A+B)\log(2+t) + c\varepsilon_1.$$

We thus arrive at the first conclusion. The second conclusion then follows from Lemmas 15.0.5 and 15.0.7. The angles φ_E, φ_I between N_E and N_I respectively and the unit tangent vectorfield N' of the geodesic congruence from O_t have been estimated in Lemmas 15.0.6 and 15.0.7. As $\varphi_{E,I} \le \varphi_E + \varphi_I$, the first estimate in part 4 of the proposition follows. The estimate for $\langle l_E, l_I\rangle$ follows immediately from the formula

$$\langle l_E, l_I\rangle = -a_E^{-1} a_I^{-1}(1 - \cos\varphi_{E,I}).$$

To estimate $\langle l, l\rangle$, we note that

$$l^\mu = (1 - f_1)l_E^\mu + f_1 l_I^\mu + (u_I - u_E)\mathbf{D}^\mu f$$

and

$$\langle l,l\rangle = 2(1-f_1)f_1\langle l_E,l_I\rangle + (u_I-u_E)\left[(1-f_1)\mathbf{D}_{l_E}f_1 + f_1\mathbf{D}_{l_I}f_1\right] + (u_I-u_E)^2\langle \mathbf{D}f_1,\mathbf{D}f_1\rangle .$$

On the other hand,

$$\nabla f_1 = h_1' a_E^{-1}\frac{N_E}{1+t}$$
$$\mathbf{D}_T f_1 = -\frac{h_1'}{1+t}\left(a_E^{-1} + \frac{\phi^{-1}u_E}{1+t}\right)$$

which imply

$$|\nabla f_1|, |\mathbf{D}_T f_1| \le c(2+t)^{-1}.$$

The desired estimate is now immediate.

The estimates for $|a-1|, |b-1|$ in part 5 follow from the formulas

$$a^{-1} = |\nabla u|$$
$$\nabla u = (1-f_1)a_E^{-1}N_E + f_1 a_I^{-1}N_I + (u_I-u_E)\nabla f_1$$
$$b^{-1} = -\langle l,T\rangle = (1-f_1)a_E^{-1} + f_1 a_I^{-1} + (u_I-u_E)\mathbf{D}_T f_1 .$$

To derive the remaining parts of the proposition, we express

$$\mathbf{D}_\mu(l_E)_\nu = (\chi_E)_{\mu\nu} - (\zeta_E)_\mu(l_E)_\nu - (\zeta_E)_\nu(l_E)_\mu + (\omega_E)(l_E)_\mu(l_E)_\nu ,$$
$$\mathbf{D}_\mu(l_I)_\nu = (\chi_I)_{\mu\nu} - (\zeta_I)_\mu(l_I)_\nu - (\zeta_I)_\nu(l_I)_\mu + (\omega_I)(l_I)_\mu(l_I)_\nu$$

and

$$(\chi_E)_{\mu\nu} = \frac{1}{2}(\mathrm{tr}\chi_E)(\gamma_E)_{\mu\nu} + (\hat{\chi}_E)_{\mu\nu},$$
$$(\chi_I)_{\mu\nu} = \frac{1}{2}(\mathrm{tr}\chi_I)(\gamma_I)_{\mu\nu} + (\hat{\chi}_I)_{\mu\nu},$$

while

$$(\gamma_E)_{\mu\nu} = \mathbf{g}_{\mu\nu} + T_\mu T_\nu - (N_E)_\mu(N_E)_\nu ,$$
$$(\gamma_I)_{\mu\nu} = \mathbf{g}_{\mu\nu} + T_\mu T_\nu - (N_I)_\mu(N_I)_\nu .$$

On the other hand, we have

$$\langle \mathbf{D}_A l, e_B\rangle = a^{-1}\chi'_{AB} + (a^{-1}-b^{-1})\eta_{AB},$$
$$\langle \mathbf{D}_A l, N\rangle = -a^{-1}\zeta'_A + (a^{-1}-b^{-1})\epsilon_A,$$
$$\langle \mathbf{D}_N l, N\rangle = a^{-1}\omega' + (a^{-1}-b^{-1})\delta .$$

Using the previously established conclusions and the assumptions of the proposition, we deduce

$$osc_{S_{t,u}} tr\chi' \leq c(A + B + \varepsilon_1)\frac{\log(2+t)}{(2+t)^2}.$$

The estimate for $\sup_{S_{t,u}} |tr\chi' - \frac{2}{r}|$ *follows as in the proof of Lemma 15.0.2. Part 3 is then obtained by integrating the equation*

$$\frac{\partial r}{\partial u} = \frac{r}{2}\overline{\mathrm{atr}\theta}$$

from the outer boundary of the matching region as in the proof of Lemma 15.0.5. The remainder of part 6 and part 7 follow similarly. The estimate in part 8 is obtained by taking into account the fact that we can express

$$\begin{aligned}
\langle \mathbf{D}_A l, T\rangle &= -b^{-2}\nabla\!\!\!/_A b + a^{-1}\epsilon_A + b^{-1}\phi^{-1}\nabla\!\!\!/_A\phi, \\
\langle \mathbf{D}_N l, T\rangle &= -b^{-2}\nabla_N b + a^{-1} + a^{-1}\delta + b^{-1}\phi^{-1}\nabla_N\phi.
\end{aligned}$$

The last part is then derived with the help of the formulas

$$\begin{aligned}
(\mathbf{D}_{l_I} l_E)_\nu &= (N_I^\mu - N_E^\mu)((\chi_E)_{\mu\nu} - (\zeta_E)_\mu (l_E)_\nu) \\
&\quad - \langle l_E, l_I\rangle(\zeta_E)_\nu + (\omega_E)(l_E)_\nu, \\
(\mathbf{D}_{l_E} l_I)_\nu &= (N_E^\mu - N_I^\mu)((\chi_I)_{\mu\nu} - (\zeta_I)_\mu (l_I)_\nu) \\
&\quad - \langle l_E, l_I\rangle(\zeta_I)_\nu + (\omega_I)(l_I)_\nu, \\
|N_E^\mu - N_I^\mu|^2 &= (1 - \cos\varphi_{E,I}).
\end{aligned}$$

Finally, using the results of Chapter 13, Section 2, and Chapter 9, Section 2, together with those of the above proposition, we obtain:

Proposition 15.0.2 *Let the hypotheses of Proposition 15.0.1 be satisfied and let*

$${}^{1}\mathcal{R}_{[2]}, {}^{1}\mathcal{K}_{[3]}, {}^{1}\mathcal{L}_{[3]} \leq c.$$

Then the global optical function u *satisfies in the matching region* M_1 *the estimates*

$$\begin{aligned}
\|r^{5/2}(\nabla\!\!\!/ \mathrm{tr}\chi', \nabla\!\!\!/ \hat{\chi}')\|_{L^4(M_1)} &\leq \mathcal{E}(t) \\
\|r^{5/2}\nabla\!\!\!/ \zeta'\|_{L^4(M_1)} &\leq \mathcal{E}(t) \\
\|r^{5/2}(\nabla\!\!\!/ \omega', \nabla_N \omega')\|_{L^4(M_1)} &\leq \mathcal{E}(t) \\
\|r^{5/2}(\nabla\!\!\!/^2 b, \nabla\!\!\!/ \nabla_N b, \nabla_N^2 b)\|_{L^4(M_1)} &\leq \mathcal{E}(t)
\end{aligned}$$

where $\mathcal{E}(t)$ is the expression

$$\mathcal{E}(t) = c(2+t)^{1/4}\log(2+t) \cdot \left({}^{e}\mathcal{O}_{[2]} + {}^{\imath}\mathcal{R}_{[2]} + {}^{\imath}\mathcal{K}_{[3]} + {}^{\imath}\mathcal{L}_{[3]} + \varepsilon_1\right).$$

Also

$$\|r^3(\nabla\!\!\!/^2 \mathrm{tr}\chi', \nabla\!\!\!/^2 \hat{\chi}')\|_{L^2(M_1)} \leq c(2+t)^{1/4}\mathcal{E}(t)$$
$$\|r^3 \nabla\!\!\!/^2 \zeta'\|_{L^2(M_1)} \leq c(2+t)^{1/4}\mathcal{E}(t)$$
$$\|r^3(\nabla\!\!\!/^2 \omega', \nabla\!\!\!/ \nabla_N \omega')\|_{L^2(M_1)} \leq c(2+t)^{1/4}\mathcal{E}(t)$$
$$\|r^3(\nabla\!\!\!/^3 b, \nabla\!\!\!/^2 \nabla_N b, \nabla\!\!\!/ \nabla_N^2 b)\|_{L^2(M_1)} \leq c(2+t)^{1/4}\mathcal{E}(t).$$

Finally, for each $q > 0$,

$$\|r^3 \nabla_N^2 \omega'\|_{L^2(M_1)} \leq c(2+t)^{1+q}\log(2+t) \cdot \left[{}^{e}\mathcal{O}_{[2]} + \varepsilon_1^{-(1+q)/2}({}^{\imath}\mathcal{R}_{[2]} + {}^{\imath}\mathcal{K}_{[3]} + {}^{\imath}\mathcal{L}_{[3]}) + \varepsilon_1\right]$$
$$\|r^3 \nabla_N^3 b\|_{L^2(M_1)} \leq c(2+t)^{1+q}\log(2+t) \cdot \left[{}^{e}\mathcal{O}_{[2]} + \varepsilon_1^{-(1+q)/2}({}^{\imath}\mathcal{R}_{[2]} + {}^{\imath}\mathcal{K}_{[3]} + {}^{\imath}\mathcal{L}_{[3]}) + \varepsilon_1\right].$$

We remark that the quantities $\|r^3 \nabla_N^2 \omega'\|_{L^2(M_1)}$, $\|r^3 \nabla_N^3 b\|_{L^2(M_1)}$ can be made as small as we wish by first choosing ε_1 sufficiently small and then choosing a common bound for the quantities ${}^{\imath}\mathcal{R}_{[2]}$, ${}^{\imath}\mathcal{K}_{[3]}$, ${}^{\imath}\mathcal{L}_{[3]}$ that is suitably small relative to ε_1.

CHAPTER 16

The Rotation Vectorfields

In this chapter we construct the rotation vectorfields ${}^{(a)}\Omega$ and estimate their deformation tensors. We assume that the estimates derived in the previous chapters for the exterior optical function u_E, interior optical function u_I, and time function t hold true. We recall that the global optical function u was defined by matching u_E with u_I according to the following formula:

$$u = (1 - f_1)u_E + f_1 u_I$$

where $f_1 = h_1 \circ \vartheta$ and

$$\vartheta = \frac{u_E + (1 + t)}{1 + t}$$

and $h_1(t)$ is a smooth function of t defined such that,

$$h_1 = \begin{cases} = 1; & t \leq \frac{7}{20} \\ = 0; & t \geq \frac{8}{20}. \end{cases} \tag{16.0.1a}$$

In the matching region M_1, defined as the set of points where $\mathbf{D} f_1$ is supported, the properties of the two optical functions are mixed. In particular, in that region u inherits the weaker differentiability properties of u_I in the direction transversal to its level hypersurfaces. Since the directions are also blurred, we conclude that u has weaker differentiability properties in M_1 in all directions. Thus, noting that the exterior optical function u_E has five bounded derivatives locally on any region of Σ_t where it is defined, we find by contrast that u has only four derivatives in the matching region M_1. This would lead to a loss of differentiability of the angular momentum operators, if we were to define them following the integral curves of $l = -\mathbf{D}u$. For this reason we define a new vectorfield L by matching directly the generators $l_E = -\mathbf{D}u_E$ and $l_I = -\mathbf{D}u_I$ of the exterior and interior optical functions respectively. We set

$$L = (1 - f_2)l_E + f_2 l_I \tag{16.0.1b}$$

where the new matching function f_2 is defined by $f_2 = h_2 \circ \vartheta$ with $h_2(t)$ a smooth function of t defined such that

$$h_2 = \begin{cases} = 1; & t \leq \frac{6}{20} \\ = 0; & t \geq \frac{7}{20}. \end{cases} \tag{16.0.1c}$$

Now in the new matching region M_2, defined as the set of points where $\mathbf{D}f_2$ is supported, L still inherits the weak differentiability properties of u_I, but we have the advantage that, since $u = u_I$ there, the 2-surfaces $S_{t,u}$ coincide with level surfaces S_{t,u_I}. We can thus benefit from the knowledge we have about the directional properties of u_I. Indeed, recall that the components $\chi(u_I), \zeta(u_I), \omega(u_I)$ have all three derivatives, tangential to the level hypersurfaces of u_I, locally bounded in $L^2(\Sigma_t)$. Moreover, the first transversal derivative of ω, $\mathbf{D}_3\omega(u_I)$ possesses two tangential derivatives and one transversal derivative, $\mathbf{D}_3^2\omega(u_I)$, locally bounded in $L^2(\Sigma_t)$. As far as the transversal derivatives of $\chi(u_I), \zeta(u_I)$ are concerned we recall that the first transversal derivative of $\chi(u_I)$ behaves precisely as the first tangential derivatives of $\zeta(u_I)$, while the first transversal derivative of $\zeta(u_I)$ behaves precisely as the first tangential derivatives of $\omega(u_I)$. With this in mind we are going to define the angular momentum operators ${}^{(a)}\Omega$, in the space-time slab, by requiring that

$$\begin{aligned} \not{\mathcal{L}}_L {}^{(a)}\Omega &= 0 \\ \langle {}^{(a)}\Omega, l \rangle &= 0 \\ \langle {}^{(a)}\Omega, T \rangle &= 0 \end{aligned} \tag{16.0.2a}$$

where $\not{\mathcal{L}}_L {}^{(a)}\Omega$ is the projection of $\mathcal{L}_L {}^{(a)}\Omega$ to the level surfaces $S_{t,u}$ of u. We also require that on the last slice Σ_{t_*} of the slab the ${}^{(a)}\Omega$ coincide with ${}^{(a)}\Omega_*$, which are defined as follows:

$$\not{\mathcal{L}}_{L_*} {}^{(a)}\Omega_* = 0 \tag{16.0.2b}$$

$$\langle {}^{(a)}\Omega_*, l_* \rangle = 0 \tag{16.0.2c}$$

where l_*, L_* are the projections of l and L on Σ_{t_*} respectively. More precisely, the angular momentum operators ${}^{(a)}\Omega$ are defined as follows:

1. In view of the general construction described in Chapter 3, we can define vectorfields ${}^{(a)}\Omega_*$ on Σ_{t_*} verifying the following properties:

$$[{}^{(a)}\Omega_*, \partial_u] = 0 \tag{16.0.2d}$$

$$\langle {}^{(a)}\Omega_*, N \rangle = 0$$

where N is the unit normal to the radial foliation on Σ_{t_*} given by the level surfaces of u. This definition agrees with 16.0.2c in the exterior region E.

2. Starting with an arbitrary surface $S_{t_*,u}$ in the exterior region E and the restrictions of ${}^{(a)}\Omega_*$ to it, we extend them to all of Σ_{t_*} according to 16.0.2c.
3. Starting with ${}^{(a)}\Omega_*$ on the 2-surfaces $S_{t_*,u}$, we extend them to the whole space-time slab according to the formulas 16.0.2a.

16.1. Estimates in the Exterior

We recall that in the exterior region $r \geq \frac{r_0}{2}$ the optical function u coincides with the exterior optical function u_E and L coincides with the generator $l = -\mathbf{D}u$.

In view of the general construction described in Chapter 3, the vectorfields ${}^{(a)}\Omega_*$, where defined in the exterior region of Σ_{t_*} by

$$[{}^{(a)}\Omega_*, \partial_u] = 0$$
$$\langle {}^{(a)}\Omega_*, N\rangle = 0,$$

also satisfy

$$[{}^{(a)}\Omega_*, {}^{(b)}\Omega_*] = \in_{abc}{}^{(c)}\Omega_*.$$

Moreover, combining the results of Propositions 3.3.3 and 3.3.4 and taking into account conditions3.3.1–3.3.3, we derive the following:

Proposition 16.1.1 *Given any S-tangent covariant tensor f on Σ_{t_*}, there exists a constant c_0, which depends only on $\mathcal{K}^{\infty}_{[1]}, \mathcal{O}^{\infty}_{[1]}, \mathcal{L}^{\infty}_{[1]}$, and $\mathcal{R}^{\infty}_0$, such that, for every $u \geq u_1(t_*)$,*[1]

$$c_0^{-1}\int_{S_{t_*,u}} r^2|\nabla\!\!\!/ f|^2 d\mu_\gamma \leq \int_{S_{t_*,u}} |\mathcal{L}_O f|^2 d\mu_\gamma$$
$$\leq c_0 \int_{S_{t_*,u}} (|f|^2 + r^2|\nabla\!\!\!/ f|^2) d\mu_\gamma$$

where γ is the metric induced on $S_{t_*,u}$, $d\mu_\gamma$ is the corresponding area element on $S_{t_*,u}$, and $|\mathcal{L}_O f|^2 = \sum_{i=1,2,3} |\mathcal{L}_{{}^{(i)}\Omega_*} f|^2$. Moreover, if f is either a 1-form or a symmetric 2-covariant traceless tensor tangent to the surfaces $S_{t_*,u}$,

$$c_0^{-1}\int_{S_{t_*,u}} |f|^2 d\mu_\gamma \leq \int_{S_{t_*,u}} |\mathcal{L}_O f|^2 d\mu_\gamma.$$

We also deduce from Propositions 3.3.1 and 3.3.2 the following bounds:

[1] Recall that $u_1(t)$ is the value of u corresponding to $\frac{r_0}{2}$.

Proposition 16.1.2 *There exists a constant c depending only on $\mathcal{O}^{\infty}_{[1]}, \mathcal{K}^{\infty}_{[0]}, \mathcal{L}^{\infty}_{[1]}$ such that, for every $u \geq u_1(t_*)$,*

$$|r^{-1\,(a)}\Omega|_{\infty,S}(t_*,u) \leq c$$
$$|\nabla\!\!\!/^{(a)}\Omega_{\infty,S}(t_*,u)| \leq c.$$

Moreover, the component ${}^{(a)}\pi_{AN}$ of deformation tensor ${}^{(a)}\pi$ of ${}^{a}\Omega_{(\infty)}$ vanishes identically and ${}^{(a)}H_{AB} = \frac{1}{2}{}^{(a)}\pi_{AB}$ verifies the bound

$$|r^{(a)}H|_{\infty,S}(t_*,u) \leq c(\mathcal{O}^{\infty}_{[1]} + \mathcal{K}^{\infty}_{0} + \mathcal{L}^{\infty}_{[1]}).$$

The angular derivatives of ${}^{(a)}H$ can be estimated by

$$|r^{2-2/p}\nabla\!\!\!/^{(a)}H|_{p,S}(t_*,u) \leq c(\mathcal{O}^{p,S}_{[2]} + \mathcal{K}^{p,S}_{[1]} + \mathcal{L}^{p,S}_{[1]}).$$

Also, in local L^2 norms ${}^{(loc)}\|\ \|(t,\kappa)$, as introduced before Lemma 13.1.1, we have as a consequence

$${}^{(loc)}\|r\nabla\!\!\!/^{(a)}H\|(t_*,\kappa) \leq c\left(\mathcal{O}_{[2]} + \mathcal{K}_{[1]} + \mathcal{L}_{[2]}\right).$$

Finally, the second angular derivatives of ${}^{(a)}H$ can be estimated by[2]

$${}^{(loc)}\|r^2\nabla\!\!\!/^{2(a)}H\|(t_*,\kappa) \leq c\left(\mathcal{O}_{[3]} + \mathcal{K}_{[2]} + \mathcal{L}_{[3]}\right).$$

We now extend the rotation vectorfields ${}^{(a)}\Omega_*$ to the entire exterior region of the space-time slab. Recall that the vectorfields ${}^{(a)}\Omega_*$ are the infinitesimal generators of an action of $SO(3)$ on Σ_{t_*}. To extend this action to the whole exterior region of the space-time slab, we consider the vectorfield l'' proportional to l such that $l''(t) = 1$. Thus $l'' = \phi e'_4 = \phi(T+N)$. Let $\Phi_{t_*,t;u}$ be the diffeomorphisms of $S_{t_*,u}$ onto $S_{t,u}$ generated by the integral curves of l''. Given p, a point on $S_{t,u}$, and O, an element of $SO(3)$, we define the action of O on p by

$$(O,p) \longrightarrow \Phi_{t_*,t,u}(O\Phi^{-1}_{t_*,t;u}(p)).$$

The corresponding rotation vectorfields are defined by the formula

$${}^{(a)}\Omega = (\Phi_{t_*,t,u})_*\ {}^{(a)}\Omega_*.$$

Clearly,

$$\langle{}^{(a)}\Omega, l\rangle = 0$$
$$\langle{}^{(a)}\Omega, T\rangle = 0 \qquad (16.1.1a)$$

[2] This follows by repeating the same argument as in Proposition 3.3.2.

$$[{}^{(a)}\Omega, l''] = 0. \tag{16.1.1b}$$

Also,

$$[{}^{(a)}\Omega, {}^{(b)}\Omega] = \in_{abc} {}^{(c)}\Omega.$$

If we denote by γ_{AB} the metric on $S_{t,u}$ and by χ' the null second fundamental form of $S_{t,u}$ relative to $e_4' = T + N$, we have along C_u

$$\partial_t \gamma_{AB} = 2\phi\chi'_{AB}. \tag{16.1.1c}$$

Recalling that the function r was defined such that the area of $S_{t,u}$ is equal to $4\pi r^2$, we find along C_u

$$\frac{\partial r}{\partial t} = \phi\frac{r}{2}\mathrm{tr}(\chi').$$

Therefore, denoting by $m(t,u)$ the normalized metric on $S_{t,u}$ defined by

$$m = r^{-2}\gamma,$$

we find along C_u

$$\partial_t m_{AB} = 2r^{-2}\phi\chi'_{AB} + r^{-1}\kappa_u\gamma_{AB} \tag{16.1.1d}$$

where

$$\kappa_u = r\phi(\mathrm{tr}(\chi') - \overline{\mathrm{tr}(\chi')}). \tag{16.1.1e}$$

We now consider the metrics $m(t_*,u)$ and $m(t,u)$ as defined on $S_{t_*,u}$ and let Λ, λ be, respectively, the largest and smallest eigenvalues of $m(t,u)$ relative to $m(t_*,u)$. Then we define as in Chapter 3

$$\mu(t,u) = \lambda\Lambda$$
$$\nu(t,u) = \sqrt{\frac{\Lambda}{\lambda}}.$$

We replace the conditions 3.3.1, 3.3.2, and 3.3.3 with the following:

$$\int_t^{t_*} r^{-1}\kappa_u(\tau)d\tau \leq c$$
$$\int_t^{t_*} |\chi'|_{\gamma_\tau} d\tau \leq c \tag{16.1.1f}$$
$$\int_t^{t_*} \kappa_u(\tau)\sup_{S_\tau}|\nabla\!\!\!/\phi|_{\gamma_\tau} d\tau \leq c$$
$$\int_t^{t_*} r|\chi'|_{\gamma_\tau}\sup_{S_\tau}|\nabla\!\!\!/\phi|_{\gamma_\tau} d\tau \leq c$$

$$\int_t^{t_*} \sup_{S_\tau} |\nabla \kappa_u(\tau)|_{\gamma_\tau} d\tau \leq c$$

$$\int_t^{t_*} r \sup_{S_\tau} |\nabla \chi'|_{\gamma_\tau} d\tau \leq c \tag{16.1.1g}$$

$$r^2 K(t, u) \to r^2 K(t_*, u) \tag{16.1.1h}$$

as $t \to t_*$ with $K(t, u)$ the Gauss curvature of $S_{t,u}$. These conditions allow us to conclude that

$$\mu(t, u) = \mu(t_*, u) e^{-\int_t^{t_*} r^{-1} \kappa_u(\tau) d\tau}$$
$$\nu(t, u) \leq \nu(t_*) e^{2 \int_t^{t_*} \phi |\chi'|_{\gamma_\tau} d\tau}$$

and prove the analogues of Propositions 3.3.3 and 3.3.4. Taking into account the estimates we have for $a, \phi, \underline{\chi}', \chi', K$, we easily derive the following extension of Proposition 16.1.1:

Proposition 16.1.3 *Given any S-tangent covariant tensor f on Σ_t, there exists a constant c_O, which depends only on $\mathcal{K}_{[1]}^\infty, \mathcal{O}_{[1]}^\infty, \mathcal{L}_{[1]}^\infty$, and $\mathcal{R}_0^\infty$, such that, for all $u \geq u_1(t)$,*

$$c_O^{-1} \int_{S_{t,u}} r^2 |\nabla f|^2 d\mu_\gamma \leq \int_{S_{t,u}} |\mathcal{L}_O f|^2 d\mu_\gamma \leq c_O \int_{S_{t,u}} (|f|^2 + r^2 |\nabla f|^2) d\mu_\gamma$$

where γ is the metric induced on $S_{t,u}$, $d\mu_\gamma$ is the corresponding area element on $S_{t,u}$, and $|\mathcal{L}_O f|^2 = \sum_{i=1,2,3} |\mathcal{L}_{{}^{(i)}\Omega_*} f|^2$. Moreover, if f is either a 1-form or a symmetric 2-covariant traceless tensor tangent to the surfaces $S_{t,u}$,

$$c_O^{-1} \int_{S_{t,u}} |f|^2 d\mu_\gamma \leq \int_{S_{t,u}} |\mathcal{L}_O f|^2 d\mu_\gamma.$$

We now proceed to estimate ${}^{(a)}\Omega$, as defined previously, and its deformation tensor ${}^{(a)}\pi$ in the exterior region. Since $l = a^{-1} e_4'$, $l'' = \phi e_4'$, we conclude from 16.1.1b that

$$[{}^{(a)}\Omega, {}^{(b)}\Omega] = \epsilon_{abc} \, {}^{(c)}\Omega \tag{16.1.2a}$$
$$[{}^{(a)}\Omega, l] = {}^{(a)}F l \tag{16.1.2b}$$

and

$$\mathbf{g}({}^{(a)}\Omega, T) = \mathbf{g}({}^{(a)}\Omega, l) = \mathbf{g}({}^{(a)}\Omega, \underline{l}) = 0. \tag{16.1.2c}$$

Warning: Throughout the remainder of this section we will use as a null pair the vectorfields $l, \underline{l}$ and thus we refer the reader to the frame coefficients formulas 13.1.1b and 13.1.1c.

From 16.1.2b and 16.1.2c we obtain

$$^{(a)}F = -(\underline{\zeta}_A + \zeta_A)^{(a)}\Omega_A \tag{16.1.3a}$$

and

$$\frac{d}{ds}{}^{(a)}\Omega_A = \chi_{AB}{}^{(a)}\Omega_B. \tag{16.1.3b}$$

We recall that for any vectorfield X we denote $^{(X)}\hat{\pi} = \mathcal{L}_X \mathbf{g}$. In the following we shall denote $^{(^{(a)}\Omega)}\hat{\pi}$ simply as $^{(a)}\pi$. Equations 16.1.2b and 16.1.2c imply

$$\begin{aligned}
\mathbf{D}_4{}^{(a)}\Omega_4 &= 0\\
\mathbf{D}_4{}^{(a)}\Omega_A &= \chi_{AB}{}^{(a)}\Omega_B\\
\mathbf{D}_A{}^{(a)}\Omega_4 &= e^\mu_A l^\nu \mathbf{D}_\mu{}^{(a)}\Omega_\nu = -e^\mu_A{}^{(a)}\Omega_\nu \mathbf{D}_\mu l^\nu\\
&= -\chi_{AB}{}^{(a)}\Omega_B\\
\mathbf{D}_3{}^{(a)}\Omega_4 &= -2\zeta_A{}^{(a)}\Omega_A\\
\mathbf{D}_4{}^{(a)}\Omega_3 &= -2\underline{\zeta}_A{}^{(a)}\Omega_A\\
\mathbf{D}_3{}^{(a)}\Omega_3 &= -2\underline{\xi}_A{}^{(a)}\Omega_A\\
\mathbf{D}_A{}^{(a)}\Omega_3 &= -\underline{\chi}_{AB}{}^{(a)}\Omega_B.
\end{aligned} \tag{16.1.3c}$$

Hence,

$$\begin{aligned}
&^{(a)}\pi_{44} = 0, \qquad {}^{(a)}\pi_{4A} = 0\\
&^{(a)}\pi_{34} = -2(\underline{\zeta}_A + \zeta_A)^{(a)}\Omega_A = 2^{(a)}F\\
&^{(a)}\pi_{33} = -4\underline{\xi}_A{}^{(a)}\Omega_A = 8^{(a)}W.
\end{aligned} \tag{16.1.4a}$$

In the following we shall derive ordinary differential equations for the remaining components

$$^{(a)}\pi_{AB} = 2^{(a)}H_{AB}, \qquad {}^{(a)}\pi_{A3} = 4^{(a)}Z_A. \tag{16.1.4b}$$

We note that in view of 16.1.2b the following commutation relation also holds:

$$[{}^{(a)}\Omega, \underline{l}] = -4^{(a)}Z_A e_A + 2^{(a)}Wl. \tag{16.1.5}$$

Lemma 16.1.1 *Let U be a 2-covariant symmetric tensorfield in space-time such that $U_{44} = U_{4A} = 0$. For any such tensorfield we denote:*

$$\begin{aligned}
U_{AB} &= 2H_{AB}(U), \qquad U_{34} = 2F(U)\\
U_{A3} &= 4Z_A(U), \qquad U_{33} = 8W(U).
\end{aligned}$$

Then $\mathcal{L}_l U$ and $\mathcal{L}_{{}^{(a)}\Omega} U$ are tensorfields of the same type and we have

$$(i)\quad \frac{d}{ds} H_{AB}(\mathcal{L}_{{}^{(a)}\Omega} U) + \chi_{AC} H_{CB}(\mathcal{L}_{{}^{(a)}\Omega} U) + \chi_{BC} H_{CA}(\mathcal{L}_{{}^{(a)}\Omega} U)$$
$$= (\mathcal{L}_{{}^{(a)}\Omega} - {}^{(a)}F) H_{AB}(\mathcal{L}_l U)$$

$$(ii)\quad \frac{d}{ds} Z_A(\mathcal{L}_{{}^{(a)}\Omega} U) + \chi_{AB} Z_B(\mathcal{L}_{{}^{(a)}\Omega} U)$$
$$- (\underline{\zeta}_B - \zeta_B) H_{AB}(\mathcal{L}_{{}^{(a)}\Omega} U) - \frac{1}{2}(\underline{\zeta}_A + \zeta_A) F(\mathcal{L}_{{}^{(a)}\Omega} U)$$
$$= (\mathcal{L}_{{}^{(a)}\Omega} - {}^{(a)}F - 2{}^{(a)}W) Z_A(\mathcal{L}_l U)$$
$$+ 2{}^{(a)}Z_B H_{AB}(\mathcal{L}_l U) - \frac{1}{2} F(U) \nabla\!\!\!/^{(a)}_A F.$$

Proof 16.1.1 *Let X, Y be two arbitrary vectorfields. Then*

$$\mathcal{L}_X \mathcal{L}_Y U - \mathcal{L}_Y \mathcal{L}_X U = \mathcal{L}_{[X,Y]} U.$$

From the basic commutation relation 16.1.2b it then follows that

$$\mathcal{L}_l \mathcal{L}_{{}^{(a)}\Omega} U_{\mu\nu} - \mathcal{L}_{{}^{(a)}\Omega} \mathcal{L}_l U_{\mu\nu} = -{}^{(a)}F \mathcal{L}_l U_{\mu\nu} - \mathbf{D}^{(a)}_\mu F U_{4\nu} - \mathbf{D}^{(a)}_\nu F U_{4\mu}.$$

The AB and $A3$ components of this relation are

$$H_{AB}(\mathcal{L}_l \mathcal{L}_{{}^{(a)}\Omega} U) - H_{AB}(\mathcal{L}_{{}^{(a)}\Omega} \mathcal{L}_l U) = -{}^{(a)}F H_{AB}(\mathcal{L}_l U)$$
$$Z_A(\mathcal{L}_l \mathcal{L}_{{}^{(a)}\Omega} U) - Z_A(\mathcal{L}_{{}^{(a)}\Omega} \mathcal{L}_l U) = -{}^{(a)}F Z_A(\mathcal{L}_l U) - (1/2) \nabla\!\!\!/^{(a)}_A F(\mathcal{L}_l U).$$

We now observe that

$$H_{AB}(\mathcal{L}_{{}^{(a)}\Omega} \mathcal{L}_l U) = (\mathcal{L}_{{}^{(a)}\Omega} H(\mathcal{L}_l U))_{AB}$$

and, in view of commutation relation 16.1.2a,

$$Z_A(\mathcal{L}_{{}^{(a)}\Omega} \mathcal{L}_l U) = \mathcal{L}_{{}^{(a)}\Omega} Z(\mathcal{L}_l U)_A + 2{}^{(a)}Z_B H_{AB}(\mathcal{L}_l U) - 2{}^{(a)}W Z_A(\mathcal{L}_l U).$$

On the other hand,

$$H_{AB}(\mathcal{L}_l \mathcal{L}_{{}^{(a)}\Omega} U) = \frac{d}{ds} H_{AB}(\mathcal{L}_{{}^{(a)}\Omega} U) + \chi_{AC} H_{CB}(\mathcal{L}_{{}^{(a)}\Omega} U) + \chi_{BC} H_{CA}(\mathcal{L}_{{}^{(a)}\Omega} U)$$

and

$$Z_A(\mathcal{L}_l \mathcal{L}_{{}^{(a)}\Omega} U) = \frac{d}{ds} Z_A(\mathcal{L}_{{}^{(a)}\Omega} U) + \chi_{AB} Z_A(\mathcal{L}_{{}^{(a)}\Omega} U)$$
$$- (\underline{\zeta}_B - \zeta_B) H_{AB}(\mathcal{L}_{{}^{(a)}\Omega} U) - (1/2)(\underline{\zeta}_B + \zeta_B) F(\mathcal{L}_{{}^{(a)}\Omega} U).$$

The results then follow. Setting $U_{\mu\nu} = \mathbf{g}_{\mu\nu}$ in Lemma 16.1.1 we deduce the following:

Corollary 16.1.1.1 *Let* ${}^{(a)}F$, ${}^{(a)}W$, ${}^{(a)}H$, ${}^{(a)}Z$ *be the null decomposition of* ${}^{(a)}\pi$. *Then:*

$$(i) \quad \frac{d}{ds}{}^{(a)}H_{AB} + \hat{\chi}^{(a)}_{AC}H_{CB} + \hat{\chi}^{(a)}_{BC}H_{CA} = (\mathcal{L}_{{}^{(a)}\Omega} - {}^{(a)}F)\hat{\chi}_{AB} + \frac{1}{2}\delta_{AB}(\mathcal{L}_{{}^{(a)}\Omega} - {}^{(a)}F)tr\chi$$

$$(ii) \quad \frac{d}{ds}{}^{(a)}Z_A - \frac{1}{2}tr\chi^{(a)}Z_A - \hat{\chi}_{AB}{}^{(a)}Z_B = (\mathcal{L}_{{}^{(a)}\Omega} - {}^{(a)}F - 2{}^{(a)}W)\zeta_A + (\underline{\zeta}_B - \zeta_B)^{(a)}H_{AB} + \frac{1}{2}(\nabla\!\!\!\!/\,{}^{(a)}F - (\underline{\zeta}_A + \zeta_A)^{(a)}F).$$

Writing equation 16.1.3b in the form

$$\frac{d}{ds}{}^{(a)}\Omega_A - \frac{1}{2}tr\chi^{(a)}\Omega_A = \hat{\chi}_{AB}{}^{(a)}\Omega_B$$

and applying Lemma 13.1.1 in the case $p = \infty$ and $\lambda_0 = -1/2, \lambda_1 = -1$, we deduce

$$|r^{-1(a)}\Omega|_{\infty,S}(t,u) \le c|r^{-1(a)}\Omega|_{\infty,S}(t_*,u). \tag{16.1.6}$$

Differentiating equation 16.1.3b tangentially to $S_{t,u}$, we obtain

$$\frac{d}{ds}\nabla\!\!\!\!/_B{}^{(a)}\Omega_A = \hat{\chi}_{AC}\nabla\!\!\!\!/_B{}^{(a)}\Omega_C - \hat{\chi}_{BC}\nabla\!\!\!\!/_A{}^{(a)}\Omega_C +{}^{(a)} E^1_{AB} \tag{16.1.7a}$$

where

$$\begin{aligned}{}^{(a)}E^1_{AB} = {}^{(a)}\Omega_C \Big[&\nabla\!\!\!\!/_B\hat{\chi}_{AC} + (\tfrac{1}{2})\nabla\!\!\!\!/_B tr\chi\delta_{AC} - \epsilon_{AC}{}^*\beta_B \\ &+ (\tfrac{1}{2})tr\chi(\underline{\zeta}_C\delta_{AB} - \underline{\zeta}_A\delta_{BC} + (\underline{\zeta}_B + \zeta_B)\delta_{AC}) \\ &+ \hat{\chi}_{AB}\underline{\zeta}_C - \hat{\chi}_{BC}\underline{\zeta}_A + (\underline{\zeta}_B + \zeta_B)\hat{\chi}_{AC}\Big].\end{aligned} \tag{16.1.7b}$$

Applying Lemma 13.1.1 with $p = \infty$ and $\lambda_0 = 0, \lambda_1 = 0$ yields

$$|\nabla\!\!\!\!/\,{}^{(a)}\Omega|_{\infty,S}(t,u) \le c\left(|\nabla\!\!\!\!/\,{}^{(a)}\Omega|_{\infty,S}(t_*,u) + \int_t^{t_*}|{}^{(a)}E^1|_{\infty,S}(t',u)dt'\right),$$

while, assuming that $\mathcal{O}_1^{\infty,S}(\chi)$, $\mathcal{O}_0^{\infty}(\chi,\zeta)$ as well as $|r^3\beta|_{\infty,S}$ and $|r^2\underline{\zeta}|_{\infty,S}$ are all bounded by a constant c, we have

$$|{}^{(a)}E^1|_{\infty,S} \le cr^{-2}.$$

Taking into account Proposition 16.1.2, we conclude the following:

Lemma 16.1.2 *There exists a constant c, which depends only on $\mathcal{O}^{\infty}_{[1]}$, $\mathcal{K}^{\infty}_{[0]}$, $\mathcal{L}^{\infty}_{[0]}$, and $\mathcal{R}^{\infty}_{0}$, such that,*

$$|r^{-1(a)}\Omega|_{\infty,S}(t,u) \leq c$$
$$|\nabla\!\!\!/^{(a)}\Omega_{\infty,S}(t,u)| \leq c.$$

We are now ready to estimate ${}^{(a)}H$, ${}^{(a)}Z$ in space-time:

Proposition 16.1.4

(i.) There exists a constant c, depending only on ${}^{e}\mathcal{O}^{\infty}_{[1]}$, $\mathcal{K}^{\infty}_{[0]}$, ${}^{e}\mathcal{L}^{\infty}_{0}$, and $\mathcal{R}^{\infty}_{0}$, such that

$$|r^{(a)}H|_{\infty,S}(t,u) \leq c({}^{e}\mathcal{O}^{\infty}_{[1]} + {}^{e}\mathcal{K}^{\infty}_{0} + {}^{e}\mathcal{L}^{\infty}_{0})$$

throughout the exterior region.

(ii.) There exists a constant c, depending only on ${}^{e}\mathcal{O}^{\infty}_{[1]}$, $\mathcal{K}^{\infty}_{[0]}$, ${}^{e}\mathcal{L}^{\infty}_{0}$, and $\mathcal{R}^{\infty}_{0}$, such that

$$|r^{(a)}Z|_{\infty,S}(t,u) \leq c({}^{e}\mathcal{O}^{\infty}_{[1]} + \mathcal{K}^{\infty}_{[1]} + {}^{e}\mathcal{L}^{\infty}_{0})$$

throughout the exterior region.

Proof 16.1.2 *According to conclusion (i) of Corollary 16.1.1.1, we have*

$$\frac{d}{ds}{}^{(a)}H_{AB} + \hat{\chi}_{AC}{}^{(a)}H_{CB} + \hat{\chi}_{BC}{}^{(a)}H_{CA} = {}^{(a)}E^{2}_{AB} \tag{16.1.8a}$$

where

$$^{(a)}E^{2}_{AB} = (\mathcal{L}_{^{(a)}\Omega} - {}^{(a)}F)\hat{\chi}_{AB} + \frac{1}{2}\delta_{AB}(\mathcal{L}_{^{(a)}\Omega} - {}^{(a)}F)tr\chi. \tag{16.1.8b}$$

Now

$$\mathcal{L}_{^{(a)}\Omega}\hat{\chi}_{AB} = {}^{(a)}\Omega_{C}\nabla\!\!\!/_{C}\hat{\chi}_{AB} + \nabla\!\!\!/_{A}{}^{(a)}\Omega_{C}\hat{\chi}_{CB} + \nabla\!\!\!/_{B}{}^{(a)}\Omega_{C}\hat{\chi}_{CA}.$$

Thus by Lemma 16.1.2

$$|\mathcal{L}_{^{(a)}\Omega}\hat{\chi}|_{\infty,S} \leq cr^{-2}\mathcal{O}^{\infty,S}_{[1]}(\chi). \tag{16.1.9a}$$

Similarly,

$$|\mathcal{L}_{^{(a)}\Omega}tr\chi|_{\infty,S} \leq cr^{-2}\mathcal{O}^{\infty,S}_{[1]}(\chi). \tag{16.1.9b}$$

Also, from equations 16.1.4a,

$$|r^{(a)}F|_{\infty,S} \leq c(\mathcal{K}^{\infty}_{0} + {}^{e}\mathcal{L}^{\infty}_{0} + {}^{e}\mathcal{O}^{\infty}_{0}), \tag{16.1.10a}$$

$$|r^{(a)}W|_{\infty,S} \leq c({}^{e}\mathcal{K}_0^\infty + {}^{e}\mathcal{L}_0^\infty + {}^{e}\mathcal{O}_0^\infty). \tag{16.1.10b}$$

It follows that

$$|{}^{(a)}E^2|_{\infty,S} \leq cr^{-2}(\mathcal{O}_1^{\infty,S}(\chi) + {}^{e}\mathcal{O}_0^\infty + {}^{e}\mathcal{K}_0^\infty + {}^{e}\mathcal{L}_0^\infty).$$

Applying Lemma 13.1.1 to 16.1.8a in the case $p = \infty$ *and* $\lambda_0 = 0, \lambda_1 = 0$, *we obtain*

$$|{}^{(a)}H|_{\infty,S}(t,u) \leq c\Big(|{}^{(a)}H|_{\infty,S}(t_*,u) + (r(t,u))^{-1}(\mathcal{O}_1^{\infty,S}(\chi) + {}^{e}\mathcal{O}_0^\infty + {}^{e}\mathcal{K}_0^\infty + {}^{e}\mathcal{L}_0^\infty)\Big)$$

from which, in view of Proposition 16.1.2, part (i) of the proposition follows.

According to part (ii) of Corollary 16.1.1.1, we have

$$\frac{d}{ds}{}^{(a)}Z_A - \frac{1}{2}tr\chi^{(a)}Z_A - \hat{\chi}_{AB}{}^{(a)}Z_B = {}^{(a)}E_A^3 \tag{16.1.11a}$$

where

$${}^{(a)}E_A^3 = (\mathcal{L}_{{}^{(a)}\Omega} - {}^{(a)}F - 2{}^{(a)}W)\zeta_A + (\underline{\zeta}_B - \zeta_B){}^{(a)}H_{AB} + \frac{1}{2}(\nabla\!\!\!/{}^{(a)}F - (\underline{\zeta}_A + \zeta_A){}^{(a)}F). \tag{16.1.11b}$$

We have

$$\mathcal{L}_{{}^{(a)}\Omega}\zeta_A = {}^{(a)}\Omega_B\nabla\!\!\!/_B\zeta_A + \nabla\!\!\!/_A{}^{(a)}\Omega_B\zeta_B.$$

Hence by Lemma 16.1.2

$$|\mathcal{L}_{{}^{(a)}\Omega}\zeta|_{\infty,S} \leq cr^{-2}\mathcal{O}_{[1]}^{\infty,S}(\zeta). \tag{16.1.11c}$$

Also,

$$|r^2\nabla\!\!\!/{}^{(a)}F|_{\infty,S} \leq c({}^{e}\mathcal{K}_{[1]}^\infty + {}^{e}\mathcal{L}_{[1]}^\infty + {}^{e}\mathcal{O}_{[1]}^\infty). \tag{16.1.11d}$$

In view of these estimates, as well as 16.1.10a, 16.1.10b, and the first part of the proposition, we obtain

$$|r^{-1(a)}E^3|_{\infty,S}(t,u) \leq c(r(t,u))^{-3}({}^{e}\mathcal{K}_{[1]}^\infty + {}^{e}\mathcal{L}_{[1]}^\infty + {}^{e}\mathcal{O}_{[1]}^\infty).$$

Applying Lemma 13.1.1 to 16.1.11a in the case $p = \infty$, $\lambda_0 = -1/2$, $\lambda_1 = -1$, yields, in view of Proposition 16.1.2,

$$|r^{-1(a)}Z|_{\infty,S}(t,u) \leq c(r(t,u))^{-2}\left({}^{e}\mathcal{K}_{[1]}^\infty + {}^{e}\mathcal{L}_{[1]}^\infty + {}^{e}\mathcal{O}_{[1]}^\infty\right),$$

from which part (ii) of the proposition follows.

Differentiating equations 16.1.4a we obtain the following expressions for the first derivatives of the deformation tensors of the rotation vectorfields:

$$\mathbf{D}_4{}^{(a)}\pi_{44} = \mathbf{D}_A{}^{(a)}\pi_{44} = \mathbf{D}_3{}^{(a)}\pi_{44} = 0 \tag{16.1.12a}$$

$$\begin{aligned} \mathbf{D}_4{}^{(a)}\pi_{A4} &= 0 \\ \mathbf{D}_B{}^{(a)}\pi_{A4} &= -\chi_{AB}{}^{(a)}F - 2\chi_{BC}{}^{(a)}H_{AC} \\ \mathbf{D}_3{}^{(a)}\pi_{A4} &= -2\zeta_A{}^{(a)}F - 4\zeta_B{}^{(a)}H_{AB} \end{aligned} \tag{16.1.12b}$$

$$\begin{aligned} \mathbf{D}_4{}^{(a)}\pi_{34} &= 2\mathbf{D}_4{}^{(a)}F \\ \mathbf{D}_A{}^{(a)}\pi_{34} &= 2\nabla\!\!\!/_A{}^{(a)}F - 4\chi_{AB}{}^{(a)}Z_B \\ \mathbf{D}_3{}^{(a)}\pi_{34} &= 2\mathbf{D}_3{}^{(a)}F + 8\zeta_A{}^{(a)}Z_A \end{aligned} \tag{16.1.12c}$$

$$\begin{aligned} \mathbf{D}_4{}^{(a)}\pi_{33} &= 4\mathbf{D}_4{}^{(a)}W - 16\underline{\zeta}\cdot{}^{(a)}Z \\ \mathbf{D}_A{}^{(a)}\pi_{33} &= 8\nabla\!\!\!/_A{}^{(a)}W - 8\underline{\chi}_{AB}{}^{(a)}Z_B - 16\zeta_A{}^{(a)}W \\ \mathbf{D}_3{}^{(a)}\pi_{33} &= 8\mathbf{D}_3{}^{(a)}W - 16\underline{\xi}\cdot{}^{(a)}Z - 32\omega^{(a)}W \end{aligned} \tag{16.1.12d}$$

$$\begin{aligned} \mathbf{D}_4{}^{(a)}\pi_{AB} &= 2d^{(a)}H_{AB}/ds \\ \mathbf{D}_C{}^{(a)}\pi_{AB} &= 2\nabla\!\!\!/_C{}^{(a)}H_{AB} - 2\chi_{AC}{}^{(a)}Z_B - 2\chi_{BC}{}^{(a)}Z_A \\ \mathbf{D}_3{}^{(a)}\pi_{AB} &= 2\mathbf{D}\!\!\!/_3{}^{(a)}H_{AB} - 4\zeta_A{}^{(a)}Z_B - 4\zeta_B{}^{(a)}Z_A \end{aligned} \tag{16.1.12e}$$

$$\begin{aligned} \mathbf{D}_4{}^{(a)}\pi_{A3} &= 4d^{(a)}Z_A/ds - 2\underline{\zeta}_A{}^{(a)}F - 4\underline{\zeta}_B{}^{(a)}H_{AB} \\ \mathbf{D}_B{}^{(a)}\pi_{A3} &= 4\nabla\!\!\!/_B{}^{(a)}Z_A - 4\chi_{AB}{}^{(a)}W - \underline{\chi}_{AB}{}^{(a)}F \\ &\quad - 2\underline{\chi}_{BC}{}^{(a)}H_{AC} - 4\zeta_B{}^{(a)}Z_A \\ \mathbf{D}_3{}^{(a)}\pi_{A3} &= 2\mathbf{D}\!\!\!/_3{}^{(a)}Z_A - 8\zeta_A{}^{(a)}W - 2\underline{\xi}_A{}^{(a)}F \\ &\quad - 4\underline{\xi}_B{}^{(a)}H_{AB} - 8\omega^{(a)}Z_A. \end{aligned} \tag{16.1.12f}$$

Remark that the only components of $\mathbf{D}^{(a)}\pi$ that cannot be estimated directly in terms of the quantities that we control already are

$$\mathbf{D}_C^{(a)}\pi_{AB}, \mathbf{D}_B^{(a)}\pi_{A3}, \mathbf{D}_3^{(a)}\pi_{AB}, \mathbf{D}_3^{(a)}\pi_{A3}.$$

That is, we need to estimate

$$\mathbf{D}_C{}^{(a)}H_{AB}, \mathbf{D}_B{}^{(a)}Z_A, \mathbf{D}_3{}^{(a)}H_{AB}, \mathbf{D}_3{}^{(a)}Z_A.$$

As usual, to estimate these quantities, we need to differentiate the equations for ${}^{(a)}H, {}^{(a)}Z$ of Corollary 16.1.1.1. Before doing this, we observe that we don't need to calculate $\mathbf{D}_3{}^{(a)}H_{AB}$, and correspondingly $\mathbf{D}_3^{(a)}\pi_{AB}$, as it can be expressed in terms of $\nabla\!\!\!\!/_B{}^{(a)}Z_A$ according to

$$\begin{aligned}\mathbf{D}_3^{(a)}\pi_{AB} - \mathbf{D}_B^{(a)}\pi_{A3} - \mathbf{D}_A^{(a)}\pi_{B3} = & -2\mathrm{tr}\underline{\chi}^{(a)}H_{AB} \\ & - 2(\hat{\underline{\chi}}_{AB}\nabla\!\!\!\!/_B{}^{(a)}\Omega_C + \hat{\underline{\chi}}_{BC}\nabla\!\!\!\!/_A{}^{(a)}\Omega_C) \\ & - (\in_{AC}{}^*\underline{\beta}_B - \in_{BC}{}^*\underline{\beta}_A)^{(a)}\Omega^C \\ & + {}^{(a)}\Omega^C(\nabla\!\!\!\!/_A\chi_{BC} + \nabla\!\!\!\!/_B\chi_{AC} \\ & \qquad - \zeta\underline{\chi}_{BC} - \zeta_B\underline{\chi}_{AC} \\ & \qquad - 2\underline{\chi}_{AB}\xi_C - 2\chi_{AB}\underline{\zeta}_C).\end{aligned} \tag{16.1.13}$$

To differentiate the equations for ${}^{(a)}H, {}^{(a)}Z$ of Corollary 16.1.1.1, we need to recall the following commutation lemma:[3]

Lemma 16.1.3 *Let U be an arbitrary p-covariant tensorfield and X an arbitrary vectorfield. Then:*

$$\mathbf{D}_3\mathcal{L}_X U_{\alpha_1\ldots\alpha_p} = \mathcal{L}_X\mathbf{D}_3 U_{\alpha_1\ldots\alpha_p} + \sum_{i=1}^{p}{}^{(X)}\Gamma^{\gamma}_{\alpha_i 3}U_{\alpha_1\ldots\gamma\ldots\alpha_p}$$

where

$${}^{(X)}\Gamma^{\gamma}_{\alpha 3} = \frac{1}{2}(\mathbf{D}_\alpha{}^{(X)}\pi^\gamma{}_3 + \mathbf{D}_3{}^{(X)}\pi^\gamma{}_\alpha - \mathbf{D}^{\gamma(X)}\pi_{\alpha 3}).$$

Corollary 16.1.3.1 *If U is S-tangential, then*

$$\nabla\!\!\!\!/_B\mathcal{L}_{{}^{(a)}\Omega}U_{A_1\ldots A_p} = \mathcal{L}_{{}^{(a)}\Omega}\nabla\!\!\!\!/_B U_{A_1\ldots A_p} + \sum_{i=1}^{p}{}^{(a)}\Gamma^{C}_{A_i B}U_{A_1\ldots C\ldots A_p}$$

where

$${}^{(a)}\Gamma^C_{AB} = \nabla\!\!\!\!/_A{}^{(a)}H_{CB} + \nabla\!\!\!\!/_B{}^{(a)}H_{CA} - \nabla\!\!\!\!/_C{}^{(a)}H_{AB}.$$

Corollary 16.1.3.2 *If U is an S-tangential 1-form, then*

$$\begin{aligned}\mathbf{D}\!\!\!\!/_3\mathcal{L}_{{}^{(a)}\Omega}U_A = & \mathcal{L}_{{}^{(a)}\Omega}\mathbf{D}\!\!\!\!/_3 U_A + {}^{(a)}\Gamma^B_{A3}U_B + \mathbf{D}\!\!\!\!/_3{}^{(a)}\Omega_B\nabla\!\!\!\!/_B U_A \\ & + {}^{(a)}\Omega_B\left[(\chi_{AB}\underline{\xi}_C + \underline{\chi}_{AB}\zeta_C)U_C - \zeta_B\mathbf{D}\!\!\!\!/_3 U_A - \underline{\chi}_{BC}\nabla\!\!\!\!/_C U_A\right].\end{aligned}$$

We are now in a position to estimate the angular derivatives.

Proposition 16.1.5

[3] See Chapter 7, page 141.

(i.) *There exists a constant c, depending only on* ${}^{e}\mathcal{O}^{\infty}_{[1]}, {}^{e}\mathcal{K}^{\infty}_{[0]}, {}^{e}\mathcal{L}^{\infty}_{[0]}$, *and* $\mathcal{R}^{\infty}_{0}$, *such that*

$$|r^{2-2/p}\nabla\!\!\!/^{(a)}H|_{p,S}(t,u) \leq c\left(\mathcal{O}^{p,S}_{[2]} + \mathcal{K}^{p,S}_{[1]} + \mathcal{L}^{p,S}_{[1]} + \mathcal{R}^{p,S}_{0}\right)$$

throughout the exterior region.

(ii.) *Also, in norms* ${}^{(loc)}\|\ \|(t,\kappa)$ *as introduced before Lemma 13.1.1, we have*

$${}^{(loc)}\|r\nabla\!\!\!/^{(a)}H\|(t,\kappa) \leq c\left({}^{e}\mathcal{O}_{[2]} + {}^{e}\mathcal{K}_{[1]} + {}^{e}\mathcal{L}_{[0]} + {}^{e}\mathcal{R}_{0}\right)$$

throughout the exterior region.

(iii.) *There exists a constant c, depending only on* ${}^{e}\mathcal{O}^{\infty}_{[1]}, {}^{e}\mathcal{K}^{\infty}_{[1]}, {}^{e}\mathcal{L}^{\infty}_{[0]}$, *and* $\mathcal{R}^{\infty}_{0}$, *such that*

$$|r^{2-2/p}\nabla\!\!\!/^{(a)}Z|_{p,S}(t,u) \leq c\left(\mathcal{O}^{p,S}_{[2]} + \mathcal{K}^{p,S}_{[2]} + \mathcal{L}^{p,S}_{[2]} + \mathcal{R}^{p,S}_{0}\right)$$

throughout the exterior region.

(iv.) *Also, in norms* ${}^{(loc)}\|\ \|(t,\kappa)$,

$${}^{(loc)}\|r\nabla\!\!\!/^{(a)}Z\|(t,\kappa) \leq c\left({}^{e}\mathcal{O}_{[2]} + {}^{e}\mathcal{K}_{[2]} + {}^{e}\mathcal{L}_{[2]} + {}^{e}\mathcal{R}_{0}\right)$$

everywhere in the exterior region.

Proof 16.1.3 *Differentiating the equation 16.1.8a tangentially to* $S_{t,u}$ *and using 16.1.2b, we find*

$$\begin{aligned}\frac{d}{ds}\nabla\!\!\!/_{C}{}^{(a)}H_{AB} + \frac{1}{2}tr\chi\nabla\!\!\!/_{C}{}^{(a)}H_{AB} &= \hat{\chi}_{CD}\nabla\!\!\!/_{D}{}^{(a)}H_{AB}\\ &+ \hat{\chi}_{AD}(\nabla\!\!\!/_{B}{}^{(a)}H_{CD} - \nabla\!\!\!/_{D}{}^{(a)}H_{BC})\\ &+ \hat{\chi}_{BD}(\nabla\!\!\!/_{A}{}^{(a)}H_{CD} - \nabla\!\!\!/_{D}{}^{(a)}H_{AC})\\ &+ \mathcal{L}_{{}^{(a)}\Omega}(\nabla\!\!\!/_{C}\hat{\chi}_{AB}) + (\frac{1}{2})(\mathcal{L}_{{}^{(a)}\Omega}\nabla\!\!\!/_{C}tr\chi)\delta_{AB}\\ &+ {}^{(a)}\not{F}_{ABC} \qquad (16.1.14a)\end{aligned}$$

where

$$\begin{aligned}{}^{(a)}\not{F}_{ABC} = &-\nabla\!\!\!/_{C}\hat{\chi}_{AD}{}^{(a)}H_{DB} - \nabla\!\!\!/_{C}\hat{\chi}_{BD}{}^{(a)}H_{DA}\\ &- {}^{(a)}F\nabla\!\!\!/_{C}\hat{\chi}_{AB} - (\frac{1}{2}){}^{(a)}F(\nabla\!\!\!/_{C}tr\chi)\delta_{AB} - \chi_{AB}\nabla\!\!\!/_{C}{}^{(a)}F\\ &+ (\underline{\zeta}_{C} + \zeta_{C})(-\hat{\chi}_{AD}{}^{(a)}H_{DB} - \hat{\chi}_{BD}{}^{(a)}H_{DA} + {}^{(a)}E_{AB})\\ &+ (\chi_{AC}\underline{\zeta}_{D} - \chi_{CD}\underline{\zeta}_{A} - \epsilon_{AD}\,{}^{*}\beta_{C}){}^{(a)}H_{DB}\\ &+ (\chi_{BC}\underline{\zeta}_{D} - \chi_{CD}\underline{\zeta}_{B} - \epsilon_{BD}\,{}^{*}\beta_{C}){}^{(a)}H_{DA}. \qquad (16.1.14b)\end{aligned}$$

Taking into account 16.1.2a, we have

$$|r^{3-2/p}(\mathcal{L}_{{}^{(a)}\Omega}\nabla\hat{\chi}, \mathcal{L}_{{}^{(a)}\Omega}\nabla tr\chi)|_{p,S} \leq c\mathcal{O}_{[2]}^{p,S}(\chi).$$

Also, in view of

$$|r^{(a)}H|_{\infty,S} \leq c,$$

we can estimate

$$|r^{3-2/p(a)}\not{E}|_{p,S} \leq cr^{-1}(\mathcal{O}_2^{p,S}(\chi) + \mathcal{K}_{[1]}^{p,S} + \mathcal{L}_{[1]}^{p,S} + \mathcal{R}_0^{p,S}).$$

Applying Lemma 13.1.1, on page 354, to 16.1.14a with $\lambda_0 = 1/2$, $\lambda_1 = 1-2/p$, *and using these estimates, we obtain*

$$|r^{1-2/p}\nabla^{(a)}H|_{p,S}(t,u) \leq c\Big[|r^{1-2/p}\nabla^{(a)}H|_{p,S}(t_*,u)$$
$$+ r^{-1}\big(\mathcal{O}_2^{p,S}(\chi) + \mathcal{O}_{[1]}^{p,S} + \mathcal{K}_{[1]}^{p,S} + \mathcal{L}_{[1]}^{p,S} + \mathcal{R}_0^{p,S}\big)\Big],$$

which, in view of Proposition 16.1.2, yields the conclusions of part (i) of the proposition. The conclusion of part (ii) of the proposition is an immediate consequence of part (i), $p = 2$, *by a simple integration.*

Differentiating equation 16.1.11a tangentially to $S_{t,u}$ *and using Corollary 16.1.2b we find*

$$\frac{d}{ds}\nabla_B{}^{(a)}Z_A = \hat{\chi}_{AC}\nabla_B{}^{(a)}Z_C - \hat{\chi}_{BC}\nabla_C{}^{(a)}Z_A$$
$$+ \mathcal{L}_{{}^{(a)}\Omega}\nabla_B\zeta_A + {}^{(a)}\Gamma_{AB}^C\zeta_C + (\frac{1}{2})\nabla_A\nabla_B{}^{(a)}F$$
$$+ (\underline{\zeta}_C - \zeta_C)\nabla_B{}^{(a)}H_{AB} - (\frac{1}{2})(\underline{\zeta}_A - \zeta_A)\nabla_B{}^{(a)}F$$
$$- 2\zeta_A\nabla_B{}^{(a)}W - {}^{(a)}\not{E}_{AB}^2 \tag{16.1.15a}$$

where

$${}^{(a)}\not{E}_{AB}^2 = (\frac{1}{2}){}^{(a)}Z_A\nabla_B tr\chi + {}^{(a)}Z_B\nabla_B\hat{\chi}_{AC}$$
$$- {}^{(a)}H_{AC}\nabla_B(\underline{\zeta}_C - \zeta_C) + (\frac{1}{2}){}^{(a)}F\nabla_B(\underline{\zeta}_A - \zeta_A)$$
$$+ (\underline{\zeta}_B + \zeta_B)(\chi_{AB}{}^{(a)}Z_B + {}^{(a)}E_A)$$
$$+ (\chi_{AB}\underline{\zeta}_C - \chi_{BC}\underline{\zeta}_A - \epsilon_{AC}{}^*\beta_B){}^{(a)}Z_C \tag{16.1.15b}$$

with ${}^{(a)}E_A$ *given by 16.1.11b. Now*

$$|r^{3-2/p}\mathcal{L}_{{}^{(a)}\Omega}\nabla\zeta|_{p,S} \leq c\mathcal{O}_{[2]}^{p,S}(\zeta).$$

Applying the Corollary 16.1.3.1 of Lemma 16.1.3 to the manifold $S_{t,u}$ and the vectorfield ${}^{(a)}\Omega$, we easily check that

$$\nabla\!\!\!/_A \nabla\!\!\!/_B {}^{(a)}\Omega^C = {}^{(a)}\Gamma^C_{AB} - K(\delta_{AB}{}^{(a)}\Omega_C - \delta_{BC}{}^{(a)}\Omega_A),$$

which, in view of

$$|r^{2-2/p}\nabla\!\!\!/^{(a)}H|_{p,S} \le c\left(\mathcal{O}^{p,S}_{[2]} + \mathcal{K}^{p,S}_{[1]} + \mathcal{L}^{p,S}_{[1]} + \mathcal{R}^{p,S}_0\right),$$

implies

$$|r^{1-2/p}\nabla\!\!\!/^{2(a)}\Omega|_{p,S} \le c\left(\mathcal{O}^{p,S}_{[2]} + \mathcal{K}^{p,S}_{[1]} + \mathcal{L}^{p,S}_{[1]} + \mathcal{R}^{p,S}_0\right). \tag{16.1.16a}$$

Using 16.1.16a, we can estimate

$$|r^{3-2/p}\nabla\!\!\!/^{2(a)}F|_{p,S} \le cr\left(\mathcal{O}^{p,S}_{[2]} + \mathcal{K}^{p,S}_{[2]} + \mathcal{L}^{p,S}_{[2]} + \mathcal{R}^{p,S}_0\right). \tag{16.1.16b}$$

We can also estimate

$$|r^{3-2/p(a)}\not{E}^2|_{p,S} \le cr^{-1}(\mathcal{O}^{p,S}_{[1]} + \mathcal{K}^{p,S}_{[1]} + \mathcal{L}^{p,S}_{[1]})$$

using

$$|r^{(a)}Z|_{\infty,S} \le c.$$

Applying Lemma 13.1.1, on page 354, to 16.1.14a with $\lambda_0 = 0$, $\lambda_1 = -2/p$, and using these estimates as well as the results of the first part of the proposition, we obtain

$$|r^{-2/p}\nabla\!\!\!/^{(a)}Z|_{p,S}(t,u) \le cr^{-2}\left(\mathcal{O}^{p,S}_{[2]} + \mathcal{K}^{p,S}_{[2]} + \mathcal{L}^{p,S}_{[2]} + \mathcal{R}^{p,S}_0\right),$$

from which the conclusions of part (iii) follow. The conclusions of part (iv) follow then easily by integration.

It remains to estimate $\not{D}_3{}^{(a)}Z$. We achieve this in the following:

Proposition 16.1.6 *There exists a constant c, depending only on the quantities ${}^e\mathcal{O}^{\infty}_{[1]}$, ${}^e\mathcal{K}^{\infty}_{[0]}$, ${}^e\mathcal{L}^{\infty}_{[0]}$, and $\mathcal{R}^{\infty}_0$, such that*

$$|r^{1/2-2/p}w_{3/2}\not{D}_3{}^{(a)}Z|_{p,S}(t,u) \le c\left(\mathcal{O}^{p,S}_{[2]} + \mathcal{K}^{p,S}_{[2]} + \mathcal{L}^{p,S}_{[2]} + \mathcal{R}^{p,S}_0\right)$$

where, we recall, $w_{3/2} = \min\{\tau_-^{3/2}r_0^{1/2}, r^{3/2}\}$.

Moreover, in norms ${}^{(loc)}\| \ \|(t,\kappa)$,

$${}^{(loc)}\|\tau_-\not{D}_3{}^{(a)}Z\|(t,\kappa) \le c\left({}^e\mathcal{O}_{[2]} + {}^e\mathcal{K}_{[2]} + {}^e\mathcal{L}_{[2]} + {}^e\mathcal{R}_0\right).$$

Proof 16.1.4 *We first differentiate the equation 16.1.11a with respect to* e_3 *according to the results of Lemma 13.1.2 to derive*

$$\frac{d}{ds}\not{D}_3{}^{(a)}Z_A - \frac{1}{2}tr\chi\not{D}_3{}^{(a)}Z_A = \hat{\chi}_{AB}\not{D}_3{}^{(a)}Z_B + \tilde{E}_A \quad (16.1.17a)$$

$$\begin{aligned}\tilde{E}_A = {} & \not{D}_3{}^{(a)}E^3 + 2\omega{}^{(a)}E^3 \\ & + (\not{D}_3\chi_{AB} + 2\omega\chi_{AB})^{(a)}Z_B \\ & + 2(\underline{\zeta}_B - \zeta_B)\not{\nabla}_B{}^{(a)}Z_A \\ & - 2(\zeta_A\underline{\zeta}_B - \underline{\zeta}_A\zeta_A - \sigma\epsilon_{AB})^{(a)}Z_B \quad (16.1.17b)\end{aligned}$$

where ${}^{(a)}E^3$ *is the error term 16.1.11b appearing in equation 16.1.11a. Using Corollary 16.1.3.2, we calculate* $\not{D}_3{}^{(a)}E^3$ *as follows:*

$$\begin{aligned}\not{D}_3{}^{(a)}E^3_A = {} & \mathcal{L}_{{}^{(a)}\Omega}\not{D}_3\zeta_A + {}^{(a)}\Gamma^C_{A3}\zeta_C + \not{D}_3{}^{(a)}\Omega_B\not{\nabla}_B\zeta_A \\ & + {}^{(a)}\Omega_B\left[(\chi_{AB}\underline{\xi}_C + \underline{\chi}_{AB}\zeta_C)\zeta_C - \zeta_B\not{D}_3\zeta_A - \underline{\chi}_{BC}\not{\nabla}_C\zeta_A\right] \\ & + \not{D}_3({}^{(a)}E^3 - \mathcal{L}_{{}^{(a)}\Omega}\zeta_A) \quad (16.1.17c)\end{aligned}$$

$$\Gamma^C_{A3} = \frac{1}{2}(\mathbf{D}^{(a)}_A\pi_{C3} + \mathbf{D}^{(a)}_3\pi_{AC} - \mathbf{D}^{(a)}_C\pi_{A3}). \quad (16.1.17d)$$

Taking into account 16.1.13 to express $\mathbf{D}^{(a)}_3\pi_{AC}$ *in terms of* $\mathbf{D}^{(a)}_C\pi_{A3}$ *and* $\mathbf{D}^{(a)}_A\pi_{C3}$, *using 16.1.12f to express* $\mathbf{D}^{(a)}_C\pi_{A3}$, $\mathbf{D}^{(a)}_A\pi_{C3}$ *in terms of* $\not{\nabla}_C{}^{(a)}Z_A, \not{\nabla}_A{}^{(a)}Z_C$, *and using the estimates we already have for* $H, Z, \not{\nabla}H, \not{\nabla}Z$, *we infer that*[4]

$$|r^{3/2-2/p}\tilde{E}w_{3/2}|_{p,S}(t,u) \leq c(\mathcal{O}^{p,S}_{[2]} + \mathcal{K}^{p,S}_{[2]} + \mathcal{L}^{p,S}_{[2]} + \mathcal{R}^{p,S}_0). \quad (16.1.17e)$$

Thus, applying once again the basic Lemma 13.1.1, we deduce

$$\begin{aligned}|r^{-1/2-2/p}\mathbf{D}_3{}^{(a)}Z|_{p,S}(t,u) \leq {} & c|r^{-1/2-2/p}\mathbf{D}_3{}^{(a)}Z|_{p,S}(t_*,u) \quad (16.1.17f) \\ & + r^{-1}w^{-1}_{3/2}(\mathcal{O}^{p,S}_{[2]} + \mathcal{K}^{p,S}_{[2]} + \mathcal{L}^{p,S}_{[2]} + \mathcal{R}^{p,S}_0),\end{aligned}$$

which, in view of Proposition 16.1.2, ends the proof of the proposition.

The following theorem contains all the estimates for the first derivatives of the deformation tensors of rotation vectorfields in a form that was used as an assumption in the proof of Theorem 8.2.1. It is an immediate consequence of Propositions 16.1.4, 16.1.5, 16.1.6, and the formulas of Corollary 16.1.1.1.

[4] The presence of the weight $w_{3/2}$ is due to the contribution of $\hat{\eta}$ to $\hat{\underline{\chi}}$.

Theorem 16.1.1 *There exists a constant c, depending only on the quantities* ${}^e\mathcal{O}^\infty_{[1]}, {}^e\mathcal{K}^\infty_{[0]}, {}^e\mathcal{L}^\infty_{[0]}$, *and* $\mathcal{R}^\infty_0$, *such that the components* ${}^{(a)}F, {}^{(a)}W, {}^{(a)}H, {}^{(a)}Z$ *of* ${}^{(a)}\pi$ *verify the following estimates in the exterior region of the space-time:*

$$\|r({}^{(a)}H, {}^{(a)}Z, {}^{(a)}F, {}^{(a)}W)\|_{\infty,e} \leq c\left({}^e\mathcal{O}^\infty_{[1]} + {}^e\mathcal{K}^\infty_{[0]} + {}^e\mathcal{L}^\infty_{[0]}\right) \tag{16.1.18a}$$

$$\begin{aligned}
{}^{(loc)}\|r\nabla\!\!\!/({}^{(a)}H, {}^{(a)}Z, {}^{(a)}F, {}^{(a)}W)\|_{2,e} &\leq c\left({}^e\mathcal{O}_{[2]} + {}^e\mathcal{K}_{[2]} + {}^e\mathcal{L}_{[2]} + {}^e\mathcal{R}_0\right)\\
{}^{(loc)}\|\tau_-\mathbf{D}\!\!\!\!/_3({}^{(a)}H, {}^{(a)}Z, {}^{(a)}F, {}^{(a)}W)\|_{2,e} &\leq c\left({}^e\mathcal{O}_{[2]} + {}^e\mathcal{K}_{[2]} + {}^e\mathcal{L}_{[2]} + {}^e\mathcal{R}_0\right)\\
{}^{(loc)}\|r\mathbf{D}\!\!\!\!/_4({}^{(a)}H, {}^{(a)}Z, {}^{(a)}F, {}^{(a)}W)\|_{2,e} &\leq c\left({}^e\mathcal{O}_{[2]} + {}^e\mathcal{K}_{[2]} + {}^e\mathcal{L}_{[2]} + {}^e\mathcal{R}_0\right).
\end{aligned} \tag{16.1.18b}$$

Moreover,

$$\begin{aligned}
|||r^{3/2}\nabla\!\!\!/({}^{(a)}H, {}^{(a)}Z, {}^{(a)}W, {}^{(a)}F)|||_{4,e} &\leq c\left({}^e\mathcal{O}_{[3]} + {}^e\mathcal{K}_{[3]} + {}^e\mathcal{L}_{[3]} + {}^e\mathcal{R}_{[1]}\right)\\
|||r^{1/2}\tau_-\mathbf{D}\!\!\!\!/_3({}^{(a)}H, {}^{(a)}Z, {}^{(a)}W, {}^{(a)}F)|||_{4,e} &\leq c\left({}^e\mathcal{O}_{[3]} + {}^e\mathcal{K}_{[3]} + {}^e\mathcal{L}_{[3]} + {}^e\mathcal{R}_{[1]}\right)\\
|||r^{3/2}\mathbf{D}\!\!\!\!/_4({}^{(a)}H, {}^{(a)}Z, {}^{(a)}W, {}^{(a)}F)|||_{4,e} &\leq c\left({}^e\mathcal{O}_{[3]} + {}^e\mathcal{K}_{[3]} + {}^e\mathcal{L}_{[3]} + {}^e\mathcal{R}_{[1]}\right).
\end{aligned} \tag{16.1.18c}$$

In the next theorem we state the results about the second derivatives of deformation tensors of the rotation vectorfields that we need in the proof of Theorem 8.2.1. These results depend on proving estimates for

$$\nabla\!\!\!/^{2\,(a)}H,\ \nabla\!\!\!/^{2\,(a)}Z,\ \nabla\!\!\!/\mathbf{D}\!\!\!\!/_3{}^{(a)}Z,$$

which we state in the following proposition:

Proposition 16.1.7

(i.) There exists a constant c, depending only on $\mathcal{O}^\infty_{[1]}, {}^e\mathcal{K}^\infty_{[0]}, {}^e\mathcal{L}^\infty_{[0]}$, *and* $\mathcal{R}^\infty_0$, *such that*

$${}^{(loc)}\|r^2\nabla\!\!\!/^{2\,(a)}H(t)\|_{2,e} \leq c\left({}^e\mathcal{O}_{[2]} + {}^e\mathcal{K}_{[2]} + {}^e\mathcal{L}_{[2]} + {}^e\mathcal{R}_{[1]}\right)$$

for all $t \leq t_*$ *in the exterior region.*

(ii.) There exists a constant c, depending only on $\mathcal{O}^\infty_{[1]}, {}^e\mathcal{K}^\infty_{[1]}, {}^e\mathcal{L}^\infty_{[2]}$, *and* $\mathcal{R}^\infty_0$, *such that*

$${}^{(loc)}\|r^2\nabla\!\!\!/^{2\,(a)}Z(t)\|_{2,e} \leq c\left({}^e\mathcal{O}_{[3]} + {}^e\mathcal{K}_{[3]} + {}^e\mathcal{L}_{[4]} + {}^e\mathcal{R}_{[1]}\right)$$

for all $t \leq t_*$, *in the exterior region.*

(iii.) There exists a constant c, depending only on $\mathcal{O}^{\infty}_{[1]}, {}^e\mathcal{K}^{\infty}_{[1]}, {}^e\mathcal{L}^{\infty}_{[1]}$, *and* $\mathcal{R}^{\infty}_0$, *such that*

$$(loc)\|r\tau_-\nabla\!\!\!/\mathbf{D}\!\!\!\!/_3{}^{(a)}Z(t)\|_{2,e} \leq c\left({}^e\mathcal{O}_{[3]} + {}^e\mathcal{K}_{[3]} + {}^e\mathcal{L}_{[3]} + {}^e\mathcal{R}_{[1]}\right)$$

for all $t \leq t_*$.

The proof of the proposition follows by differentiating the equations verified by $\nabla\!\!\!/{}^{(a)}H, \nabla\!\!\!/{}^{(a)}Z, \mathbf{D}\!\!\!\!/_3{}^{(a)}Z$ once more in the angular directions and following the same ideas as in the proof of Propositions 16.1.5 and 16.1.6.

Finally, we state the following:

Theorem 16.1.2 *There exists a constant c depending only on* $\mathcal{O}^{\infty}_{[1]}, {}^e\mathcal{K}^{\infty}_{[0]}, {}^e\mathcal{L}^{\infty}_{[0]}$, *and* $\mathcal{R}^{\infty}_0$, *such that the components* ${}^{(a)}F, {}^{(a)}W, {}^{(a)}H, {}^{(a)}Z$ *of* ${}^{(a)}\pi$ *verify the estimates*

$$(loc)\|r^2\nabla\!\!\!/^2({}^{(a)}H, {}^{(a)}Z, {}^{(a)}F, {}^{(a)}W)\|_{2,e} \leq c\left({}^e\mathcal{O}_{[3]} + {}^e\mathcal{K}_{[3]} + {}^e\mathcal{L}_{[3]} + {}^e\mathcal{R}_{[1]}\right)$$
$$(loc)\|r\tau_-\nabla\!\!\!/\mathbf{D}\!\!\!\!/_3({}^{(a)}H, {}^{(a)}Z, {}^{(a)}F, {}^{(a)}W)\|_{2,e} \leq c\left({}^e\mathcal{O}_{[3]} + {}^e\mathcal{K}_{[3]} + {}^e\mathcal{L}_{[3]} + {}^e\mathcal{R}_{[1]}\right)$$
$$(loc)\|r^2\nabla\!\!\!/\mathbf{D}\!\!\!\!/_4({}^{(a)}H, {}^{(a)}Z, {}^{(a)}F, {}^{(a)}W)\|_{2,e} \leq c\left({}^e\mathcal{O}_{[2]} + {}^e\mathcal{K}_{[2]} + {}^e\mathcal{L}_{[3]} + {}^e\mathcal{R}_{[1]}\right). \tag{16.1.19a}$$

Moreover,

$$(loc)\|r^3\nabla\!\!\!/^2\mathbf{D}\!\!\!\!/_4({}^{(a)}H, {}^{(a)}F)\|_{2,e} \leq c\left({}^e\mathcal{O}_{[3]} + {}^e\mathcal{K}_{[3]} + {}^e\mathcal{L}_{[3]} + {}^e\mathcal{R}_{[1]}\right). \tag{16.1.19b}$$

The last part of the theorem follows immediately for ${}^{(a)}F$, and it is a consequence of the formula (i) of Corollary 16.1.1.1 for ${}^{(a)}H$.

16.2. Estimates in the Interior

In this section we estimate ${}^{(a)}\pi$ in the interior region of the space-time. We recall that in the matching region $\mathcal{M}_1 \bigcup \mathcal{M}_2$ the vectorfields l, L may be different from each other and non-null. Moreover, L may not be orthogonal to the surfaces $S_{t,u}$ in the matching region $\mathcal{M}_2$. Relative to the standard pair $e'_4 = (T+N), e'_3 = (T-N)$ and and an orthonormal frame $(e_A)_{A=1,2}$, we write

$$l = {}^{(l)}\underline{p}e'_4 + {}^{(l)}pe'_3 \tag{16.2.1a}$$
$$L = {}^{(L)}\underline{p}e'_4 + {}^{(L)}pe'_3 + L_Ae_A. \tag{16.2.1b}$$

According to the definition of ${}^{(a)}\Omega$ (see 16.0.2a) we have

$$[L, {}^{(a)}\Omega] = {}^{(a)}\underline{q}e'_4 + {}^{(a)}qe'_3$$
$$\langle{}^{(a)}\Omega, e'_4\rangle = 0 \tag{16.2.1c}$$
$$\langle{}^{(a)}\Omega, e'_3\rangle = 0.$$

We recall the definition of the Ricci coefficients relative to the standard null frame:

$$
\begin{aligned}
\langle \mathbf{D}_A e_3, e_B\rangle &= \underline{\chi}'_{AB} & \langle \mathbf{D}_A e_4, e_B\rangle &= \chi'_{AB} \\
\tfrac{1}{2}\langle \mathbf{D}_3 e_3, e_A\rangle &= \underline{\xi}'_A & \tfrac{1}{2}\langle \mathbf{D}_4 e_4, e_A\rangle &= \xi_A \\
\tfrac{1}{2}\langle \mathbf{D}_4 e_3, e_A\rangle &= \underline{\zeta}'_A & \tfrac{1}{2}\langle \mathbf{D}_3 e_4, e_A\rangle &= \zeta'_A \\
\tfrac{1}{2}\langle \mathbf{D}_3 e_3, e_4\rangle &= \underline{\nu}' & \tfrac{1}{2}\langle \mathbf{D}_4 e_4, e_3\rangle &= \nu' \\
\tfrac{1}{2}\langle \mathbf{D}_A e_4, e_3\rangle &= \epsilon_A
\end{aligned}
\tag{16.2.2a}
$$

where, calculating in terms of our basic geometric quantities,

$$
\begin{aligned}
\chi'_{AB} &= \theta_{AB} - k_{AB} \\
\underline{\chi}'_{AB} &= -\theta_{AB} - k_{AB} \\
\underline{\zeta}'_A &= \phi^{-1}\nabla_A\phi - \epsilon_A - \xi'_A \\
\zeta'_A &= a^{-1}\nabla_A a + \epsilon_A + \xi'_A \\
\underline{\xi}'_A &= \phi^{-1}\nabla_A\phi - a^{-1}\nabla_A a - \xi'_A \\
\nu' &= -\phi^{-1}\nabla_N\phi + \delta \\
\underline{\nu}' &= \phi^{-1}\nabla_N\phi + \delta
\end{aligned}
\tag{16.2.2b}
$$

and

$$
\xi'_A = -a^2 e_A(\underline{\Delta}) \tag{16.2.2c}
$$

with $\underline{\Delta} = -\frac{1}{2}\langle l, l\rangle$.

Warning: In the remaining part of this section we will drop the primes from the Ricci coefficients of the standard frame.

We also introduce the following:

$$
\langle \mathbf{D}_L e_3, e_A\rangle = 2\underline{B}_A \tag{16.2.3a}
$$

$$
\langle \mathbf{D}_L e_4, e_A\rangle = 2B_A \tag{16.2.3b}
$$

$$
\langle \mathbf{D}_A L, e_3\rangle = 2\underline{G}_A \tag{16.2.3c}
$$

$$
\langle \mathbf{D}_A L, e_4\rangle = 2G_A \tag{16.2.3d}
$$

$$
\langle \mathbf{D}_4 L, T\rangle = 2U \tag{16.2.3e}
$$

$$
\langle \mathbf{D}_3 L, T\rangle = 2\underline{U} \tag{16.2.3f}
$$

$$\langle \mathbf{D}_L T, e_A \rangle = V \tag{16.2.3g}$$
$$\langle \mathbf{D}_T L, e_A \rangle = W \tag{16.2.3h}$$
$$\langle \mathbf{D}_L T, N \rangle = R. \tag{16.2.3i}$$

Also,

$${}^L\chi_{AB} = \langle \mathbf{D}_A L, e_B \rangle. \tag{16.2.3j}$$

Remark that, in view of the fact that in the matching region M_2, L may not be orthogonal to $S_{t,u}$, ${}^L\chi_{AB}$ may not be symmetric there.

Using these notations, we find, from 16.2.1c,

$${}^{(a)}\underline{q} = (\underline{G} + \underline{B}) \cdot {}^{(a)}\Omega \tag{16.2.4a}$$
$${}^{(a)}q = (G + B) \cdot {}^{(a)}\Omega. \tag{16.2.4b}$$

Also,

$$\not{\mathbf{D}}_L {}^{(a)}\Omega = {}^L\chi \cdot {}^{(a)}\Omega. \tag{16.2.4c}$$

This equation allows us to propagate ${}^{(a)}\Omega$. More precisely, taking τ, a parameter of L, $L(\tau) = 1$, we can propagate ${}^{(a)}\Omega$ from its value on the last slice or its value on the boundary of the interior region, according to the differential equation

$$\frac{d}{d\tau} {}^{(a)}\Omega_A = {}^L\chi_{BA} {}^{(a)}\Omega_B.$$

In view of the definition of L, ${}^L\chi$ has the same differentiability properties as l_E outside $M_2 \bigcup I_0$, with I_0 the inner interior region, and the differentiability properties of l_I inside the region $M_2 \bigcup I_0$. Since in that region $u = u_I$, it follows that ${}^L\chi$ has the same differentiability properties as the $\chi(u_I)$. Taking into account the asymptotic properties of the components of the Hessian of u_E and those of $\chi(u_I)$, we easily deduce, as in Lemma 16.1.2, that

$$|r^{-1} {}^{(a)}\Omega|_{\infty,S}(t, u) \le c \tag{16.2.5a}$$
$$|\nabla\!\!\!/ {}^{(a)}\Omega_{\infty,S}(t, u)| \le c, \tag{16.2.5b}$$

provided that these estimates are verified on the last slice.

We can also calculate the following:

$$\langle \mathbf{D}_A {}^{(a)}\Omega, L \rangle = \nabla\!\!\!/_A ({}^{(a)}\Omega \cdot L) - \langle {}^{(a)}\Omega, \mathbf{D}_A L \rangle \tag{16.2.6a}$$
$$= \nabla\!\!\!/_A ({}^{(a)}\Omega \cdot L) - {}^L\chi_{AB} {}^{(a)}\Omega_B. \tag{16.2.6b}$$

Hence

$${}^{(a)}\pi_{LA} = {}^L\chi_{[BA]} {}^{(a)}\Omega_B + \nabla\!\!\!/_A ({}^{(a)}\Omega \cdot L) \tag{16.2.6c}$$

where ${}^L\chi_{[BA]} = \frac{1}{2}({}^L\chi_{BA} - {}^L\chi_{AB})$. The following components of ${}^{(a)}\pi$ can be also computed in a straightforward manner:

$$
\begin{aligned}
{}^{(a)}\pi_{44} &= -4{}^{(a)}\Omega \cdot \xi \\
{}^{(a)}\pi_{33} &= -4{}^{(a)}\Omega \cdot \underline{\xi} \\
{}^{(a)}\pi_{34} &= -2{}^{(a)}\Omega \cdot (\underline{\zeta} + \zeta)
\end{aligned}
\tag{16.2.6d}
$$

All the remaining components of ${}^{(a)}\pi$ can be calculated provided we can calculate ${}^{(a)}\pi_{AB} = 2{}^{(a)}H_{AB}$, ${}^{(a)}\pi_{AT} = 4{}^{(a)}Y_A$. In order to calculate them, we derive propagation equations, following the same ideas as in the exterior region.

First, in view of the commutation formula 16.2.1c, we have

$$(\mathcal{L}_L\mathcal{L}_{{}^{(a)}\Omega} - \mathcal{L}_{{}^{(a)}\Omega}\mathcal{L}_L)\mathbf{g} = \mathcal{L}_{{}^{(a)}\underline{q}e_4}\mathbf{g} + \mathcal{L}_{{}^{(a)}qe_3}\mathbf{g}$$

or, since

$$\mathcal{L}_{fX}\mathbf{g}_{\alpha\beta} = f^{(X)}\pi + \mathbf{D}_\beta f X_\alpha + \mathbf{D}_\alpha f X_\beta,$$

hence

$$\mathcal{L}_L{}^{(a)}\pi_{\alpha\beta} = \mathcal{L}_{{}^{(a)}\Omega}{}^{(L)}\pi_{\alpha\beta} + {}^{(a)}\underline{q}{}^{(e_4)}\pi + {}^{(a)}q{}^{(e_3)}\pi + C_{\alpha\beta} \tag{16.2.7a}$$

where

$$
\begin{aligned}
C_{\alpha\beta} &= \mathbf{D}_\alpha{}^{(a)}\underline{q}(e_4)_\beta + \mathbf{D}_\beta{}^{(a)}\underline{q}(e_4)_\alpha && (16.2.7b) \\
&\quad + \mathbf{D}_\alpha{}^{(a)}q(e_3)_\beta + \mathbf{D}_\beta{}^{(a)}q(e_3)_\alpha. && (16.2.7c)
\end{aligned}
$$

We remark that C involves the derivatives of ${}^{(a)}\underline{q}$, ${}^{(a)}q$, which contain derivatives of the potentially bad terms $\underline{B}, B$. Indeed, we recall from 16.2.3a and 16.2.3b that $\underline{B}, B$ involve derivatives of e_3, e_4, therefore derivatives of l. Thus, C depends in principle of the second derivatives of l. Since we need to estimate two derivatives of ${}^{(a)}\pi$ in L^2 on slices, this requires four derivatives of l, which, as we explained at the beginning of the chapter, we simply don't have in the matching region M_1. We make, however, the crucial observation[5] that the purely tangential components of C are zero while

$$C_{AT} = \mathbf{D}_A({}^{(a)}\underline{q} + {}^{(a)}q).$$

Now, recall that

$${}^{(a)}\underline{q} + {}^{(a)}q = {}^{(a)}\Omega \cdot (\underline{G} + G) + {}^{(a)}\Omega \cdot (\underline{B} + B)$$

[5] This is to be expected in view of our definition of ${}^{(a)}\Omega$' s, in particular due to the requirement that they act tangentially to $S_{t,u}$.

and, from 16.2.3a and 16.2.3b,

$$\underline{B} + B = \langle \mathbf{D}_L T, e_A \rangle \tag{16.2.8a}$$
$$\underline{G} + G = \langle \mathbf{D}_A L, T \rangle. \tag{16.2.8b}$$

Hence, C_{AT} involves only the sum of the potentially bad terms $\underline{B}$, B, which depends only on the derivatives of T, not l. On the other hand, the sum $\underline{G} + G$ depends on the derivatives of L and thus on the Hessian of u_I in the matching region M_2. Here, however, we can take advantage on the fact that, since the surfaces $S_{t,u}$ coincide with S_{t,u_I}, the Hessian of u_I has better differentiability in the directions tangential to $S_{t,u}$. More precisely, $\underline{G} + G$ depends on the Hessian of u_I through the component $\zeta(u_I)$.

We now remark that

$$\begin{aligned}\mathcal{L}_L{}^{(a)}\pi_{AB} &= (\mathbf{D}_L^{(a)}\pi)_{AB} + {}^L\chi_{CA}^{(a)}\pi_{CB} + {}^L\chi_{CB}^{(a)}\pi_{AC} \\ &\quad - G_A^{(a)}\pi_{3B} - G_B^{(a)}\pi_{3A} - \underline{G}_A^{(a)}\pi_{3B} - \underline{G}_B^{(a)}\pi_{3A}\end{aligned} \tag{16.2.9a}$$

$$\begin{aligned}\mathcal{L}_L{}^{(a)}\pi_{AT} &= (\mathbf{D}_L^{(a)}\pi)_{AT} + {}^L\chi_{BA}^{(a)}\pi_{BT} - G_A^{(A)}\pi_{3T} - \underline{G}^{(a)}\pi_{4T} \\ &\quad + (G_B + \underline{G}_B)^{(a)}\pi_{AB} - U^{(a)}\pi_{A3} - \underline{U}^{(a)}\pi_{A4}.\end{aligned} \tag{16.2.9b}$$

Also, since

$$\mathbf{D}_L e_A = \not{\mathbf{D}}_L e_A + B_A e_3 + \underline{B} e_4 \tag{16.2.9c}$$
$$\mathbf{D}_L T = V_A e_A + RN, \tag{16.2.9d}$$

we write

$$\begin{aligned}(\mathbf{D}_L^{(a)}\pi)_{CD} &= 2\mathbf{D}_L{}^{(a)}H_{CD} - B_C^{(a)}\pi_{3D} - B_D^{(a)}\pi_{3C} \\ &\quad - \underline{B}_C^{(a)}\pi_{4D} - \underline{B}_D^{(a)}\pi_{4C}\end{aligned} \tag{16.2.9e}$$

$$\begin{aligned}(\mathbf{D}_L^{(a)}\pi)_{AT} &= 4\not{\mathbf{D}}_L{}^{(a)}Y_A - B_A^{(a)}\pi_{3T} - \underline{B}_A^{(a)}\pi_{4T} \\ &\quad - V_B^{(a)}\pi_{AB} - R^{(a)}\pi_{AN}.\end{aligned} \tag{16.2.9f}$$

Hence,

$$\begin{aligned}\mathcal{L}_L{}^{(a)}\pi_{AB} &= 2(\mathbf{D}_L{}^{(a)}H_{AB} + {}^L\chi_{CA}{}^{(a)}H_{CB} + {}^L\chi_{CB}{}^{(a)}H_{AC}) \\ &\quad - (G_A + B_A)^{(a)}\pi_{3B} - (G_B + B_B)^{(a)}\pi_{3A} \\ &\quad - (\underline{G}_A + \underline{B}_A)^{(a)}\pi_{4B} - (\underline{G}_B + \underline{G}_B)^{(a)}\pi_{4A} \\ \mathcal{L}_L{}^{(a)}\pi_{AT} &= 4(\not{\mathbf{D}}_L{}^{(a)}Y_A + {}^L\chi_{BA}{}^{(a)}Y_B) + 2(G_B + \underline{G}_B - V_B)^{(a)}H_{AB} \\ &\quad - (G_A + B_A)^{(a)}\pi_{3T} + (\underline{G}_A + \underline{B}_A)^{(a)}\pi_{4T} \\ &\quad - U^{(a)}\pi_{A3} - \underline{U}^{(a)}\pi_{A4} - R^{(a)}\pi_{AN}.\end{aligned}$$

Combining these with 16.2.7a and noting that $\mathcal{L}_{{}^{(a)}\Omega}{}^{(L)}\pi_{AB} = 2\mathcal{L}_{{}^{(a)}\Omega}{}^{L}\chi_{(AB)}$, where ${}^{L}\chi_{(AB)} = \frac{1}{2}({}^{L}\chi_{AB} + {}^{L}\chi_{BA})$, we derive the desired propagation equations for ${}^{(a)}H$:

$$
\begin{aligned}
\not{D}_L {}^{(a)}H_{AB} = {}& {}^{L}\hat{\chi}_{CA}{}^{(a)}H_{CB} + {}^{L}\hat{\chi}_{CB}{}^{(a)}H_{AC} + (\mathcal{L}_{{}^{(a)}\Omega}{}^{L}\hat{\chi})_{(AB)} \\
& + \frac{1}{2}\delta_{AB}\mathcal{L}_{{}^{(a)}\Omega}(\mathrm{tr}^{L}\chi) + ({}^{(a)}\underline{q}\chi_{AB} + {}^{(a)}q\underline{\chi}_{AB}) \\
& + \frac{1}{2}((G_A + B_A){}^{(a)}\pi_{3B} - (G_B + B_B){}^{(a)}\pi_{3A}) \\
& + \frac{1}{2}((\underline{G}_A + \underline{B}_A){}^{(a)}\pi_{4B} - (\underline{G}_B + \underline{G}_B){}^{(a)}\pi_{4A}).
\end{aligned}
\tag{16.2.10a}
$$

Also, taking into account the commutation formula

$$
[{}^{(a)}\Omega, T] = -\phi^{-1}({}^{(a)}\Omega \cdot \nabla\!\!\!/\,\phi)T + (\xi - \underline{\xi}) \cdot {}^{(a)}\Omega N - 4{}^{(a)}Y e_a
$$

and noting that ${}^{(L)}\pi_{AT} = W + G + \underline{G}$, we deduce

$$
\begin{aligned}
(\mathcal{L}_{{}^{(a)}\Omega}{}^{(L)}\pi)_{AT} = {}& \mathcal{L}_{{}^{(a)}\Omega}(W + G + \underline{G}) + 8{}^{(a)}Y_B{}^{L}\chi_{(AB)} \\
& + \phi^{-1}({}^{(a)}\Omega \cdot \nabla\!\!\!/\,\phi){}^{(L)}\pi_{AT} - (\xi - \underline{\xi}) \cdot {}^{(a)}\Omega{}^{(L)}\pi_{AN}.
\end{aligned}
$$

Hence,

$$
\begin{aligned}
\not{D}_L {}^{(a)}Y_A - \frac{1}{2}\mathrm{tr}\,{}^{L}\chi_{AB}\,{}^{(a)}Y_B = {}& \mathcal{L}_{{}^{(a)}\Omega}(W + G + \underline{G}) + {}^{L}\hat{\chi}_{AB}{}^{(a)}Y_B \\
& - \frac{1}{2}(G_B + \underline{G}_B - V_B){}^{(a)}H_{AB} \\
& + \frac{1}{4}\left((G_A + B_A){}^{(a)}\pi_{3T} - (\underline{G}_A + \underline{B}_A){}^{(a)}\pi_{4T}\right) \\
& + \frac{1}{4}\left(U{}^{(a)}\pi_{A3} + \underline{U}{}^{(a)}\pi_{A4} + R{}^{(a)}\pi_{AN}\right) \\
& + \frac{1}{4}(\phi^{-1}({}^{(a)}\Omega \cdot \nabla\!\!\!/\,\phi){}^{(L)}\pi_{AT} \\
& \qquad - (\xi - \underline{\xi}) \cdot {}^{(a)}\Omega{}^{(L)}\pi_{AN}) \\
& + \frac{1}{4}\left({}^{(a)}\underline{q}{}^{(e_4)}\pi_{AT} + {}^{(a)}q{}^{(e_3)}\pi_{AT}\right) + C_{AT}.
\end{aligned}
\tag{16.2.10b}
$$

Remark that ${}^{(a)}\pi_{3A}, {}^{(a)}\pi_{4A}$ can be expressed in terms of ${}^{(a)}\pi_{AL}, {}^{(a)}\pi_{AT}$ with the help of the following formulas (see 16.2.1b):

$$
\begin{aligned}
{}^{(a)}\pi_{AL} &= {}^{(L)}\underline{p}{}^{(a)}\pi_{A4} + {}^{(L)}p{}^{(a)}\pi_{A3} + L_B^{(a)}\pi_{AB} \\
{}^{(a)}\pi_{AT} &= \frac{1}{2}({}^{(a)}\pi_{4A} + {}^{(a)}\pi_{3A}).
\end{aligned}
$$

Also, ${}^{(a)}\pi_{3T}, {}^{(a)}\pi_{4T}$ can be expressed as linear combinations of ${}^{(a)}\pi_{33}$, ${}^{(a)}\pi_{34}$, ${}^{(a)}\pi_{44}$. Thus equations 16.2.10a, 16.2.10b form a coupled system in ${}^{(a)}H$, ${}^{(a)}Y$.

Remark that, as far as differentiability is concerned, the highest-order terms on the right-hand side of 16.2.10a, 16.2.10b are $\mathcal{L}_{{}^{(a)}\Omega}{}^{L}\chi$ and $\mathcal{L}_{{}^{(a)}\Omega}(W + G + \underline{G})$. These involve the first tangential derivatives of $\chi(u_I), \zeta(u_I)$ in the region $M_2 \bigcup I_0$, where $u = u_I$. All the other terms on the right-hand side of 16.2.10a, 16.2.10b, and also 16.2.6c, 16.2.6d, require only estimates for the Hessian of u_I and first derivatives of the Hessian of u_E. Thus the first derivatives of ${}^{(a)}\pi$ would require, at worst, $\nabla\!\!\!\!/ \mathbf{D}_3\zeta(u_I)$, hence $\nabla\!\!\!\!/^2\omega(u_I)$ in $M_2 \bigcup I_0$. Finally, since at the second-derivative level we need to take only tangential derivatives, it follows that we need, at worst, information for the third tangential derivatives of $\omega(u_I)$ in $M_2 \bigcup I_0$. Thus, in view of our results for the interior and exterior optical functions, we conclude that all the components of the second derivatives of ${}^{(a)}\pi$, which we need in the error estimates of Chapter 2, can be estimated locally in $L^2(\Sigma_t)$, provided that they can also be estimated on the last slice. This can be shown, in precisely the same fashion, from the definition of ${}^{(a)}\Omega_*$ in 16.0.2c.

It thus only remains to establish the asymptotic properties of ${}^{(a)}\pi$. This is now considerably easier than in the exterior and can be done in precisely the same manner, from propagation equations 16.2.10a and 16.2.10b, and equations 16.2.6c and 16.2.6d.[6]

We state the main results of this section in the following:

Theorem 16.2.1 *The components of the deformation tensor of ${}^{(a)}\Omega$ verify the following estimates:*

(i) There exists a constant c, depending only on $\mathcal{O}^\infty_{[1]}, \mathcal{K}^\infty_{[1]}, \mathcal{L}^\infty_{[1]}$, and $\mathcal{R}^\infty_0$, such that

$$\|{}^{(a)}\pi\|_{\infty,\imath} \leq c r_0^{-1/2}\left(\mathcal{O}^\infty_{[1]} + \mathcal{K}^\infty_{[0]} + \mathcal{L}^\infty_{[1]}\right) \tag{16.2.11a}$$

$$\|\mathbf{D}^{(a)}\pi\|_{2,\imath} \leq c\left(\mathcal{O}_{[2]} + \mathcal{K}_{[2]} + \mathcal{L}_{[2]} + \mathcal{R}_0\right). \tag{16.2.11b}$$

Also

$$\|r^{1/2}\mathbf{D}^{(a)}\pi\|_{4,\imath} \leq r_0^{-1/2}\left(\mathcal{O}_{[3]} + \mathcal{K}_{[3]} + \mathcal{L}_{[3]} + \mathcal{R}_{[1]}\right). \tag{16.2.11c}$$

(ii) There exists a constant c, depending only on $\mathcal{O}^\infty_{[1]}, \mathcal{K}^\infty_{[1]}, \mathcal{L}^\infty_{[1]}$, and $\mathcal{R}^\infty_0$, such that

$$\|r\nabla\!\!\!\!/\mathbf{D}\,{}^{(a)}\pi\|_{2,\imath} \leq c\left(\mathcal{O}_{[3]} + \mathcal{K}_{[3]} + \mathcal{L}_{[3]} + \mathcal{R}_{[1]}\right). \tag{16.2.11d}$$

[6] And the corresponding equations on the last slice.

CHAPTER 17

Conclusions

As we have shown in the proof of the main theorem given in Chapter 10, the exterior optical function ${}^{(t_*)}u$ defined on the slab $\bigcup_{t\in[0,t_*]}\Sigma_t$ converges as $t_* \to \infty$ to a global exterior optical function u. For each $t_* \geq 0$, the 0-level set of ${}^{(t_*)}u$ is the part of C_0, the 0-level set of u, contained in the slab $\bigcup_{t\in[0,t_*]}\Sigma_t$. We define for each $t \geq 0$ a diffeomorphism $\phi_{t,0}$ of S^2 onto $S_{t,0}$ such that $\phi_{t_2,0}\circ\phi^{-1}_{t_1,0}$ is the diffeomorphism of $S_{t_1,0}$ onto $S_{t_2,0}$ given by the generators of C_0. Given any $t_* \geq 0$, we consider the exterior optical function ${}^{(t_*)}u$ and define first the family ${}^{(t_*)}\phi_{t_*,\lambda}$ of diffeomorphisms of S^2 onto ${}^{(t_*)}S_{t_*,\lambda}$, the level sets of ${}^{(t_*)}u$ on Σ_{t_*}, by the requirement that ${}^{(t_*)}\phi_{t_*,\lambda}\circ\phi^{-1}_{t_*,0}$ is the diffeomorphism of $S_{t_*,0}$ onto ${}^{(t_*)}S_{t_*,\lambda}$ given by the gradient flow of ${}^{(t_*)}u$ on Σ_{t_*}. We then define the family ${}^{(t_*)}\phi_{t,\lambda}$, $t\in[0,t_*]$ of diffeomorphisms of S^2 onto ${}^{(t_*)}S_{t,\lambda}$, the level sets of the function ${}^{(t_*)}u$ on the hypersurfaces Σ_t, by the requirement that ${}^{(t_*)}\phi_{t_*,\lambda}\circ{}^{(t_*)}\phi^{-1}_{t,\lambda}$ is the diffeomorphism of ${}^{(t_*)}S_{t,\lambda}$ onto ${}^{(t_*)}S_{t_*,\lambda}$ given by the generators of ${}^{(t_*)}C_\lambda$, the λ-level set of ${}^{(t_*)}u$ in the slab $\bigcup_{t\in[0,t_*]}\Sigma_t$. These diffeomorphisms have been discussed in Chapter 16. We shall now show that as $t_* \to \infty$, ${}^{(t_*)}\phi_{t,\lambda}$ converges for each t and λ to a diffeomorphism $\phi_{t,\lambda}$ of S^2 onto $S_{t,\lambda}$, the λ-level set of the global exterior optical function u on Σ_t, where, for each λ, t_1 and t_2 $\phi_{t_2,\lambda}\circ\phi^{-1}_{t_1,\lambda}$ is the diffeomorphism of $S_{t_1,\lambda}$ onto $S_{t_2,\lambda}$ given by the generators of C_λ, the λ-level set of u in space-time.

For each $t_* \geq 0$ and $t\in[0,t_*]$, let ${}^{(t_*)}\psi_{t,\lambda}$ be the diffeomorphism of ${}^{(t_*)}S_{t,0}$ onto ${}^{(t_*)}S_{t,\lambda}$ given by the gradient flow of the function ${}^{(t_*)}u$ on the hypersurface Σ_t. In view of the convergence of the function ${}^{(t_*)}u$ to u as $t_* \to \infty$, the problem reduces to showing that for each λ_1, λ_2 and $\varepsilon > 0$ there is a T_* such that for $t_{*2} \geq t_{*1} \geq T_*$ the diffeomorphism

$$
{}^{(t_{*2})}f_{t_{*1},\lambda_1,\lambda_2} = {}^{(t_{*2})}\psi_{t_{*1},\lambda_1}\circ{}^{(t_{*2})}\psi^{-1}_{t_{*1},\lambda_2}\circ{}^{(t_{*2})}\phi_{t_{*1},\lambda_2}\circ{}^{(t_{*2})}\phi^{-1}_{t_{*1},\lambda_1}
$$

of ${}^{(t_{*2})}S_{t_{*1},\lambda_1}$ onto itself is within ε of the identity, that is,

$$
\sup_{p\in{}^{(t_{*2})}S_{t_{*1},\lambda_1}} \operatorname{dist}(p, {}^{(t_{*2})}f_{t_{*1},\lambda_1,\lambda_2}(p)) \leq \varepsilon.
$$

Proof 17.0.1 *Let* ${}^{(t_{*2})}U$ *be the vectorfield* ${}^{(t_{*2})}a\,{}^{(t_{*2})}N$ *and let* ${}^{(t_{*2})}L$ *be the vectorfield* $\phi(T + {}^{(t_{*2})}N)$, *defined by the function* ${}^{(t_{*2})}u$. *We note that*

${}^{(t_{*2})}U({}^{(t_{*2})}u) = {}^{(t_{*2})}L(t) = 1$. *Let* ${}^{(t_{*2})}\psi_\lambda$ *and* ${}^{(t_{*2})}\phi_t$ *be the 1-parameter groups of diffeomorphisms of space-time generated by* ${}^{(t_{*2})}U$ *and* ${}^{(t_{*2})}L$ *respectively and let for each t and* λ

$$ {}^{(t_{*2})}\chi_{t,\lambda} = {}^{(t_{*2})}\psi_{-\lambda} \circ {}^{(t_{*2})}\phi_{-t} \circ {}^{(t_{*2})}\psi_\lambda \circ {}^{(t_{*2})}\phi_t. $$

Then ${}^{(t_{*2})}f_{t_{*1},\lambda_1,\lambda_2}$ *is the restriction of* ${}^{(t_{*2})}\chi_{t_{*2}-t_{*1},\lambda_2-\lambda_1}$ *to the surface* ${}^{(t_{*2})}S_{t_{*1},\lambda_1}$. *To estimate the distance of the points p and* ${}^{(t_{*2})}\chi_{t_{*2}-t_{*1},\lambda_2-\lambda_1}(p)$ *on* ${}^{(t_{*2})}S_{t_{*1},\lambda_1}$ *we estimate the arc length of the curve* ${}^{(t_{*2})}\Gamma_{\lambda_2-\lambda_1} : t \to {}^{(t_{*2})}\chi_{t,\lambda_2-\lambda_1}(p)$ *on* ${}^{(t_{*2})}S_{t_{*1},\lambda_1}$ *joining these points.*

The tangent vector ${}^{(t_{*2})}X_{\lambda_2-\lambda_1,t}$ *of* ${}^{(t_{*2})}\Gamma_{\lambda_2-\lambda_1}$ *at parameter value t is given by*

$$ {}^{(t_{*2})}X_{\lambda_2-\lambda_1,t} = - {}^{(t_{*2})}\psi'_{\lambda_1-\lambda_2} \, {}^{(t_{*2})}\phi'_{-t} \, {}^{(t_{*2})}\psi'_{\lambda_2-\lambda_1} \, {}^{(t_{*2})}V_{{}^{(t_{*2})}\phi_t(p)}(\lambda_2 - \lambda_1) $$

where at each $q \in \bigcup_{t\in[t_{*1},t_{*2}]} \Sigma_t$ ${}^{(t_{*2})}V_q(\lambda)$ *is the 1-parameter family of vectors at q given by*

$$ {}^{(t_{*2})}V_q(\lambda) = {}^{(t_{*2})}\psi'_{-\lambda} \, {}^{(t_{*2})}L_{{}^{(t_{*2})}\psi_\lambda(q)} - {}^{(t_{*2})}L_q. $$

We have

$$ \frac{\partial \, {}^{(t_{*2})}V_q}{\partial\lambda} = {}^{(t_{*2})}\psi'_{-\lambda}[\, {}^{(t_{*2})}U, \, {}^{(t_{*2})}L]_{{}^{(t_{*2})}\psi_\lambda(q)}. $$

Now the commutator

$$ [\, {}^{(t_{*2})}U, \, {}^{(t_{*2})}L] = \phi \, {}^{(t_{*2})}a \, {}^{(t_{*2})}(\underline{\zeta} - \zeta). $$

Since by the main theorem we have on $\Sigma_t : |[\, {}^{(t_{*2})}U, \, {}^{(t_{*2})}L]| \leq ct^{-2}$ *and for fixed* λ_1 *and* λ ${}^{(t_{*2})}\psi'_{-\lambda}$ *is a linear map of* $T_q \, {}^{(t_{*2})}S_{t,\lambda_1+\lambda}$ *onto* $T_q \, {}^{(t_{*2})}S_{t,\lambda_1}$ *the norm of which is bounded by a constant independent of t and* t_{*2}, *we obtain*

$$ |\, {}^{(t_{*2})}V_{{}^{(t_{*2})}\phi_t(p)}(\lambda_2 - \lambda_1)| \leq c(t_{*1} + t)^{-2}. $$

On the other hand, ${}^{(t_{*2})}\phi'_{-t}$ *is a linear map of* $T_{{}^{(t_{*2})}\phi_t(q)} \, {}^{(t_{*2})}S_{t_{*1}+t,\lambda_2}$ *into* $T_q \, {}^{(t_{*2})}S_{t_{*1},\lambda_2}$ *that contracts lengths (see Chapter 16). Hence*

$$ |\, {}^{(t_{*2})}X_{\lambda_2-\lambda_1,t}| \leq c(t_{*1} + t)^{-2} $$

and

$$ \text{Arc Length}(\, {}^{(t_{*2})}\Gamma_{\lambda_2-\lambda_1}) = \int_0^{t_{*2}-t_{*1}} |\, {}^{(t_{*2})}X_{\lambda_2-\lambda_1,t}| dt \leq ct_{*1}^{-1}. $$

We conclude that

$$ \sup_{p\in \, {}^{(t_{*2})}S_{t_{*1},\lambda_1}} dist\Big(p, \, {}^{(t_{*2})}\chi_{t_{*2}-t_{*1},\lambda_2-\lambda_1}(p)\Big) \leq ct_{*1}^{-1}. $$

Now let U be the vectorfield aN and let L be the vectorfield $\phi(T+N)$ defined by the gradient flow of the global exterior optical function u and let ψ_λ and ϕ_t be the 1-parameter groups of diffeomorphisms of space-time generated by U and L respectively. Then we have $\phi_{t_2,\lambda} \circ \phi_{t_1,\lambda}^{-1} = \phi_{t_2-t_1}|_{S_{t_1,\lambda}}$, and if $\psi_{t,\lambda}$ is the diffeomorphism of $S_{t,0}$ onto $S_{t,\lambda}$ given by the gradient flow of u on Σ_t, then $\psi_{t,\lambda_2} \circ \psi_{t,\lambda_1}^{-1} = \psi_{\lambda_2-\lambda_1}|_{S_{t,\lambda_1}}$. Taking the limit $t_{*2} \to \infty$ in this argument, we conclude that the diffeomorphism $\phi_{t,\lambda_2} \circ \phi_{t,\lambda_1}^{-1}$ of S_{t,λ_1} onto S_{t,λ_2} converges as $t \to \infty$ to the restriction to S_{t,λ_1} of $\psi_{\lambda_2-\lambda_1}$. Hence if w is a p-covariant tensorfield in space-time we have

$$\frac{\partial}{\partial u}(\phi_{t,u}^* w) = \phi_{t,u}^*(\mathcal{L}_U w).$$

Also,

$$\frac{\partial}{\partial t}(\phi_{t,u}^* w) = \phi_{t,u}^*(\mathcal{L}_L w).$$

If w is a p-covariant tensorfield in space-time that is $S_{t,u}$-tangent, we define

$$\tilde{w} = \phi_{t,u}^*(r^{-p} w).$$

We say that on C_u w tends to a limit W as $t \to \infty$, and we write

$$\lim_{C_u, t\to\infty} w = W$$

if

$$\lim_{u=const., t\to\infty} \tilde{w} = W$$

on S^2. It then follows that

$$\frac{\partial W}{\partial u} = \lim_{C_u, t\to\infty} \not{\mathcal{L}}_U w. \tag{17.0.1}$$

By the results of Chapter 16, the induced metric γ on $S_{t,u}$ tends, in this sense, for each u as $t \to \infty$ to a metric $\overset{\circ}{\gamma}$ on S^2 of Gauss curvature 1. Furthermore, since $|\not{\mathcal{L}}_U \gamma| = O(r^{-1})$, the metric $\overset{\circ}{\gamma}$ is independent of u. Therefore $\overset{\circ}{\gamma}$ can be considered to be the standard metric on S^2. We have

$$\lim_{C_u, t\to\infty} \phi = 1, \qquad \lim_{C_u, t\to\infty} a = 1$$

and

$$\lim_{C_u, t\to\infty} (r \operatorname{tr}\chi) = 2, \qquad \lim_{C_u, t\to\infty} (r \operatorname{tr}\underline{\chi}) = -2.$$

Now let $(e_A : A = 1, 2)$ be an orthonormal basis for the tangent space to $S_{t,u}$ at $p = \phi_{t,u}(\omega), \omega \in S^2$, propagated along the generators of C_u according to

$$\frac{de_A}{ds} = \underline{\zeta}_A l$$

(see Chapter 9). We define a basis $(\tilde{E}_A : A = 1, 2)$ at the point $\omega \in S^2$ by

$$re_A = \phi'_{t,u}\tilde{E}_A.$$

Then $(\tilde{E}_A : A = 1, 2)$ is orthonormal relative to the metric $\tilde{\gamma} = \phi^*_{t,u}(r^{-2}\gamma)$. We note that $\tilde{E}_A$ depends on t and u; in fact,

$$\mathcal{L}_L(re_A) = \phi'_{t,u}(\partial\tilde{E}_A/\partial t).$$

Since

$$\mathcal{L}_L(re_A) = -(\frac{1}{2})(a\phi\mathrm{tr}\chi - \overline{a\phi\mathrm{tr}\chi})re_A - a\phi\hat{\chi}_{AB}re_B,$$

we obtain

$$|\partial\tilde{E}_A/\partial t|_{\tilde{\gamma}} = r^{-1}|\mathcal{L}_L(re_A)|_\gamma = O(r^{-2}).$$

Hence, at each u, $(\tilde{E}_A : A = 1, 2)$ converges as $t \to \infty$ to a basis $(E_A : A = 1, 2)$ that is orthonormal relative to $\overset{\circ}{\gamma}$. If now w is a p-covariant tensorfield in spacetime that is $S_{t,u}$-tangent, the components $\tilde{w}_{A_1 ..A_P}$ of $\tilde{w}$ relative to the basis $(\tilde{E}_A : A = 1, 2)$ coincide with the components $w_{A_1...A_P}$ of w relative to the basis $(e_A : A = 1, 2)$. Thus the statement that on C_u w tends to W as $t \to \infty$ is equivalent to

$$\lim_{u=const.,t\to\infty} \phi^*_{t,u}(w_{A_1...A_P}) = W_{A_1...A_P}$$

where $W_{A_1...A_P}$ are the components of W relative to the basis $(E_A : A = 1, 2)$. We note finally that 17.0.1 can alternatively be expressed as

$$\frac{\partial W}{\partial u} = \lim_{C_u,t\to\infty} \nabla\!\!\!/_N w.$$

The main theorem, in its complete version stated in Chapter 10, allows us to derive certain important consequences. They all concern the limiting behavior of various quantities along the null hypersurfaces C_u as $t \to \infty$. The first one follows immediately from the Main Theorem.

Conclusion 17.0.1 On any null hypersurface C_u, the normalized curvature components $r\underline{\alpha}$, $r^2\underline{\beta}$, $r^3\rho$, $r^3\sigma$ have limits as $t \to \infty$, that is,

$$\lim_{C_u,t\to\infty} r\underline{\alpha} = A(u, \cdot), \qquad \lim_{C_u,t\to\infty} r^2\underline{\beta} = B(u, \cdot)$$

$$\lim_{C_u,t\to\infty} r^3\rho = P(u, \cdot), \qquad \lim_{C_u,t\to\infty} r^3\sigma = Q(u, \cdot)$$

where A is a symmetric traceless covariant 2-tensor, B is a 1-form, and P and Q are functions on S^2 depending on u and having the following decay properties:

$$|A(u,\cdot)| \leq C(1+|u|)^{-5/2} \qquad |B(u,\cdot)| \leq C(1+|u|)^{-3/2}$$
$$|P(u,\cdot) - \overline{P}(u)| \leq (1+|u|)^{-1/2} \qquad |Q(u,\cdot) - \overline{Q}(u)| \leq (1+|u|)^{-1/2}$$

and

$$\lim_{u \to -\infty} \overline{P}(u) = 0, \qquad \lim_{u \to -\infty} \overline{Q}(u) = 0.$$

Here and in the following the pointwise norms $|\ |$ of tensors on S^2 are with respect to the standard metric $\overset{\circ}{\gamma}$.

Proof 17.0.2 *The existence of the limits $A(u,\cdot)$, $B(u,\cdot)$, $P(u,\cdot)$, $Q(u,\cdot)$ follows from the estimates for $\underline{\alpha}_4$, $\underline{\beta}_4$, ρ_4, σ_4 given by the main theorem. The other results are immediate.*

The second conclusion concerns the normalized shear $\hat{\chi}'$.[1]

Conclusion 17.0.2 On any null hypersurface C_u, the normalized shear $r^2\hat{\chi}'$ has a limit as $t \to \infty$, that is,

$$\lim_{C_u, t \to \infty} r^2 \hat{\chi}' = \Sigma(u,\cdot)$$

where Σ is a symmetric traceless covariant 2-tensor on S^2 depending on u.

Proof 17.0.3 *The proof is a simple consequence of the main theorem applied to the equation*

$$\frac{d}{ds}\hat{\chi}_{AB} + \mathrm{tr}\chi\hat{\chi}_{AB} = \alpha_{AB}.$$

Remark that this equation is considered relative to the l-pair $l = a^{-1}(T+N)$, $\underline{l} = a(T-N)$.

Since

$$\frac{d}{ds}r = \frac{r}{2}a^{-1}\phi^{-1}\overline{a\phi\mathrm{tr}\chi},$$

we find

$$\frac{d}{ds}(r^2\hat{\chi}) = r^2\left(\alpha - (\mathrm{tr}\chi - a^{-1}\phi^{-1}\overline{a\phi\mathrm{tr}\chi})\hat{\chi}\right).$$

Since the right-hand side of the last equation is of order $O(r^{-3/2})$, hence integrable, we conclude that $r^2\hat{\chi}$ has a limit as $s \to \infty$, which, in view of the fact that $\frac{dt}{ds} = a^{-1}\phi^{-1}$, proves the existence of the limit $\lim_{t\to\infty} r^2\hat{\chi} = \lim_{t\to\infty} r^2\hat{\chi}'$ and the conclusion.

[1] Namely, the traceless part of the null second fundamental form of $e_4' = a^{-1}(T+N)$.

Conclusion 17.0.3 On any null hypersurface C_u, the limit of $r\hat{\eta}$ exists as $t \to \infty$. In other words we have

$$\lim_{C_u, t \to \infty} r\hat{\eta} = \Xi(u, \cdot)$$

where Ξ is a symmetric traceless 2-covariant tensor on S^2 depending on u and having the decay property

$$|\Xi(u, \cdot)|_{\overset{\circ}{\gamma}} \leq C(1 + |u|)^{-3/2}.$$

Moreover,

$$\lim_{C_u, t \to \infty} r\hat{\theta} = -\frac{1}{2} \lim_{C_u, t \to \infty} r\underline{\hat{\chi}}' = \Xi.$$

Furthermore, Ξ is related to B by the equation

$$\overset{\circ}{\not{d}\mathrm{iv}}\, \Xi = -\frac{1}{2} B.$$

Finally, Ξ is related to Σ and A by the formulas

$$\begin{aligned} \frac{\partial \Sigma}{\partial u} &= -\Xi \\ \frac{\partial \Xi}{\partial u} &= -\frac{1}{4} A. \end{aligned}$$

Proof 17.0.4 *The proof of the first part follows from equation 11.4.6c. Multiplying it by r and proceeding as above, we deduce that $\frac{d}{ds}(r\hat{\eta}) = O(r^{-2})$ is integrable, hence $r\hat{\eta}$ has a limit.*

The second part of Conclusion 17.0.3 follows from the first part together with Conclusion 17.0.2 and the formulas

$$\begin{aligned} \hat{\chi}' &= \hat{\theta} - \hat{\eta} \\ \underline{\hat{\chi}}' &= -\hat{\theta} - \hat{\eta}. \end{aligned}$$

The third part follows from the normalized conjugate null Codazzi equation

$$\not{d}\mathrm{iv}\, \underline{\hat{\chi}}' - \underline{\hat{\chi}}' \cdot \epsilon = \frac{1}{2} \not{\nabla} \mathrm{tr} \underline{\chi}' - \frac{1}{2} \mathrm{tr} \underline{\chi}' \epsilon + \underline{\beta}$$

(standard frame), upon multiplying by r^2 and taking the limit $t \to \infty$ on C_u, in view of Conclusion 17.0.1 and the second part.

Finally the last part of the conclusion, which connects Ξ to the asymptotic shear Σ and to A, is a consequence of the following equations (see Chapter 11):

$$\nabla\!\!\!/_N\hat{\chi} = -\frac{1}{2}\mathrm{tr}\chi\hat{\eta} - \left(\mathrm{tr}\chi + \frac{1}{2}\delta\right)\hat{\chi} - \frac{1}{2}\nabla\!\!\!/\widehat{\otimes}\zeta - \frac{1}{2}\zeta\widehat{\otimes}\zeta - \frac{1}{2}\alpha,$$
$$\nabla\!\!\!/_N\hat{\eta} = \left(-\frac{1}{2}\mathrm{tr}\chi + 2\delta\right)\hat{\eta} + \frac{3}{2}\delta\hat{\chi} + \frac{1}{2}\nabla\!\!\!/\widehat{\otimes}\epsilon + \epsilon\widehat{\otimes}\zeta - \epsilon\widehat{\otimes}\epsilon - \frac{1}{4}\underline{\alpha} + \frac{1}{4}\alpha.$$

Indeed, multiplying the first of these equations by r^2 and the second by r, we observe that

$$\nabla\!\!\!/_N(r^2\hat{\chi}) = -r\hat{\eta} + O(r^{-1}),$$
$$\nabla\!\!\!/_N(r\hat{\eta}) = -\frac{1}{4}r\underline{\alpha} + O(r^{-1}).$$

The next conclusion has to do with the Bondi mass. To define, it we first introduce the Hawking mass enclosed by a 2-surface $S_{t,u}$ to be

$$m(t,u) = \frac{r}{2}(1 + \frac{1}{16\pi}\int_{S_{t,u}} \mathrm{tr}\chi\mathrm{tr}\underline{\chi}). \tag{17.0.2}$$

To establish a propagation equation for m, we introduce the following:

$$\underline{\mu} = \mathrm{d}\!\!\!/\mathrm{iv}\,\zeta + \frac{1}{2}\hat{\chi}\cdot\underline{\hat{\chi}} - \rho. \tag{17.0.3}$$

In view of the null structure equations relative to the l-pair, we have

$$\frac{d}{ds}\mathrm{tr}\underline{\chi} + \frac{1}{2}\mathrm{tr}\chi\mathrm{tr}\underline{\chi} = -2\underline{\mu} + 2|\zeta|^2$$
$$\frac{d}{ds}\mathrm{tr}\chi + \frac{1}{2}(\mathrm{tr}\chi)^2 = -|\hat{\chi}|^2.$$

Hence,

$$\frac{d}{ds}\mathrm{tr}\chi\mathrm{tr}\underline{\chi} + \mathrm{tr}\chi(\mathrm{tr}\chi\mathrm{tr}\underline{\chi}) = -2\underline{\mu}\mathrm{tr}\chi + 2\mathrm{tr}\chi|\zeta|^2 - \mathrm{tr}\underline{\chi}|\hat{\chi}|^2$$

and

$$\frac{\partial}{\partial t}\int_{S_{t,u}} \mathrm{tr}\chi\mathrm{tr}\underline{\chi} = -2\int_{S_{t,u}} a\phi\underline{\mu}\mathrm{tr}\chi + \int_{S_{t,u}} a\phi(-\mathrm{tr}\underline{\chi}|\hat{\chi}|^2 + 2\mathrm{tr}\chi|\zeta|^2). \tag{17.0.4}$$

On the other hand, in view of the Gauss equation,

$$\underline{\mu} = \mathrm{d}\!\!/\mathrm{iv}\,\zeta + K + \frac{1}{4}\mathrm{tr}\chi\mathrm{tr}\underline{\chi}.$$

Hence, applying the Gauss-Bonnet formula, we infer that

$$\int_{S_{t,u}} \underline{\mu} = \int_{S_{t,u}} (\frac{1}{2}\hat{\chi}\cdot\hat{\underline{\chi}} - \rho)$$
$$= 4\pi(1 + \frac{1}{16\pi}\int_{S_{t,u}} \mathrm{tr}\chi\mathrm{tr}\underline{\chi}) = \frac{8\pi}{r}m. \tag{17.0.5}$$

Taking into account the formula

$$\frac{d}{dt}r = \frac{r}{2}\overline{\phi a\mathrm{tr}\chi},$$

together with 17.0.4 and 17.0.5, we infer that

$$\frac{\partial}{\partial t}m(t,u) = -\frac{r}{16\pi}\int_{S_{t,u}} (a\phi\mathrm{tr}\chi - \overline{a\phi\mathrm{tr}\chi})\underline{\mu} \tag{17.0.6}$$
$$+ \frac{r}{8\pi}\int_{S_{t,u}} a\phi(\frac{1}{2}\mathrm{tr}\chi|\zeta|^2 - \frac{1}{4}\mathrm{tr}\underline{\chi}|\hat{\chi}|^2).$$

Now, since $K + \frac{1}{4}\mathrm{tr}\chi\mathrm{tr}\underline{\chi} = O(r^{-3})$, we have $\underline{\mu} = O(r^{-3})$, and, taking into account the asymptotic behavior of all terms on the right-hand side of 17.0.6, we deduce that

$$\frac{\partial}{\partial t}m(t,u) = O(r^{-2}).$$

Hence, for every fixed u, $m(t,u)$ has a limit as $t \to \infty$, which is the *Bondi mass* of the null hypersurface C_u and is denoted by $M(u)$. Clearly,

$$m(t,u) = M(u) + O(r^{-1})$$

as $t \to \infty$ on C_u.

We next differentiate $m(t,u)$ with respect to u, keeping t fixed. With the help of formula 3.1.1g we write,

$$\frac{\partial}{\partial u}m(t,u) = \frac{1}{2}\overline{a\mathrm{tr}\theta}m + \frac{r}{32\pi}\int_{S_{t,u}} a\left(\nabla_N(\mathrm{tr}\chi\mathrm{tr}\underline{\chi}) + \mathrm{tr}\theta(\mathrm{tr}\chi\mathrm{tr}\underline{\chi})\right).$$

Now, since $l = a^{-1}(T+N)$, $\underline{l} = a(T-N)$,

$$\nabla_N(\mathrm{tr}\chi\mathrm{tr}\underline{\chi}) + \mathrm{tr}\theta(\mathrm{tr}\chi\mathrm{tr}\underline{\chi}) = \frac{1}{2}a^2\left(\mathbf{D}_4(\mathrm{tr}\chi\mathrm{tr}\underline{\chi}) + \mathrm{tr}\chi(\mathrm{tr}\chi\mathrm{tr}\underline{\chi})\right) - \frac{1}{2}\left(\mathbf{D}_3(\mathrm{tr}\chi\mathrm{tr}\underline{\chi}) + \mathrm{tr}\underline{\chi}(\mathrm{tr}\chi\mathrm{tr}\underline{\chi})\right)$$

and

$$\mathbf{D}_4(\mathrm{tr}\chi\mathrm{tr}\underline{\chi}) + \mathrm{tr}\chi(\mathrm{tr}\chi\mathrm{tr}\underline{\chi}) = -2\underline{\mu}\mathrm{tr}\chi + 2\mathrm{tr}\chi|\zeta|^2 - \mathrm{tr}\underline{\chi}|\hat{\chi}|^2$$
$$\mathbf{D}_3(\mathrm{tr}\chi\mathrm{tr}\underline{\chi}) + \mathrm{tr}\underline{\chi}(\mathrm{tr}\chi\mathrm{tr}\underline{\chi}) = -2\mu\mathrm{tr}\underline{\chi} + 2\mathrm{tr}\underline{\chi}|\zeta|^2 - \mathrm{tr}\chi|\hat{\underline{\chi}}|^2$$

where

$$\mu = -\mathrm{d}\!\!/\mathrm{iv}\,\zeta + \frac{1}{2}\hat{\chi}\cdot\hat{\underline{\chi}} - \rho.$$

Therefore,

$$\frac{\partial}{\partial u}m(t,u) = \frac{r}{64\pi}\int_{S_{t,u}} \mathrm{tr}\chi|\hat{\underline{\chi}}|^2 d\mu_\gamma + O(r^{-1}). \tag{17.0.7}$$

Hence, in view of the fact that at each u the metric $\tilde{\gamma} = \phi^*_{t,u}(r^{-2}\gamma)$ converges as $t \to \infty$ to the standard metric $\overset{\circ}{\gamma}$ of the unit sphere S^2, and moreover $\frac{r}{2}\mathrm{tr}\chi$ converges to 1 and $r\hat{\underline{\chi}}$ converges to -2Ξ, we derive the *Bondi mass formula*

$$\frac{\partial}{\partial u}M = \frac{1}{8\pi}\int_{S^2} |\Xi(u,\cdot)|^2 d\mu_{\overset{\circ}{\gamma}}.$$

Finally, since the right-hand side of the Bondi mass formula is positive and integrable in u, we conclude that the Bondi mass $M(u)$ is a nondecreasing function of u with finite limits $M(-\infty)$ for $u \to -\infty$ and $M(\infty)$ for $u \to \infty$. Moreover, in view of the first three conclusions, we infer from 17.0.5 that $M(-\infty) = 0$. On the other hand, $M(\infty)$ is the total mass.

We have thus established the following:

Conclusion 17.0.4 The Hawking mass $m(t,u)$ tends to the Bondi mass $M(u)$ as t tends to ∞ on any null hypersurface C_u. More precisely,

$$m(t,u) = M(u) + O(r^{-1}).$$

Moreover, $M(u)$ verifies the Bondi mass formula,

$$\frac{\partial}{\partial u}M = \frac{1}{8\pi}\int_{S^2} |\Xi(u,\cdot)|^2 d\mu_{\overset{\circ}{\gamma}}$$

where $d\mu_{\overset{\circ}{\gamma}}$ is the area element of the standard unit sphere S^2.

Also, $M(u)$ converges to zero as $u \to -\infty$ and converges to the total mass $M(\infty)$ as $u \to \infty$.

The proof of the next two conclusions rely on the following lemma:

Lemma 17.0.1 *On any null hypersurface C_u we have relative to the standard frame*

$$\lim_{C_u, t\to\infty} r^2\bar{\delta} = 2M(u) \text{ and } \lim_{C_u, t\to\infty} r(r\overline{\mathrm{tr}\chi'} - 2) = 0.$$

More precisely,

$$r^2\bar{\delta} - 2M(u) = O(r^{-1})$$
$$r(r\overline{\mathrm{tr}\chi'} - 2) = O(r^{-1})$$

as $t \to \infty$ on any C_u.

Proof 17.0.5 *To prove the first limit we appeal to Proposition 4.4.4 applied to the divergence equation $\mathrm{div}\, k = 0$. Thus, the mean value of δ on any 2-surface $S_{t,u}$ is given by the formula*

$$4\pi r^3\bar{\delta} = \int_{u_0}^{u} du' \left(\int_{S_{t,u}} ar\hat{\theta}\cdot\hat{\eta} - \frac{1}{2}\kappa(\delta - \bar{\delta}) - ra^{-1}\nabla a \cdot \epsilon \right),$$

where u_0 is the value of u corresponding to $r = 0$ on Σ_t.

Taking into account 17.0.7, we remark that

$$\int_{S_{t,u}} r\hat{\theta}\cdot\hat{\eta} = 8\pi r \frac{\partial}{\partial u} m(t,u) + O(1).$$

Hence

$$r\bar{\delta} = \frac{2}{r^2}\int_{u_0}^{u} r\frac{\partial}{\partial u} m(t,u) + O(r^{-1}),$$

and therefore $r^2\bar{\delta}$ converges to $2M(u)$ as $t \to \infty$ on each C_u.

To prove the second part, we consider the Hawking mass $m(t,u)$, expressed relative to the standard pair $e'_4 = T + N, e'_3 = T - N$, and expand the product $\mathrm{tr}\chi'\mathrm{tr}\underline{\chi}'$ by taking into account that

$$\mathrm{tr}\chi' = \frac{2}{r} + O(r^{-2})$$
$$\mathrm{tr}\underline{\chi}' = 2\delta - \mathrm{tr}\chi'.$$

Thus

$$\mathrm{tr}\chi'\mathrm{tr}\underline{\chi}' = -\frac{4}{r^2} - \frac{4}{r}(\mathrm{tr}\chi' - \frac{2}{r}) + \frac{4\delta}{r} + O(r^{-4}).$$

Hence,

$$m(t,u) = \frac{1}{8\pi}\int_{S_{t,u}} \left(\delta - (\mathrm{tr}\chi' - \frac{2}{r})\right) + O(r^{-1})$$
$$= \frac{r^2}{2}\left(\bar{\delta} - (\overline{\mathrm{tr}\chi'} - \frac{2}{r})\right) + O(r^{-1}).$$

The second part of the lemma follows now, in view of the fact that, as $t \to \infty$ along C_u, both $m(t,u)$ and $\frac{r^2}{2}\bar{\delta}$ converge to $M(u)$.

Conclusion 17.0.5 Let

$$h = r(r\mathrm{tr}\chi' - 2).$$

On any null hypersurface C_u we have

$$\lim_{C_u, t\to\infty} h = H,$$

where H is a function on S^2 that is independent of u and of vanishing mean, that is:

$$\frac{\partial H}{\partial u} = 0, \qquad \overline{H} = 0.$$

Proof 17.0.6 *By Lemma 17.0.1*

$$\lim_{C_u, t\to\infty} \overline{h} = 0.$$

Therefore, to show that the limit $\lim_{C_u,t\to\infty} h$ exists, it suffices to show that the limit $\lim_{C_u,t\to\infty} r\nabla\!\!\!/ h$ exists. We recall the 1-form $\not{\kappa}$ introduced in Chapter 13:

$$\not{\kappa} = \nabla\!\!\!/\mathrm{tr}\chi + \mathrm{tr}\chi\zeta.$$

Consider the 1-form

$$\not{\kappa}' = \nabla\!\!\!/\mathrm{tr}\chi' + \mathrm{tr}\chi'\epsilon.$$

Then, since $\nabla\!\!\!/ h = r^2(\not{\kappa}' - \mathrm{tr}\chi'\epsilon)$ and we have $\not{\kappa}' = a\not{\kappa}$, the existence of the limit $\lim_{C_u,t\to\infty} r\nabla\!\!\!/ h$ reduces to the existence of the limits

$$\lim_{C_u,t\to\infty} r^3\not{\kappa}, \qquad \lim_{C_u,t\to\infty} r^2\epsilon.$$

The existence of the first limit will be demonstrated at this point, while the existence of the second limit will be demonstrated as part of Conclusion 17.0.7. By 13.1.6c

$$\frac{d}{ds}\not{\chi}_A + \frac{3}{2}\mathrm{tr}\chi\not{\chi}_A = F_A$$

where

$$F_A = -\hat{\chi}_{AB}\not{\chi}_B - 2\hat{\chi}_{BC}\not{\nabla}_A\hat{\chi}_{BC} - \mathrm{tr}\chi\beta_A + \mathrm{tr}\chi\hat{\chi}_{AB}\underline{\zeta}_B - 2|\hat{\chi}|^2\zeta_A,$$

and all quantities are with respect to the l-null frame. As an immediate consequence of the main theorem, on each C_u we have

$$\not{\chi}_A = O(r^{-3}), \qquad \not{\nabla}_A\hat{\chi}_{BC} = O(r^{-3}).$$

Hence

$$F_A = O(r^{-9/2})$$

and we obtain

$$\frac{d}{ds}(r^3\not{\chi}_A) = F'_A$$

where

$$F'_A = r^3F_A - \frac{3}{2}a^{-1}\phi^{-1}(a\phi\mathrm{tr}\chi - \overline{a\phi\mathrm{tr}\chi}) = O(r^{-3/2}).$$

Therefore, integrating, we conclude that on each C_u, $r^3\not{\chi}$ tends to a limit $\not{X}$ as $t \to \infty$ and

$$r^3\not{\chi} - \not{X} = O(r^{-1/2}).$$

To prove the remaining part of the conclusion, we recall from Chapter 11 the equation

$$\nabla_N\mathrm{tr}\chi' + \frac{1}{2}(\mathrm{tr}\chi')^2 = G,$$

$$G = -\frac{1}{2}\mathrm{tr}\chi'\delta - \hat{\chi}'\cdot\hat{\eta} - |\hat{\chi}'|^2 - \not{d}\mathrm{iv}\,\zeta' - |\zeta'|^2 - \rho$$

on each Σ_t. Here all quantities are relative to the standard frame. This yields

$$\nabla_N h = 2(a^{-1}\lambda - 1)(1 + r^{-1}h) - \frac{1}{2}r^{-2}h^2 + r^2G.$$

The main theorem gives $G = O(r^{-3})$, $a^{-1}\lambda - 1 = O(r^{-1})$, and $h = O(1)$. Hence

$$\nabla_N h = O(r^{-1})$$

and, integrating along an integral curve of N from a point p_1 on S_{t,u_1} to a point p_2 on S_{t,u_2}, we obtain

$$h(p_2) - h(p_1) = O(r_0^{-1}).$$

It follows, taking the limit $t \to \infty$, that

$$H(u_2, \cdot) - H(u_1, \cdot) = 0.$$

Conclusion 17.0.6 On any null hypersurface C_u we have

$$r = t - 2M(\infty)\log t + O(1)$$

as $t \to \infty$.

Proof 17.0.7 *We first write, relative to the standard frame, taking into account Lemma 17.0.1,*

$$\begin{aligned}\frac{dr}{dt} &= \frac{r}{2}\overline{\phi \mathrm{tr}\chi'}\\ &= \frac{r}{2}\overline{(1+\phi')\left(\frac{2}{r} + (\mathrm{tr}\chi' - \frac{2}{r})\right)}\\ &= 1 + \overline{\phi'} + O(r^{-2})\end{aligned}$$

where $\phi' = \phi - 1$. Now, integrating the lapse equation $\Delta\phi = |k|^2\phi$ on Σ_t in the interior of $S_{t,u}$, we derive

$$\int_{S_{t,u}} \nabla_N \phi' = \int_{u_0}^{u} du' \int_{S_{t,u'}} a\phi|k|^2.$$

Remark that $\int_{S_{t,u'}} a\phi|k|^2 \to \int_{S^2} |\Xi|^2$. Hence, in view of the Bondi mass formula established in Conclusion 17.0.4,

$$\int_{S_{t,u}} \nabla_N \phi' - 8\pi M(u) = O(r^{-1}) \tag{17.0.8}$$

as $t \to \infty$ on each C_u. Now we can write,

$$\begin{aligned}\bar{\phi}' &= \frac{1}{4\pi r^2}\int_{S_{t,u}} \phi' = -\frac{1}{4\pi}\int_{ExtS_{t,u}} div(r^{-2}N\phi')\\ &= -\frac{1}{4\pi}\int_u^\infty (r(t,u'))^{-2}\left(\int_{S_{t,u'}} a\nabla_N\phi' + (a\mathrm{tr}\theta - \overline{a\mathrm{tr}\theta})\phi'\right)du'\\ &= -\frac{1}{4\pi}\int_u^\infty (r(t,u'))^{-2}\left(\int_{S_{t,u'}} \nabla_N\phi'\right)du' + O(r^{-2}).\end{aligned}$$

Thus by 17.0.8 we have

$$\bar{\phi}'(t,u) = -2\int_u^\infty (r(t,u'))^{-2}M(u')du' + O(r^{-2}) = -\frac{2}{r}M(\infty) + O(r^{-2})$$

as $t \to \infty$ *on* C_u. *Hence on any* C_u

$$\frac{dr}{dt} = 1 - \frac{2}{r}M(\infty) + O(r^{-2})$$

as $t \to \infty$.

We recall from Chapter 11 the functions ψ and ψ', which come up in the decomposition of δ. These are defined to be the solutions, vanishing at infinity, of the equations

$$\Delta\psi = r|\hat{\eta}|^2, \ \Delta\psi' = -ra^{-1}\lambda(|\hat{\eta}|^2 - \overline{|\hat{\eta}|^2})$$

on Σ_t. Note that ψ and ψ' depend on the choice of extension of the exterior optical function to a global one. To derive the remaining conclusions we shall make use of the following lemma:

Lemma 17.0.2 *On any null hypersurface* C_u *we have*

$$\lim_{C_u, t\to\infty} \psi = \Psi \qquad \lim_{C_u, t\to\infty} \psi' = \Psi'$$

where Ψ *and* Ψ' *are functions on* S^2, *independent of* u, *which depend only on the exterior optical function. More precisely,*

$$\begin{aligned}
\Psi &= -\frac{1}{2^{1/2}4\pi}\int_{-\infty}^{\infty} du'\left(\int_{S^2} \frac{|\Xi|^2(u',\omega')}{(1-\omega\cdot\omega')^{1/2}}d\omega'\right), \\
\Psi' &= \frac{1}{2^{1/2}4\pi}\int_{-\infty}^{\infty} du'\left(\int_{S^2} \frac{(|\Xi|^2(u',\omega') - \overline{|\Xi|^2(u')})}{(1-\omega\cdot\omega')^{1/2}}d\omega'\right).
\end{aligned} \tag{17.0.9}$$

Futhermore,

$$\lim_{C_u, t\to\infty} r\nabla_N\psi = \Omega(u, \cdot) \qquad \lim_{C_u, t\to\infty} r\nabla_N\psi' = \Omega'(u, \cdot)$$

where Ω, Ω' *are functions on* S^2, *depending on* u, *that depend only on the exterior optical function. To be precise,*

$$\begin{aligned}
\Omega = &\frac{1}{2^{3/2}4\pi}\int_{-\infty}^{\infty} du'\left(\int_{S^2} \frac{|\Xi|^2(u',\omega')}{(1-\omega\cdot\omega')^{1/2}}d\omega'\right) \\
&+ \frac{1}{2}\int_{-\infty}^{\infty} du' sgn(u-u')|\Xi|^2(u',\omega), \\
\Omega' = &-\frac{1}{2^{3/2}4\pi}\int_{-\infty}^{\infty} du'\left(\int_{S^2} \frac{(|\Xi|^2(u',\omega') - \overline{|\Xi|^2(u')})}{(1-\omega\cdot\omega')^{1/2}}d\omega'\right) \\
&- \frac{1}{2}\int_{-\infty}^{\infty} du' sgn(u-u')(|\Xi|^2(u',\omega) - \overline{|\Xi|^2(u')}).
\end{aligned}$$

Proof 17.0.8 *The proof is by comparing the metric g of Σ_t to a certain flat metric $\overset{\circ}{g}$. The choice of a central line P determines, according to Chapter 9, an interior optical function that by Chapter 15 defines an extension of the exterior optical function to the interior region. The integral curves of N, the unit normal vectorfield to the level surfaces $S_{t,u}$ of this function on Σ_t, make $\Sigma_t \setminus P_t$ into the product $(u_0, \infty) \times S^2$ with the metric*

$$g = a^2 du^2 + \gamma$$

where $\gamma(u, \cdot)$ is a metric on S^2. Using r in place of u as the radial coordinate, the metric is given by

$$g = a^2 \lambda^{-2} dr^2 + \gamma.$$

Consider now the comparison flat metric

$$\overset{\circ}{g} = dr^2 + r^2 \overset{\circ}{\gamma}.$$

In terms of local coordinates ω^A, $A = 1, 2$ on S^2 we have

$$\begin{aligned}\Delta\psi &= a^{-2}\lambda^2 \frac{\partial^2\psi}{\partial r^2} \\ &\quad + a^{-1}\lambda \left(\mathrm{tr}\theta - \nabla_N \log a + \nabla_N \log \lambda\right) \frac{\partial\psi}{\partial r} \\ &\quad + \not{\Delta}\psi + \gamma^{AB} \frac{\partial \log a}{\partial \omega^A} \frac{\partial\psi}{\partial \omega^B},\end{aligned}$$

while

$$\overset{\circ}{\Delta} = \frac{\partial^2\psi}{\partial r^2} + \frac{2}{r}\frac{\partial\psi}{\partial r} + \frac{1}{r^2} \overset{\circ}{\not{\Delta}} \psi.$$

Hence, since

$$\frac{\partial\psi}{\partial r} = \lambda^{-1} a \nabla_N \psi,$$

we obtain

$$\begin{aligned}\Delta\psi - \overset{\circ}{\Delta} \psi &= (1 - \lambda^{-2}a^2)\nabla_N^2\psi + (\mathrm{tr}\theta - 2r^{-1}a^{-1}\lambda)\nabla_N\psi \\ &\quad + \lambda^{-2}a^2(\nabla_N \log\lambda - \nabla_N \log a)\nabla_N\lambda \\ &\quad + \not{\nabla} \log a \cdot \not{\nabla}\psi + \not{\Delta}\psi - r^{-2} \overset{\circ}{\not{\Delta}} \psi.\end{aligned}$$

We write the equation for ψ in the form

$$\overset{\circ}{\Delta} \psi = f \qquad (17.0.10a)$$

where

$$f = r|\hat{\eta}|^2 - (\Delta\psi - \overset{\circ}{\Delta}\psi). \tag{17.0.10b}$$

In the exterior region E_t on each Σ_t we have

$$|r\hat{\eta}|^2 - |\Xi|^2 = O(r^{-1}\tau_-^{-3/2}) \tag{17.0.10c}$$

and by the results of Chapter 11

$$\Delta\psi - \overset{\circ}{\Delta}\psi = O(r^{-2}\tau_-^{-3/2} + r^{-3}). \tag{17.0.10d}$$

In the interior region I_t we shall only use the rough estimate

$$\int_{I_t} |\Delta\psi - \overset{\circ}{\Delta}\psi\,| = O(t^{1/2}) \tag{17.0.10e}$$

and

$$\int_{I_t} r|\hat{\eta}|^2 = O(1). \tag{17.0.10f}$$

Identifying S^2 with the unit sphere in $\Re^3$, $(0,\infty)\times S^2$ is identified with $\Re^3 \setminus 0$ by $(r,\omega) \to x = r\omega$. Therefore the solution of 17.0.10a, which vanishes at infinity on Σ_t, is given by

$$\psi(t,x) = -\frac{1}{4\pi}\int_{\Re^3} \frac{f(t,x')}{|x-x'|} d^3x'. \tag{17.0.11a}$$

The interior region I_t is identified with the ball $B(\frac{1}{2}r_0(t))$ and the exterior region E_t with is identified with its complement $B(\frac{1}{2}r_0(t))^c$. Since by 17.0.10e and 17.0.10f

$$\int_{B(\frac{1}{2}r_0(t))} |f(t,x)| d^3x = O(t^{1/2}),$$

the contribution of the interior region for $x \in B(\frac{3}{4}r_0(t))^c$ to the integral in 17.0.11a is bounded by $Ct^{-1/2}$. Thus for $x \in B(\frac{3}{4}r_0(t))^c$ we can write

$$\psi(t,x) = -\frac{1}{4\pi}\int_{B(\frac{1}{2}r_0(t))^c} \frac{f(t,x')}{|x-x'|} d^3x' + O(t^{-1/2}). \tag{17.0.11b}$$

Using 17.0.10d, we find that the contribution of the term $\Delta\psi - \overset{\circ}{\Delta}\,\psi$ to the integral in 17.0.11b is bounded by Ct^{-1}. Writing $r|\hat{\eta}|^2 = r^{-1}|\Xi|^2 + (r|\hat{\eta}|^2 - r^{-1}|\Xi|^2)$ and using 17.0.10c we also find the contribution of $r|\hat{\eta}|^2 - r^{-1}|\Xi|^2$ to be bounded by Ct^{-1}. We conclude that for $x \in B(\frac{3}{4}r_0(t))^c$ we have

$$\psi(t,x) = -\frac{1}{4\pi}\int_{B(\frac{1}{2}r_0(t))^c} \frac{|\Xi|^2(u',\omega')}{r'|x-x'|} d^3x' + O(t^{-1/2}) \tag{17.0.11c}$$

where u' is the value of u that at given t corresponds to r'. Similarly, we obtain for $x \in B(\frac{3}{4}r_0(t))^c$,

$$r\frac{\partial\psi}{\partial r}(t,x) = \frac{1}{4\pi}\int_{B(\frac{1}{2}r_0(t))^c} |\Xi|^2(u',\omega')\frac{r(r-r'\omega\cdot\omega')}{r'|x-x'|^3}d^3x' + O(t^{-1/2}). \tag{17.0.11d}$$

Letting

$$\varepsilon = \frac{r'}{r} - 1, \qquad \varepsilon = \varepsilon(t,u,u'),$$

we can express these formulas in the form

$$\psi(t,u,\omega) = -\frac{1}{4\pi}\int_{u_1(t)}^{\infty}(1+\varepsilon)\lambda(t,u')du' \cdot\left(\int_{S^2}\frac{|\Xi|^2(u',\omega')}{[2(1-\omega\cdot\omega')(1+\varepsilon)+\varepsilon^2]^{1/2}}d\omega'\right), \tag{17.0.11e}$$

$$(r\nabla_N\psi)(t,u,\omega) = \frac{1}{4\pi}\int_{u_1(t)}^{\infty}(1+\varepsilon)\lambda(t,u')du' \cdot\left(\int_{S^2}\frac{[1-(1+\varepsilon)\omega\cdot\omega']|\Xi|^2(u',\omega')}{[2(1-\omega\cdot\omega')(1+\varepsilon)+\varepsilon^2]^{3/2}}d\omega'\right), \tag{17.0.11f}$$

where $u_1(t)$ is the value of u that at given t corresponds to $\frac{1}{2}r_0(t)$. Now

$$\varepsilon = \frac{r_0^{-1}\int_u^{u'}\lambda du}{1+r_0^{-1}\int_0^u \lambda du} = O(t^{-1})$$

and $u_1(t) \to -\infty$ as $t \to \infty$. It follows that

$$\psi(t,u,\omega) \to -\frac{1}{2^{1/2}4\pi}\int_{-\infty}^{\infty}du'\left(\int_{S^2}\frac{|\Xi|^2(u',\omega')}{(1-\omega\cdot\omega')^{1/2}}d\omega'\right),$$

as $t \to \infty$. Also, writing

$$\frac{[1-(1+\varepsilon)\omega\cdot\omega']}{[2(1-\omega\cdot\omega')(1+\varepsilon)+\varepsilon^2]^{3/2}} = \frac{1}{2[2(1-\omega\cdot\omega')(1+\varepsilon)+\varepsilon^2]^{1/2}} - \frac{(1+\frac{1}{2}\varepsilon)\varepsilon}{[2(1-\omega\cdot\omega')(1+\varepsilon)+\varepsilon^2]^{3/2}}$$

and using the fact that for any continuous function f on S^2

$$\lim_{\varepsilon\to 0^+_-}\int_{S^2}\frac{\varepsilon f(\omega')d\omega'}{[2(1-\omega\cdot\omega')(1+\varepsilon)+\varepsilon^2]^{3/2}} = {}^+_-2\pi f(\omega),$$

we obtain

$$(r\nabla_N\psi)(t,u,\omega) \to \frac{1}{2^{3/2}4\pi}\int_{-\infty}^{\infty} du' \left(\int_{S^2} \frac{|\Xi|^2(u',\omega')}{(1-\omega\cdot\omega')^{1/2}} d\omega'\right) + \frac{1}{2}\int_{-\infty}^{\infty} du' sgn(u-u')|\Xi|^2(u',\omega).$$

The results regarding ψ' follow in precisely the same fashion.

Conclusion 17.0.7 On any null hypersurface C_u we have

$$\lim_{C_u, t\to\infty} r^2\delta = \Omega + \Psi'$$

and

$$\lim_{C_u, t\to\infty} r^2\epsilon = E$$

where E is a 1-form on S^2, depending on u, which satisfies at each u the Hodge system

$$\begin{aligned} \overset{\circ}{c\!\!/url}\; E &= Q + \Sigma \wedge \Xi \\ \overset{\circ}{d\!\!/iv}\; E &= P - \overline{P} + \Sigma\cdot\Xi - \overline{\Sigma\cdot\Xi} \\ &\quad + \overset{\circ}{\Delta\!\!\!/}\,\Psi - \Psi' - \Omega' \end{aligned}$$

on S^2.

Proof 17.0.9 *We recall from Chapter 11 (Lemma 11.2.1) that δ decomposes into*

$$\delta = r^{-2}(q' + \psi') + r^{-1}(\nabla_N q + \nabla_N\psi).$$

By the results of Chapter 11 (Propositions 11.2.2 and 11.2.3),

$$q, q' = O(r^{-1/2}), \qquad \nabla q, \nabla q' = O(r^{-3/2}).$$

In view of Lemma 17.0.2 the conclusion regarding δ then follows. We now recall from Chapter 11 the Hodge system for ϵ:

$$c\!\!/url\epsilon = \sigma + \hat{\chi}\wedge\hat{\eta} \qquad (17.0.12a)$$

$$d\!\!/iv\epsilon = -\nabla_N\delta + \hat{\theta}\cdot\hat{\eta} - \frac{3}{2}\mathrm{tr}\theta\delta - 2a^{-1}\nabla\!\!\!/ a\cdot\epsilon. \qquad (17.0.12b)$$

According to 11.2.26d we can express

$$\begin{aligned} \nabla_N\delta - \hat{\theta}\cdot\hat{\eta} = &-\hat{\chi}\cdot\hat{\eta} - r^{-1}\Delta\!\!\!/\psi - r^{-2}(r\mathrm{tr}\theta + a^{-1}\lambda)\nabla_N\psi \\ &- r^{-1}a^{-1}\nabla\!\!\!/ a\cdot\nabla\!\!\!/\psi + r^{-2}\nabla_N p - 2r^{-3}a^{-1}\lambda p \end{aligned} \qquad (17.0.12c)$$

where

$$p = r\nabla_N q + q' + \psi'.$$

Now by Lemma 11.2.1

$$\begin{aligned}\triangle q &= \nabla_N^2 q + \mathrm{tr}\theta\nabla_N q + \not{\Delta} q + a^{-1}\not{\nabla} a \cdot \not{\nabla} q \\ &= -r(\rho + \hat{\chi}\cdot\hat{\eta} - \overline{\rho} - \overline{\hat{\chi}\cdot\hat{\eta}}) + r\hat{\chi}\cdot\hat{\eta} - 2ra^{-1}\not{\nabla} a \cdot \epsilon - \kappa\delta,\end{aligned}$$

where

$$\kappa = ra^{-1}(a\mathrm{tr}\theta - \overline{a\mathrm{tr}\theta}) = r\mathrm{tr}\theta - 2a^{-1}\lambda.$$

Substituting for $\nabla_N^2 q$ from this equation into 17.0.12c and substituting in turn the resulting expression into 17.0.12c, we bring the Hodge system for ϵ to the form

$$c\not{u}rl\,\epsilon = \sigma + \hat{\chi}\wedge\hat{\eta} \tag{17.0.13a}$$

$$\begin{aligned}\not{d}iv\,\epsilon = \rho - \overline{\rho} + \hat{\chi}\cdot\hat{\eta} - \overline{\hat{\chi}\cdot\hat{\eta}} \\ + r^{-1}\not{\Delta}\psi - r^{-2}\nabla_N\psi' - r^{-3}a^{-1}\lambda\psi' + J\end{aligned} \tag{17.0.13b}$$

where

$$\begin{aligned}J = r^{-1}\not{\Delta} q - r^{-2}\nabla_N q' - r^{-3}a^{-1}\lambda q' \\ + \frac{1}{2}r^{-2}\kappa(\nabla_N(\psi + q) - r^{-1}(\psi' + q')) + r^{-1}a^{-1}\not{\nabla} a \cdot \not{\nabla}(\psi + q).\end{aligned} \tag{17.0.13c}$$

Consider this system on the sections $S_{t,u}$ of a given C_u and let $t \to \infty$. By the main theorem in conjunction with the results of Chapter 11,

$$J = O(r^{-7/2}).$$

The first three conclusions together with Lemma 17.0.2 then imply that

$$\begin{aligned}r^3 c\not{u}rl\,\epsilon &\to Q + \Sigma\wedge\Xi \\ r^3 \not{d}iv\,\epsilon &\to P - \overline{P} + \Sigma\cdot\Xi - \overline{\Sigma\cdot\Xi} \\ &\quad + \mathring{\not{\Delta}}\,\Psi - \Psi' - \Omega'\end{aligned}$$

as $t \to \infty$. Since $\tilde{\gamma} \to \mathring{\gamma}$ and $\tilde{\not{\nabla}}$, the covariant derivative of $\tilde{\gamma}$, converges to $\mathring{\not{\nabla}}$, the covariant derivative of $\mathring{\gamma}$, if we define E to be the solution of the Hodge system

$$\begin{aligned}\mathring{c\not{u}rl}\;E &= Q + \Sigma\wedge\Xi \\ \mathring{\not{d}iv}\;E &= P - \overline{P} + \Sigma\cdot\Xi - \overline{\Sigma\cdot\Xi} \\ &\quad + \mathring{\not{\Delta}}\,\Psi - \Psi' - \Omega'\end{aligned}$$

on S^2, then on C_u, $r^2\epsilon$ tends to E as $t \to \infty$. We note that the first of the previous equations when integrated on S^2 yields

$$\overline{Q} = -\overline{\Sigma \wedge \Xi}.$$

Conclusion 17.0.8 The tensorfield Σ satisfies, at each u, the following equation on S^2:

$$\overset{\circ}{\text{div}}\ \Sigma = \frac{1}{2}\overset{\circ}{\nabla} H + E.$$

Proof 17.0.10 *This follows immediately from the normalized null Codazzi equation,*

$$div\,\hat{\chi}' + \hat{\chi}' \cdot \epsilon = \frac{1}{2}\nabla \mathrm{tr}\chi' + \frac{1}{2}\mathrm{tr}\chi'\epsilon - \beta$$

(standard frame), upon multiplying by r^3 and taking the limit $t \to \infty$ on C_u, in view of Conclusions 17.0.5 and 17.0.7 and the fact that $\beta = O(r^{-7/2})$.

Conclusion 17.0.9 On any null hypersurface C_u we have

$$\lim_{C_u, t\to\infty} r \log \phi = \Psi, \qquad \lim_{C_u, t\to\infty} r^2 \nabla_N \log \phi = \Omega.$$

Also,

$$\lim_{C_u, t\to\infty} r \log a = \Psi'.$$

Finally,

$$\lim_{C_u, t\to\infty} r^2 \zeta' = E + \overset{\circ}{\nabla}\Psi'$$

and

$$\lim_{C_u, t\to\infty} r^2 \omega' = P.$$

Proof 17.0.11 *The proof of the conclusion regarding ϕ is entirely analogous to the proof of Lemma 17.0.2. To prove the conclusion regarding a, we consider the following equation (see Chapter 9):*

$$\frac{da}{ds} = -\nu, \qquad \nu = -\nabla_N \log \phi + \delta$$

By Conclusion 17.0.7 and the first part,

$$\lim_{C_u, t\to\infty} r^2 \nu = \Psi'.$$

The conclusion regarding a then follows by integration. The conclusion regarding ζ' then follows since $\zeta' = \epsilon + \nabla\!\!\!/ \log a$. Finally, the conclusion regarding ω' follows by integrating the following equation (see Chapter 9):

$$\frac{d\omega}{ds} = 2\underline{\zeta} \cdot \zeta - |\zeta|^2 - \rho$$

and taking into account Conclusion 17.0.1 as well as the fact that

$$\lim_{C_u, t \to \infty} \omega = 0.$$

Conclusion 17.0.10 Let

$$\Sigma^+(\cdot) = \lim_{u \to \infty} \Sigma(u, \cdot), \qquad \Sigma^-(\cdot) = \lim_{u \to -\infty} \Sigma(u, \cdot).$$

Also

$$F(\cdot) = \int_{-\infty}^{\infty} |\Xi(u, \cdot)|^2 du$$

and let Φ be the solution, of vanishing mean, of the equation

$$\overset{\circ}{\Delta\!\!\!/}\, \Phi = F - \overline{F}$$

on S^2. Then $\Sigma^+ - \Sigma^-$ is determined by the following equation on S^2:

$$\overset{\circ}{d\!\!\!/iv}\, (\Sigma^+ - \Sigma^-) = \overset{\circ}{\nabla\!\!\!/}\, \Phi.$$

We note that, by Conclusion 17.0.4, $F/8\pi$ is the total energy radiated to infinity in a given direction, per unit solid angle.

Proof 17.0.12 *By Conclusion 17.0.3, $\Sigma(u, \cdot)$ tends to limits $\Sigma^+(\cdot)$ and $\Sigma^-(\cdot)$ as $t \to \infty$ and $t \to -\infty$ respectively. By Lemma 17.0.2, $\Omega'(u, \cdot)$ tends to limits $\Omega'^+(\cdot)$ and $\Omega'^-(\cdot)$ as $t \to \infty$ and $t \to -\infty$, respectively, and*

$$\Omega'^+(\cdot) - \Omega'^-(\cdot) = \int_{-\infty}^{\infty} (|\Xi(u, \cdot)|^2 - \overline{|\Xi|^2(u)})du. \tag{17.0.14}$$

Then by Conclusions 17.0.1, 17.0.3, and 17.0.7, $E(u, \cdot)$ also tends to limits $E^+(\cdot)$ and $E^-(\cdot)$ as $t \to \infty$ and $t \to -\infty$, respectively, and the difference of these limits satisfies the Hodge system

$$\overset{\circ}{c\!\!\!/url}\, (E^+ - E^-) = 0$$
$$\overset{\circ}{d\!\!\!/iv}\, (E^+ - E^-) = -\Omega'^+ + \Omega'^-.$$

It follows that

$$E^+ - E^- = \overset{\circ}{\nabla\!\!\!/}\, \Phi$$

where Φ *is the solution, of vanishing mean, of the equation*

$$\overset{\circ}{\triangle\!\!\!/}\, \Phi = -\Omega'^+ + \Omega'^-$$

on S^2*. On the other hand, by Conclusions 17.0.5 and 17.0.8 the difference* $\Sigma^+ - \Sigma^-$ *satisfies*

$$\overset{\circ}{\mathrm{d\!\!\!/iv}}\, (\Sigma^+ - \Sigma^-) = E^+ - E^-.$$

The conclusion thus follows.

We note that the integrability condition of the equation for $\Sigma^+ - \Sigma^-$ is that F is L^2-orthogonal to the 1st eigenspace of $\overset{\circ}{\triangle\!\!\!/}$ on S^2. Now the L^2- inner products of $F/8\pi$ with the three Cartesian coordinate functions $x^i, i = 1, 2, 3$, on $S^2 \subset \Re^3$, which form an orthogonal basis for the first eigenspace of $\overset{\circ}{\triangle\!\!\!/}$, represent the components $P^i, i = 1, 2, 3$ of the total linear momentum radiated to infinity. Since the initial and final states both have zero linear momentum, the vanishing of P^i manifests the law of conservation of linear momentum.

Bibliography

[Ba1] R. Bartnik. Existence of maximal surfaces in asymptotically flat space-times. *Comm. Math. Phys.* 94, 1984, 155–175.

[Ba2] R. Bartnik. The mass of an asymptotically flat manifold. *Comm. Pure appl. Math.* 39, 1986, 661–693.

[Be-Ro] L. Bel. Introduction d'un tenseur du quatrieme ordre. *C. R. Acad. Sci. Paris* 247, 1959, 1094.

[Bo] H. Bondi. *Nature* 186, 1960, 535.

[Bo-Bu-Me] H. Bondi–M.G.J. van der Burg–A.W.K. Metzner. Gravitational waves in General Relativity vii. *Proc. Roy. Soc. Lond.* A269, 1962, 21–52.

[Br1] C. Bruhat. Theoreme d'existence pour certain systemes d'equaitions aux deriveés partielles nonlinaires. *Acta Matematica* 88 (1952), 141–225.

[Br2] C. Bruhat. Un theoreme d'instabilité pour certain equations hyperboliques nonlinaires. *C. R. Acad. Sci. Paris* 276A, 1973, 281.

[Ch] D. Christodoulou. Global solutions for nonlinear hyperbolic equations for small data. *Comm. Pure appl. Math.* 39, 1986, 267–282.

[Ch-Br] C.Bruhat–D. Christodoulou. Elliptic systems in $H_{s,\delta}$ spaces on manifolds which are Euclidean at infinity. *Acta Math.* 145, 1981, 129–150.

[Ch-Eb] J. Cheeger–D. Ebin. Comparison Theorems in Riemannian Geometry. *North-Holland Publ. Comp.* 1975.

[Ch-Kl] D. Christodoulou–S. Klainerman. Asymptotic properties of linear field equations in Minkowski space. *Comm. Pure Appl. Math.* 43, 1990, 137–199.

[Ch-Mu] D. Christodoulou–N. O' Murchadha. The boost problem in General Relativity. *Comm. Math. Phys.* 80, 1981, 271–300.

[Eisen] L. P. Eisenhart. Riemannian Geometry. *Princeton University Press* 1926.

[Gi-Tr] D. Gilbarg–N.S. Trudinger. Elliptic Partial Differential Equations of Second Order. *Springer-Verlag* 1977.

[Ho] L. Hörmander. On Sobolev spaces associated with some Lies algebras. *Report N0:4, Inst. Mittag–Leffler* 1985.

[John] F. John. Formation of singularities in elastic waves. *Lecture Notes in Phys.* Springer-Verlag 1984 pages 190–214.

[Kl1] S. Klainerman. The null condition and global existence to nonlinear wave equations. *Lect. Appl. Math.* 23, 1986. 293–326.

[Kl2] S. Klainerman. Remarks on the global Sobolev inequalities in Minkowski space. *Comm. Pure appl. Math.* 40, 1987, 111–117.

[Kl3] S. Klainerman. Uniform decay estimates and the Lorentz invariance of the classical wave equation. *Comm. Pure appl. Math.* 38, 1985, 321–332.

[Ne-To] E.T. Newman–K.P. Todd. Asymptotically flat space-times. *General Relativity and Gravitation, vol. 2, A. Held, Plenum* 1980.

[Oss] R. Osserman. The isoperimetric inequality. *Bull A.M.S.* 84, Nov. 1978, 1182–1238.

[Pe1] R. Penrose. Structure of space-time. *Batelle Rencontre, C.M. DeWitt and J.A. Wheeler* 1967.

[Pe2] R. Penrose. Zero rest mass fields including gravitation: asymptotic behaviour. *Proc. Roy. Soc. Lond.* A284, 1962, 159–203.

[Sa] R.K. Sacks. Gravitational waves in General Relativity viii. *Proc. Roy. Soc. Lond.* A270, 1962, 103–126.

[Sc-Ya] R. Schoen–S.T. Yau. On the proof of the positive mass conjecture in general relativity. *Comm. Math. Phys.* 65, 1979, 45–76.

[Si] T. Sideris. Formation of singularities in 3-D compressible fluids. *Comm. Math. Phys.* 101, 1985, 475–485.

[Wi] E. Witten A new proof of the positive energy theorem. *Comm. Math. Phys.* 80. 1981, 381–402.

Milton Keynes UK
Ingram Content Group UK Ltd.
UKHW021024140924
448309UK00006B/273